VISUAL OPTICS
AND REFRACTION

VISUAL OPTICS AND REFRACTION

a clinical approach

DAVID D. MICHAELS, M.D.

Assistant Clinical Professor of Ophthalmology,
University of California School of Medicine,
Los Angeles, California;
Chairman, Department of Ophthalmology,
San Pedro and Peninsula Hospital,
San Pedro, California

WITH 391 ILLUSTRATIONS

SAINT LOUIS

The C. V. Mosby Company

1975

Printed in the United States of America

Distributed in Great Britain by Henry Kimpton, London

Library of Congress Cataloging in Publication Data

Michaels, David D 1925-
Visual optics and refraction.

Bibliography: p.
Includes index.
1. Eye—Accommodation and refraction.
2. Optics, Physiological. I. Title. [DNLM:
1. Optics. 2. Refraction, Ocular. WW300 M621v]
RE925.M46 617.7'55 75-22215
ISBN 0-8016-3424-5

GW/CB/B 9 8 7 6 5 4 3 2 1

To my children

Larry, Mark, Leslie, and Michelle

for the many hours there was
no time to play

Preface

The purpose of this book is to teach the principles and techniques of refraction not as a mechanical routine but as a clinical discipline. Visual physiology and psychology have therefore received more than usual attention, whereas the optics is condensed to what is pertinent for practical work. Since refractive errors also occur in diseased eyes, the discussions on pathophysiology may not be inappropriate. Mathematics has been generally avoided, and the beginner with little preparation and less experience will find sufficient direction to get started. The more important formulas are boxed for easy reference.

The subject matter is divided into three parts. The first part deals with optical and physiologic principles basic to interpreting refractive tests. It is a maxim of pedagogy that facts are best anchored to a rigging of theory, perhaps buoyed by a smile or a simile. Although practice transcends theory, the "art" of refraction is not in insulating layers of technology but in cutting through the irrelevant to reach a correct diagnosis rapidly and efficiently. The second part attempts a fair evaluation of present methods and techniques of refraction, their advantages, and limitations. The emphasis is on those tests most likely to give useful information in the restricted time of a busy practice. Certain alternative techniques are included that are suitable for special situations. The last part deals with specific clinical areas and suggests a course of action based on the analysis of signs and symptoms as well as on optical measurements. Although it may be rash, even reckless, to assess not only the how and why but also the why not, readers are entitled to an evaluation, even if they ultimately arrive at differing conclusions. It is differences that make horse races, sometimes red faces. For those requiring a quick survey, Chapters 8, 9, 11, 13, 14, and 17 provide a self-contained core program.

An elementary treatment must necessarily be positive yet shun magisterial pronouncements. Extensive text documentation would serve no purpose; the pertinent references are given at the end of each chapter. The latter include papers of historical as well as current interest. Historical sidelights help enliven the subject and pay tribute to the pioneers fast fading in the flood of new research.

Written in spare moments between full-time practice, part-time teaching, and little time for anything else, this book would not have been possible without the cooperation of my patients, the indulgence of my publisher, and the forebearance of my family.

I should like to acknowledge the encouragement of my friends Drs. George Zugsmith, Arthur Linksz, and Bradley Straatsma; the editorial assistance of Ms. V. Pettijohn; and the secretarial help of Ms. I. McLaughlin, N. Millet, D. Donahugh, B. Brayman, C. Myers, K. Nagle, and L. Alexander.

David D. Michaels

Contents

PART III Evaluation

PART I

BASICS

1

The nature of light

More than any other sense organ the eyes feed the brain with information by converting light into coded neural activity, although what the code is, no one is yet prepared to say. Study optics, advised Voltaire, and you will see that it is impossible for objects to appear other than you see them. Nevertheless, the eyes miss a great deal of what is going on because they respond to only a narrow band, a single octave, of the radiations that reach them. The bent stick in the water, the image behind the mirror, and the mirage in the desert illustrate, sometimes painfully, that the eyes can be deceived. We speak of green light and red apples, but the green turns gray in moonlight, and, likely as not, the apple is a shade of yellow to the color blind. To what extent our senses can be trusted is a question of unending fascination to the epistemologist. Indeed the paradox of modern physics is that we are forever separated from a good part of the universe by a wall of indeterminacy. Truth is consistent sensation, not absolute but relative, and as Einstein proved, one cannot always choose one's relatives.

Light travels in straight lines most of the time,* like a beam of sunlight passing through the dusty air of a darkened room. If an opaque body is held in its path, a shadow is cast on the wall. In textbook diagrams of this simple experiment the shadow is delineated by lines, called "light rays," and the shadow edges are perfectly sharp. Experimental shadow edges, however, are diffused. This difference illustrates the two approaches to optics (Fig. 1-1): geometric and physical optics. Geometric optics deals with a scaffold of light rays held together by four basic postulates: rectilinear propagation, the independence of individual rays, the law of reflection, and the law of refraction. This geometric mansion serves admirably for clinical purposes, although there are some vacancies. For example, the limits of resolution, the glow of a fluorescent bulb, or the workings of a laser beam cannot be explained by geometry. So we also take a brief peek in the basement at waves and quanta, interference and diffraction, the domain of physical optics.

*A special branch called "Schlieren optics," deals with the behavior of light in inhomogeneous media. The most common example is the variations of the earth's atmosphere with altitude. The "shadow" test for lens glass quality is a practical application.

WAVES AND CORPUSCLES

The central concept of the wave theory is strange in that it concerns a kind of motion in which nothing definite appears to move. If a stone is poked with a stick, a wave propagates down the stick. A pebble thrown into a pool of water produces ripples; yet individual water particles move only up and down. The pebble and the stick illustrate that a wave is a propagation of motion rather than substance and that it requires some sort of medium.

Christiaan Huygens (1629-1695) reasoned that light unlike sound, which cannot survive a vacuum, is a wave in an all-pervasive and imaginary ether with the hardness of the

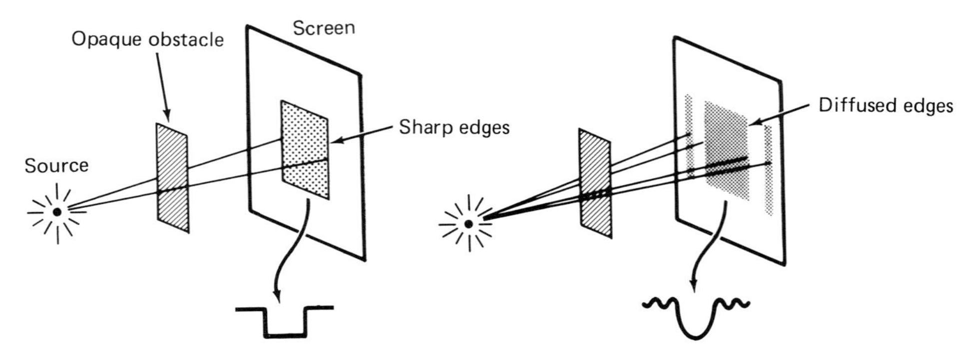

Fig. 1-1. The formation of shadows illustrates the two approaches to optics. The geometric shadow *(left)* has sharp edges; the experimental shadow *(right)* has diffused edges because of diffraction.

stick and the elasticity of the water. And how, you ask, can such a wave travel over great distances, for example, from a star, without dissipating itself? One will cease to be astonished, writes Huygens addressing unborn generations of puzzled optics students, if one considers that an infinite number of wavelets originating at the same instant makes practically one single wave, which has force enough to reach our eyes. To explain how the wave travels in straight lines, Huygens invented a principle, which is named after him: Each wavelet's effect is limited to that part which touches the enveloping new wave front. By connecting equivalent points on the progressing wave with a straight line, one obtains the direction of propagation. This direction is the light ray, but why the ray has such Euclidean wisdom, Huygens did not say.

Across the channel, Huygens' contemporary, Isaac Newton, was looking through his prism (still preserved in the British Museum).

> In the sun's beam which was propagated into the room through the hole in the window shut, at the distance of some feet, I held the prism in such a posture that its axis might be perpendicular to that beam and I observed that the most refracted part thereof appeared violet, the least refracted red, the middle parts blue, green, and yellow in order.*

What Newton saw, others had seen but overlooked. If sunlight is homogeneous as was then believed, any spreading of its light by a mechanical effect of the prism would imply a varying law of refraction, and another prism should produce still more colors. On the contrary, by using a second prism, Newton showed that the part of the beam bent too much the first time was also bent too much the second and by exactly the same amount. Refraction was therefore regular and lawful for the individual parts and not for sunlight as a whole; the prism separated rather than modified the light. Since color lies in the light, Newton concluded it must be corpuscular, with a different kind of corpuscle for each color. Light corpuscles did not require a hypothetical ether, and rectilinear propagation followed directly from the laws of motion. The theory predicted a higher light velocity in water than air, but the experiment to disprove this point was still a hundred years away. Newton was fond of wording his conclusions as if daring anyone to disagree with him: "Are not the rays of light very small bodies emitted from shining substances?" And for a century to come the scientific world, bowing to his genius, answered, "Of course they are."

Thomas Young, physician, physicist, and philologist, had studied tides, energy, shipbuilding, Egyptian hieroglyphics, ocular accommodation, and color vision. He discovered astigmatism in his own eyes. Tscherning called him the "Father of Physiological Optics." In 1801 Young repeated an experiment that had been tried unsuccessfully by Grimaldi 150 years earlier.

*From Newton, I.: Opticks: or a treatise of the reflexions, refractions, inflexions and colours of light, 1704 (reprint), New York, 1952, Dover Publications, Inc.

Grimaldi had concluded that light waves could interfere with each other but erred by using two separate light sources whose frequencies are much too erratic. What he observed was physiologic border contrast (or Mach bands), which he mistook for interference. Young had the idea of placing two pinholes in front of a single source, thus obtaining two identical twin waves of the same frequency. Catching the waves on a screen after each had traveled a slightly different distance, a series of dark areas appeared on the screen, and interference was demonstrated as that peculiar reality in which two parts light add up to one part darkness.

When light rays pass the edge of an obstacle, they are bent, a phenomenon called "diffraction." Young showed diffraction to be an interference effect resulting from the interaction of the unscreened wave and the wave from the edge of the obstacle. Diffraction can be demonstrated by simply lifting two fingers and while viewing the sky, bringing the fingers together. Just before they touch, the slit of light between them widens.

Young's contemporary, a remarkable young Frenchman, Augustine Fresnel, saw in interference the solution to Huygens' puzzle. What prevented the Huygenian wavelets from traveling backward and what limited their cumulative effects exclusively to the enveloping wave was simply their mutual interference and reinforcement. Seldom in the history of science was so much theory built on so little equipment. So conclusive were Fresnel's experiments that the wave theory was not to be challenged again until the present century.

There still remained the problem of Huygens' ether. "We can scarcely avoid the inference," wrote James Clerk Maxwell in 1873, "that light consists of transverse undulations of the same medium which is the cause of electric and magnetic phenomena." Maxwell's ether was something new. It was not the elastic-solid medium proposed earlier but an ether whose physical stresses and motions were electric and magnetic fields. Light is only the visible range of a broad symphony of radiations, including the useful x rays, the entertaining radio waves, the awesome gamma rays, and the mysterious cosmic rays.

PENCILS, BEAMS, AND RAYS

We have it on Huygens' authority that light waves are spherical surfaces concentric to their point of origin. A true representation would require drawing approximately 30,000 wave fronts to the inch; thus our schematic drawing (Fig. 1-2) shows only a few representative wave fronts.

The curvature of a wave front in air depends on the radius. The further from its origin the less its curvature. At some point the radius is so long and the wave becomes so flat that it is plane for all practical purposes. In visual optics this distance is considered to be 6 meters. In the accepted sign convention, converging waves are positive, and diverging waves are negative; a lens or mirror converging light is positive, and one diverging light is negative.

The direction of wave propagation is the light ray perpendicular to its surface. If we think of the ray as a physical entity, such as passing light through a small opening, we are in a mess because diffraction forever prevents the isolation of a single ray. Only in Euclidean geometry can matter, which occupies space, be represented by lines, which do not. A light ray is noncommittal; it can represent equally well the direction of flow of luminous energy, a wave normal, or the path of a train of light corpuscles or quanta.

A group of light rays passing through a limiting aperture is termed a "pencil." The aperture may be the pupil of the eye, the mounting of a trial lens or the opening in a pinhole camera. The presence of an aperture is implicit in the notion of parallel rays. Since natural light sources always emit diverging rays, the rays from a distant object can only become parallel relative to some limiting aperture.

A collection of pencils is called a "beam." The individual pencils making up the beam

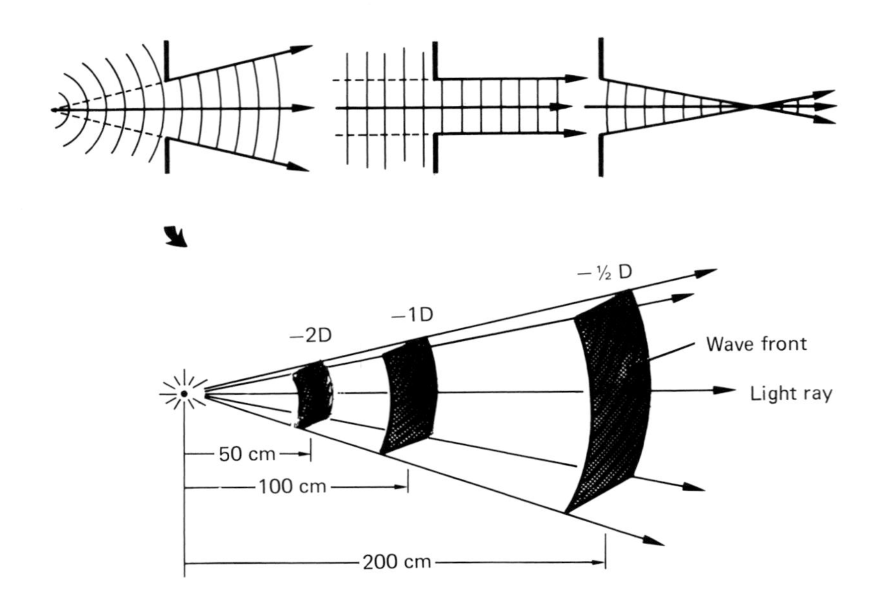

Fig. 1-2. Diverging, parallel, and converging pencils as limited by an aperture *(top)*. The progressive decrease of curvature of an advancing wave front is illustrated in the bottom figure.

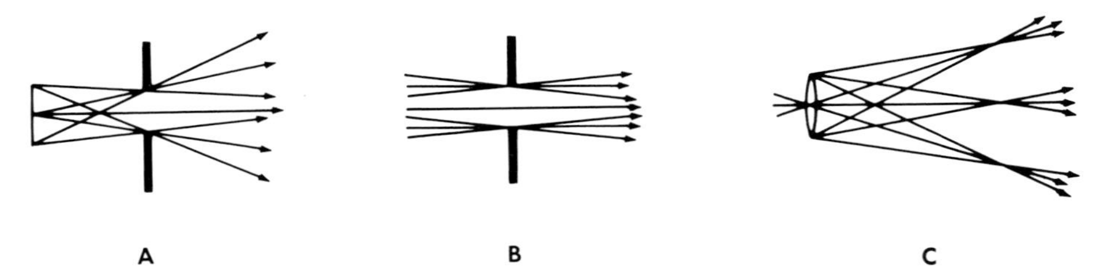

Fig. 1-3. Three light beams consisting of divergent (**A**), parallel (**B**), and convergent (**C**) pencils. In speaking of light "vergence," the term always refers to the pencils and not the beam.

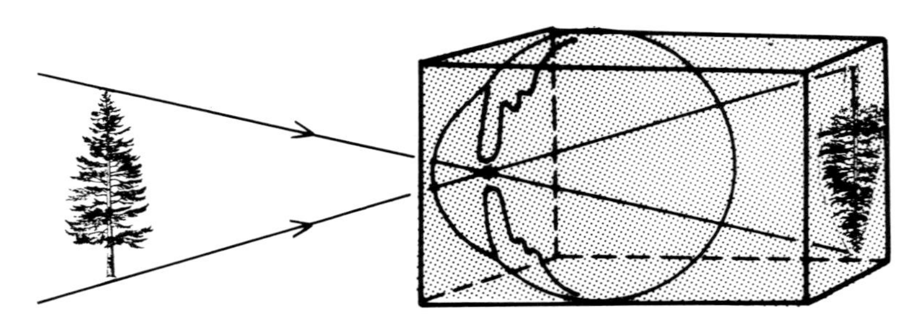

Fig. 1-4. The eye behaves in many respects like a pinhole camera. Each pencil passes through the aperture independently and forms a small patch of light on the screen (or retina) of the same shape as the opening. Since the pupil is round, retinal images are made of blur "circles." The rectilinear path of the rays can be confirmed by using a ground glass screen (or dissecting a scleral window) and observing the small inverted image.

may converge or diverge; thus vergence always refers to the pencils and not the beam (Fig. 1-3).

A pinhole camera illustrates rectilinear propagation and the formation of an optical image (Fig. 1-4). All the rays pass through the pinhole ignoring each other, those from each object point proceeding in a straight line to an equivalent image point. The image is real, that is, it can be caught on a screen, inverted, and its size is in the same ratio as the object and image distances. The image is made up of little patches of light that duplicate the shape of the aperture. Since the pupil is round, retinal images are made up of blur circles. The smaller the aperture the smaller these circles and the less they overlap, thus producing sharper images. If the aperture is too small, however, not only does the image become dim, but it also becomes blurred again—one of the effects of diffraction. Pinholes are frequently used to evaluate whether a patient's vision can be improved by optical means. If the opening is too small (less than about 0.5 mm), diffraction will make the vision worse instead of better.

VELOCITY OF LIGHT

Galileo first proposed an experiment to determine light velocity. Two observers, each with a screened lantern, were to face each other from a considerable distance. When the first observer uncovered his lantern, the second was to immediately uncover his. The distances were too small for the experiment to work, but the principle was sound, and in 1675 Romer used the idea to measure the time it took for light from one of Jupiter's satellites to travel across the earth's orbit. The approximate velocity of light is now considered to be 3×10^{10} cm/sec.

The astronomer measures light velocity in a vacuum, but the velocity in air is so similar that, for all practical purposes, they can be assumed to be identical. The velocity of light in air is thus the practical standard or the common denominator to which velocities in all other media are compared. Expressed as a ratio, it is called the "refractive index." For example, $\frac{\text{velocity in air}}{\text{velocity in glass}}$ = refractive index of glass. A glass of refractive index 1.5 means that the velocity in air is 1.5 times faster than in this glass. The symbol for refractive index is n, which represents a pure number without dimensions and is much easier to deal with than actual velocities.

In clinical optics we seldom deal with more than two media, one of greater refractive index than the other. It is convenient to indicate the refractive index of the first medium by the symbol n and then refractive index of the second medium by the symbol n′. The direction of the refracted ray depends on whether light passes into a medium of a higher or lower index; therefore it is important that the sequence be right. The terms "rare" and "dense" are often used in a relative sense in comparing media of lower or higher refractive index.

Since Newton's discovery, it has been known that light is composed of many colors, that is, many wavelengths. The refractive index varies with each wavelength, since each travels at its own velocity.

The change in refractive index with change in wavelength varies with different media and is termed "dispersion." For example, a prism bends blue light more than red, but a prism made of flint glass bends the blue light more than a prism made of crown glass. The flint prism has a higher dispersion. Achromatic optical systems can be constructed by making use of glasses of varying dispersive power. In glass manufacturing the reciprocal of dispersive power is frequently used and is termed the "constringence" of the glass.

REFLECTION AND REFRACTION

Reflections in quiet pools of water have fascinated man since time began, and mirrors are, by feminine acclaim, the most popular optical instruments. Reflection from polished surfaces is referred to as "regular," or "specular." Reflection from ordinary objects is irregular; the irregularity is in the direction and not in the character of the reflection. The apple is red because its surface absorbs all wavelengths except those which, by impinging on a light-adapted eye of an observer with normal color vision, arouse the sensation "red." Colors are sensations that may be compared, altered, or named but not described. Had Newton been color-blind, his prism would have led him nowhere. Thus when we speak of "red light," it is just loose terminology, since the same wavelength appears colorless to the dark-adapted eye; the stimulus-response connection is to be understood.

Euclid named mirror optics "catoptrics," and Alhazen stated the law of reflection in 1100 A.D.: The angles of reflection and incidence are equal and lie in the same plane. One of the quirks of optical geometry is that angles are measured relative to a perpendicular (or normal) to the surface at the point of impact and not the surface itself.

When light enters a transparent medium of greater index, it is slowed down; that is, it is refracted, and the rays emerge at a new angle. If the rays are perpendicular to the surface, the angle is unchanged. This indicates that the light is only slowed and unbent and not that there is no refraction.

Before Snell's findings, scrupulous measurements of the angles of refraction for various angles of incidence in a variety of media led to no definite connection between them. The notion that if only one gathered enough data, something would turn up, worked no better in optics than in formulating foreign policy. Snell discovered the solution but for some unknown reason did not publish it. Descartes had seen Snell's results, presenting it in 1637 in the now familiar form:

$$n \sin i = n' \sin i'$$

This rule states that the ratio of the sines of the angles is a constant for any two optical media for a particular wavelength. The symbols i and i′ and n and n′ are the angles of incidence and refraction in the first and second medium, respectively. Fig. 1-5 is Descartes' original diagram, which is as useful as any to grasp the essence of Snell's law, the basis of all clinical optical formulas.

A simple illustration of Snell's law is the critical angle (Fig. 1-6). When light passes

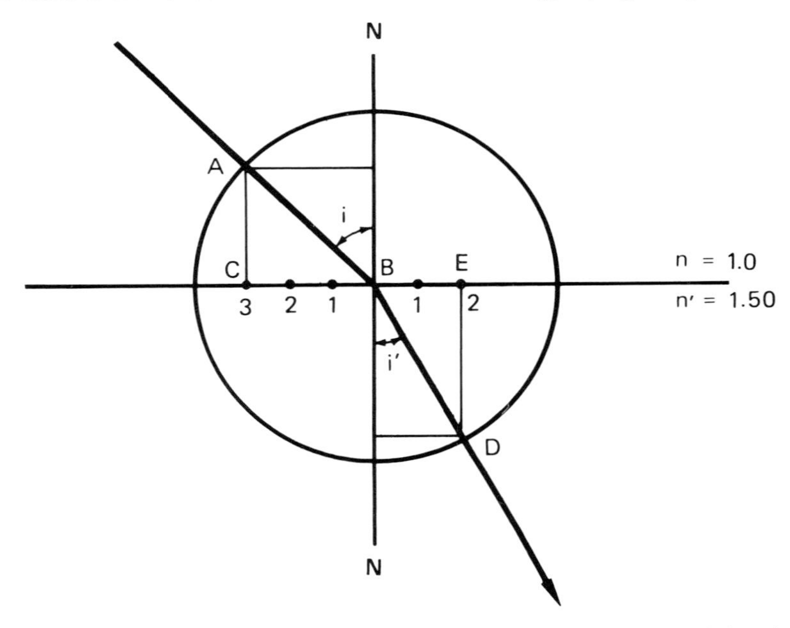

Fig. 1-5. Diagram used by Descartes to "explain" Snell's law. Descartes assumed that the ball traveling from the rarer medium *(AB)* moved half as easily in the denser medium so that *BE* = ⅔ *BC*. The geometry is correct, but the theory is wrong. The true principle, first stated by Fermat, is that light chooses the path through *B* which involves the least time. The same problem stimulated Bernoulli to found the calculus of variations. The light is actually slowed, not accelerated in the denser medium.

from a medium of lower to higher index, Snell's law predicts a bending toward the normal; that is, the angle of refraction is less than the angle of incidence. If the incident angle is zero, the ray is undeviated. When the incident ray grazes the surface, the angle of incidence is 90°, and the angle of refraction is $1/n'$. For glass of index 1.5, this turns out to be approximately 42°. One of the characteristics of light rays is that their path is reversible; that is, a ray proceeding from object to image retraces its path if we pretend the image becomes the object. It follows that when light passes from a denser to a rarer medium, such as from an object embedded in a glass block, the refracted angle is always greater than the incident angle; indeed, when the incident angle is 42°, the refracted ray will just graze the surface. This angle of incidence, which causes the emerging ray to skip along the surface, is the critical angle (i_c), and $\sin i_c = 1/n$ where n refers to the *first* (denser) medium. If the incident ray exceeds the critical angle, the light is reflected back into the glass. A critical angle only arises when rays travel from a denser to a rarer medium. Diamond, for example, has a high critical angle, which accounts for its sparkle.

The critical angle illustrates that both angles and index difference are essential ingredients of Snell's law. A goniolens allows us to see the angle of the anterior chamber of the eye not by changing the angle of incidence but by altering the index difference between cornea and air; the rays that would otherwise be internally reflected are now permitted to escape. The critical angle is also applied in reflecting prisms; the rays striking the second face of the prism exceed the critical angle and are reflected in a predictable direction. The same idea allows piping light down a solid glass rod. An image can be transmitted by a series of rods if their relative orientation is maintained regardless of how the bundle as a whole is twisted (fiber optics).

POLARIZATION

Maxwell's theory had forced a modification of the concept of light waves from geometric

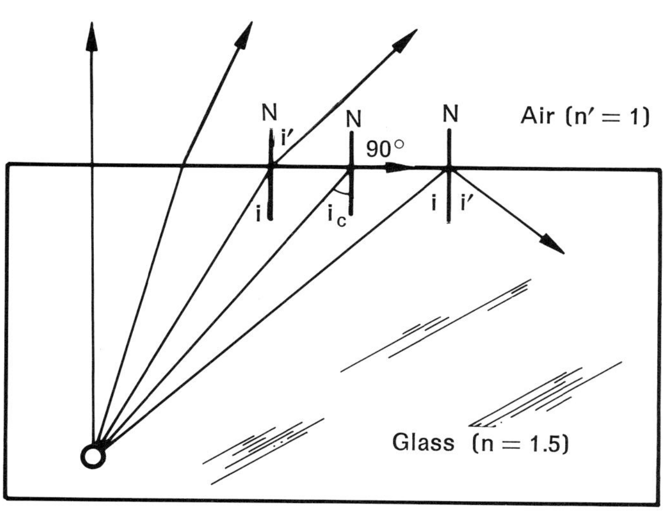

$\sin i_c = 1/1.5$

i_c for glass = 42°

Fig. 1-6. Diagram illustrating refraction from a denser medium. According to Snell's law, the angle of refraction increases faster than the angle of incidence, and at one particular angle (i_c), the refracted ray just grazes the surface. Incident rays greater than the critical angle are reflected back into the glass.

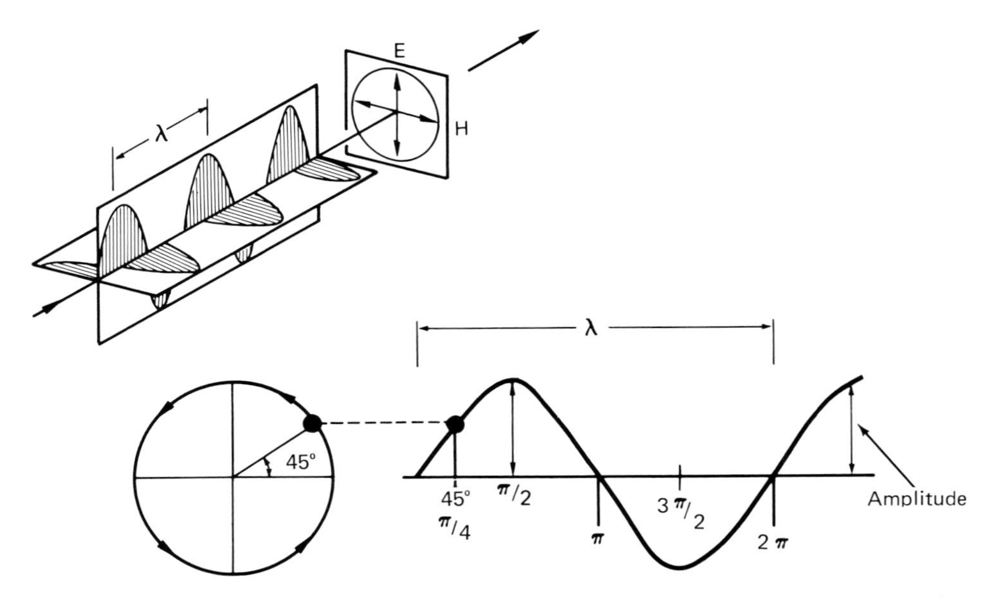

Fig. 1-7. Linearly polarized wave *(top)* showing the electrical *(E)* and magnetic *(H)* vectors. The simple harmonic wave *(bottom)* corresponds to a shadow cast on a moving screen by a continuously rotating ball. The angle of rotation is used to define various portions of the wave. The distance between two successive positions in the same phase is the wavelength (λ), the time for a complete wave motion to pass a given point is the frequency, and their product is the wave velocity.

spheres to electromagnetic fields. Fig. 1-7 attempts to show a wave in three dimensions. The electrical vector, the one that counts, oscillates only in a vertical plane. Such a wave is said to be linearly polarized. Ordinary light consists of trains of waves oscillating in all directions at random. Certain crystals such as Iceland spar, tourmaline, and quinine iodosulfate have the property of double refraction; that is, they isolate certain planes of oscillation. In 1929 Land developed a method of embedding quinine iodosulfate in cellulose acetate films. Large sheets of the plastic are stretched to unscramble its molecules in the direction of stretch, and the sheet is dipped in an iodine solution whose molecules add in a common orientation.

A linearly polarized wave illustrates some general features of wave motion. The curve resembles the effect of flipping a cord that is stationary at one end or of bouncing a beam of light from a tiny mirror on the end of a vibrating tuning fork onto a moving screen. The mathematician calls it a "sine curve" (sinus = wave). Two particles on the curve having the same displacement and direction are said to be "in phase," and the distance between two successive particles in the same phase is called the "wavelength." The wave repeats itself periodically but not indefinitely because the intra-atomic electron jumps producing it require another ticket, and the price is energy. It is therefore impossible to obtain a single pure wavelength of light. The time for a complete wave motion to pass a given point is called the "period," and its reciprocal is the "frequency." Wave velocity is the product of wavelength and frequency. When light passes into a denser medium, velocity changes, but frequency remains the same; hence it is the wavelength that changes proportionately. Although frequency has more physical reality, wavelength is easier to measure. Thus physicists prefer rulers to stopwatches.

When the walls of Jericho fell, it is rumored, at the sound of the trumpets, it was probably a concussion effect. Waves transfer energy by amplitude. Two waves of zero phase difference have four times the intensity (constructive interference), whereas a phase difference of 180°, where crest meets trough, has zero intensity as in the dark areas of Young's experiment.

Ordinary optical glass is isotropic but when subject to strain becomes doubly refractive, a characteristic useful in identifying heat-treated lenses or points of stress. Skylight is partially polarized (by scattering) as confirmed by viewing it through polarized film. Polarized sunglasses eliminate the linear waves reflected from large expanses of snow or water. Because their direction is horizontal, the spectacle polarizers are vertical. During the brief popularity of "3-D" movies, millions of polarizers found their way into the hands of movie patrons, and we make use of them clinically to measure stereopsis, malingering, and suppression and in other binocular techniques of refraction. The human eye is a poor detector of polarization (even the lowly bee does better), the exception being Haidinger brushes, an entoptic phenomenon useful as a practical test of macular integrity.

LIGHT SOURCES

The eye evolved in sunlight, its sensitivity is tied to it, and daylight is the most important natural light source. To plague the photometrist and photographer, its character varies with weather and season; hence artificial sources provide the standard.

When the first electric illuminant appeared, there was considerable concern of ocular injury; indeed London oculists petitioned for a law against its unshaded use. Although we still prefer candles for their romantic glow if not their luminous efficiency, electric sources have become universal.

Artificial light sources may be thermal or luminescent. Thermal sources dissipate most of their radiant energy as heat. Luminescent sources are more efficient, but their bluish cold colors make them less popular.

A solid heated to incandescence turns red, orange, and yellow until finally it is "white hot." The color quality of the emitted radiations depends on temperature. A high-color temperature means that there are more blue radiations and a low-color temperature means that the red ones predominate. Daylight is equivalent to a color temperature of 25,000° K. (It is, of course, meaningless to say that daylight *is* that color temperature.) The color temperature of an ordinary incandescent bulb is equivalent to 2750° K, and the candela, the new international candle, is equivalent to 2041.7° K (the temperature at which molten platinum solidifies).

When a current is passed through hydrogen, it glows a purple color. The color is due to four bright lines (there are others outside the visible range), which always occupy the same spectral position. Similarly, sodium salt thrown into a Bunsen burner vaporizes with a characteristic yellow color. Potassium glows red, and other gases, other colors. These line spectra delineate the elements—the portraiture of spectrochemical analysis.

When Newton pointed his prism at the sun, he saw consecutive colors. About 1820 Fraunhofer, aiming his spectroscope in the same direction, was startled to find several hundred dark lines. He labeled the major ones alphabetically. The lines result from the absorption (by scattering of certain frequencies) by atmospheric gases and provide a convenient method of identifying a particular spectral region. In evaluating the dispersion of glass, for example, we compare its effect on wave bands from the middle and from each end of the spectrum. The wavelengths are identified by their Fraunhofer letters C, D, and F.

In Bologna, circa 1603, a part-time alchemist, Vincenzo Cascariolo, found some unusual white stones, which in the tradition of the times, he tried to convert into gold. By pulverizing and heating with coal, he obtained a porous cake that to his surprise glowed in the dark. The glow was phosphorescence, and stones with such magic properties were eventually named phosphorus ("light bearer"). The substance now produces more magic in television tubes than Cascariolo ever dreamed.

Ordinary pigments and dyes appear colored by selective reflection; fluorescent pigments produce color by emission as well as by reflection. Clothes washed in certain detergents appear "whiter than white" because fluorescent particles convert the sun's ultraviolet light to visible rays. *Fluorescein*, a fluorescent dye, has numerous clinical applications from outlining corneal abrasions to retinal angiography. The peak excitation frequency of fluorescein is approximately 500 nm, and the emission is approximately 530 nm; thus a blue filter is used over the source, and a green filter is used for observation (or photography).

Phosphorescence depends on a more leisurely return of electrons to lower energy levels. The electrons, held in metastable states, make the radiation persist for some time—the basis of the television image. Sequential reproduction is feasible because the eye is unaware that the picture is being assembled piecemeal. Color television is a technologic tour de force; electron guns excite phospors of the primary colors, the "lenses" are electrical and magnetic fields, and the "rays" are electron trajectories.

Laser optics

Luminescence results from the spontaneous emission of many photons by many atoms. If an appropriate source of atoms, such as a ruby crystal, is powerfully irradiated by light of the right frequency, many electrons occupy a higher energy level and can release their photons simultaneously. The result is a laser beam of extraordinary intensity. The waves from ordinary light sources proceed with the inefficiency of a disorganized mob; the waves produced by lasers all march in step with destructive consequences befitting their Teutonic discipline. Lasers are used in ophthalmology for photocoagulation (retinal detachment, tumor and vascular obliteration). Because they damage skin and eye tissue, great care is needed in handling them; the hazard applies to physician and patient.

THE MEASUREMENT OF LIGHT

The official definition of light, by people who make their living measuring it, is radiant flux evaluated with respect to its capacity to evoke visual brightness. Radiometric units are easy to understand; the physicist measures radiations in ergs or joules per time (that is, in watts) and cares not at all whether there is anyone to see them. It is the photometrist's units that cause all the worry because he wants to measure brightness and relate it to cost. Instead of radiant flux, the photometrist speaks of luminous flux. Why must radiant flux be redefined before one can see it?

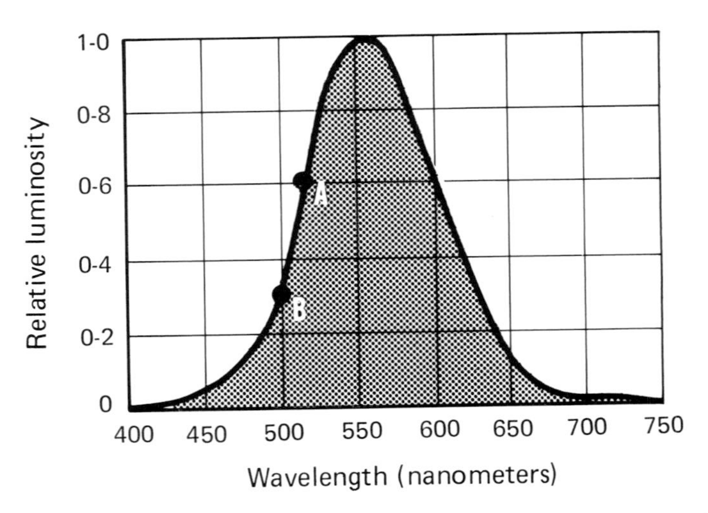

Fig. 1-8. Photopic luminosity represents the relative effectiveness of different wavelengths to produce a given brightness in a "standard" observer after the radiant flux is equalized for all wavelengths (equal energy spectrum). Note that the curve does not represent brightness variations in the visible spectrum. Thus it takes twice as much radiant energy to make *B* appear the same brightness as *A*, but *A* is not twice as bright as *B*.

Because equal amounts of radiant energy do not produce equal effects on the eye. The visual response is bell shaped, peaking at 555 nm and falling off on either side into a sea of darkness. The curve, called the "photopic luminosity curve," gives the rate of exchange between radiometric and photometric units (Fig. 1-8). Photometric units therefore are only meaningful with reference to a human observer; hence we are told the units are psychophysical.

Psychophysics was invented by Fechner, who equated stimulus and response and started a controversy that still divides psychologists. The problem is that the eye is poor at discriminating absolute brightness, although it can detect 1% differences. If one watt of wavelength A yields half the lumens of one watt of wavelength B, it is not seen half as bright. The photometrist has to obtain his measurements backward, that is, by evaluating how much energy is needed to make A equally as bright as B. The results of such data for all visible wavelengths yield the photopic luminosity curve. The least required radiant energy is the peak of the curve and corresponds to 555 nm. The number of lumens produced by 1 watt of 555 nm is based on an arbitrary standard, which was a sperm candle, is now the candela, but could just as well have been a firefly if that were reproducible.

The photometrist's observer must not only have an "average" eye, but this eye must also be light adapted; otherwise the measurements are meaningless. Unfortunately the exact adaptation level has never been specified. The photopic luminosity curve values, fixed by decree of an international commission, liberates photometry from disconcerting individual variability.* The number of lumens corresponding to peak sensitivity has changed from time to time and will probably change again; the current value as established in 1972 is 685 lumens per watt at 555 nm, which corresponds to 100% on the relative luminosity curve.

All photometric terms are preceded by the adjective luminous, with the luminosity curve providing the conversion. One merely multiplies the radiometric value by the spectral luminous efficiency for the wavelength in question. The Illuminating Engineering Society in its most recent handbook has taken some admirable steps toward simplifying the almost unbelievable confusion of photometric units.

Radiant energy is energy traveling in the form of electromagnetic waves. The term "radiation," widely used, should be reserved for a process rather than a quantity. The unit of radiant energy is the joule or erg. *Luminous energy* is the photometric equivalent of radiant energy and applies only to the visible portion of the electromagnetic spectrum. The unit of luminous energy is the lumen-second.

Radiant power (or radiant flux) is radiant energy transferred per unit time. It is measured in power units (work/time), hence joules/sec or watts. *Luminous power* is the quantity of radiant power that produces visual sensations in a human observer. The unit of luminous power is the lumen, defined as the luminous power emitted within a unit solid angle (or steradian) by a point source of luminous intensity of one candela. Thus a point source of uniform radiation with a luminous intensity of one candela emits a total of 4π lumens. At a wavelength of 555 nm, 1 lumen is equivalent to 0.00147 watts; or 1 watt at 555 nm is equivalent to 685 lumens. To convert watts to lumens for other portions of the spectrum, multiply by the appropriate luminous efficiency (from the photopic luminosity curve).

Radiant intensity is the radiant power from a point source per unit solid angle. *Luminous intensity* is the photometric equivalent of radiant intensity. The unit of luminous intensity is the candela.

Radiance is the radiant power leaving a surface per unit solid angle and unit projected area of that surface. The unit of radiance is the watt per steradian and square centimeter. *Luminance*, the photometric equivalent, is the quotient of luminous flux leaving the surface. The unit is the candela per square meter (or nit). In illuminating engineering the luminance of a surface in a given direction is expressed as lumens/unit area. Thus 1 lumen/square foot is said to have a luminance of 1 foot lambert.

*The scotopic spectral sensitivity curve is based on the known absorption characteristics of rhodopsin, but it is only a matter of time before the absorption spectrum of the conglomerate cone pigments is completely determined. The photopic curve will then have a quantum photochemical basis, and the photometrist will be able to drop the "psycho" part of his psychophysical definition. It is doubtful whether the photometric system is partly psychologic in the meantime.

Irradiance is the radiant power incident upon a surface per unit area. The unit is the watt/square cm. *Illuminance* is the luminous power incident on a surface. The unit of illuminance is lumens/square foot (or footcandle), or lumens/square meter (lux), or lumens/square centimeter (phot).

The illuminance of a surface follows the inverse square law:

$$\text{Illuminance} = \frac{\text{Luminous intensity}}{\text{Distance squared}}$$

A source of 100 candela 5 feet directly above a table gives 100/25 or 4 footcandles illuminance. If the table is tilted 60°, it varies with the cosine of the angle; hence it is reduced to half. The light entering the eye, however, also depends on reflectance, for example, whether the table is covered by a black or white cloth. Reflectance is often ignored in the standard 7 footcandles specified for perimetry without any hint as to whether the target should be matt, lacquered, or luminescent.

The illuminance of the retinal image does not follow the inverse square law, since its size diminishes by the square of the distance. The retinal illuminance is therefore independent of the distance of the source (Chapter 6).

In 1729 Pierre Bouguer described the first scientific method of measuring light. If two neighboring pieces of the same kind of paper, each normally illuminated, appear equally bright, intensity is proportional to the square of the distances. One has only to measure the distance of the unknown source, which produces the same illuminance as some standard. If the sources have different colors, the hue is ignored by a technique involving critical flicker fusion frequency, which depends only on intensity (heterochromatic photometry). Photometric measurements are now made mostly by photoelectric detectors, calibrated, of course, to read like the standard eye. Photoelectric cells do not suffer fatigue, frustration, or frontal headaches.

The function of the eyes depends on adequate lighting. Proper illumination contributes to visual comfort and efficiency. The first code on lighting issued by the Illuminating Engineering Society in 1915 is periodically revised. Intensity receives the greatest emphasis, but spectral distribution has its effects as in color discrimination. The illuminating engineer tries to control glare, diffusion, distribution, and direction of light. Criteria used to evaluate adequate lighting include photophobia, visual thresholds, speed and accuracy of performance, blink rate, heart rate, and psychologic preference. Frequently, the conclusion is that existing illumination is inadequate, and more light would be, if not absolutely necessary, at least more pleasant. There is no real evidence that poor lighting damages the eye, although it is a common belief.

Some recommended lighting*

Task	*Footcandles*
Auditoriums	15
Automobile assembly	200
Bank lobby	50
Chemical laboratory	50
Church	15
Church (with special zeal)	30
Court room	70
Dance hall	5
Engraving	200
Food service	50
Glass inspection	200
Hospital corridor	20
Autopsy table	1000
Eye examining room	50
Intensive care room	100
Close work laboratory	100
Operating table	2500
Waiting room	20
Office	150
Printing	100
Residence dining	15
Kitchen	150
Reading	70
Bedroom	±10
Entrance	5

QUANTUM OPTICS

Until the turn of the present century, physics assumed that energy was continuous, but attempts to analyze the energy distribution in the spectrum predicted an ultraviolet catastrophe. In 1900, by some trial and error juggling, Planck found the equation that solved the problem. To make theory agree with fact, Planck postulated that each oscillator gives off energy, not continuously, but in small packages called "quanta." His formula indicated that there is a minimum

*Approximate recommendations of the Illuminating Engineering Society.

amount of energy that cannot be further subdivided, just as there is an ultimate subdivision of matter. The implications were so profound that at first even Planck believed quanta represented only pulses of electromagnetic waves, not a new conception of the waves themselves. The full significance did not become clear for another five years.

In 1905 Einstein published three papers. One on the photoelectric effect proposed nothing less than an uncompromising corpuscular view of light. Radiation itself is quantized and travels as particles of energy during emission, as well as during absorption. The energy is fixed for each light frequency; the intensity regulates the number of quanta and not their energy. The particles behave like a torpedo whose explosion is independent of the distance at which it is launched. The more torpedoes the more likely the hit, but the explosive force is the same because each carries its energy with it. Some low-frequency light particles have little explosive force just as some torpedoes are duds. Red light has little effect on a photographic plate, ultraviolet light causes skin burns, and x rays are even more potent. Cosmic rays, having the highest frequency, also have the highest quantal energy. And how do these discrete photons dance the uninterrupted tune of undulation? Their behavior is somewhat like individual pulses of force applied to a child's swing, yielding a harmonic periodicity. More than fattening a statistical curve by weight of numbers, light is wavelike and particle-like at the same time, each aspect representing different categories of observation. Here was a theory that Newton would have approved. To Pope's epitaph:

> Nature and Nature's law, lay veiled in Night.
> God said "Let Newton be" and all was Light.

A modern wit now added:

> It could not last, the Devil howling "Ho"!
> "Let Einstein be," restored the status quo.

Waves or particles? The controversy remains unsettled, and nearly any dogmatic statement is likely to be proved wrong. Bohr's principle of complementarity is a modus vivendi. The principle states that an experiment that allows observation of one aspect prevents observation of the other just as one cannot study morbid pathology and physiology in the same specimen. To observe wave properties we use our wave eye, and to study photons, our corpuscular eye, but the microscope is always monocular so we can never use our binocular vision. In practical optics, we deal with large magnitudes, and the wave theory works beautifully, but if one intends to play with photochemistry, luminescence, lasers, or elementary particles, quantum theory determines the rule of the game. This rule laid down by Heisenberg and Born is called the "principle of indeterminacy."

REFERENCES

Arey, L. B., and Bickel, W. H.: The number of nerve fibers in the human optic nerve, Anat. Rec. **58**:supp., 1935.

Becker, S. C.: Unrecognized errors induced by present-day gonioprisms and a proposal for their elimination, Arch. Ophthal. **82**:160-168, 1969.

Beckman, H., Barraco, R., Sugar, H. S., Gaynes, E., and Blau, R.: Laser iridectomies, Am. J. Ophthal. **72**:393-402, 1971.

Bouma, P. J.: Physical aspects of color, Eindhoven, 1947, Philips Industries.

Bragg, W.: The universe of light, New York, 1959, Dover Publications, Inc.

Cline, B. L.: Men who made a new physics, New York, 1965, The New American Library, Inc.

Cogan, D. G.: Some ocular phenomena produced with polarized light, Arch. Ophthal. **25**:391-400, 1941.

Connes, P.: How light is analyzed, Sci. Am. **219**:72-82, 1968.

Descartes, R.: La dioptrique, Leyden, 1638.

Ditchburn, R. W.: Light, ed. 2, New York, 1963, Interscience Publishers, John Wiley & Sons, Inc.

Efron, A.: Exploring light, New York, 1969, Hayden Book Co, Inc.

Einstein, A., and Infeld, L.: The evolution of physics, New York, 1938, Simon & Schuster, Inc.

Feather, N.: Vibrations and waves, Middlesex, England, 1964, Penguin Books, Ltd.

Fresnel, A.: Complete works, 2 vols., Paris, 1868.

Frisch, O. R.: Take a photon, Contemporary Physics **7**:45-53, 1965.

Gamow, G.: Biography of physics, New York, 1961, Harper & Row, Publishers.

Guillemin, V.: The story of quantum mechanics, New York, 1968, Charles Schribner's Sons.

Guth, S.: Luminance addition: general considerations and some results at foveal threshold, J. Opt. Soc. Am. **55**:718-722, 1965.

Haidinger, W.: Ueber das direkte Erkennen des polarischen Lichts und der Lage des Polarisationbene, Ann. Phys. Chem. **63:** 29-39, 1844.

Hardy, A. C., and Perrin, F. H.: The principles of optics, New York, 1932, McGraw-Hill Book Co.

Hardy, LeGrand H., and Rand, G.: Elementary illumination for the ophthalmologist, Arch. Ophthal. **33:**1-8, 1945.

Harvey, E. N.: A history of luminescence, Philadelphia, 1957, American Philosophical Society.

Heavens, O. S.: Optical properties of thin solid films, London, 1955, Butterworth and Co., Ltd.

Hoffmann, B.: The strange story of the quantum, New York, 1959, Dover Publications, Inc.

Huygens, C.: Treatise on light, translated by S. P. Thompson, New York, 1962, Dover Publications, Inc.

Jaffe, B.: Michelson and the speed of light, Garden City, 1960, Doubleday & Co., Inc.

Kapany, N. S.: Fiber optics, principles and applications, New York, 1967, Academic Press, Inc.

Kaufman, J. E.: Illuminating Engineering Society Lighting Handbook, ed. 5, New York, 1972, Illuminating Engineering Society.

Klein, M. V.: Optics, New York, 1970, John Wiley & Sons, Inc.

Lancaster, W. B.: Lighting standards, Am. J. Ophthal. **20:**1221-1231, 1937.

LeGrand, Y.: Spectral luminosity, Handbook Sensory Physiol. **7**(4):413-433, 1972.

L'Esperance, F. A.: The ocular histopathologic effect of krypton and argon laser radiation, Am. J. Ophthal. **68:**263-273, 1969.

Luckiesh, M., and Moss, F. K.: Seeing in tungsten, mercury and sodium lights, Trans. Illum. Eng. Soc. **31:**655-674, 1936.

Lytel, A., and Buckmaster, L.: ABC's of lasers and masers, Indianapolis, 1972, Howard W. Sams & Co., Inc.

Mach, E.: The principles of physical optics; an historical and philosophical treatment, translated by J. J. Anderson and A. F. A. Young, New York, 1925, Dover Publications, Inc.

Malus, E. L.: Optique, J. de l'Ecole Polyt. **7:**1-44; 84-129, 1808.

Maxwell, J. C.: A dynamical theory of the electromagnetic field, Proc. R. Soc. London **13:**531-536, 1864.

Meyer-Arendt, J. R.: Radiometry and photometry: units and conversion factors, Appl. Optics **7:**2081-2084, 1968.

Meyer-Arendt, J. R.: Introduction to classical and modern optics, Englewood Cliffs, 1972, Prentice-Hall, Inc.

Minnaert, M.: Light and colour in the open air, London, 1959, Clarke, Irwin & Co., Ltd.

Monk, G. S.: Light, principles and experiments, New York, 1963, Dover Publications, Inc.

Newton, I.: Opticks: or a treatise of the reflexions, refractions, inflexions and colours of light, 1704 (reprint), New York, 1952, Dover Publications, Inc.

Pierenne, M. H.: Optics, painting, and photography, Cambridge, 1970, Cambridge University Press.

Planck, M.: Theory of light, London, 1932, The Macmillan Co.

Riggs, L. A.: Light as a stimulus for vision. In Graham, C. H., editor: Vision and visual perception, New York, 1965, John Wiley & Sons, Inc., pp. 1-38.

Robertson, J. K.: Introduction to physical optics, New York, 1933, D. Van Nostrand Co., Inc.

Ruechardt, E.: Light, visible and invisible, Ann Arbor, 1965, University of Michigan Press.

Schapero, M., Cline, D., and Hofstetter, H. W., editors: Dictionary of visual science, Philadelphia, 1960, Chilton Book Co.

Schawlow, A. L.: Optical masers, Sci. Am. **204:**52-61, 1961.

Schurcliff, W. A., and Ballard, S. S.: Polarized light, Princeton, 1964, D. Van Nostrand Co., Inc.

Sears, F. W.: Optics, Reading, 1958, Addison-Wesley Publishing Co.

Tinker, M. A.: Illumination standards, Am. J. Public Health **36:**963-973, 1946.

Tolansky, S.: Revolution in optics, Middlesex, England, 1968, Penguin Books, Ltd.

van Heel, A. C. S., and Velzel, C. H. F.: What is light? New York, 1968, McGraw-Hill Book Co.

Walls, G. L.: The fundamental character of the photometric system, Am. J. Physics **20:**145-151, 1952.

Walsh, J.: Photometry, ed. 3, London, 1958, Constable & Co., Ltd.

Young, T.: A course of lectures on natural philosophy and the mechanical arts, 2 vols., London, 1807.

2

Basic optics

Optics deals with the origin, behavior, and effects of light. The physicist studies its source; the dermatologist, its erythemal potency; the painter, its color; and the philosopher, its implications. To the practicing ophthalmologist, optics is a tool for the diagnosis and treatment of refractive errors. It is in terms of its clinical application that "visual optics" is used in the title of this book. The dry facts of optics are a poor hors d'oeuvre for refraction unless, flavored with ocular physiology, we apply them in interpreting clinical tests, evaluating diagnostic criteria, and planning more effective treatment.

Optics has its own vocabulary, a symbolic shorthand, transposable into transparent English. Learning a new language and thinking with it is not achieved overnight by a single reading of a single book. It takes time. Clinicians who develop an antipathy to optics have probably never become comfortable with its language. If ophthalmic optics teachers ever hold a convention, their first order of business should be an attempt to standardize the plethora of symbols that give individuality to textbooks but mask the real simplicity of the subject. The beginner, faced by what appears to be a mountain of different formulas, decides optics is not for him and takes up something more practical like electron microscopy, histochemistry, or electrophysiology. Fortunately optics does not change, and once mastered it will last for a lifetime of practical use.

REFLECTION BY PLANE MIRRORS

The measurement of interpupillary distance and corneal curvature, the optics of the slit lamp and retinoscope, locating the optical center of a lens by superimposed reflections, and checking its surface for imperfections are a few clinical illustrations of reflection. Recent innovations include measurements of retinal photopigments, optical evaluations of retinal image quality, and, on a larger scale, construction of sail-like mirrors to power spacecrafts. A mirror explosion is currently in progress to harness solar energy.

The plane mirror image is the same size as the object, as far behind the surface as the object is in front, and reversed left to right but not up and down. The reversal or perversion of the image is not an optical conspiracy against common sense but has reference to a view of ourselves compared to a real person facing us. The mirror only reflects what is presented to it; a row of letters has to be turned toward the mirror in the first place. If a camel looks in, one cannot expect a chorus girl to look out.

Unlike lenses (or curved mirrors), a plane mirror does not reassemble light rays to a focus but only displaces their direction. Thus a plane mirror increases the optical distance of a short-refracting room. A reversed acuity chart is placed above the head of the patient who views its mirror image, although one must be tolerant of occasional patients who insist on turning their heads to look at the actual chart. A plane mirror rotated through an angle, displaces the image through twice the angle, thus acting as an optical lever. When backing an automobile, objects seen by the driver in the rearview mirror approach twice as fast. The field of view of a plane mirror is that of an open window; to see one's full length, the mirror need be only half one's height.

Images formed by plane mirrors are composed not of actual light rays but of their backward projections or construction lines. It means the rays entering the observer's eye appear as if they originated at the mirror image. Such an image is called a "virtual image." Since no real light rays participate in its formation, it cannot be caught on a screen. The real rays in turn enter the eye and are brought to a real focus on the retina by the ocular optical system.

Symbols and sign convention

A thick lens is a system of two thin lenses, a thin lens is a system of two single refracting surfaces, and a single refracting surface is an elaboration of the behavior of individual light waves. Symbols that change chameleon-like from one chapter to the next obscure the simpler relations within the more elaborate. The symbol F represents focal power, whether it be a mirror, lens, or eye. Image space symbols are object space symbols primed; thus l' is the image distance corresponding to the object distance l; f′ is the image focal length that logically accompanies n′ and h′, the image index and image size, respectively. By limiting the symbol F to power, the letter D is allowed to represent diopters, and thus one need not guess whether 4 D means four diopters or four times some power D. Subscripts (and Greek letters) are generally avoided; the context will indicate whether one is talking about focal power, focal length, or focal point. In optical diagrams, virtual rays are indicated by dotted lines and real rays by solid lines. The symbols and sign convention adopted in this book are those used in the references listed at the end of this chapter.

Of the various sign conventions, the Cartesian is not only simple but logical as well. Directions to the left and down are minus, and directions to the right and up are plus. Light rays moving left to right are plus. Convergent waves and converging lenses are plus; diverging waves and diverging lenses are minus. Distances are measured *from* the optical surface *to* the point in question. The focal power of a lens has the same sign as its second focal length. This convention is fairly standard in ophthalmic optics, but it is not universal; consequently one cannot select a formula out of any book indiscriminately.

REFLECTION BY SPHERICAL MIRRORS

Spherical mirrors have less practical utility than optical simplicity, so we use them to introduce some important optical concepts. Two spherical mirrors are shown in Fig. 2-1, each with a center of curvature C, a vertex (or pole) P, and a radius, r. The curvature is the reciprocal of the radius; the longer the radius the flatter the mirror. In all optical diagrams, curvature and distances are greatly exaggerated and are to be taken schematically. The line passing through the pole and center of curvature is the optic, or principal, axis. The light direction is assumed to be from left to right. Optical distances are measured from the surface to the point in question and are considered positive or negative relative to the light direction. For example, in the concave mirror the radius is minus. In all optical diagrams in this book the sign to be attached to the distance is indicated by a unidirectional arrow.

An object O at a distance l is imaged at I, a distance l' from the mirror according to the mirror equation:

$$\frac{1}{l'} + \frac{1}{l} = \frac{2}{r}$$

An object placed 20 cm from a concave mirror of 10 cm radius is imaged at $1/-20 + 1/l = 2/-10$; $l' = -6.6$ cm. The minus sign means that the object is to the left (that is, in front) of the mirror. Note that in our sign convention, object distances generally have a minus sign.

To visualize the light rays and check computation, it is useful to draw a diagram as indicated in Fig. 2-1. Drawings are a thinking tool. Formulas are often meaningless unless we get their feel in terms of the behavior of light rays. The diagram makes use of two or three critical rays to locate the image. Ray *1*, aimed at the center of curvature, actually passes through it in the concave mirror; only its projection passes through the center of curvature in the convex mirror. Since this ray coincides with the normal, the angle of

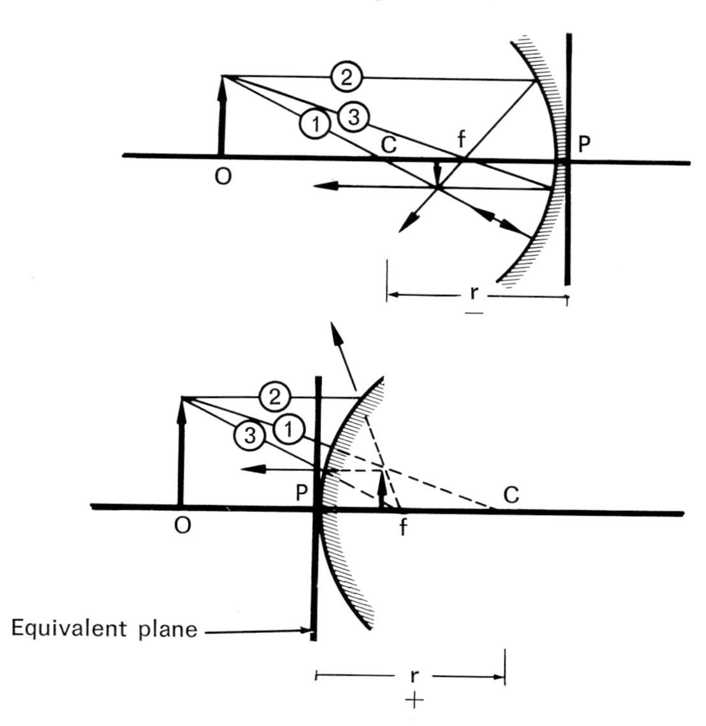

Fig. 2-1. Reflection at spherical mirrors showing the three rays most generally useful in finding the character and position of the image. The concave mirror *(top)* converges light and forms real images; the convex mirror *(bottom)* diverges light and forms virtual images. Distances are labeled from the mirror surface: plus if in the same direction as the light (always assumed to be from left to right) and minus if in the reverse direction. Note that in this diagram aberrations are ignored.

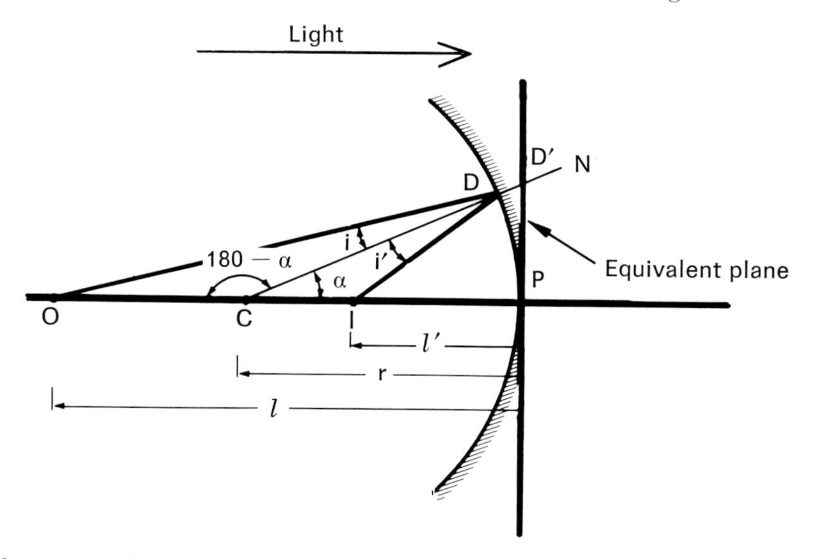

Fig. 2-2. A diagram used to derive the paraxial equation for spherical mirrors. The region from *D* to *P* is assumed to be so near the optic axis that its curvature is negligible. This leads to the fundamental simplifying concept in clinical optics—the paraxial theory—in which an equivalent plane replaces the surface and aberrations may be ignored. Although the point *D* does not appear to be very near the optic axis, this is simply an excuse for legibility. All optical diagrams are arbitrarily exaggerated, and it is supposed to be understood that all the rays shown really belong to the paraxial region.

incidence is zero, and the ray is reflected on itself. Any ray that is reflected without deviation should certainly be included among our construction rays. Ray 2, parallel to the optic axis, is reflected toward f, the focal point. The focal point is real in the concave mirror and virtual in the convex mirror. In the convex mirror a parallel incident ray is reflected as if coming from f. Since light is reversible, we can turn ray 2 around to get ray 3, which is reflected parallel to the optic axis.

If all three rays originate from the top of an object, like the arrow shown in the diagram, their intersection locates the top of the image. Since both object and image rest on the optic axis, any ray coincident with the principal axis has simultaneously the characteristics of the other three rays combined. It is reflected on itself and passes (really or virtually) through the focal point and the center of curvature.

Ray tracing is neat but tedious. The formula we started with is certainly quicker and more accurate. Let us see how it is derived—not as an exercise in high school algebra—but to understand a fundamental optical concept: the paraxial theory.

Fig. 2-2 illustrates a concave spherical mirror of radius *r*, pole *P*, and center of curvature *C*. An arbitrary ray from an axial object *O* strikes the mirror at *D* and is reflected towards the image *I*. *I* is the image of *O*. The angle of reflection i' is equal to the angle of incidence i at the normal drawn from *C* through *D*. The object distance is l, the image distance is l', and all distances in the diagram are negative.

From the law of sines:

$$\frac{\text{CO}}{\sin i} = \frac{\text{CI}}{\sin i'} \text{ and } \frac{\text{DO}}{\sin (180 - \alpha)} = \frac{\text{DI}}{\sin \alpha}$$

Since $\sin i = \sin i'$ and $\sin (180 - \alpha) = \sin \alpha$:

$$\frac{\text{CO}}{\text{DO}} = \frac{\text{CI}}{\text{DI}}$$

And since $\text{CO} = l - \text{r}$ and $\text{CI} = \text{r} - l'$:

$$\frac{l - \text{r}}{\text{DO}} = \frac{\text{r} - l'}{\text{DI}}$$

Although there is nothing in the figure by which to measure DO and DI, suppose the curvature of the mirror in the region from D to P is very small, so small it may be neglected. (It may not look small in the diagram because, like all optical diagrams, it is schematic and exaggerated.) If we pretend the point D lies in the region just above or below the optic axis, that is, the paraxial region, then D and D′ become coincident, and the problem is solved because if the arc PD equals the line PD′, then DO equals PO, and DI equals PI. Since $\text{PO} = l$, and $\text{PI} = l'$:

$$\frac{l - \text{r}}{l} = \frac{\text{r} - l'}{l'}$$

$$\boxed{\frac{1}{l} + \frac{1}{l'} = \frac{2}{\text{r}}}$$

This paraxial formula applies without further modification to either convex or concave spherical mirrors. For convex mirrors the radius is plus, and for concave mirrors the radius is minus, according to our sign convention.

Paraxial formulas eliminate unnecessary detail and simplify computations. Returning to Fig. 2-2, it is apparent that if the point D is assumed to be very close to the optic axis, then the angle of the ray must be very small. The mathematician tells us that the value in radians (the arc DP) for such small angles is equal to their tangents (the line D′P). To be even more elegant (to impress other mathematicians), he would say we neglect the higher orders of the trigonometric expansion; that is, one can round off some numbers and drop the terms beyond the first. This is why elementary optics is often called "first-order optics."

All clinical optics is first order, based on paraxial theory; hence one can add or subtract trial lenses by simple algebra and leave the eccentricities of aberrations to the lens designer. The justification for the paraxial simplification is that eyes for which lenses are prescribed have a pupil that limits rays to a central area of practical, if not of mathematic, paraxiality. But we must not be surprised if telescopic mirrors are parabolic, and problems may arise not predicted by

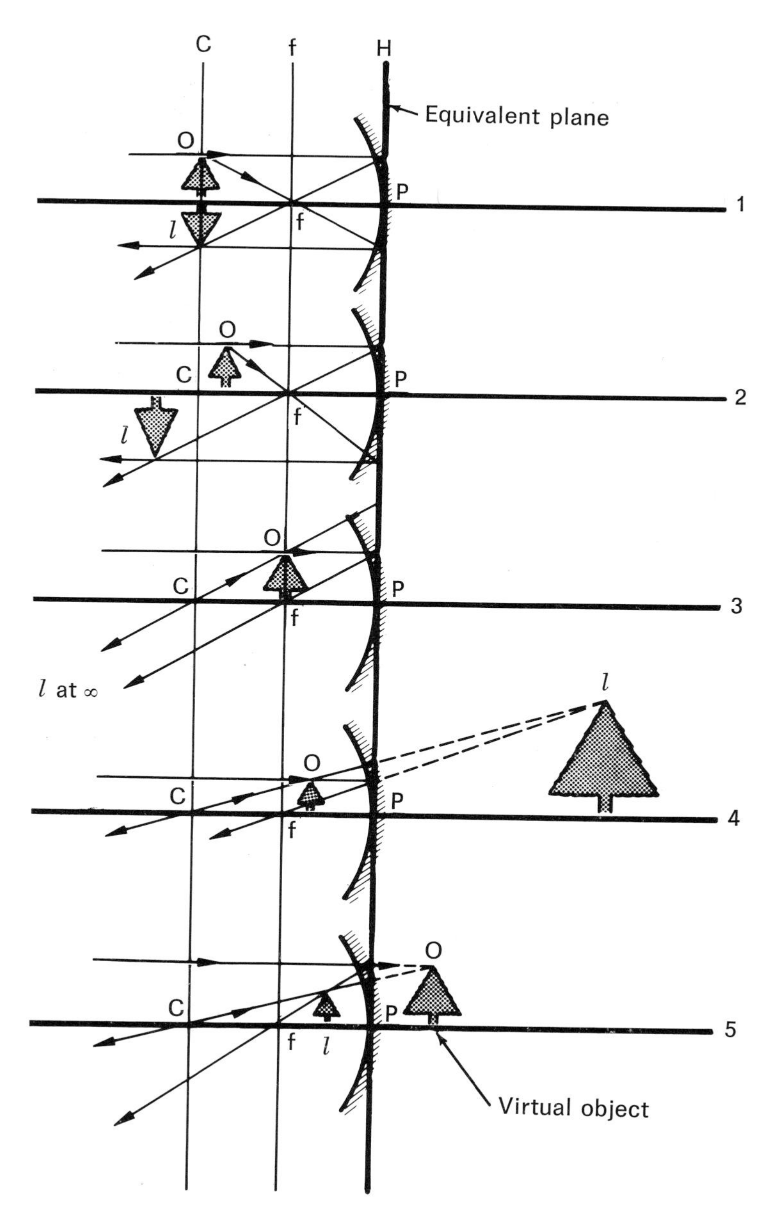

Fig. 2-3. Sequence of ray tracings for a concave spherical mirror. Constructions are based on the equivalent plane—the curved surfaces are shown only for esthetics. Note that an object at the focal point gives an image at infinity; an object inside the focal point gives a virtual image. The last sketch shows rays heading toward a virtual object.

simple formulas when patients look through the periphery of strong spectacles.

By assuming that all rays in our optical diagrams are paraxially confined by some invisible aperture, the curved surface can be replaced by a plane tangent to it. Fig. 2-3 shows such equivalent planes which are represented by lines intersecting the plane of the paper. The equivalent plane can be drawn as tall as necessary to make the diagram legible, providing we understand that all rays are really paraxial. It must not be concluded that the spherical mirror has been replaced by a plane mirror; the plane is only optically equivalent. The same applies to lenses, which are often sketched as vertical lines with an arrow to indicate whether they are plus or minus. The line represents a "thin" lens—the equivalent plane of a real one.

The paraxial mirror equation predicts that an infinitely far object (incident rays parallel) gives an image at half the radius—the focal length of the mirror. Conversely, an object at the focal point results in an image at infinity. Unlike lenses, spherical mirrors only have one focal point. The German term "Brennpunkt" literally means burning point and emphasizes that paper placed at the focal point of a concave mirror aimed at the sun will be ignited. (Of course, it will not work for a convex mirror, which has a virtual focal point.) A mirror with an infinitely long radius has, according to the paraxial equation, equal object and image distances but on opposite sides. This is what was previously concluded about plane mirrors.

Spherical mirrors magnify; witness the Palomar telescope. The linear, or lateral or transverse magnification compares image size (h′) to object size (h):

$$m = \frac{h'}{h} = -\frac{l'}{l}$$

The magnification ratio is a pure number. A plus sign indicates that the image is upright, and a minus sign indicates that it is inverted. A ratio greater than unity means magnification; a ratio less than unity, minification.

The business of a spherical mirror is to change the vergence of the incident light waves by its optical, or focal, power. Focal power, measured in diopters, is the reciprocal of focal length, measured in meters. Curvature and focal power are sometimes confused because unfortunately both are expressed in the same dioptric units. In the standard sign convention, converging optical systems are plus and diverging systems, minus. Since concave mirrors converge light, they are given a plus power; hence the formula includes a minus sign:

$$F = -\frac{2}{r} = -\frac{1}{f}$$

This formula then applies to either convex or concave mirrors. The confusion arises because of the tendency to relate concave curves with minus power; true for refracting surfaces, not for mirrors.

The cornea, for example, is a powerful convex mirror. Its radius is only 7.7 mm, and its catoptric focal power is as follows:

$$F = -\frac{2}{0.0077} = -259.7 \text{ D}$$

Notes on problem solving

Although we approach optics with a minimum of mathematics and a maximum of examples, solving an occasional quantitative problem is still the best way to confirm one's grasp of principles. A schematic ray tracing diagram (free hand will do) will indicate if the answer makes sense. Particular attention should be given to signs, fundamental in optics because they tell in which direction to measure. Answers can also be checked by dimensions; diopters can be substracted from diopters but not from inches. Proper numbers in proper places will give the answer in appropriate phrases. A convenient and handy optical laboratory is the refracting room, with its trial lenses and prisms. Almost all elementary optical theory can be verified with a homemade optical bench. Above all, one should reason out the answer rather than substitute blindly in a formula, which can always be looked up. One clear concept is worth a hundred hazy formulas.

REFRACTION AT PLANE SURFACES

The simplest and certainly the oldest example of refraction occurs at the surface of quiet pools of water. We look into the pool and see a fish displaced closer to the surface. Although patients are not examined under water, and even zoologists are indifferent to piscatorial presbyopia, plane surfaces illustrate some of the most important principles of refraction.

In Fig. 2-4 a fish *O* in a pool of water of refractive index *n* is viewed directly from above.* The rays entering the pupil are, for all practical purposes, paraxial. Since the eye is in air, the first medium (n) is water, and the second (n') is air. It follows from Snell's law that $l'/n' = l/n$; the object distance di-

*If the angle of observation is too oblique, the rays are so far displaced from the paraxial region that each is refracted a different amount, resulting in gross distortion.

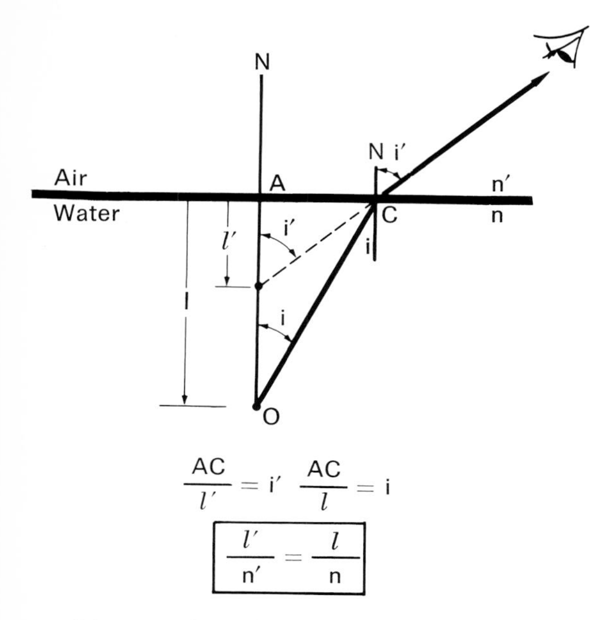

Fig. 2-4. The refraction at a water-air interface serves to develop the paraxial equation for plane surfaces. The obliquity of the angle of view is exaggerated; rays striking the eye are supposed to belong to the area adjacent to the optic axis *ON*. Although the image *I* appears to be in water, it is actually formed by backward projections of the air rays. Such an image is said to be "virtual." The object, on the other hand, is physically and optically in water. It is fundamental in optics that objects belong to object space and images to image space. In this example, object and image spaces are not only optically but also physically different. With lenses surrounded by air, object and image spaces are only optically separate.

vided by the water index equals the image distance divided by the air index. (Note that primed and unprimed values go together.) This is the paraxial equation for plane refracting surfaces.

A submerged submarine located 100 meters below the surface of the ocean (n = 4/3) will appear to be $\frac{100}{4/3}$ or 75 meters below the surface. Aside from the trivial arithmetic—or the military implications—the example illustrates two important optical concepts: object and image space and reduced curvatures and distances.

The image *I* appears to be physically in and surrounded by water. Actually the image is in air because it represents the backward projections of the air rays. The image is a kind of phantom, made-up of construction lines rather than real rays; in short, it is virtual. The real rays enter the eye, which sees them as if they were coming from I. It should be recognized that an object placed 75 meters below the waterline will not give an image at 100 meters but at $\frac{75}{4/3}$ or 56 meters. To retrace the light path, one must follow the optical rules, which apply to real rays and not their projections. One cannot assume an image 75 meters away is identical to an object at the same location. The object and all the rays before refraction belong to object space; the image and all the rays after refraction belong to image space. This example illustrates it neatly because the two spaces are both optically and physically different. In dealing with lenses, both object and image are generally in air, which is simultaneously object and image space. The optical difference exists, however, even when the media are physically identical. This is an important difference but difficult to grasp and frequently misunderstood.

The paraxial equation, by which we computed the location of the image ($l/n = l'/n'$), states that object and image distances modified by the index of their respective media are equal for plane surfaces. The actual distances divided by the index are termed "reduced distances" because they have been reduced to the common medium of optics—air. Although the actual distances (100 and 75 meters) differ, the reduced distances $\frac{100}{4/3}$ and $\frac{75}{1}$ are identical. It means light traveling 100 meters in water is equivalent to light traveling 75 meters in air. The light is slowed in water and hence must go farther to equal its proficiency in air. Indeed any optical distance in any medium is made equivalent when divided by the index of that medium. One of the first lessons of arithmetic is that apples cannot be compared to oranges; we are taught to reduce fractions to a common denominator before we are allowed to play with them. The common denominator of optics is air because refractive index is defined

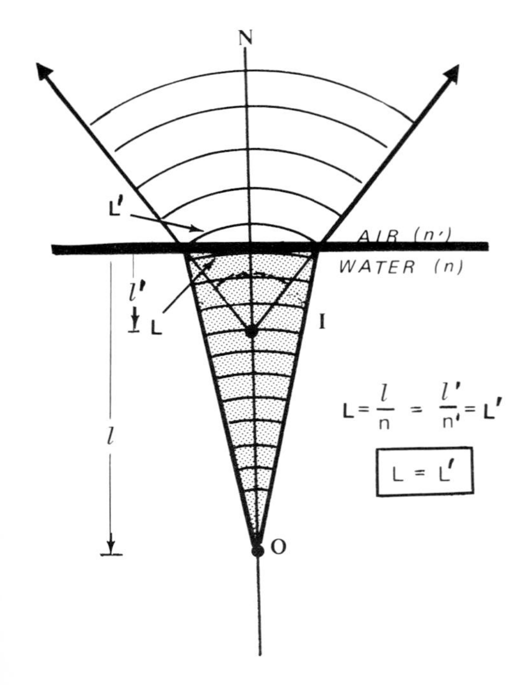

Fig. 2-5. This is a duplicate of Fig. 2-4 (not to scale), showing wave fronts rather than light rays. Note that there is a change in curvature as the wave passes from water to air; the curvature is greater in the latter. But if we modify the curvature by the index of the medium in which it travels, the two modified curvatures, or "reduced vergences," are identical; L′ = L. This modification of optical distances, curvature, and focal lengths by the index of its respective medium means that these quantities are reduced to a common denominator of optics—which is always air because that is how refractive index is defined. So universal is this reduction that we sometimes forget that a 1 D lens gives a 1 meter focal length only if the lens is in air. Although most lenses *are* in air, the one we remove in the operating room is not. When dealing with the optics of the eye, the index is usually that of water.

as the ratio of the speed of light in a given medium compared to air.

All practical optics formulas are based on reduced distances and reduced focal lengths. For example, a +1 D lens focuses rays at 100 cm but only if the lens is in air. When placed in water, its power differs. Focal power is not simply 1/f but n/f, and the index cannot always be ignored. We become so used to dealing with lenses surrounded by air that we may forget the unity in the numerator is a substitution for the index of air. When light enters the eye, whose index differs, the simple reciprocal will give wrong answers.

Fig. 2-5 is a duplicate of the previous figure in terms of waves rather than rays. The waves pass from water to air as if coming from *I*. We see the curvature of the air, and water waves are altered by the boundary. Why then do we say the boundary has no power? It is said to have no power because reduced curvature remains unchanged, the reduced curvature being the reciprocal of the reduced distance. The actual curvature of the wave in water just striking the surface is $\frac{1}{100}$, and that of the refracted wave is $\frac{1}{75}$, but the reduced curvatures (actual curvature multiplied by its respective refractive index) are $\left(\frac{1}{100}\right) \times (1.33)$ and $\left(\frac{1}{75}\right) \times (1.00)$, which are identical. Again, the definition of optical power is based on reduced curvature, or "vergence," a noncommittal term for either divergence or convergence.* Only "reduced" vergence is important in clinical optics, and the adjective is to be understood whenever the term is used. The symbols for vergence are L and L′, representing object and image space, respectively; $L = n/l$, and $L' = n'/l'$. For plane refracting surfaces the reduced vergences are equal: L = L′.

Real and virtual images

The differences between real and virtual images are often misunderstood, so let us summarize their characteristics. A virtual image can be seen as in a plane mirror but cannot be caught on a screen. It represents the intersection of construction lines that is, projections rather than light rays. A virtual image becomes a virtual object for the eye whose optical system converts it into a real retinal image. Any real image can be caught on a screen, although it need not be (an example is the aerial image of indirect ophthalmoscopy). Although the retinal image is real, it is not always clear, as in ametropia. The retinal image is trans-

*The concept of vergence, that is, reciprocals of linear units, was introduced by the astronomer Herschel in 1827. The more useful notion of "reduced vergence" we owe to our very own optical hero, Gullstrand.

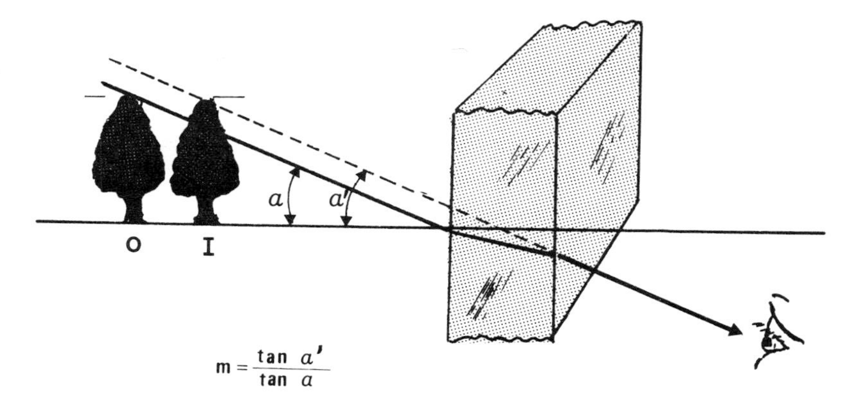

Fig. 2-6. Light rays passing obliquely through a slab of glass are displaced (not deviated), according to the thickness of the plate. The displacement results in angular magnification *(m)* because the image appears nearer than the object. The visual angle without the plate is α and with the plate α'; the ratio is the magnification. This is the reason for front surface mirrors in the refracting room and haploscope. A ray perpendicular to the plate of glass is neither deviated nor displaced but still refracted; that is, it is slowed but unbent.

mitted to the brain, which sees no images at all; it sees perceptual objects. Optical images, real or virtual, belong to image space; optical objects, real or virtual, belong to object space. Optical images are always physical; perceptions are always subjective.

When light rays pass through a slab of glass, there are two interfaces and two refractions, the second equal and opposite to the first. The emerging ray is parallel to the original but laterally displaced (Fig. 2-6). The amount of displacement depends on the angle of incidence and the thickness of the plate. The displacement for distant objects is minor, but near objects appear magnified because of the larger angle subtended at the eye. This angular magnification* results when one is looking not only through plate glass windows, car windshields, and display dials but also through plano lenses that have thickness. And if the thickness differs for the two eyes, they may induce clinical aniseikonia.

*Various kinds of magnification clutter up the optical vocabulary. The only two of clinical importance are linear magnification, which compares the lengths of image to object, and angular magnification, which compares the visual angle subtended at an observer's eye with and without a visual aid.

REFRACTION AT INCLINED PLANE SURFACES: PRISMS

A prism is essentially two angulated plane refracting surfaces. The angle between them is the apical angle, and a perpendicular dropped from the apex to the base points in the direction of the prism as specified clinically. Tracing a light ray through a prism involves only the application of Snell's law to each surface in turn, and all prism problems can be solved this way. Fig. 2-7 shows an incident ray *AB* striking the first surface at an angle i_1 and refracted at an angle i'_1. The ray proceeds to the next face, which it meets at an angle i_2 to finally emerge at an angle i'_2. The relation between these angles and the apical angle of the prism can be deduced from a fortunate geometric characteristic of the figure: the intersection of the two normals form an angle equal to the apical angle (α), which is the exterior angle of a triangle, and consequently $\alpha = i'_1 + i_2$. If we know the incident angle and the index of the glass, the emerging angle is readily computed.

Since no ray can be normal to both prism surfaces simultaneously, the final ray is deviated. The deviation is the sum of the deviations at each surface. The primary optical duty of a prism is to deviate light rays. It can

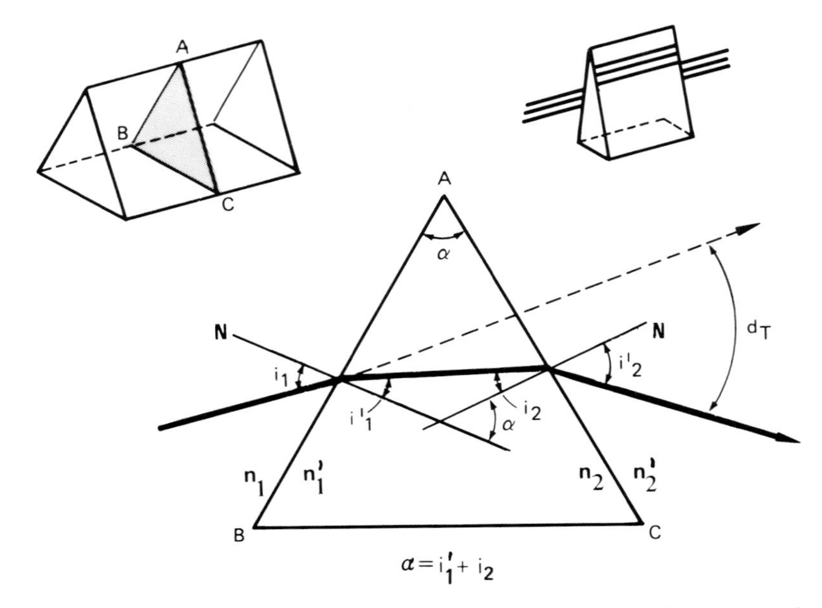

Fig. 2-7. The path of a light ray through a prism follows Snell's law applied to each surface. At the first surface, $n_1 \sin i_1 = n'_1 \sin i'_1$. Once the refracted angle at the first face (i'_1) is known, i_2 follows from the relation to the apical angle: $i_2 = \alpha - i'_1$. Applying Snell's law a second time ($n_2 \sin i_2 = n'_2 \sin i'_2$, we get the final angle i'_2. The total deviation d_T is the sum of the deviations produced at each interface: $d_T = i_1 + i'_2 - \alpha$. Since the index of the prism is generally greater than the surrounding medium, the light is deviated toward the base. For thin ophthalmic prisms, the sines are proportional to the angles, and if held so one face is normal to the line of sight, the deviation in degrees is simply $(n - 1)\ \alpha$.

be shown experimentally and proved by the calculus that the deviation is at a minimum when the ray traverses the prism symmetrically. If deviation is plotted against incident angle, the graph is U shaped, with symmetrical passage falling at the bottom of the U.

Most ophthalmic prisms are very thin; indeed the most important are not prisms at all but the prismatic deviations produced by lenses. For thin prisms, the minimum deviation is $d = (n - 1)\ \alpha$, and if of glass index 1.5, it reduces to $d = \alpha/2$. Hence the deviation of a thin prism is about half the apical angle (in degrees). The angle of incidence is not included in the previous formula, it being assumed that the prism is held before the eye so that one of the faces is perpendicular to the visual axis. Unfortunately a patient looking through a prism generally turns his eye toward various objects in the visual field; the rays now pass not only asymmetrically but eccentrically. The deviation fluctuates and so does the patient's stomach. Thus, although prisms are optically simple, they are therapeutic dynamite, and prudent practitioners only prescribe them for a precise purpose.

The apical angle was formerly the method of calibrating prisms. Jackson at one time recommended the angle of minimum deviation, which soon led to confusion between apical degrees and degrees of deviation. About 1890 Prentice proposed the much more useful prism diopter (symbol Δ). In Fig. 2-8 the angle α represents the angular deviation of a thin prism. The linear deviation is 1 cm/100 cm or 1% of the distance.

$$\tan \alpha = \frac{AB}{BO} = \frac{1}{100}$$

Hence for small angles, the angle whose tan is 0.01 is 35 minutes or about half a degree. One degree therefore represents approximately two prism diopters, and one prism diopter deviates a ray 1 cm at 1 meter.

Although prism diopters are based on angular deviations, we gratefully discard trigonometry for centimeters; in fact, we like it so much that we use prism diopters to desig-

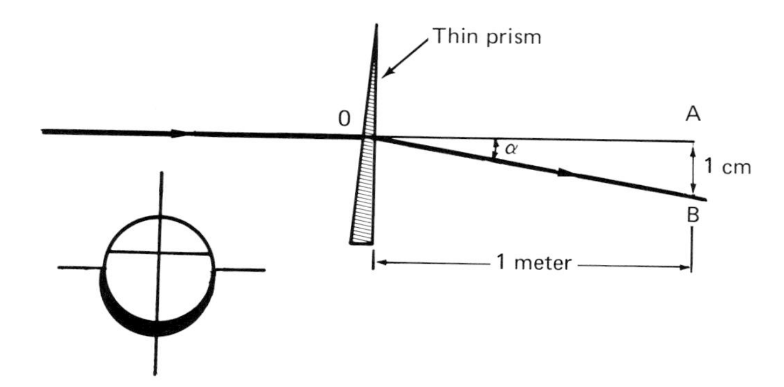

Fig. 2-8. Prisms used in ophthalmic optics are generally thin, with the most important ones not being prisms at all but the prismatic deviation resulting when the line of sight does not pass through the optical center of a lens. For such purposes, the prism diopter (Δ) has many advantages; for example, it can be directly related to diopters of focal power or ocular rotations. For thin prisms the angular deviation in degrees that corresponds to a linear deviation of 1 cm at 1 meter is tan 0.01 or about a ½°. It follows that 1° is equivalent to 1.74 Δ. The orthoptist who measures strabismus in degrees of deviation will get about half the value as the clinician using loose prisms calibrated in prism diopters—a difference of some consequence in planning a surgical correction.

nate lens decentrations and ocular rotations. Prisms do not alter light vergence, and therefore a diopter of prism power differs from a diopter of optical power. One must specify prism diopter for accuracy if not for euphony and brevity.

REFRACTION AT SINGLE SPHERICAL SURFACES

Every lens has two surfaces, with a refraction at each in succession; hence a single surface is the optics of "half" a lens. If we understand the refraction at one surface, the behavior of the lens as a whole follows. In addition to being a way station to lens theory, single refracting surfaces also help explain the optics of the eye. The human eye consists of a series of surfaces, not completely spherical, separated by nonhomogeneous media, not exactly centered on a common axis. It is much too complex for clinical computations; consequently we make use of a simplified model to explain to ourselves the optics of ametropia, the workings of refractive tests, and the effects of visual aids. In this model eye, all refracting surfaces are reduced to a single spherical surface, and all the media are reduced to a common medium, which is generally water. Logically the model is called a "reduced eye," and optically it is nothing more than a single spherical surface. For example, a contact lens acts by replacing this surface with a new one.

Finally, in building an understanding of the optics of lenses, the eye, and lenses applied to the eye, single spherical refracting surfaces are the basic construction module. Indeed the formulas for single surfaces are the most important in all clinical optics because they reappear, practically unaltered, in those for lenses, lens systems, and visual optics.

The paraxial equation for single spherical surfaces can be derived along the same lines as that for spherical mirrors (see appendix, pp. 34 and 35).

$$\frac{n'}{l'} - \frac{n}{l} = \frac{n' - n}{r}$$

The formula states that object distance *(l)* and image distance *(l′)* are related by a constant. The constant includes the radius and the *difference* in index (sometimes called "refractivity") on the two sides. Therefore an infinite number of conjugate object–image relations are connected by this constant. For example, an object placed 20 cm from a convex spherical surface of radius 5 cm, separating air from glass (index 1.5), gives an image (Fig. 2-9) shown on p. 28.

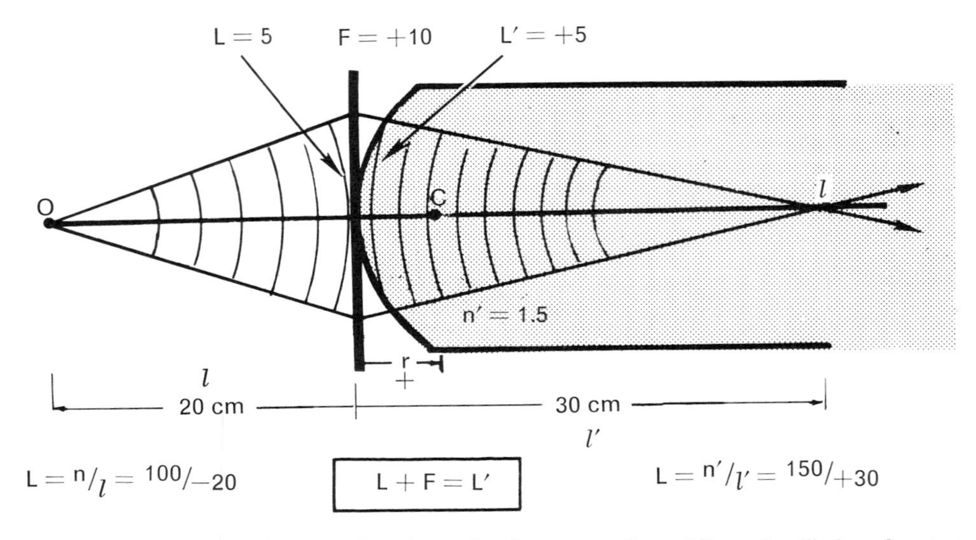

Fig. 2-9. The refraction of light at single spherical refracting surfaces follows Snell's law, but instead of using trigonometry, the position of the image can be computed much more simply from the paraxial equation. We simply assume all the rays involved belong to the region immediately surrounding the optic axis (again the diagram is arbitrarily exaggerated for legibility). Its clinical application hinges on the fact that the front surface of the eye behaves, for most practical purposes, like a single refracting surface separating air from water. Contact lenses "work" because they replace the cornea by a new optical surface.

$$\frac{1.5}{l'} - \frac{1}{-20} = \frac{0.5}{+5};\ l' = +30 \text{ cm}$$

The plus sign means the image is to the right of the surface (the standard light direction from left to right is assumed). The constant $\frac{0.5}{+5}$ dictates that an object at 5 cm will form a virtual image 15 cm to the left of the surface.

When the rays originate from a distant object (l is infinitely large and $1/l$ is zero), the image distance is then the second focal length (f′). When the refracted rays are parallel, the object must be at the first focal point (f). Hence:

$$\frac{n' - n}{r} = \frac{n'}{f'} = -\frac{n}{f}$$

"First" and "second" are better terms than "anterior" and "posterior" because the relative location of focal points depends on the sequence of indices (Fig. 2-10). For example, light passing from a denser medium, such as from glass to air, represents the sequence at the second surface of a lens.

EXAMPLE: The radius of curvature of the Donders reduced eye model is 5 mm, separating air from water (n = 4/3). What are the first and second focal lengths of this model eye?

$$\frac{n'}{f'} = -\frac{n}{f} = \frac{n' - n}{r};\ \frac{4/3}{f'} = -\frac{1}{f} = \frac{4/3 - 1}{.005}$$

$$f' = +20 \text{ mm};\ f = -15 \text{ mm}$$

The paraxial formula for single spherical refracting surfaces is similar to that for mirrors.* Unlike mirrors, however, spherical refracting surfaces have two focal points, differing in sign and numerically unequal because they are in the same ratio as the indices. Lenses, on the other hand, are generally surrounded by the same medium; thus their focal lengths are opposite in sign but numerically identical.

*The similarities between reflecting and refracting surfaces can be explained by assuming that mirrors reflect light "backward" instead of refracting it "forward." Since the second medium for mirrors is the same as the first, we can substitute (−n) for n′ in the refraction equation, the minus sign indicating reversal. The paraxial mirror equation then follows directly by substitution in the spherical refraction equation. No memorizing is necessary, and the confusion of sign differences is avoided.

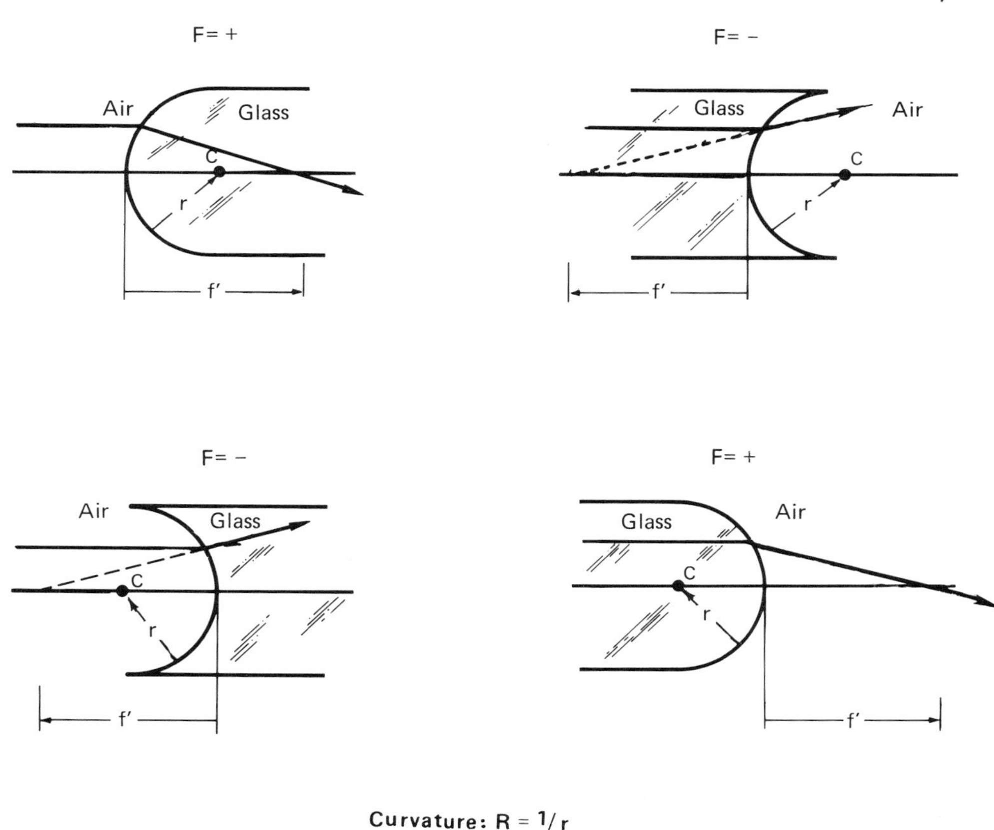

Fig. 2-10. Diagrams illustrating the importance of refractivity (i.e., index difference) as contrasted to curvature. The curvature of all four surfaces is the same, based on the radius, but the optical power—the bending effect on light—depends on whether the rays pass from a medium of lower or higher index. Thus the *rear* surface of the cornea actually has a minus power, since its index is greater than that of aqueous.

The optical power (F) of single spherical surfaces is the reciprocal of focal length:

$$F = \frac{n'}{f} = -\frac{n}{f}$$

It also depends on curvature and index difference:

$$\boxed{F = \frac{n' - n}{r}}$$

Optical power can thus be defined as the product of refractivity $(n' - n)$ and curvature (R) where $R = 1/r$:

$$F = (n' - n)\,(R)$$

The cornea is the most powerful optical component of the eye not because its radius is so small but because the index difference from air is so large. For example, Thomas Young showed that ocular accommodation was lenticular by proving it was not corneal. The approximate focal power of the cornea in air is $F = \frac{n' - n}{n}$; $\frac{1.376 - 1}{0.0077}$ $= +48.8$ D, but when the eye is immersed in water ($n = 4/3$), its power is reduced to $\frac{1.376 - 1.333}{0.077} = +5.6$ D. Since he could still focus on near objects under water, Young demonstrated that accommodation could not be corneal.

EXAMPLE: A concave refracting surface separates glass from air. The light passes *from glass* (index 1.5) to air. If the radius of curvature is 40 cm, what is the optical power?

$$F = \frac{n' - n}{r} = \frac{1 - 1.5}{-0.4} = +1.25\ D$$

If the sequence is reversed, the light passes *from air* into glass:

$$F = \frac{n' - n}{r} = \frac{1.5 - 1}{-0.4} = -1.25\ D$$

The example illustrates that although curvature (R) remains constant $\left(\frac{1}{0.4} = 2.50\ D\right)$, the focal power varies according to the sequence of media. For example, in some of the literature a 60 D convex lens is used to represent the model eye. This is misleading because it ignores the difference in media.

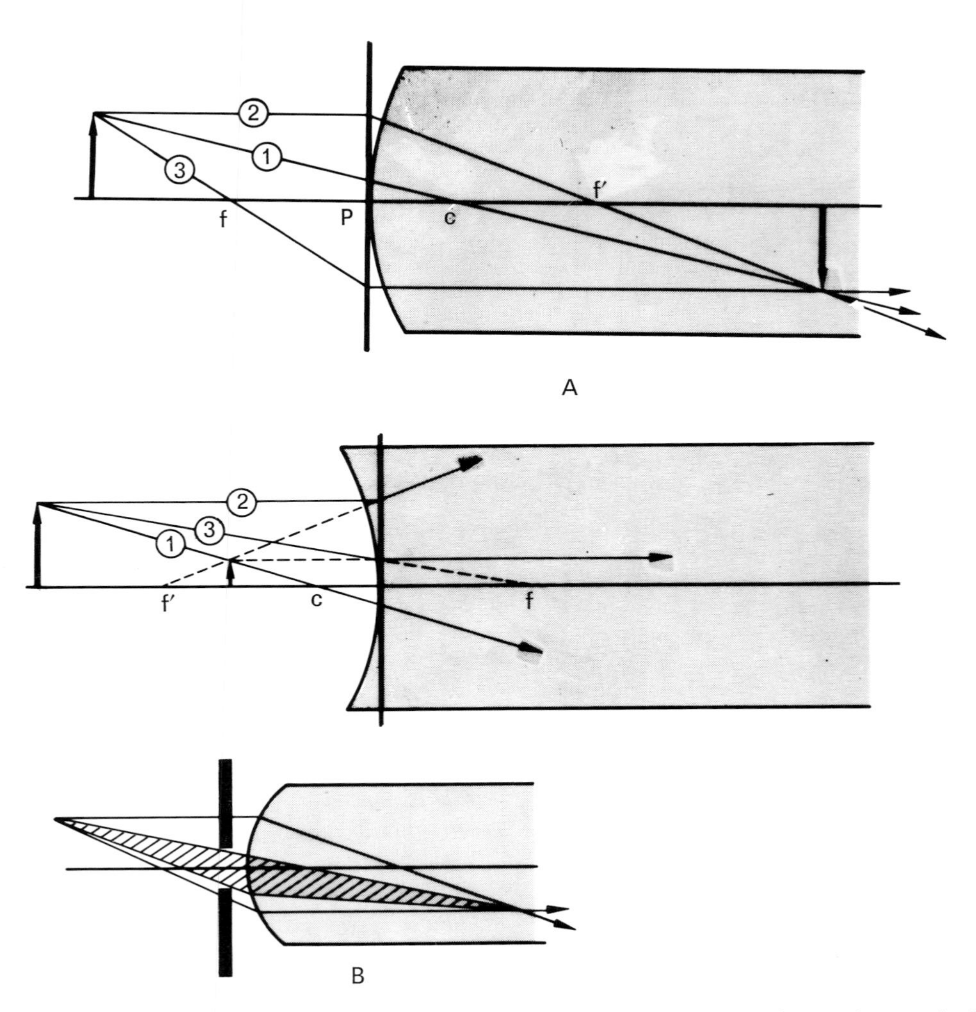

Fig. 2-11. Ray tracings for convex and concave single spherical refracting surfaces. The standard three construction rays are used to locate the position of the image. They need not even participate in forming the image, as shown in **B** where they are cut off by the aperture. The construction will still locate the image, providing all rays are assumed to be paraxial. Note that the distance from C to f' is identical to the distance from P to f. The center of curvature for single refracting surfaces is its "nodal point" because the rays through it are undeviated.

The second focal length of a 60 D lens is $\frac{1000}{60} = 16.67$ mm, but the focal length of the model eye is $\frac{1333}{60} = 22.22$ mm.

Clinically the curvature of a lens is measured with a lens clock, or lens measure; the curvature of the cornea, with a keratometer. Although both instruments really measure curvature, they are calibrated in diopters of focal power based on an assumed refractive index: 1.53 for the lens measure and 1.3375 for the cornea. To get the actual curvature one has to uncalibrate the reading by dividing it by the index.

Ray tracings are useful to diagram the passage of light through single spherical refracting surfaces (Fig. 2-11). Tracings for single spherical refracting surfaces, as for mirrors, represent only paraxial rays, even though they are arbitrarily enlarged vertically for legibility. Indeed the curved surface should really be represented by an equivalent plane tangent to the pole because the rays are confined to a region where curvature is assumed to be negligible. (We show the curve only for esthetics.) The diagram need not even represent actual light rays. For example, in Fig. 2-11, *B*, the construction rays are cut off by the aperture but still serve to locate the position and character of the (paraxial) image.

Some representative ray tracings for a convex surface are shown in Fig. 2-12. As the object approaches the surface, the image recedes from the second focal point and reaches infinity to the right when the object arrives at the first focal point on the left. When the object moves inside the first focal point, a virtual image reappears from left infinity. What happens between right and left infinity? Since the rays are parallel, they focus (or originate) in either direction; the emerging rays are convergent, become parallel, and finally diverge. The position of the image simply follows the vergence of the refracted rays. This sequence should be carefully studied, since it duplicates the movement of the far point of an ametropic eye when corrected by lenses.

The second focal point is the closest point

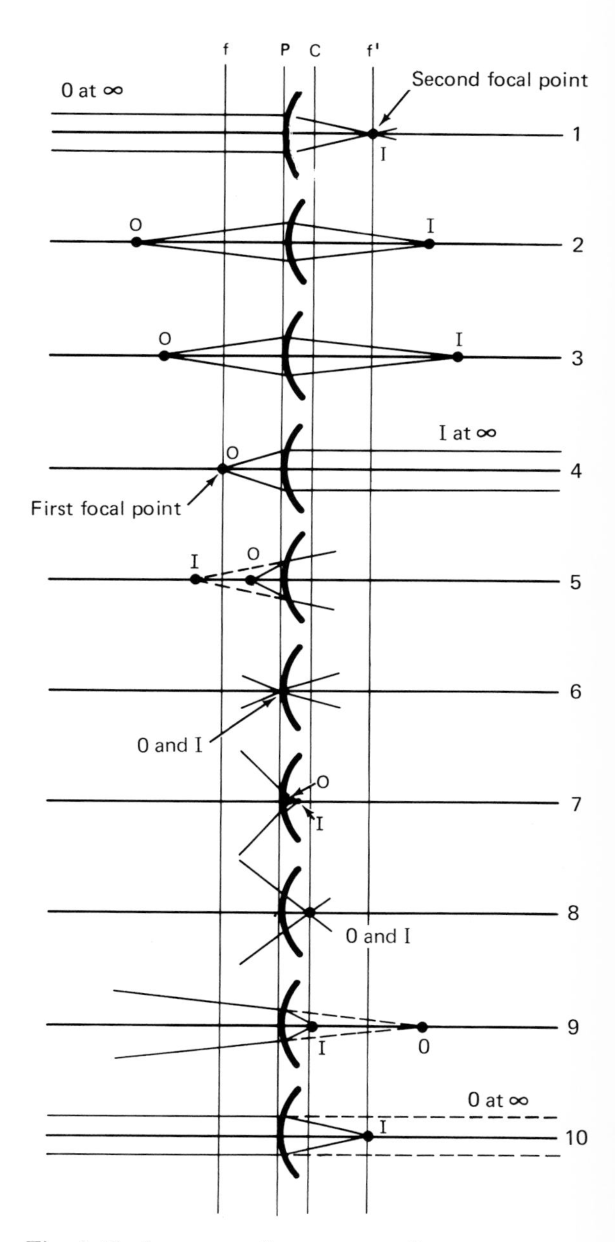

Fig. 2-12. Sequence of ray tracings for a convex single refracting surface showing the disposition of the image as the object is brought nearer to it. If the surface represents the eye, the sequence is identical to the movement of the far point when lenses are placed in front of it in subjective testing or retinoscopy. By studying sequences we get a glimpse of the optical dynamics; the ideal way to learn geometric optics is probably with moving cartoons!

a real object can be imaged. When the object reaches the surface, the image catches up with it. If the object is virtual, the image is real, and the two coincide again at the center of curvature. A virtual object at right infinity (or left infinity) forms an image at the second focal point, which exhausts all possibilities. When object and image are equidistant from the center of curvature, the magnification is −1; when both are at the surface, the magnification is +1.

THE CURVATURE OF WAVE FRONTS: VERGENCE

Vergence* is a noncommittal term for either converging or diverging light waves and includes both curvature and index. The symbols $L = n/l$ and $L' = n'/l'$ describe the vergence of the waves in the media in which they travel. The advantage of vergences as contrasted to linear distances is that vergences can be added or subtracted directly. We need only specify the vergence of the wave at the moment of impact. For example, the addition of −2 D to +3 D gives the answer +1 D, even in a semiconscious state, but a problem to find the image of an object 50 cm from a lens of 33⅓ cm focal length is a puzzle in algebraic fractions. However, they are the same problem if converted to vergences: an object 50 cm from the lens has an incident vergence of −2 D, which when combined with a +3 D lens gives a final vergence of +1 D. The wave focuses at +100 cm, which is the position of the image (Fig. 2-13). The advantage of this method in clinical optics is that waves are mostly in air, and their vergence is the simple reciprocal of their radius; $L = 1/l$ and $L' = 1/l'$. Only when waves enter the eye or a single refracting surface is it necessary to substitute an index other than air. Moreover the distances involved in clinical refraction are generally round numbers whose reciprocals are obvious from mental computation (Table 2-1).

*The term "vergence" is also used but in an entirely different sense, for binocular rotations. It is therefore necessary to distinguish between the vergence of eyes and the vergence of light waves. Generally, this is obvious, but sometimes it leads to confusion as in the phrase "vergence-induced accommodation."

The paraxial equation for single spherical refracting surfaces can be put directly into the vergence form:

Linear form: $$\frac{n'}{l'} - \frac{n}{l} = \frac{n' - n}{r}$$

Vergence form: $$L' - L = F$$

The vergence formula is used whenever lenses are added to a trial frame, in neutraliz-

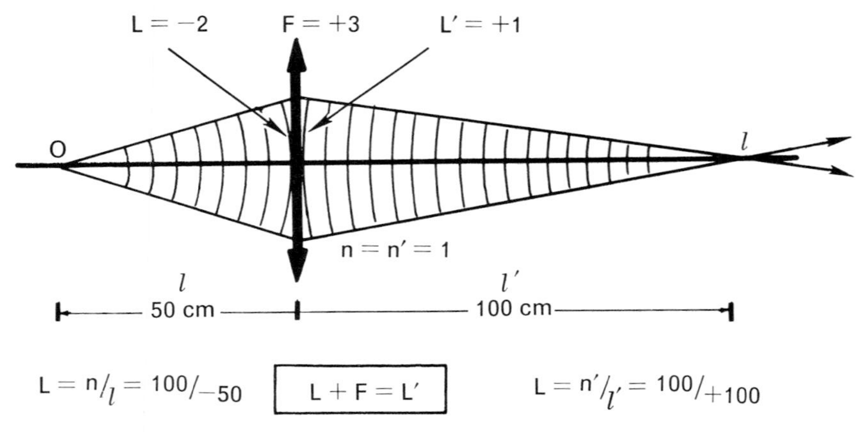

Fig. 2-13. The beauty of the (reduced) vergence method of computation is illustrated by this diagram. The position of the image is found by simply adding the incident vergence to the vergence of the optical instrument (a lens in this case) to obtain the emerging vergence. If the system is in air, the reciprocal of the final vergence gives the linear distance of the image. The same principle is used when we add a +4 D contact lens to correct an eye whose power is 56 D, making it 60 D emmetropic.

Table 2-1. Equivalent of diopters in inches and millimeters

Diopters	*mm*	*Inches*	*Diopters*	*mm*	*Inches*	*Diopters*	*mm*	*Inches*	*Diopters*	*mm*	*Inches*
0.125	8000	314.96	2.00	500	19.68	4.75	211	8.29	9.00	111	4.37
0.25	4000	157.48	2.125	471	18.53	5.00	200	7.87	9.50	104	4.14
0.375	2667	105.00	2.25	444	17.50	5.25	191	7.50	10.00	100	3.94
0.50	2000	78.74	2.375	422	16.58	5.50	182	7.16	10.50	95.4	3.75
0.625	1600	62.99	2.50	400	15.75	5.75	174	6.85	11.00	90.9	3.58
0.75	1333	52.49	2.625	383	15.00	6.00	167	6.56	11.50	88.1	3.42
0.875	1142	45.00	2.75	364	14.32	6.25	160	6.30	12.00	83.3	3.28
1.00	1000	39.37	2.875	349	13.69	6.50	154	6.06	13.00	76.9	3.03
1.125	905	35.00	3.00	333	13.12	6.75	148	5.84	14.00	71.4	2.81
1.25	800	31.50	3.25	308	12.11	7.00	143	5.62	15.00	66.7	2.62
1.375	737	28.63	3.50	286	11.25	7.25	138	5.44	16.00	62.5	2.46
1.50	666	26.25	3.75	267	10.50	7.50	133	5.25	17.00	58.8	2.32
1.625	615	24.23	4.00	250	9.84	7.75	129	5.08	18.00	55.5	2.19
1.75	572	22.50	4.25	236	9.26	8.00	125	4.92	19.00	52.6	2.07
1.875	534	21.00	4.50	222	8.75	8.50	118	4.63	20.00	50.0	1.97

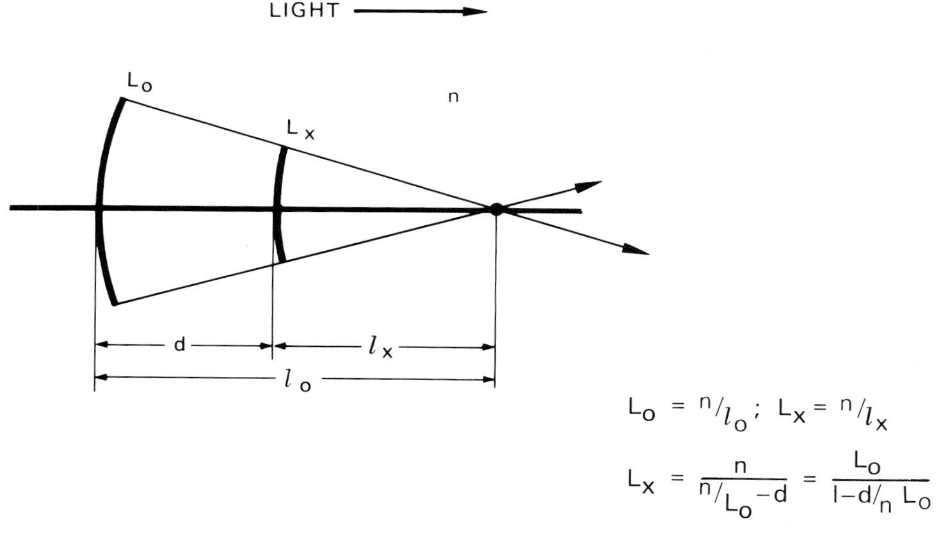

$$L_o = n/l_o;\quad L_x = n/l_x$$

$$L_x = \frac{n}{n/L_o - d} = \frac{L_o}{1 - d/n\, L_o}$$

Fig. 2-14. Diagram illustrating that the vergence of a wave in a given medium can be directly computed from the initial vergence and how far it has traveled. The principle is applied whenever a lens is placed in front of the eye at the spectacle plane; the wave leaving the back of the lens is altered by the time it reaches the cornea. Changing the "power" of a lens by moving it nearer or farther from the eye is another application of effectivity.

ing an unknown lens, or in comparing two spectacle prescriptions. It is perfectly general, being applicable to lenses, in which $L = 1/l$ and $L' = 1/l'$; to single surfaces, in which $L = n/l$ and $L' = n'/l'$; or to mirrors if $(-n)$ is substituted for n'. Whether the optical instrument is a single lens or a complex telescopic system, if the incident vergence is L and the emerging vergence is L′, the focal power of the system is the algebraic difference between them. Conversely, for a plane boundary such as the surface of a pool of water, $F = 0$, and $L' = L$, an expression previously derived.

The duty, purpose, and function of a refracting surface, or lens, is to change the incident vergence, that is, to impress a new vergence on the incident wave front. If the change is toward divergence, it has a minus optical power; if toward convergence, a positive optical power.

Instead of the vergence of wave fronts immediately, before or after refraction, we may wish to know the vergence at any arbitrary point. Fig. 2-14 shows a converging wave traveling in medium n. At position O the vergence L_o is n/l_o, and at position X the vergence L_x is n/l_x. If the distance between O and X is d, the new vergence is given by the following equation:

$$L_x = \frac{L_o}{1 - \frac{d}{n} L_o}$$

For example, it takes more accommodation to see an object clearly at 25 cm than at 50 cm. An object moved from 25 cm to 50 cm requires 2 D less accommodation because the vergences are −4 D and −2 D, respectively. This is why the beginning presbyope pushes small objects further from his eye; it can handle only the lesser vergence. Note that in computing these results from the effectivity formula, d must be in meters, since the vergence is in diopters. The index is that of air—unity.

APPENDIX

Derivation of paraxial equation for single spherical refracting surfaces

In Fig. 2-15, *A* shows the trace of a three-dimensional spherical refracting surface, which is the arc of a circle in the plane of the paper. Its center of curvature is C, and, since this is a convex surface, the radius is positive

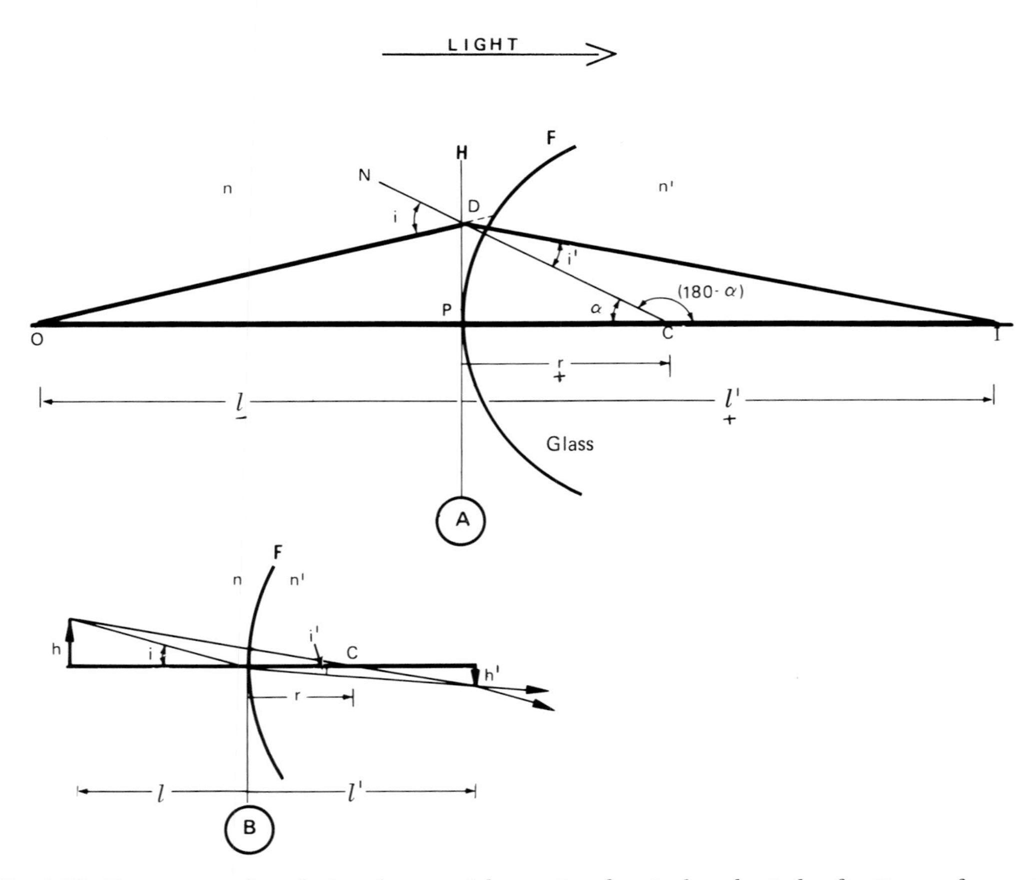

Fig. 2-15. Diagram used to derive the paraxial equation for single spherical refracting surfaces—the single most important equation in clinical optics.

measured from the vertex *P* to *C*. The surface separates two media of index *n* and *n'*; n' is greater than n in this example.

A ray from the axial object *O* strikes a point *D* on the equivalent plane of the curved surface. The equivalent plane, represented in the figure by a line, is drawn tangent to the curve at *P*. The point *D* is assumed to belong to the paraxial region; thus the difference between it and the surface proper is negligible. The refracted ray intersects the optic axis at *I*, which is the image of *O*. The object distance PO is *l*, and PI is *l'*. A radius from *C* drawn through *D* provides the normal around which we plot the angles of incidence and refraction, *i* and *i'*.

In triangle ODC

$$\frac{OC}{OD} = \frac{\sin(180 - i)}{\sin \alpha} = \frac{\sin i}{\sin \alpha}$$

and OC = $-l$ + r; and OD = $-l$ because OD = OP. Therefore:

$$\frac{-l + r}{-l} = \frac{\sin i}{\sin \alpha}$$

In triangle IDC

$$\frac{DI}{CI} = \frac{\sin(180 - \alpha)}{\sin i'} = \frac{\sin \alpha}{\sin i'}$$

and CI = l' − r, and DI = l'; therefore:

$$\frac{l'}{l' - r} = \frac{\sin \alpha}{\sin i'}$$

Multiplying the terms from each triangle:

$$\left(\frac{\sin i}{\sin \alpha}\right)\left(\frac{\sin \alpha}{\sin i'}\right) = \left(\frac{-l + r}{-l}\right)\left(\frac{l'}{l' - r}\right)$$

$$\frac{\sin i}{\sin i'} = \frac{-ll' + l'r}{-ll' + lr}$$

But sin i/sin i' = n'/n according to Snell's law; hence:

$$n'/n = \frac{-ll' + l'r}{-ll' + lr}$$

$$-nll' + nl'r = -n'll' + n'lr$$

Dividing by $-ll'r$:

$$\frac{n}{r} - \frac{n}{l} = \frac{n'}{r} - \frac{n'}{l'}$$

This may be rearranged to give the formula:

$$\frac{n'}{l'} - \frac{n}{l} = \frac{n' - n}{r} \text{ or } L' - L = F$$

This is the standard form of the paraxial equation for single spherical refracting surfaces—the most important equation in elementary optics.

The following expression for magnification (m) can be derived from Fig. 2-15, *B:* tan i = h/l and tan i' = h'/l'. For small angles, the tangents are equal to the angles in radians; hence h/l = i and h'/l' = i'. From Snell's law, ni = ni'. Substituting equivalent values:

$$n\left(\frac{h}{l}\right) = n'\left(\frac{h'}{l'}\right)$$

$$m = \frac{h'}{h} = \frac{nl'}{n'l}$$

REFERENCES

Bennett, A. G.: The use and abuse of approximations, Optician **125**:349-351, 378-381, 1953.

Bennett, A. G.: Emsley and Swaine's ophthalmic lenses, London, 1968, Hatton Press.

Bennett, A. G., and Francis, J. L.: Visual optics. In Davson, H., editor: The eye, vol. 4, New York, 1962, Academic Press, Inc.

Boeder, P.: An introduction to the mathematics of ophthalmic optics, Fall River, 1937, The Distinguished Service Foundation of Optometry.

Dennett, W. S.: A new method of numbering prisms, Trans. Am. Ophthal. Soc. **5**:422-426, 1888.

Emsley, H. H.: Visual optics, London, 1955, Hatton Press.

Fincham, E. F.: Optics, London, 1956, Hatton Press.

Fry, G. A.: Geometrical optics, Philadelphia, 1969, Chilton Book Co.

Gluck, I. D.: It's all done with mirrors, Garden City, 1968, Doubleday & Co., Inc.

Greguss, P., and Galin, M. A.: Holography, Ann. Ophthal. **4**:817-820, 1972.

Gullstrand, A.: Einfuhrung in die Methoden der Dioptrik des Auges des Menschen, Leipzig, 1911, S. Hirzel Verlag.

Herschel, J.: On light, an article in the Encyclopedia Metropolitana, London, 1840.

International Commission on Optics: Recommendations for the standardization of sign conventions and symbols in geometrical optics and for the definition of refractive indices of optical glass, J. Opt. Soc. Am. **41**:140-141, 1951.

Jalie, M.: The principles of ophthalmic lenses, London, 1967, Association of Dispensing Opticians.

Longhurst, R. S.: Geometrical and physical optics, New York, 1958, John Wiley & Sons.

Martin, L. C.: Geometrical optics, London, 1955, Pitman.

Morgan, M. W.: Distortions of ophthalmic prisms, Am. J. Optom. **40**:344-350, 1963.

Ogle, K. N.: Distortion of the image by ophthalmic prisms, Arch. Ophthal. **47**(2):121-131, 1952.

Prentice, C. F.: A metric system of numbering and measuring prisms, Arch. Ophthal. **19**:64-75, 128-135, 1890.

Riva, C.: New ocular fundus reflectometer, Appl. Optics **11**:1845-1849, 1972.

Smith, R.: A compleat system of opticks, Cambridge, 1738.

Struik, D. J.: A concise history of mathematics, ed. 3, New York, 1967, Dover Publications, Inc.

Tennant, E. R.: Workbook in geometrical and theoretical optics, Chicago, 1971, Illinois College Optometry.

Weale, R. A.: Vision and fundus reflectometry, Doc. Ophthal. **19**:252-286, 1965.

Zanker, A.: Nomograph for the calculation of the index of refraction of a prism, Appl. Optics **13**:2449, 1974.

3

Lens optics

A lens is a piece of glass or plastic whose optical behavior depends on the consecutive refraction at each surface. The light striking the first boundary is refracted, and the wave, now inside the glass, travels through its thickness to reach the second surface. The effective vergence of the wave is slightly altered by the thickness of the glass. Since most ophthalmic lenses are rather thin, we pretend this change is negligible, and the total power is simply the sum of its two surface powers.

If we assume that only paraxial rays are transmitted, each lens surface may be represented by an equivalent plane tangent to its pole (Fig. 3-1). If thickness is neglected, these two planes fuse into a single equivalent plane. This is the "thin lens," a conceptual model of a real lens, as much position as substance. It is the position at which a single refraction replaces two consecutive refractions at each surface, and into this position all the optical power is compressed. A thin lens suffers no peripheral aberrations and can be (indeed must be) represented by a single vertical line to indicate it is an infinitely thin plane. It greatly simplifies computations in practice not only because real ophthalmic lenses are rather thin but also because the pupil of the eye acts as a limiting paraxial aperture. Of course, aphakic patients may not let us forget that their lenses are not infinitely thin; so the reality of thick lenses cannot always be ignored.

THIN LENS OPTICS

The optical power of a thin lens is the sum of its two surface powers F_1 and F_2. Since the power of each surface is $F = \frac{n' - n}{r}$, the total power of the thin lens is

$$F = (n' - 1)\left(\frac{1}{r_1} - \frac{1}{r_2}\right)$$

where n′ is the index of the glass, the radii are those of the two surfaces, and the lens is assumed to be in air. The expression gives the power in terms of lens construction and is called the "lens maker's equation." It may seem strange that thin lens power should involve the radius of each surface. Both surfaces contribute to the power because they are not being replaced by plane ones any more than the equivalent plane of a spherical mirror in a plane mirror. Moreover the power is not characterized by two unique surfaces; any combination will do if they add up to the same total. Thus a +4 D lens can be constructed by +2, +2; +3, +1; +4, plano; +6, −2; or any other combination. By changing base curve, lenses are "bent" to make them approximate the rotations of the eye without altering the power needed to correct the refractive error.

EXAMPLE: The human crystalline lens, average index 1.41, is surrounded by an aqueous medium of index 1.336. If the radius of its front surface is 10 mm and the radius of its back surface is 6 mm, what is its thin lens power?

$$F = n' - n\left(\frac{1}{r_1} - \frac{1}{r_2}\right)$$

$$= 1.41 - 1.336\left(\frac{1000}{+10} - \frac{1000}{-6}\right)$$

$$= +19.7 \text{ D}$$

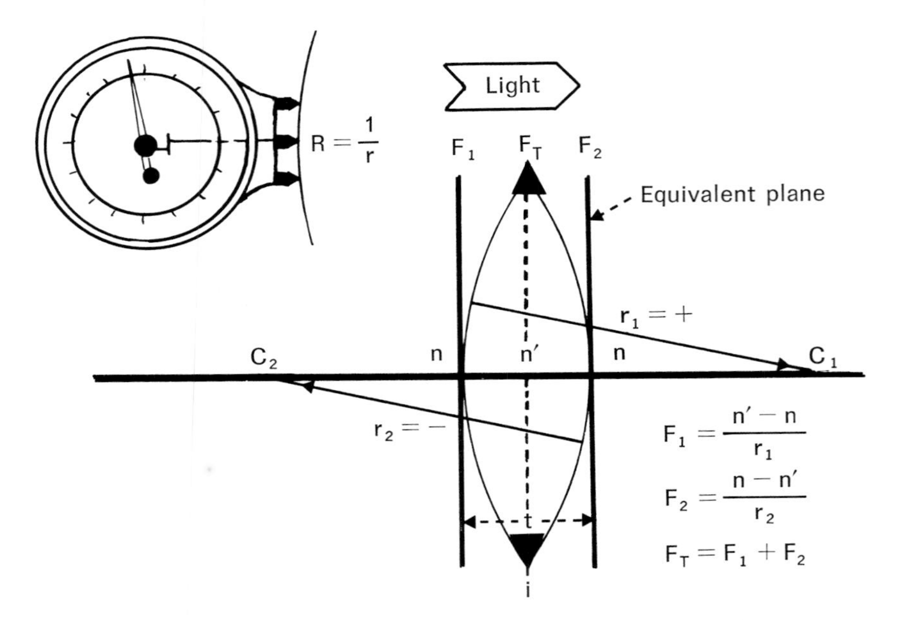

Fig. 3-1. Diagram showing a real lens of thickness t and surface powers F_1 and F_2. Each surface can be replaced by an equivalent plane tangent to its pole. If we pretend the thickness is negligible, the two equivalent planes fuse; this resultant single equivalent plane is the "thin lens" (F_T). Its thinness is not a physical property but an optical conception; hence it must always be represented by an infinitely thin line. Thin lenses have no peripheral aberrations because they only handle paraxial rays by definition.

The example illustrates the effect of the surrounding medium on focal power. If the crystalline lens were removed from the eye and hand neutralized, its absolute power would not be +19.7 D but (1.41 − 1.00) (+226) or +109 D, which is extremely different. The +19.7 D represents the air-equivalent (or reduced) power of the crystalline lens: the same power the lens exerts on a wave in aqueous humour as a +19.7 D lens would exert on a wave in air. Air is always the common denominator, being the reference index of visual optics. Clinically we seldom care about absolute power; our interest is in equivalent power. Occasionally, because of the development of intraocular lenses, it may be necessary to compute the power of a lens in air that will reproduce the power of the crystalline in a bed of aqueous.

The image formed by the first lens surface becomes the object for the second, and the final image belongs to the lens as a whole. Applying the single surface paraxial equation ($n'/l' - n/l = n'/f'$) to each surface in turn, one obtains the paraxial equation for the thin lens:

$$\frac{1}{f'} = \frac{1}{l'} - \frac{1}{l}$$

The unity in the numerator represents the index of air (assumed to surround the lens), and the index of the glass is not in the formula because thickness is ignored. Since $1/l'$ and $1/l$ are the image and object vergences and $1/f'$ the focal power, the paraxial equation can be put directly into the vergence form:

$$F = L' - L$$

This is the same basic equation derived for single surfaces in the previous chapter. The only difference is that air replaces the double media of single refracting surfaces (Fig. 3-2).

A lens images a distant object at its second focal point. If L approaches zero, then $L' = F$ and $1/l' = 1/f'$. When the image is at infinity, L' approaches zero, and the object

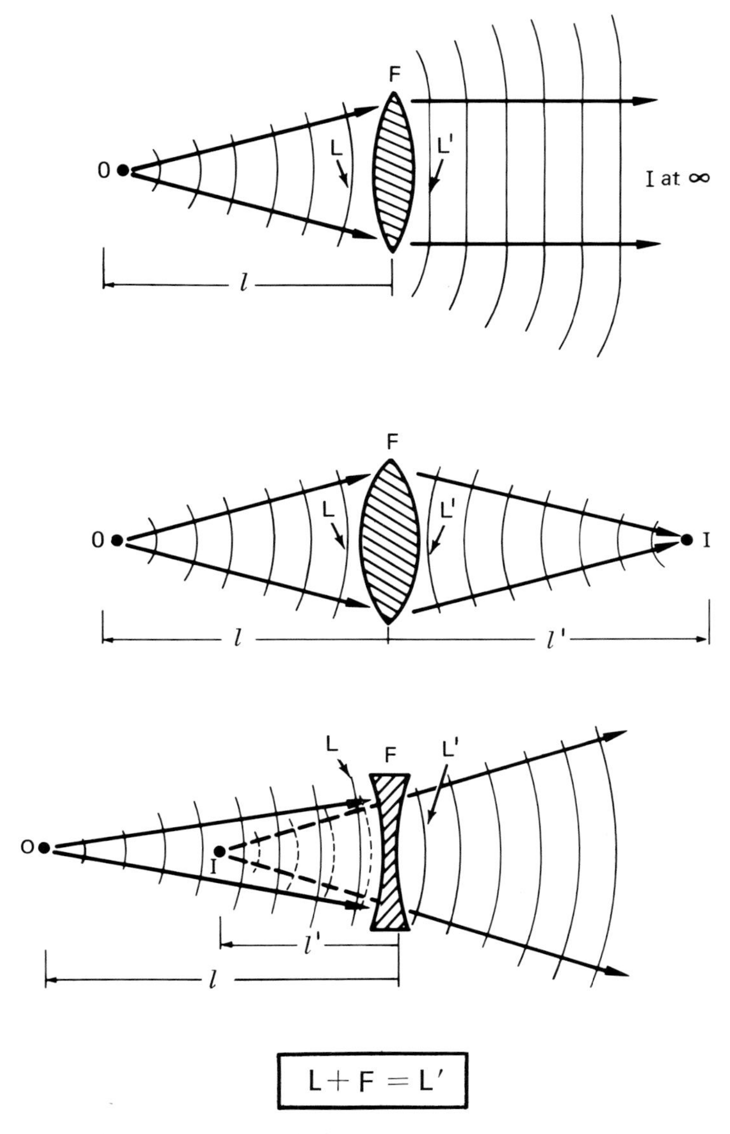

Fig. 3-2. The ability of a lens to impress a change in vergence on the incident waves is the essence of its optical power. If we pretend the waves are confined to the paraxial region, the resultant vergence can be simply computed by algebraically adding the incident vergence to the lens power in diopters.

must be at the first focal point; hence:

$$F = \frac{1}{f'} = -\frac{1}{f}$$

The second focal point lies in image space and is numerically equal but opposite in sign to the first focal point. Although focal points are paired, they are not conjugate to each other; the first focal point is conjugate to plus infinity; the second focal point is conjugate to negative infinity. Positive and negative infinity refers to our sign convention—not science fiction. A plus sign indicates "to the right," and a minus sign indicates "to the left," relative to light direction, always assumed left to right (Fig. 3-3).

The chief function of a lens is to change

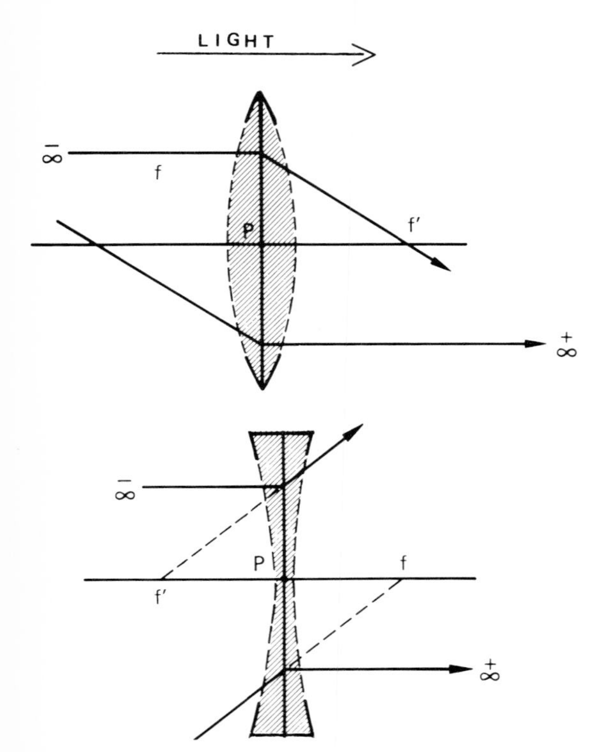

Fig. 3-3. Focal points of a thin lens are numerically equal but on opposite sides. Their position is reversed for plus and minus lenses; hence we call them "first" and "second" focal points rather than "anterior" or "posterior." Focal points are not conjugate to each other but to optical infinity (or 6 meters in practice). The sign we attach to infinity is not science fiction but to indicate direction. Plus means to the right, minus to the left—relative to the light direction (always assumed to proceed from left to right).

the vergence of the incident wave. A lens that impresses a positive vergence is plus; one that impresses a negative change is minus. We care more about optical effect than lens construction because in correcting ametropia the focal point must coincide with the far point of the eye. Consequently we generally use the focal length formula and leave the constructional equation (by index and radii) to the optician. The sign of the focal power is always that of the second focal length and opposite that of the first.*

*The only text that has it reversed is J. P. C. Southall's *Mirrors, Prisms, and Lenses*. The reason for this idiosyncrasy escapes me.

EXAMPLE: What is the second focal length of a biconvex thin lens (index 1.5) whose radii of curvatures are 10 inches? What is the effect of grinding one surface plane?

$$F = n' - 1\left(\frac{1}{r} - \frac{1}{r_2}\right)$$

(a) $1.5 - 1\left(\frac{40}{+10} - \frac{40}{-10}\right)$; F = +4 D, f′ = +10 inches

(b) $1.5 - 1\left(\frac{1}{\infty} - \frac{40}{-10}\right)$; F = +2 D, f′ = +20 inches

The example illustrates that the focal length of an equiconvex lens of index 1.5 equals the radius of either surface. When only equiradial trial lenses of this index were manufactured, a No. 10 glass meant it had been ground by a tool of 10-inch radius; the radius of each surface was 10 inches, and the focal length was 10 inches. These designations worked until new kinds of optical glass appeared and lenses acquired meniscus profiles.

The earliest designation of convex lenses was based on the estimated age of the presbyopic user. In the reign of the first Queen Elizabeth, spectacles were ordered "of sixty or sixty-five year of age." Presbyopia was not differentiated from absolute hyperopia, and eyes that could not see clearly at near were presumed prematurely aged. But the inch system of designating lenses soon led to chaos; the English inch was not the same as the Parisian inch, which differed in turn from the Prussian inch.

When Donders in 1864 propounded the clinical importance of accommodation and defined its amplitude as the difference between the far point and near point, the need for a reciprocal unit became obvious. Helmholtz favored the reciprocal Prussian inch and proposed to name it "Zolltel." A distinguished committee including Javal, Donders, Leber, Nagel, and Wells among others unanimously recommended optical power instead of focal lengths to designate trial lenses.

Monoyer in 1872 coined the term "dioptrie." The advantage of the diopter (the British spell it in the French fashion, dioptre) is that it is directly additive as 1/40 + 1/10, for example, are not. However, the diopter is not the equal unit originally planned; the linear focal difference between 10 and 11 diopters is not the same as between 1 and 2 diopters. Despite the sliding scale, the diopter has been universally adopted, and a suggestion to subdivide it into millidiopters and hectodiopters was happily ignored. We still

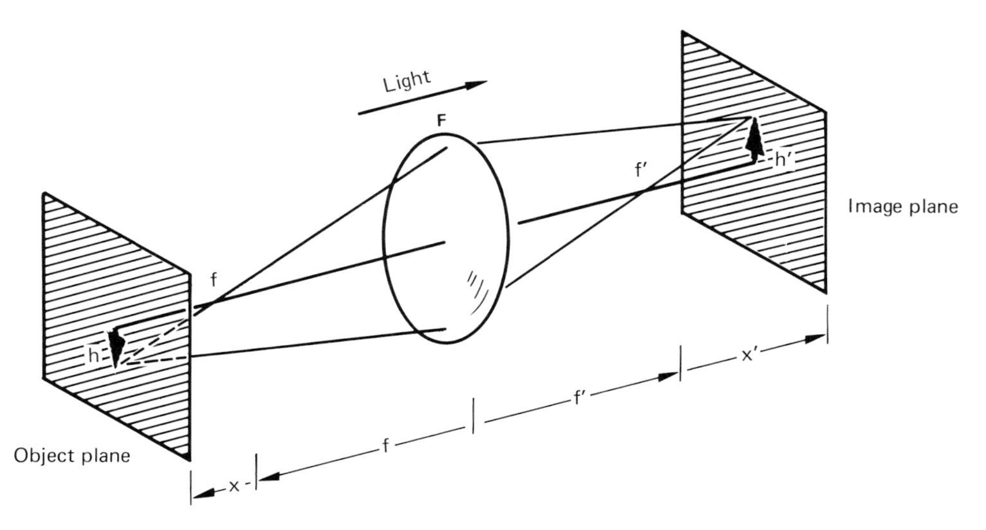

Fig. 3-4. This diagram is to remind us that objects and images lie in planes and may have three dimensions. The image plane for a real lens would be slightly bent toward it at the edges, an aberration called "curvature of field," which is ignored in thin lens theory. The steps from the focal points to the object and image planes (x and x') are sometimes useful in slide rule computations.

have fractional diopters but never more than 1/4, or at most 1/8, in clinical practice.

The lateral magnification of thin lenses compares linear dimensions of object (h) and image (h'). For a thin lens as for single refracting surfaces, object and image size are in the same ratio as their distances (or vergences) in air.

$$m = \frac{h'}{h} = \frac{l'}{l} = \frac{L}{L'}$$

For example, an object placed 15.87 mm from a +60 D lens images at −63 + 60 = −3 D; l' = −333 mm. The magnification is l'/l = +21 ×.

Positive magnification means the image is upright, and negative magnification means that it inverted with respect to the object (Fig. 3-4).*

Ray tracing illustrates the optics of thin lenses, graphically portrays the formulas, and helps confirm the results of computation (Fig. 3-5). Since only two construction rays (and the optic axis) are needed to locate the image, the following simple method requires only a ruler. Draw the thin lens perpendicular to the optic axis and a ray parallel to the axis refracted through the second focal point. The distance of the parallel ray from the axis is immaterial, since the ray is in any case assumed to be paraxial. With a straight edge, connect the object point in question through the optical center of the thin lens to the point where the line intersects the focal ray. This gives its conjugate image point.

The optical center is the unique point through which a ray passes the thin lens without deviation. It is analogous to the center of rotation of a single refracting surface and is also called a "nodal" point. The optical center of a thin lens is always located at its intersection with the optic axis and need not be computed.

For a plus lens the image of a real object is real and inverted when the object lies outside the focal point. An object at twice the focal length gives an image of the same size and distance (m = −1). An object inside the focal point gives a virtual image. Object and

*Fig. 3-4 shows some additional dimensions, which are labeled x and x', called "extrafocal distances." They are the steps from the first focal point to the object (x) and from the second focal point to the image (x'). From the similar triangles: $-xx' = (f')^2$. This expression, known as "Newton's relation," is frequently useful to check computations, especially by slide rule. Moreover the linear magnification is useful in the optics of magnifying loupes when expressed in terms of extrafocal distances: $m = -f/x = -x'/f'$.

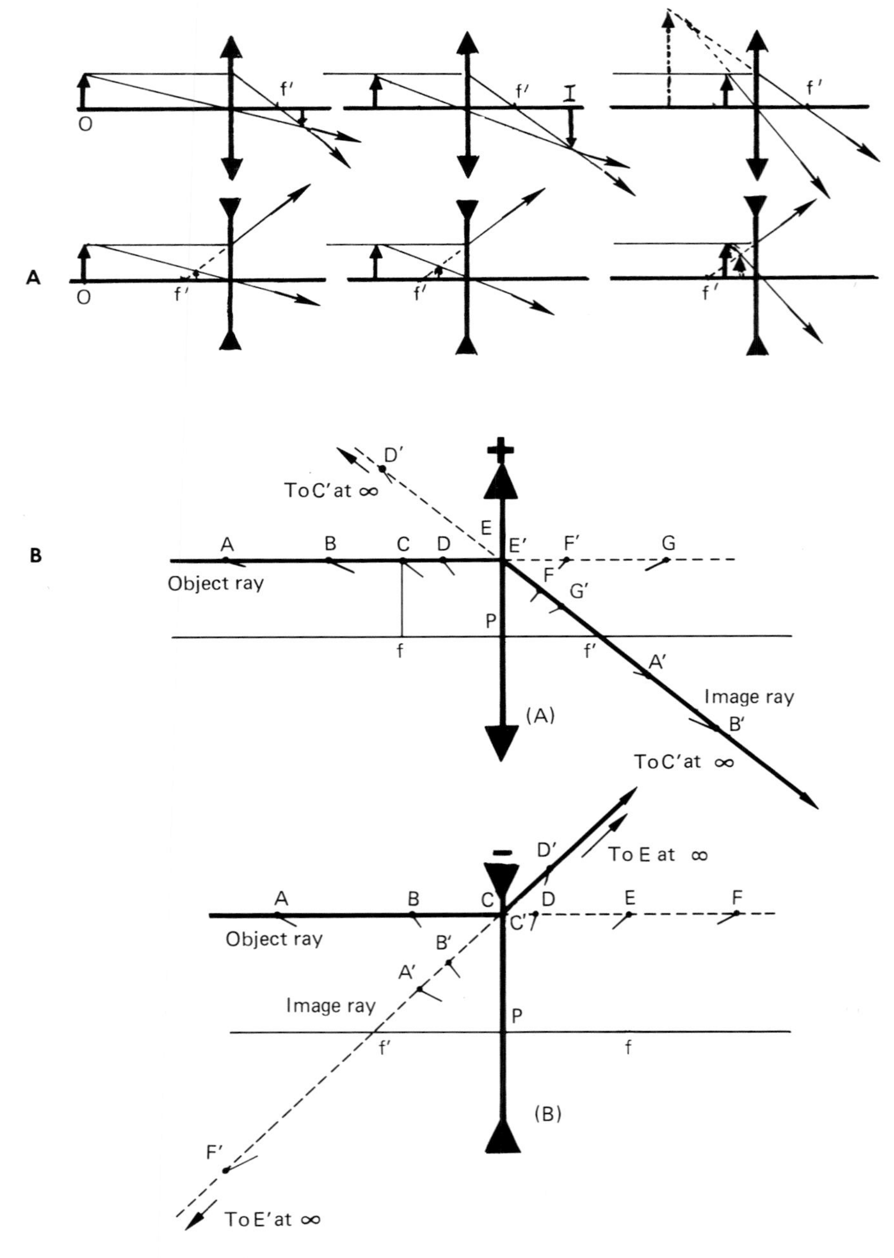

Fig. 3-5. A shows a series of ray tracings for plus and minus thin lenses using two construction lines. **B** illustrates a simple method of confirming computations by using only the ray through the second focal point and a straight edge through the optical center of the lens (for details see text).

image are the same size, and both are upright (m = +1) at the lens. If the object is virtual (converging incident rays), the image is real, erect, and between the lens and second focal plane. These results are summarized in Table 3-1.

When a distant object is viewed through a plus lens that is moved up and down across the line of vision, the target appears to move "against"; the direction of movement is "with" through a minus lens. This movement, the result of prismatic effect (Fig.

Table 3-1. Object–image relations for thin lenses

Nature of lens	*Nature and position of object*	*Nature and position of image*
Plus	Real and outside 2f	Real, inverted, diminished, and between f and 2f
Plus	Real and at 2f	Real, inverted, same size as object, and at 2f
Plus	Real and between 2f and f	Real, inverted, magnified, and outside 2f
Plus	Real and inside focus	Virtual, erect, magnified, and farther from lens than object
Plus	Virtual in any position	Real, erect, diminished, and inside focus
Minus	Real in any position	Virtual, erect, diminished, and inside focus
Minus	Virtual outside focus	Virtual, inverted, and outside focus
Minus	Virtual inside focus	Real, erect, magnified, and farther from lens than object

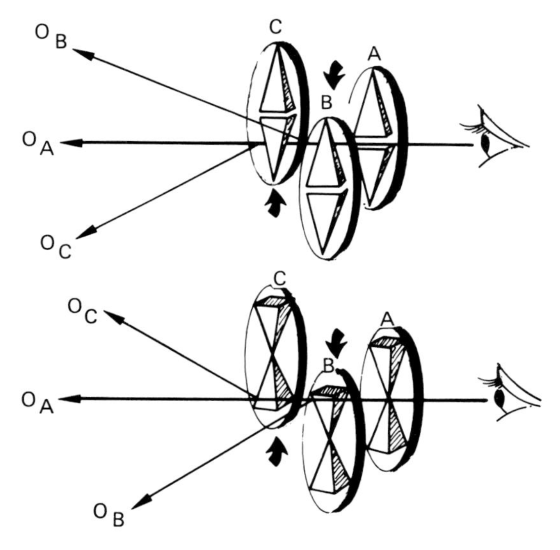

Fig. 3-6. Schematic diagram explaining the behavior of plus and minus lenses in producing apparent motion of objects as in hand neutralization. The "against" motion of a plus lens *(top)* and the "with" motion of a minus lens *(bottom)* are readily visualized by pretending the lenses are composed of two prisms, base to base and apex to apex.

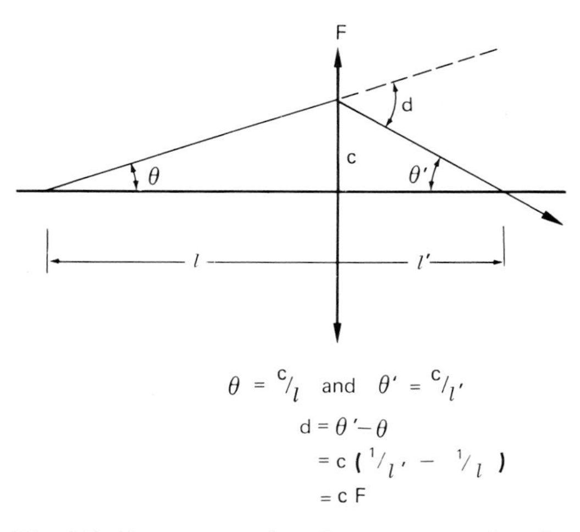

Fig. 3-7. Diagram used to derive Prentice's rule, which is fundamental in planning (or avoiding) the prismatic effects induced by lenses. It shows that the prismatic effect at a given point of a thin lens depends only on its power and the distance from the optical center. If the deviation *(d)* is expressed in prism diopters, one need only multiply the displacement *(c)* from the optical center in centimeters by the lens power *(F)* in diopters.

3-6), is the basis of hand neutralization. The lens behaves like a series of prisms; the more peripheral the incident ray the greater the prismatic effect. The direction of the prism base is relative to the optic axis. A plus lens acts as a biprism base to base, a minus lens as a biprism apex to apex.

The prismatic effect at a specified point of the lens is computed from Prentice's formula

$$d = cF$$

where d is the deviation in prism diopters, c the distance of the ray from the optic axis in centimeters, and F the focal power of the lens (Fig. 3-7). This is the only exception to the rule that any distance in a formula including diopters must be in meters. The reason for this exception is that prism diopters are based on centimeter deviations. Therefore the symbol c is appropriate to remind us that in this formula the distance of the ray from the axis is in centimeters.

EXAMPLE: What is the prismatic deviation 8 mm below the optical center of a +4 D lens?

$$d = cF;\ d = 0.8(+4) = 3.2^{\Delta} \text{ base up}$$

Whenever we adjust a microscope, focus an acuity projector, move nearer a patient's eye in retinoscopy, or displace a spectacle lens down the nose, we make use of effectivity. Effectivity means that the optical effect of a lens depends on position just as the height of a brick has a variable effect on the toe it falls on. The principle is derived directly from the effectivity of wave fronts (p. 33); indeed the equations look identical except that lens power is substituted for wave vergence:

$$F_x = \frac{F_o}{1 - dF_o}$$

F_x is the effective power of a lens F_o at some new position d meters farther or nearer (the lens is assumed to be in air). Fig. 3-8 shows a +1 D lens focusing parallel rays at 100 cm. A lens moved 20 cm nearer the screen requires an effective power of $\frac{100}{80}$ or +1.25 D. The same result is obtained directly from the formula:

$$F_x = \frac{+1}{1 - (.2)(+1)} = +1.25 \text{ D}$$

The formula method is quicker and requires no sketch, but care must be taken to give d the correct sign. The lens required to focus the light on the screen 20 cm further is 100/120 = +0.83 D.

THIN LENS SYSTEMS

There is a tendency to think of optical systems only in terms of microscopes, telescopes, and expensive cameras; however, every lens is a system of two surfaces. In thin lens theory, their separation is assumed to be negligible, but it cannot be ignored in aphakic corrections, contact lenses, or aniseikonia. Moreover, in ametropia the eye and the correcting lens form an optical system. For example, a hyperopic eye represents a minus lens, and the correcting plus lens is some distance in front of it, based on the length of the lashes and, in the nature of things, the size of the nose on which the frame rests. This constitutes a Galilean telescope, resulting in magnification that may differ for the two eyes and induce aniseikonia. So unless we come to grips with lens systems and real lenses, it is difficult to appreciate the limitations of thin ones.

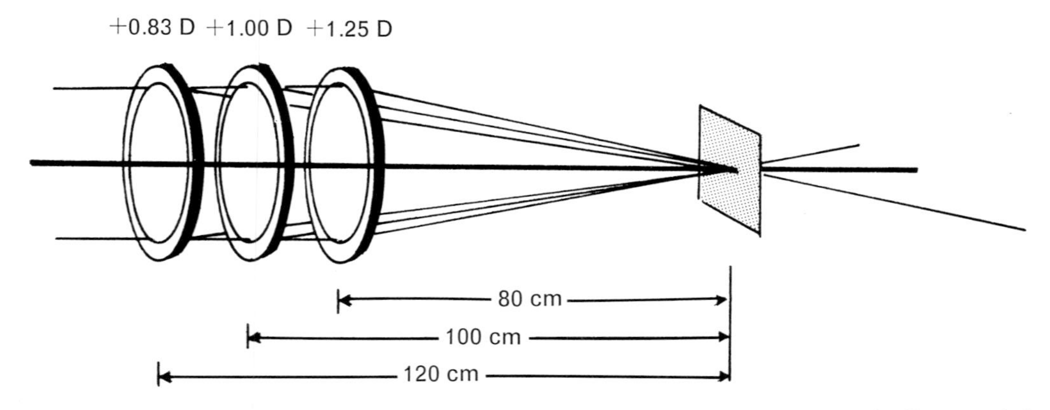

Fig. 3-8. Diagram illustrating that three different lenses can have the same optical effect. Each lens focuses a sharp image on the screen, although of different size. The principle explains why a contact lens must have a different power than a spectacle lens to correct the same refractive error.

A real lens is a system of two thin lenses (representing each surface), separated by glass. Postponing consideration of the glass, one has a simple thin lens system in air. Fig. 3-9 shows such a system composed of two thin lenses, $F_1 = +4$ D and $F_2 = +2$ D, separated 15 cm. If this system views a distant object, the incident vergence is zero, and the refracted wave (from the standard vergence formula) is $L_1 + F_1 = L'_1$; plano + 4 D = +4 D. The +4 D wave would focus 25 cm from the first lens, but after traveling 15 cm, it strikes the second lens. At this point, its vergence is +10 D. The effective power of the wave leaving the first lens has changed from +4 to +10. The +10 D can be computed two ways: we can draw it and note that the wave is 10 cm from its intended focus when it reaches the second lens, or we use the effectivity formula:

$$L_x = \frac{L_o}{1 - dL_o}$$

$$L_x = \frac{+4}{1 - 0.15(+4)} = +10\text{ D}$$

Both methods give the same result.

At the second lens the refraction ($L_2 + F_2 = L'_2$) is (+10) + (+2) = +12 D gives the final vergence, producing an image $\frac{100}{+12}$ = +8.3 cm from the second lens. The image does not belong to the second lens but to the system as a whole. Since the incident rays were parallel, we can call it the "second focal point" of the system (f'_v). The subscript is to remind us that it is the second focal point of the lens combination.

The method just described is applicable to any system of thin lenses. One simply considers the image of the first lens as the object for the next lens. But it takes too long and is too tedious. A single formula that summarizes the whole thing would be quicker and less subject to errors. The key to such a formula is the changed effectivity that the wave leaving the first lens undergoes by the time it reaches the second. As long as the system views a distant object, that is, the incident vergence is zero, the wave leaving

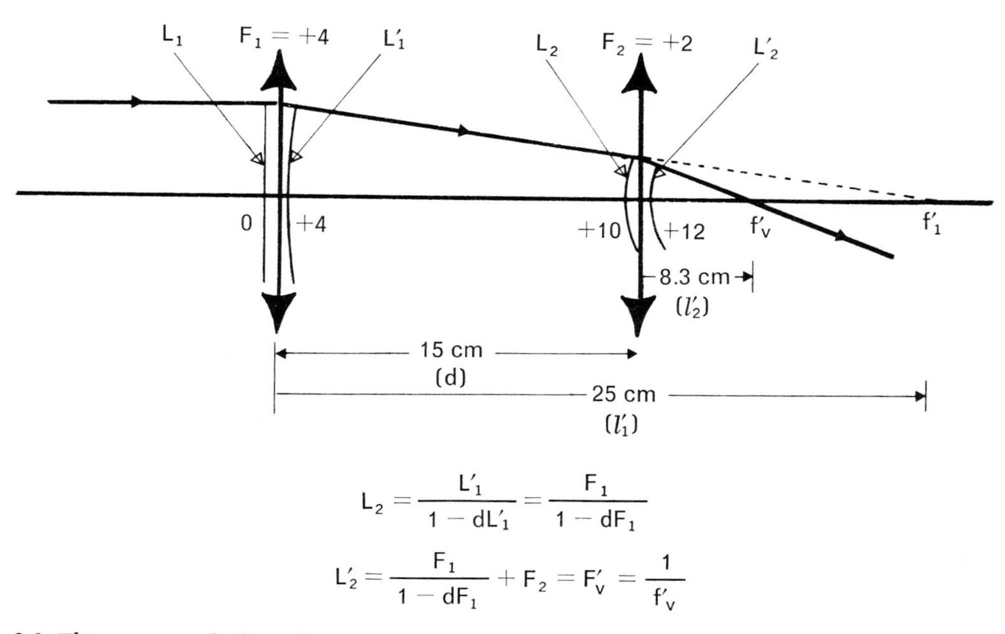

$$L_2 = \frac{L'_1}{1 - dL'_1} = \frac{F_1}{1 - dF_1}$$

$$L'_2 = \frac{F_1}{1 - dF_1} + F_2 = F'_v = \frac{1}{f'_v}$$

Fig. 3-9. This imposing-looking diagram is nothing more than a simple thin lens system in air. The numbers are used to illustrate how the rays are traced through each lens, using only the paraxial equation L + F = L′, and the change in effectivity of the wave because it must travel the distance *d* before reaching the second lens. The example is then used to derive the important and practical concept of vertex powers, which are *the* powers of clinical refraction.

the first lens must equal the power of that lens ($L'_1 = F_1$). The vergence of the refracted wave at the position of the second lens is therefore $\frac{F_1}{1 - dF_1}$. Adding this to the second lens, the final vergence is obtained:

$$L'_2 = \frac{F_1}{1 - dF_1} + F_2$$

Since the incident rays were parallel, the final image is also the second *focal point*, which means that L'_2 must be the second focal power (F'_v) of the system. Our general formula is therefore:

$$F'_v = \frac{F_1}{1 - dF_1} + F_2$$

Thus to find the final vergence add the effective power of the first lens at the plane of the second to the power of the second lens. Let us see if it is true by substituting the numerical values of the previous example in the formula: $\frac{+4}{1 - 0.15\,(+4)} + 2 =$ +12 D. The reciprocal ($F'_v = 1/f'_v$) gives the back vertex focal length; $\frac{100}{+12} = +8.3$ cm, which is the position of the image measured from the second lens.

In ophthalmic optics, F'_v is called the back vertex focal power, hence the real reason for the v subscript. There is also a front vertex power (F_v) whose reciprocal, the front vertex focal length, designates the point at which an object must be placed so that parallel rays emerge from the system. We can find it by the same reasoning, tracing the rays backward or pretending the system is turned around. Since the sequence of lenses is now reversed, the front vertex power is:

$$F_v = \frac{F_2}{1 - dF_2} + F_1$$

The two vertex powers always differ unless the two thin lenses of the system happen to be identical. Generally, the two lenses are not the same. For example, with a telescope, it makes a difference which end one looks through. In all cases, the reciprocals of the two vertex focal powers give the vertex focal lengths: $F_v = -1/f_v$ and $F'_v = 1/f'_v$. These are entirely analogous to the two focal lengths of a thin lens but apply to the lens system. The front vertex focal length is measured from the first lens, and the back vertex focal length is measured from the second lens.

THICK LENS OPTICS

A thick lens differs from a double thin lens system in that the distance between them is now the thickness of the glass (t) measured in millimeters instead of centimeters, and the index of the glass must be considered. We therefore modify our vertex formulas to include the index of the glass:

Back vertex power:

$$F'_v = \frac{1}{f'_v} = \frac{F_1}{1 - t/nF_1} + F_2$$

Front vertex power:

$$F_v = -\frac{1}{f_v} = \frac{F_2}{1 - t/nF_2} + F_1$$

The front and back vertex powers are identical for an equiconvex or equiconcave lens, but since modern ophthalmic lenses are generally meniscus shaped, the two vertex powers differ (Fig. 3-10).

The back vertex focal power is the power we write in the lens prescription, the power stamped on our trial lenses, and the power read off with the lensometer (the correct way, of course, with the ocular surface of the lens facing away from us). There is nothing esoteric about back vertex focal power—it is *the* ophthalmic power of practical refraction. No matter what the surface powers of a lens, the back vertex focal length must coincide with the far point of the eye to correct any ametropia, and the rear surface is always a tangible point from which to measure.

The front vertex focal power is the power measured by hand neutralization because the ocular surface of an ophthalmic lens is concave, and the neutralizing trial lens can only be approximated to its front surface. For strong corrections, such as aphakic prescriptions, the lensometer and hand neutralization give different results (unless one turns the lens the "wrong" way in the lensometer).

The vertex formulas involve fractions and are somewhat clumsy for practical computa-

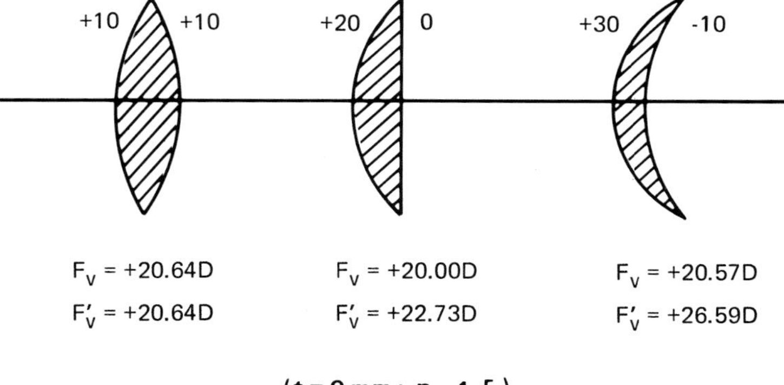

Fig. 3-10. Three lenses of the same approximate powers ($F_1 + F_2$) but of different front and rear vertex powers. All three lenses are assumed to have a constant thickness and glass index. It will be seen that if the clinician orders the biconvex lens and the patient receives the concavoconvex lens, he will be 6 D overcorrected.

tions. For the small thicknesses usually involved, the fraction can be divided out and higher terms dropped to obtain the following useful approximations:

$$F'_v = F_1 + F_2 + \frac{t}{n}(F_1)^2$$

$$F_v = F_1 + F_2 + \frac{t}{n}(F_2)^2$$

The sum ($F_1 + F_2$) is the thin lens power; the last term to be added to it is the "allowance factor." The allowance factor is the difference between a thick and a thin lens; when thickness is zero, it drops out leaving only the equivalent thin lens power. Note that the allowance factor involves only one surface power, the front one for F'_v and the back one for F_v and its value is always positive.

EXAMPLE: What is the back vertex power of an ophthalmic lens, index 1.523, whose surface powers are +12 D and −2 D and whose thickness is 6 mm?

$$F'_v = F_1 + F_2 + t/n\,(F_1)^2$$

$$= (+12) + (-2) + \frac{.006}{1.523}(144)$$

$$= +10.57 \text{ D}$$

In this example there is about 0.5 D difference between the real lens and its thin lens equivalent. The front vertex power incidentally would be +10.02 D; thus it does make a difference which way the lens is placed in a lensometer. Although the prescription specifies back vertex power, surface powers are variable depending on lens design. The optician using tabulated allowance factors picks the surface powers to match the prescription for the particular glass index, thickness, and base curve decided on.

AFOCAL SYSTEMS AND MAGNIFICATION

A telescope is a lens system that discharges parallel incident rays without change but at a new angle to produce magnification. A system that does not alter the vergence of parallel incident waves is referred to as "afocal" (without focus). To obtain an afocal system we need two lenses so positioned that the second neutralizes not the power of the first but the effective power of the first at the plane of the second:

$$F_2 = -\left(\frac{F_1}{1 - dF_1}\right)$$

There are two ways of accomplishing this (Fig. 3-11): the wave after refraction by the first lens can be allowed to come to a focus and then neutralized by a second plus lens (astronomic telescope), or the wave can be neutralized at some point while it is still converging by a second minus lens (Galilean telescope). The separation of the lenses in either telescope is the algebraic sum of their two focal lengths. Two +10 D thin lenses

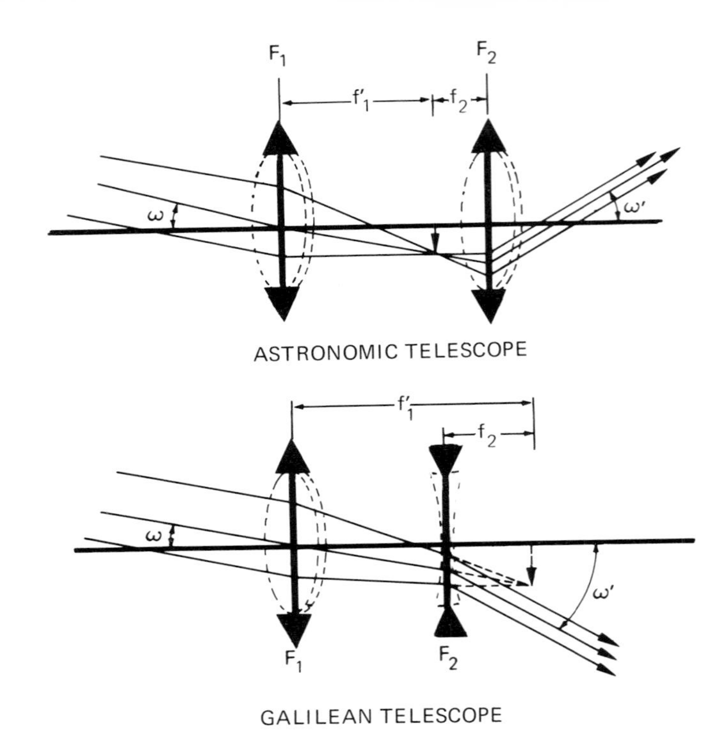

Fig. 3-11. The importance of the telescope in clinical practice has nothing to do with its astronomic function. The indirect ophthalmoscope is an astronomic telescope; even more important, when a spectacle lens is used to correct refractive errors, the lens and eye constitute a Galilean telescope. The distance between them and the powers involved determine the magnification. The hyperope looks through it "forward," the myope looks through it "backward." The optics of the telescope is simply a matter of one lens neutralizing the *effective* power of the other.

separated 20 cm would constitute an astronomic system; a +10 D and −20 D separated 5 cm, a Galilean system. The astronomic system produces an inverted image, but the astronomer does not care if his view of a star is upside down. In the indirect ophthalmoscope, also an astronomic system, it is a nuisance.

The Galilean system produces a virtual, upright image. The correction of ametropia is analogous to a Galilean system; one lens represents the ocular error and the other, the correcting lens. The magnification of a telescope is in ratio to the focal powers of the two lenses:

$$m = -\frac{F_2}{F_1}$$

Substituting the neutralizing effective power of F_1 for F_2 and simplifying, we obtain the formula:

$$m = \frac{1}{1 - dF_1}$$

The magnification of a Galilean system therefore depends only on the power of the first lens and its distance from the second. Applying this formula to a thick lens of front surface power F_1 and back surface power F_2 and including its thickness t and index n, we obtain the afocal magnification of an ophthalmic lens:

$$m = \frac{1}{1 - t/n\ F_1}$$

Consider an ophthalmic lens of front surface power +8.00 D, index 1.523 and thickness 6 mm. To make this lens afocal, its rear surface power would have to be

$$F_2 = -\frac{F_1}{1 - t/n\ F_1}$$

or −8.26 D. This lens is now a "solid glass Galilean telescope," or afocal lens. Even though it has no power, it produces a magnification of

$$m = \frac{1}{1 - \frac{0.006}{1.523}(+8)}$$

or 3%. This magnification due only to front surface power and thickness is called the "shape factor." The "shape" refers to lens profile and not circumferential configuration.

Afocal lenses are seldom prescribed as such, but all real ophthalmic lenses have some thickness and a front surface power and therefore produce some afocal magnification, in addition, of course, to any power magnification. Suppose, for example, we order a prescription +8 D to be made as a biconvex lens, but a zealous optician decides to make it in modern form with surface powers +14 D and −6 D. If the glass has an index of 1.523 and a thickness of 8 mm, the shape magnification of the first lens is

$$m = \frac{1}{1 - \frac{0.008}{1.523}(+4)}$$

or 2%, whereas the second (assuming the same thickness) has a shape magnification of

$$m = \frac{1}{1 - \frac{0.008}{1.523}(+14)}$$

or 7%. The change in base curve alone has induced a difference of 5%. Whereas 2% magnification might have been accepted relative to the other eye, 7% magnification may easily lead to discomfort and spatial distortion. So if some patients cannot tolerate their new lenses, check their surface powers (with a lens clock) and their thickness (with a lens caliper) before concluding they must be neurotic because the lensometer confirms your prescription.

The maximum practical front surface power of ophthalmic lenses is about 20 D, and the maximum thickness is about 9 mm; hence the greatest shape magnification is around 14%. This is insufficient to compensate for the ocular size difference in monocular aphakia but more than enough to induce symptomatic aniseikonia. Aniseikonia is a binocular problem; it is the small size differences that cause the trouble because large ones only result in suppression.

GAUSSIAN OPTICS

The equivalent power (F_e) is the power of a single thin lens that optically replaces and is equivalent to the optical system. It represents a single refraction and is far simpler than tracing rays through the system. Moreover the thin lens formulas then are easily applied to locate focal or image points.

The equivalent power of a lens system is given by

$$F_e = F_1 + F_2 - \frac{d}{n} F_1 F_2,$$

where F_1 and F_2 are two lenses separated by a distance d in a medium of index n.* Since the equivalent lens is a thin lens, *its* focal points are always equal and opposite like the focal lengths of any other thin lens but unlike the vertex focal lengths:

$$F_e = +\frac{1}{f'_e} = -\frac{1}{f_e}$$

Equivalent power is sometimes called "true power," which is not to imply that vertex powers are false, but to emphasize its optical character. Clinically lenses are calibrated in vertex rather than equivalent power, since the latter computations are too tedious. But equivalent power is *the* power when we speak of the eye as an optical instrument.

If one could avoid ray tracing through lens systems only by computing the equivalent power, one could dispense with sequen-

*The equivalent power (F_e) is actually already computed in finding vertex power. If one reduces fractions in the vertex power equations, it appears as the numerator. For example,

$$F'_v = \frac{F_1}{1 - \frac{d}{n}F_1} + F_2 = \frac{F_1 + F_2 - d/n\, F_1F_2}{1 - \frac{d}{n}F_1} = \frac{F_e}{1 - \frac{d}{n}F_1}$$

and similarly, $F_v = \dfrac{F_e}{1 - \frac{d}{n}F_2}$

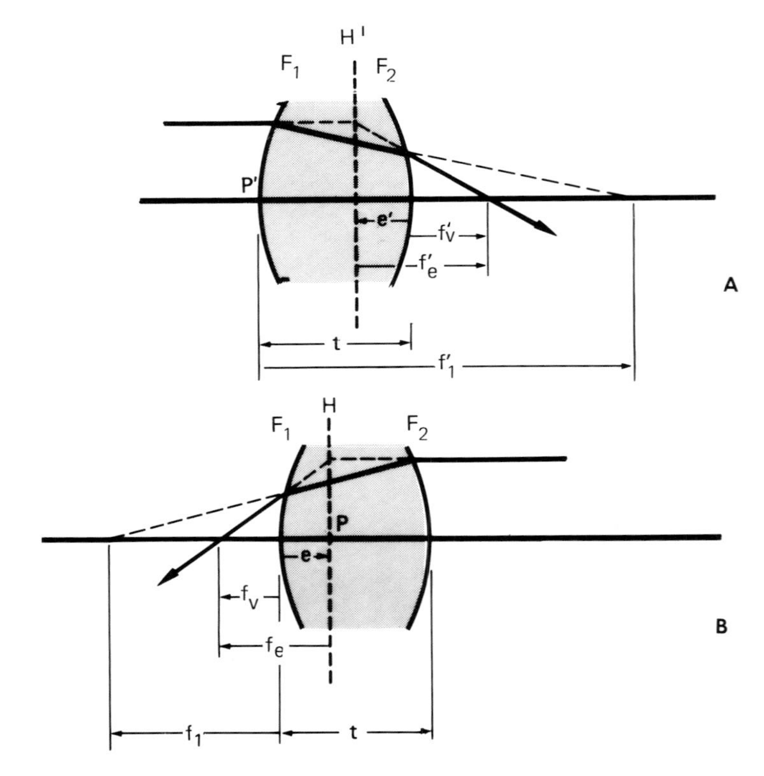

Fig. 3-12. The two diagrams illustrate the meaning of the first, **B**, and second, **A**, equivalent planes (*H* and *H'*). In this example of a thick lens, the second equivalent plane is where an equivalent thin lens must be placed so that a single refraction will have the same effect as two consecutive refractions at each surface. There are *two* equivalent planes because incident parallel rays have a different sequence than emerging parallel rays. Thus the first equivalent plane (H) is related to the first focal point of the system (f_e), and the second equivalent plane (H') is related to the second focal point (f'_e) of the system.

tial mathematics and vertex powers. It is not enough, however, to know the power of the equivalent thin lens; we must also specify its position. It can only replace the system at this position called the "equivalent plane." However, there are two equivalent planes, labeled first (*H*) and second (*H'*) (Fig. 3-12). The equivalent lens must be placed not at either one or the other nor at both simultaneously but at each in succession. The equivalent lens must be at the first equivalent plane to catch the incident rays and, by Gaussian sorcery, at the second to release them, which is the only way it will work. The beauty of this Gaussian magic is that no matter how complex the system, no more than two equivalent planes and one equivalent lens are ever needed to replace it. The first equivalent plane (H) is measured from the first lens, and the second equivalent plane (H') is measured from the second lens of the system. The distances e and e' are simply the differences between the vertex and equivalent focal lengths, which are already known:

$$e' = f'_v - f'_e \text{ and } e = f_v - f_e$$

The sign of e and e' determines in which direction to measure each equivalent plane from its respective lens. The planes need not be in sequence; the second may precede the first. Their exact position is not known in advance but by computing e and e' and noting how the sign comes out. (See appendix, pp. 52 and 53.)

The intersection of the equivalent planes with the optic axis represent the "equivalent" or "principal" points. Once the equivalent planes and power are known, the thin lens

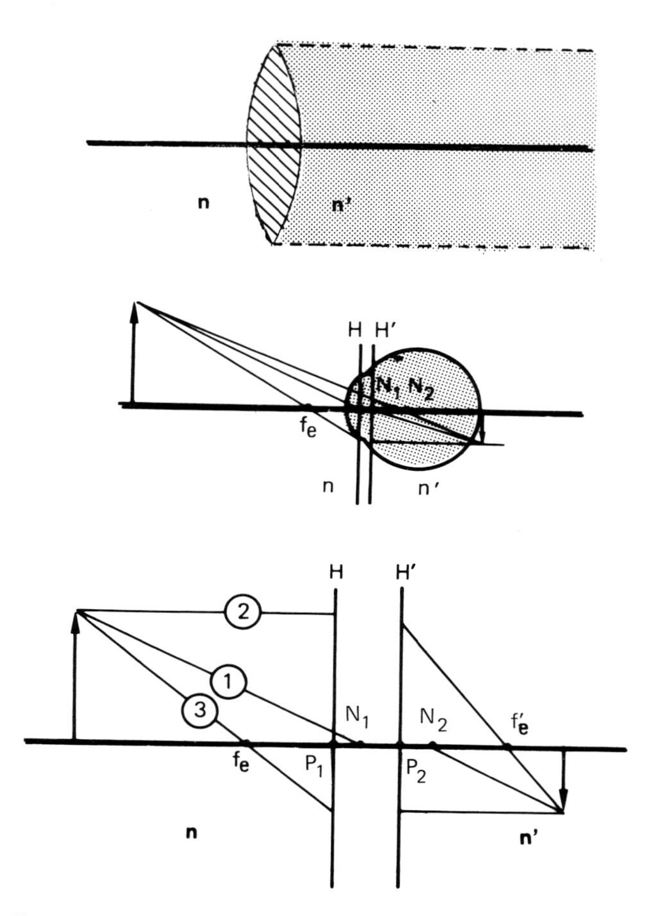

Fig. 3-13. Diagram illustrating the application of principal and nodal points in a lens system *(top)* and the eye *(bottom)*. For all ophthalmic lenses the principal (or equivalent) points and nodal points are identical. The nodal points only become displaced from the principal points when the index of refraction on one side of the system differs from the other—as in the case of the eye (or a single refracting surface).

vergence formula ($L + F_e = L'$) can be used, but instead of measuring object and image distances to the first and second lens of the system, one measures to the equivalent planes. There is no point to these computations unless one intends to use the system over and over. That is how Gullstrand won the Nobel prize. He calculated the equivalent planes and power of the eye, and we have been using them ever since. With the equivalent power of the eye and the position of its equivalent planes known, only simple thin lens formulas are necessary for computations instead of tracing the rays through individual ocular media. Since the first equivalent plane of the reduced eye is less than 1.5 mm from the cornea, object distances are clinically referred to it. The simplicity of visual optics is thus in great measure due to Gauss who invented first order optics, which is sometimes called Gaussian optics.

> I may perhaps be blamed for having inserted too many optical experiments. However, when one considers how little surgeons in general understand about optics, that they scarcely know the term, there is reason to believe that I have not told them too much for an understanding of the function of the parts of the eye and the incidents that arise in the treatment of diseases of the eye.

This little apology appeared in the first comprehensive treatise of ophthalmology by Maitre-Jan, *Traite des maladies de l'oeil,* written in 1707! (Plus ca change, plus c'est la même chose.)

Vertex powers are focal powers, that is, they apply only to parallel rays. The calibration of trial lenses is thus strictly accurate only for distant objects. Of course, that is they way we generally use them to measure and correct ametropia, but back vertex power cannot be used to locate the image unless the incident vergence is zero. To find the image of a near object we must either trace the rays through the system or know its Gaussian constants, that is, its "cardinal points."

Every optical system has three pairs of cardinal points: two (equivalent) focal points, two equivalent (or principal) points, and two nodal points. An object placed at the first equivalent point produces an upright image of the same size at the second equivalent point; hence the equivalent planes are the planes of unit *linear* magnification.

The nodal points coincide with the equivalent points except when the refracting medium behind the system has a different index than that in front, as if one end of a telescope were placed in water (Fig. 3-13). The nodal and equivalent points coincide for lenses because the surrounding medium is air. (The optician frequently speaks of the "nodal" points of a lens, but he really means the equivalent points.) For the eye, however, the second medium is not air. Hence there are both a pair of equivalent and a pair of nodal points. The distance separating the nodal points is always the same as that between the equivalent points.

The separation of nodal and equivalent points is analogous to single refracting surfaces in which the pole represents the equivalent point and the center of curvature represents the nodal point. In a thin lens, the points all coincide at its optical center. In the model eye, the paired equivalent and nodal points are fused as in single surfaces. Nodal points like equivalent points are mathematic concepts, not anatomic points through which "we project visual perceptions."

The two nodal points of a system are conjugate to each other; an object placed at one will give an image that subtends the same angle at the other. The nodal points are points of unit *angular* magnification; a ray directed toward the first emerges undeviated from the second.

Six cardinal points and equivalent power give all the necessary information about an optical system. This is the essence of Gaussian theory. It is a theory in so far as it is limited to paraxial rays and centered systems.

APPENDIX

Numerical example: *Computing a lens system*

A +5 D and a +8 D thin lens in air are separated by 12.5 mm. What are the vertex powers, equivalent power, and position of equivalent planes?

If an object is placed 40 cm from the first lens, where is the image, and what is its magnification (Fig. 3-14).

1. Equivalent power

$$F_e = F_1 + F_2 - dF_1F_2$$

$$= +5 + 8 - \frac{12.5}{1000}(+5)(+8)$$

$$= +12.5\ \text{D}$$

$$F_e = -\frac{1}{f_e} = \frac{1}{f'_e};\ f_e = -8\ \text{cm};\ f'_e = +8\ \text{cm}$$

2. Vertex powers

$$F'_v = \frac{F_e}{1 - dF_1};\quad \frac{+12.50}{1 - \frac{12.5}{1000}(+5)} = +13.3\ \text{D}$$

$$F'_v = \frac{1}{f'_v};\ f'_v = +7.5\ \text{cm}$$

$$F_v = \frac{F_e}{1 - dF_2};\quad \frac{+12.50}{1 - \frac{12.5}{1000}(+8)} = +13.9\ \text{D}$$

$$F_v = -\frac{1}{f_v};\ f_v = -7.2\ \text{cm}$$

3. Position of equivalent planes

$$e' = f'_v - f'_e;\ +7.5 - 8 = -0.5\ \text{cm}$$

$$e = f_v - f_e;\ -7.2 + 8 = +0.8\ \text{cm}$$

Note that e′ is the distance from the second lens to H′, and e is the distance from the first lens to H. This means that a thin lens of power +12.50 placed at H′ focuses parallel light at the same point as the lens system; similarly, a thin lens of +12.50 D placed at H would make light emerge parallel, as does the system.

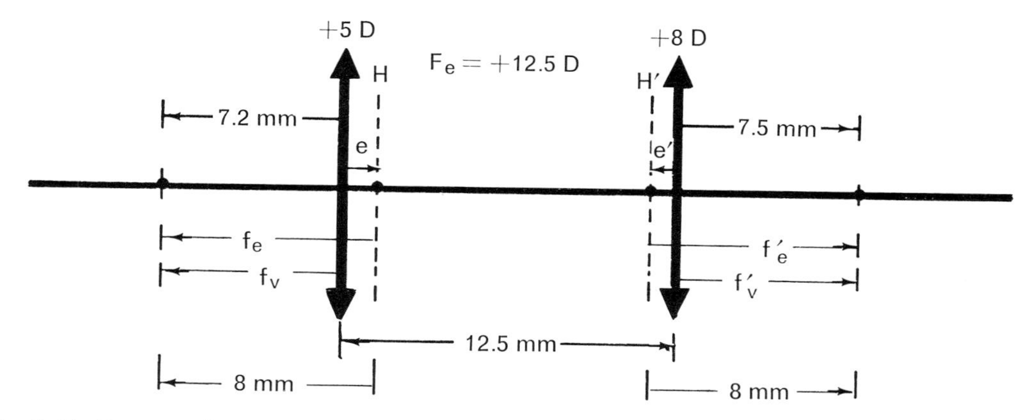

Fig. 3-14. Diagram illustrating a numerical example in which equivalent and vertex powers and the equivalent planes are computed. Note that the position of the equivalent planes is simply the difference between the respective vertex and equivalent focal lengths.

Once the cardinal points are known, the second half of the problem can be solved. An object 40 cm from the first lens is 40.8 cm from the first equivalent plane. Applying the thin lens paraxial equation:

$$\frac{1}{l'} - \frac{1}{l} = \frac{1}{f'_e}; \frac{1}{l'} - \frac{1}{-40.8} = \frac{1}{+8}$$

$l' = +9.9$ cm from H′ or 9.4 cm from F_2

The magnification $m = l'/l = +9.9/-40.8 = -0.24$

This example seems rather long and tedious considering one could have traced the rays through this system by the vergence method fairly easily. Recall, however, that the cardinal points for this system need only be computed once and then can be used over and over to determine the image position for various objects. When the system is more complicated as in the case of the eye, ray tracing for each object distance would be a nuisance indeed; thus we are grateful to Gullstrand for providing the Gaussian constants. The method Gullstrand used is identical in principle to the method just described, including only the index and taking each two ocular refractive components and combining them with the next two and so on for all the ocular surfaces to obtain an equivalent power of 58.64 D.

REFERENCES

Askovitz, S. I.: A simple nomogram for problems in optics, Arch. Ophthal. **53:**702-705, 1955.

Barnett, A.: The numbering of lenses, Optician **102:** 315-316, 1942.

Conrady, A. E.: Applied optics and optical design, New York, 1960, Dover Publications, Inc.

Fincham, W. H. A.: Optics, ed. 6, London, 1956, Hatton Press, Ltd.

Gauss, C. F.: Dioptrische Untersuchungen, Goettingen, 1841.

Gleichen, A.: Lehrbuch der geometrischen Optik, Leipzig, 1902, B. G. Teubingen.

Goldstein, D.: Optics for optometrists, Philadelphia, 1971, Pennsylvania College of Optometry.

Herzberger, M.: Modern geometrical optics, New York, 1958, Interscience Publishers, John Wiley & Sons, Inc.

Jenkins, F. A., and White, H. E.: Fundamentals of optics, ed. 3, New York, 1957, McGraw-Hill Book Co.

Laurance, L.: General and practical optics, ed. 3, London, 1920, School of Optics.

Linksz, A.: Physiology of the eye: Optics, vol. 1, New York, 1950, Grune & Shatton.

Monoyer, F.: Theorie generale des systemes dioptriques centres, Paris, 1883, Physiques et seances.

Morgan, J.: Introduction to geometrical and physical optics, New York, 1953, McGraw-Hill Book Co., Inc.

Nussbaum, A.: Geometric optics, Reading, 1968, Addison-Wesley Publishing Co.

Pascal, J. I.: Selected studies in visual optics, St. Louis, 1952, The C. V. Mosby Co.

Rubin, M. L.: The optics of indirect ophthalmoscopy, Survey Ophthal. **9:**449-464, 1964.

Sears, F. W.: Principles of physics: optics, ed. 3, Cambridge, 1948, Addison-Wesley Press.

Southall, J. P. C.: The principles and methods of geometrical optics, New York, 1913, The Macmillan Co.

Southall, J. P. C.: Mirrors, prisms, and lenses, ed. 3, New York, 1936, The Macmillan Co.

4

Cylindrical lenses

Approximately 80% of all ophthalmic lens prescriptions specify a cylinder; hence a considerable part of clinical refraction deals with the detection, measurement, and evaluation of ocular astigmatism.

Refraction at spherical surfaces is symmetrical and can be represented by either horizontal or vertical cross-sectional diagrams.* Astigmatic pencils are asymmetrical and must be visualized in three dimensions, which accounts for part of the difficulty in understanding them. The simplest astigmatic surface is a section cut from a solid glass cylinder; a plane through its axis has no curvature, whereas the perpendicular plane has maximum curvature.

In ophthalmic optics, cylinders are designated by axis. A +1.00 D × 180° cylinder means the axis is horizontal and the +1.00 D power is in the vertical meridian. The standard or "TABO" notation (after the optical committee that adopted it in 1917) is identical for *both* eyes, beginning with zero on the right, and ending with 180° on the left. The scale applies to the patient's eye as the examiner views it (Fig. 4-1). It is not necessary to specify angles larger than 180° (for example, 220° = 40°), and the degree symbol is usually omitted.

*Spherical surfaces also exhibit astigmatism if the incident rays strike it at an oblique angle. This is considered in the discussion on lens aberrations in Chapter 5.

ASTIGMATIC PENCIL

A parallel pencil* incident on a simple cylinder is shown in Fig. 4-2. The rays in the plane of the axis are undeviated; those in the plane perpendicular to it form an image line parallel to the axis. This image is located at 1/F, where F is the focal power in that meridian. Thus a +4 × 90 cylinder forms a vertical focal line 25 cm from the lens. A −4 × 90 cylinder forms a virtual vertical focal line at −25 cm; the real rays leave the cylinder as if diverging from this line.

In Fig. 4-3 a pencil diverging from a single axial object point is incident on a plus cylinder axis vertical. The rays in the three illustrated representative planes converge toward the image points I_1, I_2, and I_3 to form a vertical line, so that even a single object point results in a line image. This conversion of point objects into line images is the essence of astigmatic imagery, and the name is derived from it: "a" meaning without and "stigma" meaning point.

Although the three divergent planes of Fig. 4-3 are far from the axis, we pretend that the angles are small and the planes are paraxial. The thin lens formulas then apply to cylinders. An object 50 cm from a +4 × 90 cylinder has an incident vergence $L = \frac{100}{-50}$

*A parallel pencil is sometimes referred to in optics texts as a "cylindrical bundle", meaning the rays arrange themselves in the form of a cylinder. It has nothing to do with cylindrical lenses or astigmatism, and the term is avoided in this book.

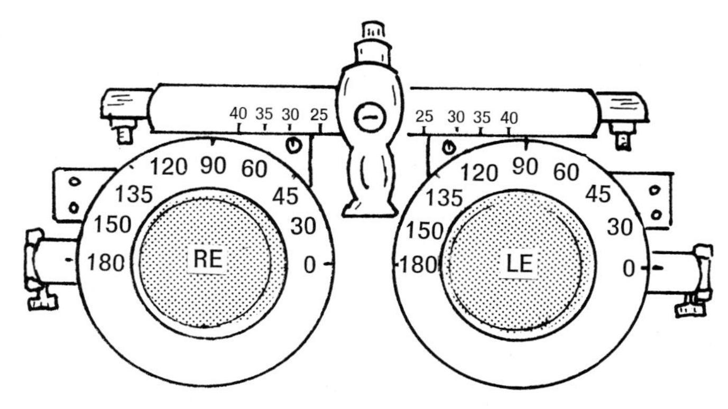

Fig. 4-1. The standard, or "TABO," notation is based on the examiner's point of view and is the *same* for each eye. All ophthalmic prescriptions are written in this notation. In using an astigmatic dial calibrated for the patient's point of view, it is necessary to transpose the axis to the standard notation.

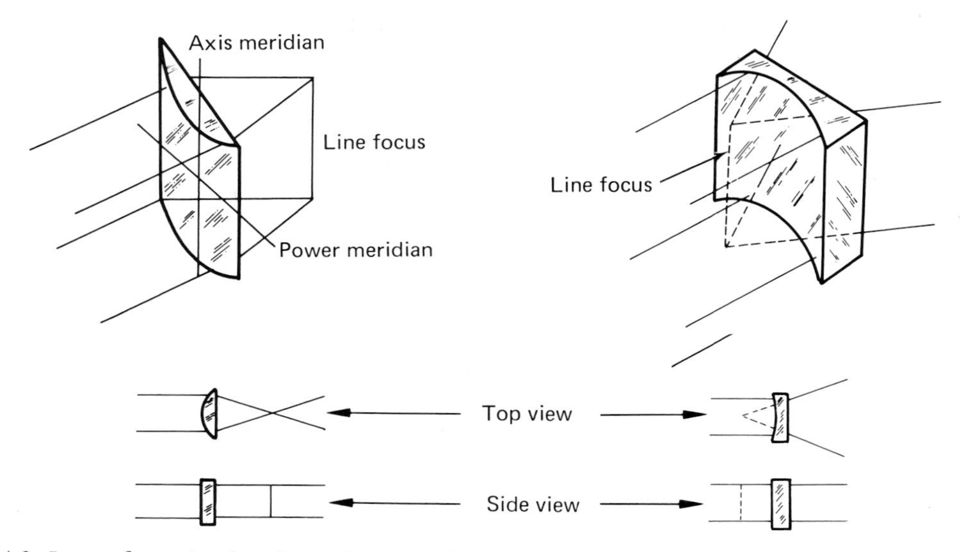

Fig. 4-2. Image formation by plus and minus plano cylinders. The plus cylinder forms a real vertical focal line; the minus cylinder, a virtual focal line. The focal line is always parallel to the axis of the cylinder.

or -2 D, which combines with the $+4$ D in the horizontal meridian to give a vertical image line at $\frac{100}{+2}$ or $+50$ cm. The light in the vertical plane encounters zero power and results in a virtual horizontal image line 50 cm to the left of the lens. We call them image rather than focal lines because the incident rays in this example are not parallel. (This verbal distinction is often ignored.) The example emphasizes that even a simple cylinder forms two image lines, a fact sometimes forgotten when one of them is at infinity. This is the principle of the Maddox rod.

Maddox rod

The original instrument invented by Maddox about 1890 consisted of a single glass rod mounted in a metal disc and illustrates its optics. When positioned before the eye, a distant light spot appears as a streak perpendicular to the axis of the glass rod. The rod, acting as a very strong cylinder, forms a real focal line parallel to the axis, but it is so close to the eye that it cannot be focused on. Instead the patient sees the un-

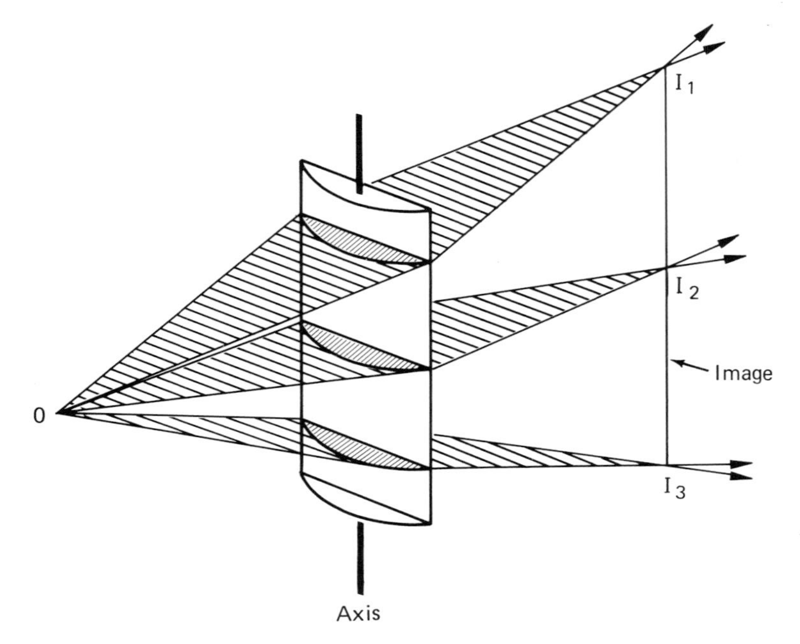

Fig. 4-3. The conversion of an object point into a line image by a plano cylinder is the essence of astigmatic imagery. The point *O* represents any series of object points, all of which form line images. In clinical optics we pretend the three illustrated planes are paraxial so that we can apply our thin lens formulas.

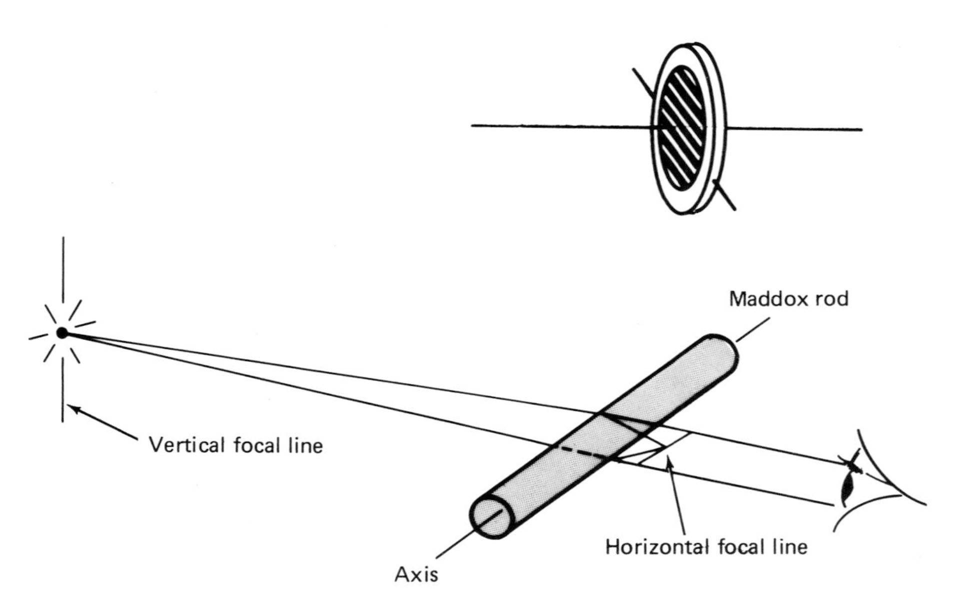

Fig. 4-4. The optics of a single Maddox rod illustrates the principle involved. The rod acting as a very strong cylinder forms a focal line parallel to its axis, but it is so close to the eye that it cannot be focused on. Instead the patient sees the virtual vertical focal line, which he usually (but not inevitably) localizes at the plane of the light source. If one views the source through a half-silvered mirror, one can "project" the vertical line to the surface of the mirror and watch the dancing displacement consequent to changes in accommodative convergence.

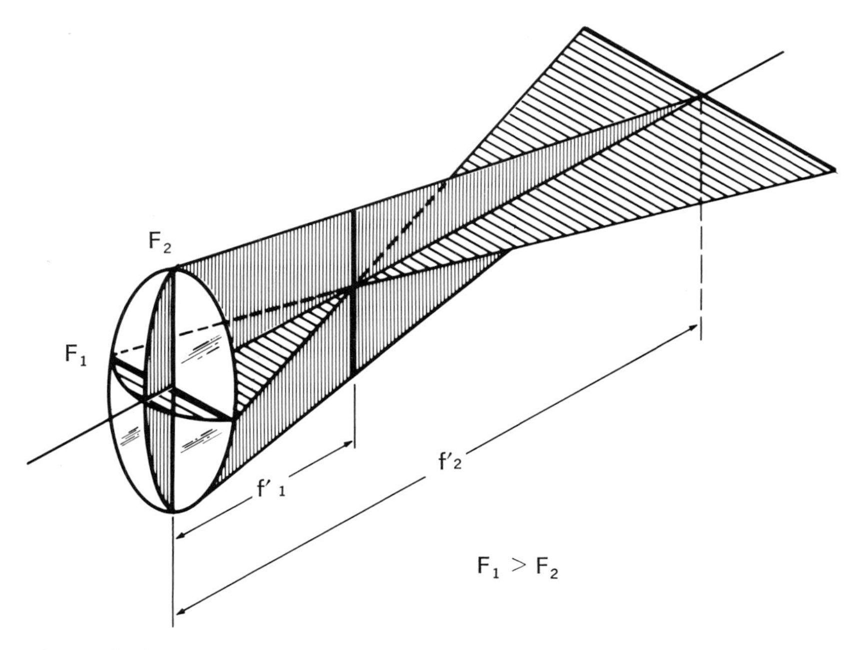

Fig. 4-5. A spherocylindrical (or "toric") surface has a different curvature in each meridian but has one maximum and one minimum meridian, called the "principal meridians." In this example, the horizontal meridian has the greatest curvature; hence the vertical focal line is focused first. Note that there are not really two separate focal lines but a series of convolutions of the same beam; what is seen depends on where the screen intersects it.

deviated rays, which form a virtual image line perpendicular to the rod at the plane of the light source. The eye focuses on this virtual image and generally, although not inevitably, sees it at that distance (Fig. 4-4).

Most ophthalmic prescriptions are combinations of spheres and cylinders called "toric surfaces" because of a resemblance to the base molding of Ionic columns. They look like a slice from a doughnut or automobile tire, with a different curvature in each meridian and a maximum and minimum perpendicular to each other (Fig. 4-5).

A negative toric surface can be visualized as a mold from which the plus toric is shaped. The saddle is a toric curve, convex in one meridian and concave in the other. In addition to their toricity, ophthalmic lenses are bent into a meniscus profile. The terms "meniscus" and "toric" are sometimes confused; meniscus means that one lens surface is not flat, whereas toric means spherocylindrical.

The astigmatic pencil produced by the spherocylindrical surface $+2 + 2 \times 90$ from a distant object forms a vertical line at 25 cm and a horizontal line at 50 cm. The lines appear consecutively, not simultaneously, and represent different cross sections of the same beam. We become so used to speaking of two focal lines that we may mistake them for two independent images. The same rays actually participate in forming each. The astigmatic pencil, consisting of lines and all the configurations in between, is named the "conoid of Sturm" after the French mathematician who described it in 1845. The distance between the focal lines is Sturm's interval. Incidentally, Sturm erroneously believed astigmatism to be a desirable kind of "depth of focus" that would make ocular accommodation unnecessary.

Representative cross sections of the conoid of Sturm (Fig. 4-6) are mostly elliptic, the long axis depending on the nearest focal line. The location and direction of the focal lines

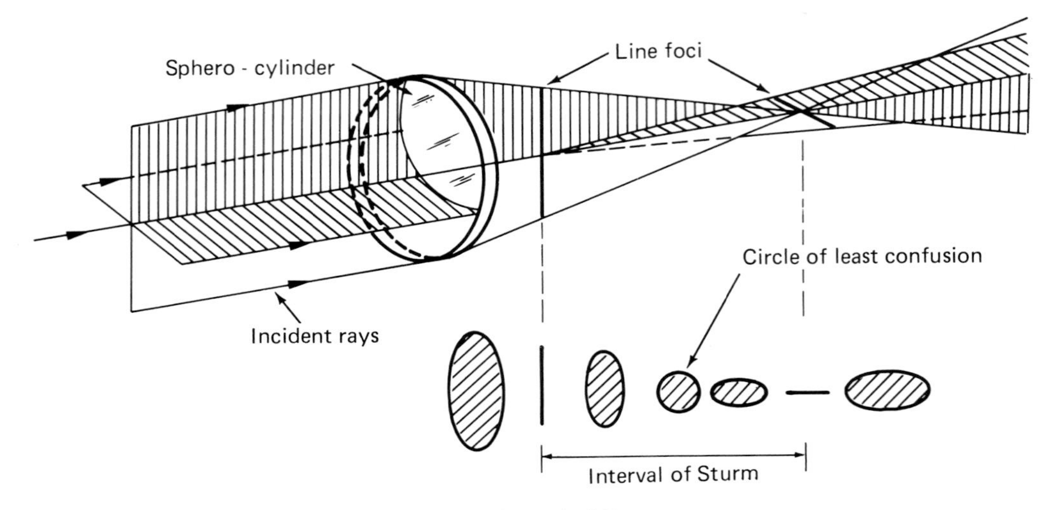

Fig. 4-6. Conoid of Sturm.

vary with the power and orientation of the primary lens meridians. The vertical meridian yields a horizontal line, and the horizontal meridian yields a vertical line. The lines may be real or virtual, the one nearest the lens belonging to the meridian of greatest plus power. At the dioptric midpoint the cross section is circular—the circle of least confusion. Its circular shape mimics the shape of the lens aperture; square openings produce squares of least confusion. The phrase "least confusion" refers to the uncorrected astigmatic eye; when the circle is on the retina the uncorrected vision is least confused.

The diameter of the circle of least confusion gets smaller when the focal lines approach each other. The principle of correcting ocular astigmatism is to collapse the interval to a point image. The confusion circle also depends on the diameter of the aperture, and therefore no matter how severe the astigmatism, the vision will always be better if the pupil is small. The circle is not the smallest area of the beam, however, as is widely taught; hence it does not produce the brightest image.

OPTIC CROSS AND TRANSPOSITION

An optic cross is a convenient diagram of a spherocylinder to indicate the powers in each primary meridian and the axis orientation. Fig. 4-7 shows the optic cross for several different spherocylinders.

The optic cross indicates total power in each primary meridian, not actual surface powers. Since surface powers determine magnification, these may be shown by adding the curved line as in Fig. 4-8. The added curve represents the spherical surface and is drawn according to whether the toric surface is in front or behind.

The term "base curve" is somewhat equivocal. To the manufacturer it means the finished curve on a semifinished blank, generally the lowest curve on the toric surface for plus cylinder forms and the sphere for minus cylinder forms. For our purposes, we define it as the weakest curve on the toric side whatever the lens form.

Every lens prescription can be written in one of three ways: plus cylinder form, minus cylinder form, and crossed cylinder form. The following represent the same prescription: $-2.00 + 5.25 \times 180$; $+3.25 - 5.25 \times 90$; $-2.00 \times 90 \supset +3.25 \times 180$. The clinician writes prescriptions according to the trial lenses used in the examinations; the optician in turn selects the form that matches the lens blanks. Conversion from one to the other is called "transposition."

It is almost impossible to make transposition errors if the prescription is first dia-

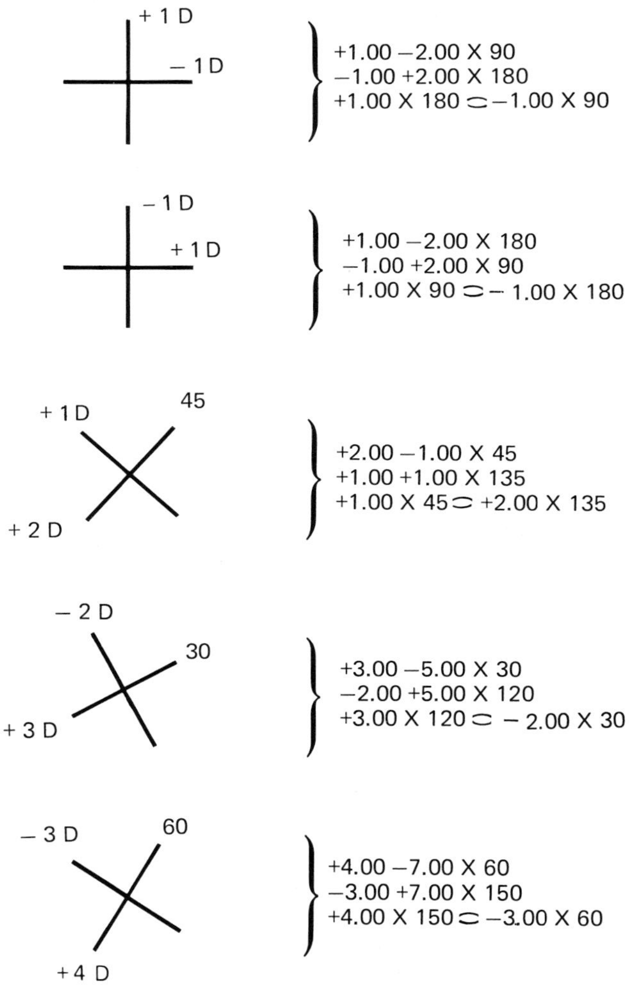

+ 1 D
− 1 D

+1.00 −2.00 X 90
−1.00 +2.00 X 180
+1.00 X 180 ⊂ −1.00 X 90

− 1 D
+ 1 D

+1.00 −2.00 X 180
−1.00 +2.00 X 90
+1.00 X 90 ⊂ − 1.00 X 180

+ 1 D
45
+ 2 D

+2.00 −1.00 X 45
+1.00 +1.00 X 135
+1.00 X 45 ⊂ +2.00 X 135

− 2 D
30
+ 3 D

+3.00 −5.00 X 30
−2.00 +5.00 X 120
+3.00 X 120 ⊂ − 2.00 X 30

− 3 D
60
+ 4 D

+4.00 −7.00 X 60
−3.00 +7.00 X 150
+4.00 X 150 ⊂ −3.00 X 60

Fig. 4-7. Optic cross diagram for several different spherocylinders, each written in one of three prescription forms.

grammed on an optic cross because we can see at once if it represents the same lens. The two prescriptions $-3 + 3 \times 90$ and $+3 + 3 \times 90$ each have the same cylinder, but the first has no power in the horizontal meridian, and the second has +6 D. The spherocylindrical form emphasizes power difference; the optic cross, total power. The spherocylinder form is a holdover of the days when the lens was made by literally cementing a flat sphere to a flat cylinder back to back. In testing for astigmatism by retinoscopy, astigmatic dial, or cross cylinder we care more about the total power in each principal meridian.

It is good practice to diagram all prescriptions on the cross until the visualization of powers in the primary meridians becomes automatic. Only then should reliance be placed on the following transposition rules. To convert one spherocylinder to another:

1. Add the original sphere and cylinder algebraically to find the new sphere.
2. Change the sign of the cylinder, but retain its power.
3. Turn the cylinder axis through 90° (that is, add or subtract 90°, whichever is appropriate).

The crossed cylinder form is not used in

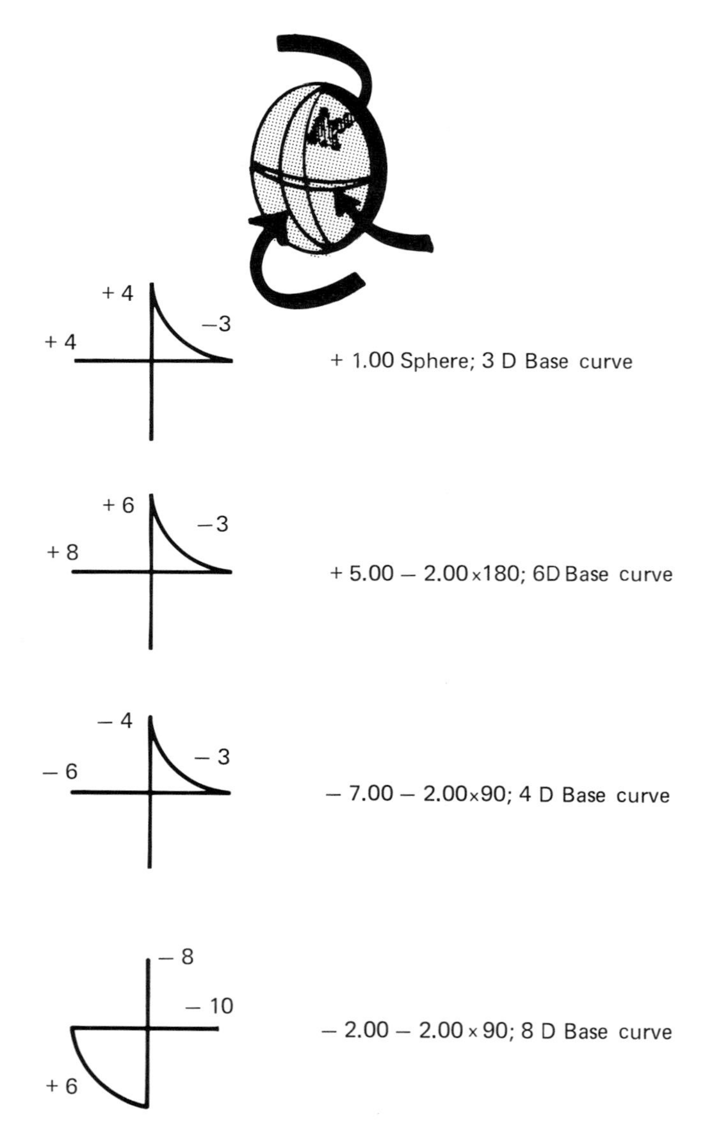

Fig. 4-8. Toric cross diagram for several different spherocylindrical prescriptions. The curved line represents the front or rear spherical surface, and the cross represents the toric surface powers. The total power in each meridian is the sum of front and rear surface powers added algebraically.

grinding ophthalmic lenses but is important in the theory of testing astigmatism and in cylinder retinoscopy. Rather than remember a rule, it is best written directly, transliterated from the optic cross. For example, $+0.50 - 1.00 \times 180$ can be written as $+0.50 \times 90 \frown - 0.50 \times 180$. The symbol $\supset$ means "combined with." The crossed cylinder form emphasizes axis as contrasted to the optic cross, which indicates power. This particular spherocylinder is a "half diopter" Jackson cross cylinder, which is calibrated only to indicate axis.

SPHERICAL EQUIVALENT

The spherical equivalent represents the dioptric midpoint of the interfocal interval and corresponds to the circle of least confusion. It can be computed by taking half the sum of the powers on the optic cross or by adding half the cylinder to the sphere algebraically. The spherical equivalent of +2

+ 2 × 180 is +3 D, and the circle of least confusion is 33.33 cm from the lens (not at the linear midpoint between the focal lines, which in this case is 37.5 cm).

A spherical equivalent does not truly match a toric surface because, being a sphere, it forms point images. But it is a kind of compromise, the best correction under the circumstances. The circumstances may be that toric lenses are unavailable (as occasionally happened during World War II), too expensive, or not advisable for other reasons. To eliminate the cylinder from the prescription +2 + 4 × 90, we may, without stopping to draw it on the cross, simply write +2.00 sphere, a lopsided prescription. The correct spherical compromise is a +4.00 sphere.

In practice, a more common situation arises when the cylindrical element is to be reduced but not eliminated. The patient may never have worn lenses, and prudence suggests prescribing the cylinder in graduated doses. Suppose the prescription is +2 + 4 × 90 and the cylinder is to be cut in half but still maintains the circle of confusion on the retina. We cannot order +2 + 2 × 90 because if the original prescription had been written in minus cylinder form (+6 − 4 × 180), the same reasoning would lead to +6 − 2 × 180, and the two "cut" cylinder prescriptions are not the same; that is, they do not represent the same optic cross. The correct way is with spherical equivalents: Add half the reduced cylinder to the sphere algebraically. This works whether the prescription is written in plus or minus cylinder form. We obtain +3 + 2 × 90 or +5 − 2 × 180, which are identical, and both maintain the circle of least confusion on the retina. Both have the same spherical equivalence as the original "uncut" prescription, and that is the point. The cut prescription not only reduces one meridian but also undercorrects one and overcorrects the other by the same amount.

Spherical equivalents also enter into the clinical testing of astigmatism. Here it is not a question of cutting cylinders but changing them in consequence of the patient's answers. For example, to allow the patient to make valid comparisons in the cross cylinder test, the sphere is proportionately modified with every cylinder change. If we add −0.50 D cylinder, the sphere is increased by +0.25 D.

ASTIGMATIC IMAGERY

Fig. 4-9 illustrates an astigmatic pencil from three representative object points, the rays from each forming a conoid of Sturm. Since the points represent samples of an infinite number of points making up the object, every one is imaged the same way. The image consists of overlapping focal lines, ellipses, or circles depending on the position of the screen. At *A*, the image is composed of overlapping vertical lines; at *B*, of overlapping circles; and, at *C*, of overlapping horizontal lines. A cross target will give an image *A* in which the vertical limb is sharp and the horizontal limb blurred; this is reversed at *C*. At *B*, neither limb is as sharp as before, but both are clearer simultaneously. This is why the spherical equivalent gives the best vision under the circumstances. Of course, if the astigmatism is 6 D, the blur circles are so large that no sphere helps much.

Stenopeic slit

The stenopeic slit is an elongated aperture—a slit cut in a disk—and works on the principle of astigmatic imagery. Placed before the eye, it limits the entering rays to that meridian. If the slit is horizontal (Fig. 4-10), the vertical focal line is reduced to a point. Hence the vision is improved if the retina is at *C*. The change is reversed if the retina is at *A*.

Each primary ocular meridian could theoretically be refracted through the slit (assuming no accommodation), and the result would be the emmetropic prescription in the form of the optic cross. But the slit must be narrower than the pupil, or it will not work; and if it is too narrow, diffraction destroys the image. Either way it reduces retinal illumination. Thus the slit is interesting but seldom useful in routine refraction, except to demonstrate astigmatism to the patient.

POWER IN OBLIQUE MERIDIANS

Since a cylinder has meridians of maximum and minimum curvature, there must also be some intermediate curvatures (Fig. 4-11). The intersection of an oblique plane with a solid cylinder forms an ellipse, not a circle, which makes for tedious mathematics. If the

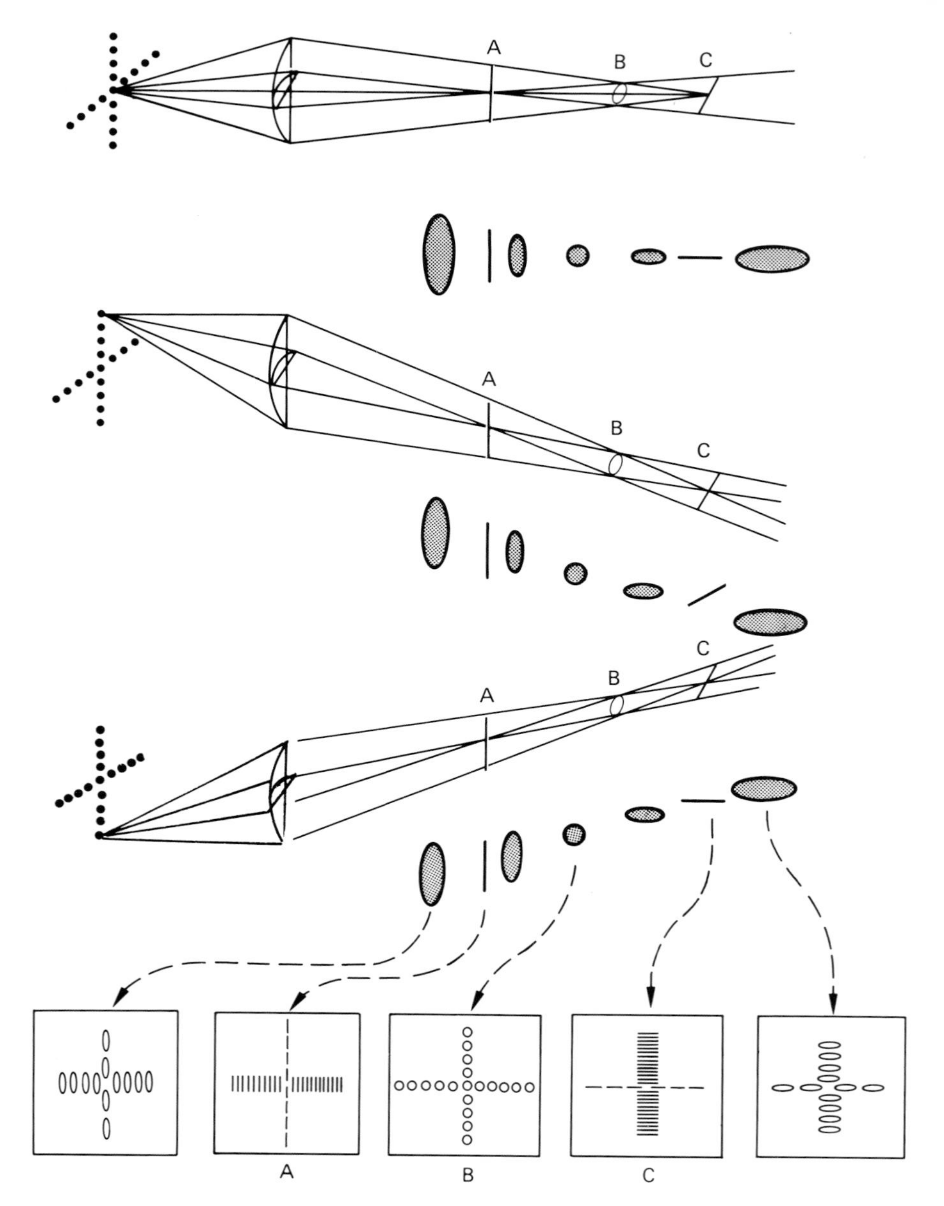

Fig. 4-9. Astigmatic imagery. The image of the object (a cross represented by a series of object points) depends on the position of the screen. Note that at *C*, the object points of both the vertical and horizontal limbs produce horizontal image lines. This is why a patient viewing an astigmatic dial sees the horizontal lines sharper and the vertical lines blurred and grayer.

incident beam is assumed to be paraxial, however, we can pretend the ellipse is practically a circle, and the oblique curvature is then related to the maximum curvature by

$$R_\theta = R \sin^2 \theta$$

where R_θ is the curvature in a meridian θ degrees from the axis and R is the maximum curvature. Adding the curvature in any oblique meridian to the one at right angles to it always equals the sum of the primary meridian curvatures. This is a special case of a more general theorem discovered in 1771 by Euler, who managed to be one of the world's

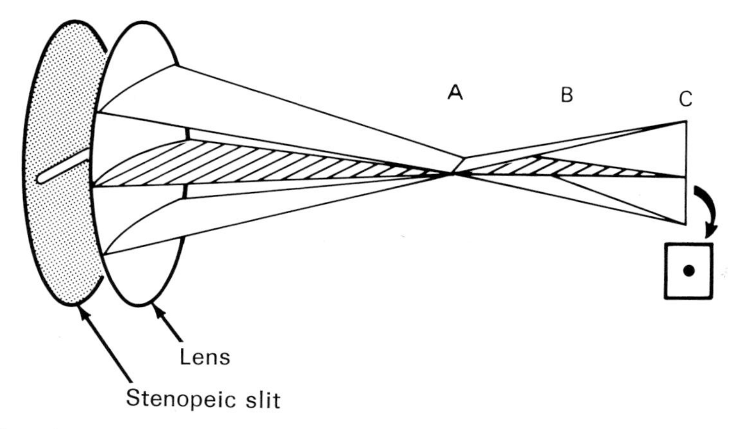

Fig. 4-10. The effect of a horizontal stenopeic slit is to pinhole the vertical focal line and produce a sharper image. Of course it only works if the retina is at position *C*. If the retina were at position *A*, the best vision would be obtained with the slit vertical.

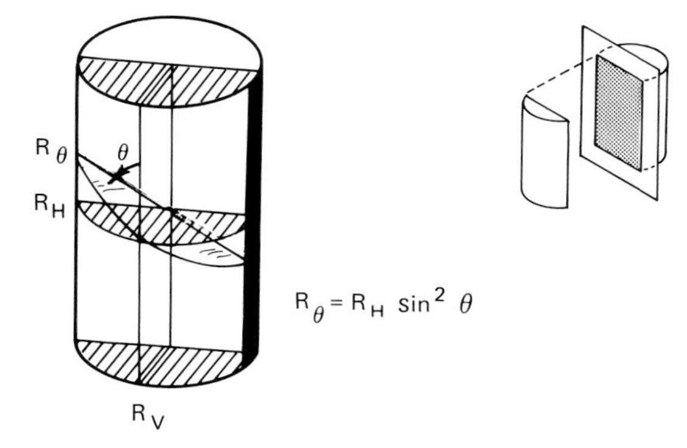

Fig. 4-11. Curvature of a plano cylinder in an oblique meridian.

greatest mathematicians despite the consecutive loss of vision in both eyes.

When a lens measure is placed on a simple cylinder $+1 \times 180$, it reads zero power in the axis meridian, maximum power in the vertical meridian, and other powers in between. In the 45° meridian, for example, it would read $\sin^2 45° = \frac{1}{2}(+1) = +0.50$ D. In the 30° meridian it would read, not 30/90 or +0.33 D, but $\sin^2 30\ (+1) = +0.25$ D. The power is simply the maximum power multiplied by the sine squared of the angle from the axis. So far so good, but if we measure the 30° meridian of this lens in a lensometer, it does not read +0.25 D but +1.00 D. In fact it never reads anything but +1.00 D no matter where we measure. And if we placed this lens on an optical bench, it would always focus at 100 cm.

The reason the lensometer and lens measure give different results is that the latter really measures curvature calibrated as focal power by an assumed index, which presupposes the lens has an optical center. Cylinders have no optical center; they have an axis, so the sine-squared formula has more curvature ingenuity than focal truth. Despite the textbooks, the sine-squared formula never really gives power but only curvature, and the rays do not actually focus where the formula predicts because they take off in skewed directions. But the formula is so simple and the correct computations so tedious that opticians continue to use it, admittedly as an approximation.

PRISMATIC EFFECT

The prismatic effect of a spherocylinder can be computed as for a thin lens by Prentice's formula: $d = cF$, where d is in prism diopters, c is the distance of the ray from the axis in centimeters, and F is the focal power in the meridian investigated. For example, if the line of vision passes 8 mm below the optical center of a $+2 + 2 \times 180$, the focal power in the vertical meridian (from the optic cross) is +4 D, and the prismatic effect is $d = 0.8\ (+4)$ or 3.2Δ base up. Similarly, a point 5 mm nasal to the optic center would have a $0.5\ (+2) = 1\Delta$ base-out effect.

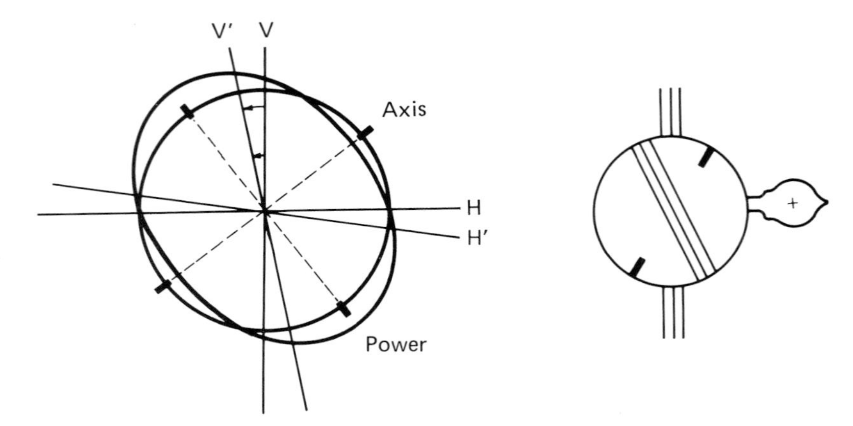

Fig. 4-12. Distortion and magnification of a plus cylinder. The magnification is in the meridian of greatest plus power.

If the primary meridians are horizontal and vertical, the prismatic effect can be computed by obtaining the distance to the respective axes. For oblique meridians, however, horizontal and vertical prismatic effects must be computed by dropping a perpendicular to each principal axis and resolving the resultant oblique prism into horizontal and vertical components. The best way is to treat the spherocylinder as a separate sphere and cylinder. Even then the process is somewhat tedious, and, for the small powers generally involved, the approximate sine-squared formula is used by most opticians. By this method the vertical prism induced at a point 8 mm below the "optic center" of a +2 + 2 × 45 spherocylinder would be d = 0.8 (+3) = 2.4Δ base up. The +3 D is the power in the vertical meridian computed from +2 + $\sin^2 45$ (+2) = +2 + 1 = +3 D.

Prismatic effect explains the displacement of a line seen through a cylinder when it is rotated as in hand neutralization. A vertical line observed through a +1.00 cylinder axis vertical rotates clockwise as the lens is rotated counterclockwise—an "against" break. The break results because the more peripheral points are farther from the axis and therefore more deviated (Fig. 4-12). So a circular target is distorted into an ellipse whose long axis is in the power meridian; that is, the magnification is in the meridian of plus power. These distortions make life for the patient with oblique astigmatism more difficult.

EFFECTIVITY

When two spherical thin lenses, +4 D and +6.6 D, are placed 25 cm and 15 cm, respectively, from a screen, they form sharp (but unequal-sized) images. Both lenses have the same effective power at the screen, although each maintains its own equivalent (thin lens) power. It takes only +4 D to form an image at 25 cm, whereas +6.6 D are necessary to bring parallel rays to a focus 10 cm nearer the screen.

To compute the power of a lens with the same effect as a +4 + 1 × 90 spherocylinder, 10 cm nearer the screen, draw an optic cross (Fig. 4-13), and use the effectivity formula. The power needed in the vertical meridian is +6.6 D, and the power needed in the horizontal meridian is +10.0 D or, in spherocylindrical form; +6.6 + 3.3 × 90. Note that the cylinder difference is now exaggerated, since each meridian is altered unequally. This is one reason the keratometer, which measures power at the plane of the cornea, gives a different reading than retinoscopy, which measures power at the spectacle plane.

When a spherical lens is tilted about a horizontal axis, the image of a point is elongated as if a cylinder had been added to the sphere. The power of the induced cylinder is approximately $\tan^2 \alpha$, where α is the angle of tilt. For a +1 D sphere tilted 15° about the horizontal, the power of the induced cylinder is 0.07 D; tilted 30°, it is 0.33 D; tilted 45°,

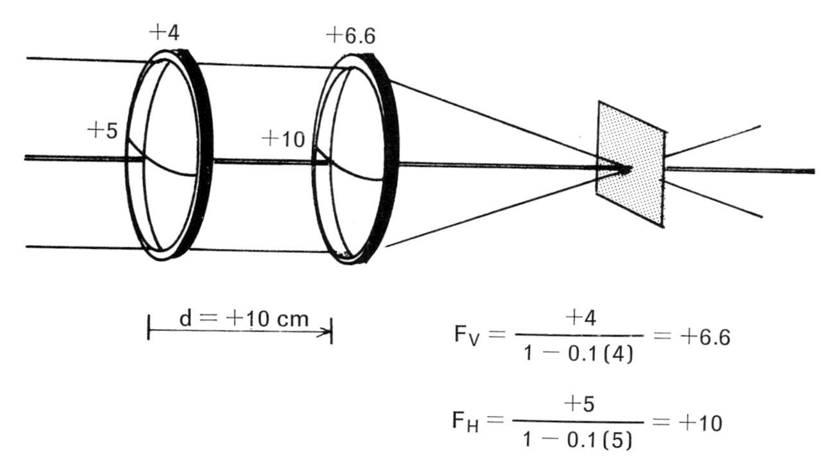

Fig. 4-13. Effectivity of spherocylinder. The lens required 10 cm nearer the screen differs because of effectivity. Since the change affects each meridian by a different amount, the cylinder changes from $+4 - 1 \times 90$ to $+10 - 3.3 \times 90$. The same principle plagues the patient wearing a high cylinder when he views a near object; the cylinder is no longer correct at the reading distance.

it is 1.0 D. The effect on stronger spheres is approximately proportional. The sign of the induced cylinder is the same as the sphere, and its axis is the axis of lens rotation.* The astigmatic effect follows from the oblique incidence of the rays. If the tilt is about 8°, the change is taken into consideration by the lens designer and is about optimum in reducing oblique astigmatism for distance and reading through a single vision lens.

Historical note on astigmatism

"My eye requires a glass whose shape does not correspond to a portion of a true sphere, but rather to a portion of a spheroid or perhaps a cylindric." Thus wrote the Rev. Mr. Goodrich to his local optician in Philadelphia, who provided him with the first American-made cylinder. Goodrich, a layman, had arrived at his own diagnosis between 1825 and 1826, thus anticipating Airy's description by a year. Airy, a British astronomer, was the first to correct his own compound myopic astigmatism with a spherocylinder. Thomas Young described his astigmatism in 1801, but nothing came of it, possibly because he added that it caused him no inconvenience.

Stokes in 1849 developed a rotary cylinder, the forerunner of the modern Jackson cross cylinder. By 1860 Fick, Helmholtz, Bowman, Listing, and Knapp had measured their own astigmatism, and after Donders, it ceased to be a laboratory curiosity. Strangely, clinical astigmatism had to wait 125 years to be discovered despite the fact that I. Barrow, Newton's predecessor at Cambridge, had described oblique astigmatism in optical systems as early as 1669.

*There is also a negligible change in spherical power due to the tilt, but the induced astigmatism is far more important.

OBLIQUELY CROSSED CYLINDERS

Two cylinders combined at an angle other than 90° are said to be crossed obliquely. For example, two +1 D cylinders with their axes superimposed, give twice the cylindrical effect; crossed at right angles, the result is a +1 D sphere; crossed at an intermediate angle, the result is some sphere and some cylinder. The induced spherical component of oblique cylinder combinations is unavoidable, which is one reason the Stokes rotating cross cylinder never became popular. The mathematics of oblique cylinders is tedious and may be avoided by duplicating the problem with trial cylinders, which are then hand neutralized or placed in a lensometer. In the clinical testing for astigmatism, however, in which one cylinder represents the ocular astigmatism and the other the correcting (glass) cylinder, this is not feasible. The situation arises whenever the axis of a correcting cylinder is displaced from the true axis of ocular astigmatism. This is the basis of the Jackson cross cylinder, cylinder retinoscopy, "rocking" the cylinder, and the astigmatic dial tests. Since one cylinder represents the eye and the other the correction, the only

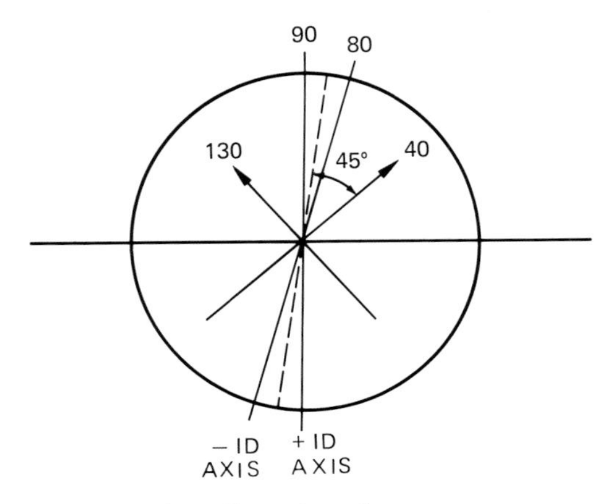

Fig. 4-14. The effect of combining two opposite but equal plano cylinders with their axes separated 10° is to create an entirely new astigmatism whose principal meridians are at 40° and 130°. The result can be confirmed by placing two trial cylinders in a lensometer or by computation as outlined in the appendix to this chapter.

kind of obliquely crossed cylinder problems we care about always have *opposite* signs.

Combine a +1 D and a −1 D cylinder from the trial set: +1 × 90 and −1 × 80. Place this combination in the lensometer, and note the new primary meridians. They are not at 90°, 80°, or in between, but at an entirely new and somewhat unexpected position: 130° and 40° (Fig. 4-14). Although the cylinder axes were displaced only 10°, the new meridians of maximum and minimum power have shifted some 45° roughly by a factor of four. When the two opposite cylinders have approximately the same power, the new astigmatism is some 45° from the bisector of the angle between them. In this case the bisector is $\frac{1}{2}$ (10°) = 85°. Add 45° to obtain 130°; subtract 45° to obtain 40°, which is the result found with the lensometer. The formula method to compute the result is shown below.

APPENDIX
Formula solution for oblique cylinders

Equations	*Numerical example: Combine +1 × 90 and −1 × 80*
Step 1. Transpose cylinders to the same sign. Set spherical component aside.	+1 × 90 −1 + 1 × 170 (S = −1)
Step 2. Choose cylinder with smaller numerical axis as C_1.	C_1 = +1 × 90
Step 3. Find the angle α from: (axis C_2 − axis C_1)	$\alpha = 170 - 90 = 80$
Step 4. Find the angle θ from the formula: $\text{Tan } 2\theta = \frac{C_2 \sin 2\alpha}{C_1 + C_2 \cos 2\alpha}$	$\text{Tan } 2\theta = \frac{1(\sin 160)}{1 + 1(\cos 160)}$ $2\theta = 260°$ $\theta = 130°$
Step 5. Determine the spherical component (S) from: $S = C_1 \sin^2 \theta + C_2 \sin^2 (\alpha - \theta)$	$S = 1 (\sin^2 130) + 1 (\sin^2 50)$ $= +1.2$ D
Step 6. Determine the resultant cylinder power from: $C_T = C_1 + C_2 - 2S$	$C_T = +1 + 1 - (2.4) = -0.4$
Step 7. The direction, in standard notation, of the resultant cylinder is given by adding θ to the axis of C_1.	Resultant axis = 90 + 130 = 220 or 40°
Step 8. Find the total spherical component by adding S to the sphere set aside in Step 1.	Total sphere = (−1) + (+1.2) = +0.2 D
Answer	+0.2 − 0.4 × 40

REFERENCES

Airy, G. B.: On a peculiar defect of the eye and a mode of correcting it, Trans. Cambridge Philosoph. Soc. **2**:267, 1827.

Askovitz, S. I.: The circle of least confusion on Sturm's conoid of astigmatism, Arch. Ophthal. **54**:691-697, 1956.

Bennett, A. G.: Prismatic effect of sphero-cylinders: a new graphical construction, Optician **119**:633-637, 1950.

Bennett, A. G.: Prismatic effects of cylinders: a recurrent fallacy, Manufact. Optician **3**:575-577, 1950.

Burian, H. M., and Ogle, K. N.: Meridional aniseikonia at oblique axes, Arch. Ophthal. **33**:293-310, 1945.

Casanovas, J.: The untoward influence of astigmatism on the statement of visual acuity, Am. J. Ophthal. **61**(2): 1059-1062, 1966.

Euler, L.: Sur la perfection des verres objectifs des lunettes, Memoires de Berlin **3**:274-296, 1747.

Goode, H.: On a peculiar defect of vision, Trans. Cambridge Philosoph. Soc. **8**:493, 1846.

Green, J.: On certain stereoscopical illusions evoked by prismatic and cylindrical glasses, Trans. Am. Ophthal. Soc. **5**:449, 1888.

Hallett, J. W.: Optics of cylinder magnification, Am. J. Ophthal. **33**:1090-1095, 1950.

Hays, L.: Lawrence's diseases of the eye, Philadelphia, 1854, Blanchard & Lea.

Kronfeld, P. C., and Devney, C.: Frequency of astigmatism, Arch. Ophthal. **4**:873-884, 1930.

Maddox, E. E.: Ophthal. Review, May, 1890.

Pascal, J. I.: Power of cylinders in oblique meridians, Arch. Ophthal. **22**:290-291, 1939.

Prangen, A. De H.: Significance of Sturm's interval in refraction, Am. J. Ophthal. **24**:413-418, 1941.

Riddell, W. J. B.: Note on spherical equivalent of sphero-cylindrical lenses, Br. J. Ophthal. **27**:302-304, 1943.

Snyder, C.: The Rev. Mr. Goodrich and his visual problem, Arch. Ophthal. **73**:587-589, 1965.

Stokes, G. C.: On a mode of measuring astigmatism of a defective eye, Br. Assoc. Adv. Sci., pp. 10-11, 1850.

Sturm, J. C. F.: Memoire sur la theorie de la vision, Compte Rendu de l'Academie des Sciences, Paris **20**:554-560, 761-767, 1238-1357, 1845.

Swaine, W.: Astigmatic lens distortion and scissors movement, Optician **87**:285-289, 1934.

5

Practical ophthalmic optics

Prescription writing is one of the oldest practices associated with the art of healing. Its effectiveness depends on the validity of the diagnosis and the choice of therapeutic agent, its mode of administration, mechanism of action, and side effects. The treatment of refractive disorders involves an optical prescription whose efficacy depends not only on the accuracy of the examination but also on the care with which it is translated into the final lens and the manner in which it is fitted to the eye. Not specifying the decentration, bifocal style, quality, or color of a lens makes as little sense as not indicating the dosage and route of administration of an antibiotic. Only the clinician who has taken the history and studied the patient has the information from which such decisions can be made. Moreover, one of the characteristics of correcting ametropia is that the lens and eye together form a new optical system in which the fit of the lens plays a major role. If the fit is wrong, not only is the prescription altered, but the entire optical behavior of the lens differs from what was intended.

The inventor of spectacles is not known, but in a sermon delivered in Florence in 1305 Brother Giordano stated that he saw and spoke with him, and "it is not yet 20 years since the art was invented." Moreover, Allessandro Spina, according to the Annals of the monastery of St. Catherine in Pisa, saw the product and duplicated it. Spina died in 1313, and although opticianry is not the oldest profession, it was plagued by enough amateurs to cause Venetian glassmakers to lay down strict rules. For example, in 1317 they granted "Francisco, son of the late surgeon Niccolo, the right to make spectacles of glass and sell them in Venice." No conflict of interest seems to have been recorded by the local surgical college. Roger Bacon, frequently quoted as the inventor of eyeglasses, was at least half right, since his knowledge extended to single lenses and magnifiers to be held near the object rather than in front of the eye. The first medical writer to mention spectacles was Guy de Chauliac (circa 1363): "If these collyria are not successful, spectacles must be used."

In the Florentine church of Santa Maria Maggiore on the tomb of Salvino Armato is inscribed "Inventor of eyeglasses. May God forgive him his sins. AD 1317." Whatever Armato's sins, inventing spectacles was not included for he, his genealogy, and his tomb, were apparently dreamed up by Del Migliore, a seventeenth century member of the local chamber of commerce pushing the city virtues. No one knows whose tomb it is; the bust is that of a Roman soldier circa 100 A.D., obviously added to impress the tourists. Why the invention of spectacles should be a sin is not clear; perhaps the epitaph was coined by some disgruntled lens grinder who failed his board exam.

LENS FORM AND LENS POWER

In elementary optics a converging lens is plus, a diverging lens is minus, and it makes little difference how the surface powers are distributed, whether biconvex, planoconvex, or concavoconvex. In ophthalmic optics, shape determines optical efficiency because the eye looks through different lens zones at different angles. In thin lens optics, focal power is the simple sum of two surface powers; for example, a thin lens whose front surface is +16 D and whose rear surface is −6 D is considered to have a power of +10 D. Real lenses, however, have some thickness, which alters the power. A lens with the same surfaces as in the previous example and with a thickness of 5 mm of crown glass (n = 1.523) has a back vertex power of $F_1 + F_2 + t/n\ (F_1)^2$ or +10.84 D, a front vertex power of $F_1 + F_2 + t/n\ (F_2)^2$ or +10.12 D, and an equivalent power of $F_1 + F_2 - t/n\ F_1F_2$ or +10.31 D. There is thus a 0.75 difference according to the way the lens is placed and the power read off in the lensometer, and a 0.31 D difference between the approximate (thin) lens power and the equivalent (or true) power.

In 1912 von Rohr suggested back vertex power* as the practical unit, and it has become so universal that when an optician speaks of diopters, or the clinician writes his prescription in diopters, back vertex power is understood. The advantage of this power unit is that whatever the lens profile, thickness, index, or distance from the eye, the power is correct for the refractive error, since the back vertex focal length then coincides with the far point. Because clinical refraction deals only with the far point, the rear lens surface always provides a tangible measuring point.

MEASUREMENT OF LENS POWER

The physicist measures lens power with an optical bench and elaborate accessories and may take days to complete the task for an involved system.† The refractionist deals with single, relatively thin, and low-powered lenses and makes use of three methods, each of which, however, measures a different power aspect.

Hand neutralization

In hand neutralization, trial lenses of known power are used to eliminate "motion" seen through the unknown lens relative to a distant target. Since most ophthalmic lenses are meniscus, the neutralizing lens can only be apposed to its front surface; hence hand neutralization measures front vertex power. A cross serves as a convenient distant target, and a sheet of paper with a ½-inch hole between the lenses acts as a baffle. The lens held at arm's length is moved perpendicular to the clinician's line of vision. Spherical lenses show the same motion in all meridians: "with" if minus, "against" if plus, and none if plano. Spherocylinders are identified by the "break" in the limbs of the target; an "against" break indicates the plus cylinder axis and a "with" break, the minus cylinder axis. When the cross limbs appear continuous, they coincide with the primary meridians. Their intersection is the optical center, which is then ink marked. The motion in each meridian is now neutralized by appropriate spherical trial lenses, and the axis is read from a protractor. The result is best recorded on an optic cross diagram from which the prescription may be transposed to spherocylindrical form.

If the lens contains a prism, the target appears displaced toward the apex of the prism. It is measured by apposing a known prism, with its base in the direction of the displacement.

Hand neutralization is clumsy and time-consuming and frequently leads to transposition errors. More important, the results may be misleading in strong prescriptions, such as in aphakia, because it does not measure back vertex power, the power of clinical interest. Neutralizing an unknown lens is seldom so urgent that it cannot wait to be placed in the lensometer.

Lensometer

The lensometer (Focimeter, Vertometer, Refractionometer, Vertexometer, etc.) is a miniature optical bench, including a light source, reticle, movable target, telescope, and lens holder. The reticle consists of concentric circles, a single spherical line, and a set of triple cylindrical lines. The eye piece is adjusted to the operator's vision, and the unknown lens is carefully placed with its *ocular* surface against the lens stop. The

*Opticians sometimes refer to back vertex focal power as "effective power," not to be confused with effectivity, which applies to any lens, thick or thin, displaced to a new position.

†Occasionally even a clinician must return to the optical bench, for example, to measure the equivalent power of subnormal visual aids whose varying constructions are not readily comparable by vertex powers (Sloan and Jablonski, 1959).

instrument now reads back vertex power directly. If front vertex power is desired, the lens is placed in the instrument "backward." The two vertex powers are similar for minus spheres, but front vertex power is always weaker in plus lenses.

Rotate the power drum until the target comes into focus. If both spherical and cylindrical lines are sharp, the lens is spherical; if only one set of lines is sharp, the lens is a spherocylinder. The prescription may be read in plus or minus cylinder form. To obtain the result in plus cylinder form, set the power drum at the lower of the two possible focusing positions, and rotate the cylinder axis wheel until the single line is aligned. Now rotate the power drum until the triple lines are in focus. The first reading is the sphere, and the *difference* between the first and second readings is the plus cylinder power whose axis is given by the protractor. The intersection of the target lines locates the optical center whose position is marked with the inking device. The distance between the optical centers of the lenses in a frame (measured with a millimeter ruler) should coincide with the patient's interpupillary separation (unless deliberately decentered). Reading glasses are, of course, centered for the near interpupillary distance. Inking optical centers is the only way of confirming the match to the pupil distance, but this does not mean that they automatically lie in front of each pupil, since eyes are often (as many as 60%) asymmetrical, with one closer to the midline than the other. This may be checked by noting that when the lenses are properly centered, their surface reflections and the corneal reflex are all in line.

Ophthalmic prisms are evaluated by placing the lens with its inked optical center at the midpoint of the lens stop and the horizontal axis coincident with the axis marker pins. Focus the target and note the displacement relative to the concentric, calibrated reticle circles. The direction of the prism base in standard notation is indicated by the scale.

A bifocal "add" is the difference between the two spherical or the two cylindrical readings, taken through the distance and near portions of the lens. Bifocal prescriptions are written as an add because they were first made by literally cementing a small reading add to the distance portion. To measure the add in fused (front surface) bifocals (especially in aphakic prescriptions), read the distance prescription in the usual way; now turn the lens with its convex surface against the lens stop to measure the reading portion; the difference is the correct add. The reason for this is that the lensometer only gives back vertex power if the incident rays are parallel. Since they are not parallel through the segment (seg), the lens must be reversed.

The tolerance of the instrument, if kept in proper working order, is ±0.12 D. Inaccurately adjusted eyepieces, improper centering of loose lenses, placing the wrong end of the lens against the lens stop, and an unfortunate habit of leaving the inkwell dry will lead to errors. Projection lensometers allow several observers to use the same instrument without individual eyepiece adjustments. Recent innovations include electronic scanning with print-out results.

Lens measure

A lens measure (lens gauge, lens clock, or Geneva lens measure) looks like a pocket watch with three legs, two fixed and one moving. When applied to a curved surface, the displacement of the central leg activates a pointer indicating the "sag," which is proportional to curvature. The instrument, usually calibrated for glass of index 1.53 (stamped on the face) must be corrected if the lens has a different index n. The correction formula is as follows:

$$\text{True power} = \frac{n - 1}{0.53}\,(\text{lens measure reading})$$

For example, if the lens measure reads +6.00 D when used on plastic of index 1.49, the true power is $\frac{1.49 - 1}{0.53}(6.00) = +5.55$ D. A higher index naturally gives higher power.

The lens measure legs should be applied

perpendicular to the lens surface. When all three are in contact, the reading is correct, and the sum of the front and rear surface powers is the approximate or thin lens power. Toric lenses give a maximum and a minimum reading for the same surface.

Lens clocks not only measure base curve, but, in addition, an odd or fractional reading usually means the lens is "corrected." Some manufacturers place a trademark on their corrected curve lenses, made visible by "breathing" on them. (An occasional patient whose glasses become steamed sees muscae "AO-tantes.") Since shape magnification varies with base curve, it must be known to evaluate induced aniseikonia. A base curve that differs from what a patient is used to may be the wrong choice if he is sensitive to magnification changes.

Lens clocks deserve the same care given to a fine watch, since they readily get out of adjustment. The zero setting should be periodically checked against a known flat. Accuracy is within 0.25 D.

Lens caliper

Lens calipers measure thickness in "points"; "5 point" means 0.2 mm and 10 points, 0.1 mm. The legs are applied tangent to the lens surface at its optical center. A vernier scale is attached to the movable arm for interpolation, and the reading should be zero when the points touch. (An adjustment is provided by a set screw.) Thickness is not only a factor in magnification, but also a minimum thickness (2.2 mm) is required for heat treatment to make the lens impact resistant.

OPTICAL GLASS

The origin of glass production is lost in antiquity. The first glass was probably formed by the action of lightning on sand and was prized, not for its optics, but for the sharp edges that could be fashioned into weapons. Modern optical glass, developed in the last century, consists of ingredients that are mixed, heated to high temperatures, cooled, stirred, and annealed. The finished sheets, cut into small pieces, are molded and supplied to the lens manufacturer who nowadays completes the operations by electronically controlled, and computerized, machinery.

Lenses are molded to the approximate shape, mounted on iron blocks in pitch, and ground with various grades of emery. The surfaces are then polished, washed, and shipped. Single vision lenses are supplied as semifinished blanks or finished uncut lenses. In the semifinished lens, toric plus cylinders are provided to the local laboratory which grinds the spherical surface. With minus cylinders and practically all bifocals, the ocular toric surface is ground by the local laboratory. Finished uncut lenses are supplied in a variety of sizes in the most frequently used prescription ranges, requiring only cutting and edging and greatly reducing the time to fill the prescription.

Common ophthalmic crown glass ($n = 1.523$) contains 70% sand, 12% lime and 18% soda, with small amounts of potassium, arsenic, borax, and antimony to improve quality. Flint glass contains lead oxide and has a higher index of refraction ($n = 1.625$), greater luster, and dispersion. Barium glass contains barium oxide and is similar to flint with less dispersion ($n = 1.621$). The specific gravity of crown glass is 2.54, and 3.66 is the specific gravity of flint, compared to plastic, 1.32. Flint and barium are commonly used for fused bifocal segments. Single vision flint lenses reduce thickness ("Thinlite") but the weight is greater and, since heat treatment is impossible, are no longer available.*

Colored glass is made by adding metallic oxides to the raw materials. Photochromic glass contains microscopic particles of silver halide, which decompose by exposure to ultraviolet rays into silver and halogen and recombine in the dark to the original colorless state.

Ophthalmic glass is required to be free of striae, bubbles, and foreign particles. It should be hard, color constant, and of reproducible index and dispersion.

*A new high-index glass ($n = 1.70$) has recently become available under the trade name High-Lite. These lenses can be chemically treated and are only 18% heavier than crown glass. Only single vision series are in production so far.

PLASTIC LENSES

An increasing proportion of ophthalmic lenses are now made of plastic because they are lighter, safer, and can be molded into aspheric shapes. Plastic has better transmission, less internal reflection, fogs less in high humidity, and absorbs about 90% of ultraviolet rays. Plastic lenses weigh about 40% less than glass ones, a considerable advantage in thick plus perscriptions.

Plastics suitable for optical lenses must be homogeneous and transparent and retain their form when molded or cast. The two principal plastics used in lenses are polymethyl methacrylate (n = 1.49), which is thermoplastic, and a material developed in 1947, CR-39, or allyl diglycol carbonate (n = 1.50) which is thermosetting. CR-39 lenses are cast and have high impact and scratch resistance.

An important advantage of plastics is the ability to manufacture low-cost aspheric curves for aphakic corrections. Even temporary plastic aspherics are now produced at reasonable prices. Another benefit of plastic technology is the Fresnel lens and prism made of thin vinyl film. Although contrast degradation and light scattering properties limit their usefulness, they are ideal for temporary aphakic corrections and temporary adds and prisms in strabismus or in evaluating potential adjustment to bifocals and trifocals. Fresnel prisms, even of high power, are cosmetically acceptable because the membrane thickness is only 1 mm. The vinyl material, cut to shape and size, is placed on the ocular surface of the lens. Two 30Δ Fresnel prisms on rotary plano (Risley) carriers can be used to evaluate occulomotor deviations, or a series may be arranged to form a prism bar.

Disadvantages of plastic lenses are relatively low refractive index, requiring thicker lenses than equivalent glass powers, and less resistance to scratching. Once finished, plastic lenses cannot be reshaped or resurfaced. They do not absorb infrared rays, and their base curve, set by the manufacturer, cannot be varied except on special order, which may cause magnification problems when switching from glass to plastic. Although plastic bifocals and trifocals are now available, large lenses (over 60 mm) are only now becoming available.

SAFETY LENSES

In January 1972 the Federal Drug Administration issued a policy statement whose essentials are as follows:

> Use of impact-resistant lenses in eyeglasses and sunglasses . . . (a) Examination of data available on the frequency of eye injuries resulting from the shattering of ordinary crown glass lenses indicate that the use of such lenses constitutes an avoidable hazard to the eye of the wearer. (b) The consensus of the ophthalmic community is that the number of eye injuries would be substantially reduced by the use in eyeglasses and sunglasses of either plastic lenses, heat-treated crown glass lenses, or lenses made impact resistant by other methods. (c) To protect the public more adequately from potential eye injury, eyeglasses and sunglasses must be fitted with impact-resistant lenses, except in those cases where the physician or optometrist finds that such lenses will not fulfill the visual requirements of the particular patient, directs in writing the use of other lenses, and gives written notification thereof to the patient. (d) The physician or optometrist shall have the option of ordering heat-treated glass lenses made impact resistant by other methods; however, all such lenses must be capable of withstanding an impact test in which a $\frac{5}{8}$ inch steel ball weighing approximately 0.56 ounce is dropped from a height of 50 inches upon the horizontal upper surface of the lens.

Safety lenses are classified as impact resistant (industrial and dress eyewear), laminated, and plastic. The impact-resistant lens is a standard lens brought close to the melting point and rapidly cooled to produce fusion of molecules. It makes the glass resistant but not impervious to fracture. When viewed through a polarizer, a characteristic strain pattern in the form of a Maltese cross appears in heat-treated lenses. Heat treatment requires that the lens have a minimal thickness of 2.2 mm, which adds to its weight and may cause some surface defacement or curvature change. Once treated, the lens cannot be resurfaced or reedged. Heat treatment is not always reliable, and impact resistance may not be demonstrable under all conditions.

Rather than look a statutory gift horse in the mouth, the drop-ball test should be used to verify the process.

A chemical strengthening process for ophthalmic lenses is currently being introduced. The technique involves a chemical ion exchange in the glass surface that has been used successfully in laboratory glassware and in automobile and aircraft windshields. It potentially strengthens a glass several times more than conventional tempering methods. The lenses are immersed in a molten salt bath; an ion exchange occurs between salt and potassium ions when heated. The larger potassium ions diffuse into the glass surface, and this exchange produces high compression. The ion exchange occurs below the strain point of the glass, there is no warping, and lenses of all sizes and thickness can be treated simultaneously. No characteristic pattern is seen in the polarizer; so the drop-ball test confirms the treatment.

Laminated lenses consist of two thin glass lenses separated by an interposed sheet of plastic that keeps them from splintering. This optical sandwich is somewhat thicker than ordinary lenses and is indicated in particularly hazardous occupations and for sports wear. Hard resin plastic lenses are safer, cheaper, and more cosmetically acceptable and have more or less replaced laminated glasses.

Other protective devices such as goggles, face masks, helmets, hoods, and shields are sometimes indicated. Safety goggles will fit snugly around the orbit; masks are designed for protection against dust, wind, fumes, and liquid splash hazards; helmets protect the head of a welder, and face shields serve a similar function in polishing and buffing operations.

QUALITY OF OPHTHALMIC LENSES

The quality of an ophthalmic lens depends both on mechanical construction and optical precision. The limits of tolerance should be practical but not so high as to unduly increase cost. The requirements for a telescope would be prohibitive and unnecessary for an ophthalmic lens because the eye has its own tolerances and built-in means of dealing with most aberrations. Although it is a good rule to use only first-quality lenses, tolerance limits vary somewhat from one laboratory to the next.

Current American standards for ophthalmic lenses are set by a subcommittee for the American Standards Institute,* acting in cooperation with the Optical Manufacturer's Association. Tolerances are set for power (±0.06 D from plano to 3.00 D; ±0.12 D up to 12.75 D), for thickness (±0.2 mm), for axis (within 3°), for centration (within ±0.25Δ), for seg size (±0.5 mm) and for physical quality and appearance (for example, lenses inspected against a dark background in light from an open-shaded, 100-watt lamp must be free from glass and surface faults and chips).

Lens quality may be evaluated by inspection carried out by reflected light, transmitted light, and "shadowing." The lens, which must be absolutely clean, is held about a foot from the eye in front of a distant source, whereupon defects such as bubbles, rings, blisters, and waves will be evident. Tarnish, scratches, and chips are made visible by holding the lens close to a bright source and viewing the reflected image of the source in various lens zones. In shadowing, the lens is held at its focal point from the eye, and, viewing a strong light source (minus lenses require an auxiliary lens), the entire lens aperture fills with light. When the lens is slowly moved to one side, surface defects such as veins will appear as bright streaks in the transition zone.

The configuration of a lens depends on the spectacle frame. New frames with new shapes are continuously introduced, and memorizing them is impossible.† In 1962 the Optical Manufacturer's Association adopted the "boxing method" of shape specification. The lens outline is boxed in a rectangle whose sides are tangent to its apex, sides, and bottom. The width of the box is the "eye size," and the difference between width and depth is the "pattern."

DECENTRATION

Light rays that pass eccentric to the optical center of a lens are deviated; the more peripheral the ray the greater the deviation.

*ANSI: Z80.1, 1964,

†A complete catalogue of currently available frames is published quarterly, including styles, sizes, colors, and prices. *Frames* is published by Zulch and Zulch, Inc., P. O. Box 4427, Sylmar, Calif.

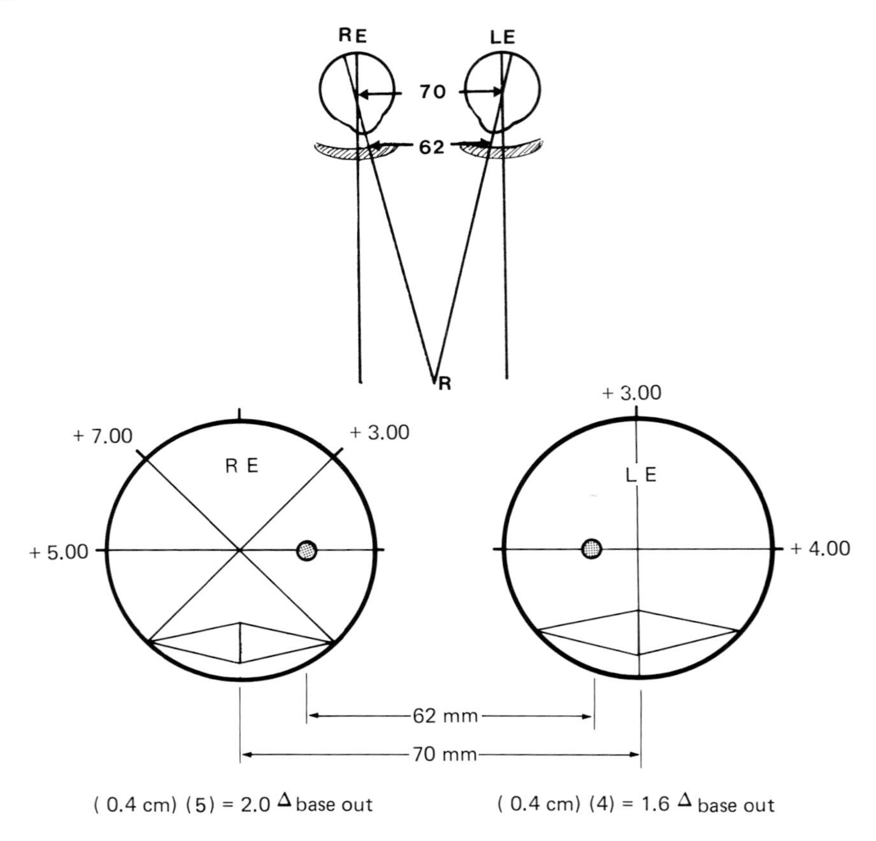

Fig. 5-1. Horizontal prismatic effect resulting when a pair of eyes with an interocular distance of 62 mm view a distant object through lenses whose optical centers are separated 70 mm. Because of the oblique cylinder axis in the right eye, there will also be a slight vertical prismatic effect, which is ignored in this illustration.

This is Prentice's rule. To align the optical center of spectacle lenses with the pupils the lens pattern is moved on the lens blank before cutting.

The effect of horizontal prisms on ocular rotations are additive when their bases are opposite and subtractive when their bases are the same. Vertical prisms are additive when their bases are the same and subtractive when opposite. Of course, binocular base-up prisms still produce displacement, but the patient only has to learn a new hand-eye coordination, not strain the fusional vergences.

EXAMPLE: What is the horizontal prismatic effect when a patient with an interpupillary distance of 62 mm looks through a pair of lenses whose optical centers are separated by 70 mm? The prescription is RE: +3.00 +4.00 × 45; LE: +3.00 + 1.00 × 90 (Fig. 5-1).

The power in the horizontal meridian of the right lens is +5.00 (from the sine-squared formula given in Chapter 4), and the prismatic effect is $0.4 \times 5.00 = 2\Delta$ base out. The power in the horizontal meridian of the left lens is +4.00, and the prismatic effect is $0.4 \times 4.00 = 1.6\Delta$ base out.

The total prismatic effect is therefore 3.6Δ base out.

When a prismatic effect is desired, for example, to correct a muscle imbalance, the problem comes up in reverse. The rule is that a positive lens is decentered in the same direction as the required prism base, and a negative lens is decentered in the opposite direction.

In reading the eyes drop below the optic center, and a vertical prismatic effect results.

For example, when the visual axis passes 8 mm below the optic center of a −8.50 D lens, the result is 6.8Δ base down. If the other eye looks through an identical lens, the only consequence is a displacement of the target. In anisometropia, however, there is an induced vertical imbalance, which varies with the power difference. Patients with single vision lenses can compensate for it by dropping their heads so that the visual axes pass through the optical centers. With bifocals, this is no longer possible since patients would look above the segs, and optical compensation for the imbalance is required. The vertical imbalance of anisometropia is only one of the problems in prescribing bifocals.

One may wish to prescribe both a vertical and a horizontal prism. The optician replaces them by a single oblique prism computed by

$$\text{New prism power} = \sqrt{(H)^2 + (V)^2}$$

$$\tan\theta = \frac{V}{H}$$

where V and H represent the respective prisms and θ is the angle between the base to apex line of the resultant prism and the horizontal meridian.

EXAMPLE: Resolve the following prismatic combination to be worn before the right eye; 3.5Δ base in and 2Δ base up:

$$\text{Resultant} = \sqrt{(3.5)^2 + (2)^2} = 4.03\Delta$$

$$\tan\theta = \frac{2}{3.5} \colon \tan\theta = 30°$$

Answer: 4Δ base up and in at 30°

Conversely an oblique prism may be resolved into its horizontal and vertical components from the following simple relation:

Horizontal component = prism power ($\cos\theta$)
Vertical component = prism power ($\sin\theta$)

The rotary prism, invented by Herschel and adapted for clinical use by Cretes and Risley consists of two prisms, base to apex, producing a smoothly changing deviation. If the base to apex lines are vertical at the zero setting, one can plot the vectors to show that as the prism rotates in opposite directions the vertical components neutralize and the horizontal components add to a maximum. The combination is rotated as a whole to vary the directions in and out or up and down.

Prisms cause distortion; with base down the floor appears concave, objects appear taller, and one has the impression of walking uphill; base up makes the floor appear convex and gives the feeling of standing on a hill; horizontal prisms make objects appear larger toward the base.

Some induced prismatic effects frequently overlooked are caused by improperly aligned refractors, tilted frames, and facial asymmetries. Sometimes the old lenses are off center, but the patient has adapted to them, and the new, correctly aligned prescription cannot be tolerated.

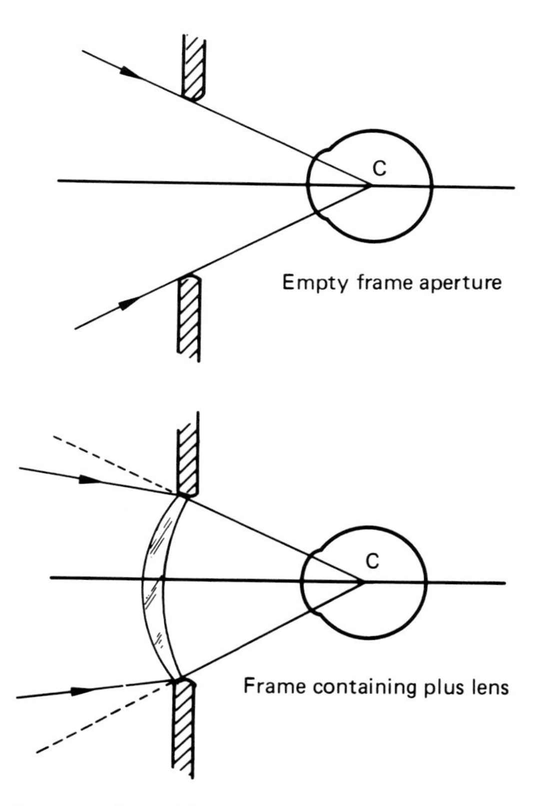

Fig. 5-2. The field of view is reduced by a convex lens; the stronger the power the greater the resultant scotoma. With minus lenses a zone of double vision is created. It follows that all spectacle lenses, especially aphakic ones, should be fitted as close to the cornea as feasible.

FIELD OF VIEW

The opening of a spectacle frame acts like a window and limits the field of view (Fig. 5-2). When the eye is close to the frame, the effect is negligible, but with a strong convex lens, the field may be restricted by as much as 2.5% per diopter for a fixed eye and head position. Minus lenses increase the real field of view, although there may be a diplopic zone in the periphery.

The most important limiting factor, however, is the distance of the lens from the eye. The closer the lens is fitted the greater the field of view. The field of view of a reading lens or magnifier is best evaluated empirically by a linear scale at the indicated distance with and without its aid.

ANTIREFLECTIVE COATINGS

Lenses generally reflect about 4% of the incident light at each surface, thus resulting in an 8% transmission loss. In 1892 a British designer noted that old, tarnished lenses made faster photographic objectives. The improved transmission was due to chemical changes at the lens surface. A film had formed that eliminated surface reflections by destructive interference. The coating is now produced artificially with magnesium or aluminum fluoride. The films are deposited to a thickness of one-fourth wavelength by controlled evaporation in a vacuum chamber. Since the waves reflected from the anterior and posterior surface of the film have a 180° phase difference, they cancel each other. Cancellation of wave amplitude is best when the film index is the square root of the glass index. Antireflective coatings only work for selected wavelengths (generally yellow-green), and consequently long and short waves continue to be reflected and give the coated lens a purple tint. Coated lenses should not be confused with tinted ones; the former improve light transmission, and the latter reduce it.

Clinical indications for antireflective coating depend on the symptoms, the prescription, and the illumination, as well as cosmetic factors. Ghost images are particularly annoying to myopes and aphakes, to those subject to glare from large expanses of water or snow, and to actors or speakers who require less visible lenses. Weak minus prescriptions frequently cause ghost images because bright objects are reflected toward the optic axis both from above and below. A shallower base curve or a slight change in head or lens position often alleviates the problem, but little can be done to eliminate it. In multifocals, reflections from the top of the segments are reduced by a gold pigment coating that absorbs unwanted light.

ABSORPTIVE LENSES

Absorptive lenses reduce or eliminate potentially dangerous ocular radiations. At the turn of the century Sir William Crookes investigated protective filters for furnace workers, and the resultant lens is still produced to absorb radiations below 350 nm. Special absorptive lenses can reduce ultraviolet, infrared, or visible light transmission. Wavelengths shorter than 300 nm are absorbed by the cornea, and those longer than 300 nm (up to 400 nm) are absorbed by the crystalline lens; thus the chances of damaging rays reaching the retina are normally slight.

Sunglasses, in an almost infinite variety of shapes and sizes with or without correction, are often made larger than regular lenses; hence their magnification varies according to the thickness. Color, although a personal preference, should not be so dark as to reduce transmission less than 10% to 20%. Colored glass lenses contain metallic oxides, but their absorption curve is somewhat lumpy, with convex lenses too dense in the middle and concave lenses too colorless in the center. Strong prescriptions are best made by color coating white lenses. Color coatings are thin films of metal oxides, usually gray, blue, green, pink, or brown. Since the colors are not standardized, broken lenses should be sent to the laboratory for an exact match. Colors are available for both plastic and glass lenses. The density is independent of lens thickness but can be artificially graded, (e.g., maximum color at the top). Color coating a spare pair of white lenses is an economic way to convert them to sunglasses.

It should be noted that manufacturer's data on lens transmission for various colors are for 2 mm thick plano lenses and do not include the added prescription thickness.

In selecting an appropriate filter, consideration is given to its intended use, whether sport or street wear, industrial or high altitudes, snow or water, or for tasks that may or may not require faithful color rendering. Yellow-absorbing filters are considered to decrease dazzle, and blue-absorbing filters are supposed to enhance contrast, for example, by reducing the long wavelengths that induce fluorescence in cataracts. Polarized filters are useful on water and snow but also reduce visible transmission. Photochromatic filters darken with direct ultraviolet exposure, (which does not always help the driver behind a windshield). They are available in both lighter (photogray) and darker (photosun) shades.

All absorption glasses reduce retinal illumination to some extent and may cause difficulties with night vision. For example, yellow lenses combined with green-tinted car windshields are a distinct night driving hazard. The requirements of night vision must also be kept in mind when prescribing slightly tinted lenses for constant wear.

LENS ABERRATIONS

Ophthalmic lenses suffer aberrations, an impediment to both form and clarity of image reproduction. Unlike thin lenses, which only handle theoretical paraxial rays, ophthalmic lenses have a real aperture, that is, the diameter of the frame, and peripheral rays enter the eye. When the eye rotates to view eccentric objects, the spectacle lens does not move with it, and although the incident rays are limited by the pupil, they pass the lens obliquely. Moreover the theoretical thin lens deals only with monochromatic rays, whereas the real world is polychromatic, which also contributes to a spoiled image.

Chromatic aberrations

Newton observed that a prism disperses polychromatic light into its component colors and concluded correctly that different wavelengths travel through the same medium at different velocities. Each medium thus has a different refractive index for different wavelengths. Lenses act like a series of prisms and focus the short wavelengths sooner than the long ones, resulting in a chromatic difference in focus and magnification (Fig. 5-3). A quantitative expression for this difference is obtained by comparing the effect on a representative wavelength from each end and the middle of the spectrum (identified by their Fraunhofer lines: C (656.3 nm), D (587.6 nm) and F (486.1 nm). The extent to which the lens breaks up white light is its dispersive power $= \frac{n_F - n_C}{n_D - 1}$, usually expressed as the reciprocal Abbe number (symbol V). The V value for crown glass is about 60, compared to flint glass of about 40. The chromatic aberration of crown glass is approximately $\frac{F}{V}$; hence a focal difference of 0.067 D for a +4 D lens means the power is that much stronger for blue than for red light. For a −10 D lens, the difference is −0.17 D, meaning the focus is virtual with the blue closer to the lens.

By combining a plus crown glass with a weaker minus flint glass, the long and short wavelengths may be focused at the same plane, a combination called an "achromatic system," or "doublet," providing $\frac{F_1}{V_1} = -\frac{F_2}{V_2}$. For thin lenses in contact, the sum is the power of the desired system. For example, to obtain a +10 D achromatic system of crown and flint, F_1 must be $\frac{+10}{60 - 40}(60) = +30$ D, and F_2 must be $\frac{+10}{60 - 40}(30)$ or −20 D.

Chromatic aberration is an inevitable but minor annoyance because the eye itself is not achromatic. A myope sees some blue fringes when looking above the optical center of the lens, with reddish fringes below. The hyperope gets the reversed effect. These are due to the base-up or base-down prismatic effect of the lens. In bifocals the seg acts like an additional prism, and the chromatism increases toward the bottom. When the prismatic effect of distance and reading portions are neutralized, the chromatic effect dimin-

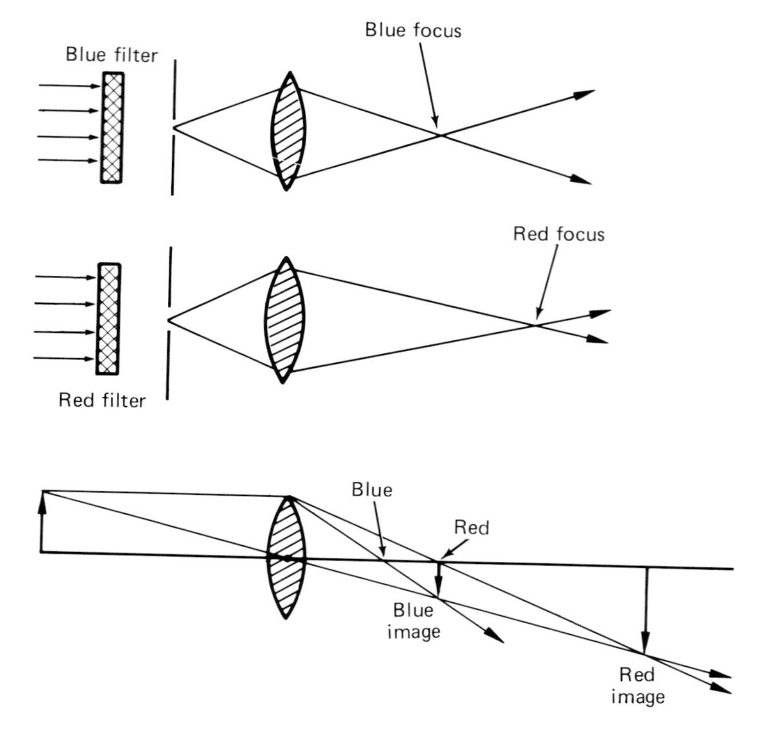

Fig. 5-3. The effect of chromatic aberration is to produce not only a difference in focus but a chromatic magnification as well. Chromatic aberrations are inevitable in spectacle lenses but become particularly annoying with certain types of bifocals.

ishes, and if the seg is made of different glass, the chromatic effect can be greatly reduced.

Monochromatic aberrations

Monochromatic aberrations occur even for single wavelengths and add to the chromatic aberrations. Their optics was first worked out by von Seidel; hence they are also called "Seidel aberrations." Some impair image quality (spherical aberration, coma, and radial astigmatism), and some deform the image (curvature of field and distortion) (Fig. 5-4). Optical aberrations are the playground of the lens designer and beyond the scope of this book, but a qualitative insight into their effects can be obtained without mathematics.

Spherical aberration is the inevitable result of axial rays striking a spherical surface at a greater peripheral angle, hence focusing earlier than central rays. Peripheral and central angles can be equalized by paraboloid surfaces, such as those used in reflecting mirrors. Spherical aberration is of little consequence in ophthalmic lenses because the pupil limits the rays to a small central area. For the wider fields of camera and microscope lenses, their elimination requires aplanatic designs.

Coma is due to spherical aberration of an oblique pencil. Since the angle of refraction varies for each ray, the image consists of overlapping variable circles of diminishing brightness. The image of a point is therefore comet shaped, hence the name.

Radial, marginal, or oblique astigmatism results when light rays strike the lens obliquely, or the eye views an extra-axial target. (Recall that tilting a sphere creates astigmatism for the same reason.) The oblique rays are refracted to form a conoid of Sturm whose interval increases with the angle of obliquity. For a series of eccentric points, these focal lines form surfaces that diverge like a teacup and saucer, and such surfaces are termed "astigmatic image shells." When

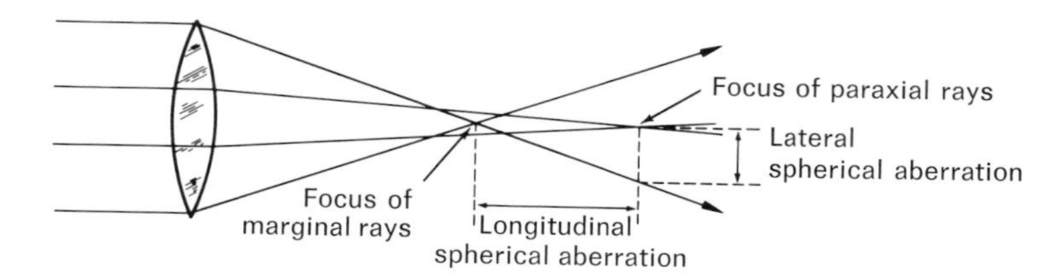

Spherical aberration

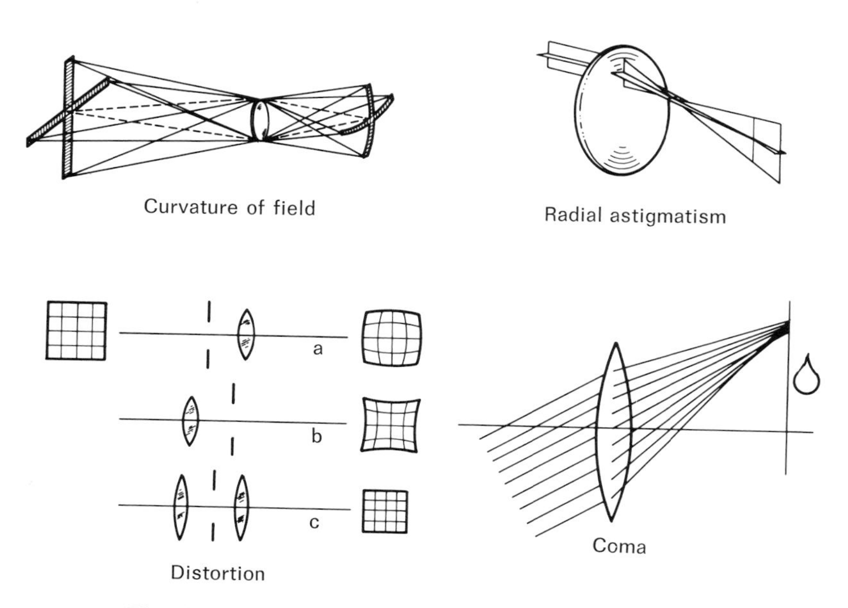

Fig. 5-4. Monochromatic aberrations. For details see text.

the eye turns toward the top of a plus sphere, the astigmatism is as if a plus cylinder axis 180 had been added (similar to tilting a plus sphere about its horizontal axis). The horizontal line focuses first; consequently horizontal objects are blurred. The vertical focal line is little affected so the horizontal axis error is of greater consequence. When the eye turns horizontally right or left behind a plus sphere, a plus cylinder axis 90 is created, which blurs vertical objects. Since the eye may also turn obliquely, it is conventional to call one focal line "tangential" and the other "sagittal." Tangential means that lines tangent to the circular field of view are blurred; sagittal means that lines parallel to the direction of view are blurred. The astigmatic interval for an eccentricity of 40° may reach 0.33 D for a +3.00 D sphere of 6.25 base curve. A hyperopic eye may vary accommodation and compensate for it by bringing the circle of confusion on the retina; the myopic or fully corrected hyperopic eye cannot, since accommodation is fully relaxed. There is something to be said therefore for permitting some residual accommodative play in ametropic prescriptions.

Curvature of field results because all parts of an extended linear object cannot be equally distant from a spherical surface. The peripheral rays focus sooner than the central rays, and the image is curved toward the edges in a plus lens and away from the edges in a minus lens. The surface on which the image lies is named after its discoverer, Petzval. The retina acts as a kind of Petzval surface for the optics of the eye. If there is radial astigmatism, the distance from the sagittal focus to the Petzval surface is one half the distance from the sagittal to the tangential focus, and the two aberrations can be treated concurrently by the lens designer.

Distortion results from unequal magnification; square objects appear as pincushions through a plus lens and as barrels through a minus lens. It is usually because of inappropriately positioned stops. For example, if the stop is between the lens and the object, the peripheral rays must travel farther than if the stop were absent; hence the image edges are cramped into a barrel effect. If the stop is between the lens and the image, the peripheral rays strike the lens sooner, and the image is stretched into a pincushion. An aperture between lenses of a system reduces distortion (making it orthoscopic), hence the advantageous location of the pupil between cornea and crystalline lens. Distortion may cause difficulties in lenses over 3.00 D; hence the lens designer tries to compensate for it.

Ophthalmic lens aberrations of greatest significance are radial astigmatism, curvature of field, distortion, and chromatism in that order.

ELEMENTS OF LENS DESIGN

The goal of ophthalmic lens design is to eliminate or at least alleviate those aberrations most detrimental to acuity as the eye turns behind the lens. Since excursions without head movements are small, effective designs are usually limited to eccentricies of about 30°. To understand the principle it is necessary to introduce the concept of the far point of the eye, which is discussed more fully in Chapter 7. The far point represents that point toward (or from) which incident rays must be directed to focus on the retina. The far point in 1 D hyperopia is 1 meter behind the eye; in 1 D myopia, 1 meter in front of the eye. A spectacle lens corrects ametropia when its second focal point coincides with the far point—by definition. A +1 D lens corrects 1 D of hyperopia because its second focal length is at the far point. The rays never actually reach it but enter the eye to focus on the retina.

Fig. 5-5, *A*, illustrates the far point in 1 D hyperopia. As the eye turns, the far point moves with it, describing a circle concentric to its center of rotation. For the single refracting surface of a model (reduced) eye, with its center of curvature at *c*, the far point circle is also the Petzval surface, a sphere in three dimensions. The line joining the center of the pupil and the fovea always passes through the center of rotation for every direction of gaze, so that one can assume a fixed stop at *c*. The exact position of the center of rotation is not known but is considered to be 13 to 15 mm from the cornea based on experimental averages. Fig. 5-5, *B*, shows a series of little hypothetical lenses so arranged about the eye that they image all eccentric object points on the far point sphere. Although these image points are still subject to aberrations, if the stop is small, spherical aberration and coma are minimal, and only radial astigmatism, curvature, and distortion must be considered.

The designer's problem is to duplicate with a real lens the optical behavior of the series of hypothetical lenses, and although Fig. 5-5, *C*, illustrates that no single lens can accomplish this, some lens profiles are better than others. To make the Petzval and far point surfaces coincide, the base curve would need to be very steep (about −20 D, for example). For practical base curves there is a difference called the "image shell error," which increases for increasing angles of obliquity. Astigmatic intervals can be decreased by proper selection of lens curvatures; hence the final design may produce a lens either free from oblique astigmatism but with a residual shell error or a lens in which the tangential error approximates the back vertex power of the lens. Such lens forms are termed "corrected curve."

By assuming an average center of rotation and vertex distance, lens designers compute the effect of various curvature combinations for different prescriptions at various angles of view, usually up to 30°. Early designs were primarily concerned with astigmatic error. For example, the von Rohr "Punktal" lens had specific curvatures for almost every prescription, resulting in prohibitive price and

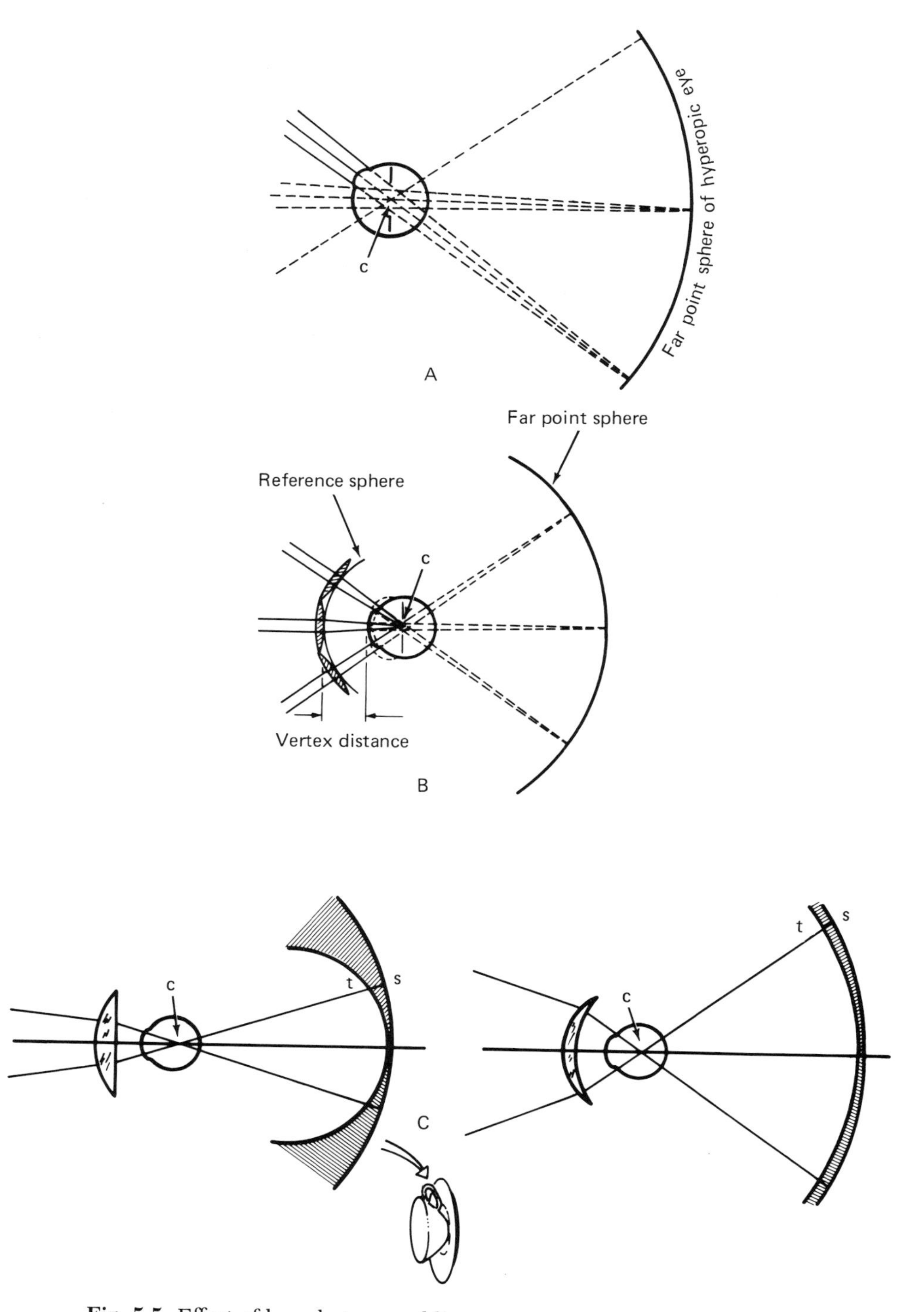

Fig. 5-5. Effect of lens design on oblique astigmatism. For details see text.

production problems. Modern lens designs compromise on power and astigmatic errors on the reasonable assumption that there is no point in exceeding the eye's own discrimination. Prescriptions are grouped in series, each with a particular base curve. These groupings, using only base curve as a variable (as contrasted to bitoric grinding), is the basis of modern corrected curve lenses.

Special design considerations enter into the engineering of multifocals, cataract and high minus corrections, aspherics, low-vision aids, and iseikonic and variable focus lenses. Modulation transfer function techniques have been used to evaluate resolution, haze, and surface defects.

CORRECTED CURVE LENSES

The term "corrected curve" is equivocal, since most lenses are corrected for something, and the term often brings to mind only a photograph of a sparkling lens on a velvet surface. The designer thinks of a corrected curve lens as one which gives optimum optical performance under certain controlled conditions specifically by eliminating those aberrations deemed most detrimental to acuity, for a series of spherical and spherocylindrical prescriptions, for different directions of view, for controlled magnification, at various vertex distances, and for eyes whose center of rotation changes with axial length. Moreover it is limited by the available glass index, thickness, and frame sizes and shapes and must be produced at reasonable cost. Clearly the task is not simple, and the final lens is generally a compromise that represents a particular designer's thinking and point of view.

Early ophthalmic lenses were equiradial or flat on one side. In 1804 Wollaston observed by trial and error that peripheral vision improved when lenses were curved. Such lenses, which he called "periscopic" (or "looking about glasses"), were not to be produced commercially for another fifty years. In 1898 Ostwald noted that oblique astigmatism could be eliminated by certain combinations of lens curvatures, and in 1904 Tscherning demonstrated that the Ostwald and Wollaston designs represented the two forms, shallow and steep, for which the tangential and sagittal errors switched places. Percival in 1901 emphasized the importance of placing the circle of confusion on the far point sphere, especially for reading vision. It is noteworthy that each of these pioneers in lens design was a physician, the last three practicing ophthalmologists.

Modern corrected curve lenses include the American Optical "Tillyer," Baush and Lomb "Orthogon," Titmus "Normalsite," and Shuron/Continental "Kurova." In the past decade, corrected curve lenses with minus cylinders have become available such as the American Optical "Masterpiece" and the Shuron/Continental "Shursite." Rear or minus cylinders reduce meridional magnification because the toric surface is closer to the eye. The thickness differences are also less visible, and since bifocals are generally rear torics, the transition from single vision lenses is made more palatable. Rear surface torics are usually flatter than plus spherocylinders, and consequently these lenses cannot be worn if the lashes are long and the bridge of the nose flat. Details of lens construction and the base curve of the series for each prescription may be obtained from the manufacturer.

Base curves, even when part of a corrected curve lens, should not be arbitrarily switched in high corrections, especially if the patient is sensitive to magnification differences. An attempt should also be made to maintain similar base curves for each pair of spectacles (sometimes difficult with bifocals and oversized sunglasses). Corrected curve lenses of the highest quality may cause more problems than they solve if fitted improperly.

COSMETIC FACTORS IN SPECTACLE WEAR

Yesterday's fad is today's fashion and will provide tomorrow's nostalgia. While all the rules seem to be changing before the winds of individuality and the onslaught of mass production, the dispenser is privileged to rush in where optical science fears to tread. But some precautions are timeless; oversized lenses are best avoided in high prescriptions, flimsy frames are a poor choice for youngsters, frames that cannot be readily adjusted will mean extra expense for an aphakic patient, a bifocal will not work for close work above eye level, and a reading glass will not focus at arm's length. The patient's cosmetic frame fantasies must also be tempered by the anatomic realities of lashes, nose, cheeks, and ears.

Cosmetic lenses are frequently useful to improve appearance, such as a balancing lens for a blind eye, a minifying or magnifying lens for a prosthesis or enophthalmos, a ptosis crutch for myasthenia, and a prism for ocular or orbital deformities. Prisms are also recommended for the cosmetic improvement of strabismus, but they seldom work.

PSYCHOLOGIC FACTORS IN SPECTACLE WEAR

Practically all patients have mistaken ideas about the curative properties of spectacles, and most will have forgotten the instructions concerning when and how to wear them by the time they are delivered. It is therefore useful to reevaluate and to discuss the purpose of the lenses and to explain the prognosis of ametropia, astigmatism, or amblyopia. Parents should be made aware that some conditions progress with growth and some improve; changes are thus anticipated. Patience and empathy are always taxed by the new aphake, the older presbyope, and the partially seeing. Failure to anticipate difficulties with reading range, magnification, and field of view invariably lead to complaints later.

Few people enjoy wearing glasses, but even fewer prefer the alternative. By emphasizing positive aspects, the acceptance of a visual aid can be made more rational if not more pleasant. The value of lenses that increase acuity is obvious, but when prescribed for safety, efficiency, or comfort, a more detailed explanation is needed to improve motivation.

OPHTHALMIC ASPECTS OF DISPENSING

A correcting lens becomes an integral part of a new optical system in which an improper fit may turn a first-rate refraction into a third-rate visual result. A poorly centered lens that touches the lashes and fits at the wrong distance and angulation does not provide the quality of seeing that the clinician or lens designer intended, and the blame is usually placed on the refraction. Nor does the fault lie with the optician, given insufficient information, asymmetrical eyes, a nose that will not support the frame selected by an uninformed patient, or an arbitrary lens design chosen only on the basis of the latest advertisement.

Interocular distance (PD)

Of the dozen different ways of measuring interocular distance, ranging from the simple millimeter ruler to the most elaborate self-illuminated pupillometer, the simplest and most accurate utilizes corneal reflections. A penlight is held at the reading distance, and the examiner, sighting with one eye, measures the separation of the reflections with a millimeter ruler. If the eyes are asymmetrical, the measurement is taken from the midline of the crest of the nose to each eye. Although this measures only the near PD, the distance PD is more accurately calculated by adding 4 mm to the near measurement. The interocular distance will be somewhat smaller if the measurement is taken further from the cornea because the lines of sight converge (the 4 mm difference is based on a spectacle plane 15 mm from the cornea). An allowance is made for monocular patients and those who work predominantly toward one side (e.g., a violinist). In strabismus the ruler may be alternately tilted to cover one eye, aligning the zero mark with the corneal reflection of the fixing eye.

The interocular distance must be known for proper adjustment of trial frame or refractor, to compute or avoid prismatic effects, and to evaluate the ratio of accommodative convergence to accommodation (AC/A ratio) and the extent of ocular deviations.

Vertex distance

The vertex distance is the distance between the back surface of the correcting lens and the cornea and locates the spectacle plane. It is advisable to fit lenses as close to the eye as the lashes permit to increase the field of view. The vertex distance is measured with a millimeter ruler or special caliper (Distometer, Lenscorometer). Trial frames and refractors often have an attached gauge, which is usually not accurate enough because of parallactic sighting.

An accurate method is a stenopeic slit and trial frame; a metal millimeter ruler is passed through the slit to the closed lid (allow 1 mm for lid thickness). Vertex distances vary, ranging from 7 to 22 mm, and should be measured, not assumed.

One of the peculiarities of spectacles is that the best lens design may not work if the base curve is too shallow for the lashes and the lens cannot be fitted at the vertex distance that the designer intended in computing the curves. It follows that one cannot automatically assume a particular optical result, and a working relationship with the optician is necessary to resolve such optical fitting problems, which are often compounded by an injudicious frame selection.

Vertical alignment

Although reasonable care is used in measuring interocular distance, the potentially more troublesome vertical alignment is often ignored. When a single vision lens is used for reading, the visual axis passes below the optical center so that a pantoscopic tilt is necessary to place the optical center about 5 mm below the line of sight (8° of tilt is most common). If the lens is angulated too much, however, the line of sight may pass more than 9 mm above the optical center. The larger the vertex distance the greater the discrepancy. Close-fitting frames often tend to sit too high, whereas distant-fitting frames fit too low, increasing radial astigmatism with loss of optical efficiency.

One way of estimating the correct vertical fit is to look at the patient's face in profile as he views a distant object. With the temples horizontal, pass an imaginary perpendicular line from the 180° axis of the lens and see if it intersects the external canthus (the approximate position of the center of rotation) (Fig. 5-6). Most designers agree that lens performance depends as much on poor fit as on poor design.

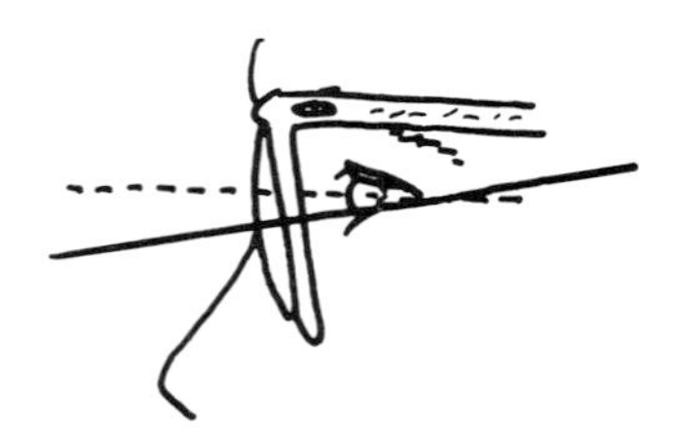

Fig. 5-6. Vertical positioning of a spectacle lens depends on the pantoscopic tilt, the size of the nose, and the length of the lashes. A single vision lens fitted for both distance and reading should approximate the position shown. A line perpendicular to the eyewire intersects the external canthus and gives the optical behavior the lens designer intended.

Writing the lens prescription

To avoid misunderstanding and misinterpretation, the prescription is written in appropriate symbols with a copy in case of loss or breakage.

1. The sphere, written first, may be abbreviated sph. to indicate there is no cylinder. It is not necessary to add D or Diopt. Three digits are specified; thus it is 2.00 and not 2 and 0.25, not $\frac{1}{4}$. Do not omit the sign.
2. The cylinder may be written in plus or minus form. The symbol ○ is unnecessary, a capital X serves for "axis," and the degree mark is omitted.
3. If a prismatic element is to be included, specify clearly degrees or prism diopters and the direction of the prism base. For oblique prisms it is conventional to use the TABO notation, adding whether the base is up or down or in or out in the axis given.
4. Include the distance and near interpupillary distance and the vertex distance in strong corrections or contact lens orders.
5. For multifocals specify type, style, size, height, inset, and decentration.
6. General specifications may include safety features, base curve, front or

rear surface cylinders, coating, tint, glass, or plastic.

7. Agreement should be reached with the laboratory as to mechanical and optical tolerances.

All lenses should be checked after delivery. Even the best opticians make mistakes. It is tragic to have a good refraction ruined by poorly fitting frames, incorrect decentrations, or transposition errors.

REFERENCES

Adams, A. J., Kapash, R. J., and Barkan, E.: Visual performance and optical properties of Fresnel membrane prisms, Am. J. Optom. **48:**289-297, 1971.

Barnard, E. E. P.: Visual problems under water, Proc. R. Soc. Med. **54:**755-756, 1961.

Bechtold, E. W., and Langsen, A. L.: The effect of pantoscopic tilt on ophthalmic lens performance, Am. J. Optom. **42:**515-524, 1965.

Bennett, A. G.: Graphical methods of solving optical problems, London, 1948, The Hatton Press, Ltd.

Bennett, A. G.: British standards relating to ophthalmic lenses, Optician **139:**507-511, 535-539, 563-568, 1960.

Bryant, R. J.: Ballistic testing of spectacle lenses, Am. J. Optom. **46:**84-95, 1969.

Chase, G. A., Kozlowski, T. R., and Krause, R. P.: Chemical strengthening of ophthalmic lenses, Am. J. Optom. **50:**470-476, 1973.

Clark, B. A. J.: The luminous transmission factor of sunglasses, Am. J. Optom. **46:**362-378, 1969.

Cridland, N.: The British standard for spectacle lenses, Br. J. Ophthal. **40:**611-618, 1956.

Davis, J. K.: Corrected curve lenses and lens quality, Am. J. Optom. **39:**135-152, 1962.

Davis, J. K.: A study of lens construction and use, Int. Ophthal. Clin. **11:**169-259, 1971.

Davis, J. K., Fernald, H. G., and Rayner, A. W.: The design of a general purpose single vision lens series, Am. J. Optom. **42:**203-236, 1965.

Davis, J. K., and Rich, J. B.: Spectacle lenses for the myopic patient, Am. J. Optom. **44:**424-427, 1967.

Dowaliby, M.: Practical aspects of ophthalmic optics, Chicago, 1972, The Professional Press, Inc.

Dowaliby, M., Griffin, J., Palmer, B., and Voorhees, L.: A study involving glass safety lenses, Am. J. Optom. **49:**128-136, 1972.

Drew, R.: Professional ophthalmic dispensing, Chicago, 1970, The Professional Press, Inc.

Emsley, H. H., and Swaine, W.: Ophthalmic lenses, ed. 2, London, 1932, The Hatton Press, Ltd.

Epting, J. B., and Morgret, F. C.: Ophthalmic mechanics and dispensing, Philadelphia, 1964, Chilton Books.

Faraday, M.: On the manufacture of glass for optical purposes, Philosoph. Trans. R. Soc. London, **120:** 1-57, 1830.

Fry, G. A.: Lens shapes for zylonite and metal frames, J. Am. Optom. Assoc. **31:**716-722, 1960.

Fry, G. A.: Geometrical optics, Philadelphia, 1969, Chilton Books.

Grolman, B.: The sighting center, Am. J. Optom. **40:**666-675, 1963.

Hirsch, M. J.: Prism in spectacle lenses for cosmesis, Am. J. Optom. **45:**409-413, 1968.

Humphries, M. K.: Survey of injuries caused by crown glass lenses, Richmond, 1971, Society for Prevention of Blindness.

Jalie, M.: The thickness of spectacle lenses, Optician **152:**239-242, 1966.

Jalie, M.: The principles of ophthalmic lenses, London, 1970, Association of Dispensing Opticians.

Kamellin, S.: Catmin lenses, Am. J. Ophthal. **28:**993-998, 1945.

Keeney, A. H.: Safety lens materials, Survey Ophthal. **5:**404-412, 1960.

Keeney, A. H.: New ophthalmic standards and their development, Trans. Am. Acad. Ophthal. Otolaryngol. **76:**1286-1288, 1972.

Kors, K., and Peters, H. B.: Absorption characteristics of selected commercially available ophthalmic lenses, Am. J. Optom. **49:**727-735, 1972.

Levy, J. D., and Kerr, K. E.: Analysis of methods for improving the appearance of convex spectacle lenses, Am. J. Optom. **47:**690-701, 1970.

Miles, P. W.: The importance of corrected curve lenses, Am. J. Ophthal. **35:**1320-1328, 1952.

Morgan, M., and Peters, H. B.: The optics of ophthalmic lenses, Berkeley, 1948, University of California Press.

Neumueller, J.: The correction lens, Am. J. Optom. **25:**247-261, 326-339, 370-429, 1948.

Newton, A. W.: Industrial eye protection: appraisal of some currently safety lens materials, J. Inst. Eng. **39:**163-170, 1967.

Obrig, T. E.: Modern ophthalmic lenses and optical glass, Philadelphia, 1944, Chilton Publishing Co.

Ostwald, F.: Ueber periscopic Glaser, Arch. Ophthal. **46:**475, 1898.

Pacey, D. J.: Antireflection coatings for ophthalmic lenses, Optician **147:**633-637; **148:**2-5, 55-61, 1964.

Percival, A. S.: The prescribing of spectacles, ed. 3, New York, 1928, William Wood & Co.

Peters, H. B.: The fracture resistance of industrially damaged safety glass lenses, Am. J. Optom. **39:**33-35, 1962.

Petzval, J.: Bericht über die Ergebnisse einiger dioptrischer Untersuchungen, Sitzungsber. d. math.-naturwiss., Vienna, **26:**50-75, 92-105, 129-145, 1857.

Richards, O. W.: Sunglasses for eye protection, Am. J. Optom. **48:**200-203, 1971.

Richards, O. W., and Grolman B.: Avoid tinted contact lenses when driving at night! J. Am. Optom. Assoc. **34:**53-55, 1962.

Risley, S. D.: A new rotary prism, Trans. Am. Ophthal. Soc. **5:**412-413, 1888.

Roberts, W. E.: Roentgenographic demonstration of

glass fragments in the eye, Am. J. Roentgen. **66:**44-51, 1951.
Rosen, E.: The history of eyeglasses, J. Hist. Med. **11:**13-46, 183-218, 1956.
Sasieni, L. S.: The principles and practice of optical dispensing and fitting, ed. 2, London, 1962, Hammond & Co.
Seidel, L.: Zur Dioptrik, Astron. Nachr. **37:**105-120, 1853.
Slater, P. N., and Weinstein, W.: Light transmitted by very small pinholes, J. Opt. Soc. Am. **48:**146-149, 1958.
Sloan, L. L., and Jablonski, M.D.: Reading aids for the partially blind; classification and measurements of more than two hundred devices, Arch. Opthal. **62:**465-484, 1959.
Smith, F. D.: Optical image evaluation and the transfer function, Appl. Optics **2:**335-350, 1963.
Stimson, R. L.: Ophthalmic dispensing, Springfield, Ill., 1971, Charles C Thomas, Publisher.
Volk, D.: Conoid refracting surfaces and conoid lenses, Am. J. Ophthal. **46**(II):86-95, 1958.
von Rohr, M., editor: Die Theorie der optischen Instrumente, Berlin, 1904, Carl Zeiss, Inc.
von Rohr, M.: Die Brille als optisches Instrument, Berlin, 1921, Julius Springer Verlag.
Ward, B., and Davis, J. K.: The modulation transfer function as a performance specification for ophthalmic lens and protective devices, Am. J. Optom. **49:**234-259, 1972.
Waters, E. H.: Ophthalmic mechanics, Ann Arbor, 1947, Edwards Brothers, Inc.
Whitney, D. B.: An automatic focusing device for ophthalmic lenses, Am. J. Optom. **35:**182-190, 1958.
Wigglesworth, E. C.: A comparative assessment of eye protective devices and a proposed system of acceptance testing and grading, Am. J. Optom. **49:**287-304, 1972.
Wollaston, W. H.: On an improvement in the form of spectacle glasses, Philosoph. Magazine **17:**327-329, 1803.
Wyszecki, G.: Theoretical investigation of colored lenses for snow goggles, J. Opt. Soc. Am. **46:**1071-1074, 1956.
Zugsmith, G. S., and Michaels, D. D.: Use of plastics in ophthalmic surgery, Eye Ear Nose Throat Digest **24:**1-6, 1962.

6

Physiologic optics

It is platitudinously true that the eye is an optical instrument. Indeed it is often compared to a camera, since the components in both serve the same function. But the camera picture is the end result, whereas the retinal image is only the beginning of the transmitted message. More than a passive screen, the retina performs important preliminary processing of the visual message so as to reflect the sharpness of the optical image. Unfortunately we have little to offer at the neurophysiologic level, so we concentrate on improving the optics.

All the giants of physiologic optics of the nineteenth century, including Listing, Helmholtz, Donders, Knapp, Gullstrand, and Tscherning, measured and calculated the optical constants of the eye, convinced that the keratometer and phakometer would accomplish for ophthamology what microscopy had achieved for pathology. Ametropia was to be conquered by formula. But instead of optical precision they found aberrations, irregularities, and disconcerting variances in optical components—biologic probability instead of mathematical certainty. The performance of the eye as an optical virtuoso turned out to be only so-so, compensating with biologic elegance for its dioptric aberrance.

THE EYE AS A STATIC OPTICAL INSTRUMENT

The two major optical components of the eye are the cornea and lens. The cornea is the heavyweight, contributing some +43 D; the optical punch of the lens is blunted by the surrounding aqueous to a mere +19 D. The total power of the eye is not the simple sum of the two components but the equivalent power of these two components separated by 5.4 mm of aqueous:

$$F_E = F_C + F_L - t/n\ F_C L_L \text{ or } +58.7\ D$$

Cornea

The principle of measuring corneal curvature is based on an idea of Scheiner who as early as 1619 compared the corneal reflections to those in different-sized glass marbles. "Keratometry," or "ophthalmometry," (the first term seems more appropriate) measures the image reflected by the cornea acting as a brilliant convex mirror. If the object is relatively distant (about $\frac{1}{3}$ meter), the image is for all practical purposes at the mirror's focal point, that is, half the radius of curvature. Once the radius is known, the refracting power can be computed by $F = \frac{n' - n}{r}$. Since the refractive index of the cornea is not measured, an index of 1.3375 is assumed.* This is the essence of the optics of the keratometer.

*This assumed value requires some explanation, since it in no way resembles the real index of the cornea, which is 1.376. Instead the index is purely arbitrary, chosen so that the instrument reads 45 D when the radius is 7.5 mm. The assumed index thus allows for the contribution of the posterior corneal surface (about −5.8 D), which is never actually measured.

EXAMPLE: The keratometer reads 44 D. What is the corneal radius?

$$r = \frac{n' - n}{F} = \frac{337.5}{44} = 7.67 \text{ mm}$$

(This is the formula by which to select a trial contact lens if one loses the tables supplied by the manufacturer.)

Although the keratometer measures corneal power indirectly, it is the only clinical instrument that measures any kind of ocular power. So the keratometer gives us a clue whether, for example, a myopic eye is optically too strong or anatomically too long.

Crystalline lens

The curvature of the lens is measured on the same principle as the cornea by a "phakometer," which uses reflected lenticular images. Since these are a hundred times less bright than the corneal reflections, it is not adaptable to clinical use. The crystalline lens, which is composed of sheets like the layers of an onion, is even less homogeneous than the cornea. Even maximum curvature changes do not account for the accommodative amplitude. Consequently, Gullstrand postulated a cortex and nucleus with differing indices (1.386 and 1.406). As a result the usually quoted average index of the lens (1.41) does not work, a matter still not fully resolved.

The equivalent power of the crystalline lens, based on Gullstrand's average index, is +19.7 D, which means it exerts the same power in situ that a +19.7 D trial lens would on a wave in air.

Axial length

Until four decades ago all axial length measurements were based exclusively on anatomic specimens, subject to the vagaries of postmortem changes. Optical measurements were based on ophthalmoscopy. X rays made possible in vivo measurements. In low dosage they pass through the eye without refraction and, aimed perpendicular to the optic axis, intersect the retina twice unless they just graze the posterior pole. The patient sees a disclike phosphene and only reports when it is as small as possible. The distance between the x-ray beam and a telescope sighted on the anterior corneal pole gives the axial length. Unfortunately the dangers of irradiation make this technique unsuitable for clinical testing.

Ultrasonography is a more recent innovation, based on the reflection of sound energy from each ocular surface. When these are converted to length and added together, they give axial length with results that agree well with x-ray and optical methods. Of course, ultrasound does not measure refractive power but only axial length and that piecemeal. Moreover the echo is reflected at different rates through different media and from the back of the eye, not the focal plane of the retina.

Refractive index

Each of the refractive components of the eye has a different index, varying with its constituent layers. The refractive index of the cornea as a whole, for example, is 1.376; however, its surface is bathed by a precorneal tear film of index 1.33 (depending on mucous content), and its stroma is made up of collagen fibrils (index 1.55) and ground substance (index 1.345). The hydration of the fibrils changes the index and may be excessive in glaucoma or when a contact lens fits too tight, resulting in irregular refraction with halos and hazy vision. Corneal transparency is thus intimately related to a stable index and a regularity of structure to prevent scattering by constructive interference.

The discontinuities of the crystalline lens are made visible by the slit lamp beam, representing fibers formed throughout life. Their number increases with age like the rings of a tree trunk, causing scattering and light dispersion and perhaps the yellowish pigmentation that disturbs the vision of old age. The refractive index changes from one layer to the next probably progressively toward the interior.

The index of refraction of the aqueous and vitreous is homogeneous, but the vitreous may degenerate, and the condensation of collapsed fibrils give rise to annoying floaters or even retinal detachment.

Our optical computations assume the refractive index of the eye is stable, although we know better in diabetic myopia. All clinical instruments designed to measure optical power of the eye (e.g., the keratometer), are arbitrarily calibrated.

The pupil

The pupil regulates the light entering the eye, and the delicate iris muscles regulate the pupil according to the prevailing illumination, target distance, age, refractive error, and emotional state. Pupil size can range from 2 to 8 mm, with approximately 3.5 mm considered average (an assumption used in computing spectacle lens designs). But the pupil in older people is often smaller, the myopic pupil is generally larger, the glaucomatous pupil is fixed, and the illumination is seldom average. When looking at a near target, the pupil constricts, and it dilates in fright. A small pupil reduces aberrations, thus increasing depth of focus; a large pupil allows more light but increases aberrations.

The pupil is usually round unless the iris muscle is selectively paralyzed by disease or a blow on the eye or if parts of it are removed by an iridectomy. Normally the pupils are equal and react symmetrically. The most common cause of inequality is not disease but the inadvertent use of drugs. Inequality is never due to oblique illumination, some texts notwithstanding.

When one eye is exposed to light, the homolateral and contralateral pupils constrict; this is the consensual reflex, an important diagnostic clue to the integrity of retinal function. It is present in all animals with semidecussating optic nerve fibers. With near fixation, the pupil constricts as part of the triad reflex (accommodation, convergence, and miosis). An isolated direct and consensual reflex paralysis to light but not accommodation constitutes the Argyll-Robertson pupil. It differs from amaurotic paralysis, in which stimulation of the affected eye causes no response and illumination of the normal eye contracts both pupils. An unusual sluggish, delayed light reflex with prolonged iris and ciliary contraction and loss of tendon reflexes constitutes Adie's syndrome.

Pharmacologically the iris resembles the ciliary muscle; all mydriatics produce some cycloplegia, and all cycloplegics produce mydriasis, but a lower drug concentration is required to dilate the pupil. Individuals differ in drug responsiveness, and consequently the pupil is a poor indicator of cycloplegia.

The pupil is ideally situated between the two major refracting components and helps exclude the rays that might disturb the optical perfection of the retinal image. Optically one distinguishes between the entrance and exit pupils. The clinician viewing the pupil sees its enlarged image through the cornea—the entrance pupil; the retina in turn views the pupil through the lens—the exit pupil. The ray directed to the center of the entrance pupil and emerging from the center of the exit pupil is the "chief ray" and defines the center of each retinal blur circle. The chief ray from the center of the exit pupil to the fovea is of special importance and is called the "primary line of sight." The diameter of the pupil is most readily measured by comparison to calibrated apertures (pupillometer). A contact lens, for example, must be large enough to cover the pupil, not an easy task when it is eccentric or irregular. An eccentric pupil need not interfere with acuity, however, if the rays can be brought to a focus; hence it is more important to choose a clear rather than a central area for a successful optical iridectomy.

If the fogging lens in subjective refraction is reduced too much, the patient accommodates and says that the letters now appear not only clearer but smaller and darker. The minification is due to increased optical power; the darkening is due to enhanced contrast of smaller blur circles from pupil constriction.

Depth of focus

The greatest distance a screen can be moved without a noticeable blur is the depth of focus; conversely, the greatest distance an object can be moved without changing the focus on a fixed screen is the depth of field.

The depth of focus of a camera depends on the "grain" of the film, as well as the optics of the lens and size of the stop. Similarly, the depth of focus of the eye varies with pupil diameter, refractive power, and perhaps the anatomic length of foveal cones. There are therefore no absolute values for the depth of focus; the range is greater for throwing a beach ball in sunlight than for threading a needle by candlelight. For "average" pupil diameters it is on the order of 0.75 D. Patients whose resolution has been destroyed (e.g., by macular degeneration) have a larger depth of focus because the change in blurredness is poorly discriminated. This is one reason (among others) for giving the weakest magnifiers that will accomplish the task.

Although the depth of focus allows a range of clear vision for a constant dioptric focus, it is a nuisance in clinical testing. It gives the impression that the patient has more accommodation than is really used. Thus a reading glass deemed adequate with good illumination in the refracting room proves to be a poor choice when the patient returns to a poorly lit workbench. Older people with small pupils and patients using miotics require larger lens changes in refractive tests because the increased depth of focus reduces their discrimination.

One of the most persistent, perpetual, and prevalent problems in advanced presbyopia is the reduced range of near vision as the biofocal add is increased. A useful expedient is to substitute a better depth of focus. By increasing illumination, it permits a better near range, as well as some intermediate vision. Increased illumination may also allow a patient with scarred corneas or irregular astigmatism to achieve useful vision. Patients with posterior subcapsular cataracts, however, generally do better with a dilated pupil.

Aberrations

Like most optical systems the eye suffers aberrations, which prompted Helmholtz, according to a frequently quoted statement, to exclaim that he would not only return such a poorly made instrument to the inept designer but would scold the optician for his carelessness. The statement must not be taken seriously because no designer has yet achieved a device that, with negligible effort, adjusts its own focus, regulates its own sensitivity, adapts to a range of illumination from noon light to moonlight, and manages to discriminate brightness, color, and contours at a quantum efficiency approaching unity.

Helmholtz's view is typical of nineteenth century physiology, which, peeking behind the biologic curtain, had discovered a machine. There was the digestive furnace, the muscular engine, and the osseous levers and ligamentary pulleys strung on the neural wires, with the mind somehow ensconced among the wheels and levers. Philosophers soon began to refer to their wives as ingeniously constructed mechanisms, and anatomists searched assiduously for the little creature that sits in the brain and looks through the eye. How disappointing it suffered aberrations.

In fact most ocular aberrations go unnoticed, and special conditions must be devised to demonstrate them. Unlike a camera film, only the foveal image must be sharp; peripheral distortions just fall on nondiscriminating rods. Because of its mobility, the eye foveally scans objects, the pupil excludes peripheral rays, the spherical eyeground minimizes curvature of field, and neurophysiologic mechanisms facilitate and inhibit. More impressive than the retinal wiring is the master switchboard that manages to enhance, suppress, emphasize, and express useful information even from a poorly focused picture (an analogy that cannot fail to comfort all shareholders in the telephone company).

Spherical aberration of the eye depends on pupil diameter, refractive power, and blur criterion and ranges from 0.5 D to 3.0 D, even becoming negative (central rays focusing first) with strong accommodation. Spherical aberration causes the rays to form not point images but a caustic. In elementary optics, one assumes the dioptric midpoint lies on the retina of the emmetropic eye; Gullstrand in fact believed the best focus to be the point of maximum light concentration and made his schematic eye hyperopic accordingly. Clinically only patients can tell us, by subjective testing, which point of the caustic suits them best and gives the most serviceable image. This is one reason subjective tests differ from objective ones, and the notion that tests are more discriminating in dim light should be accepted with reservations.

The visual axis is displaced relative to the optic axis, and the light is dispersed; that is, there is chromatic aberration. The long and short wavelengths do not focus simultaneously on the retina; in fact they differ by 1.00 to 1.50 D with the red last, blue first, and yellow-green in the middle. The emmetropic eye viewing a distant polychromatic target will be strictly in focus only for the middle wavelengths (the only ones allowed in thin lens optics theory). The emmetropic eye has to accommodate less to see the blue, which appears out of focus to the myope. This principle is utilized in the clinical "duochrome" test—a refinement technique in subjective refraction.

Retinal illumination

The amount of light reaching the photoreceptors determines whether a signal is generated. Only one or two quanta are required to activate the rods; a greater number are required to activate the cones. The light distribution in the retinal image can be studied through a scleral window or, more practically, by measuring the light reflected from the retina with a modified ophthalmoscope. The amount of light reaching the retina depends on target luminance (but not its distance), pupil diameter, transmittance of ocular media, and the angle of the light rays. A difference in relative luminous efficiency for central and peripheral pupillary rays is the Stiles-Crawford effect. The effect is as if the pupil were covered by a gradient density filter, progressively more dense radially. The cause lies not in the pupil but in the angulation of photoreceptors, which channel the rays like a funnel and therefore catch more quanta when struck head on.

For a constant pupil size, the ocular transmittance and angle of view remaining invariant, the retinal illuminance (in trolands) is the product of target luminance (in nits) and pupil area (in square millimeters). These strange-sounding units were not invented by Dr. Seuss but to provide physiologic optics with more appropriate photometric dimensions. A nit is a unit of luminance equal to one candela per square meter. A troland (named after a psychophysiologist) is a unit of retinal illuminance and is what one gets after multiplying nits by pupil area.

The image of a luminous point on the retina is a mound of light peaking in the center and falling off on each side in a series of maxima and minima. The light distribution at a border corresponds to half a mound and can be measured by photoelectric detectors. By means of a series of slits the photocell output may be tabulated for various parts of the retinal image. Once the distribution of illuminance for a single point is known, the retinal illuminance can be computed for any pattern. But the focus is distorted by aberrations, scattering, and irregularities; hence real retinal images lack the chorus girl symmetry and proportions of first-order, optical theory.

When a patch of light is presented to the eye intermittently, and the recurring cycles are high enough to fuse, the effect is as if the stimulus had the average luminance of the cycle (Talbot-Plateau law). When the frequency is reduced, an annoying flicker appears (well known to home movie watchers). The flicker varies with intensity and voltage and is a practical problem with fluorescent bulbs. It can be eliminated by using two or more bulbs in fluorescent fixtures. The most common example of visual fusion of discontinuous stimulation is the television screen. Fortunately no known pathologic mechanisms interfere with this kind of fusion, and although prolonged televiewing may harm the psyche, it seems to have no detrimental effects on the eye.

Ocular transmission

The limits of the visible spectrum vary between 390 nm and 760 nm.* The range can be extended in the infrared by increasing intensity and in the ultraviolet by removing the crystalline lens—indeed recent aphakes may complain that objects have a bluish tint. The retina is thus sensitive beyond the average spectral limits, but the rays do not reach it because of corneal and lens absorption. Additional ocular losses result from scatter,

*The "nanometer" (nm) is identical to and replaces the older term "millimicron" ($m\mu$, or 10^{-9} meter).

internal reflections, absorption by macular pigment, and progressive lenticular pigmentation. Diffusion from the sclera, flare in the optical media, reflections from the cornea and pigment epithelium, light scatter, and lens fluorescence all combine to produce stray light within the eye. These may become clinically important in the presence of peripheral glare sources.

When the eye is exposed to a source of light that is substantially greater than the luminance to which it is adapted, there is loss of visibility and discomfort. Glare sources set up stray light within the eye that veils the retinal image and reduce contrast. The luminance of the veiling patch may be used as a measure of the stray light. To maintain visibility, either the size of the object or its contrast must be increased—a point to remember in illuminating Snellen wall charts or plastic-coated reading cards.

Resolution

The ability of the eye to discriminate two closely spaced targets is termed "resolution." The image of a luminous point is a diffraction pattern and is called an "Airy disc" after its discoverer (Fig. 6-1).

The size of the disc is inversely proportional to the diameter of the entrance pupil and directly proportional to wavelength. Its exact size is indeterminate because the illumination fades without sharp demarcation. In 1879 John Strutt, Lord Rayleigh, first proposed that two points are resolvable if their Airy discs are separated by the distance between the center of one and the first minimum of the other. Smaller separations fuse into a single larger blur. For a circular pupil of 3 mm diameter and wavelength of 555 nm, the resolving power according to this criterion is about 47 seconds, a fair approximation to the 1-minute angle found clinically. Current methods favor closely spaced bars or gratings rather than points, but clinicians justifiably continue to use letters and numbers as the most practical criterion of resolving power.

Magnification and resolution are sometimes confused. If two Airy discs are projected on an imaginary expanding balloon, there is magnification, but since the Airy discs swell proportionately, they are no further separated than before. Such enlargement unaccompanied by improved resolutions is called "empty magnification." A patient with macular degeneration obtains limited benefit from magnifiers because his resolving power is lost, and newspaper print enlarged indefinitely leads to a meaningless jumble of black and white dots. The physiologic basis of resolution is the ability of the eye to detect brightness differences, that is, contrast gradients.

Contrast is seldom measured clinically; yet it makes a difference whether one views

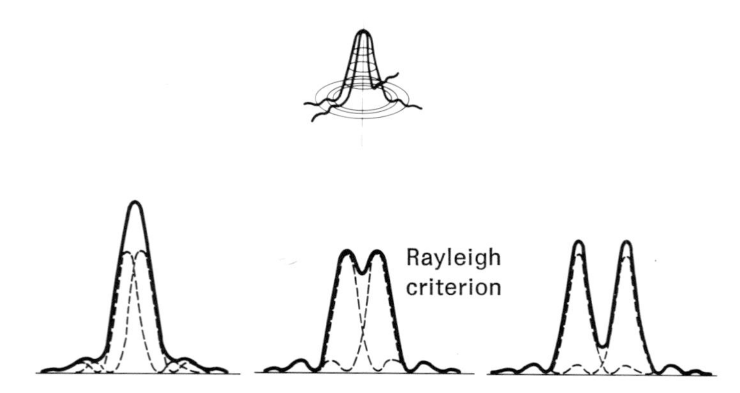

Fig. 6-1. The resolution of two points of light depends on the separation of their Airy discs, one of which is shown in the top figure. The separation between the two Airy discs must be a certain minimum, or dip, in the retinal illumination of at least 1%.

a white or a black thread against a black velvet background.

Ocular fixation is not perfect. The eye undergoes a series of minute oscillations with a frequency of 30 to 70 cycles per second superimposed on slower drifts and jerks, which are believed to enhance contrast. A classic illustration of contrast enhancement are the bands first described by Ernst Mach in 1865. At the borders of alternate black and white bars one sees a brighter stripe next to the dark border and a darker stripe next to the bright border. Although the intensity in a given bar is uniform, perceived brightness is not because it is influenced by the adjacent border.*

Optics of the visual pathway

In the retina the distribution of cones is highest at the fovea, and this is the area with the greatest resolving power. Critical seeing requires the extraocular muscles to point the fovea toward the object of interest. When motility is disturbed as in eccentric fixation, acuity drops in proportion to the cone density of the new fixation area and frequently more.

The ratio of ganglion cells to cones is highest at the fovea, which has not only a private wire system to the cortex but also diffuses interconnecting horizontal cells to excite and inhibit adjacent areas. Whereas photoreceptors give rise to steady potentials, the ganglion cell responds with an all-or-none spike, with a frequency proportional to light intensity. These responses are particularly active at the onset and end of stimulation and practically silent during continuous exposure. On-and-off activity is believed to be one of the retinal border enhancing mechanisms. By electrophysiologic recording from single ganglion cells, it is possible to map out that area of the retina whose stimulation contributes signals to it. Such areas are termed "receptor fields."

The simplest level of visual perception probably involves detection of movement, and one of the major neurophysiologic insights of the past decade has been gained from an analysis of spatial signals. Retinal cells of cats respond equally to movement in all directions, although over a limited spatial frequency. Like ganglion cells, the geniculate cells also respond to movements in all directions. Cells of the striate visual cortex of cats and monkeys, however, respond to the orientation of the edge or line as it is moved across the receptive field—a specific response code.

In the "optics" of this system, the retinal receptor fields (recorded by microelectrodes from ganglion or geniculate cells) consist of doughnut-shaped areas. A small spot of light confined to the center has a stimulating effect; a spot presented simultaneously in the periphery has an inhibitory effect on the center. When the retinal receptor field is plotted for cortical cells in area 17, the doughnut is elongated. A line (or edge) passing such a field will, depending on its orientation, produce more or less activation of the periphery and hence a variable inhibition of the center. This is how the signal is coded. Based on psychophysical studies of high-contrast, masking gratings, it seems likely that the human visual system is similarly organized.

SCHEMATIC AND REDUCED EYES

The schematic eye is a Gaussian blueprint, a mathematical model of a real eye in which all surfaces are perfectly centered and all rays are perpetually limited to the paraxial region. The best known schematic eye is based on Gullstrand's average measurements, but, of course, neither Gullstrand nor anyone else can tell us if the next patient's eye matches these averages. Thus the schematic eye is a relative and not a celestial blueprint, which is useful to explain to ourselves the optics of ametropia, correcting lenses, and refractive tests. Since Gullstrand's measurements far exceed practical accuracy, there is every reason to simplify. It is silly to reckon retinal image size to the fourth decimal when the models—much less clinical methods—have not this accuracy. Instead of 58.64 D for ocular power, a rounded value of

*The effect can also be produced by pasting a black polygon in the center of a circular white field and rotating the pattern on a child's top. The phenomenon has also been termed "simultaneous contrast."

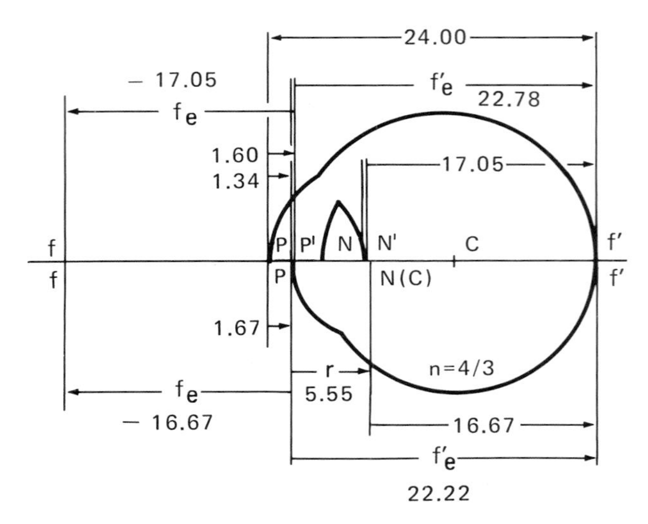

Fig. 6-2. Comparison of Gullstrand's schematic eye *(top)* and the simplified "reduced" eye *(bottom)*. The reduced eye has only one surface, one equivalent plane, and one nodal point, which is also the center of curvature of the surface.

60 D will do as well; and instead of a series of surfaces, a single refracting surface separating air from a common medium (such as water) will serve all our needs. This simplified or "reduced eye" model is the most practical of the schematic eyes.

The first reduced eye, invented by Huygens, consisted of a single refracting surface separating air from water. Listing, a pupil of Gauss, resurrected it 200 years later, and all modern variants are based on it. Donders, with a passion for round numbers and a mistaken idea of corneal curvature, preferred a model of 66.6 D and 5 mm radius (which gives focal lengths of −15 mm and +20 mm). The reduced eye of Fig. 6-2 has a single refracting surface of radius 5.5 mm separating air from water, with a focal power $F = \frac{n' - n}{r} = +60$ D and focal lengths $-n/f = -16.66$ mm and $n'/f' = +22.22$ mm. Since only a single refracting surface is involved, its center of curvature is also its nodal point, whereas the anterior pole represents the equivalent point.* Although the reduced eye looks like an eye, it would be wrong to call the front surface a "cornea" because it is equivalent to the power of the eye as a whole. Its curve is only a concession to our sense of esthetics; optically it should be drawn as an equivalent plane. All the rays that will ever enter its "pupil" are paraxial. Our reduced eye has no real pupil because retinal illumination is not part of the definition of emmetropia.

Emmetropia and ametropia

An eye is considered to be emmetropic if parallel rays focus on the retina with accommodation relaxed. The reduced eye of 60 D is emmetropic when the retina is 22.22 mm behind the refracting surface. The definition says nothing about its biologic health—the eye may be emmetropically blind. Indeed the power need not be 60 D; any other power will do, providing the retina straddles the focal point. If the retina is not at the focal plane, the eye is ametropic (accommodation assumed to be inactive). An eye of optical power 60 D and axial length longer than 22.22 mm is myopic; the optical image falls in front of the retina and is blurred. An eye of 60 D with axial length less than 22.22 mm is hyperopic.* Its retinal image is also blurred unless it accommodates, but that is forbidden, at least in this chapter.

*The distance from the nodal point to the retina numerically equals the anterior focal length. Many texts use the more exact 17.05 mm (based on a schematic eye power of +58.64 D), but the difference is negligible.

Ametropia and axial length

It is readily shown that for a reduced eye of 60 D, 0.3 mm represents about 1 D of axial error. In ophthalmoscopy, 1 mm of fundus elevation is considered to correspond to 3 D of defocus, such as in following the course of papilledema. A 60 D eye with the retina at 21.85 mm represents 1 D hyperopia.* But the same degree of hyperopia could also occur if the power were 59 D and the retina were at 22.22 mm. In the first case—axial ametropia—the power was "normal" but the retina malplaced. In the second case—refractive ametropia—the retinal position was "normal" but the refractive power too weak (Fig. 6-3). This is not just playing with numbers; it makes a difference in retinal magnification as we see later. Of course the "normal" is only an assumed Gullstrand average, since there is nothing intrinsically emmetropic about 60 D; an eye could be emmetropic with a power of 70 D and a retina at 19 mm.* But we must start somewhere, so the combination 60 D and 22.22 mm is *arbitrarily* taken to represent emmetropia (some texts call it "normal emmetropia"). An eye whose power differs will by definition have refractive ametropia; an eye whose axial length differs will by definition have axial ametropia. About two thirds of all ametropias are estimated to be axial, and one third are attributed to variations in curvature or index.

EXAMPLE: The length of the fetal eye increases from 14 to 17 mm during the last trimester. The theoretical change in power is

$$F = \frac{n'}{f'};\ \frac{1.33}{.014} = +95\text{ D};\ \frac{1.33}{.017} = +78\text{ D}$$

or about 17 D.

The astigmatic eye can be visualized in

*The term "hypermetropia" was coined by Donders. "Hyperopia" is shorter, sounds better, and, if precedent is needed, was adopted by Helmholtz. This is no reflection on Donders—we do not use his "brachymetropia" for myopia either.

*For example, E. J. Tron found that an emmetropic eye may have powers ranging from 57.47 to 72.10 D and axial variations from 20.46 to 25.45 mm. Corneal emmetropic powers may range from 37 to 48 D, and lenticular powers from 20.19 to 34.09 D.

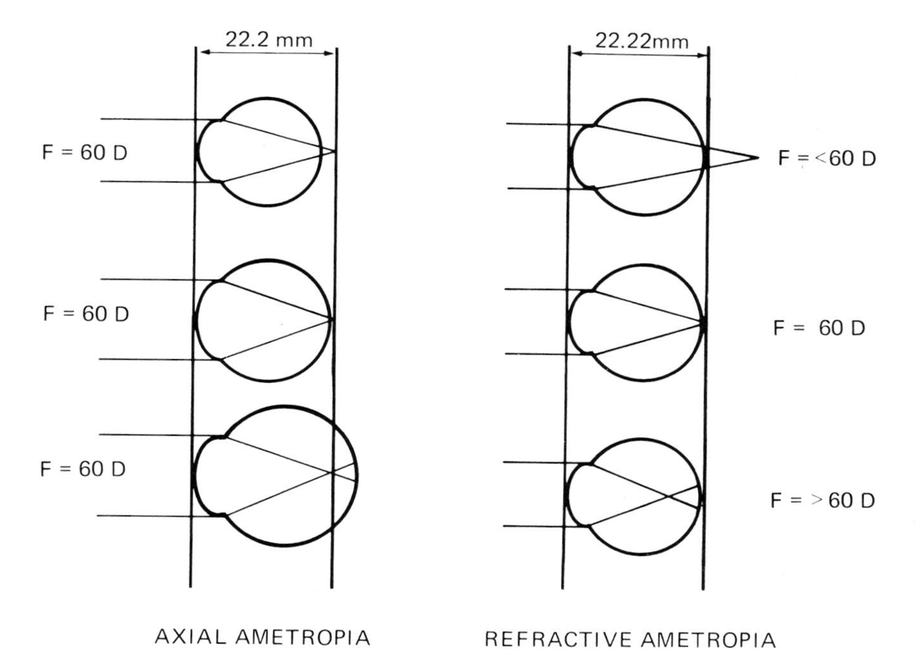

Fig. 6-3. Comparison of axial and refractive ametropia. In axial ametropia the ocular power remains the same, but the length of the globe varies; in refractive ametropia, axial length remains constant, and the ocular power varies.

terms of the optic cross (accommodation assumed inactive). If the power in the vertical meridian deviates from emmetropia by +2 D and in the horizontal meridian by −3 D, the horizontal line is 2 D anterior to the retina, and the vertical line is 3 D behind. But the distances are not 50 cm and 33.33 cm because the 60 D emmetropic power must be added. The horizontal line is thus $\frac{1.33 \times 1000}{62} = 21.45$ mm or 0.77 mm in front, and the vertical line is $\frac{1.33 \times 1000}{57} = 23.33$ mm or 1.11 mm behind the retina.

In the aphakic eye the lens component is missing, and the surface of the reduced eye now represents the cornea. The aphakic eye thus has a power of +43 D, and is 58 − 43 or 15 D hyperopic. Its retinal image is magnified by $\frac{58}{43} = 34\%$. It requires a correcting lens of +10.50 D at a spectacle plane 15 mm from the cornea.

Visual angle

A light ray through the nodal point remains undeviated, and since the nodal point is as far from the retina as the *first* focal length (−16.66 mm), the size of a sharp retinal image (h′) can be computed from the visual angle ω:

$$h' = f_e \tan \omega$$

For example, the limb of a Snellen E subtending 1 minute will give a retinal image diameter of (−16.66) (0.00029) or −0.0048 mm. The minus sign indicates the image is inverted, as most retinal images are. For reasonably distant objects the exact point at which the visual angle is measured need not be specified—the corneal pole will do as well as the nodal point (Fig. 6-4).

Optically the size of a visual object is determined only by the size of the retinal image, that is, by the visual angle. If a coin held at arm's length subtends the same angle as the moon, both are optically the same size.

The psychologist knows better and indeed has known since Berkeley's essay (1709) that retinal image size is only one factor that determines perceived size. For aside from its dimensions, the retinal image contains a whole host of stimulus correlates (referred to as "cues" in the older literature). These monocular cues are what the painter incorporates on canvas to give us the illusion of a three-dimensional scene. So much are they a part of our common visual experience that artists must train themselves to ignore what they perceive so as to duplicate the cues correctly. This is why "primitive" masters as well as amateurs generally get their perspective wrong.

Judgment of size thus depends not only on visual angle but on perceived distance. Our previous knowledge of objects affects it; a person viewed through binoculars appears not larger but nearer.

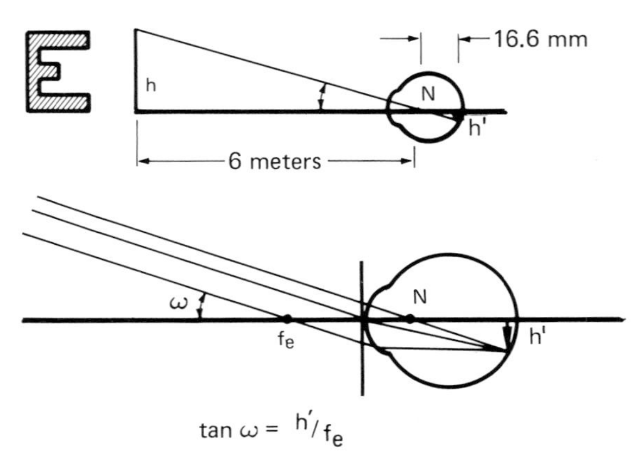

Fig. 6-4. The visual angle subtended by a distant object can be measured at the nodal point, pole, or first focal point of the reduced eye. *Top,* If the object size is known, the size of the retinal image is determined from the similar triangles. *Bottom,* If only the visual angle is known, the retinal image can be computed from the tangent relationship.

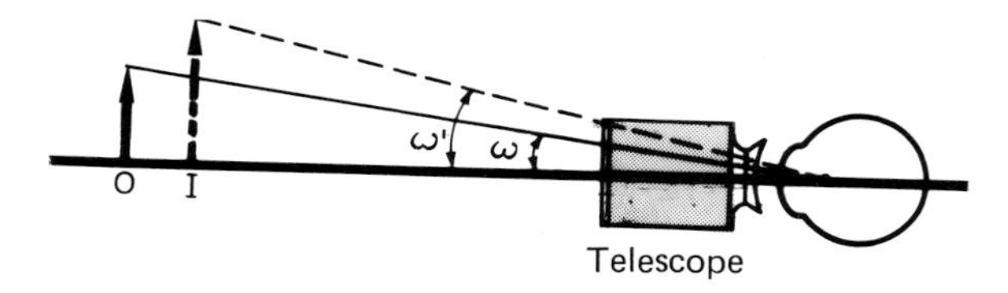

Fig. 6-5. The increased visual angle (ω') produced by a telescope accounts for its angular magnification: $m = \omega'/\omega$.

In optics, all this is ignored; to the model eye, coin and moon are the same size because they subtend the same visual angle. If a telescope is now interposed, the ratio of the visual angles with and without the instrument, that is, the angular magnification, is increased (Fig. 6-5). A telescope magnifies because it alters this angular ratio; the incident vergence is not changed.

$$\text{Angular magnification} = \frac{\text{Visual angle with telescope}}{\text{Visual angle without telescope}}$$

A plus lens magnifies because it increases the visual angle. Since objects at different distances may all subtend the same angle, a ratio gives insufficient data, and a reference distance called the "conventional least distance of distinct vision" is assumed. The arbitrary distance, set by some unknown partially presbyopic optics teacher, is taken to be 25 cm despite periodic attempts to have it changed to a more reasonable 40 cm. A person with an adequate amplitude of accommodation would not use either distance but simply bring a small object as close as possible to achieve maximum magnification. Most patients with subnormal vision, however, no longer have good accommodation, so the reference distance of 25 cm provides the standard for comparison. If the lens is held at its focal distance from the object, the angular magnification for the 25 cm reference distance is $\frac{F}{4}$, where F is the power of the lens. A +8 D lens therefore gives 2× magnification by this criterion.

Retinal image size

The size of a sharply focused retinal image may be computed from the similar triangles subtended at the nodal point of the reduced eye:

$$\frac{\text{Size of image}}{\text{Size of object}} = \frac{\text{Nodal point distance}}{\text{Object distance}}$$

EXAMPLE: What is the size of the blind spot measured on a tangent screen 1 meter from the eye? Assume the optic disc diameter is 1.5 mm.

$$\frac{-1.5}{h} = \frac{16.6}{-1000};\ h = +90.0 \text{ mm}$$

Computations of this type are frequently made to calibrate a tangent screen, and we see the results are based on assumed emmetropic values. If the eye in the example is 2 mm longer, the blind spot diameter is 80 mm; if 4 mm longer, 73 mm—an error of 11% and 19%. The actual power or length of a given eye is not generally known; hence such computations have only relative significance. To take absolute blind spot size seriously is to ignore optics, which may be the reason so many "focal infections" were diagnosed in the older literature. Incidentally these elongated eyes are not theoretical; the standard deviation of emmetropic axial length (from Stenstrom's data) is ±1.09 mm, and an axial length of 26 mm for a myopic eye is not unusual.

In 1668 Mariotte discovered the blind spot. Mistaking nerve fibers for photoreceptors, he concluded that the choroid must be the organ where "vision is made." Mariotte's error is excusable, but some current explanations that the blind spot is invisible because it is covered by the other eye are not—no hole appears in the monocular field either. The blind spot cannot be perceived because it has no cortical representation. We are only aware of it by inference; a previously seen object disappears when its image falls on the disc. No psychic or physiologic process fills it in any more than the space between cerebral hemispheres. This is why there is no vertical seam in our monocular visual field even though each half is represented in a different hemisphere. It has

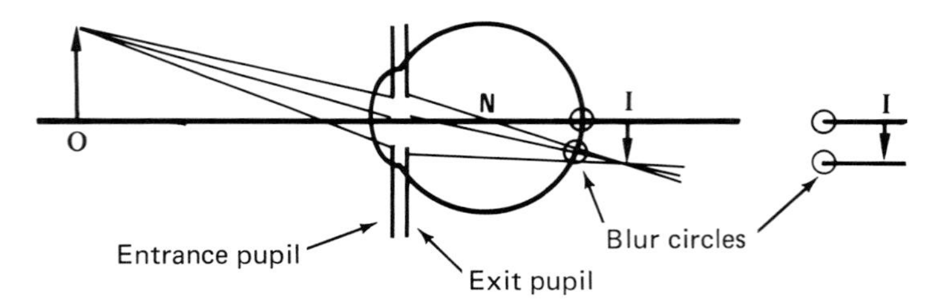

Fig. 6-6. Diagram illustrating that blurred retinal images are determined not by the nodal point but by the diameter of the limiting blur circles, which depend on the rays through the entrance and exit pupil.

little connection with such adaptations as Troxler's phenomenon (the disappearance of objects seen in indirect vision with steady central fixation) or Arago's spot (the nonseeing area in scotopic vision due to the absence of rods at the fovea) or the absence of the retinal vessel shadows (which can, however, be made visible by moving an oblique light beam). The blind spot is more akin to Anton's disease in which, as the result of cortical destruction, the patient is blind to his blindness.

When an eye views a near object without accommodating or a myope looks at a distant target, the retinal image is blurred. The blurred image is made up of blur circles (not to be confused with blurred images). The diameter of the blur circles varies with the diameter of the pupil and how much the eye is out of focus. If the pupil constricts, the blur circles collapse about a central ray that passes not through the nodal point but through the entrance and exit pupils. The size of a blurred retinal image depends on the distance between the centers of the limiting blur circles (Fig. 6-6) and cannot be computed from the nodal point. For example, a contact lens disturbs the retinal image size less than a spectacle lens simply because it is nearer the pupil. For any eye the size of the retinal image, clear or blurred, is proportional to the visual angle subtended at the entrance pupil. In elementary visual optics the nodal point and entrance pupil are assumed to coincide, and neither one is generally known except on the average. Finally, the nodal point is a mathematical concept, not an anatomic point in the vitreous. Nor is it a perceptual point through which we "project" visual sensations. Mixing mathematics, anatomy, and perception only leads to a psychophysical mishmash.

A sharp optical image is formed on the retina of the reduced emmetropic eye. In axial ametropia the optical power is 60 D, but the globe length is altered, and the retinal image is blurred. Since the retina is further in myopia and closer in hyperopia, the blurred myopic image is larger and the blurred hyperopic image smaller than the sharp emmetropic image.

In refractive ametropia the axial length is constant, but the power varies. The radius of the reduced surface will be smaller in myopia and longer in hyperopia, and thus the blurred ametropic image is the same size as the sharp emmetropic image.

In all ametropias, whether axial or refractive, objects that subtend the same angle generally are not perceived larger or smaller because they are judged by habitual visual standards. Only when these are altered by spectacles do spatial difficulties arise.

Visual direction

The limits of the sensitive retina is the visual field, first measured by Thomas Young. Its diagnostic value was first appreciated by Von Graefe and depends on precise rectilinear projections of the visual field on the retina and of the retina in the visual cortex.

Although rectilinear light propagation was known to Euclid, sight was believed to result from the emanation of pictures from the eye. This naive view, under the name of the "projection theory," still influences our conceptions of binocular vision, more impressive for its simplicity than for its profundity.

Alhazen (1100 A.D.) is credited with getting the light direction right, but it was Kepler (1604) who first pointed out that the ocular image must be "painted" on the retina not by the intercrossing of light rays as in a pinhole camera but by the collecting and refracting power of the cornea and lens. Scheiner soon demonstrated the small inverted retinal image by a scleral window, and the question of why we see objects right side up has been asked by students ever since. Strangely they seldom ask why objects are not seen diminished or reversed left to right.

The notion that we learn to see upright stems from an almost universal misinterpretation of an experiment by Stratton, performed about 1896, in which it is claimed that he learned to see right side up after wearing a small astronomic telescope in front of one eye. He wore the device for approximately 8 days (the other eye was covered) and, with a strong sense of body image, spoke of "upright appearance." Although he never said he saw objects erect, he did learn a new visuotactual coordination. Stratton's feeling of normalcy after wearing the device has never been captured by others who have repeated the experiment with more elaborate optics and progressively longer wearing times.

The inversion of the retinal image is not to be dismissed by any relativistic dodge—already suggested by Kepler, its discoverer. Since the branches of the tree are in the sky and its roots in the ground, the whole matter is supposed to be a pseudoproblem because these relations are preserved in the image. In fact, the inversion of the retinal image is a *necessary* prerequisite to upright vision. The matter is of more than academic interest because if such a major restructuring of the visual apparatus is possible, then all kinds of minor orthoptic miracles should be easily achievable. But if upright vision is based on an inverted retinal image tied to the visual cortex by genetically predetermined and maturationally fixed pathways, the system will allow for little tampering.

THE EYE AS A DYNAMIC OPTICAL INSTRUMENT

The remarkable faculty of the eye by which it adjusts its focus automatically to view objects at various distances is called "accommodation." It distinguishes the eye from a static camera by giving range to its resolution. Wound up in infancy, the accommodative mechanism gradually and inexorably runs down to impotence. The age at which it produces clinical symptoms cannot be exactly specified and is roughly that at which we cease to indulge and begin to preach.

Optics of accommodation

The closest point of distinct vision is the near point of accommodation (punctum proximum). The amplitude of accommodation is the dioptric difference between the far point and near point—the flexibility of the ocular focus.

The far point of an emmetropic eye is at infinity, and therefore its accommodative amplitude is simply the reciprocal of the near point. The amplitude of an ametropic eye is computed from the famous Donders formula best expressed in diopters:

Ametropia + amplitude = Nearest point of distinct vision (in diopters)

In this formula the amplitude is always positive, myopia is taken as a plus, and hyperopia as a negative quantity. For example, a 2 D myope with an 8 D amplitude will be able to see clearly as close as 10 cm. A 2 D hyperope with an 8 D amplitude will be able to see a target clearly at 16.6 cm. A 10 D hyperope with an amplitude of 8 D cannot see objects clearly at any distance; his nearest point of distinct vision is virtual, 50 cm behind the eye.

A theoretical problem is to compute the reading prescription while leaving a proportion of the accommodative amplitude in reserve. For example, if a 1 D myope with an amplitude of 3 D is to read print at 25 cm and keep one third of the ampliutde in reserve, the required reading lens is +1 D. The principle is that the power needed at 25 cm is +4 D of which 1 D will be supplied by the myopia, leaving 3 D to be contributed by the accommodation. Since the available accommodation is to be two thirds of 3 D or 2 D, the required lens is +1 D. It is best to reason these problems through and not depend on a formula. The steps for finding the reading prescription are as follows: (1) what power is needed for the distance in question; (2) how much is added or subtracted because of ametropia; (3) what accommodation is available so as to leave the required reserve; (4) the difference between (step 1 and step 2) and step 3 is the required lens. In these simple calculations the vertex distance between the spectacle lens and the eye is ignored.

EXAMPLE: What reading lens is required for a 4 D hyperope to read at 33.33 cm if one half of his 5 D amplitude of accommodation is to remain in reserve?

1. Optical power required for 33.33 cm equals 3 D.
2. Hyperopia of 4 D is added to the demand; hence total power required at this distance equals 7 D.
3. Available amplitude is $\frac{1}{2}$ (5) or +2.50 D.
4. Reading lens required is 7 − 2.50 = +4.50 D.

In practice one would almost never prescribe reading lenses on such an arbitrary basis, but these problems frequently arise on board examinations, and therefore no book with optics in its title would dare omit them completely.

Accommodation is normally binocular, and no aniseikonia is induced; since the entrance pupil is unchanged, there is no difference in retinal image size.

Accommodation and ametropia

When a minus lens is placed in front of an emmetropic eye observing a distance object, accommodation is activated to maintain clear vision. The minus lens had induced an optical hyperopia. A plus lens will not relax accommodation less than zero; so the distance vision remains blurred, an induced optical myopia. One of the theories of myopia that comes to us from Germany, that land of fairy tales, is that excessive accommodation somehow becomes permanently frozen into the ciliary mechanism—a kind of rusted pseudomyopia.

The ability of hyperopes to compensate for their refractive errors and their inability to do so when the hyperopia exceeds their accommodative amplitude led to endless confusion between hyperopia and presbyopia. James Ware (1812) a British surgeon (and the first to devote his practice exclusively to ophthalmology), first clarified the difference.

The amount of clinically correctable hyperopia is termed "manifest hyperopia." Since there is a certain tonus of accommodation that cannot be further relaxed without cycloplegia, some "latent" hyperopia remains masked. The total hyperopia is the latent plus the manifest hyperopia. If some hyperopia remains outstanding even after maximum accommodative effort, the uncompensated portion is termed "absolute hyperopia"; the compensated portion is the "facultative hyperopia." With advancing age, the absolute hyperopia increases at the expense of the facultative.

Although the myope needs to accommodate less in viewing a near object than does the emmetrope, the correcting lens restores a more normal accommodative convergence relation at the reading distance. In the presbyopic age group, on the other hand, myopia may act as a built-in reading glass.

In uncorrected astigmatism, accommodation may improve vision by placing the circle of confusion on the retina. The circle of least confusion is about half the diameter of an equivalent amount of myopia; consequently, all other factors being equal, one would expect the vision to be twice as good. Accommodation can also be stimulated by a minus cylinder. Although the cylinder moves only one focal line, it displaces the circle of confusion by changing the interval of Sturm. The patient then accommodates to restore it to the retina.

Anisometropic patients are in trouble most of the time, floundering between aniseikonia and amblyopia and swamped by induced heterophorias. Even their accommodation is against them because it is binocularly symmetrical. The job of the refractionist is to get these patients on dry land or at least throw them a rope by eliminating the refractive difference promptly. Only if a patient has already drowned in a sea of suppression would one ever consider correcting one eye for distance and the other for near.

Night myopia

When illumination is greatly reduced, the steadiness and accuracy of accommodation diminish, but instead of falling to zero, it becomes positive and produces myopia. A similar effect is observed in empty visual fields or whenever contrast is greatly reduced. The amount of myopia ranges up to 1.50 to 1.75 D, depending on the baseline emmetropic criterion. Known for some 150 years, night or empty-field myopia has recently received renewed interest in connection with space flights.

Night myopia is partly the result of spherical aberration—the dilated pupil causes a shift in the caustic anterior to the retina. A contributing factor is the change in peak spectral sensitivity from yellow to green—the Purkinje shift. There is, however, a residual degree of myopia that can be abolished by cycloplegics. The origin of this accommodative stimulus is still debated, but the effect has clinical implications in testing under reduced illumination and with fogging lenses.

Light and dark adaptation

One property of the eye that distinguishes it from a photographic plate is the ability to adapt to sunlight and moonlight, an operating efficiency of several millionfold. The major feature in this process is the changes in the concentration of photopigments in the rods and cones. The light bleaches the photopigment, and the decomposition products reassemble to form new photopigment in the dark. The rate of dark adaptation is rapid for the first 10 minutes and is almost completed in 30 minutes. In light adaptation the sensitivity falls much more rapidly and is almost completed in 75 seconds. Thus in night driving the problem is not seeing the approaching car but the road after the car has passed. An outstanding characteristic of the dark-adapted eye is not only its extreme sensitivity but the absence of color vision as well. The monochromatic nature of the scotopic eye indicates that rods do not participate in color perception.

When dark adaptation is measured over a range of illumination, the resultant curve exhibits a break at the transition point between photopic and scotopic vision. The photopic branch lasts 10 minutes and is followed by a slower scotopic process, indicating that both cones and rods are involved. The shift indicates that absorption characteristics of the cone photochemical differ from that of the rods. With long wavelengths, only one part of the curve is evident, one reason for wearing red goggles in preadaptation.

The clinical evaluation of dark adaptation is essentially simple in theory. The eye is preadapted to a standardized test field for a specified time, and the patient identifies the visibility of a low-intensity, test target as a function of the duration in the dark. Practical difficulties arise from the fact that the pupil dilates; the rate of adaptation is influenced by the region of the retina stimulated, variations in the wavelength composition, the area of the adapting and test light, the time it is exposed, and whether the recognition criterion includes texture, form, and depth.

Although most of our vision is carried out under photopic conditions, such activities as night driving and flying, astronomic observations, and a variety of military duties are performed under low levels of illumination. Night vision is also influenced by absorption lenses, and even ordinary glasses reduce transmission 8%. Acuity is influenced by illumination and contrast, and the ability of the eye to adapt from one photopic level to another has an annoying inertia in older people with miotic pupils or in patients on antiglaucoma therapy. Night vision is also characteristically reduced in retinal diseases such as retinitis pigmentosa, extensive choroidal atrophy, some vitreoretinal degenerations, and in severe vitamin deficiencies. The optical myopia in dim light has already been described.

COLOR VISION

Color perception increases the range of visual discrimination from a series of "just noticeably different" brightness alone to a similar series for each spectral hue. This multiplication of steps that can be discriminated is its chief biologic value. Seeing two different colors does not require particularly high resolution.

Color sensations can be analyzed by introspection into three attributes: hue, saturation, and brightness, which are related psychophysically to wavelength, admixture of white light, and intensity, respectively. (Hue and saturation are sometimes grouped together as "chromaticity".) The three color attributes are interrelated; a change in one generally results in a change in the others. For example, a desaturated red appears pink, and all hues are lost at scotopic illumination levels. Color depends on the state of adaptation, the region of retina stimulated, simultaneous and successive contrast, and an awareness of the overall direction and kind of illumination. The colors of color theories are film colors; the colors of objects are surface colors and are influenced by texture and background. Color vision depends on a normal sensory receiving mechanism, which is deficient in some 2% of the male population and somewhat peculiar in another 6%. Color vision disturbances may also be acquired (and temporary) in choroideremia, macular degeneration, central serous retinopathy, and retrobulbar neuritis.

Physiology of color vision

Normal color vision is trichromatic; that is, all colors can be reproduced (assuming negative coefficients) by the appropriate mixture of three primary colors: red, green, and blue. There may be more than one set of primaries, as long as a mixture of any two does not reproduce the third. When mixed in equal proportions, the primaries yield white: blue + green + red = white or red + green = yellow and yellow + blue = white. Yellow and blue and similar combinations are termed "complementary colors." Expressed algebraically, any color C can be reproduced by mixing three primaries, R, G, and B, in luminosity proportions α, β, and γ as follows:

$$C = \alpha R + \beta G + \gamma B$$

For example, 2R + 2G + 2B = 2 white (W) since the luminosities are additive, and 1R + 3G + 6B = 2G + 5B + 1W by extracting one unit of each primary to form 1W. The last color is said to be desaturated by 1W (that is, the lowest primary proportion expresses the color saturation).

The theory of color vision, first proposed by Thomas Young and resurrected by Helmholtz, assumes the presence of three receptors, each activated in a ratio α, β, or γ when stimulated by a particular combination of wavelengths.* The receptor response is a physiologic duplicate of the experimental, color mixing equation; the luminosity proportions are neuron discharge frequencies; and the three primaries are different patterns of responses. Fig. 6-7 shows a graphic representation of these responses for the range of spectral wavelengths. An ordinate at a particular wavelength on the graph intersects with the three curves in the proportions α, β, and γ needed to reproduce or match it. At the ends of the spectrum, the ordinate intersects only two curves, which means one of the primary proportions is zero. The

*Modern evidence is rather convincing that the triple response is based on three different photochemicals. Red-, green-, and blue-responsive cones have been identified by microspectrophotometry. Retinal densitometry can measure the rate of bleaching for the red and green (the amount of blue-sensitive pigment is still too small to be measured), and the bleaching rates are proportional to the brightness response of the cones, adding up to the photopic luminosity curve.

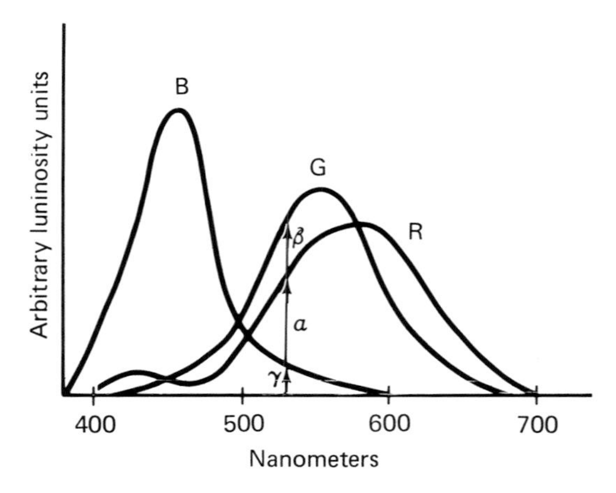

Fig. 6-7. The three curves represent the amounts (luminosities) of each primary color required to match the color sensation aroused in a normal photopic observer by the wavelength on the abscissa. The color indicated by the ordinate (about 530 nm) would require α red, β green, and γ blue luminosities, or approximately 2R + 3G + 1B. Since equal amounts of the three primaries give white, the same result could be expressed by 1R + 2G + 1W. The lowest proportion for any ordinate represents the saturation; the uppermost curve, the dominant color. The curves also represent the theoretical responses of three primary color receptors in a trichromatic observer.

curves are constructed from experimental matching data for each spectral color. The lowest curve at any particular wavelength represents the color saturation; the uppermost curve represents the dominant color. It might be expected that at those wavelengths at which the dominant colors shift (where the top curves intersect), the ability to detect a slight change in wavelength would be maximal. This is indeed found in the hue discrimination (or more correctly, the wavelength discrimination) curve, which is roughly W shaped, the minimum falling at the intersections of red-green and blue-green.

Color blindness

"Color blindness" is usually a misnomer; few people have truly monochromatic vision. What is meant is "color deficiency." The usual is red-green confusion, sometimes called "Daltonism" after its most illustrious patient. Color deficiency can be classified as follows:

1. Dichromatic vision
   a. Protanopia (red deficient) } Daltonism
   b. Deuteranopia (green deficient) } Daltonism
   c. Tritanopia (blue deficient)
2. Anomalous trichromatic vision
   a. Protanomalous
   b. Deuteranomalous
   c. Tritanomalous
3. Monochromatic vision
   a. Rod monochromat
   b. Cone monochromat

The characteristic of dichromatic vision is that all colors can be reproduced by mixing two instead of three primaries; let us call them A and B. The mixture of these two in various luminosity proportions reproduce all other colors:

$$C = \alpha A + \beta B$$

Equal amounts of A and B yield white. The protanope and deuteranope both confuse red and green, but in protanopia the visible spectrum is shortened in the long region, and the neutral point (an area in the spectrum that appears colorless) is around 495 nm. The deuteranope has a neutral point at 505 nm, but the visible spectrum is not shortened. The tritanope has either one or two neutral points (depending on whether the green cones copy the blue or vice versa) and confuses blue and green.

Various theories have been proposed to explain these peculiarities of dichromatic vision. The simplest theory and one which may serve as a foundation for further study of the more recent literature was first proposed by Fick (of ocular axis fame) in 1878. Fick assumed that in protanopia, or red deficiency, the red receptors are not truly absent but aberrant. They still respond as "red" (pattern of discharge) but the frequency of their response exactly duplicates the green receptors, which remain normal. The result of this imitation is that red and green receptors always respond with the identical frequency, producing a curve that looks like Fig. 6-8, *A*. Since red plus green is yellow, we might as well label the superimposed curves yellow. The two colors the protanope sees are therefore yellow and blue. We know this to be actually true from a few rare cases of unilateral color blindness. Fick's theory, although based on the Young-Helmholtz idea, preserves the dignity of yellow as the perceptual primary insisted on by Goethe and Hering.

Since the red receptors copy the green, the shift results in a shortened spectrum and a neutral point where the two intersect (at 495 nm). The protanope has no difficulty distinguishing wavelengths on the left of the neutral point (the yellows) from those on the right (the blues). But he can only distinguish the various yellows (which appear as red-orange-yellow-green to the normal trichromat) by differences in brightness and saturation. A given spectral color *a* is more saturated (has less white) than spectral color *b* because the lower curve indicates less blue to be mixed with it. Since color blindness is practically always bilateral and congenital, the protanope learns to call one particular shade of yellow "red" and another shade of yellow "green." The names attached to colors are learned, and dichromats name colors normally. But if the usual brightness and saturation of these yellows are altered, confusion

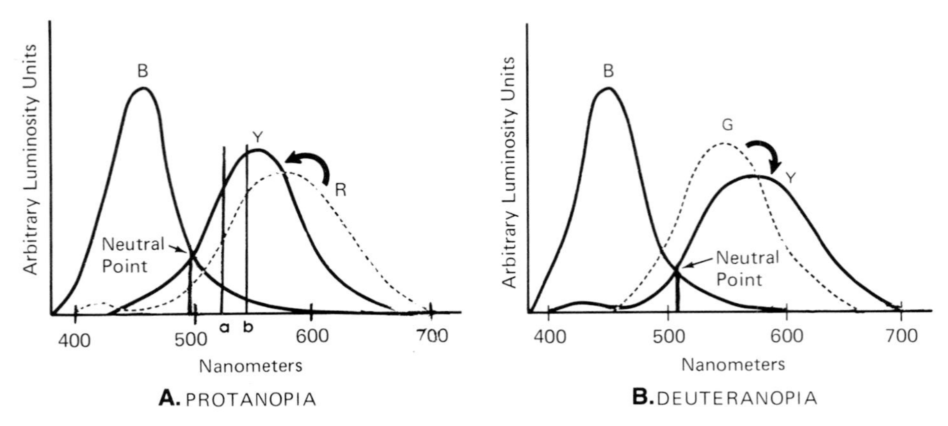

Fig. 6-8. Diagram explaining the physiology of the two most common forms of dichromatic vision. **A,** The red photoreceptors of the protanope are aberrant, responding by a pattern of red but at exactly the same frequency as the green photoreceptors. For any wavelength, the red response is the same as the green, or, in effect, the red response curve is superimposed on the green giving a sensation of yellow (equal amounts of red and green give yellow). The protanope therefore sees all wavelengths to the left of the neutral point (495 nm) as blue and those on the right as yellow. The neutral point is colorless (equal amounts of red, green, and blue give white). He distinguishes the different yellows, *a* and *b*, because *a* is less bright and more saturated, and his visible spectrum is shortened at the long end. **B,** The deuteranope's green photoreceptors copy the red, again resulting in two dichromatic colors, blue and yellow, but the neutral point differs, and the spectrum is not shortened.

results. This is the principle of color confusion tests.

In deuteranopia the green receptors respond abnormally; they still mediate green but duplicate the frequency of the red receptors, which remain normal. This is equivalent to superimposing the green on the normal red curve (Fig. 6-8, *B*). Thus the deuteranope also sees only yellows and blues, but his spectrum is not shortened, and his neutral point differs (505 nm). Indeed the location of the neutral point is the scientific way of differentiating them. Although all colors on the left of the deuteranope's neutral point appear yellow, their varying saturation and brightness allow him to distinguish between them. The yellows differ in some minor respects from those seen by the protanope, but it is not difficult to see why both protanope and deuteranope confuse red and green.

The rare tritanope has his blue response curve shifted and superimposed on the green or vice versa. He sees blue-greens and reds and, since the curves may intersect twice, two neutral points. Naturally he confuses the blues with the greens, but neither blues nor greens are confused with the reds.

Anomalous trichromats also confuse colors but still require three primary colors to match each wavelength—only their proportions differ from the normal. Anomalous trichromatic vision is so rare that in practice one need only differentiate between the anomalous protanope and deuteranope.

The diagnosis is made by the proportion of red and green used to match a particular yellow (chosen by Lord Rayleigh because of a handy candle flame). The instrument called an "anomaloscope" has split field with the Rayleigh yellow on one side and a mixture of red and green on the other. Patients are asked to combine the red and green to match the yellow. If they use more green than normal, they are deuteranomalous; if more red, protanomalous.

Monochromatic vision is rare; the rod monochromat not only sees all spectral colors as shades of gray but has photophobia, nystagmus, and poor acuity as well, and the electroretinogram shows characteristic findings, despite normal eye grounds. Failure to make the diagnosis has been known to result in committing children as feeble-minded. The cone monochromat is even more rare; although he is truly color blind, there is no nystagmus, acuity is good, and the electroretinogram is normal.

Color vision tests

Most clinical tests of color vision are screening rather than diagnostic tests. The defect is classified as mild or severe, but the type is not specifically identified. Since verbal descriptions must be avoided and locating the neutral point is too tedious, clinical tests are usually the pigment-matching type (the Holmgren wool test), pigment-confusion type (the Stilling, Ishihara, Dvorine, or Hardy-Rand-Rittler plates), or the ordering of colors into a continuous array (the Farnsworth D-15 test and the Farnsworth 100 hue test). Each test must be administered under the illumination for which they were standardized. The Hardy-Rand-Rittler and Farnsworth D-15 tests have blue-yellow plates for detecting acquired color defects. Lantern tests are designed to evaluate the ability to detect color signals (such as ribbed or frosted glass to simulate weather conditions) and are used as screening tests for railroad and transportation workers.

Although color vision defects are neither curable nor preventable, their identification is of importance in a variety of occupations (printers, pilots, railroad engineers, military academy candidates, etc.) and may be corroborative in the diagnosis of certain ocular diseases in which color blindness is an accompanying feature. Color vision screening is such a simple rapid test, requiring minimal equipment, that it should be part of every routine eye refraction. Knowledge of the presence or absence of a color defect may be of considerable importance in planning a vocation or career. Conversely, the rejection of an applicant for a particular job because of an erroneously interpreted or administered color test may have tragic consequences. For those occupations that require

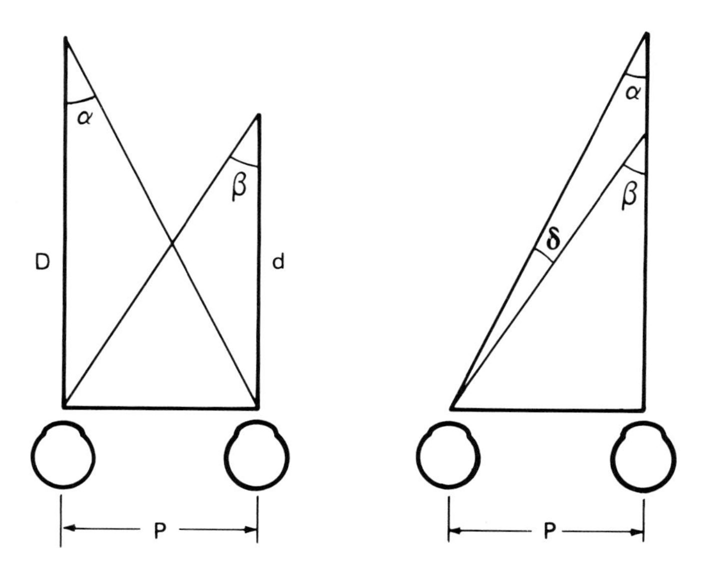

$$\tan \alpha = P/D\ ;\ \tan \beta = P/d$$

$$\delta \text{ radians} = \frac{P}{d} - \frac{P}{D} = \frac{P\,(D - d)}{Dd}$$

Fig. 6-9. Geometry of stereoacuity.

a high degree of color aptitude, a more extensive hue discrimination test may be administered.

THE EYE AS A BINOCULAR INSTRUMENT

Most people have two eyes, which provides not only for a handy spare but makes possible certain visual capacities not readily achieved by monocular individuals. The two eyes permit viewing the world from slightly different points of view simultaneously.

There is a common misconception that only binocular individuals see in three dimensions, as if monocular vision were flat. With few exceptions, all objects are solid, and their relative distances are recognized even in monocular vision. Indeed it is hard to "see" a perspectively drawn figure in other than three dimensions (e.g., the Necker cube). In ordinary vision, all monocular cues operate for the binocular as well as the monocular individual, and they prevent the flat vision that would be totally incompatible with real behavior. Only in artificial laboratory situations could one ever be misled into thinking that binocularity is a prerequisite to three-dimensional vision. Binocularity confers a quantitative and not a qualitative advantage.

Only three binocular factors have been considered of any use in distance judgments: convergence, which is limited to about 2 meters; physiologic diplopia, which probably plays no role at all; and horizontal retinal disparity—the correlate of stereopsis. "Stereopsis" means seeing solidly, an unfortunate term, since solidity may be lessened but not eliminated by closing one eye. But it has become tied to disparity by long usage.

The geometric principles of stereoacuity are illustrated in Fig. 6-9. Stereoacuity is no more synonymous with depth perception than a patient is blind because his resolution is not perfect. Squinters, who have no stereoacuity worth mentioning, manage to thread needles, park their cars, or even pilot an aircraft. Indeed the binocular pilot attempting a landing on water or snow had better have something more than disparity to depend on. Stereoacuity is also not the primary factor in space perception or even the primary factor in stereopsis—except, perhaps, to an artillery observer.

REFERENCES

Alpern, M.: Metacontrast, J. Opt. Soc. Am. **43**:648-657, 1953.

Aubert, H.: Physiologische Optik. In Graefe-Saemisch Handbuch der Gesammte Augenheilkunde, ed. 1, 1876.

Baker, H. D.: Initial stages of dark and light adaptation, J. Opt. Soc. Am. **53**:98-103, 1963.

Barlow, H. B., Dark and light adaptation: psychophysics, Handbook Sensory Physiol. **7**(4):1-28, 1972.

Barlow, H. B., Blakemore, C., and Pettigrew, J. D.: The neural mechanism of binocular depth discrimination, J. Physiol. (London) **193**:327-342, 1967.

Bartley, S. H.: Vision: a study of its basis, New York, 1963, D. Van Nostrand Co.

Baumgardt, E.: On direct scaling methods, Vision Research **7**:679-681, 1967.

Bergmans, J.: Seeing colors, New York, 1960, The Macmillan Co.

Berkeley, G.: A new theory of vision and other writings, New York, 1922, Everyman's Library.

Bishop, P. O., and Henry, G. H.: Receptive fields of simple cells in the striate cortex, J. Physiol. (London) **231**:31-60, 1973.

Brindley, G. S.: Physiology of the retina and visual pathway, London, 1970, E. Arnold and Co.

Brown, J. L.: Afterimages. In Graham, C. H., editor: Vision and visual perception, New York, 1965, John Wiley & Sons, Inc., pp. 479-503.

Burde, R. M.: The pupil, Int. Ophthal. Clin. **7**:839-855, 1967.

Campbell, D. J., Koester, C. J., Rittler, M. C., and Tackaberry, R. B.: Physiological optics, New York, 1974, Harper & Row Publishers, Inc.

Campbell, F. W., and Gubisch, R. W.: Optical quality of the human eye, J. Gen. Physiol. **186**:558-578, 1966.

Cornsweet, T. N.: Visual perception, New York, 1970, Academic Press, Inc.

Cowan, A.: The role of the pupil in ametropia, Am. J. Ophthal **40**:481-485, 1955.

Dalton, J.: Extraordinary facts relating to vision of colours, Mem. Manchester Lit. Phil. Soc. **5**: 28, 1798.

Davson, H.: The physiology of the eye, New York, 1972, Academic Press, Inc.

Deller, J. F. P., O'Connor, A. D., and Sorsby, A.: X-ray measurement of diameters of the living eye, Proc. R. Soc. Lond. **134**:456-466, 1947.

Ditchburn, R. W., and Ginsborg, B. L.: Vision with a stabilized retinal image, Nature **170**:36-37, 1952.

Donders, F. C.: Accommodation and refraction of the eye, London, 1864, The New Sydenham Society.

Dowling, J. E.: The organization of vertebrate visual

receptors. In Allen, J. M., editor: Molecular organization and biologic function, New York, 1967, Harper & Row Publishers, Inc., pp. 186-210.

DuCroz, J. J., and Rushton, W. A. H.: The separation of cone mechanisms in dark adaptation, J. Physiol. **195**:263-271, 1968.

Enoch, J. M.: Optical properties of the retinal receptors, J. Opt. Soc. Am. **53**:71-85, 1963.

Fick, A.: Lerhe von der Lichtempfindung. In Hermann, L., editor: Handbuch der Physiologie, vol. III, Leipzig, 1879, Vogel.

Fry, G.: The optical performance of the human eye, Progr. Opt. **8**:53-131, 1970.

Gernet, H., and Francerschetti, A.: Ultrasonic measurements in calculation of refraction and magnification expected after Fukala's operation in high myopia, Ophthalmologica **148**:393-404, 1964.

Gibson, J. J.: The perception of the visual world, Boston, 1950, Houghton Mifflin Co.

Goldmann, H., and Hagen, R.: Zur direkt Messung der total Brechkraft des lebenden menschlichen Auges, Ophthalmologica **104**:15-22, 1942.

Graham, C. H., editor: Vision and visual perception, New York, 1965, John Wiley & Sons, Inc.

Granit, R.: Sensory mechanisms of the retina, New York, 1947, Oxford University Press.

Gregory, R. L.: The intelligent eye, New York, 1970, McGraw-Hill Book Co.

Gullstrand, A.: Die optische Abbildung, ed. 3, Helmholtz Handbuch der physiologischen Optik, Hamburg, 1909.

Hartline, H. K., and Graham, C. H.: Nerve impulses from single receptors in the eye, J. Cell. Comp. Physiol. **1**:227-295, 1932.

Hecht, S.: Rods, cones, and the chemical basis of vision, Physiol. Rev. **17**:239-290, 1937.

Hubel, D. H., and Wiesel, T. N.: Receptive fields of single neurons in the cat's striate cortex, J. Physiol. (London) **148**:574-591, 1959.

Hyslop, J. H.: Upright vision, Psychol. Rev. **4**:71-73, 142-163, 1897.

Ivanoff, A.: About the spherical aberration of the eye, J. Opt. Soc. Am. **46**:901-903, 1956.

Jansson, F.: Determination of the axis length of the eye roentgenologically and by ultrasound, Acta Ophthal. **41**:236-246, 1963.

Jones, L. A.: The science of color, New York, 1953, Thomas Y. Crowell Co.

Kalmus, H.: Diagnosis and genetics of defective colour vision, Oxford, 1965, Pergamon Press.

Kepler, J.: Paralipomena ad Vitellonem, 1604.

Lakowski, R.: A critical evaluation of colour vision tests, Br. J. Physiol. Opt. **23**:286-309, 1966.

LeConte, J.: An exposition of the principles of monocular and binocular vision, New York, 1881, D. Appleton & Co.

LeGrand, Y.: Form and space vision, Bloomington, 1967, University of Indiana Press.

LeGrand, Y.: Light, colour, and vision, ed. 2, London, 1968, Chapman & Hall, Ltd.

Linksz, A.: Physiology of the eye: vision, vol. 2, New York, 1952, Grune & Stratton.

Linksz, A.: An essay on color vision and clinical color-vision tests, New York, 1964, Grune & Stratton.

Linksz, A.: A short primer on color vision and its defects, J. Pediatr. Ophthal. **5**:183-190, 1968.

Listing, J. B.: Dioptrik des Auges. In Wagner's Handworterbuch der Physiologie, vol. 4, 1884, p. 473.

Lowenstein, O., and Loewenfeld, I. E.: The pupil. In Davson, H., editor: The eye, vol. 3, London, 1969, Academic Press, Inc.

Ludvigh, E., and McCarthy, E. F.: Absorption of visible light by the refractive media of the human eye, Arch. Ophthal. **20**:37-51, 1938.

Mariotte, P. E.: A new discovery touching vision, translated by Justel, R. Soc. Lond. Philosoph. Trans. **3**:668-671, 1668.

Maurice, D. M.: The cornea and sclera. In Davson, H., editor: The eye, vol. 1, New York, 1969, Academic Press, Inc., pp. 489-600.

Michaels, D. D.: The nature of the photoreceptor process, Am. J. Optom. **33**:59-76, 1956.

Moses, R. A.: Adler's physiology of the eye, clinical application. ed. 5, St. Louis, 1970, The C. V. Mosby Co.

Muller, J.: Elements of physiology, translated by W. Baly, London, vol. 1, 1838, vol. 2, 1842, Taylor & Walton.

Nikara, T., Bishop, P. O., and Pettigrew, J. D.: Analysis of retinal correspondence by studying receptive fields of binocular single units in cat striate cortex, Exp. Brain Res. **6**:353-372, 1968.

Norren, D. V., and Vos, J. J.: Spectral transmission of the human ocular media, Vision Res. **14**:1237-1244, 1974.

Ogle, K. N.: Some aspects of the eye as an image-forming mechanism, J. Opt. Soc. Am. **33**:506-512, 1943.

Ogle, K. N.: Binocular vision, Philadelphia, 1950, W. B. Saunders Co.

Park, R. S., and Park, G. E.: The center of ocular rotation in the horizontal plane, Am. J. Physiol. **104**:545-552, 1933.

Pettigrew, J. D., Nikara, T., and Bishop, P. O.: Responses to moving slits by single units in cat striate cortex, Exp. Brain Res. **6**:373-390, 1968.

Pirenne, M. H.: Vision and the eye, ed 2, London, 1967, Chapman & Hall, Publishers.

Polyak, S. L.: The retina, Chicago, 1941, The University of Chicago Press.

Polyak, S.: The vertebrate visual system, Chicago, 1957, The University of Chicago Press.

Pomerantzeff, O., Fish, H., Govignon, J., and Schepens, C.: Wide angle optical model of the human eye, Ann Ophthal. **3**:815-819, 1971.

Purkinje, J.: Beobachtungen und Versuche zur Physiologie der Sinne, Berlin, 1825, G. Reimer.

Rubin, M. L., and Walls, G. L.: Fundamentals of visual science, Springfield, Ill., 1969, Charles C Thomas, Publisher.

Rushton, J.: Clinical measurement of axial length in the living eye, Trans. Ophthal. Soc. U. K. **58**:136-140, 1938.

Rushton, W. A. H.: The intensity factor in vision. In McElroy, W. D., and Glass, B., editors: Light and life, Baltimore, 1961, Johns Hopkins University Press.

Rushton, W. A. H.: Visual adaptation, Proc. R. Soc. **162**:20-46, 1965.

Rushton, W. A. H., and Gubisch, R. W.: Glare: its measurement by cone thresholds and by bleaching of cone pigments, J. Opt. Soc. Am. **56**:104-110, 1966.

Safir, A., and Hyams, L. A.: Distribution of cone orientation as an explanation of the Stiles-Crawford effect, J. Opt. Soc. Am. **59**:757-766, 1969.

Scheiner, C.: Oculus hoc est; fundamentum opticum, Innsbruck, 1619.

Schouten, J. F., and Ornstein, L. S.: Measurements on direct and indirect adaptation by means of a binocular method, J. Opt. Soc. Am. **29**:168-182, 1939.

Schwartz, J. T., and Ogle, K. N.: The depth of focus of the eye, Arch Ophthal. **61**:578-588, 1959.

Smith, K. U., and Smith, W. M.: Perception and motion, Philadelphia, 1962, W. B. Saunders Co.

Sorsby, A., and O'Connor, A. D.: Measurement of the diameter of the living eye by means of x-rays, Nature **156**:779-780, 1945.

Southall, J. P. C.: Introduction to physiological optics, London, 1937, Oxford University Press.

Sperry, R. W.: Visuomotor coordination in the newt (Triturus viridescens) after regeneration of the optic nerve, J. Comp. Neurol. **79**:33-55, 1943.

Sperry, R. W.: The eye and the brain, Sci. Am. **194**:48-52, 1956.

Stenstrom, S.: Optics and the eye, London, 1964, Butterworth & Co.

Stiles, W. S.: Color vision: the approach through increment threshold sensitivity, Proc. Nat. Acad. Sci. **45**:100-114, 1959.

Stiles, W. S., and Crawford, B. H.: The luminous efficiency of rays entering the eye pupil at different points, Proc. R. Soc. Lond. **112**:428-450, 1933.

Straatsma, B. R., Landers, M. B., Kreiger, A. E., and Apt, L.: Topography of the adult human retina. In The retina, Berkeley, 1969, University of California Press, pp. 379-410.

Stratton, G. M.: Vision without inversion of the retinal image, Psychol. Rev. **4**:341-381, 466-471, 480-481, 1897.

Stratton, G. M.: The spatial harmony of touch and sight, Mind **8**:492-505, 1899.

Strutt, J. W.: Notes, chiefly historical, on some fundamental propositions in optics, Philosoph. Mag. **21**:466-476, 1886.

Tait, E. F.: Intraocular astigmatism, Am. J. Ophthal. **41**:813-825, 1956.

Trendelenburg, W.: Der Gesichtsinn, ed. 2, Berlin, 1961, Springer Verlag.

Troland, L. T.: The principles of psychophysiology, II. Sensations, New York, 1930, D. Van Nostrand Co.

Tscherning, M.: Physiologic optics, translated by C. Weiland, Philadelphia, 1924, The Keystone Publishing Co.

von Helmholtz, H.: Helmholtz's treatise on physiological optics, ed. 3, 1909-1911 translated by J. P. C. Southall, New York, 1925, Optical Society of America.

von Hippel, A.: Ein Fall von einseitiger, kongenitaler rot-grün Blindheit, Arch. Ophthal. **26**:176, 1880.

von Tschermak-Seysenegg, A.: Introduction to physiological optics, translated by P. Boeder, Springfield, Ill., 1952, Charles C Thomas, Publisher.

Wainstock, M. A., editor: Ultrasonography in ophthalmology, Int. Ophthal. Clin., vol. 9, 1969.

Wald, G.: The receptors of human color vision, Science **145**:1007-1917, 1964.

Wald, G.: The molecular basis of visual excitation, Science **162**:230-239, 1968.

Waley, S. G.: The lens: function and macromolecular composition. In Davson H., editor: The eye, vol. 1, New York, 1969, Academic Press, Inc., pp. 299-379.

Walls, G. L.: The problem of visual direction, Am. J. Optom. **28**:55-83, 115-146, 173-220, 1951.

Weale, R. A.: Limits of human vision, Nature **191**: 471-473, 1961.

Weale, R. A.: From sight to light, Edinburgh, 1968, Oliver & Boyd, Ltd.

Westheimer, G.: Spatial interaction in human cone vision, J. Physiol. (London) **190**:139-154, 1967.

Westheimer, G.: Dependence of the magnitude of the Stiles-Crawford effect on retinal location. J. Physiol. (London) **192**:309-315, 1967.

Westheimer, G., and Campbell, F. W.: Light distribution in the image formed by the living human eye, J. Opt. Soc. Am. **52**:1040-1045, 1962.

Woodworth, R. S.: Experimental psychology. New York, 1938, Henry Holt & Co.

Wright, W. D.: The measurement of color, London, 1958, Hilger & Watts.

Wyszecki, G., and Stiles, W. S.: Color science, New York, 1967, John Wiley & Sons, Inc.

Yoss, R. E., Moyer, N. J., and Hollenhorst, R. W.: Hippus and other spontaneous rhythmic pupillary waves, Am. J. Ophthal. **70**:935-941, 1970.

Young, T.: On the theory of light and colors, Philos. Trans. R. Soc., Nov. 12, 1801.

Zoethout, W. D.: Physiological optics, ed. 4, Chicago, 1947, Professional Press, Inc.

7

Optical principles in the correction of ametropia

The emmetropic eye, wrote Donders, presents both in its structure and function the standard by which anomalies of refraction must be estimated. This, in a sentence, is the theory of correcting ametropia. The statistician would not hold with Donders' standard for a moment because emmetropia is an optical ideal, not a biologic average, and optical exactitude does not always lead to perceptual aptitude.

Emmetropia is that static ocular condition in which refractive power is proportional to axial length. Of course, power and length are theoretical problems for textbooks and not clinical skill. Clinically, we evaluate the discrimination of brightness, color, contrast, shape, and size, which in turn depend on pupil diameter, adaptation, accommodation, fusional status, emotional stress, and ocular disease. A static eye means one thing under cycloplegia, another under "fog," and something else in casual seeing. In practice we measure the far point, not the focal point; acuity, not retinal blur; and refractive error, not refractive power. This is not to disparage a good optical image. That a sharp focus is merely geometric dogma and that we have gone as far as we can with lenses and must now concentrate on exploring (and exploiting) the psychologic and integrative aspects of vision is the theme of the fringe group of refractionists hitting on all cylinders.

Although a punctiliously emmetropic eye is hard to find, most people are in fact nearly emmetropic. It is still a puzzle why the optical components arrange themselves with the symmetry of a Hollywood cavalry charge answering a genetic bugle.

THE FAR POINT

When a myope moves nearer the chart to see the letters or a retinoscopist approaches the eye to find neutral, both are hunting for the far point. The far point (punctum remotum or PR) is the point conjugate to the sharply focused retinal image—the key problem of clinical refraction.

The classification of ametropia by the relative position of the optical image is a convenient fiction, since it cannot be measured. For example, the information that an optical image is 1 mm in front of the retina would tell us little more than the eye is myopic. The degree of myopia would have to be computed by assuming certain values for optical power and emmetropic axial length.* But if we knew that the incident vergence must be −4 D to produce a sharp retinal image, the correcting trial lens could immediately be selected. The reciprocal of this vergence is the far point. Although the far point provides no clue as to axial length, refractive power, or the distance of the optical image relative to the retina, it gives all the practical

*For example, if the emmetropic standard is assumed to be 60 D and 22.22 mm, 1 mm defocus represents 2.8 D of myopia, but for an emmetropic eye of 70 D and 19 mm, the same linear distance would represent 3.9 D of myopia.

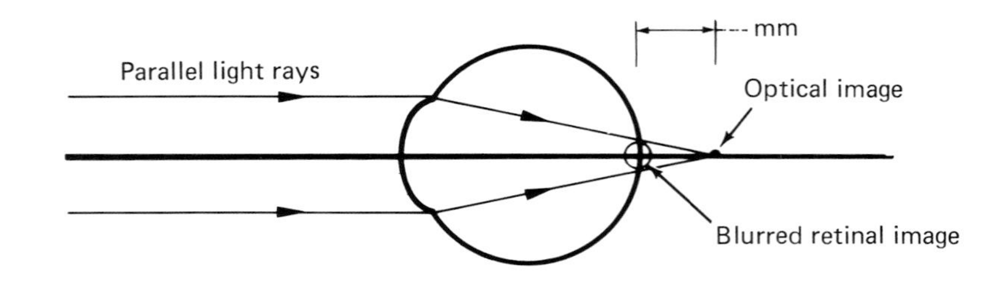

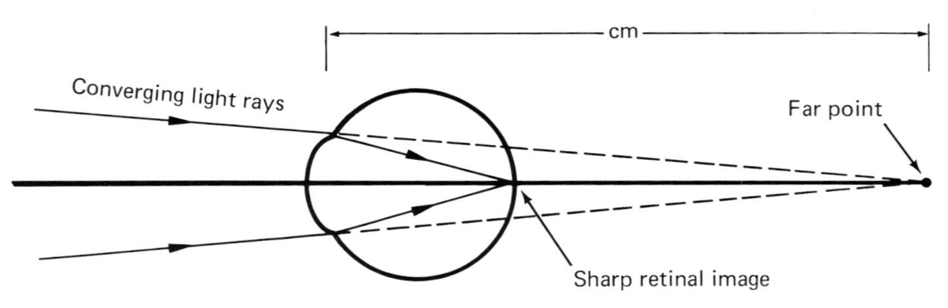

Fig. 7-1. Two methods of defining refractive error: by position of the optical image relative to the retina *(top)*, and by vergence required to place the optical image on the retina *(bottom)*. The first method is measured in millimeters and concerns the rays inside the eye; the second method is measured in centimeters and deals with the direction of rays outside the eye—the first is theoretical; the second, practical.

information we care about in clinical refraction.

The far point defines ametropia in terms of incident rays; the optical image represents the intersection of rays inside the eye. The far point is measured in meters or fractions thereof; the optical image, in millimeters or fractions thereof. With the emmetropic correction, the far point shifts to infinity, and the optical image moves on to the retina. It is difficult to understand how the two can be confused, but they frequently are (Fig. 7-1).

The far point of a myopic eye is in front of the eye; the optical image is in front of the retina. Optically the hyperopic far point and optical image are "behind" the retina; anatomically, only orbital fat is behind the eye.

The far point expresses the vergence of the incident light by the linear distance of their intersection rather than trigonometry. The more nearly parallel the rays the further off they cross.

The optical image is formed by the intersection of the refracted rays inside the eye. In hyperopia these rays (barring accommodation) are intercepted by the retina, resulting in a blurred retinal image. The optical image, ignoring aberrations, is always sharp but not necessarily on the retina. The retinal image is always on the retina but not necessarily clear.

When a −10 D lens is appropriately positioned before a 10 D myopic eye, it is optically corrected; anatomically it remains myopic as before. Emmetropia is created by optically shifting the far point from 10 cm to infinity or at least 20 feet. This shift is measured by retinoscopy or subjective refraction. Instead of actually locating the far point (impossible when virtual) we move it with lenses to the peephole of the retinoscope or to the Snellen chart in subjective testing.

For example, the far point of a 4 D hyperopic eye is 25 cm behind the cornea. With a +1 D lens, it moves to 33.33 cm, to 50 cm with a +2 D lens, and to infinity with a +4 D lens. The addition of another +1 D (for a total of +5 D) would move the far point to 100 cm in front of the eye. This far point voyage has been described as proceeding around the world, from plus to minus infinity, or even beyond infinity, but we should now under-

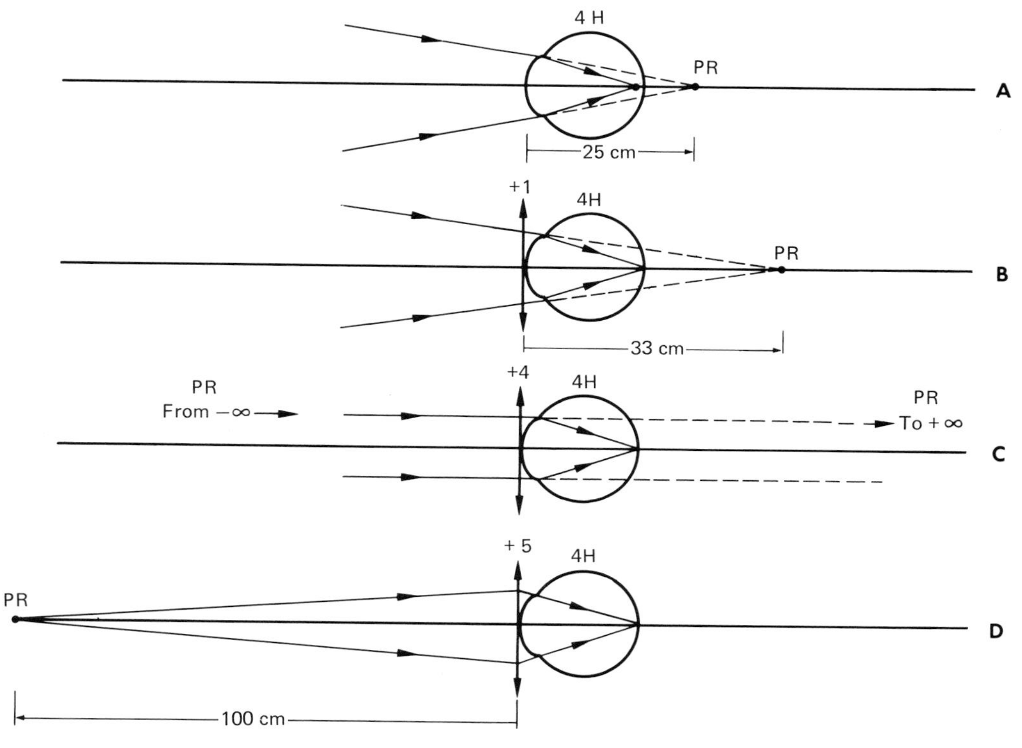

Fig. 7-2. Optical sequence illustrating the displacement of the far point of a 4 D hyperopic eye when trial lenses are placed in front of it (vertex distance is here ignored). The far point shifts from 25 cm behind the eye (**A**), to 33 cm (**B**), to positive infinity from negative infinity (**C**), and finally to 100 cm in front of the eye (**D**). This is also the sequence in clinical retinoscopy as the far point is brought to the instrument held at 1 meter.

stand what these descriptions mean in terms of light rays (Fig. 7-2).

MEASURING THE FAR POINT

The far point of any myopic eye could be measured with a sliding acuity target, and this is the basis of the simplest optometer, such as +10 D lens and a calibrated ruler. The farthest distance the target is seen clearly represents the far point (Fig. 7-3). Note that the calibration illustrates a peculiarity of focal power reciprocals; the stronger the power the less spacing between scale markings. For example, the separation between 4 and 5 D on the myopic side is only 0.48 cm, whereas the same hyperopic difference is 3.34 cm. A small error in judging sharpness, easily possible as a result of depth of focus, causes a marked difference in the measured ametropia, which is one reason we prefer to shift the far point rather than find it with an optometer.

The first "optometer" was invented by Porterfield (1759), who coined the term. He got the idea from de la Hire, who had set out to prove, using the Scheiner double pinholes, that accommodation did not exist. The more astute Porterfield adapted it to measure ametropia and incidentally to disprove de la Hire's conclusions (Fig. 7-4).

A half century later, Thomas Young resurrected the same optometer to discover his own astigmatism. He records his monocular diplopia disappeared at 7 inches when the pinholes were horizontal and at 10 inches with the pinholes vertical. Young's correction must therefore have been −5.71 + 1.71 × 180, and it is not hard to see why, compared to his myopia, his astigmatism caused little inconvenience. Had it troubled him more, he might have recognized its clinical importance sixty years before Donders.

Von Graefe adapted the Galilean telescope and Hirschberg adapted the astronomic

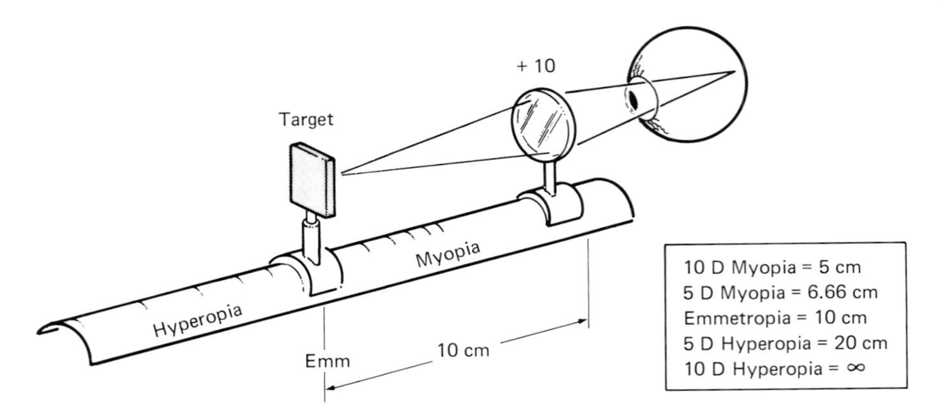

Fig. 7-3. Simple optometer. The purpose of the +10 D lens is to inhibit accommodation and bring the far point within reach of the scale. Note that the hyperopic scale is more widely spaced than the myopic scale, a consequence of converting to reciprocal diopters. For all subjective optometers, at such close distances a small linear error leads to large dioptric mistakes.

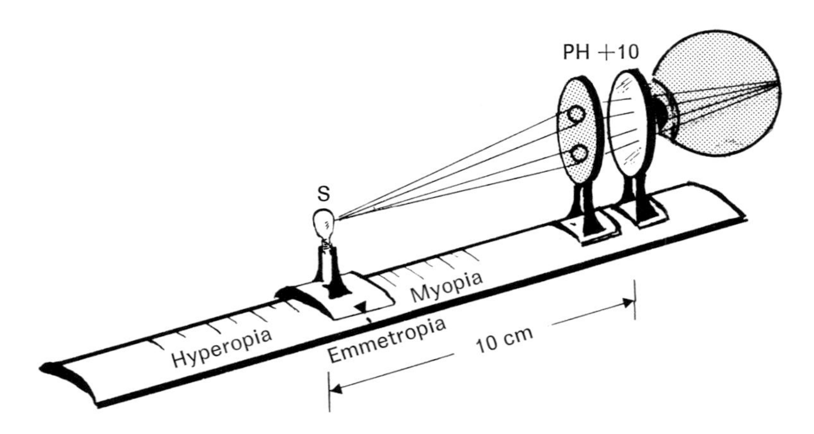

Fig. 7-4. The Porterfield optometer (often attributed to Thomas Young) incorporates a double-pinhole Scheiner disc and a strong convex lens. The pinholes create a monocular diplopia unless the optical image is exactly focused on the retina. The chief clinical difficulty is keeping one or the other pinhole from slipping off the pupil.

telescope to oculometry by calibrating the distance between the lenses. The myope requiring a minus vergence decreases the lens separation, and the hyperope increases the distance between them.

Similar to the double pinhole and suffering the same handicap of crucial alignment in front of the patient's pupil is the double prism. If two 3Δ prisms base to base are carefully centered before the pupil and a target 3 cm in diameter is viewed at 1 meter, the emmetrope will see two discs just touching; the myope sees them overlapping, and the hyperope sees them separated.

One problem with most optometers is the increasing visual angle as the target approaches the eye. The magnification tips the odds even when the image is blurred. Badal, a French ophthalmologist, recognizing this inconsistency, designed an optometer whose focal point coincided with the nodal point of the eye (Fig. 7-5). The target now subtends the same visual angle whatever its distance. It also works if the focal point of the optom-

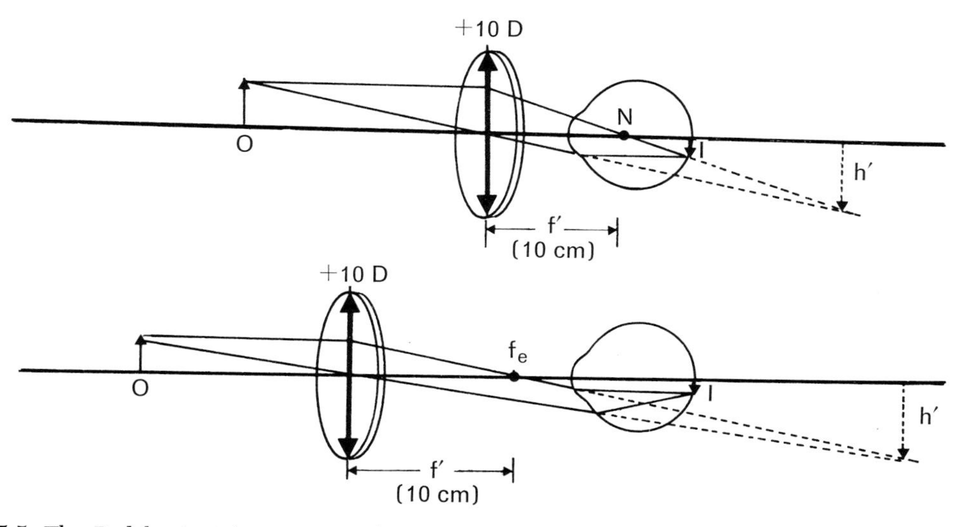

Fig. 7-5. The Badal principle is commonly used in research optometers to maintain a constant retinal image size as the target approaches the eye. *Top,* Original design in which the second focal point of the lens coincides with the nodal point. *Bottom,* The principle also works if the second focal point of the lens coincides with the anterior focal point of the eye. The second optometer is more practical because it gives the spectacle refraction (very nearly). In both optometers, the retinal image is blurred until the object occupies the artificial far point, but it retains substantially the same size because the image is determined by the ray through the nodal point in the first case and by the ray through the anterior focal point of the eye in the second. These ocular points are not actually known; their location is assumed according to Gullstrand's constants.

eter lens is made to coincide with the anterior focal point of the eye, a principle used in research haploscopes of the physiologic optics laboratory. Of course, neither the nodal or anterior focal point of the eye is generally known, and the values are assumed, based on Gullstrand's schematic eye.

All subjective optometers, however mechanically dressed or optically flavored, are clinically hard to digest. The patient must judge the sharpness of a near target without accommodating, and a small linear error will lead to large dioptric mistakes. Moreover the examiner has no control over the degree of blur the patient selects as an end point, and the answers cannot be independently confirmed. Astigmatic eyes have two far points, so the optometer must also allow for axis variations, which is clumsy and compounds the difficulties.

Objective optometers are independent of patient answers—the end point is judged by the examiner or a photoelectric scanner (neither of which is infallible). A sharp retinal image is created by a beam passing through one part of the pupil, and the emerging vergence is measured. Since the sharp retinal image and the far point are optically conjugate, the emerging vergence characterizes the ametropia. The earliest objective optometer was the ophthalmoscope. An emmetropic examiner sees a sharp fundus picture if the patient is emmetropic; the ametropic fundus is blurred until focused by the Rekoss lenses. If the examiner does not accommodate, if he selects a point reasonably near the fovea, if there are no significant aberrations, and if the patient maintains his distant fixation, the Rekoss lens represents the emmetropic correction. In astigmatism the difficulties are doubled, and a horizontal or vertical retinal vessel of the disc is selected to evaluate the meridians. The disc itself appears distorted (e.g., a vertical oval in myopic "with-the-rule" astigmatism) but only if the cylinder is large. (Some modern ophthalmoscopes such as the Keeler Projectoscope incorporate graticule targets in the focal plane of the condensing lens.) A small peephole increases depth of focus; so ophthalmoscopy is a poor method of measuring refractive error. It is surprising that its use persisted even after the invention of the retinoscope.*

The latest optometers utilize infrared rays (so as not to stimulate accommodation) and computerized photoelectronic scanning (so as not to disturb the clinician). One simply pushes a button, and the machine grinds out a "prescription." We can now, the ads assure us, turn in our retinoscopes, turn these devices over to the office technician and turn the refracting room into an optical bazaar while we do something more important like taking tensions (which, little do they know, is being taken by another office technician with an electronic tonometer).

Now no one wants to disparage accurate measurements, especially if no effort need be expended to make them. It is only that objective optometers will not tell us what the patient sees, how he sees, or if he sees. The prescription comes out just the same if the eye is blind. This may be disappointing—after so much technology one expects an infallible revelation. Optometers do have their place in mass screening, in epidemiologic studies, and in research on ametropia. But they give no clue as to comfort, efficiency, or binocular cooperation and no distinction between the focal plane measured and that actually selected by the patient or the axis tested and the visual axis. Fluctuation in accommodation, aberrations, depth of focus, poor fixation, and uncooperative patients all trouble the instrument as much as the examiner in conventional refraction. One would no more prescribe lenses on such a precarious basis than allow the patient to select them out of a box; indeed the latter is more practical.

*Marcus Gunn in the 1880's would not permit his students to use the retinoscope and insisted on the "direct" (ophthalmoscopic) method because it provided "better training." All statistical studies of the incidence of ametropia (by sex, age, occupation, race, geography, etc.) before the turn of the century were based on such tenuous "objective" ophthalmoscopic measurements.

SPECTACLE AND OCULAR REFRACTION

Any lens whose second focal point coincides with the far point of the eye corrects ametropia (Fig. 7-6). Consequently there are an infinite number of different lenses that correct the same refractive error (Fig. 7-7). Although lenses differ in equivalent power, they all have the same effective power at the cornea. Even a convex lens will correct myopia if placed at the appropriate distance. Of course it is not practical, since lenses must be worn in a frame that sits on the nose, and the field of view would become disastrously small.

Convex lenses were actually suggested as a kind of treatment to prevent myopia by Hooke in 1681. But this was going too far even for such a thoroughgoing empiricist as Molyneux (1692) who quickly pointed out that not only would this spectacle distance be incommodious, one would need to write inverted and retrograde, and "what is yet more inconvenient, from the bottom towards the top of the page, which is hardly practical on account of blotting the wet writing."

In geometric optics, object distances are measured to the optical center of the lens; in visual optics the far point is measured to the equivalent plane of the eye. The equivalent plane is not generally known but is about 3 mm from the cornea of Gullstrand's eye and 1.5 mm behind the surface of the reduced eye. These distances are small enough to be neglected and thus the corneal pole serves as the clinical point of reference. The dioptric reciprocal of the far point (measured from the cornea) is the "ocular refraction"—the power of the contact lens prescription. The power at the spectacle plane differs from the ocular refraction by the induced effectivity (Fig. 7-8). If F_S is the spectacle refraction and F_O the ocular refraction, the relation between them is given by the effectivity formula:

$$F_O = \frac{F_S}{1 - dF_S}$$

Clinically the correcting lens is invariably expressed in terms of spectacle refraction and depends on the vertex distance d. For purposes of computation, one can assume

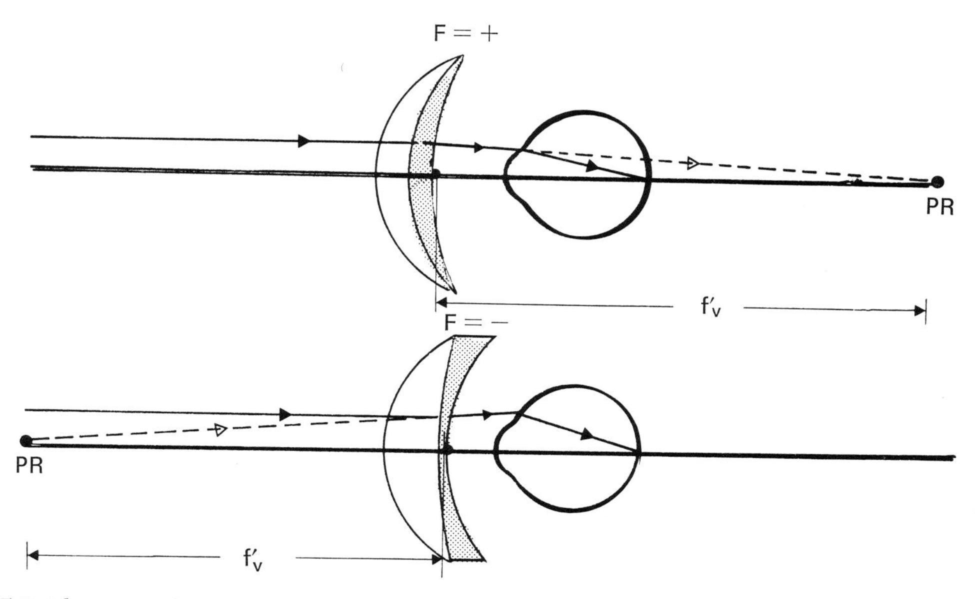

Fig. 7-6. The principle of spectacle correction is that the back vertex focal length (f'_v) coincides with the far point of the eye. In hyperopia *(top)* a parallel pencil of light is converged toward its second focal point, which coincides with the far point. These rays enter the eye and form a sharp retinal image. The far point of the myopic eye *(bottom)* is to the left; hence parallel rays must be diverged as if coming from this point so that they will focus on the retina.

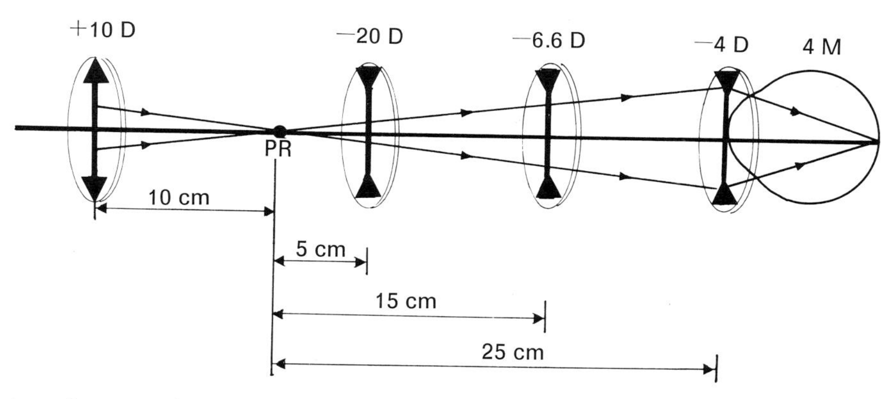

Fig. 7-7. An infinite number of different lenses may correct the same refractive error, providing only that their second focal points coincide with the far point. Conversely, the same spectacle lens displaced to some new position will now no longer be adequate.

some arbitrary distance (from 7 to 22 mm). The best way is to measure, not assume—at least in strong prescriptions. For example, if the ocular refraction is +10 D, the spectacle refraction 10 mm from the cornea is $\frac{+10}{1 + (0.1)(+10)} = +9.2$ D, but the spectacle refraction 20 mm from the cornea is $\frac{+10}{1 + (.02)(+10)} = +8.3$ D, almost a 1 D difference (Fig. 7-9). So a small dispensing error can make a considerable difference in the vision of an aphakic patient. In prescribing contact lenses, the ocular refraction is specified by computing it from the spectacle refraction.

EXAMPLE: What is the contact lens prescription for an eye corrected by a spectacle lens (a) −5 D and (b) +5 D located 16.75 mm from the cornea?

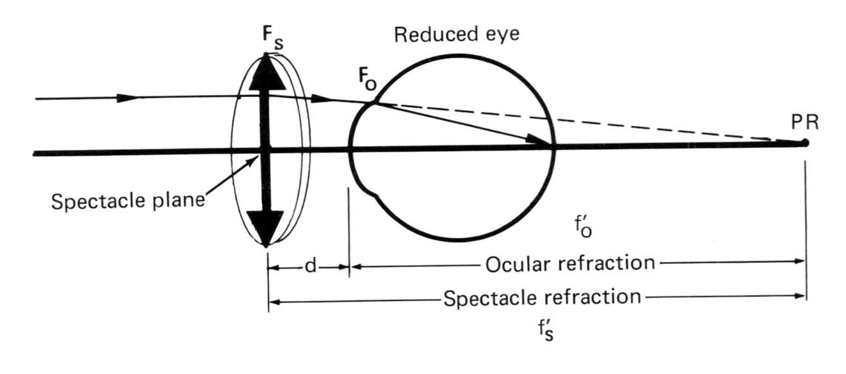

Fig. 7-8. The ocular refraction (F_o) in the reciprocal of the far point measured from the cornea (more technically, the equivalent plane of the eye). The spectacle refraction (F_s) differs from it by virtue of the effectivity added by the vertex distance (d). By applying the effectivity formula, one can be computed from the other. For example, the contact lens prescription represents the ocular refraction as distinct from the spectacle prescription. In determining the ocular refraction from the spectacle refraction, d has a plus sign; it is given a minus sign for the reverse calculation (according to the light direction, which is always assumed left to right).

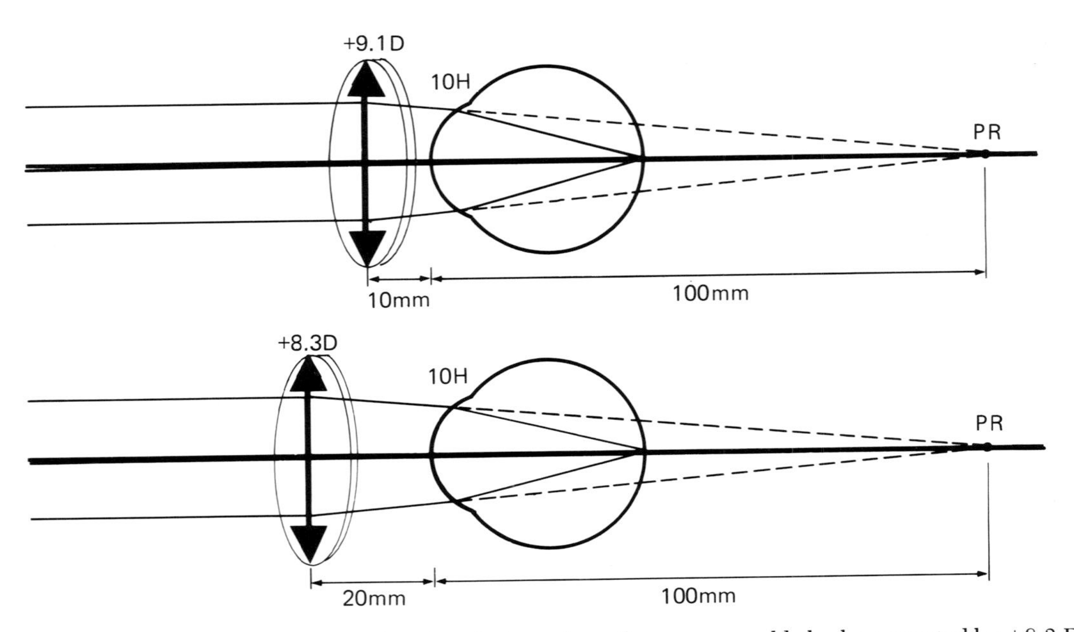

Fig. 7-9. An eye requiring a +9.1 D lens fitted 10 mm from the cornea would also be corrected by +8.3 D lens at 20 mm. Conversely, if the refraction were performed at 20 mm and the final spectacles are dispensed at 10 mm, the patient would be almost 1 diopter undercorrected.

$$\text{(a) } F_O = \frac{-5}{1 - (0.01675)(-5)} = -4.62 \text{ D}$$

$$\text{(b) } F_O = \frac{+5}{1 - (0.01675)(+5)} = +5.45 \text{ D}$$

The example illustrates that it takes a stronger convex or weaker concave power to correct the eye at the plane of the cornea. The sign of d is taken according to the light direction—positive when computing the ocular from the spectacle refraction and negative when computing the spectacle from the ocular refraction. Thus a frame may be fitted a few millimeters nearer or farther from the eye; the stronger the correction the more critical the displacement.

For small vertex distances, the effectivity formula can be simplified to $F^2/1000$, which gives the dioptric change per millimeter displacement. For example, a +8 D prescription at 10 mm would require 0.064 D per millimeter or 0.64 D less plus at 20 mm. Conversely, if a patient wearing a +8 D lens moved the same lens 10 mm further, it would become 0.64 D stronger—which is why beginning presbyopes push their distance glasses down their nose. If an aphakic patient cannot adjust to bifocals, he can dispense with them by slipping his +10 spheres some 2.5 cm further down his nose and achieve not only a satisfactory reading add as a temporary expedient but get some magnification besides. A separate pair of single vision reading lenses can be ordered for prolonged near work.*

Care must be taken in solving problems of this type to differentiate between (1) the *new* prescription required at some new distance and (2) the change in power resulting when the *old* lens is displaced to a new position. The rule is best memorized by recalling that when presbyopes push their glasses further down their noses, both plus and minus lenses acquire more plus power.

The effectivity change for astigmatic eyes varies with the power in each primary meridian. For example, if the ocular refraction is +1 +5 × 180, the spectacle correction at 15 mm would need to be 0.54 D less plus in the vertical and 0.15 D less plus in the horizontal or approximately +0.87 + 4.62 × 180. In this example the spectacle cylinder is smaller than the corresponding ocular cylinder.

*Another advantage of single vision lenses for aphakes is that aspheric curves can be maintained. Aphakic bifocals are not aspheric in the segment.

ASTIGMATIC CORRECTION

Astigmatic eyes have two far points, one for each primary meridian. The position of the focal lines relative to the retina provides a convenient classification (Fig. 7-10). If one focal line is on the retina, the astigmatism is simple; if both are in front, the astigmatism is compound myopic; if both are behind, the astigmatism is compound hyperopic; and if one is anterior and the other posterior, it is mixed. The term "mixed" is not precise un-

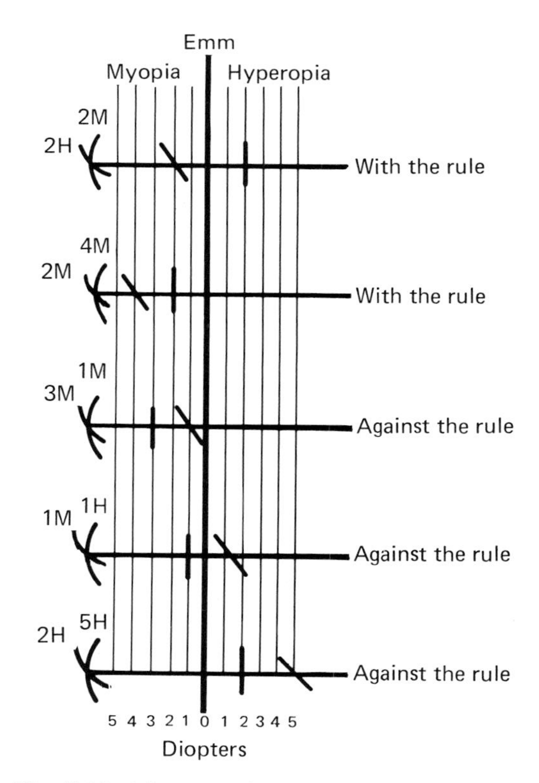

Fig. 7-10. Schematic diagram of a series of eyes showing the optical power and the position of the focal lines in dioptric displacements from the emmetropic retina. The conventional clinical classification of astigmatism is based on the position and sequence of these focal lines.

less one focal line is as far behind as the other is in front, that is, equal-mixed astigmatism.

If the more myopic ocular meridian is vertical or near vertical, the astigmatism is said to be "with the rule"; if the more myopic meridian is horizontal or near horizontal, it is "against the rule." If the primary meridian is in between, the astigmatism is oblique. The rule can be remembered by noting that minus cylinder axis 180 means with the rule. The classification has little diagnostic significance—it is a constant memory claim; one either forgets the rule, or one cannot remember the name.

Fig. 7-11, *A*, illustrates an eye with compound hyperopic astigmatism with the rule. Both focal lines are behind the retina, although the horizontal line is closer (accommodation assumed inactive). Fig. 7-11, *B*, shows compound myopic astigmatism with the rule. Both focal lines are now anterior, but this time the vertical lies nearer the retina. The first patient sees the horizontal limb of a clock dial clearer; whereas the second reports the vertical line clearest. If the hyperopic patient is allowed to accommodate, he brings the circle of confusion on to the retina, and all the astigmatic dial lines appear equal (although somewhat blurred). This illustrates the importance of accommodative control in all subjective dial tests for astigmatism. Unless accommodation is inhibited by cycloplegia or a fogging lens, the answers are invalid. With a fogging lens, all astigmatic eyes are converted to compound myopic astigmatism; both focal lines now lie anterior to the retina, and accommodation only creates more blur.

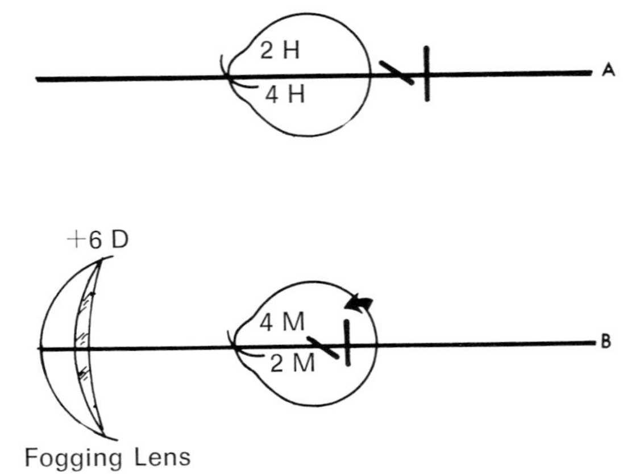

Fig. 7-11. A, Schematic drawing of an eye with compound hyperopic astigmatism, the horizontal focal line falling nearer the retina. The patient's responses on the astigmatic dial would be invalidated by accommodation. **B,** If a +6 D fogging lens is placed before this eye, the ametropia is converted to compound myopic astigmatism. Accommodation is inhibited, and the refractive correction proceeds reliably by collapsing the conoid of Sturm with minus cylinders.

The principle of correcting ocular astigmatism is to collapse the interval of Sturm with cylinders. Since the eye has been converted to compound myopic astigmatism by the fogging lens, we can use either plus cylinders to advance the focal line nearest the retina or minus cylinders to displace the more blurred line closer to the retina. In the first case the patient must judge when all the dial lines appear equally blurred; in the second case the end point is when they appear equally clear. Obviously the second is easier and more reliable; so minus cylinders are practically mandatory in the fogging technique of refraction.

For example, if the horizontal focal line is 1 D anterior and the vertical focal line is 2 D anterior to the retina, the patient naturally reports the horizontal dial line clearer (Fig. 7-12, *A*). To displace the vertical line toward the horizontal line we add minus cylinder axis vertical (Fig. 7-12, *B*). (This adds minus *power* to the horizontal ocular meridian, which is responsible for the vertical focal line.) The minus correcting cylinder axis is thus always perpendicular to the line reported clearest on the dial. Of course, we could have added +1 axis vertical and moved the horizontal focal line forward to the level of the more blurred vertical line (Fig. 7-12, *C*). To equalize the astigmatic dial requires −1 D axis vertical. In this example, the final prescription would be −1 − 1 × 90, the sphere serving to move both focal lines on to the retina (Fig. 7-12, *D*). The retinal image is now composed of points and not focal lines.

Although minus cylinders move only one

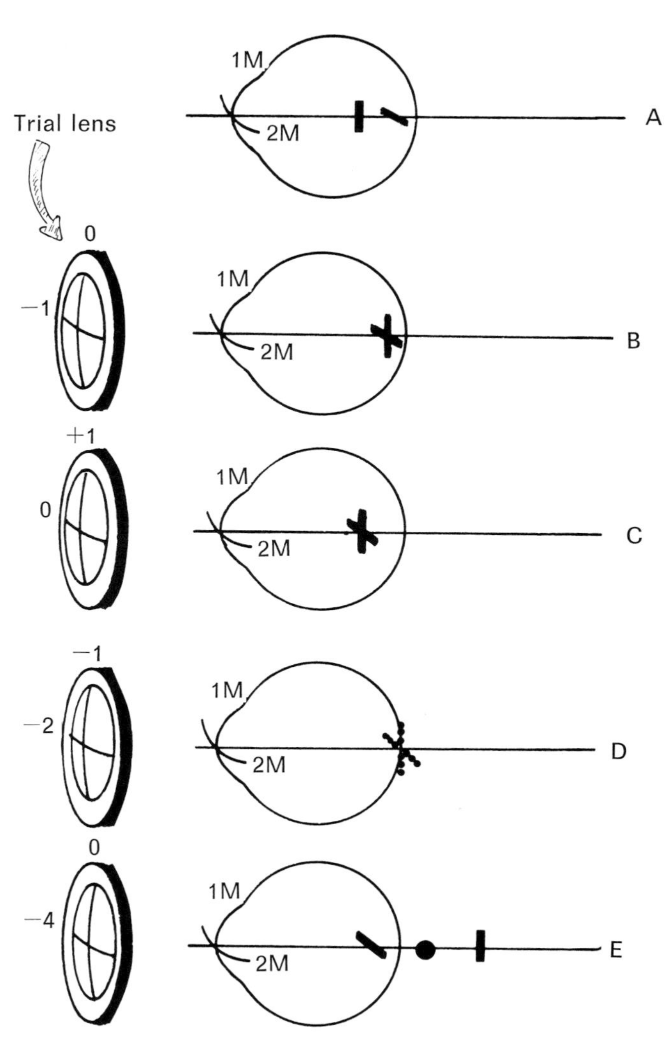

Fig. 7-12. Schema of the optical state of affairs when an eye with compound myopic astigmatism is corrected by plus or minus cylinders, a spherocylinder, and the effect of minus cylinder overcorrection. For details see text.

focal line, they alter the interval of Sturm and therefore displace the circle of least confusion. Suppose in the previous example we had added −4 D axis vertical. The vertical focal line is displaced 2 D behind the retina while the horizontal line remains immobile 1 D in front (Fig. 7-12, *E*). But the circle of confusion is now at the new midpoint of Sturm's interval, 0.5 D behind the retina—the patient accommodates, and the test is invalid. So care must be taken (1) to make the initial fogging lens strong enough and (2) not to overshoot by adding more minus cylinder than necessary to equalize the lines. Conversely, the fog must not create so much blur that the judgment of equality is made unreliable.

The axis of the minus correcting cylinder is always given in standard (TABO) notation. If the sharpest line on the clock dial is vertical, the minus cylinder axis is 180; if the horizontal line is clearest, the minus cylinder axis is 90. If the axis is oblique, the situation becomes confusing. Suppose the patient reports the 11-5 clock line clearest. The clock dial is numbered from the patient's point of view, but the cylinder axis notation is from the examiner's point of view. If one holds a pencil in front of one's eye in the position of the 11-5 line, and imagines oneself looking at the eye from in front with the pencil as positioned, its orientation will be in the 60° meridian according to the standard notation. The correcting minus cylinder axis is perpendicular to it, hence at 150°.

A simple rule of thumb by which to remember how to position the minus cylinder axis from the clock dial is to multiply the smaller "clock" number by 30°. Thus for the 11-5 line the smaller number is 5; hence $5 \times 30° = 150°$. If the sharpest reported line were 1-7, the minus cylinder axis is to be placed at $1 \times 30° = 30°$. Of course not all astigmatic dials are clock dials; some have radiating lines at 10° intervals, so one must be able to reason out the answer.

The earliest astigmatic dial, a cross grid, was invented by Colonel Goulier, a professor of topography at the Metz Military Academy. Goulier, impressed by the difficulty of some of his students in reading their instruments, independently diagnosed and corrected astigmatism approximately eight years before Donders. The clock dial was invented in 1866 by a Boston ophthalmologist, John Green. Before Green, astigmatic testing consisted of locating the two far points with an optometer as Young had done. The fogging method was first described by J. Z. Laurence in 1863, and the standard axis notation by H. Knapp in 1886. Astigmatism as a significant clinical cause for asthenopia was proposed by a neurologist, S. Weir Mitchell, in 1876. His patient's delight with his new glasses is understandable, considering that the previous treatment consisted of, among other things, cauterization of the spine. The cross cylinder technique did not appear until the end of the century when Jackson mentioned its use, almost in passing, in a paper of 1887.

When astigmatism should be corrected and how much cylinder should be prescribed is a clinical judgment that depends on symptoms, acuity, previous correction, and associated features (amblyopia, strabismus, aniseikonia, etc.). The astigmatic error that is here today and gone tomorrow or found with one test and not another can usually be ignored. Conversely, a large cylinder cannot always be prescribed in the first pair of glasses because distortion and meridional aniseikonia may make it intolerable.

SPECTACLE MAGNIFICATION

Although the chief purpose of ophthalmic lenses is to sharpen the retinal focus and improve vision, they also cause magnification. The latter is a mixed blessing, particularly when it differs for the two eyes. The magnification of a spectacle lens is a function of its power, its construction (shape and thickness), and its vertex distance from the eye. Let us ignore construction and consider first the effect of an infinitely thin lens. In Fig. 7-13, a thin convex correcting lens is placed in front of a reduced hyperopic eye viewing a distant object. The lens is at the spectacle plane S, a distance d from the cornea. The object subtends an angle ω at the lens, and forms an image h' at its second focal plane. This image in turn becomes the object for the eye and subtends an angle ω' at the pole P (actually the entrance pupil). The ratio ω/ω' is the spectacle magnification (SM). Since $\tan\,\omega = h'/f'$ and $\tan\,\omega' = \frac{h'}{f' - d}$, we obtain by substitution:

$$SM = \frac{1}{1 - dF_S}$$

This may be simplified to $SM = 1 + dF_S$ or, since magnification is usually expressed in percent, $SM_{\%} = dF_S$, where d is in centimeters.

EXAMPLE: What is the spectacle magnification of each of the following thin lenses placed 13 mm from the cornea?

(a) +2 D; SM = 1.3 (+2) = +2.6%
(b) −2 D; SM = 1.3 (−2) = −2.6%
(c) +5 D; SM = 1.3 (+5) = +6.5%
(d) +10 D; SM = 1.3 (+10) = +13%

The example illustrates that a thin lens by virtue of its optical power produces magnification; the greater the power the greater the

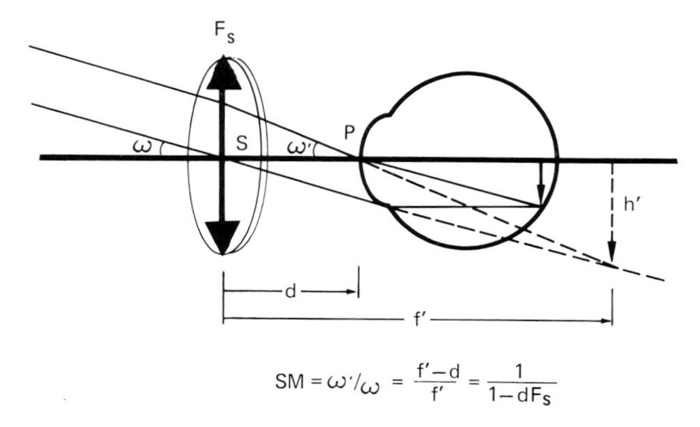

Fig. 7-13. Schematic diagram illustrating the spectacle magnification *(SM)* of an infinitely thin lens (also called the "power" factor). The visual angle (ω) formed by a distant object at the cornea (more technically, the entrance pupil) is altered when the lens is interposed. The image h' formed by the lens becomes the virtual object for the eye, subtending a new visual angle (ω'). Since for any eye the basic size of the retinal image is directly proportional to the visual angle, the spectacle magnification is the ratio of the two. It depends only on the power of the lens and its distance from the eye; hence any plus spectacle lens magnifies, and any minus spectacle lens minifies by virtue of the inevitable vertex distance. Spectacle magnification can be eliminated only by fitting a contact lens or, in theory, an intraocular lens.

magnification. Moreover, magnification is a function of vertex distance; if the distance is doubled, the +10 D lens produces 26% magnification. We also see that plus lenses magnify and minus lenses minify; to avoid magnification the lens must be placed in contact with the cornea.* It follows that spectacle lenses invariably produce some magnification because of the inevitable vertex distance to clear the lashes. The statement frequently made in textbooks that lenses placed at the anterior focal point of the eye never change retinal image size is obviously incorrect.

The magnification of a thin lens, due to its power, is called the "power factor." Since real ophthalmic lenses are not infinitely thin, even plano lenses produce some afocal magnification (described in Chapter 3). The afocal magnification is the result of front surface power (F_1), thickness (t), and the glass index (n) and is called the "shape factor":

$$m = \frac{1}{1 - t/n\ F_1}$$

The total spectacle magnification of an ophthalmic lens is the product of the power and shape factors:

$$SM = \left(\frac{1}{1 - dF'_v}\right)\left(\frac{1}{1 - t/n\ F_1}\right)$$

Note that the power in the power factor is now the back vertex focal power of a real lens instead of the equivalent power of a thin one. In the shape factor, t is the thickness of the glass; d in the power factor is the distance of the lens from the cornea. The complete formula does not contain equivalent power (which is not generally known) and is entirely practical. All the variables can be evaluated in the refracting room: the back vertex power with a lensometer, the front surface power with a lens measure, the thickness with a lens caliper, the vertex distance with a ruler, and the glass index from the manufacturer's specifications. If the lens is concave or less than +4 D, the shape factor is negligible and may be ignored. For example, the shape

*A contact lens does not actually produce unit magnification because the corneal pole is only an approximation to the entrance pupil, the true point of reference. Since the cornea in Gullstrand's eye is 3 mm from the entrance pupil, a +10 D contact lens produces at least 3% magnification. Since the 3 mm would also have to be added to the vertex distance to compute the spectacle magnification, the relative difference between contact and spectacle lens magnification still holds true. In an aphakic eye the iris plane falls further back to about 5 or 6 mm, and the residual aniseikonia of a "thin" contact lens is 5% or 6%.

factor magnification for a plano convex +4 D lens, 2 mm thick, is only 0.5%.

An approximate formula for total spectacle magnification is

$$SM = (1 + dF'_v)(1 + t/n\ F_1)$$

or directly in percent

$$SM_{\%} = dF'_v + t/n\ F_1$$

where d and t are in *centimeters*. Note that spectacle magnification compares the retinal image on a before-and-after basis—with and without the lens in the same eye. It applies only to distant objects because back vertex focal power and afocal magnification assume incident parallel rays.

EXAMPLE: What is the spectacle magnification of a +10 D lens, 15 mm from the eye if the front surface power is +12 D, thickness is 4 mm, and refractive index is 1.523?

$$SM_{\%} = dF'_v + t/n\ F_1$$
$$= (1.5)(+10) + \frac{0.4}{1.523}(+12)$$
$$= 18\%$$

Although it is never actually computed, spectacle magnification compares retinal image size with and without the lens in the same eye.

$$SM = \frac{\text{Retinal image size in corrected ametropic eye}}{\text{Retinal image size in uncorrected ametropic eye}}$$

Convex lenses magnify and minus lenses minify the retinal image compared to its size before the lens was introduced in the *same* eye.

In anisometropia, it may be desirable to equalize the magnification as much as possible. A change in spectacle magnification can be produced by altering either the power or shape factor. The first is accomplished by moving the spectacle frame closer or nearer the eye, that is, by altering the vertex distance. Obviously little change can be effected this way unless the power is high. Alternatively one can switch to a contact lens, a common expedient in monocular aphakia. The approximate, anticipated change in magnification of a contact lens compared to a spectacle lens is $m_{\%} = -dF_S$, where F_S is the spectacle power and d is in centimeters. For example, an eye requiring a −8 D spectacle lens 15 mm from the cornea will obtain approximately 12% magnification with a contact lens; an eye wearing a +10 D spectacle lens at the same distance would achieve about 15% minification by switching to a contact lens.

A change in magnification by altering the shape factor is limited by the power required to correct the ametropia at the spectacle plane, and consequently only small changes are possible. For example, the base curve may be altered (to change front surface power), the thickness varied, or a glass ordered of high refractive index. Although the resultant magnification change is small, it may be sufficient to correct an induced aniseikonia.

RELATIVE SPECTACLE MAGNIFICATION

Spectacle magnification compares retinal image size with and without the lens. The ratio of retinal image size in the two eyes, one of which is emmetropic and the other ametropic, is called "relative spectacle magnification" (RSM).

$$RSM = \frac{\text{Retinal image size in "standard" emmetropic eye}}{\text{Retinal image size in corrected ametropic eye}}$$

Since there is no "standard" emmetropic eye, one must be assumed. We use the standard 60 D equivalent power previously adopted for the reduced eye; call it F_E. Relative spectacle magnification compares the equivalent power of the standard eye (F_E) to the equivalent power of the corrected ametropic eye; call it F_A. Hence $RSM = \frac{F_E}{F_A}$.

The equivalent power of the corrected ametropic eye is that of the (spectacle lens) + (eye) system

$$F_A = F_L + F_O - dF_L F_O$$

where F_L is the spectacle lens power, F_O is the uncorrected ocular power, and d is the vertex distance between them. Spectacle power and the vertex distance can be measured, but ocular power is not generally known. It is presumably 60 D in axial ametropia, more than 60 D in refractive myopia, and less than 60 D in refractive hyperopia.

Axial ametropia

Most large ametropias such as progressive myopias are axial; the ocular power can be assumed to remain constant at 60 D while the length changes. If the correcting lens is now placed at the anterior focal point $1/F_O$ (or -16.66 mm), the equivalent power of the lens plus eye system (F_A) remains 60 D. For example, if a -6 D lens is placed -16.66 mm from an axially myopic eye ($F_O = 60$ D), the equivalent power of the system is $F_A = (-6) + (60) - 0.0166\ (-6)\ (60) = 60$ D; hence $F_A = F_E$. Since the relative spectacle magnification is defined as the ratio $\frac{F_E}{F_A}$, any lens placed at the anterior focal point of the eye gives unit RSM. This is "Knapp's rule."

Knapp's rule states that if ametropia is assumed to be axial, and if the axially ametropic eye has the same power as the standard emmetropic eye, and if one knows this standard power so that the lens can be placed at the anterior focal point, then no magnification will result compared to the emmetropic eye. This is a rather long list of assumptions, but all are necessary for Knapp's rule to work.

The whole purpose of such computations is to compare the magnification of lenses in two different eyes as in anisometropia. Here we do not care about the magnification in the same eye with and without the lens (SM) but the effect on retinal image size with the lens in place compared to another eye with its lens in place (RSM). The common denominator, that is, the standard of comparison, is logically the emmetropic eye. The logic is dimmed because there is no standard emmetropic eye—one must be assumed. Moreover one must assume, without the possibility of measuring, that the ametropic eye is not only axially ametropic but also that its power is the same as that of the standard emmetropic eye. It is unlikely that any ametropia is purely axial or purely refractive. Nevertheless, Knapp's rule has a curious fascination for those who want to correct aniseikonia by calculation rather than measurement, even though it has little practical value.

Refractive ametropia

Passing from the questionable logic of axial ametropia, we come to the more certain muddle of refractive ametropias. Which ametropias are refractive? The eye in aphakia, diabetic myopia, drug-induced ametropias, and accommodative spasm are probably refractive. As for the remainder, we cannot measure and hence cannot be sure.

The eye in refractive ametropia has a "normal" length, but the power is larger than 60 D in myopia and less than 60 D in hyperopia. When a correcting lens is placed before such an eye, the lens makes up for the excess or deficient power so that the equivalent power of the spectacle lens plus eye system is restored to 60 D and the rays focus on the retina. The equivalent power of the spectacle lens at the plane of the eye is $\frac{F_S}{1 - dF_S}$, and the equivalent power of the eye is F_O. The sum of the two must equal 60 D or F_E.* Substituting this combination for F_E in the formula for relative spectacle magnification gives:

$$\text{RSM}_{\text{refractive}} = \frac{1}{1 - dF_S}$$

We see that the relative spectacle magnification in refractive ametropias is identical to the spectacle magnification; a plus lens always produces a larger retinal image, and a minus lens always produces a smaller retinal image compared to the standard emmetropic eye.

EXAMPLE: What is the relative spectacle magnification in (a) axial and (b) refractive ametropia, corrected by a $+10$ D lens placed 16.66 mm from the cornea? Assume the standard emmetropic eye to be 60 D.

$$\text{(a) RSM}_{\text{axial}} = \frac{60}{+10 + 60 - 0.0166\,(+10)\,(+60)} = \frac{60}{60} = +1.00 \text{ or } 0\%$$

$$\text{(b) RSM}_{\text{refractive}} = \frac{1}{1 - 0.0166\,(+10)} = 1.20 \text{ or } +20\%$$

In theory at least, aniseikonia can be avoided in axial ametropia by placing the spectacle lens at the anterior focal point of

*The optical principle is analogous to computing the power of the second lens in a Galilean telescope that will neutralize the effective vergence of the first.

the eye; in refractive ametropia an induced aniseikonia is unavoidable. It would require a correcting lens inside the eye (or in practice, a corneal contact lens) to give unit relative spectacle magnification. Since ocular power cannot be clinically measured, and we never know what proportion of ametropia is due to curvature or axial length, these computations are rather speculative. For example, an aniseikonia may exist in two emmetropic eyes if one has a power of 60 D and the other 70 D, each with its appropriate retinal focal plane. Conversely, aniseikonia may be induced in two identical eyes by plano lenses if their thickness and front surface powers differ markedly. These computations therefore indicate a trend, a relative rather than an absolute comparison. Since most ametropias are axial, and most aphakic eyes are refractive (ignoring presurgical ametropia), an empirical adjustment in the shape and power factors can be attempted to reduce a computed, induced aniseikonia. The only way to be sure of what is really happening is to measure with an instrument like the eikonometer. Unfortunately it is tedious, impractical in children, and presupposes normal binocular vision because the measurement is based on spatial distortions induced by unequal ocular images.

Size differences are also induced by astigmatic corrections—not between the two eyes, but between the two meridians of the same eye. Since astigmatism is usually the result of curvature changes (i.e., refractive ametropia), the RSM is identical to the SM. For example, in the prescription $+1.00 + 4.00 \times 90$ worn 15 mm from the cornea, the magnification will be 0.15% in the vertical meridian and 8% in the horizontal meridian. This "meridional aniseikonia" will cause symptoms if the patient has not previously worn cylinders or if the astigmatism developed rapidly.* Perhaps the most remarkable thing about aniseikonia is that patients often compensate for it even when the diagnosis is missed or ignored.

SPECTACLE AND OCULAR ACCOMMODATION

Just as the far point in practice is measured to the cornea (the equivalent ocular plane in theory) to give the "ocular refraction," the near point measured to the cornea gives the "ocular accommodation." It is obtained by bringing letters nearer the uncorrected eye until they blur. Another method of measuring the near point is adding minus lenses for a constant target distance. Here the accommodation is measured at the spectacle plane, and the result is the spectacle accommodation. Clinically we usually determine spectacle accommodation because ametropia is first corrected by lenses. The near point is conventionally referred to the spectacle plane. Ocular and spectacle accommodation thus differ by the effective power.

Fig. 7-14 shows a reduced eye corrected by a +10 D thin lens placed 15 mm in front of it. The spectacle ametropia is 10 D, but the ocular ametropia is $F_O = \frac{F_S}{1 - dF_S} = \frac{+10}{1 - .015\,(+10)} = +11.76$ D. When this eye views an object at 40 cm, the incident vergence at the lens is −2.50 D, and the emerging vergence is +7.50 D, but by the time the wave reaches the eye 15 mm away, it has changed to $\frac{+7.50}{1 - .015\,(+7.50)} = +8.43$ D. Since the ocular hyperopia is 11.76 D, the actual stimulus to accommodation is 11.76 D − 8.43 D = −3.33 D instead of the theoretical −2.50 D. The spectacle-corrected hyperopic eye must therefore accommodate 0.83 D more than a natural emmetrope. If the hyperopia had been corrected by a contact lens, the necessary accommodation would have been 2.50 D.* The hyperope wearing a contact lens can be expected, all other factors being equal, to require reading glasses later than if he were wearing spectacles. The reverse

*One of the advantages of the recently introduced, rear surface corrected curve lenses is that the cylinder is nearer the eye and these meridional differences are minimized.

*This optical effectivity has been advanced as an indication for prescribing contact lenses in hyperopes with accommodative esophoria-tropia. Adding more plus for reading would seem a simpler alternative.

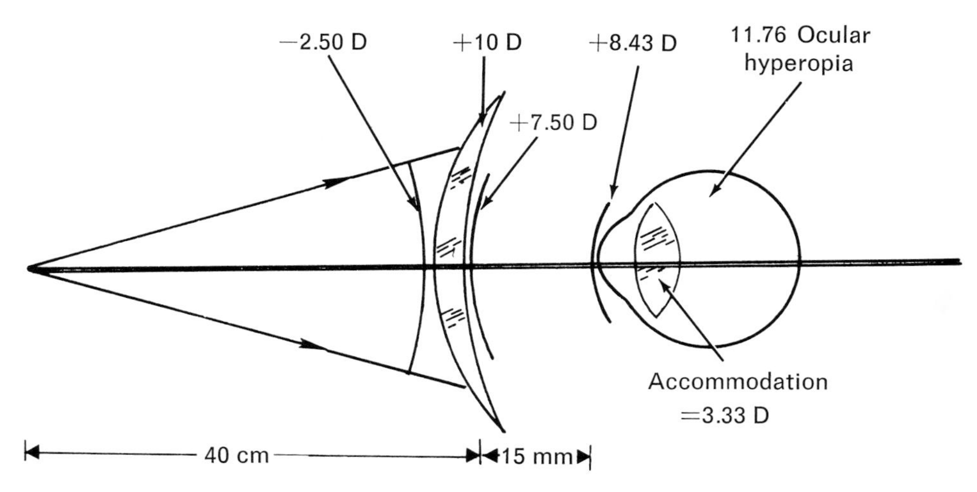

Fig. 7-14. Diagram illustrating the effect of a distance correcting lens on accommodation. The patient has a spectacle refraction of +10 D and an ocular refraction of 11.76 D hyperopia. When he reads at 40 cm, the vergence of the wave leaving the lens is +7.50 D but changes to +8.43 D by the time it reaches the cornea 15 mm further. The patient must therefore accommodate 11.76 − 8.43 or 3.33 D instead of the 2.50 D required of a natural emmetrope. A similar computation will show that a myope wearing his spectacle correction requires less accommodation than the natural emmetrope. If the vertex distance were eliminated by prescribing a contact lens, the hyperope would now use less accommodation and the myope more accommodation than with spectacles.

situation obtains in myopia. A prepresbyopic myope changing from spectacles to contact lenses now finds himself unable to read fine print at his usual reading distance.

EXAMPLE: Contrast the ocular and spectacle accommodation of an eye corrected by a −10 D thin lens placed at a spectacle plane of 15 mm, viewing an object at 40 cm.

The ocular refraction F_O is $\frac{-10}{1 - .015\,(-10)} =$ −8.69 D. The incident vergence at the spectacle plane is −2.50 D; the incident vergence at the eye is $\frac{-12.50}{1 - .015\,(-12.50)} = -10.53$ D.

The stimulus to accommodation is 10.53 − 8.69 = 1.84 D. This represents 2.50 − 1.84 or 0.66 D less ocular accommodation than that required of the emmetrope (or contact lens corrected myope).

The example illustrates that with a spectacle correction, the myope needs to accommodate less and the hyperope must accommodate more than the emmetrope. The greater the ametropia and the closer the viewing distance, the larger the difference between ocular and spectacle accomodation.

A simplified formula (for "average" vertex distances) by which to compute ocular accommodation without going through the previously described steps is given by Pascal (1951). For hyperopia, the "unit of ocular accommodation" is simply 1 + 4% F_S; and for myopia, 1 − 3% F_S, where F_S is the spectacle thin lens power taken numerically (without respect to sign). In the 10 D myope of the previous example, the unit ocular accommodation is 1 − .03 (10) = 0.7 D; hence the total ocular accommodation required for 40 cm is (2.50) (0.7) = 1.75 D, which is near enough the exact result.* The advantage of Pascal's formula is that once the accommodative unit is computed, it can be used to determine the ocular accommodation for any read-

*A discrepancy in Pascal's formula occurs when the correcting lens has less plus power than the divergence of the wave from the near object. For a corrected hyperope of 2 D viewing an object at 33 cm, a −1 D vergence leaves the spectacle lens, but only −.98 D enters the eye—a decreased rather than increased effect. This is a minor point but that it is a "physiologic unit" of accommodation is something else. The difference between ocular and spectacle accommodation is a matter of optical effectivity and not physiology. Whether a unit of ocular accommodation is the consequence of an invariant unit of ciliary muscle contraction is still debated.

ing distance by simply multiplying it by the dioptric equivalent of that distance. Thus the 10 D myope whose accommodative unit is 0.7 D would need to accommodate 3.5 D at 20 cm, 7 D at 10 cm, and 1.4 D at 50 cm.

The difference between ocular and spectacle accommodation in an eye with astigmatism is to exaggerate the cylindrical error at the reading distance. For example, an eye wearing the correction $+2 + 4 \times 90$ at a distance of 20 mm from the cornea reads print at 25 cm. Pascal's accommodative unit in the vertical meridian is 1.08, and the required ocular accommodation is $4 \times 1.08 = 4.16$ D. A similar computation for the horizontal meridian gives $4 \times 1.24 = 4.96$ D. Thus a 0.8 D astigmatic difference remains uncorrected at the reading distance for which the eye cannot compensate because accommodation is symmetrical. Since both meridians cannot be placed on the retina simultaneously, the eye focuses on the circle of least confusion. There is no remedy for these near effectivity changes in astigmatism short of a separate pair of reading glasses. But it should come as no surprise that astigmatism measured at the reading distance (e.g., by a miniature astigmatic dial) differs from that measured at 20 feet. Many cases of so-called astigmatic accommodation reported in the literature are only the consequence of this effectivity change.

The difference between ocular and spectacle accommodation also applies to amplitude. If the emmetropic near point is measured to the cornea, the amplitude will always be less than that measured from the spectacle plane. For example, if the near point is 10 cm measured to the spectacle plane, the spectacle amplitude is 10 D. But if the eye is corrected by a +5 D lens worn at 15 mm, the accommodative unit is 1.2 D and the ocular amplitude is $10 \times 1.2 = 12$ D. Donders in constructing his famous table measured the near point from the equivalent plane; Duane's table, almost as famous, is based on measurements to the spectacle plane (14 mm). Donders' figures for a given age are therefore lower than Duane's, whereas the latter are more nearly representative of the amplitudes as measured clinically with the emmetropic spectacle correction in place.

The contrasting figures for ocular and spectacle accommodation must be considered in changing from spectacles to contact lenses or vice versa. They can be ignored in prescribing presbyopic adds if ametropia, add, and near point are all referred to the same plane.

Effectivity at near

We have seen previously that when a spectacle lens is moved further from the cornea, both plus and minus lenses acquire more plus power. This is strictly true only when the eye views distant objects. It is also true at the reading distance for minus lenses, but it is more complicated for plus lenses.

A plus lens gains plus power when moved away from the eye only if the near object is further than twice the focal length of the lens. For the average reading distance of 40 cm, this means a hyperope cannot increase the effectivity of the plus lens unless the correction exceeds +5 D. The reason is that the negative vergence from the near target increases faster than the plus power effectivity of the displaced lens. The low hyperope can still obtain more plus power, however, by moving the object further away at the same time as he displaces his glasses down his nose. And in all cases, a convex lens moved further from the cornea will increase magnification; thus the patient gets the benefit of a larger image even if slightly blurred.

SPECTACLES AND EYE MOVEMENTS

When the visual axis does not pass through the optical center of a thin lens, there is, according to Prentice's rule, a prismatic effect. An eye looking 0.8 cm below the optic center of a +8 D lens sees objects displaced 6.4Δ downward, that is, toward the apex of the induced prism. The eye rotates in the direction of the apex to maintain foveal fixation—the prism has induced an ocular rotation. Since the lens is spherical, an increased rotation is required for all objects in the visual field that do not lie on the principal axis. The more eccentric the line of fixation

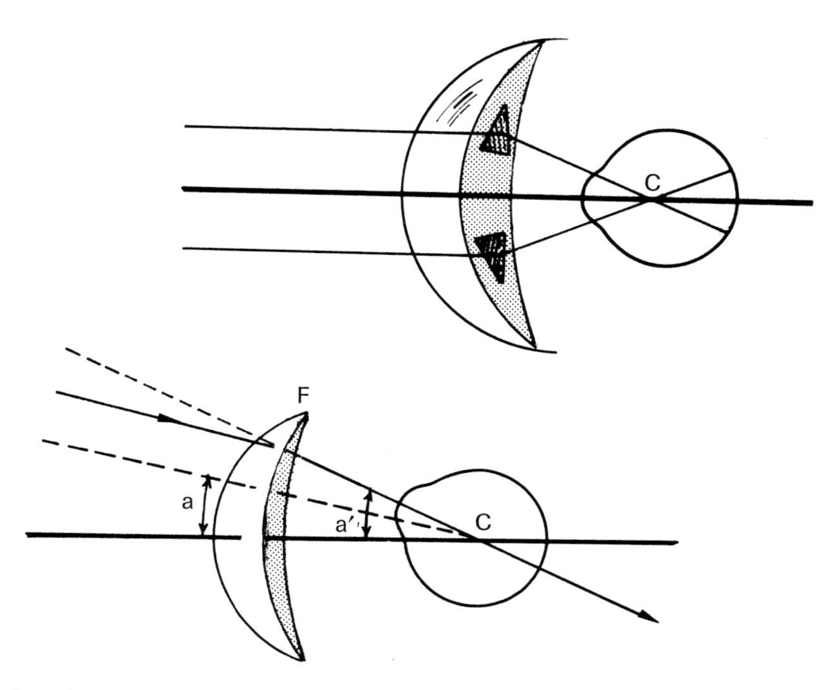

Fig. 7-15. When the eye turns to view a distant object through the periphery of a convex spherical lens, the chief ray passing through its center of rotation (*C*) subtends a larger angle than if the lens were absent. The eye must therefore turn a greater amount as might be expected from the base-down effect in the upper portion of a plus sphere.

the more the prismatic effect and the greater the induced rotation. Looking below the optical center of a minus sphere results in a base-down prismatic effect, the apparent displacement is up, and the eye must rotate less to view the object than without the lens (Fig. 7-15).

A patient with normal binocular vision viewing a near object drops his eyes below the optical center. If the lenses are identical, there is an induced symmetrical binocular ocular rotation. But if there is anisometropia, the prismatic effect and the induced rotations differ for the two eyes. For example, a patient wearing OD: +8 D and OS: +4 D looking 8 mm below the optic center will have 3.2Δ base up OD; that is, the right eye must turn down 3.2Δ more than the left to maintain single binocular vision. If the eyes are now dissociated, the right eye will turn up 3.2Δ more than the left—the same prism has induced a right hyperphoria.

If the induced prism exceeds the vertical fusional vergences, diplopia results. Since the amplitude of vertical vergences is considerably less than the horizontal, a small, induced vertical prismatic effect has greater consequences. Even if there is no diplopia, the induced heterophoria may lead to discomfort.

The effect of a prism on ocular rotations also depends on the purpose for which it is prescribed. For example, a base-out prism may be ordered to reduce an esophoria; the prism produces an inward rotation of the eyes under binocular conditions, thus reducing the degree of in-turning when they are dissociated. (A "phoria" is defined as a tendency of the eyes to deviate that manifests itself when binocular fusion is disrupted.)

A base-out prism placed before eyes that initially measure orthophoria produces convergence to maintain fusion. When the eyes are dissociated, the examiner observes an outward movement—an exophoria. Thus a base-out prism relieves esophoria and induces exophoria. Conversely, a base-in prism (providing it is not so large as to break fusion) stimulates divergence and results in an inward rotation on dissociation—an induced esophoria.

On the other hand, if the eyes are already dissociated and we wish to measure the existing phoria, a prism placed before the eye does not stimulate a vergence movement but simply redirects the ray toward the fovea; therefore base-out prism measures esophoria, and base-in prism measures exophoria.

Whether a phoria should be prismatically corrected depends on its amount, the amplitude of compensating fusional vergences, and the symptoms. Thus the orthoptist may treat esophoria with a base-in prism to increase fusional divergence, whereas the refractionist may order a base-out prism to relieve it.

If the patient does not have binocular vision, the rotational effect of a prism varies with the eye. For example, a prism placed before the nonfixing eye has no effect unless the fixing eye is covered. If the prism is placed before the fixing eye, the nonfixing one executes a yoked version movement. The result may be that a nonsuppressing area is now stimulated, which is intolerable, and in a matter of days or weeks, the tropia increases by an amount that exactly matches the prism to restore the status quo. This principle is sometimes useful to test the anticipated effects of a surgical correction. Another approach is a reverse prism, for example, a base-in prism over the nonfixing esotropic eye with partial occlusion of the fixing eye. This gets the rays away from the suppression area and decreases the need for inhibition, with the goal of restoring its sensitivity. The purpose of the partial occlusion of the fixing eye is to induce the nonfixating eye to take up fixation. The result is a disruption of established vergence, or version reflexes, so that a more correct pattern can be learned.

The effect of a prism on light rays is always the same; it deviates them toward the base. The effect of a prism on ocular rotations depends on whether there is binocular vision or strabismus, fusion or dissociation, and amblyopia or suppression. The kind of prism to be used in a given clinical situation depends on whether one intends to induce or measure a heterophoria, reduce the phoria or increase the fusional vergence, or displace the light into or out of a suppression area in strabismus. These variable effects of a prism on the eyes illustrate the difference between theoretical optics and its clinical application.

Center of rotation

Ocular movements take place around a hypothetical center of rotation, the significance of which is that it constitutes the effective stop of the mobile eye, important in the design and fitting of lenses. This stop, also called the "sighting center," is the key point in ocular kinematics; Listing's plane passes through it, and all ocular rotations may be analyzed as if they occurred about an axis that lies in this plane. Experimentally the sighting center is located by triangulation of corneal reflections for various fixation angles. Results indicate that it is not a single point but that it varies with the axial length of the globe, translatory movements, and angle of view. For clinical purposes these complexities are ignored, and the center of rotation is taken as a single point 14.5 mm behind the corneal pole.

Prismatic deviation for near objects

Trial prisms are calibrated in prism diopters, but the calibration assumes the eye is observing a distant object. When the object is near, the result is an overestimation of ocular rotation—the eye turns less than the prism number indicates.

To demonstrate, hold a pencil horizontally superimposed on some distant horizontal line (e.g., the horizon) at the reading distance. Now place a 6Δ base-down prism before your eye, and note that the distant horizontal line is displaced upward considerably more than the near one. The explanation is shown in Fig. 7-16. The chief ray passing through the center of rotation of the eye from the near object intersects the prism nearer the base. From similar triangles it can be shown that the actual deviation for the near target is

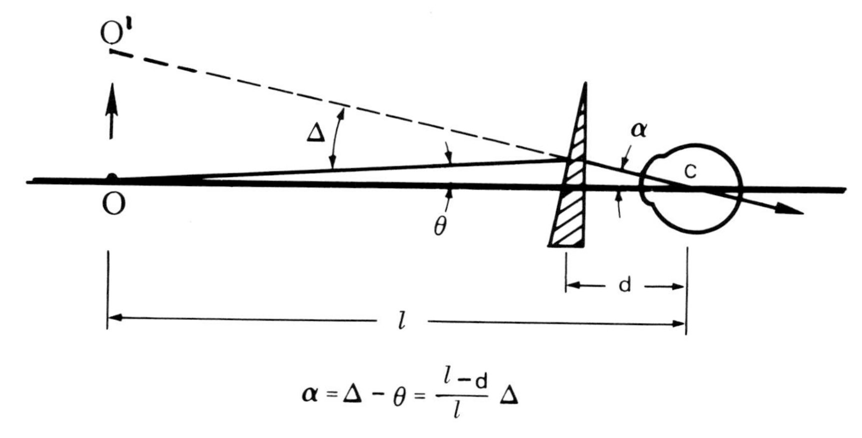

Fig. 7-16. The image O' of a near object O is displaced by the prism through an angle Δ (according to the calibration on the prism handle). To receive this image on the fovea, the eye must rotate through an angle α, which is less by an amount θ compared to its rotation if the prism were absent. The induced ocular rotation in prism diopters (α) is therefore less than the deviation (Δ) in prism diopters by the ratio of the object distance and the distance of the prism from the eye.

$\frac{l - d}{l}$ (Δ), where l represents the target distance, d is the distance of the prism (or lens in case of a prismatic effect) from the center of rotation, and Δ is the prismatic deviation for a distant object (i.e., the prism label). For example, the Risley prism of a refractor can be as much as 5 cm from the center of rotation. Hence in measuring a deviation at 40 cm, when it reads 10Δ, the eye really turns only $\frac{40 - 5}{40}$ (10) or 8.75Δ. In measuring strabismus by holding loose prisms over the eyes of a squirming infant, it is easy to achieve a 25% error on the near deviation. Moreover, if the prism is not held perpendicular to the line of sight, the error is compounded because the ray no longer passes the prism symmetrically. The assumption that the prism cover test has an accuracy of 2Δ is rather optimistic.* Finally, orthoptists not infrequently measure squint angles in degrees —a planned surgical procedure, based on the mistaken assumption that the angle is in prism diopters, will undercorrect by a factor of 2×.

*This limit, frequently misquoted, actually applies to the clinician's ability to detect a 2Δ ocular movement and not his ability to measure it with this accuracy by hand-held prisms.

REFERENCES

Alvarez, L.: Two-element variable power spherical lens, Feb. 21, 1967, U. S. Patent No. 3,507,565.

Armaly, M. F.: The size and location of the normal blind spot, Arch. Ophthal. **81**:192-201, 1969.

Badal, A. J.: Nouvel optomètre, Ann. Ocul. **75**:5-13, 1876.

Bennett, A. G., and Francis, J. L.: Visual optics. In Davson, H., editor: The eye, New York, 1962, Academic Press, Inc.

Carleton, E. H., and Madigan, L. F.: Relationship between aniseikonia and ametropia, Arch. Ophthal. **18**:237-247, 1937.

Cowan, A.: Emmetropia, Am. J. Ophthal. **34**:1021-1024, 1951.

de la Hire, P.: De quelques faits d'optique, et de la manière dont se fait la vision, Mem. Hist. Acad. R. Sci. Paris, 1709.

Diamond, S.: Conjugate and oblique prism correction, Am. J. Ophthal. **58**:89-95, 1964.

Donders, F. C.: On the anomalies of accommodation and refraction of the eye, translated by W. D. Moore, London, 1864, The New Sydenham Society.

Duke-Elder, S.: The practice of refraction, ed. 8, St. Louis, 1969, The C. V. Mosby Co.

Duke-Elder, S., and Abrams, D.: System of ophthalmology; ophthalmic optics and refraction, vol. 5, St. Louis, 1970, The C. V. Mosby Co.

Elliott, R. H.: Errors of refraction, Br. J. Ophthal. **2**:313-322, 1918.

Emsley, H. H.: Visual optics, ed. 3, London, 1944, Hatton Press, Ltd.

Fry, G. A., and Hill, W. W.: The center of rotation of the eye, Am. J. Optom. **39**:581-595, 1962.

Gettes, B. C.: Practical refraction, New York, 1957, Grune & Stratton.

Guyton, D. L.: Automated refraction, Invest. Ophthal. **13**:814-818, 1974.

Hartinger, H.: Neuzeitliche Gerate für die subjective Sehprüfung, Z. Ophthal. Optik **15**:129-139, 1931.

Hartstein, J.: Review of refraction, St. Louis 1971, The C. V. Mosby Co.

Hooke, R.: Micrographia, or some physiological descriptions of minute bodies made by magnifying glasses, New York, 1961, Dover Publications, Inc.

Jackson, E.: Trial set of small lenses and a modified trial case, Trans. Am. Ophthal. Soc., pp. 595-598, 1887.

Javal, E.: Histoire et bibliographie de l'astigmatisme, Ann. Ocul. **55**:105-127, 1866.

Knapp, H.: The influence of spectacles as the optical constants and visual acuteness of the eye, Arch. Ophthal. Otolaryngol. **1**:2, 1870.

Lancaster, W. B.: Refraction correlated with optics and physiological optics; motility limited to heterophoria, Springfield, Ill., 1952, Charles C Thomas, Publishers.

Landolt, E.: Refraction and accommodation of the eye, translated by C. M. Culver, Edinburgh, 1886, Y. J. Pentland.

Laurance, L., and Wood, H. O.: Visual optics and sight testing, London, 1936, The School of Optics, Ltd.

Laurence, J. Z.: Optical defects of the eye, London, 1865.

Ludvigh, E.: Amount of eye movement objectively perceptible to the unaided eye, Am. J. Ophthal. **32**:649-650, 1949.

Mason, F. L.: The principles of optometry, Berkeley, 1940, University of California Press.

Mayer, E.: Memoire sur les ordonnances des lentilles spheriques, Ann. Ocul. **142**:333-372, 1909.

Neumueller, J.: Effect of ametropia upon correction for corneal astigmatism, Am. J. Optom. **7**:201-212, 1930.

O'Brien, J. M., and Bannon, R. E.: Accommodative astigmatism, Am. J. Ophthal. **30**:289-296, 1947.

Ogle, K. N.: Optics: an introduction for ophthalmologists, ed. 2, Springfield, Ill., 1968, Charles C Thomas, Publisher.

Pascal, J. I.: Lens efficiency—clinical concept, Br. J. Ophthal. **30**:291-298, 1946.

Pascal, J. I.: Scope and significance of the accommodative unit, Am. J. Optom. **29**:113-128, 1952.

Percival, A. S.: The prescribing of spectacles, ed. 3, New York, 1928, William Wood & Co.

Porterfield, W.: Treatise on the eye, Edinburgh, 1759, G. Hamilton & J. Balfour.

Rubin, M. L.: The sliding lens paradox, Survey Ophthal. **17**:180-195, 1972.

Rubin, M. L.: Optics for clinicians, ed. 2, Gainesville, 1974, Triad Scientific Publishers.

Sheard, C.: Ophthalmic optics with applications to physiological optics. In Glasser, O., editor: Medical physics, Chicago, 1950, Blakiston Co.

Sheard, C.: Visual and ophthalmic optics. Philadelphia, 1957, Chilton Book Co.

Thorington, J.: Refraction of the human eye and methods of estimating the refraction, Philadelphia, 1930, P. Blakiston's Son & Co., Inc.

Tour, R. L.: Astigmatism, Int. Ophthal. Clin. **5**:369-388, 1965.

Volk, D.: Aspheric ophthalmic lenses, Int. Ophthal. Clin. **5**:471-494, 1965.

Westheimer, G.: Optical and motor factors in the formation of the retinal image, J. Opt. Soc. Am. **53**:86-93, 1963.

Young, T.: On the mechanism of the eye, the Bakerian lecture, Philosoph. Trans. **17**:23, 1801.

PART II

TECHNIQUE

8

Principles of refraction

Clinical refraction, at least since Donders, is based on a solid foundation of optics, physiology, and the psychology of perception. Without optics, the refractionist is impotent; without application, optics is only an idle exercise in geometry. The physiology, based mostly on the research of the present century, has established the biologic character of ametropia and the significance of oculomotor and accommodative anomalies in producing visual symptoms.

Although refraction is currently out of proponents, physiologic optics out of fashion, and geometric optics out of the question, the ophthalmologist will treat more refractive problems than any other kind. And when diligently applied, it can still provide a modicum of professional satisfaction by restoring useful and comfortable vision to the vast majority of patients.

REFRACTION, COMMON SENSE, AND PSYCHOPHYSICS

It has been said that whatever exists, exists in some amount and consequently can be measured. The precise rules are provided by psychophysics. It tells us how to evaluate a stereoscopic threshold or the beauty of a sunset and the significance to be attached to each. In clinical refraction, human responses are measured under conditions of fluctuating attention, conflicting motivations, and limited intelligence, with tests of imperfect design. There is no time nor is the patient trained for methods of limits, paired comparisons, random presentations, or multiple testing, so we must substitute common sense and clinical judgment for the precision of the laboratory. An irrational response may mean a witless patient; more likely the test was poorly chosen, the conditions inept, or the instructions misinterpreted. Ask a leading question and most patients will oblige with an invalid answer.

The symptoms determine that of two patients exhibiting the same refractive error, one receives spectacles and the other does not. On the other hand, an asymptomatic patient will discourage the tendency to order an optical prosthesis merely for the sake of doing something positive.

Refraction is part of a complete ophthalmologic examination and cannot be performed in a vacuum. The status of the chamber angle and intraocular tension must be known before a cycloplegic is used, retinoscopy will be useless in keratoconus, subjective tests require a different interpretation in the presence of subnormal vision, and patching the only good eye of a child with macular toxoplasmosis is a tragic deprivation to improve a phantom amblyopia.

Precision is gained by giving clear instructions in words that have a common meaning. Children may not understand what is wanted, tense patients are afraid to give the wrong answer, malingerers the right one, and the deaf patient may not have heard the question. Most people do better on tests that present a clear choice of alternatives, such as clear-blur, light-dark, or single-double. "Which is clearer?" signifies little to the patient who wonders whether the examiner means the first lens, the second lens, neither lens, or the vision before the drops were put in; it is better to use a sphere, a cylinder, or a prism to demonstrate what is wanted. Each patient requires a different approach, just as some compositions are scored for a brass band and others for a string quartet.

Every refractive test has its optical basis and its limits of accuracy. To make rational lens changes in response to the patient's answers means visualizing what is happening on the retina. The astigmatic dial is used with the eye fogged; the cross cylinder works best without fog; reverse the conditions, and the validity goes out the window. Attempts to exceed the accuracy of the test—or the accuracy of the patient's discrimination—are futile and only leads to boredom and frustration for all concerned. Since refraction deals with people (not "organisms"), allowance must be made for individual differences in temperament and motivation. One person is soothed by the violin; another is infuriated by a piccolo. There is no substitute for the experience of actual contact with patients. Despite the apparently improvised chaos of a busy office, this clinical sixth sense is a valuable if intangible asset.

The clinician without reliable data is sooner or later without a reliable diagnosis, basis for treatment, or freedom of action. Tests must provide reasonably consistent results when repeated under similar conditions. Patients seldom try to deliberately mislead the examiner, but patients are human and subject to errors, and children not infrequently give the response they believe is wanted.

Facts, even quantitative facts, are not always what they appear. In an experiment to evaluate the effect of eye exercises on chickens, 33⅓% improved, 33⅓% became worse, and no data could be obtained on the other 33⅓% because that one ran away. Refractive tests must also be valid, that is, measure what we intend to measure. The clinical significance of 20/200 vision hinges on the fact that from the human point of view, most people have 20/20 vision. Our culture, our education, even our lives are based on it. The patient with 20/200 vision cannot read a price tag, a book, a phone dial, or safely drive a car. If retinal resolution is destroyed or the transparency of the ocular media disturbed, the inability to restore the standard has a different practical meaning.

In Donders' day the far point was located with subjective optometers, and instrumental accommodation was a variable nuisance. The refractionist, told to aim for optical emmetropia, found cycloplegia a convenient answer and supposed he had achieved a more accurate result. But patients could not wear the prescription because when the effect wore off, they were almost invariably overcorrected. Cycloplegic refraction provides different diagnostic information that supplements but does not replace the examination of the nonparalyzed eye.

All measurements are ultimately subjective; someone must read the dial, the scale, or the end point. The examiner's reading may be more reliable, but this does not necessarily make it more representative of the function tested. We can expedite the examination, obtain more valid and reliable answers, and make the experience pleasant for the patient by a pinch of patience, a dash of discipline, and a sprinkling of humor. Keep calm; we shall all be fables bye and bye.

REFRACTIVE EQUIPMENT

Diagnostic efficiency depends on accurate instruments, easily manipulated and comfortably adjustable. Professional time is best spent on examining patients, not polishing lenses, focusing instruments, or cleaning equipment. Refracting rooms should provide longitude, not lassitude. Have controls at arm's reach, with foot switches for lighting and instruments, interoffice communication, and dictating facilities and counterbalanced centralized instruments that are clean, focused, and periodically calibrated in a pleasant, quiet, and uninterrupted setting.

Trial lenses and frame

A set of trial lenses is indispensable if only to confirm a prescription found with the refractor. The set should include both plus and minus cylinders and if possible, with color-coded power markings on the handle rather than engraved on the glass and axis markings easily visible. Lens powers are best calibrated in back vertex powers in 0.25 D steps with a pair of ±0.12 D spheres and cylinders for interpolation. The number of dioptric steps is optional, but the fewer the lenses in front of the eye the less the reflections, aberrations, and chance for transposition errors.

Accessories include circular prisms, occluders, frosted glass, pinholes (single and multiple), a stenopeic slit, a Maddox rod, a red glass, a double prism, and several cross cylinders. A Scheiner disc, red-green glasses, polarizing lenses, and graduated, neutral density filters are sometimes useful. A tape measure, PD ruler, and vertex caliper should be handy.

Trial frames should be sturdy and light, with adjustments for interocular distance, temple length, nose support, bridge size, angulation, and vertex distance. At least four lens cells (sphere, cylinder, occluder, and prism or add) are required. Manufacturer's instructions should be followed in positioning the trial lenses; the cylinder is generally anterior so that the axis is visible. Most modern trial lenses are directly additive; lens position and air spacing are compensated.

Refractors

Refractors (phoroptors) are time-saving trial frames with self-contained lenses in discs that revolve in front of each eye. There are separate discs for spheres, cylinders, and accessories. Cylinders may be plus or minus (unfortunately never both), and accessories are optional. In addition, self-centering cross cylinders, rotating prisms, Maddox rods, and an attachment for holding near-point targets are incorporated.

Refractors provide many lens choices and variations, for example, to maintain spherical equivalents. Its disadvantages include a telescope effect (minimum of 30 mm tube thickness) with limitation of field, inability to see the patient's face, reading tests with eyes in primary rather than down position, and an occasional temptation to twirl the lenses faster than the patient's ability to respond. Unless the unit is sealed, the lenses become smudged and require expensive cleaning. All factors considered, however, refractors are a tremendous convenience, and it is difficult to conceive of a modern refracting room without one.

Acuity targets

Different acuity targets are available to suit various requirements such as illiterate E's, unlearnable letters; broken rings; ingenious pictographs; graduated bars, lines, or numbers calibrated in fractions (imperial or metric); decimals; percentages (efficiency or loss); visual angles; and some scales understood only by their inventors. Acuity charts may be white, gray, colored, printed, engraved, projected, or polarized for both direct and reversed viewing. They may be illuminated externally, internally, variably, or improperly. No precise level has ever been agreed on. There are special targets for flyers, military candidates, malingerers, drivers, and one in several languages for diplomats—the test is to misunderstand as many as possible. There are targets for far, near, and intermediate distances in Snellen, Jaeger, and printer's points for normal and subnormal vision in letters foreign and domestic.

Astigmatic dials

Astigmatic dials were introduced by Javal in 1866 and the well-known clock dial by John Green, two years later. A wide selection of dials is available, composed of fixed, rotating, or projected lines, arrows, or patterns. Some are simpler to understand, and some are easier to use; one clinician swears by the clock, another by the cross, and patients not infrequently at all.

Equipment for lens evaluation

An optical prescription will not work unless translated into the proper lens and the proper fit, used under the proper circumstances. To check the final prescription a lensometer, lens measure, lens caliper, radiuscope (for contact lenses), protractor, and a lens marking device will be required. A polariscope is useful to confirm that lenses have been heat treated.

Equipment for oculomotor evaluation

The items required for motility examination of a young child include little more than a flashlight, some interesting toys, loose prisms, or a prism bar. Testing older strabismics depends on the clinician's philosophy: mechanistic or sensorial. It may involve only motor tests, such as the cover test, or the gamut of sensorial techniques from a simple red filter to elaborate synoptophores. The following will be helpful until the orthoptist arrives: Worth's four dot test (distance and near); stereoscope or vectographic slides for stereoacuity, fixation disparity, and latent nystagmus; red-green (Lancaster) projectors and Hess screen for phoria testing and evaluation of paralytic squints; a projection ophthalmoscope, Bagolini striated lenses, afterimage tester, and Haidinger brushes for evaluating amblyopia; and a simple amblyoscope (Worth) to measure the subjective angle of deviation. A Stevens phorometer, red glass, a Maddox rod, rotary prisms, and neutral density filters are useful accessories to the trial set or refractor.

Equipment for the evaluation of subnormal vision

In addition to the standard trial set, a strong reading light, and a selection of reading material, there should be available some hand-held and stand magnifiers, head-borne (clip-on) loupes, strong bifocal trial lenses, and a selected sampling of variable focus distance magnifiers. It is best to have two of each of the more commonly used visual aids so that one can be given to the patient for home trial.

Miscellaneous equipment

In addition to conventional refractive equipment, special instrumentation is required for contact lens work, including trial sets for standard and aphakic corrections. Other useful items are an optokinetic drum or tape, Amsler grid, Howard-Dolman or Verhoeff stereotester, a visual skills test battery, a screening color vision test, and, depending on interest and inclination, various so-called tests of visual perception.

Tests designed to give some insight into the nonoptical factors in visual disorders (e.g., cortical dysfunction) are part of the routine neuro-ophthalmologic examination. They include tests for form, color, contour, and spatial relations, the recognition and naming of simple objects, the execution of fine movements on command, and memory for the written or spoken word. Unlike peripheral lesions, with their precise topographic localization, central lesions show considerable overlap with disorders of speech, hearing, reading, and writing.

Tests for the recognition of form and pattern have also been used in human engineering, for example, in designing road signs, instrument dials, work layout, camouflage, advertising copy, and television commercials. More controversial are tests to identify the poor school achiever, in general, and the poor reader, in particular. Clinical attitude ranges from accepting such tests as a mild form of alchemy to uncritical enthusiasm. Briefly, so-called perception tests* fall into one or more of the following categories (generally graded for specific age groups).

The Gestalt completion tests involve the identification of incomplete pictures or patterns, mutilated words, and the recognition of patterns within designs (hidden pictures).

Tachistoscopic tests consist of brief exposure of letters, digits, pictures, words, or sentences. A simple instrument can be constructed by adapting a camera shutter to the acuity projector. The results are used in the diagnosis and treatment of reading difficulties. For example, fast readers have a larger span per single fixation than poor ones.

Hand-eye coordination tests take various forms, such as cheiroscopic tracing, mirror tracing, pointing, following, and copying simple figures. Accuracy and timing are evaluated for each eye and under binocular conditions. Form and color dominance tests evaluate preference for grouping by shape, color, pattern, size, or brightness. Retinal rivalry for colors or patterns, Necker cube reversal, and various visual illusions also belong in this group.

Miscellaneous tests include autokinetic movement, critical fusion frequency, puzzle solving, afterimages, and tests for sensory and motor eye dominance. Will changing eye dominance improve reading? Will perceptual training im-

*Most of these tests are described and reviewed in Buros, O. K.: Mental measurements yearbook, Highland Park, N. J., 1965, Gryphon Press.

prove school performance? Will the preference for certain colors make better citizens? If past experience is any guide, they will not. What they will do is silence criticism that the ocular problems have been ignored.

THE ETIOLOGY OF AMETROPIA

Clinical refraction deals with the diagnosis and treatment of ametropia. However, considering its prevalence, remarkably little is known of its etiology. Obviously eyes differ. What is it that makes them differ?

The earliest views considered the ametropic eye a diseased eye, especially if myopic. Ametropia was the result of inflammation or metabolic, endocrine, abiotrophic, and nutritional insults. Feed a man the right amount of calcium, parathyroid, Vitamin D, and protein, and he will be emmetropic; feed him wrong, and he ends up hyperopic—an optical cretin, or myopic—with lenticular obesity and axial elephantiasis.

The biologic view holds that ametropia, at least of moderate amount, is a normal variant of a Gaussian distribution, an accident of birth and genetic probability, and no more a disease than height or sex or the length of one's nose. The myopic eye is not elongated by pathologic stretch but by physiologic growth. If any exogenous factor is involved, it is in the Darwinian sense, an evolutionary adaptation found useful by the race, not by the individual.

The environmentalist pointing to the progression of myopia during the school years is ready to declare the entire educational process a public calamity. Myopia is the result of prolonged near work, which presumably causes its effects like a dripping faucet, not by intensity but by sustained repetition. The solution is to send people back to the prairie; tie a person to a stenographer's job or to the stock market quotations and distance vision deteriorates.

The psychologist looking within views ametropia as the consequence of human needs, impulses, and desires, which impel attempts to master and thus adapt to the environment. Nearsighted bookworms are not nearsighted because they read, but because their personalities meander between valleys and peaks of autonomic activity. Treatment in this view may dispense with lenses and concentrate instead on tranquilizers, relieving psychic tensions, and training perceptual skills.

The statistician, given enough data, can hang any theory to an extrapolated hook because he performs no experiments, contemplates no causes, and draws no etiologic conclusions. If there is a higher incidence of myopia among Chinese than Frenchmen, and poor Chinese like rotten fish, and rich Frenchmen like rotten cheese, he will allow that fresh fish prevents myopia, and Camembert deserves a therapeutic trial.

Thus postulated mechanisms of ametropia are numerous, with an ingenuity inversely proportional to the possibility of experimental verification. It is hard to distinguish between the drunken theory and the sober hypothesis. (And although I have caricatured the theories a little, you ought to see some of the theories.)

The eye is made of fourteen individual optical components, none of which have ever been measured simultaneously.* All statistical studies are based to some extent on assumed or secondarily computed values. Until the appearance of the x-ray method, no one had ever measured axial length in vivo, and to this day we depend on Gullstrand's corneal index (measured on two cadavers) and an assumed homogeneous refractive index for the obviously heterogeneous crystalline lens. The range of emmetropic eyes can vary considerably from 22.3 to 26.0 mm, the power of the cornea can vary from 39.0 D to 47.6 D, and the power of the lens can vary from 15.5 D to 23.9 D. The respective averages and standard deviations for 107 eyes (clinical refraction between zero and 0.5 D) were 24.4 mm ± 0.85, 43.1 D ± 1.62, and 19.7 D ± 1.62 (Sorsby and others, 1957). Goldmann and Hagen (1942) studied 6 emmetropic eyes by the x-ray method and

*Three studies are available in which the major optical components were measured (rather than secondarily computed): Tron (1929), Stenstrom (1948), and Sorsby and associates (1957).

found a range of 22.2 to 24.4 mm. Stenstrom (1948) studied 1,000 eyes and found an axial length of 24.00 ± 0.035 mm for the series. Jansson (1963) compared the ultrasound and x-ray methods in 36 eyes and found a mean axial length of 24.43 mm ± 0.168 mm by ultrasound compared to 23.30 mm ± 0.158 mm by the roentgenologic method with about the same error (0.15 mm).

Similar measurements for ametropic eyes show that, with few exceptions, values fall within the emmetropic range. Only when the clinical refraction exceeds 6 D of hyperopia or 4 D of myopia does the axial length fall outside the emmetropic limits. Ametropia less than 4 D therefore is in most cases a matter of disproportionate rather than excessive growth (correlation ametropia). A smaller percentage (about 5%) of ametropias (over 4 D) is due to an abnormal component, abnormal in the sense that the curvature or axial length falls outside the emmetropic range (component ametropia). Thus the myopia of prematurity belongs to the second category, just as acromegaly is not a normal growth variant.

What can we extract from this extracted mass of measurements? Only three of the eye's optical components are generally agreed to have any great consequence in ametropia: axial length, corneal curvature, and the power of the crystalline lens. In 1913 a Swiss ophthalmologist, A. Steiger, first suggested that each component is normally distributed and proposed that refractive error is simply the result of their chance association, a point on a normal curve of physiologic variation. Steiger's theory was simple, reasonable, and biologic but did not explain why there were so many emmetropes or near-emmetropes. The distribution of refractive error in the general population is disproportionately peaked around the zero value (the technical term is "leptokurtic"), a small difference but, as statistics go, an important one.

To account for the high number of people without large refractive errors, Straub (1909) postulated a process of emmetropization. The optical components are not associated by chance but are correlated to fit together. Nature incapable of producing only punctiliously perfect eyes has provided for the maximum number of nearly emmetropic ones. But if a great deal may be said about the statistical game of measuring ocular dimensions, considerably less can be asserted about why they match so neatly. Is it a hereditary predisposition, the work of a genetic organizer, some mysterious biologic force such as "harmonious growth," a cortical-subcortical servomechanism, the result of environmental blessing or dietary bliss, or simply a mathematic artifact?

The concept of hereditary predisposition has been studied by comparing uniovular twins and tracing familial incidence. The refraction in twins is statistically more similar than chance predicts, and there is a higher incidence in families, suggesting either a dominant or recessive inheritance. Knowledge of the exact genetics is still lacking because large human pedigrees are difficult to obtain, and the dioptric character of the phenocopies do not differ diagnostically.

The eye, almost full size at birth (axial length 18 mm), grows to its adult value (about 24 mm) by age 14 years. There is a rapid infantile phase to age 3 years, with a slower growth thereafter. Elongation is balanced by proportional changes in the cornea and lens so that overall refractive changes are slight. The development of ametropia is in most cases a matter of disproportionate rather than excessive growth. Thus myopia results when axial elongation exceeds the compensatory reduction of curvature, but the period during which the most dramatic growth changes occur—the first three years—is not easily accessible to precise optical measurements. Chromosomes are no more optically knowledgeable than beginning eye residents, and there is no inherent reason why a different gene should control each component. Harmonic growth is not an anthropomorphic adjustment of an out-of-focus camera; it is genes acting on biologic variables, not necessarily optical components. The normal distribution of optical components is no assurance of homogeneity or of intrinsic biologic independence. By the laws of probability there are fewer high ametropes for the same reason that there are fewer idiots or geniuses.

The concept of a genetic organizer was suggested by Sorsby. Just as the optic vesicle is the organizer for the anatomic eye, so the retina is the genetic organizer for the optical eye. Thus the size of the retina would determine the size of the overlying sclera (recall that the sclera is formed by a mesodermal condensation around the optic cup). The sclera in turn determines the curvature of the cornea and from this it follows by simple mechanics that a small retina tends to produce a more curved cornea, balancing short axial length with increased corneal power. Conversely, a large retina will not only lead to the manufacture of a deflated cornea but also, by an expanded ciliary ring, lead to a flatter crystalline lens. It is not yet clear why the retina is small in one case and large in another, but the mechanism is presumably genetic rather than environmental.

Some consider emmetropia and ametropia as biologic servomechanisms, in which macular output influences ciliary tonus by way of the autonomic nervous system. The relaxation of parasympathetic tone from higher centers produces axial elongation, choroidal stretch, and traction on the optic nerve.

During the growth period there is a greater incidence of myopia than can be accounted for by chance variation, and practically all clinical studies of ametropia have concentrated on myopia. Cohn of Breslau (1886) first emphasized that near work might cause myopia, a position subsequently supported by such an all-star cast as Javal, Landolt, Tscherning, Stilling, Bishop Harmon, and Risley among others. An equally outstanding team, including Donders, Priestley-Smith, Hess, Steiger, Tron, Brown, and Kronfeld, propounded the biologic or hereditary view. But most of us are still inclined to hedge our bets, if only because the stakes are high and simple hygienic measures can do no harm. Two generations have grown up with their noses glued to the television set, with no apparent increase in the incidence of myopia; yet I still advise patients to watch (if they must) from across the room. Primum non nocere. The notion that poor posture, poor reading habits, and poor illumination contribute to refractive errors is hard to discard.

If monkeys are placed in a restricted visual environment, some become myopic, but monkeys are not humans, and humans will not be restricted. The conclusion that more Chinese are myopic because of their reading habits sounded reasonable until someone pointed out that 90% of the population studied was illiterate. And the incidence of myopia among literate white settlers and native tribesmen in equatorial Africa was found to be practically identical. If prolonged accommodation results in myopia, there should be no hyperopes left, since they would be the first and the most affected. Yet such careful observers as Tscherning and Lancaster, not given to crackpot theories, recommended and used atropine in the more myopic eye, whereas more recently van Alphen suggests parasympathomimetics and encourages near work to prevent myopic progression. Small wonder, as Hirsch remarks, that those who work with refractive errors are in no immediate danger of losing their humility.

Although the cause of myopia remains unclear, genetic influences are undoubtedly a major factor. The question remains whether environmental causes trigger or aggravate it.

Finally, even a chance association of optical components leads to more emmetropes or near emmetropes. In comparing axial length and corneal curvature, for example, there are three emmetropic possibilities: high curvature, short length; average curvature, average length; and weak curvature, long length.

The technical problems of studying the incidence and distribution of ametropia are, it must be admitted, considerable. The very definition of emmetropia must be decided a priori. That is, should emmetropia be measured with or without cycloplegia by objective or subjective methods; and with which targets, what state of light adaptation, how large a pupil, level of illumination, and contrast ratio; and will the dioptric result be referred to the spectacle, corneal, equivalent, or nodal point plane? If astigmatism is present as it generally is, will the record show the power in the vertical or horizontal meridian or will it be the spherical equivalent? How large a deviation from this assumed emmetropic standard will be required to categorize the eye as ametropic? After all, 0.25 D is only a convenient practical unit for our trial lenses; it could just as easily be 0.1 D or even 0.01 D. Diopters are arbitrary units of refraction based on the metric system. There is no assurance that genetic organizers work in 0.25 D steps or have ever heard of the metric system. Clearly the more refined the measurement the more precise the criteria, and the more discriminating the patient the fewer the number of emmetropes one will find by any definition.

CLINICAL PATHOLOGY OF AMETROPIA

Eyes with low degrees of refractive error do not differ in structure from emmetropic ones, and even the changes in high ametropia are not pathognomonic, although they are fairly characteristic. And an ametropic eye may participate in the disturbances of any other ocular disease. By virtue of its optical construction, the ametropic fundus appears magnified or minified in ophthalmoscopy, and the fundus details may be distorted in severe astigmatism.

In direct ophthalmoscopy the fundus is an object viewed through the ocular optical system, which acts as a simple magnifier. Comparing the angle under which the fundus is viewed with the angle under which it would be viewed if placed at the conventional distance of distance vision (25 cm), the magnification is $\frac{250 \text{ mm}}{16.67 \text{ mm}} = 15\times$. The myopic eye having a stronger optical system is magnified; the hyperopic fundus is minified compared to the emmetropic eye. In indirect ophthalmoscopy, conditions are reversed; the hyperopic fundus is magnified and the myopic fundus minified. The magnification of the emmetropic fundus is about 3× in indirect ophthalmoscopy, depending on the power of the condensing lens.

Hyperopia

The significant changes in the hyperopic eye are anatomic rather than pathologic. The small eye predisposes to a narrow chamber angle, which, coupled with a relatively large lens, may lead to an abrupt and marked rise in intraocular tension when the pupil dilates. A combination of higher pressure in the posterior chamber, pupillary block to the forward circulation of aqueous, relaxation of peripheral iris folds, and congestion of vessels may precipitate an acute glaucoma attack, characterized by pain, decreased vision, halos, corneal edema, and a fixed dilated pupil.

The hyperopic fundus reflexes are bright, a reflection of the fact that it usually characterizes the younger eye. The disc may appear hyperemic and, if surrounded by a prominent nerve fiber pattern that obscures its margins, may be mistaken for papilledema or optic neuritis (pseudoneuritis). This congenital pattern may be seen in other members of the family and can be differentiated from a true neuropathy by the presence of good vision, normal visual fields, and the absence of progression.

Hyperopia has been reported to occur secondary to trauma, possibly a consequence of retinal edema. A secondary hyperopia occurs after scleral resection or extensive diathermy for retinal detachment. A retro-ocular orbital tumor may cause axial hyperopia. A change in ametropia is characteristic of diabetes; usually the trend is toward myopia with a high blood sugar a reversal when blood sugar returns to normal. These changes are attributed to variations in the hydration of the crystalline lens because they are not abolished by cycloplegia.

The hyperopic macula may rarely be displaced farther from the disc, giving rise to an apparent divergent strabismus. Despite the fact that the hyperope must exercise accommodation more than the emmetrope or myope, there is no significant change in the onset of presbyopia if the refractive error is adequately corrected and the effectivity of the spectacle lens taken into account. Nor does lenticular activity predispose the hyperopic eye to cataract formation.

Myopia

With progressive elongation of the globe, a number of alterations occur, confined mostly to the posterior segment and exaggerated in older patients. Whether these alterations are due to a biomechanical stretch or to a failure of the inner tunics to grow apace of the sclerocorneal envelope is unknown. The first seems the more logical or at least, the simpler alternative. At one time these changes were attributed to inflammation, a kind of chronic uveitis, but since Salzmann (1902), most clinicians accept them as atrophic manifestations.

In the highly myopic eye (generally over 7 D), the sclera may be thinned and the choroid atrophic, with the entire posterior fundus giving the appearance of diffuse atrophy, which is less marked toward the periphery. Focal areas of atrophy, consisting of areas of depigmentation or pigmentation, may be scattered throughout and are characteristic of advanced degeneration. They are attributed to vascular occlusion and secondary hemorrhages. In advanced stages, splits or clefts form in the lamina vitrea (lacquer cracks), which appear clinically as whitish to yellowish irregular stripes resembling striae gravidarum. They are mostly confined to the paramacular area and seldom give rise to visual symptoms.

Tears in Bruch's membrane with secondary hemorrhage at the macula are also rare. The patient may report a sudden decrease of vision with metamorphopsia. The eventual result is a darkly pigmented, sharply circumscribed area smaller than the disc, containing areas of gray, red, or white (Fuchs spot). Fuchs described fifty such cases in 1901 and followed them over a period of years, noting that the spots tend to become larger and gray or white, with a surrounding zone of atrophy. The spots never completely disappear, and vision does not return to normal. The average degree of myopia in Fuchs' cases was about 12 D, but there is no invariable relation between axial length and Fuchs spots or lacquer cracks.

As the result of posterior elongation, the sclerochoroidal canal becomes more oblique from the nasal to the temporal side, and a conus becomes visible ophthalmoscopically (Fig. 8-1). The conus consists of a pale scleral crescent or a more reddish choroidal crescent. With increasing atrophy of the choroid, the scleral crescent grows larger at the expense of the choroidal crescent and later may surround the disc. A myopic conus is more prevalent on the temporal side (75%) and is directly related to the degree of axial elongation. It varies in size, and the retinal vessels pass undisturbed over the atrophic areas. Myopic crescents should not be confused with congenital crescents (usually inferior), which occur in nonmyopic eyes.

The vitreous may show liquefaction, detachment, or opacities, and although not invariably related to myopia, these are more common in myopic eyes and a source of con-

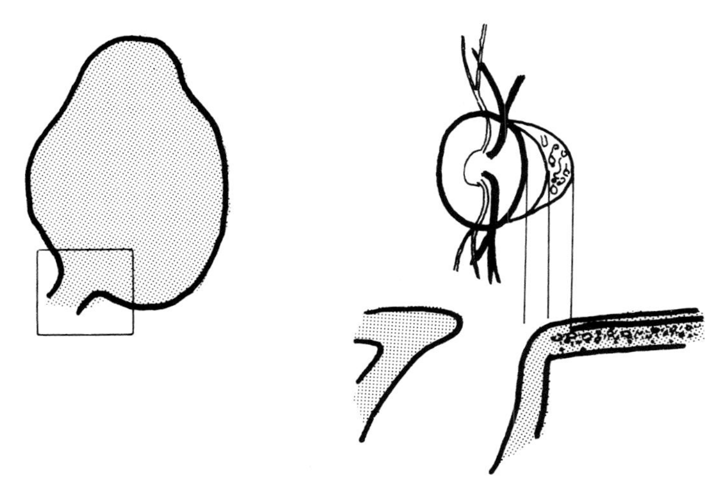

Fig. 8-1. Myopic conus.

siderable annoyance in reading, especially under bright light. Posterior vitreous detachment appears as a circular, sometimes doughnut-shaped opacity, representing the former attachment to the optic disc.

There is no better door to the posterior vitreous than the biomicroscope with a contact or preset Hruby lens. The preset lens has more reflections and less magnification, but the simplicity of its use, especially with the new model biomicroscopes (Haag-Streit, American Optical, Zeiss), permits its application to any well-dilated eye. No topical anesthetic agent or special preparation is required. The Hruby concave lens has a power of −58 D, is about 19 mm in diameter with a flat anterior surface, and forms an erect virtual image of the fundus at the level of the iris, which can be focused by the biomicroscope. In high myopia, a +60 D convex lens (Bayadi) may be used instead to form an inverted, real image, about 2 cm in front of the cornea.

The angle between the lamp and microscope is maintained within 17°, which permits binocular stereoscopic vision. Using 10× magnification, focus on the fundus and identify the landmarks. Proceeding anteriorly and using the optical section and sagittal movements of the coupled system, note the typical fibrillar or gossamer vitreous structure. In vitreous detachment, more common above than below, note an optically empty space between vitreous and retina. In localized posterior detachment, a circular opacity is seen at or near the disc, which actually represents a hole in the vitreous. The hole may be broken, giving an L shape. A vitreous detachment is present in many older patients, reaching 90% in those with retinal detachment.

The most serious complication afflicting the highly myopic eye is retinal detachment. The incidence is variously estimated at 5% to 8%. The mechanism is presumed to be peripheral retinal degeneration with lattice degeneration and hole formation. Retinal detachment is also more likely after cataract extraction and trauma. Unfortunately there are no effective methods of prevention.

CLINICAL ASPECTS OF AMETROPIA

The etiology of ametropia remains a puzzle, an historical succession of contradictions, with geneticists and environmentalists arage with a fractricidal mania to destroy each others' claims. The clinician, unable to unravel these learned contradictions, must satisfy the daily requirements of patients with refractive errors, and somehow translate what no one understands into a knowledgeable lens prescription.

Hyperopia

Biologically hyperopia is a point on a Gaussian curve, which changes during the growth years. Unlike the myopic eye, it is considered to be imperfectly developed. Anatomically it resembles the emmetropic eye.

A hyperopic "habitus," so striking to earlier writers, will strike the modern reader more droll than diagnostic. It consisted of a brachycephalic cranium, a flat nose, flat face, flat eyelids, small pupils and small mobile eyes, a large interpupillary distance and large angle kappa, an ellipsoidal form of the equator when viewed through the separated lids with the eye rotated toward the nose, a lack of orbital depth, and an asymmetry of the face. Not to be outdone, modern writers have added a psychologic habitus. Thus the farsighted boy is described as lazy, mischievous, dumb, inattentive, and "motor minded." He is bright but does not want to study, is more interested in girls or athletics, is out for a good time, and is nearly always a jolly good fellow. What jolly good psychology, where explanations are replaced by adjectives as if people could be typed by one criterion.

Since there are many more young hyperopes, one might have predicted a positive correlation with just about any major or minor learning, reading, or behavioral problem. Indeed such periodic reports appear regularly in the literature, accompanied, of course, by suitable statistics to translate what everyone knows into language that no one can understand.

Most younger children (about 80%) are hyperopic, the amount decreasing with elongation of the globe and flattening of the cornea and lens. The growth pattern resembles that of the central nervous system, rather than the skeletal system, and is genetically determined.

Not all children become less hyperopic; some become more so (about 9% as measured in one longitudinal study), especially if already markedly hyperopic. Most of them, however, become less hyperopic by about 0.07 D per year, whereas 10% become myopic at the rate of 0.25 D a year, and 2% become myopic at an even more rapid rate.

These statistics do not replace the actual examination of individual children but indicate a trend. Growth is not linear and therefore more changes can be anticipated in some years than in others. These accelerated or decelerated periods may have a fortuitous relation to school vacation, to puberty, to change in habits or environment, or to incidental optical or pharmacologic therapy. It is easy to read into such correlations evidence for both vice and virtue, the conquest of mind over matter, or the triumph of statistical inference over common sense.

Hyperopia may be axial, curvature, index, and positional or a combination of these. The axial character of the microphthalmic eye, the curvature of an aphakic eye, positional effects of a subluxated lens, or index changes in diabetes are fairly obvious. For the rest, the precise mechanism of hyperopia can seldom be identified by clinical refraction.

More useful is a classification by accommodation required to compensate for it: latent or manifest and facultative or absolute. The uncorrected hyperope, being obliged to use accommodation at all distances, may develop blurred vision, headaches, discomfort, fatigue, intermittent diplopia, and other symptoms classified as "eyestrain." Characteristically they follow close work, are exacerbated by the use of the eyes, and are relieved by discontinuing the visual task. Symptoms may be aggravated in children by a tendency to hold objects nearer because of short arms, to gain magnification, and to mobilize convergence. In adults, symptoms are sometimes induced by a change in occupation or avocation that requires more prolonged near seeing under less than optimum conditions. As the hyperopic patient approaches the presbyopic age, the increased accommodative demand makes itself evident sooner than in the emmetrope, especially under poor light, in which depth of focus fails to make up the deficit. Even the spectacle-corrected hyperope can be expected to manifest near symptoms earlier because the effectivity of the lens induces a greater accommodative stimulus.

The diagnosis of hyperopia is fairly simple; the decision whether it should be corrected is somewhat more difficult. If symptoms are associated with the use of the eyes, the minimum plus lens, which alleviates them and is compatible with clear distance vision, is indicated. The prescription is therefore based on trial and error. To Donders belongs the credit of appreciating that treatment consists not of gradually reducing plus but increasing it as more latent hyperopia becomes manifest. The physiologic basis of this reasoning is that prolonged use of the ciliary muscle leads to hypertonicity, which is only gradually relaxed by wearing correcting lenses. Additional plus is given if symptoms recur. Donders' frequently quoted rule to correct all the manifest plus one fourth of the latent hyperopia was never meant for general application. He suggested it only in such cases in which little hyperopia was found or symptoms were not relieved by the manifest correction or for a dilated patient who could only be seen once. The blanket application of the rule covers too many variables: age, amplitude, symptoms, visual requirements, depth of focus, and resolving power.

The definition of "latent hyperopia" varies with different authors, some considering it the amount of hyperopia masked by normal ciliary tonus, some, considering it the amount masked by excessive ciliary tonus, and some considering it synonymous with accommodative spasm. Much of the refraction literature of a few decades ago was concerned with the diagnosis latent hyperopia without the use of cycloplegia (comparing positive to negative relative accommodation, dynamic retinoscopy, base-in prism, cross cylinder, cyclodamia, etc.). The easiest way to measure it is by comparing cycloplegic to postcycloplegic findings (average difference about 0.75 D depending on age).

If hyperopia is unequal so that the patient cannot obtain simultaneous clear vision, a correction which equalizes vision should be given to prevent suppression and amblyopia.

To temper these therapeutic imperatives, only a small percentage of young hyperopes require correction. The older hyperope may first require a distance prescription for near, then eventually for constant use, with an additional reading add later. An attempt to differentiate between symptoms of hyperopia per se, associated anisometropia or aniseikonia, and astigmatism or heterophoria may require an empirical trial, just as in selecting the most effective antibiotic or the resection to straighten a squint. There is no statute that every lens correct every optical defect or alleviate all symptoms on every occasion; there is a code that suggests the possibility of failure. "Treat the patient, not the eye" is a rule worth recalling in hyperopia.

Myopia

The incidence of myopia depends on the age and population sample. About 25% to 35% of adults are myopic, 50% of these have myopia under 1 D, 10% to 15% exceed 1 D, but less than 3% reach the degenerative stage. Because myopia is more prevalent with increasing years of schooling, Germany declared it a public calamity at the turn of the century and initiated a new order of school hygiene. To the effects of extraocular muscle compression, intraocular tension, postural hyperemia, and dietary deficiency were added the dire consequences of reading by poor light and writing with pale ink in gloomy classrooms. We get rid of such pronouncements by admitting their half truth. Myopia develops during the growth phase not because children attend classes but because they are old enough to attend school. Better an educated child that wear glasses than an uneducated one—that wears glasses. There is little evidence that reading changes myopia one iota.

Myopia is more likely to occur in a child whose parents are myopic, whose hyperopia has diminished to zero by 6 years of age, who does rather well in school, who enjoys reading, and who seldom complains of discomfort or poor vision. The genetic pattern is more evident in high degrees of myopia, although there is no way of predicting which eyes will develop degenerative myopia, and it is doubtful that overall refraction as measured clinically is inherited as a unigenetic entity. Evidence from twins indicates less difference between uniovular than binovular twins.

The clinical feature of myopia is reduced acuity, generally without other symptoms. Since the condition is slowly progressive, the child seldom complains, and it is picked up by the school nurse, the pediatrician, or the father who notes the youngster cannot see which team is winning on the scoreboard. Because the myope's vision is clear at some distance, suppression and amblyopia is less likely in anisometropic myopes than anisometropic hyperopes. The association of myopia with better than average school performance has been both affirmed and denied. The relation to intelligence seems to depend on how much reading is part of the IQ test.

There is some evidence that myopic eyes are more susceptible to glaucoma, at least to that induced by topical steroids. In any case, reduced rigidity may mask increased pressure if taken by a Schiøtz tonometer, a myopic conus may obscure a pathologic cupping, and glaucomatous field changes may be erroneously attributed to chorioretinal degeneration.

The diagnosis of myopia involves little difficulty and can generally be suspected from the reduced distance vision. There is no linear relation between the drop in acuity and the degree of myopia because of other variables, which preclude predicting one from the other; these variables include contrast, illumination, type of target, depth of focus, blur interpretation, and ocular aberrations, not to mention vanity and whether the child has learned the trick of squinting the lids together. The differential diagnosis mainly involves ruling out pseudomyopia.

When Nero watched Rome burn, the chances are he was seeing a blurred picture through narrowed lids. Contemporary historians tell us his eyes were "dull and weak," and since he committed suicide at 31 years of age, it is not likely that the weakness was presbyopia; thus Nero is the most infamous of myopes. Like his games, the question of myopia prevention always attracts an audience. Unfortunately, of much that is useful, little has been written. Treatment depends on one's view of the etiology of myopia and the mechanism of its progression. It ranges from undercorrection, cycloplegia, and bifocals to reduce ciliary contraction; base-in prism to reduce convergence; parasympathomimetics to increase ciliary tone and prevent the choroid from stretching; systemic vitamins, hormones, and diet to enhance scleral resistance; avoiding stooping, bending, and reading to eliminate the compression effects of the extraocular muscles (occasionally supplemented by rectus or oblique tenotomy); flat-fitting contact lenses or radial surgical incisions to alter corneal curvature; and eye exercises, palming, staring, and hypnosis to improve acuity and blur interpretation. Not only does myopia beget more myopia, but every treatment generates its opposite with the inevitability of Newton's law. Most of the results can be dismissed with a shrug of the shoulders. Some surgical procedures to correct ametropia seem rather extravagant, more dangerous than elegant, and less innovant than litigant.

The treatment of myopia is easily solved by considering each case on its own merits. By neglecting further specification, it leaves the clinician free to nibble at every therapeutic notion without having to swallow its entire philosophy. Such common sense recommendations as good light, good contrast, good posture, good diet, and good general health are virtuous and go without saying. For the rest, it is more difficult to arrive at a consensus in an area where dogma and tradition compete with a myriad post hoc facts.

A reasonable therapeutic program is a correction adequate for the patient's visual needs. This permits considerable leeway on cutting the minus. There is no real evidence that deliberate undercorrection prevents myopic progression; overcorrected or undercorrected it will stop—and cannot be made to progress further!

Undercorrection may be indicated in early presbyopia or when the full correction induces an esophoria or intermittent esotropia.

Newly corrected myopes may note some difficulty with near vision because of the unaccustomed accommodation, which can lead to temporary spasm with esophoria—an occasional cause for refractive amblyopia. Maximum acuity may not be obtained until the prescription is worn for some days or weeks. Whether the glasses should be worn full time or only for distance depends on associated anisometropia, astigmatism, or heterophoria. It is not reasonable to deprive patients of their astigmatic correction or binocular vision or make them adapt to a new accommodative-convergence relationship just to satisfy a pet philosophy.

In the absence of secondary degenerative changes, the visual prognosis in myopia is excellent. Patients receiving minus lenses will usually be concerned with their appearance, the annoying reflections, the possibility of breakage, and the minification (which quickly disappears). All these features can be alleviated by proper choice of lens, frame, and fit.

Although myopia is not a disease, wrote Tscherning, it always causes a disagreeable feeling, which it is our duty to prevent as much as possible. He did not elaborate whether he meant the patient or the physician. The primary concern is to avoid the secondary degenerative changes and complications of high myopia. Fortunately such cases are rare (1% to 3%). The diagnosis of pathologic (or degenerative) myopia is based not on the dioptric error but on fundus changes. The term "malignant myopia" is not only inaccurate but inappropriate.

The pathologic physiology of high myopia depends on degenerative changes. The acuity may not be correctable because of scarring, hemorrhage, atrophy, photorecepter disorientation, glaucoma, or amblyopia. The light threshold may be elevated, the dark adaptation curve may be anomalous, visual fields may be altered, the vitreous may be detached, and the peripheral retina may show lattice degeneration and holes.

The treatment of high myopia should in theory not differ from other forms. If certain methods are believed to be helpful with these cases, there is no reason not to apply them to all. It is probably the better part of valor to avoid body contact sports, which might result in trauma to the eye. Suitable visual aids may be tried when conventional spectacles are ineffective.

Myopia of prematurity

A rare form of myopia (incidence less than 1% of myopes) occurs at birth in premature babies, especially those subject to oxygen therapy. Whether the mechanism is an incomplete retrolental fibroplasia or a mesenchymal defect is not clear. A significant proportion show other mesodermal defects, and about half have strabismus. The myopia is generally fairly high (8 to 12 D) and seldom progresses. It is often associated with a myopic conus, but the vision and prognosis are good. There appears to be no hereditary factor, and full correction to improve vision is indicated.

Astigmatism

It is rare to find an eye completely free from astigmatism; its incidence varies with the clinician's enthusiasm for refraction. Of all lens prescriptions, 80% or more contain a cylinder up to 1.5 D, and probably there are many astigmatic errors for which no correction is ordered. Only 2% to 3% exceed 3 D. Astigmatism is thus the most common of all refractive errors.

The primary source of astigmatism is the anterior surface of the cornea, as evidenced by the correlation between the total astigmatism measured by retinoscopy or subjective techniques. The correlation is not perfect, however, because the crystalline lens makes a small contribution, which varies with age (generally against the rule, reaching 1.00 to 1.25 D in senility). The contribution of other optical components is negligible, the posterior corneal surface, for example, tending to parallel the anterior. In measuring corneal astigmatism by keratometry, recall that the spectacle lens will differ from it by the effectivity of the vertex distance. Anatomically astigmatism is chiefly due to toroidicity of the cornea, with tilting and eccentricity about a common optic axis as minor factors. Ocular astigmatism, at least for the optical zone of the cornea, is almost invariably regular; that is, the primary meridians are perpendicular to each other. Irregular astigmatism may consist of either primary meridians that are not at right angles or variations in curvature in the same meridian. The diagnosis is readily made by keratometer, slit lamp, placido disc, or simply by observing the variations in corneal reflection of a light source. The most common causes of irregular astigmatism are corneal scarring and corneal disease (e.g., keratoconus).

The optical classification into simple or compound myopic, hyperopic, or mixed has already been considered. Clinically, compound hyperopic astigmatism is usually masked by an active accommodation that places the circle of least confusion on the retina. In such cases, one can anticipate symptoms without any accompanying decrease in acuity. In compound myopic astigmatism the acuity drops because accommodation is ineffective; in hyperopic astigmatism, acuity drops only if accommodation is unavailable.

The extent to which astigmatism reduces visual acuity depends not only on the amount of astigmatism but also on the size of the pupil, the flexibility of accommodation, the position of the principal meridians, and the type of target used to measure resolution. The deleterious effects of uncorrected oblique astigmatism may only become evident in attempts to read small print and scales closely set together under conditions of reduced contrast and minimal depth of focus. Because the vertical and horizontal strokes predominate with alphabetic letters, astigmatic errors with primary meridians in the vertical or horizontal positions cause substantially smaller reduction in legibility.

The pathogenesis of discomfort in astigmatism is not known, although a variety of factors have been implicated. Most commonly mentioned is the need to shift the circle of confusion on to the retina. Actually the ciliary strain would be no different than in hyperopia and absent in compound myopic astigmatism. Indeed, when the accommodative response (measured objectively) to small spherocylinders is compared to their spherical equivalents, a statistically significant lower response is obtained with spherocylinders.

Another factor frequently mentioned is squinting the lids to produce a stenopeic slit, leading to strain of the orbicularis muscle. Since this does not happen in myopia, it probably does not occur in astigmatism. Most important but least understood are the means whereby patients compensate for even high astigmatic errors and why some patients adapt and others fail to adapt to the proper cylindrical correction.

More often than not, astigmatism is symmetrical in the two eyes, one being frequently the mirror image of the other. It is interesting that primary meridians at oblique axes can often be traced in other members of the same family, and the similarities speak for more than coincidence. Large changes in astigmatism do not occur with age. There may be slight changes (0.25 to 0.75 D) during the school years, approaching 1.00 to 1.25 D in older age, generally toward against the rule. These changes are attributed to the crystalline lens, although the corneal toricity may also change in some cases. Unpredictable changes may occur with the development of cataract. The postoperative astigmatism of cataract extraction tends to be against the rule, following the scarring of the incision. Indeed in the surgical correction of astigmatism, an incision is made tangent to the cornea in "direct" astigmatism and radially in "inverse" astigmatism (Sato, 1953). The surgical treatment of astigmatism is, of course, in the experimental stage—and likely to remain so. Astigmatism can also result from a chalazion, pterygium, strabismus, or retinal detachment surgery.

Astigmatic errors are more often associated with hyperopia than with myopia, which provides some evidence of the ballooning effect of ocular growth.

There is no convincing evidence that sectional ciliary contraction can induce (or compensate for) astigmatism. Such a mechanism would lead to predictable changes in astigmatism at the near point. Although changes are found, they are due to the effectivity of the spectacle cylinder and vertex distance.

The diagnosis of astigmatism is made by objective and subjective techniques. Chasing the last 0.12 D of cylindrical error only leads to quiet desperation for the clinician and troubled anguish of indecision for the patient. The astigmatic error that the patient ruminates, vacillates, and speculates about will probably be absent or altered on another test and another day.

Volumes can and have been written about the treatment of astigmatism. When should it be corrected, how should it be corrected, should it be ordered for constant or intermittent wear, should all or part of it be corrected, and should the cylinder be tilted toward vertical, horizontal, or the previously worn axis? Opinions are plentiful, but facts are few because there is so much individual variation. What works well for one patient is a disaster for another.

In general, it seems reasonable to correct astigmatic errors especially in children, which interfere with acuity or the successful binocular cooperation of the eyes. The older patient who has never worn glasses is approached more judiciously. If the axis is vertical or horizontal, the cylinder can usually be ordered, providing the patient notes a definite improvement in vision. If the axis if oblique or if the previously worn cylinder axis differs from what is found, the problem becomes more complicated.

Oblique astigmatism not previously corrected in older patients means that they have compensated for the distortion in the same way we compensate for the color of sunglasses. The mechanism of compensation is presumed to be perceptual, and a new correction requires a new compensation. Some patients manage well; others do not. A good way to evaluate the problem is to place the correction in a trial frame and have the patient walk (not sit) about the office. Movement exaggerates untoward effects. If the distortion is considerable, an alternative is the equivalent sphere or reducing the cylinder. (Do not omit a proportionate change in the sphere if the cylinder is cut.) The reduced cylinder is then again placed in the trial frame, and the procedure is repeated.

If the patient is already wearing a cylinder but its axis differs from what is found on the current refraction, repeat the examination and recheck all the findings. Note particularly whether the new cylinder provides significantly better acuity than the old. If there is no real change in acuity and the patient is otherwise asymptomatic, there is little justification for change. If there is a noticeable improvement in acuity, the question arises whether the new cylinder should be ordered at the correct axis, the old axis, or somewhere in between. If the primary meridians are near vertical or horizontal, the change will cause less difficulty than when the new axis is near 45°-135°. The reason will be at once apparent if a strong cylinder is held up at an oblique axis and one compares not the acuity but one's spatial orientation. The effect is entirely monocular, although it is compounded by stereoscopic distortions with binocular vision. The best way to check is to place the new prescription in a trial frame and note the distortion for various axis orientations monocularly and binocularly. The method is empirical, but it will save considerable grief later.

Low refractive errors

Perceptual proficiency does not always keep company with optical efficiency; one patient with low ametropia complains bitterly, and another wears the wrong prescription with serenity. How and when such symptoms should be treated requires insight and empathy and a modest view of one's professional infallibility.

Ocular discomfort also arises from such nonoptical causes as blepharitis, chronic conjunctivitis, a dry cornea, a drugged ciliary body, a dropsical aqueous, or a drudging personality. Blurred vision may be the price of a chalazion, the depreciation of ocular transmission, or the bankruptcy of the receiving or transmitting mechanisms. Inefficient binocular vision may be due to suppression, amblyopia, aniseikonia, or heterophoria. Perceptual disturbances may arise from the fear of failing vision or from a search for security, sympathy, and workmen's compensation. Any or all may be associated with a low refractive error, and the clinician must choose between the science of prescribing with dignity and the art of doing nothing gracefully.

Every treatment has its accompanying psychologic effect, depending on the personality of the prescriber, the expectations of the patient, and the rapport between them. The placebo effect is evident in the pharmacologically impotent sugar pill, the impoverished absorption of a minimally tinted lens, or the imperative command of a Germanic bedside manner. To what extent an optical placebo should be used for its beneficial effect in psychosomatic visual disorders versus its detrimental effect on the patient's pocketbook are ethical judgments that the clinician must make on an individual basis. In this connection there is no harm, it may do good, and at any rate it costs little to prescribe a rest, a change of scene, a diet, or a diversion.

OCULAR HYGIENE AND PREVENTIVE CARE

The eye is the dominant organ of perception, and all the terror, anguish, and finality of blindness comes through in Milton's line: "O dark, dark, dark, amid the blaze of noon." Fundamental to any program of preventive care are public education with respect to the proper use of the eyes, vision screening for early diagnosis, periodic examination for prompt treatment, the formulation of programs to facilitate visual tasks and prevent ocular injuries, and rehabilitation of the visually handicapped.

SOME MEDICOLEGAL ASPECTS

When a patient alleges damages or injury as the result of examination or treatment, the clinician is being accused of negligence. Unexpected and untoward results are unavoidable, but the patient who is convinced that everything necessary was done and nothing useful was omitted is not likely to sue. The best way to avoid medicolegal difficulties is to establish good rapport, follow some simple rules of conduct, and take time to listen to the patient's complaints.

A permanent, legible, accessible record is kept for each patient, showing dates of examinations, diagnosis, treatments, and follow-up results. A complete copy of the lens prescription should be kept on file, showing not only the prescription but also the type of lens, style of bifocal, decentration, tint, base curve, etc., for later identification or duplication. Although impact-resistant lenses are now mandatory, they are not immune to breakage. If for some reason a nonimpactresistant lens is ordered, the patient's written consent must be obtained. Frames should be nonflammable and sturdy, especially for children. Prescriptions that are anticipated to produce spatial distortions should first be worn only at home under proper supervision.

In patching the eye of an amblyopic child (or following the removal of a foreign body), make sure the acuity in the remaining eye is adequate for safe daily activities. Common sense and sympathy must be balanced when vision cannot be fully corrected in older patients with respect to the hazards of driving, occupation, or recreation.

It is generally assumed that physicians possess a reasonable degree of learning and skill, although they are not responsible for failure to cure. In interpreting reasonable care and skill, the courts have applied the rule that it must be of such quality and diligence as are ordinarily exercised by thoroughly educated surgeons in such localities. In the doctrine of res ipsa loquitur (the thing speaks for itself), the burden of proof is placed on the physician to explain the inference of negligence. If there is any doubt about an action, it is best to obtain legal advise.

Finally, the prescription for good practice by Guy de Chauliac a few centuries ago is still worth taking:

> Let the surgeon be bold in all sure things, and fearful in dangerous things; let him avoid all faulty treatments and practices. Let him be gracious to the sick, considerate to his associates, and cautious in his prognostications. Let him be modest, dignified, gentle, dutiful, and merciful. Let his reward be according to his work, to the means of the patient, to the quality of the issue, and to his own dignity.

APPENDIX

Sources of ophthalmic instruments

Albert Aloe Co., 805 Locust St., St. Louis, Mo. 63101
American Optical Co., Southbridge, Mass. 01550
Austin Belgard, 109 N. Wabash Ave., Chicago, Ill. 60602
Baush & Lomb Optical Co., Rochester, N. Y.
Benson Optical Co., Minneapolis, Minn.
Bernell Corp., 316 South Eddy St., South Bend, Ind. 46617
Clement Clarke, Ltd., 16 Wigmore St., London, England
Curry Paxton, Ltd., 101 Park Ave., New York, N. Y. 10017
Hamblin, Ltd., 15 Wigmore St., London, England
House of Vision, Inc., 30 N. Michigan Ave., Chicago, Ill. 60602
Keeler Optical Products, Inc., Philadelphia, Pa. 19143
Keystone View Co., Meadville, Pa. 16335
Kollmorgen Optical Co., 2 Franklin Ave., Brooklyn, N. Y. 11211
MacBeth Daylighting Corp., P.O. Box 950, Newburgh, N. Y. 12550
Matalene Surgical Instruments, 125 E. 46th St., New York, N. Y. 10017
V. Mueller & Co., 330 S. Honore St., Chicago, Ill. 60612
National Electric Instrument Co., 9221 Corona Ave., Elmhurst, Long Island, N. Y.
National Society for the Prevention of Blindness, Inc., 16 E. 40th St., New York, N. Y. 10016
New Era Optical Co., 17 North Wabash, Chicago, Ill. 60602
New York Association for the Blind, Low Vision Services, 111 E. 59th St., New York, N. Y. 10022
Parsons, A. H. Laboratories, 520 Powell St., San Francisco, Calif. 94108
Poll, A. H. Co., 44 W. 55th St., New York, N. Y. 10019
Ritter Co., Ritter Park, Rochester, N. Y.
Rodenstock Co., 29/43 Isartal St., Munich, Germany
Shuron Optical Co., Geneva, N. Y. 14456
Sparta Instrument Corp., 305 Fairfield Ave., Fairfield, N. J. 07006
Storz, 4570 Audubon Ave., St. Louis, Mo. 63110
Univis Lens Co., Dayton, Ohio
Welch Allyn Inc., Auburn, N. Y. 13021
Zeiss (Opton & Werkel), 435 Fifth Ave., New York, N. Y. 10016

REFERENCES

Alfano, J. E.: Myopia caused by prematurity, Am. J. Ophthal. **46**:45-49, 1958.

Aquavella, J. V.: Thermokeratoplasty, Ophthal. Surg. **5**:39-47, 1974.

Bard, L. A.: Transient myopia associated with promethazine therapy, Am. J. Ophthal. **58**:682-686, 1964.

Barraquer, J. L.: A new technique for surgical correction of myopia (keratomileusis), Ann. Inst. Barraquer **5**:206-229, 1964.

Baum, W. W.: Muscle imbalance in myopia, Am. J. Ophthal. **24**:291-295, 1941.

Beasley, F. J.: Transient myopia and retinal edema during hydrochlorthiazide therapy, Arch. Ophthal. **65**:212-213, 1961.

Belmont, O.: Management of myopia, Int. Ophthal. Clin. **5**:395-402, 1965.

Birge, H. L.: Myopia caused by prematurity, Am. J. Ophthal. **41**:292-298, 1956.

Brown, E. V. L.: Apparent increase of hyperopia up to age of nine years, Am. J. Ophthal. **19**:1106, 1936.

Brown, E. V. L.: Net average yearly changes in refraction of atropinized eyes from birth to beyond middle life, Arch. Ophthal. **19**:719-734, 1938.
Brown, E. V. L.: Use-abuse theory of changes in refraction versus biologic theory, Arch. Ophthal. **28**: 845-850, 1942.
Brown, E. V. L.: Comparison of refraction of strabismic eyes with that of nonstrabismic eyes from birth to the twenty-fifth year, Arch. Ophthal. **44**:357-361, 1950.
Brown, E. V. L., and Kronfeld, P. C.: The refraction curve in the U. S. with special reference to the first two decades. In Proceedings of the Thirteenth International Congress of Ophthalmology, vol. 13, 1929, pp. 87-98.
Bucklers, M.: Changes in refraction during life, Br. J. Ophthal. **37**:587-592, 1953.
Cambiaggi, A.: Myopia and retinal detachment: statistical study of some of their relationships, Am. J. Ophthal. **58**:642-650, 1964.
Cohn, H.: The hygiene of the eye in schools, translated by W. P. Turnbull, London, 1886, Simpkin, Marshall & Co.
Coleman, D. J., and Carlin, B.: A new system for visual axis measurements in the human eye using ultrasound, Arch. Ophthal. **77**:124-127, 1967.
Cook, R. C., and Glasscock, R. E.: Refractive and ocular findings in the newborn, Am. J. Ophthal. **34**:1407-1413, 1951.
Coulombre, A. J.: Experimental embryology of the vertebrate eye, Invest. Ophthal. **4**:411-419, 1965.
Cowan, A.: Myopia, Am. J. Ophthal. **25**:844-853, 1942.
Curtin, B. J.: The pathogenesis of congenital myopia; a study of 66 cases, Arch. Ophthal. **69**:166-173, 1963.
Curtin, B. J., and Karlin, D. B.: Axial length measurements and fundus changes of the myopic eye, Am. J. Ophthal. **71**:42-53, 1971.
Duke-Elder, W. S.: Changes in refraction in diabetes mellitus, Br. J. Ophthal. **9**:167-187, 1925.
Duke-Elder, W. S.: An investigation of the effect upon the eyes of occupations involving close work, Br. J. Ophthal. **14**:609-620, 1930.
Dunstan, W. R.: Variation and refraction, Br. J. Ophthal. **18**:404-421, 1934.
Durham, D. G.: Office efficiency, Int. Ophthal. Clin. **3**:349-358, 1963.
Ewalt, H. W.: The Baltimore myopia control project, J. Am. Optom. Assoc. **17**:167-185, 1945.
Exford, J.: A longitudinal study of refractive trends after age forty, Am. J. Optom. **41**:685-692, 1964.
Feldman, J. B.: Myopia, vitamin A, and calcium, Am. J. Ophthal. **33**:777-784, 1950.
Filatov, V. P., and Volokitensho, A. E.: Osmotherapy in high myopia without macular changes, Vestn. Oftalmol. **17**:515-519, 1940.
Fletcher, M. C., and Brandon, S.: Myopia of prematurity, Am. J. Ophthal. **40**:474-481, 1955.
Fox, D. H., Smith, P. M., and Hirsch, M. J.: Patient preference for cylinder location in spectacle lenses, Am. J. Optom. **48**:412-415, 1971.
Franceschetti, A., and Luyckx, J.: Study of emmetropization effect of the crystalline lens by ultrasonic echography, Am. J. Ophthal. **61**:1096-1100, 1966.
Francois, J. L.: Heredity in ophthalmology, St. Louis, 1961, The C. V. Mosby Co.
Frank, E.: Bodily build and refraction, Klin. Monatsbl. Augenheilkd. **101**:184-197, 1938.
Friedman, B. B.: Acute myopia induced by sulfonamides, Am. J. Ophthal. **24**:935-942, 1941.
Fuchs, E.: Der centrale schwarze Fleck bei Myopie, Z. Augenheilkd. **5**:171, 1901.
Gardiner, P. A.: The relation of myopia to growth, Lancet **266**:476-479, 1954.
Gardiner, P. A.: Dietary treatment of myopia in children, Lancet **281**:1152-1155, 1958.
Gernet, H.: Compensatory behavior of corneal refraction and length of the globe in buphthalmos, Klin. Monatsbl. Augenheilkd. **144**:429-431, 1964.
Giles, G. H.: The distribution of visual defects, Br. J. Physiol. Opt. **7**:3-32, 1950.
Godard, P.: Influence de la myopia sur la formation de la personalité, Clin. Ophthal. **31**:512-519, 1927.
Goldmann, H., and Hagen, R.: Zur direkt Messung der total Brechkraft des lekenden menschlichen Auges, Ophthalmologica **104**:15-22, 1942.
Goldstein, J. H., Vukevich, W. M., Kaplan, D., Paolino, J., and Diamond, H. S.: Myopia and dental caries, J.A.M.A. **218**:1572-1573, 1971.
Good, J. M.: Office ophthalmology equipment, Int. Ophthal. Clin. **3**:359-371, 1963.
Gould, G. M.: Biographic clinics, Philadelphia, 1909, Blakiston.
Graether, J. M.: Retinal changes in degenerative myopia, Int. Ophthal. Clin. **2**:109-132, 1962.
Grossman, E. E., and Hanley, W.: Transient myopia during treatment for hypertension with autonomic blocking agents, Arch. Ophthal. **63**:853-855, 1960.
Grosvenor, T.: Refractive error distribution in New Zealand's Polynesian and European children, Am. J. Optom. **47**:673-679, 1970.
Haines, H.: An evaluation of the visual status and academic achievement of a selected group of elementary school children over a period of seven years, Am. J. Optom. **32**:279-288, 1955.
Halpern, A., and Kulvin, M.: Transient myopia during treatment with carbonic anhydrase inhibitors, Am. J. Ophthal. **48**:534-535, 1959.
Hartridge, H.: The causes of errors of refraction, Br. J. Physiol. Opt. **7**:143-149, 1950.
Hayden, R.: Development and prevention of myopia at the United States Naval Academy, Arch. Ophthal. **25**:539-547, 1941.
Henderson, T.: The constitutional factor in myopia, Trans. Ophthal. Soc. U. K. **54**:451-459, 1934.
Herm, R. J.: Refraction of children, Int. Ophthal. Clin. **5**:413-422, 1965.
Hirsch, M. J.: An analysis of inhomogeneity of myopia in adults, Am. J. Optom. **27**:562-571, 1950.
Hirsch, M. J.: The changes in refraction between the

ages of 5 and 14—theoretical and practical considerations, Am. J. Optom. **29**:445-459, 1952.
Hirsch, M. J.: Sex differences in the instances of various grades of myopia, Am. J. Optom. **30**:135-138, 1953.
Hirsch, M. J.: The relationship between measles and myopia, Am. J. Optom. **34**:289-297, 1957.
Hirsch, M. J.: Changes in refractive state after the age of forty-five, Am. J. Optom. **35**:229-237, 1958.
Hirsch, M. J.: The longitudinal study in refraction, Am. J. Optom. **41**:137-141, 1964.
Hirsch, M. J.: Predictability of refraction at age 14 on the basis of testing at age 6—interim report from the Ojai longitudinal study of refraction, Am. J. Optom. **41**:567-574, 1964.
Hirsch, M. J.: The prevention and/or cure of myopia, Am. J. Optom. **42**:327-336, 1965.
Hirsch, M. J., editor: Refractive state of the eye, a symposium, American Academy of Optometry series, Minneapolis, 1967, Burgess Publishing Co.
Hirsch, M. J.: Summary of current research on refractive anomalies. In Refractive anomalies of the eye, Public Health Service publication No. 1687, National Institute of Neurologic Diseases and Blindness, monograph No. 5, U. S. Dept. Health, Education and Welfare, 1967.
Hirsch, M. J., and Levin, J. M.: Myopia and dental caries, Am. J. Optom. **50**:484-488, 1973.
Hirsch, M. J., and Weymouth, F. W.: Notes on ametropia—a further analysis of Stenstrom's data, Am. J. Optom. **24**:601-608, 1947.
Hofstetter, H. W.: Some interrelationships of age, refraction, and rate of refractive change, Am. J. Optom. **31**:161-169, 1954.
Hofstetter, H. W.: Emmetropization—biological process or mathematical artifact? Am. J. Optom. **46**:447-450, 1969.
Hynes, E. A.: Refractive changes in normal young men, Arch. Ophthal. **60**:761-767, 1956.
Jackson, E.: Changes in refraction with age, Am. J. Ophthal. **3**:228-230, 1920.
Jackson, E.: Control of myopia, Am. J. Ophthal. **14**:719-725, 1931.
Jackson, E.: Norms of refraction, J.A.M.A. **98**:132-133, 1932.
Jaeger, E.: Ueber die Einstellungen des dioptrischen Apparates im menschlichen Auge, Vienna, 1861, Seidel u. Sohn.
Jampolsky, A., and Flom, B.: Transient myopia associated with anterior displacement of the crystalline lens, Am. J. Ophthal. **36**:81-85, 1953.
Jansson, F.: Determination of the axis length of the eye roentgenologically and by ultrasound, Acta Ophthal. **41**:236-246, 1963.
Jansson, F., and Kock, E.: Determination of the velocity of ultrasound in the human lens and vitreous, Acta Ophthal. **40**:420-428, 1962.
Jones, C. P.: A study of one hundred refraction cases in Indians fresh from the plains, J.A.M.A. **51**:308-402, 1908.
Keller, J. T.: A comparison of the refractive status of myopic children and their parents, Am. J. Optom. **49**:206-211, 1972.
Kettesy, A.: The stabilization of the refraction and its role in the formation of ametropia, Br. J. Ophthal. **33**:39-47, 1949.
Kimura, T.: Analysis of errors in ultrasound biometry. In Gitter, K. A., Keeney, A. H., Sarin, L. K., and Meyer, D., editors: Ophthalmic ultrasound, St. Louis, 1969, The C. V. Mosby Co.
Knapp, A. A.: Vitamin D complex in progressive myopia, Am. J. Ophthal. **22**:1329-1336, 1939.
Kronfeld, P. C., and Devney, C.: The frequency of astigmatism, Arch. Ophthal. **4**:873-884, 1930.
Kronfeld, P. C., and Devney, C.: Ein Beitrag zur Kenntnis der Refractionskurve, Arch. Ophthal. **126**:487-501, 1931.
Laval, J.: Vitamin D and myopia, Arch. Ophthal. **19**:47-54, 1938.
Law, F. W.: Calcium and parathyroid therapy in progressive myopia, Trans. Ophthal. Soc. U. K. **54**:281-290, 1934.
Lebensohn, J. E.: Visual charts, Int. Ophthal. Clin. **5**:347-367, 1965.
Loring, E. G. J.: Is the human eye changing its form and becoming near sighted under the influence of modern education? N. Y. Med. Rec. **12**:732-734, 1877.
Lyle, W. M., Grosvenor, T., and Dean, K. C.: Corneal astigmatism in American Indian children, Am. J. Optom. **49**:517-524, 1972.
Mackenzie, W.: Practical treatise on diseases of the eye, Boston, 1833, Carter, Hender & Co.
Mark, H. H.: Emmetropization; physical aspects of a statistical phenomenom, Ann. Ophthal. **4**:393-401, 1972.
Merin, S., Rowe, H., Auerback, E., and Landau, J.: Syndrome of congenital high myopia with nyctalopia, Am. J. Ophthal. **70**:541-547, 1970.
Miles, P. W.: Children with myopia treated with bifocal lenses, Mo. Med. **54**:1152-1155, 1957.
Miller, H.: Is myopia a deficiency disease? Am. J. Ophthal. **23**:296-305, 1940.
Miller, W. M., and Borley, W. E.: Surgical treatment of degenerative myopia; scleral reinforcement, Am. J. Ophthal. **57**:796-804, 1964.
Morgan, M. W.: The nature of ametropia, Am. J. Optom. Monograph No. 27, 1947.
Morgan, M. W.: Changes in refraction over a period of twenty years in a non-visually selected sample, Am. J. Optom. **35**:281-299, 1958.
Morgan, M. W.: Relationship of refractive error to bookishness and androgyny, Am. J. Optom. **37**:171-185, 1960.
Morgan, O. G.: Some cases of traumatic myopia, Br. J. Ophthal. **24**:403-406, 1940.
Mundt, G. H., and Hughes, W. F.: Ultrasonics in ocular diagnosis, Am. J. Ophthal. **41**:488-498, 1956.
Nadell, M. C., and Hirsch, M. J.: The relationship of the birthplace of parents and grandparents to the

refractive state of the child, Am. J. Optom. **32**:137-141, 1955.
Nugent, O. B.: The use of base-in prisms in the treatment of school myopia, W. Va. Med. J. **45**:103-106, 1949.
Ochi, S.: Relation of ocular muscles and sclera in the etiology of myopia, Am. J. Ophthal. **3**:675-678, 1919.
Olurin, W.: Refractive errors in Nigerians: a hospital clinical study, Ann. Ophthal. **5**:971-976, 1973.
Pischel, D. K.: Detachment of the vitreous as seen by slitlamp examination, Am. J. Ophthal. **36**:1497-1507, 1953.
Pistocchi, P., and Lamberti, O.: Further statistical investigation on the correlation of refraction, ocular motility, and amblyopia, Arch. Ottal. **66**:253-258, 1962.
Podos, S. M., Becker, B., and Morton, W. R.: High myopia and primary open angle glaucoma, Am. J. Ophthal. **62**:1039-1043, 1966.
Rasmussen, O. D.: Incidence of myopia in China, Br. J. Ophthal. **20**:359-360, 1936.
Rasmussen, O. D.: Myopia in England and Scotland, Opt. J. Rev. Optom. **85**:42, 1948.
Reinecke, R. D., and Herm, R. J.: Refraction, a programmed text, New York, 1965, Appleton-Century-Crofts.
Rengstorff, R. H.: Myopia induced by ocular instillation of physostigmine, Am. J. Optom. **47**:221-227, 1970.
Rice, T. B.: Physical defects in character. I. Farsightedness, Hygeia **8**:536-538, 1930.
Rice, T. B.: Physical defects in character. II. Nearsightedness, Hygeia **8**:644-646, 1930.
Rotter, H.: Technique of biomicroscopy of the posterior pole, Am. J. Ophthal. **42**:409-415, 1956.
Salzmann, M.: Die Atrophie der Aderhaut in Burzsichtigen Auge, Arch. Ophthal. **54**:337-350, 1902.
Sato, T.: The causes and prevention of acquired myopia, Tokyo, 1957, Kaneharu Shuppan Co., Ltd.
Sato, T., Akiyama, K., and Shibata, H.: A new surgical approach to myopia, Am. J. Ophthal. **36**:823-829, 1953.
Schepens, C. L.: Retinal detachment and myopia, Am. J. Ophthal. **58**:695, 1964.
Schmerl, E.: Higher refractional anomalies; their frequency in different groups and counties, Am. J. Ophthal. **32**:551-565, 1949.
Slataper, F. J.: Age norms of refraction and vision, Arch. Ophthal. **43**:466-481, 1950.
Sorsby, A.: School myopia, Br. J. Ophthal. **16**:217-224, 1932.
Sorsby, A.: The pre-myopic state; its bearing on the incidence of myopia, Trans. Ophthal. Soc. U. K. **54**:459-467, 1934.
Sorsby, A., Benjamin, B., Davey, J. B., Sheridan, M., and Tanner, J. M.: Emmetropia and its aberrations, a study in the correlation of the optical components of the eye, Med. Res. Council Report No. 293, London, 1957, H. M. Stationary Office.
Sorsby, A., Benjamin, B., and Sheridan, M. L.: Refraction and its components during the growth of the eye from the age of three, Med. Res. Council Report No. 301, London, 1961, H. M. Stationary Office.
Sorsby, A., and Fraser, G. R.: Statistical note on the components of ocular refraction in twins, J. Med. Genet. **1**:47-49, 1964.
Sorsby, A., and Leary, G. A.: A longitudinal study of refraction and its components during growth, London, 1970, H. M. Stationery Office.
Spivey, B. E.: A review of ocular abnormalities associated with the x chromosome, Survey Ophthal. **10**:223-231, 1965.
Stansbury, F. C.: Pathogenesis of myopia, Arch. Ophthal. **39**:273-299, 1948.
Steiger, A.: Beitrag zur Physiologie und Pathologie der Hornhaut-refraktion, Wiesbaden, 1895, Bergmann.
Steiger, A.: Die Entstehung der spharischen Refraktionen des menschlichen Auges, Berlin, 1913, S. Karger.
Stenstrom, S.: Investigation of the variation and the correlation of the optical elements of human eyes, translated by D. Woolf, Am. J. Optom. Monograph No. 58, 1948.
Stenstrom, S.: Ueber den Astigmatismus des brechenden Systems des Auges und den seiner Komponenten, Acta Ophthal. **27**:455-474, 1949.
Stevens, S. S.: The psychophysics of sensory functions, Am. Sci. **48**:226-252, 1960.
Stevens, S. S.: The surprising simplicity of sensory metrics, Am. Psychol. **17**:29-39, 1962.
Stilling, J.: Zur Anatomie des myopischen Auges, Z. Augenheilkd. **14**:21-23, 1905.
Stocker, F. W.: Pathologic anatomy of myopic eye with regard to newer theories of etiology and pathogenesis of myopia, Arch. Ophthal. **56**:476-493, 1943.
Straub, M.: Ueber die Aetiologie der Brechungsanomalien des Auges und den Ursprung der Emmetropie, Graefe's Arch. Ophthal. **70**:130-199, 1909.
Stromberg, E.: Ueber Refraktion und Achsenlange des menschlichen Auges, Acta Ophthal. **14**:281, 1936.
Sugar, H. S.: Late refractive changes following various operations for angle-closure glaucoma, Am. J. Ophthal. **53**:43-45, 1962.
Tassman, I. S.: Frequency of various kinds of refractive errors, Am. J. Ophthal. **15**:1044-1053, 1932.
Thurstone, L. L.: The measurement of values, Chicago, 1959, The University of Chicago Press.
Tron, E.: Variation statistische Untersuchungen über Refraktion, Graefe's Arch. Ophthal. **122**:1-33, 1929.
van Alphen, G. W. H. M.: On emmetropia and ametropia, Ophthalmologica (supp.) **142**:1-92, 1961.
van Alphen, G. W. H. M., Lely, C., Nass, C. A. G., and van Leeuwen, H.: A comparative psychological investigation in myopes and emmetropes, Proc. R. Netherlands Acad. Sci. **55**:689-696, 1952.
Walton, W. G.: Refractive changes in the eye over a period of years, Am. J. Optom. **27**:267-286, 1950.
Ware, J.: Observations relative to near and distant sight of different persons, Philosoph. Trans. R. Soc. London **103**:31, 1812.
Wibaut, F.: Ueber die Emmetropisation und den

Ursprung der spharischen Refraktions anomalien, Graefe's Arch. Ophthal. **116**:596-612, 1925.

Wilmer, H. A., and Scammon, R. E.: Growth of the components of the human eyeball, Arch. Ophthal. **43**:599-619, 620-637, 1950.

Wold, K. C.: Hereditary myopia, Arch. Ophthal. **42**: 225-239, 1949.

Wong, W. W.: Semantic awareness in refractive procedures, Ann. Ophthal. **3**:832-838, 1971.

Woods, A. C.: Report from the Wilmer Institute on the results obtained in the treatment of myopia by visual training, Am. J. Ophthal. **29**:28-57, 1946.

Yamanaka, T.: Correlation between tonus of ciliary muscle and astigmatism of cornea, Jap. J. Ophthal. **7**:18-29, 1963.

Young, F. A.: An estimate of the hereditary component of myopia, Am. J. Optom. **35**:337-345, 1958.

Young, F. A.: The development and retention of myopia by monkeys, Am. J. Optom. **38**:545-555, 1962.

Young, F. A.: The effect of restricted visual space on the primate eye, Am. J. Ophthal. **52**:799-806, 1961.

Young, F. A.: Myopia and personality, Am. J. Optom. **44**:192-201, 1967.

9

Clinical history

Every diagnosis begins with a history, which provides the clues that give direction to the examination. The rationale of the history is to solve the diagnostic puzzle, a time-honored concept in the practice of medicine but sometimes forgotten in clinical refraction. Lenses are prescribed, withheld, or modified only on the basis of retinoscopy or astigmatic dial. Refraction may even be delegated to a machine or technician. But no mathematic exactitude defines pain, headache, or visual discomfort, and no formulas characterize subjective complaints. We must rely on patients to recount their past responses to stress, disease, or treatment and to tell us what they feel or see or, indeed, if they can see at all. What refracting machine will give us this information? To provide clear comfortable and efficient vision, we must know the precision, distance, size, contrast, illumination, and timing of the visual task. What instrument will provide these specifications? The inability to function in the visual environment changes our point of view. Every malady has its mental spectres, especially when the shadows are visible. Fear is often hidden, sometimes overcome, but seldom extinguished. What technician will probe these psychic subtleties?

Some believe that the presence of symptoms can be established by answers to printed questions, their variations programmed, and their significance evaluated by a computer. Step by step the art of medicine will disappear before the machine, quality before quantity, empathy before computation; soon the patient will disappear and only switches and buttons remain. But I doubt it. And the Cassandrian experts who predict that it is only a matter of time before the physician is replaced by an electronic Delphian oracle are probably just making extrapolating fools of themselves.

The history is not only a diagnostic but a therapeutic tool. For a long time it was the only tool, and the physician's empathy at the bedside promoted hope if not cure. Patients now demand less ingenuity and quicker results, but rapport is still vital to gain the patient's trust. The more serious the problem and the more prolonged and dangerous the treatment, the greater this need for belief. What computer will inspire this confidence? Clinical refraction is at a historic standstill because we have been led to believe that ametropia can be beaten down with formulas, but the realities of practice are the patient and his symptoms, not impersonal optical equations.

TAKING THE HISTORY

Within limits of polishing the phraseology, the history will probably follow a standard outline: identifying information, chief complaint, past ocular history, medical history, family history, and social, environmental, and occupational history, concluding with a review of ocular symptomalogy.

The outline is not to parody a pattern but to prevent omissions, and printed record forms help serve as reminders until the habit is established. Ultimately it is the patient's symptoms that determine the direction, the sequence, and the emphasis of the examination. There is no point in testing everything to find something wrong; it will only end in a detour with significant landmarks buried in a mass of irrelevant information, not to mention the time, effort, and expense involved.

Identifying information

An initial note will include such basic information as name, address, age, sex, race, and occupation. This information should be accurate for prompt retrieval, permanent and private for medicolegal protection, and legible for bookkeeping.

Identifying data help direct the examination to the most likely causes. Each age has attributes and characteristic ailments that determine the norms to which we compare refractive test results and the prognosis of treatment. The greater our knowledge of the incidence of visual disturbances the more fruitful these suspicions; a pinch of perhaps may save us a pound of puzzlement.

Chief complaint

Some patients describe their symptoms with a bubbling spontaneity, some with cautious reticence, and some with a challenging obscurity; we must funnel the first, syphon the second, and endure the third. We hear of tearing, itching, and pain but seldom of suppression, scotomas, or squint. Since patients do not complain of uveitis and convergence insufficiency but of photophobia and double vision, the history deals with symptoms and not the neat etiologic entities categorized in textbooks.

Those who are sick see a doctor, and those who become ill summon a physician; the patient's choice of words selects the channel of communication. Some complain of dizziness when they mean lightheadedness, diplopia when they mean blur, blur when they mean a floater, and floaters when they mean a scotoma. Symptom constellations or syndromes must be assembled, since they have greater diagnostic significance. Headache by itself means little, but coupled with bitemporal hemianopia it is something else again.

The chief complaint may not be the dominant diagnostic feature, but is is the one that concerns the patient most. If a man complains that he cannot see his morning bus and is given atropine, orthoptics, base-in prism, or thyroid to "improve" his myopia, he had better be told why he will continue to be late for work—or the waiting room will soon be filled with purple myopes waving their glasses with a vengeance.

An elaboration of the chief complaint might proceed along the following traditional lines:

1. Initial onset—sudden or gradual
2. Duration—acute, chronic, or recurrent
3. Location—ocular, orbital, facial, or cranial
4. Character—pain, tearing, discharge, itching, scratching, etc.
5. Severity—in terms of interference with function
6. Frequency—relation to body rhythms, work, season
7. Precipitating factors—visual task, light, glare, lack of sleep
8. Progression—getting better or worse, or unchanged
9. Related features—whether the symptom is part of a syndrome
10. Previous treatment—date, nature, results

Within the wailing of the ailing, we pursue the detail of the failing: from if to then, and from how to when; we track the crafty where —into the inner diagnostic lair. If the chief complaint is the main actor in the diagnostic mystery, not infrequently there is a subplot, such as a cupped disc, an old amblyopia, or poor night vision. "Disease," says Harrison, "often tells its secrets in casual parentheses." Other problems and other questions may arise during the examination, and the time set aside for the refraction serves nicely to get acquainted with the patient.

Past ocular history

A history of strabismus may throw light on an otherwise unexplained amblyopia, and a patient's prior response to three unsuccessful bifocals may convince us that a separate pair of reading glasses is a better choice.

Some information pertinent to the history may only become available later, including data from the family physician, pediatrician, parents, siblings, school authorities, and, alas, lawyers for the plaintiff, the defendant, the insurance company, and the legal aid societies.

Medical history

If the ocular problem is the drama, the body is the stage on which it is played. The scene is set by the cerebral director, the hormonal orchestra, and the metabolic impressario. Disease does not select the physician who will treat it, so the medical history may provide a clue for sudden changes in ametropia, amplitude, or pupil size. Steroids, antibiotics, tranquilizers, diuretics, muscle relaxants, and autonomic drugs have their toxic ocular consequences, even if the systemic disease for which they are used has no direct ocular effect. Indeed the pharmacopia is expanding so fast that even the ladies' magazines cannot keep up, along with some illegal variants with which the authorities cannot keep up.

Family history

A history of strabismus in a sibling, of myopia in a father, or of rubella in a mother may confirm a diagnosis and help in patient counseling. In choroideremia or retinoblastoma the family history may require an extensive investigation of biologic relatives. A suspected papilledema in a hyperopic child is not excluded by a similar fundus picture in the mother, but it is reassuring enough to avoid dangerous diagnostic procedures.

Social, environmental, and occupational history

The consequences of visual disability are profound and may require major readjustments in life-style. Professional recommendations regarding school work, jobs, driving, or flying are required daily. Every task has its visual criteria. After prescribing a strong reading glass to a piano teacher, it is too late to call on Donders to explain why the unhappy patient cannot see at arm's length. All patients cannot be treated by the same methods; better an approximate reading glass on the nose than a theoretically perfect bifocal in the dresser drawer.

The inspector of small parts, the crane operator, the electrician tracing a color-coded diagram, the high-speed machine operator, the pilot, the bus driver, the signal man, and even the bartender have fairly exact visual demands. Only patients can tell us their precise visual needs, which influence what tests we choose to include in the clinical refraction.

Hazards like ultraviolet rays for arc welders, infrared rays for melters and furnace men, noxious fumes for chemical workers, splattering for sandblasters, impact effects for riveters, and the splash for caustic chemical handlers determine the choice of filters and safety features such as masks, helmets, and goggles. Safety lenses are now a universal legal requirement. Questions about optimum lighting and the potential hazards of sunlight, reading, and television are frequently asked, and patients with partial vision need particular reassurance.

A review of major visual symptoms helps prevent omission; the length of the list depends on the clinical problem and the patient's suggestibility. One can readily make up a list of bad humors and doleful lamentations.

A brief summary pinpoints the main difficulty and saves reading time later. "Hysteria" and "neurosis" are not usually ophthalmic diagnoses and should not be used to pigeonhole problems not fully explored. To state that a problem is psychologic is to define it, not to offer a solution. Conversely, we must not write "noncontributory" when we mean "not interested."

EVALUATING THE PATIENT

Disease leaves its mark on the lustre of the cornea, the droop of the lid, the symmetry of the pupil, the flexibility of the focus, the turbidity of the lens, and the mottling of the disc. It changes the timbre of the voice, the texture of the skin, the tide of the breathing, the temper of the bones, the totter of the gait, and the tension of the nerves. "You mention your name as if I should recognize it," says Sherlock Holmes, "but beyond the obvious facts that you are a bachelor, a solicitor, a Freemason, and an asthmatic, I know nothing whatever about you." The trained senses perceive much at a glance if we teach them what to look for.

Unraveling the account of mysterious aches and pains and weighing their significance requires some insight into practical patient psychology. More important than the wish to play psychiatrist is the will to anticipate the patient's ability to follow directions, to discriminate target differences, and to cooperate with recommended treatment. This determines when certain techniques are to be avoided, how often tests must be repeated, what reliance to place on the result, and the limits of practical treatment. Admittedly, not every investigation leads to a happy solution. Even Sherlock Holmes failed on occasion (once so gravely he had to be resurrected by popular demand).

Rapport means gaining the patient's confidence by individual attention and a passion for patience. The rewards have never been stated more eloquently than by Emerson:

> We mark with light in the memory the few interviews we have had, in the dreary years of routine, with souls that made our souls wiser; that spoke what we thought; that told us what we knew; that gave us leave to be what we inly were. Discharge to men that priestly office, and, present or absent, you shall be followed with their love as by an angel.

EVALUATING THE SYMPTOMS

Skill in analyzing symptoms is useful and practical and well worth cultivating. Symptom etiology can be approached in traditional terms of congenital, inflammatory, degenerative, vascular, or neoplastic mechanisms. Whatever the approach, a single examination only reveals a stationary tableau in the drama of disease. It is by serial examination and the history that we reconstruct a three-dimensional picture of the clinical problem.

If there are symptoms but no manifest signs, the disturbance may not have progressed far enough or lasted long enough to produce them. Functional changes frequently precede structural ones in time and severity and provide an opportunity for preventive treatment. The presence of both signs and symptoms allows comparison for consistency.

Headache

Next to constipation, headache is the most common complaint and is severe enough for one out of ten patients to consult a physician. It may be as painful as a toothache, as benign as a forgotten breakfast, as malignant as a brain tumor, or as trivial as a nagging spouse, but it is seldom imaginary. Headache, says Wolff, always means something wrong.

Headache is referred pain, and the mechanisms are all subsurface. They include intracranial or extracranial traction, pressure, displacement, dilation and distention of vessels, inflammation, sustained muscular contraction, and metabolic disturbances ranging from hypothyroidism to hangover.

Ocular symptoms are a leading feature in most varieties of headaches, and the ophthalmologist is frequently the first to be consulted. Ocular manifestations include scotomas, hemianopia, ophthalmoplegia, amaurosis, the ocular congestion of cluster headache, the ophthalmic pain of trigeminal neuralgia, the third nerve palsy of aneurysm, the field changes in brain tumor, and the palsy of head trauma. Chances are that frontal pressure only mimics the pressures of life, the cluster headache will be relieved by histamine desensitization, the throb of migraine will be aborted by prophylactic ergot, the elevated sedimentation rate of temporal arteritis will respond to massive steroids, and the anisometropic headache will improve with the correct spectacles. The pain may also mean glaucoma, herpes zoster, nasopharyngeal tumor, or a tooth abscess. No diagnostic shortcuts separate the trivial from the life-threatening disease. Clinical evaluation of headache may involve time, expense, and several specialties.

Tension, the most common cause of headache, may be triggered by an ocular problem. The headache of refractive error is poorly understood and is attributed in part to persistent contraction of periorbital and neck muscles and to the intimate connections of the spinal tract of the fifth nerve with cervical segments. Practically the entire anterior half of the cranium is supplied by the ophthalmic division of the fifth nerve so that a neurologic pathway for referred pain is readily available. Since pain thresholds vary, no definite type of head pain is characteristic of a particular refractive disorder. More likely the symptomatic struggle for clear and efficient vision is won or lost in the association areas of the brain.

What can we do? "If I wished to show a student the difficulties of medical practice," wrote Oliver Wendell Holmes, "I should give him a headache to treat." Although that was before major and minor tranquilizers, energizers, muscle relaxants, and psychomotor stimulants, the aphorism still holds if we substitute "diagnose" for "treat." The ophthalmologic work-up will include tonometry, fundus study, palpation and auscultation of the temporal and carotid arteries, perimetry, refraction, muscle balance, and perhaps ophthalmodynamometry. The results may be as unrewarding as skull films, sinus films, electroencephalography, or angiography, but they are part of our responsibility to the headache patient. If the medical treatment of headache is not simple, it is because life is not simple. There will always be people caught between hates and hostilities, frustration and futilities, who wander from physician to clinic searching for a pill, a spectacle, an operation, or a diet that will relieve their symptoms and who will carry their headaches to the grave.

Asthenopia

If someone complained of eyestrain, or asthenopia, in England a hundred years ago, he would have been advised to stop reading and do gymnastics or emigrate to Australia and take up sheepherding. Perhaps spectacles might have been offered (after urethral cauterization, rectus tenotomy, and a prolonged rest in the country), but they were believed to aggravate the condition and to result at best in amblyopia or at worst in amaurosis. The etiology was attributed to such interesting causes as spermatorrhea, masturbation, atonic retinas, or nervous and muscular fatigue. If some daring soul insisted on lenses, he was told to try a series of them at an optician's shop. But what is eyestrain?

Eyestrain has been said to include symptoms as diverse as tearing, itching, burning, and scratching; increased sensitivity to light, glare, motion, and tension; decreased acuity, efficiency, and equanimity; flushing of the eye, quivering of the lids, and tingling of the brow; double vision, distorted vision, and dazzled vision; eye ache, headache, neck ache, and bellyache; and local and general strain, weakness, and tiredness. It has been attributed in part or in whole to weak or excessive accommodation, to low hyperopia and high astigmatism, to an imbalance of the muscles or the illumination, to the small size or great precision of the visual task, and to neurosis or neurasthenia. It is variously considered to be characteristic of, the cause of, and the result of "visual fatigue."

What can we do? Drop the term. It is a fig leaf concept that cannot conceal the semantic profusion, copious assumptions, and dangling theories. When Donders published his great work on refraction and accommodation in 1864, it was hailed the following year at the first scientific session of the American Ophthalmological Society as having "demonstrated with mathematic certainty the treatment of myopia, hyperopia, asthenopia, and astigmatism." But the thrill of recognition inspired the forecasting of more benefits than the theory could deliver, and learned writers sometimes view in Donders more than Donders knew. The nature of visual fatigue is still a question, low refractive errors often cause more trouble than high ones, early presbyopes seldom have eyestrain, subnormal accommodation is rare, and hyperopia is considerably more prevalent than the patients who complain of it. More than weak accommodation or insufficient convergence symptoms may result from the failure of binocular cooperation of the eyes; the glare, frustration, and monotony of the visual task; induced heterophorias, aniseikonia, and inadequate fusional reserves; optical aberrations and perceptual permutations; the imperfect centration, quality, or design of lenses; and perhaps inaccurate measurements or invalid interpretation. Let us record the symptom and postulate a possible mechanism; if it is relieved by lenses, prisms, orthoptics, or surgery, we will make note of it but not assume causality. Perhaps this will lead to some searching inquiries into an area in which the only recent advances consist of paraphrasing Donders.

Visual fatigue

Visual fatigue is a semantic chameleon. Watch it change color in such definitions as "asthenopia is eye fatigue caused by tiring" or it is "a visual symptom caused by excess effort—the eyeballs feel tired, full and ache." Although we understand by common experience a patient's statement "I am tired," there is a difference between the fatigue after a game of tennis, a game of chess, and a game of love. Muscular effort, the expenditure of energy, and the exhaustion of glycogen differ between: "I am too tired to read this book" and "Reading this book makes me tired." Boredom, conflict, frustration, and motivation enter into a realistic concept of fatigue. Soldiers with "combat fatigue" are rarely "tired" and, under stress, may perform extraordinary feats despite prolonged muscular activity or lack of sleep.

Patients may tire not of seeing but of their work, their surroundings, their employers, or their failure to achieve the goals they have set for themselves. If symptoms arise during some visual activity, they readily equate the two and assume one is the cause for the other. The clinician in turn finds an accommodative problem and concludes he has isolated the cause of the difficulty.

Measurement of repeated visual performance shows a decrement, which is attributed to fatigue—logical; the experiment is repeated, and an increase is found, which is attributed to training and practice—equally logical. Either the logic or the physiology is wrong because subjective fatigue occurs when performance shows no decrement or when it is actually improving. Ergographic studies of the near point of accommodation and convergence show both decrement and improvement. Visual tasks requiring a high degree of alertness and attention are easily susceptible to distraction, faulty illumination, irregular habits, as well as refractive errors. Indirect criteria of fatigue such as blink rate, changes in phoria or fusional vergences, critical fusion frequency, reaction time, and accuracy and speed of performance are equivocal and nonspecific. Fatigue is a generalized experience and is rarely localized to the eyes. It may express itself in blunted perception, disagreeable sensations, and a desire to sleep. It is not so much ocular overuse as overplayed conflict, overdone monotony, and uncoordinated action.

What can we do? Stop defining our terms like a revolving door. Separate fatigue from the conditions under which it occurs and differentiate between subjective fatigue and objective decrement of performance. Modify simplistic notions that visual fatigue (or visual efficiency) can be defined by input-output formulas (implicit in such terms as "accommodative asthenopia," "exhaustion exophoria," and "weak fusion"). Realize that the "human machine" may compensate for failure by adaptation mechanisms (well known to the neurologist), which can themselves produce symptoms.

Ocular pain

The pain of acute glaucoma, uveitis, a corneal foreign body, or a hordeolum is clear-cut; the ocular referred pain of migraine, trigeminal neuralgia, sinusitis, and otitis has a rational neuroanatomic basis; but the ill-defined aches, pressures, smarting, cramping, drawing, gnawing, or burning attributed to refractive disorders are not so well understood.

Pain may be superficial or deep; the deep pain is frequently felt in areas other than those stimulated, where indeed it may be more acute because of associated muscle contraction. For example, blepharospasm occurs in both direct and referred pain. Conversely, traction on extraocular muscles or the instillation of eserine produces fairly characteristic ocular pain—which incidentally is not at all similar to eyestrain. Eyestrain is not always abolished by cycloplegics; indeed it can occur concurrently with the pain induced by eserine. Moreover, few complain of pain despite the constant use of ciliary and extraocular muscles. Tractions on medial or lateral recti produce localized pain, but it does not resemble that found in convergence and divergence insufficiency. Here are two extremes: television viewing requires almost unchanging fixation, whereas in reading the eyes undergo countless saccades. Yet symptoms are no more frequent in one than the other.

When no direct cause can be implicated, it is not unusual to fall back on functional manifestations. Spiral or tubular visual fields are not necessarily confirmations but prove only that the patient was tired, bored, or inattentive. And thus a small change is made in the sphere here, the cylinder there, the axis is rotated 2°, and the frame receives a twist, all in the secret hope that it will get, if not a cure, the patient out of the office.

What can we do? Use less logic uncontaminated with experience. Although the correction of refractive errors may bring complete relief, it does not mean every small refractive error must be corrected. Treatment depends on the symptoms, which is why the history is so important. Conversely, large phorias or low fusional vergence reserves can sometimes be treated with low corrections if the AC/A ratio is high; even small vertical imbalances frequently cause problems that can be corrected by minimal decentration. Small differences in magnification, intermittent tropias, and moderate hyperopia in early presbyopes illustrate that quantity of lens power can give disproportionate qualitative relief.

Blurred vision

The smooth course of human affairs hinges on the fact that most people have good vision. We stake our lives when we assume that the other driver sees us coming. So universal is 20/20 acuity that we consider the average the normal despite variations in axial length, receptor density, foveal topography, neural integration, and cortical interpretation. What other mass-produced instrument can boast of such quality control?

Poor vision is seldom ignored, although it may be neglected. Amblyopic children rarely complain of poor vision; conversely, young myopes hesitant about their answers may be misdiagnosed as amblyopes. If vision is improved by lenses (or by a pinhole), the problem is likely to be optical, but ametropia also occurs in diseased eyes, and visual aids may significantly improve vision. Amblyopia cannot always be prevented, but it can be treated effectively if diagnosed early. Distance and near vision often are affected unequally, and a strong reading glass may permit the patient with cataract or macular disease to read, a not inconsiderable achievement.

What can we do? Every patient deserves a careful refraction as part of a complete ocular examination. The visual result cannot be predicted until the ametropia is corrected —it may even take some months in refractive amblyopia. The correction depends on the patient's visual needs; an acceptable blur to a ditchdigger may be intolerable to a truck driver—hence one cannot push plus arbitrarily in all prescriptions. The anisometrope sometimes prefers happy monocular to uncomfortable binocular vision, and some cosmetically conscious matrons will sacrifice acuity rather than submit to bifocals. Reduced near vision in the middle years is universal, but the indistinct intermediate vision, especially with the second presbyopic add, is almost as common and must be anticipated. Amblyopia is treated by early occlusion, total occlusion, and constant occlusion of the better eye.

Photophobia

Sensitivity to light is a common complaint. It is to be expected in anterior segment inflammations, albinism, febrile diseases, headache, drug-induced mydriasis, beginning contact lens wearers, and recent aphakes with large iridectomies. Associated symptoms may include lacrimation, blepharospasm, frowning, increased blink rate, or pain, which persist even without intense illumination. Glare, on the other hand, is the obvious and offensive result of excess or improper illumination, and the discomfort is due to ocular stray light. Some patients insist on tinted lenses because they assume light is harmful and a filter will act as a preservative like an eyewash. It is doubtful that this means "conversion-hysteria"; more likely it is a quest for protection, especially if there is impaired vision.

What can we do? A complete ocular examination will rule out the obvious causes of photophobia. True photophobia depends on the integrity of the trigeminal nerve, and a therapeutic trial with anesthetics, vasoconstricters, mydriatics, or cycloplegics may be necessary. In the absence of organic disease, a request for tinted lenses can be honored for "functional" photophobia, providing they are not too dark for safe night vision.

Floaters

"Spots" and "dots," "globs" and "blobs," "spiders" and "gliders," "cobwebs" and "octopeds," "flying fleas" and "small debris" are a few of the descriptive terms patients apply to annoying floaters. The symptoms are due to opacities in the ocular media, which may cast shadows on the retina. A sudden shower of such floaters can be a frightening experience. They are generally mobile, with a quick component in the direction of fixation; the closer to the retina the more prominent they appear. If totally opaque, the shadow is dark; semitransparent floaters may have a bright or a dark center depending on refractive index. Physiologic debris in the vitreous and condensation of vitreous fibrils, called "muscae volitantes," are the most common entoptic phenomena. More serious (and more controversial) are the potential effects of vitreous degeneration, for retinal detachment or macular edema. Accompanying symptoms may include flashes of light (the lightning streaks of Foster Moore).

What can we do? To describe is easy, but to prescribe is difficult. The opacity may be impossible to identify either with ophthalmoscope or slit lamp, and, in the absence of vitreous or retinal disease, the most important treatment is reassurance. No treatment effectively promotes absorption, but placebo eye drops may keep the patient busy until the floater disappears of its own accord. The correction of refractive error sometimes makes the floater less noticeable by optically defocusing the shadow.

Diplopia

The patient with true double vision comes into the examining room with one eye closed or patched. Diplopia is incompatible with ordinary activities and thus is one of the most incapacitating visual symptoms. A patient with both eyes open walking about and complaining of diplopia is either malingering, has misinterpreted the symptom, or has intermittent diplopia. Occasionally, as in minor strokes, the diplopia has resolved by the time the patient appears in the office. Despite the textbooks, I have never seen a patient complain of physiologic diplopia.

Diplopia follows a sudden loss of binocular coordination in which one object is seen in two directions. Foveal integrity is not a necessary prerequisite; a patient with macular degeneration will experience diplopia if the peripheral fusion mechanism is disrupted. An obvious squint without diplopia, means a strabismus of long-standing. Diplopia may also result from mechanical displacements as in blow-out fracture of the orbit, tumor, hemorrhage, and inflammation; from affections of the nerves as in orbital apex syndrome and diabetic neuropathy; or from the brain as in aneurysm, tumor, and vascular accidents. The diplopia of decompensated heterophoria may follow the prescription of reading glasses, anisometropic bifocals, improperly centered lenses, lack of sleep, boredom, or alcohol ingestion.

A diplopia that persists with one eye closed has probably been mistaken for blurred vision—a reasonable error undeserving the label of "hysteria." True monocular diplopia is unusual and may result from double pupils, anterior and posterior lenticonus, or lens subluxation or opacities. It rarely occurs after strabismus surgery (and then only when the other eye is open) and is even more rare in malingering.

What can we do? The correct evaluation of diplopia is a serious matter not only in its incapacitating character but also because its causes are potentially life threatening. Many clinical tests are available, ranging from the objective cover test to the subjective synoptophore. Recent developments permit electrophysiologic study of the extraocular muscles directly. Diplopia is conveniently classified as crossed or uncrossed by the position of the double images. (The "x" in "exo" reminds us which is crossed.) Neurologically, crossed diplopia generally means third nerve involvement; uncrossed diplopia means sixth nerve involvement. Diagnosis depends on eliciting associated signs, including head turning, vertigo, deafness, and nystagmus, and the involvement of other cranial nerves or adjacent anatomic areas (skin, sinuses, nose, and ear). Sites of old scars from previous trauma are easily overlooked. An area of periorbital anesthesia may confirm an orbital fracture. Corneal sensitivity can be checked with a wisp of cotton. There is little value to muscle balance tests in the acute stage, but they may help in the diagnosis of intermittent diplopia. Visual fields, both central and peripheral, detailed intraocular examination, and skull and orbital x-rays studies are part of the diagnostic work-up.

Treatment of diplopia is directed at the underlying cause. Relieving prisms are occasionally indicated, but if prescribed by instinct, with reason they will be returned.

Dyslexia

Primary reading disability has been defined as "a congenital reading disability—characterized by the presence of dyslexia, dysgraphia, dyscalculia, right-left disorientation, and agnosia."* It sounds important, it is undeniably learned, and it may give a feeling of comfort to the authors if not to the patient. Under the label of "dyslexia," slow readers, poor writers, emotional calculators, immature learners, and cerebrally impaired children, caught between an anxious mother and a conscientious pediatrician, are brought to the ophthalmologist in the vain hope that a pair of spectacles will solve all the difficulties.

Unquestionably it is necessary to see the letters before one can read them; yet myopes are seldom poor readers. The reading process, says one writer, involves so much mental energy that little is left over for comprehension. If only we could measure mental energy, we could leave two thirds of it in reserve! And how shall we interpret the teacher's report that a child "fails to work to capacity" in the absence of any clear standard and in a schoolroom culture in which conformity is an asset and individuality a handicap. The reasonable theory that children enter school before their eyes are mature enough to cope with the demands of the curriculum does not hold up, at least for accommodation and acuity. Nor are the results of reading tests—sometimes improperly administered, inaccurately interpreted, and gainfully disseminated—always a reliable index, when based on every self-styled reading expert's own ladder of neurologic development. And when we come to "cross dominance," what cerebral pandemonium must reign in the uncoordinated hemispheres of the Chinese.

Reading difficulties are complex syndromes involving the interaction of ocular, perceptual, associative, and motor factors. The cause in most cases is neither an ocular nor a neurologic deficit, and the exact mechanism is unknown. Most likely there are multiple causes, with an abundance of unsupported theories for both etiology and treatment.

At some time in the diagnostic work-up, the child is referred for an eye examination. Many visual functions have been studied (as early as 1826 by Muller), but invariably no specific visual abnormality including ametropia, binocular coordination, eye tracking

*From Kolson, C. J., and Kaluger, G.: Clinical aspects of remedial reading, Springfield, Ill., 1963, Charles C Thomas, Publisher, p. 19.

and fixations, perceptual span, eye dominance, and perceptual cognitive processes (closure, grouping, reversals, constancy, and rivalry) are implicated.

What can we do? The role of the ophthalmologist is early recognition and exclusion of correctable visual defects. It seems the best way to teach reading is by word recognition, not training perceptual span, ocular dominance, motor coordination, and ancillary skills. How dyslexic children should be taught to read is a problem for the educator, who needs no amateur's advice. As primary physician, the ophthalmologist can assist in bringing the child's potential to the best level be referring him to the appropriate remedial agency. We follow Hamlet's advice to the actors: Suit the examination to the history and the history to the problem—nor saw the air too much with theories—but use all gently.

Refractive failure

The characteristic syndrome of refractive failure is a choleric patient, clinking his spectacles and caustically commenting "I cannot wear these glasses."

What can we do? First, maintain our poise. Second, correlate the lenses with the prescription. Third, evaluate the symptoms. Is the difficulty with the vision, the distance, the spatial orientation, the comfort, or the payment? Any or all may have etiologic significance and may suggest appropriate steps. We recheck the P.D., the base curve, the centration, the thickness, the position, the balance, and the fit of the frame. We compare the old spectacles to the new and old complaints to the current ones. We repeat the refraction and find it adequate. *Then* we tell him: "Wear the glasses, you'll get used to them."

REFERENCES

Adams, G. L., and Pearlman, J. T.: Prevention of mental disorders in ophthalmic patients, Ann. Ophthal. **4:** 555-560, 1972.

Allan, F. N.: The clinical management of weakness and fatigue, J.A.M.A. **127:**957-960, 1945.

Alvarez, W. C.: Psychosomatic medicine that every physician should know, J.A.M.A. **135:**704-708, 1947.

Anton, M.: La migrana oftalmica, Rev. Med. Cir. Habana, **34:**285-296, 1929.

Apt, L.: Headaches in children, Int. Ophthal. Clin. **2:**859-872, 1962.

Bahn, C. A.: The psychoneurotic factor in ophthalmic practice, Am. J. Ophthal. **26:**369-378, 1943.

Bannon, R. E.: Symptoms and case history—the patient as a person, Am. J. Optom. **29:**275-285, 1952.

Bar, W.: Psychologic problems in ophthalmologic diagnosis, Am. J. Ophthal. **25:**321-329, 1942.

Bartley, S. H., and Chute, E.: Fatigue and impairment in man, New York, 1947, McGraw-Hill Book Co., Inc.

Belmont, O.: Refraction troubles, Int. Ophthal. Clin. **1:**261-275, 1961.

Bender, I. E., Imvs, H. A., Rothney, J. W. M., Kemple, C., and England, M. R.: Motivation and visual factors, Hanover, 1942, Dartmouth College Publications.

Benson, D. F., Brown, J., and Tomlinson, E. B.: Varieties of alexia: word and letter blindness, Neurology **21:**951-957, 1971.

Blatt, N.: Weakness of accommodation, Arch. Ophthal. **5:**362-373, 1931.

Bowen, S. F.: Retinal entoptic phenomena, Arch. Ophthal. **69:**551-561, 1963.

Browning, C. W., Quinn, L. H., and Crasilneck, H. B.: The use of hypnosis in supression amblyopia of children, Am. J. Ophthal. **46:**53-67, 1958.

Chambers, R., and Cinotti, A. A.: Functional disorders of central vision, Am. J. Ophthal. **59:**1091-1095, 1965.

Clauson, J., and Karrer, R.: Phosphene threshold as related to age and sex, J. Psychol. **47:**189-198, 1959.

Cogan, D. G.: Medical progress: Popular misconceptions pertaining to ophthalmology, N. Engl. J. Med. **224:**462-466, 1941.

Dandy, W. D.: Intracranial aneurysms, New York, 1947, Comstock Publishing Co., Inc.

Drews, R. C.: Organic versus functional ocular problems, Int. Ophthal. Clin. **7:**665-696, 1967.

Eckhardt, L. B., McLean, J. M., and Goodell, H.: Experimental studies on headache: genesis of pain from eye, Proc. Res. Nerv. Dis. **23:**209-227, 1943.

Eyles, M.: Functional home exercises in cases of eyestrain, Am. J. Ophthal. **31:**45-48, 1948.

Ferree, C. E., and Rand, G.: Intensity of light in relation to the near point and the apparent range of accommodation, Am. J. Ophthal. **18:**307-318, 1935.

Fonda, G. E.: Refraction problems, Rochester, 1969, American Academy of Ophthalmology and Otolaryngology.

Francois, J., and Neetens, A.: Tear flow in man, Am. J. Ophthal. **76:**351-358, 1973.

Friedman, B.: Observations on entoptic phenomena, Arch. Ophthal. **28:**285-312, 1942.

Fugate, J. M., and Fry, G. A.: Relations of changes in pupil size to visual discomfort, Illum. Eng. **51:**537-549, 1956.

Goldberg, H. K.: The ophthalmologist looks at the reading problem, Am. J. Ophthal. **47:**67-74, 1959.

Goldor, H.: Headache and eye pain, Int. Ophthal. Clin. **7:**697-705, 1967.

Harrington, D. O.: Psychosomatic interrelationships

in ophthalmology, Am. J. Ophthal. **31**:1241-1250, 1948.

Heaton, J. M.: The pain in eyestrain, Am. J. Ophthal. **61**:104-112, 1966.

Hedges, T. R.: Ophthalmologic view of headache, Headache **11**:31-34, 1971.

Howe, L.: The fatigue of accommodation, J.A.M.A. **67**:100-104, 1916.

Humphrey, C. E., and Murgolo, W. J.: The use of phophenes in detecting visual loss, Eye Ear Nose Throat Mon. **36**:170, 1956.

Ironside, R., and Batchelor, I. R. C.: The ocular manifestations of hysteria in relation to flying, Br. J. Ophthal. **29**:88-95, 1945.

Jenkins, D. R.: Problems of computer application in medical research, Trans. N. Y. Acad. Sci. **28**:439-447, 1966.

Keeney, A. H., and Keeney, V. T., editors: Dyslexia, diagnosis and treatment, St. Louis, 1968, The C. V. Mosby Co.

Kolson, C. J., and Kaluger, G.: Clinical aspects of remedial reading, Springfield, 1963, Charles C Thomas, Publisher.

Kunkle, E. C.: Mechanisms of headache, with particular reference to vascular headache, Trans. Am. Acad. Ophthal. Otolaryngol., **67**:758-765, 1963.

Lancaster, W. B.: The story of asthenopia, Arch. Ophthal. **30**:167-178, 1943.

Lebensohn, J. E.: Photophobia mechanisms and implications, Am. J. Ophthal. **34**:1294-1300, 1951.

Lowenstein, O., Feinberg, R., and Loewenfeld, I. E.: Pupillary movements during acute and chronic fatigue, Invest. Ophthal. **2**:138-157, 1963.

Luckiesh, M., and Moss, F. K.: Fatigue of convergence induced by reading as a function of illumination intensity, Am. J. Ophthal. **18**:319-323, 1935.

Mason, E. A.: The hospitalized child—his emotional needs, N. Engl. J. Med. **272**:406-414, 1965.

Miles, P. W.: Refractive treatment of asthenopia, Am. J. Ophthal. **32**:111-121, 1949.

Miller, D.: A review of speed-reading theory and techniques for the ophthalmologist, Am. J. Ophthal. **62**:334-338, 1966.

Moore, R. F.: Subjective "lighting streaks," Br. J. Ophthal. **19**:545-547, 1935.

Nebel, B.: The phosphene of quick eye motion, Arch. Ophthal. **58**:235-243, 1957.

Orton, S. T.: Reading, writing, and speech problems in children, New York, 1937, W. W. Norton & Co., Inc.

Pino, R. H., and Hultin, G. L.: Treatment of asthenopia —nonpathologic and nonrefractive in origin, Am. J. Ophthal. **27**:520-523, 1944.

Ponder, E., and Kennedy, W. P.: On the act of blinking, Q. J. Exp. Physiol. **18**:89-110, 1927.

Priestley, B., and Foree, K.: Clinical significance of some entoptic phenomena, Arch. Ophthal. **53**:390-397, 1955.

Roper, K. H.: Headache, ophthalmological aspects, Q. Bull. Northwestern Univ. Med. School **30**:29-34, 1956.

Schapero, M., and Hirsch, M. J.: The relationship of refractive error and Guilford-Martin temperament test scores, Am. J. Optom. **29**:32-36, 1952.

Souders, B. F.: Hysterical convergence spasm, Arch. Ophthal. **28**:361-365, 1942.

Stillerman, M.: The refraction failure, Int. Ophthal. Clin. **5**:555-567, 1965.

Veasey, C. A.: The dissatisfied refraction patient, Am. J. Ophthal. **30**:1286-1293, 1947.

Verhoeff, F. H.: Are Moore's lighting streaks of serious potent? Am. J. Ophthal. **41**:837-840, 1956.

Wong, W. W.: Personality patterns in ocular discomfort, Arch. Ophthal. **42**:443-450, 1949.

Woolf, H. G.: Headache and other head pain, New York, 1948, Oxford University Press.

10

Visual acuity

When Alice complained that she could see nobody down the road, the indignant King replied, "I only wish I had such eyes, to be able to see nobody—and at that distance too." It seems that even in Wonderland, the definition of "acuity" depends on who gives it. The Snellen chart at 20 feet is the standard clinical method of testing acuity. Missing from the standard are specifications with respect to the kind and number of letters per line, their separation from each other, their contrast with the background, the gradation of difficulty from one line to the next, the overall illumination, the patient's adaptation and pupil size, whether the 20 feet must be real or may be virtual, and in what units the results should be recorded. Although an official definition of visual acuity exists, few researchers pay much attention to it, and practically no clinician uses it. But before we resolve to revolutionize the definition, consider that people differ, visual tasks differ, requirements differ, and refracting rooms differ, and these differences are not likely to be eliminated by more rigid criteria. To impose the conditions of the laboratory on clinical testing would be impractical, and the results would not necessarily be more useful. If the definition of acuity is somewhat vague, it is probably best left so. What must not be left vague are the factors that influence it, since acuity is *the* criterion of the adequacy of our refractive techniques; if results do not come up to standard, we should be able to separate poor technique from poor resolution.

ACUITY TASKS

Asked for a definition of acuity, the psychophysicist speaks of quantal thresholds; the physiologist, of photochemistry, receptor fields, and neuron recruitment; the lens designer, of modulation transfer functions; the psychologist, of "form" perception; and the ophthalmologist recites a Snellen fraction. To approach this elephant from a clinical point of view, we will have to penetrate the technical babel and grasp some of its individual appendages. For didactic convenience, we can speak of several different tasks: visibility, resolution, recognition, and localization (Fig. 10-1). These capacities are reciprocals of operationally defined thresholds. When the threshold is an intensity, we have a "sensitivity", and when the threshold is an extensity, we have an "acuity," but sometimes it is difficult to say when intensity leaves off and extensity begins.

Visibility

The term "visibility" refers to seeing whether an object is present or not in an otherwise empty visual field. It is not required to recognize, resolve, or localize detail (e.g., a star barely visible against the night sky).* Although stars appear of different brightness, we must not conclude like ancient astronomers that they are of different size. Their

*The hopeless inadequacy of these definitions will be evident. A star once visible will inevitably be localized in some direction and recognized as different from a Zeppelin. The suggested substitution of "detection" for the well-established term "visibility" does not appear to offer any advantages.

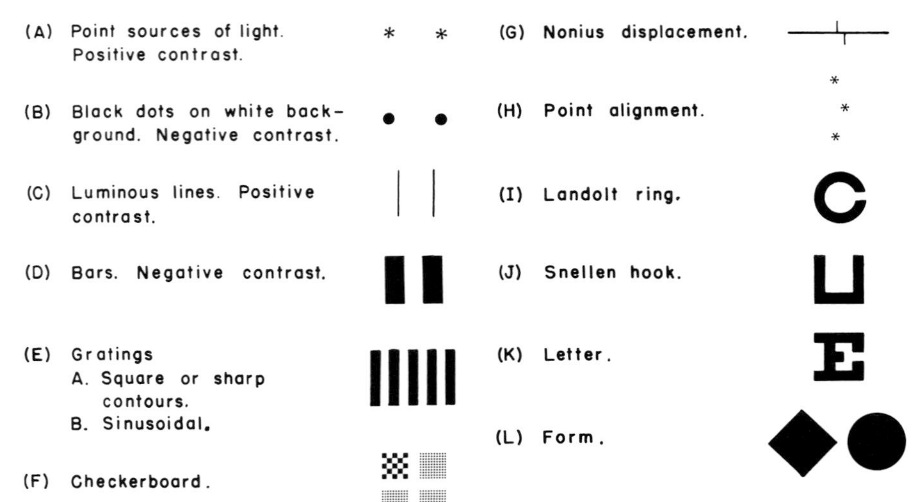

Fig. 10-1. Various types of acuity tasks.

extensity, the angle they subtend at the nodal point of the eye, is the same and, for all practical purposes, negligible. The visibility limit is intensity, not extensity. Indeed, if the sky is perfectly dark, their visibility is a measure of the absolute threshold of vision. Given adequate exposure time, adaptation, and fixation, a luminous target theoretically remains visible even if infinitely small, providing intensity is increased enough.

The reciprocal relation between intensity, area, and time are expressed by several laws of physiologic optics. Thus Ricco's law ($I \times A = C$) applies to central fixation and Piper's law ($I \times \sqrt{A} = C$) to the peripheral retina (where I = luminance, A is a restricted stimulus area, and C is a constant). The Bunsen-Roscoe law ($I \times t = C$) applies to time. The laws are empirical generalizations and only approximately true. Unlike a photographic plate, the eye does not collect light indefinitely; if intensity is too low, nothing is seen no matter what the exposure because rhodopsin regenerates faster than the light can break it down. Of course, thresholds are always a matter of probability. The observer may report seeing a target when none is present or does not see it when it is. The value is arbitrary, conventionally taken as that for which the observer is right as often as he is wrong—hardly a practical criterion for clinical work.*

*The cause of these random oscillations in threshold is still unclear. They have been attributed to fluctuations in photochemical concentrations or in the number of cones, bipolars, or ganglion cells momentarily in a nonlatent phase. Most plausible is the known fluctuation in the number of quanta a light source sends out each time it is flashed.

As the sky lightens, the star becomes invisible. Obviously its size has not changed; hence it must be intensity that no longer makes itself felt above the surround. The visibility of a bright target on a less bright background thus depends on contrast discrimination, in which contrast is expressed as $\frac{I_T - I_G}{I_G}$ or $\frac{\Delta I}{I}$, where I is the luminance of the target (T) or ground (G), and ΔI is the luminance difference. Although brightness discrimination is best under photopic conditions, we cannot always choose the conditions. The intensity of a star or a projector bulb cannot be increased, so we wait until it gets dark or draw the shutters to watch a movie. A faint star is made visible with a telescope not by increasing angular size but by gathering and concentrating the light on the retina.

The visibility of a star against the less than totally dark sky has its clinical counterpart in perimetry. We measure visual field isopters by the size of the target and the distance at which the patient fixates. Thus 2/1000 means a 2 mm target at 1 meter. This ratio should not, however, be called an "acuity," and isopters in no sense measure resolution. Target visibility depends on brightness discrimination. It is only by historic accident that we use targets of different area rather than of different intensity—the pioneer perimetrists without self-luminous targets had to substitute size. We still follow suit out of habit even though modern projection perimeters make it unnecessary.

It is understood that contrast, illumination, and reflectance remain constant during serial examinations, or perimetric fields would have little diagnostic value. The size of isopters varies for the same reason that brightness discrimination varies with different regions of the retina. There are less cones in the periphery, and more must transmit their information on the same party line. But we must not reduce brightness by blurring the image in the private line region of the fovea; thus a lens correction should be worn when measuring central fields.

Brightness discrimination is also the limiting factor in the visibility of a dark object on a bright field. Here the bright background has the practical testing advantage of photopic vision without shifting retinal adaptation. When Hecht and Mintz (1939) performed their now classic experiment on the threshold visibility of a black thread on a white field and found it to be an amazing ½ second (not minute) of arc, the background luminance was an optimal 30.2 millilambert. (The black thread must also be of minimum length, about 30 minutes.) If this luminance was altered, brightness discrimination fell markedly. Although the shadow cast by the black thread on the retina is much larger than its geometric image because of diffraction, aberrations, and scattering, it produces a sufficient dip in the overall luminance to stimulate some cones less than others. Hecht and Mintz computed the required dip and found it to be about 1% of background luminance. This figure came as no surprise, since it had been known for over a century that the least noticeable difference for vision is 1/100 at least for average illumination (about 2 log units). This value has come to be called the "psychophysical" or "Weber fraction."

Psychophysics was founded in Germany about 1820 by E. H. Weber, a professor of anatomy and physiology at the Leipzig medical school. Weber had studied two-point discrimination of the skin and discovered the muscle sense in the discrimination of weights. He concluded that the just noticeable difference (jnd) could be expressed as a fraction that was constant for a given sense modality.

Fechner following Weber is often (unfairly) considered only an echo of his predecessor. Noting that instead of a one-to-one correspondence, sensation followed an arithmetic series, whereas the stimulus was characterized by a geometric series, Fechner assumed that sensations could not only be measured by jnd, but also that the latter might provide a kind of scale unit to quantify them. He invented practically all the psychophysical methods used to this day, summarizing his results by the mathematic expression, $S = K \log I$, which he believed related the physical and spiritual world. The philosophy is typical of the obscure romanticism of Schelling and Fichte, but his magnum opus *Elemente der Psychophysik* (1860) laid the cornerstone of quantitative psychology. His last years were devoted to the evaluation of art and beauty, the foundation of experimental esthetics, which has now degenerated into the pollster's canvas for mediocrity.

Resolution

Discrimination of two (or more) spatially separated targets is called "resolution." Visibility is assumed to be above threshold; that is, each target is of sufficient size and intensity to be visible if presented by itself. But if we expect to discriminate two black threads of ½-second angle separated by another ½ second, we shall be disappointed. Resolution thresholds are always much larger than the minimum visible (average 1 minute).* Sometimes they reach less than 30 seconds under ideal conditions.

The simplest resolution targets are also the oldest—the double star in the night sky. Hooke, a contemporary of Newton, stated that "the sharpest eye cannot well distinguish a spot on the moon, or the distance of two stars, which subtend a less angle at the eye than a half a minute; and hardly half a hundred men can distinguish it when it subtends one minute." One cannot be sure whether Hooke was measuring moon spots, a visibility threshold (the canals of Mars were also favored), or the separation of stars, a resolution threshold. When Kollicker a hundred years later found a foveal cone to subtend 1 minute at the nodal point, everything seemed to fit the notion that two stars could be resolved if their images fell on two cones, leaving an unstimulated one between them. And Snellen and Landolt proceeded to construct their optotypes on this basis.

*An angle of 1 minute at the nodal point of our reduced eye ($f = 16.66$ mm) corresponds to approximately 5 μ. Based on Polyak's data, this encompasses about 4 cones, somewhat less by recent estimates. The exact cone diameter is hard to compute, since no one is sure which part is most important in resolution or if the larger base acts as a funnel.

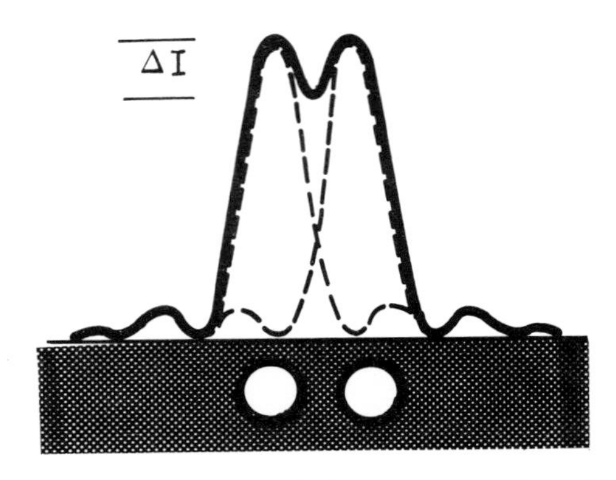

Fig. 10-2. Dip in retinal illumination produced by diffraction patterns of two bright points.

Hooke had actually underestimated acuity, and Kollicker had overestimated cone diameter, but Snellen's unit happens to be the most clinically useful just the same. Incidentally, Euclid anticipated Hooke by a few centuries. Porterfield (1759) coined the term "minimum visible" and Giraud-Teulon (1864), the term "minimum separabile."

The image of a point of light is a diffraction pattern, a mound of light on the retina. The spread is apparent when using a sharply focused ophthalmoscope spot to illuminate the fundus. The mound resembles a normal curve (or point spread function). With two bright points, there are two mounds. To distinguish them it is necessary for only a just noticeable dip in the illumination to stimulate some cones less than their neighbors (Fig. 10-2). Under ideal conditions, (the Stiles-Crawford effect helps), the required dip is about 5%, which is considerably better than the 20% assumed in Rayleigh's criterion. Resolution ultimately depends on both contrast and cone density, varying with the area of the retina stimulated and the state of adaptation. If the cone threshold is low or the intensity of the targets high, the mounds of light fuse into a single, apparently larger object. Because this irradiation effect is absent with dark-on-light targets, their resolution is somewhat better. With gratings, the difference between light on dark and dark on light becomes academic.

The minimum separable can be measured by points, lines, bars, gratings, checkerboards, or Landolt rings and some (but not all) Snellen letters. Imagine two black bars expanding and swelling until they occupy the entire field, leaving the white stripe between them unchanged. At some point the task of resolving two black bars switches to the visibility of a single white line; the extensity threshold becomes an intensity threshold.

Resolution of horizontal bars or a Snellen E depends on the separation of the centers of the black (or white) bars, that is, the crests of the light mounds on the retina (line spread function).

Two black bars are more likely to be resolved on a white background than a gray background, even for the same angular subtense. The image of a grating (evaluated by a photodetector) is a waveform of intensity distribution whose spatial frequency is a function of the magnification and modulation amplitude, that is, the intensity difference between crest and trough. This transfer function is an expression of optical performance:

$$\text{Transfer function} = \frac{\text{Image contrast}}{\text{Object contrast}}$$

This ratio tells how well or how poorly an optical system transfers contrast from object to image. A transfer factor of zero indicates that the system did not resolve the grating or that it blurred it out. A transfer function of 0.5 means the image contains half the contrast of the object. By presenting gratings of similar amplitude but different frequencies, the ability of the system to resolve progressively finer gratings can be plotted. The plot of transfer function versus spatial frequency (lines per millimeter) is the "spatial modulation transfer function." The eye differs from a lens in that is is attenuated at both high and low spatial frequencies, and one of the more exciting areas of current research in human resolution is whether the eye acts as a frequency analyzer.

Recognition

Clinical measurement of recognition is based on letters, numbers, and forms. Letters are more likely to be recognized by the literate, in the context of words, and in a familiar language. But their meaning may also be imperfect, distorted, or stereotyped. However meaning becomes attached to symbols, it must doubtless be learned. The question is how best to go about it.

It is as difficult for teachers to agree on how children should learn to read as for theologians to agree on a definition of sin. Do the poor readers fall by their own impediment, or are they frustrated by poor pedagogy, parental indifference, and cultural indolence? Surveying the dyslexia literature, one gets the impression nothing is happening every minute.

In the "look and see" method children are taught to see words whole; they must remember everything and learn nothing. In the phonetic approach, they learn sets of sounds that enable them to decipher new words. Its principles are discussed with wit and insight by Linksz (1973), and should be read by every clinician who wishes to discuss the subject intelligently with parent or teacher.

Children undoubtedly could profit by a more intelligent phonetic approach. For example, here is a child who has spent several years speaking sentences and untold hours watching television. He can recite commercials retail and has followed cartoons wholesale. And the first sentences he is taught are "Sally saw the dog. Watch the dog run." Any self-respecting child soon comes to hate both Sally and her dog. The good teacher can create interests, encourage motivations, and recognize difficulties. The trick is not to turn up more elaborate reading tests but to turn out more enlightened educators.

Early notions of form perception were based on attempts to build shape out of lines as lines are built of points. The Gestaltists, appreciating that "squareness" was more than four lines, derived empirical principles such as figure emerging from ground, closure, symmetry, inclusion, and "goodness," which stimulated much research, although their brain-field isomorphism is no longer taken seriously. Current interests center around physiologic mechanisms of contour formation and the psychophysics of computer-generated patterns. Still undecided is how form should be defined, whether in terms of a projection on a flat surface, in three dimensions, or by geometric properties such as perimeter or perimeter/area.

Localization

Three visual tasks are grouped under localization: vernier (or nonius) acuity, stereoscopic acuity, and motion acuity. Vernier acuity is discrimination of lateral misalignment in the frontal plane, stereoscopic acuity is a misalignment in depth, and motion acuity is detectable movement. All three have in common amazingly low thresholds on the order of a few seconds of arc, illustrating the inadequacy of any anatomic cone separation theory.

Vernier acuity is used in the slide rule, the range finder, the micrometer, or in any other fine linear measuring stick. Its interest is both theoretical and practical, since any theory of acuity must somehow explain the high order of discrimination involved, and if it accounts for one, it will automatically account for the others. The similarity of vernier and motion acuity is apparent when one of the misaligned lines is allowed to oscillate; the similarity of vernier acuity and stereoacuity is apparent if we let the misalignment represent the disparity of one eye's view of a stereogram.

Vernier acuity does not depend on contrast, cone interspaces, or the width of the target. There is no space for light irradiation to obliterate, and it is not even necessary to have a line or border—it also works for a central misaligned point between two vertical points. The underlying factor is the precision of oculocentric directions in which each retinal point (not each receptor) maintains its independence. How this is accomplished is still a mystery, since the threshold is less than the diameter of any single cone (or row of cones). Hering (1899) postulated that each of the misaligned borders, for example, would chiefly stimulate two separate rows of cones. Since this would limit the threshold to a single cone diameter (still insufficient to explain the few seconds observed experimentally), Andersen and Weymouth (1923) added physiologic micronystagmus as an averaging process, whereas Marshall and Talbot (1942) replaced the geometric edges with diffraction gradients of illumination. Yet vernier acuity is measurable even without contours and when all eye movements are eliminated (by a contact lens mirror technique). Even a contact lens is unnecessary; the sharply outlined stabilized retinal blood vessels can be made to appear in one's own eye if their shadow is thrown on an area not usually stimulated. Stabilization does not reduce acuity, although there is fading with adaptation. Thus

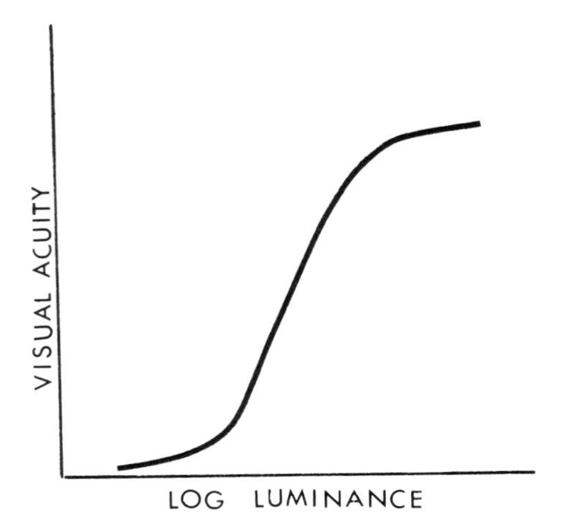

Fig. 10-3. Relation between visual acuity and target luminance.

the problem is back in the lap of the neurophysiologist still hunting for mechanisms of retinal interaction to explain it.

SOME FACTORS AFFECTING VISUAL ACUITY

Although visual acuity conveys the idea of sharpness of vision, its ramifications are so broad as to encompass the entire gamut of seeing. Looking at the dawn sky, we see the visibility of stars, the contrast of the moon, the shape and texture of clouds; we resolve the airplane and recognize its motion; and we discriminate the manifold hues of the sunrise. All forms of acuity tasks above and below threshold are thus involved. For didactic purposes, the variables affecting acuity may be classified as physical, physiologic, and psychologic, and we pretend each acts independently. Clinically this is impossible; increased illumination not only sends more light to the retina, but it also constricts the pupil, alters adaptation, and changes depth of focus, color temperature, and accommodation, or it may even induce glare and closing of the eye.

Physical factors

König, a pupil of Helmholtz, showed that if two black bars of good contrast on a white field are viewed in a dark room whose illumination is slowly increased, discrimination improves along a sigmoid-shaped curve (Fig. 10-3). Obviously finer detail or less separation can be discriminated under better illumination. The curve is flat to the threshold, rises linearly with the logarithm of light intensity, then flattens out again at average illumination with little further improvement (unless it induces glare). Contrast is altered in both directions; the black bars appear blacker, and the white field, brighter. With clinical targets of black on white letters (a difference of twenty times in luminance), 10 to 20 footcandles are sufficient to reach the upper plateau.

Many theories have been proposed to account for the improved acuity with increasing intensity: irradiation; random variations in light threshold among cones; variations in the steady state of photochemical concentration; or neurophysiologic interaction. Although they explain resolution, none account for localization, so no general theory of acuity is yet in sight.

When contrast is reduced, more illumination is required. Just how much more depends on whether one listens to the illuminating engineer or the psychophysicist.* The fact remains that optimum contrast is not always obtained in everyday seeing. The radiologist, microscopist, photographer, meteorologist, radar operator, television director, and astronomer seldom work with maximum contrast.

Luminance is the product of illuminance and the diffuse reflection of the surface. If 100 footcandles illuminate white paper, gray cloth, and black velvet, whose reflections are 80%, 8%, and 0.8%, respectively, the luminance will be 80, 8, and 0.8 footlamberts. The approximate reflectance of white paper is 80%; newsprint, 65%; black ink, 2%; outdoor grass, 10%; clouds or snow, 80%; and wood surfaces, 40%. To discriminate a black thread against a velvet surface requires twenty times more illumination than if seen against a white surface.

Reducing the contrast of clinically presented Snellen letters turns out to have surprisingly little effect (Ludvigh, 1941). Reducing contrast from 90% to 5% (at 23 footcandles of general illumination) reduced vision by only one line.

Contrast does make a difference in the protective coloration of animals. And the principles

*The contrasting views are argued in two papers: Tinker (1947) and Luckiesh and Taylor (1944).

of camouflage were laid down by a Chinese military writer about 500 B.C.: "To beat the enemy we must seem unable to attack when able, far away when near, near when far away, active when inactive. By making him see what is not there, and not see what there is, to entice and then to crush him."

A black and white photograph of a well-executed painting shows consistent tone values for near and far, brights and shadows. Thus the amateur artist learns to separate brightness from hue by squeezing the eyelids together. Hues differ in brightness, and pigments differ in reflectance. Clearly the three types of cone responses are not limited to primary colors, or acuity under green light, for example, would be only one third as good.

Unlike the monochromic acuity chart or the printed page the natural environment is polychromatic. Color contrast adds to brightness contrast. Discrimination is possible for two hues even though they have identical brightness (an interesting problem for the designer of highway signs). Discrimination may therefore be improved by color coding—or obscured by camouflage, a poor filter, or a glare source. Mercury lamps, for example, are highly efficient sources, but people looked so ghastly that they were given up for indoor use. And the butcher who painted his shop walls red found his sales drop off when his customers projected the green afterimage onto the meat counter.

Although exposure time is rarely a factor in ordinary seeing, it becomes significant in dealing with fast moving machinery, movies, or in detecting a blip on a radar screen or a tank on the gound by a jet pilot flying a reconnaissance mission. Every visual task, on the other hand, takes time. If one attempts to read this page under reduced contrast or illumination or by blurring the optical image, the speed of performance decreases. The child with high astigmatism, anisometropia, or high anisophoria is going to take considerably more time interpreting the printed symbols in the reading primer. Indeed one characteristic of efficient reading is a natural rhythm of fixation, the symmetry of which is evident if one has a vitreous floater and observes its motion. It also takes time to accommodate for targets at different distances, and it takes time to react to what is seen, a factor of importance in driving and flying.

Visual acuity for a line or grating is better when oriented vertically or horizontally than obliquely, a kind of "retinal astigmatism." It accounts for the greater confusion of certain letters (e.g., P and R, N and H). The reason is not known.

Acuity for moving targets is worse than for stationary ones if the movement is more rapid than the ability of the eye to maintain fixation. Dynamic acuity decreases as a cubic function of target velocity, and correlations between so-called dynamic and static acuities are low. Despite its obvious practical importance in driving, it is seldom measured clinically.

A number of purely optical problems enter into the measurement of acuity through corrective spectacles. The lenses reduce light transmission and induce reflections, magnification, and variations in focus if not properly positioned. A fairly common error is measuring distance vision through a near prescription.

When acuity is expressed as a visual angle, it is assumed that size and distance are interchangeable. There is evidence, however, that acuity varies with distance, even though the visual angle remains constant, an effect first demonstrated by Aubert and Forster (1857) on two perimeters at equivalent distances. We ignore it when we compare acuity in equivalent refracting rooms. But distance cannot be neglected by an artillery gunner, ship's lookout, mountain climber, and pilot for whom atmospheric haze and weather conditions modify the intervening space. So just to be sure, we should always indicate at what actual distance our measurement was made.

Physiologic factors

There is psychophysical, optical, neurophysiologic, and perimetric evidence that resolution is highest at the fovea and decreases with eccentric stimulation. The threshold of resolution, like the body politic, is as good as the population permits. The population of cones decreases toward the periphery, but acuity falls not in proportion to cone density but rather in proportion to the cone to ganglion cell ratio (roughly as a straight line for eccentricities up to 20° to 30°). The decrease also depends on lateral interaction by horizontal and amacrine cells and on the effects of neurophysiologic facilitation and inhibition.

A foveal area of high resolution requires a highly mobile eye for fixation. Acuity is thus tied to motor function, and vision is blanked out during saccades because of motion blur. Fortunately fixation is remarkably accurate (within ±5 minutes of arc), involving servomechanisms of the highest precision.

The standard set of six oculorotary muscles are present in all animals unless the eye is microscopic. Compensatory automatic movements to keep the eye still during head or body motion are present in all eyes that can turn within the orbit. Voluntary or exploratory movements are intimately related to the presence of a fovea. Fishes with spontaneous movements have a fovea generally in the temporal retina as do most reptiles. Birds have little eye mobility, relying on flexibility of the neck and their temporal foveas for accurate distance judgments. In mammals the more frontal the eye position the more fixed the pattern of conjugate movements. Voluntary movements are correlated with acuity, which happen to correspond fairly well well with intelligence (Walls, 1942). Curious is the correlation with lid opening; one can estimate an animal's ocular motility by how much sclera shows.

The peripheral retina does not have better motion acuity as is commonly stated; rather its resolution is so poor that it only sees disembodied motion and not the moving target, hence the misconception.

A plot of acuity for different states of adaptation follows the expected pattern, showing a rod and cone branch. Visibility of bright on dark is better under scotopic conditions; resolution improves under photopic conditions. High resolution targets such as fine gratings exhibit only a cone branch like the adaptation curve to red light. Acuity is thus influenced by the immediate past history of light stimulation, hence the reason for delaying (for a few seconds or minutes) the subjective refraction examination after direct or indirect ophthalmoscopy and retinoscopy.

Variations in pupil size change acuity by altering retinal illumination, increasing depth of focus, and modifying the diameter of the blur circles. Thus the eye is in focus from real to optical infinity (6 meters). If the pupil were infinitely small and there was no diffraction, the eye could focus any object whatever its optical defect or accommodative amplitude. Actually acuity decreases rapidly for pupils smaller than 2 mm and more so if illumination is low. Optimal acuity is found for physiologic pupil size (about 3 mm). With larger pupil diameters, aberrations increase; with smaller ones, diffraction increases. The effect can be confirmed by monochromatic or coherent light.

The pinhole disc is a practical and diagnostically profitable little instrument. Because it decreases blur circles, acuity improves in an optically defective eye and thus differentiates between those conditions likely to benefit by lens correction. But the size must not be much less than 1 mm, or diffraction will make the vision worse, and target illumination should be increased to compensate. Occasionally in older patients the results are misleading, such as in nuclear sclerosis or central corneal opacities, and in functional amblyopia, the vision fails to improve.

By increasing depth of focus, the pinhole also acts as a magnifier, permitting a presbyope to hold targets nearer. It may be used to diagnose the type of ametropia, to exaggerate floaters, to check the final correction, or as a therapeutic device (multiple pinhole spectacles, pinhole contact lenses).

Every clinician knows that binocular acuity is better than monocular; it is not twice as good but about 5% to 10% better even on rough clinical measurements. A reasonable interpretation is that monocular acuity is not the lower limit to which the other eye adds; rather binocular acuity is the baseline from which the covered eye subtracts. Monocular acuity is worse than binocular acuity because of Fechner's paradox: the seeing eye is inhibited by covering the other.

Formation of a sharp optical image on the retina depends on precise accommodation. This factor is probably involved in the improved acuity when the patient makes an "effort to see," coupled with the associated miosis, which decreases blur circle diameter.

Changes in acuity with age are better understood at the end than the beginning. The measurement of acuity in infants depends on objective techniques. For example, Dayton and associates (1962), using optokinetic nystagmus, found resolution of 7.5 minutes or better in 1- to 5-day-old infants. Generally, acuity is probably better than the conventionally accepted values of 20/30 by 3 years of age and 20/20 by age 5 years. Wavelength discrimination and depth perception appear to follow a parallel development. Unequivocal data of course would be extremely useful in the diagnosis of amblyopia. In its absence, if there is an anomalous fixation pattern or anisometropic hyperopia, it is wise to begin treatment.

The best known method of measuring acuity objectively is optokinetic nystagmus inhibition by a variable-sized fixation point or stationary grating. Acuity is then the visual angle subtended by the smallest fixation point that arrests the optokinetic nystagmus. The method may be coupled with oculography, and the correlation with Snellen acuity is high (0.85). Also worth trying in children is the subjective identification of spherical candy cake decorations (Bock, 1960). The child is allowed to taste one and then selects from an arrangement held in the examiner's hand (about 12/24 vision at 1 foot).

In older patients, acuity may decline for a variety of reasons. Miosis, lens pigmentation, and media opacities limit the light reaching the retina. But dim vision is by no means inevitable, and many a gleam in an oldster's eye is due to more than lens reflections.

Psychologic factors

Seeing involves discrimination not only of detail, size, and position but also shape, pattern texture, motion, and color. All this is in the context of meaning, expectations, and past experience, modified by other senses, and varying with general health, fatigue, boredom, drugs, or emotional state. Seeing is selective, of evolutionary survival value, and limited only by our response capabilities. In this respect, evolution has not kept pace with technology; we have no receptors for electricity or ionizing radiations.

The distinction between sensation and perception was first made by Thomas Reid about 200 years ago; perception is a conception of the object perceived and an immediate and irresistible conviction of its present existence. No one has yet disproved his theory that it is a faculty with which the human race is endowed by the Creator. Associationists produce perceptions by pulling sensations out of a "meaningful" hat; gestaltists are sure that meaning is given immediately; behaviorists ignore the problem, and psychophysicists confine themselves to measuring it, optimistic that sooner or later someone will find a physiologic explanation in the little black box. We perceive first and make sense out of it later—if at all, by analysis and abstraction. If there is a hierarchy, it is more likely from the general to the specific and from the global to the detail. The problem is still with us; there is an uncomfortable similarity between visibility (sensation) and recognition (perception).

The influence of past experience is evident in the interpretation of blurred images, letters, multilated words, and familiar versus nonsense syllables. Numerous targets have been designed in an attempt to eliminate the factor of familiarity, such as Landolt rings, squares, polygons, triangles, dots, lines, hooks, vernier displacement, and brightness gradients. An extensive comparative study done by The United States Army (1948) showed that although all tests had fairly good reliability (0.80 or higher), they do not measure the same capacity. Factor analysis, for example, revealed that Landolt rings and checkerboards measured resolution; gradient tests, brightness discrimination; figure tests, form perception; and Snellen letters, a combination. Checkerboards were found the purest test of resolution (with Landolt rings second), and resolution targets are naturally influenced most by optical blur. Little variance was due to the examiner, order of test presentation, or time of day the subjects were tested. Significantly the subjects liked letter tests better than nonletter tests.

Shape tests do not measure resolution, and resolution tests do not measure the practical visual requirements of patients.

The perpetual question of whether letters should be replaced by pure resolution targets had been decided long ago by years of practical experience. Letters are better, the official adoption of the Landolt ring by the 1909 International Congress of Ophthalmology notwithstanding.

Acuity performance like any other human performance is subject to impairment by ocular and general health, emotional stress, boredom, and a variety of drugs acting both peripherally and centrally. It also improves with practice even for unfamiliar target bars and improves more for familiar letters or shapes. (Trained observers, for example, can improve their scores by reporting when the bars are separated enough to produce a dumbbell-shaped single target.) The set (Einstellung) is evident in clinical testing when the patient stumbles over the score number at the end of a line of letters (or calls the number 3 a "backward E").

Motivation is of importance in practical testing, since patients may refuse to guess for fear of embarrassment. How many letters must be missed before we deny credit? And out of how many per line? The decision is somewhat arbitrary; statistical studies suggest 5% to 10% based on average errors of normal subjects. The experienced clinician

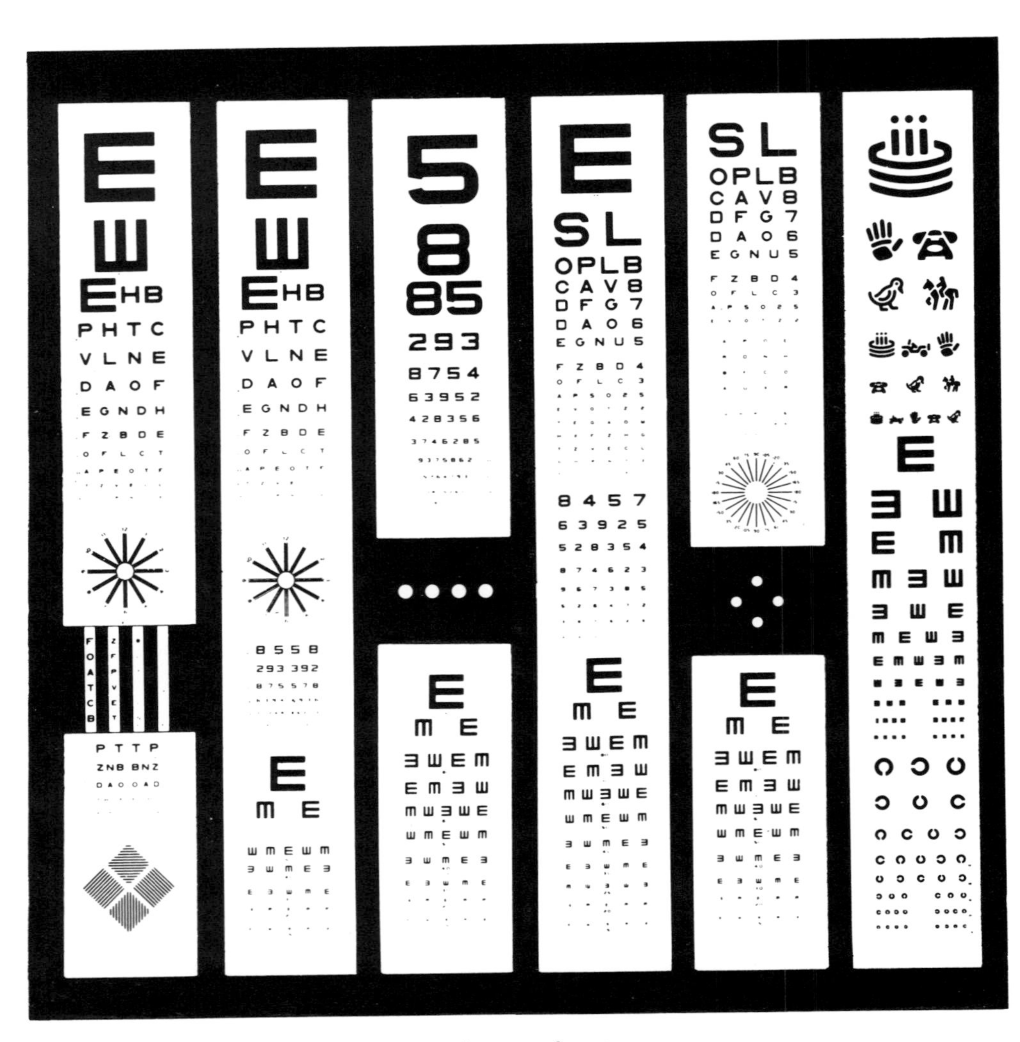

Fig. 10-4. Selection of acuity targets.

will also note how they are missed—the confusing interconversions of the amblyope, the intermittent disappearance of suppression, the total absence of the central letters in a line with a central scotoma. Conversely, we sometimes push and coax our patients into more effort on the clinical test than they are willing to expend at home or on the job. The difference between "easy" and "hard" 20/20 may tip the balance toward an unhappy patient. The effort to see involves attention, accommodation, miosis, and fixation and can produce significant improvement. Less clear is the influence of extraocular muscles or lids in deforming corneal curvature, either directly or indirectly, by raising intraocular tension. To what extent these effects are involved in "palming," staring, the simultaneous stimulation of other sense modalities, hypnosis, and eye "exercises" is not known. What is known is that such improvements are not related to changes in refractive error with which they are frequently confused.

CLINICAL MEASUREMENT

Visual acuity is tested monocularly and binocularly with a variety of targets, usually a species of Snellen chart, although not necessarily the original design (Fig. 10-4). Most acuity charts are printed with maximum contrast, about 84% reflectance for the white background and 4% for the black letters. The precise level of illumination has never been standardized; approximately 5 to 20 foot-candles evenly distributed are recommended. Although contrast on projected charts is not as good as on engraved ones, the difference is negligible above these illumination levels, and the variety, simplicity, and flexibility of projected targets would be difficult to give up.

It is conventional to record the visual angle of the smallest target recognized in fractional form. The advantage of the fraction is that it leaves no doubt as to the distance at which the test was made. Fractions speak louder than words. If target detail of 1 minute is resolved, acuity is expressed as 20/20 or 6/6. If vision is better than 20/20 (not uncommon), it is recorded as 6/4 or 20/10. If a certain number of letters are missed, it may be noted as 6/9 − 2, indicating two letters were miscalled on the 6/9 line. This is a purely qualitative "footnote," applicable for one particular chart on which the number of letters per line remains constant for subsequent (serial) testing. It means little when compared to other charts with different letters. If the patient cannot discriminate the largest target (usually 20/200 or 6/60), he is allowed to approach the chart until the target is resolved, and the distance is noted (for example, 5/60 means that the 20/60 target was seen at 5 feet). For vision less than this, the distance is noted at which the patient can count fingers (e.g., FC at 4 feet), recognize hand movements (e.g., HM at 1 foot), or perceive light (LP). If there is only light perception, the examiner should note whether it is correctly or incorrectly localized and if color vision is present.

In the seventeenth century Dr. Jurin described the method of measuring acuity using test charts constructed of parallel, alternate black and white lines. The patient was to "remove himself backwards" until he could no longer distinguish them, and then "let him stand and measure the distance, and by calculation under what angle each of the black and white spaces appear to his eye—which being known, he hath a standard." Three hundred years later, although we still have no standard, the method has changed little. Jurin's is one of two techniques of measuring acuity. The patient approaches a target of constant size until it is resolved; or the target distance remains fixed, but its size is increased until resolved.

The first method, still used by the armed services, allows better quantification, since scores need not be fitted into some predetermined pigeonhole limited by the number of lines on a chart. The entire chart can consist of standard (20/20) letters, with enough of them to be unlearnable, and the distance at which they are discriminated is recorded as the numerator of the visual fraction. Thus 12/20 means the 20/20 letter was read at 12 feet. The result can be converted by dividing into 400; thus 400/12 = 33 or 20/33.

The second method is more applicable to refraction in which it is not convenient to change target distance. The usual testing distance is 20 feet (610 cm), selected on the optical basis of light vergence. The waves incident at the eye have a vergence of −0.166 D, a negligible amount in terms of both discrimination and lens correction.

Office space being what it is, the question arises whether a smaller testing distance might do as well. In fact, Snellen used 5 meters, and the International Congress of Ophthalmology in 1909 accepted it as an alternative. For distances shorter than 5 meters, a mirror system is generally recommended, but even this has recently been questioned (Hofstetter, 1973). For example, 4 meters would not only require less space but directly relate to the 40 cm near-testing distance (40 cm/50 would be equivalent to 4/5 instead of 6/7.5).

Acuity targets

Landolt rings consist of "C" targets subtending an overall angle of 5 minutes, with a stroke width and gap of 1 minute. The patient is required to localize the position of the gap, which in practice must be confined to four directions. There is thus an element of guessing, and there will be a size at which the chance of a correct guess is one out of four (such as an identation in the intensity distribution of the black ring even when the gap is not resolved). Target blur is also influenced by the position of the primary meridians of an existing astigmatism; in fact these targets have been recommended to identify the primary meridians (by adding an additional gap perpendicular to the first). Astigmatism is most effective when the major axis of the blur ellipse is perpendicular to the gap, least effective when it is parallel, and intermediate when it is oblique. It is not difficult to see why Landolt's rings have never become popular with clinicians. The 1916 and 1930 committees of the American Medical Association recognized the inevitable by recommending 18 letters of comparable difficulty, although upholding the "scientific" accuracy of Landolt rings.

The checkerboard pattern, such as in Bausch & Lomb's Ortho-Rater, consists of four squares, one of which differs from the other three. The patient is required to identify which one differs and its position (one of four directions). The test square consists of a series of small alternate black and white squares subtending the minimum angle of resolution. When squares cannot be resolved, they appear uniformly gray and, if properly constructed, of the same contrast. It apparently makes no difference whether size variation is achieved by decreasing dimensions of the small squares or by reducing the size of the overall test target. Either way, the patient need not resolve the detail in the test target, providing it is recognized as different from the others. Moreover, in astigmatism the small squares are distorted, but the patient will be able to identify them correctly even though they form a kind of grating image.

Letter targets have the advantage of more nearly representing practical visual tasks. Although they do not measure pure resolution, are influenced by literacy and past experience, can be recognized if moderately blurred or incomplete, and vary in legibility, contrast, and typography, they do provide an index of "all-around seeing ability." Identification of letters is immediate and unequivocal. It is not unusual, for example, for a patient to skip some targets or even start over on a line. This is easily recognized with letters but leads to confusion with Landolt rings or checkerboards unless individually presented. Moreover, long years of clinical experience have established associations between letter acuity and ocular disease, such as in cataract progression. Most adults have grown up with letter charts in school, industrial, or screening tests, and the ability to check recognition for themselves is no small part of letter charts' appeal. For these reasons, letters are likely to continue to be the choice for clinical acuity testing.

Some letters are easier to recognize than others. Numerous investigations to grade their relative difficulty lead to as many different results. The variance depends on the criterion used, such as resistance to blurring, short exposure, peripheral presentation, contrast variations, illuminance variations, color variations, letters of differing types, external or internal lighting, with or without control of pupil size, accommodation, ametropia, etc., and the selection of scoring methods. Some letters (C, D, O, G, and Q) are readily confused, some (A, V, L, T, O) involve little resolution, some (A, T, M, V, O, X, and W) are symmetrical for mirror viewing, some are reinforced by crossing strokes, and some are recognizable by shape (A and V). If the selected group of letters is too restricted, guessing enters into the scoring. The choice of letters used in the experiment will also influence the likely confusion (thus "T" scores easy in some investigations but difficult in others because "Y" is included as an alternate). A reasonable compromise is the selection of ten nonserified letters with a variety of vertical, horizontal, oblique, and curved contours: Z, N, H, R, V, K, D, C, O, and S, in the estimated order of difficulty (Sloan, 1951).

Near point acuity

Testing near vision with letters is complicated by three distinct systems of notation: Snellen, Jaeger, and the point system (printer's type). The Snellen notation is simply an equivalent reduction for the near distance, maintaining the same visual angle. Industrial and compensation forms still use the impractical 14-inch test distance instead of the more reasonable 16 inches (40 cm). Strictly speaking, one cannot specify a near acuity as 20/40 but rather as 14/28, 16/32, 40/80, or 2 minutes at 16 inches, to indicate the actual distance used.

The Jaeger notation is an historic enigma that appears to have diplomatic immunity from change. It correlates with nothing, and Jaeger would never commit himself to the distance at which the print should be used. Not much better is the point system, which doubtless is useful to the printing trade from which it was arbitrarily adopted but is not useful for clinical much less for scientific measurement. The interconversion from one to the other is only approximate and, considering the variety of print, an abdication of uniformity. Thus 6-point type is approximately equivalent to Jaeger 3, which is approximately equivalent to 14/28, all of which means little unless one knows that this is about the size of the print in the telephone directory. Either acuity measurements are based on visual angles or they are not; if they are, we should abandon targets that have only an accidental relation to them. Type is unquestionably useful in evaluating a presbyopic add, for example, of a telephone operator (or any other patient who brings in print as a work sample)—but this is perpetually confused with *measuring* near acuity. It leads to difficulties in correlating distance and near acuity, which are diagnostically important.

The history of print type is a fascinating story in itself. The earliest printers were jewelers who tried to duplicate script. Books were supposed to be recognizable by their print: one kind for religious subjects (Gothic), another for classics (Roman), and still another for popular reading material. Modifications were added with changes in printing technique and the price of paper and to attract attention in advertising. The use of serifs, the height of upstroke or downstroke, and their compression and thickness all add individuality to the printed page but confound standardization in measuring acuity. Whether serifs should be added to test letters depends on how much importance one attaches to overall contrast versus contour masking.

Binocular acuity

A binocular (stereoscopic) method of testing acuity is incorporated in various screening techniques and may be used in the refracting room by means of polarizers and anaglyphs. The patient is expected to fuse the peripheral field, and in the case of monocular testing, the other eye sees dummy targets. As discussed in Chapter 15, there are certain diagnostic advantages to binocular refraction such as in evaluating suppression, aniseikonia, fixation disparity, astigmatism, and heterophoria, but the measurement of acuity is another matter. The stereoscope is not only a poor method of measuring monocular acuity but also a poor method of measuring binocular acuity. Not everyone fuses peripheral fields, the patient must suppress the empty field seen by the untested eye, and there is a reciprocal effect because of ocular dominance. According to Linksz, the clinical value of occlusion rests not on naturalness or simplicity but on the fact that accumulated clinical experience has established this measure as the standard.

Nomenclature and notation

The best known and most widely used system of acuity notation is the Snellen fraction: V = d/D, where d is the distance at which a given letter can just be discriminated and D is the distance at which the same letter subtends 1 minute. Thus 20/40 means that at a viewing distance of 20 feet, the patient can just discriminate an object subtending 1-minute detail at a distance of 40 feet. The patient's minimum separable is thus 2 minutes.

The geometry of the Snellen fraction is illustrated in Fig. 10-5. Since the angles are small, the tangents are equal to the sines, and the linear distances may be substituted.

The decimal system of acuity notation, first suggested by Monoyer in 1875, represents the reciprocal of the visual angle (in minutes) or the numerical value of the fractional Snellen notation. Thus 20/40 represents a 2-minute visual angle and a decimal notation of 0.5. This system is still popular on the continent (presumably because it allows easy interconversion between Snellen fractions expressed in feet or meters). It is much more likely to be misinterpreted as a percentage of vision loss, it gives no indication of the actual distance at which the test was made, and it is not even good algebra. The Snellen fraction is not a proper fraction and cannot be reduced because the denominator varies. A uniform decimal system is impossible with a uniform geometric progression of letter size. The only advantage of the decimal is to mentally compare one poor acuity to another. Finally, the decimal notation of 1.0 (20/20) gives the impression that this is maximum acuity, whereas many patients can do better.

It is true that 20/40 might also be misinterpreted as a 50% visual loss. This seems to worry chart designers more than patients. Actually, it is an unlikely possibility when the entire chart is exposed. There is no use pretending that patients can ever be made to understand our medical jargon, and no one is proposing to change the system of recording blood pressure, blood counts, or blood sugar simply because the numbers might be misinterpreted. The Snellen fraction is in fact neither so complex that its meaning cannot be explained, nor does it preclude interpretation for the patient in practical terms of visual function—the ability to see the blackboard, drive a car, or carry out some particular reading task. 20/40 means the minimum visual angle is doubled, and the resolving power is one half; it is up to us to make sure the patient does not get the idea that his vision is halved.

A subcommittee of the American Committee on Optics and Visual Physiology

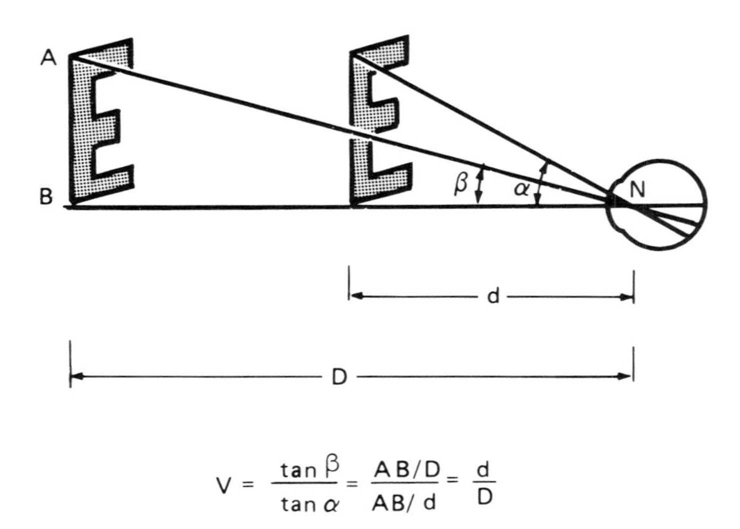

$$V = \frac{\tan \beta}{\tan \alpha} = \frac{AB/D}{AB/d} = \frac{d}{D}$$

Fig. 10-5. Geometry of Snellen fraction.

(Ogle, 1953) studied the problem of acuity nomenclature and recommended, among other things, that a decimal system should not be adopted and that "for the designation of visual acuity serious attention should be given at once to the adoption of the visual angle of resolution in minutes of arc of the component parts of the test letter of the chart." The conclusion is not in accord with the statement that "the members of the committee almost unanimously and strongly felt that visual acuity should be stated in terms of the quantity actually measured," which would require an expression of both the visual angle and the test distance (e.g., 3 minutes at 20 feet) in which case one might just as economically write 20/60, or 6/9. We have seen that visual angles are not directly comparable at different distances. Not only are there contradictory Aubert-Forster effects, but more important in refraction, there is also magnification and convergence induced by spectacle lenses, and there are different consequences of such diseases as nuclear sclerosis, congenital nystagmus, and corneal opacities for distance and near acuity.

In addition to fractional Snellen, decimal, and visual angle notation, a number of other systems have been proposed, some arbitrary and some empirical. An arbitrary system from one to ten, one to fifteen, or any other number would be reasonable if everyone used the same charts, that is, the same number of intermediate steps. For clinical purposes, the range from 1 to 10 minutes of visual angle at 20 feet is generally sufficient, but there is no agreement on how the steps should be subdivided. The "oxyopters" of Blascovics was an admirable attempt to make acuity notation analogous to lens diopters. Oxyopters were to be defined by dividing the minimum visual angle into 60 (e.g., 20/60, minimum angle 3 minutes = 20 oxyopters). Although the system attaches a higher number to better acuity and is free of any connotation of visual efficiency, it suffers from the handicap of not specifying the test conditions. Javal, noting that block letters subtend four times the area when the visual angle is halved, wanted to grade acuity by the square of the decimal notation. (No one has, as far as I know, used the more logical steradian.) Another alternative suggested by Linksz is a specific unit of human visual acuity (he proposed to name it "snellen") in which 1 minute (or 30 seconds, depending on the choice of standard acuity) would be equivalent to 100 snellen, either graduated in arithmetic or geometric progression. This system would work well providing we all used the same chart at equivalent distances (Linksz had the Lancaster chart in mind).

Obviously there has been no great rush to change the present system. Not only are clinicians attached to the Snellen fraction for good reasons of clinical familiarity and simplicity, but also too many insurance, driver's license, employment, accident, and compensation forms are around specifying results to be recorded in Snellen fractions. The 20/20 concept is so entrenched in the public mind that it is doubtful that 6/6 will be acceptable when we switch to the metric system. Perhaps the matter will eventually be settled by a representative international body (another glass menagerie that will prepare for every contingency and fail at the first practicality).

One problem in constructing acuity charts is how much more difficult (how much larger) a line of targets should be compared to the previous one. A geometric progression is in accord with the psychophysical rule that sensations change arithmetically with a logarithmic (i.e., geometric) change in stimulus.* The difference is best illustrated by comparing the arithmetic chart of Monoyer (Table 10-1), in which each letter size decreases by one tenth the size of the largest letter, with the geometric chart of Green (the one commonly used), in which each line is about 25% larger than the previous one (Table 10-2). It will be seen that the decimal notation provides poor accuracy in the range of poorer

*An arithmetic progression consists of quantities that increase or decrease by a common difference; for example, the series 1, 5, 9, 13, 17 has the common difference +4. A geometric progression is one in which quantities change by a constant factor; for example, the series 3, 6, 12, 24, 48 has the common factor 2.

acuities and falsely high accuracy for good acuities.

The refractionist cares little about how the steps are divided, providing his patient can be corrected to 20/20; and the researcher would not be satisfied by any system of pigeonholing, preferring continuously variable sizes or distances. For clinical measurement of acuity, however, it is necessary to have a certain number of steps, the choice being balanced between time available for testing (40 steps would take forever), the average distribution of acuities, precision, and the desirability to express the resultant fraction as a whole number. Most clinicians prefer a geometric sequence. It follows that any practical system is a compromise, usually involving ten or eleven steps with a little fudging to get whole number fractions.

To achieve the requisite steps, the usual progression is one line 25% larger than the preceding (or 20% smaller than the letter size above). This happens to be $\sqrt[3]{2}$ or 1.25, a figure first adopted by Green in 1905. Other values have also been suggested (1.41 by Blascovics and 1.19 by the United States Army). A reasonable basis on which to decide the issue was worked out by Dreyer (1964).* He computed the value so that the number of letters (equalized for legibility) missed plotted a linear change against the logarithm of the visual angle. His conclusion was that the Green system (1.25) fits the data best in terms of acceptable ranges of measurement errors.

A 20/200 letter takes up more space than a 20/20 one; it follows that more small letters than large ones can be presented on any given line of the acuity chart. Is this fair? The nature of printing being what it is, the printer must crowd a certain number of letters and words per line. Thus there is practical justification for presenting at least the 20/20 letters at a similar separation. Now it has been known for some time that an amblyopic eye does considerably better when letters are presented individually than when crowded together. This masking has been termed the "crowding phenomenon" or "separation difficulty." These terms imply something more mysterious than is actually involved. First, the crowding phenomenon is characteristic not only of amblyopia but also of any decrease in resolution. Second, probably the effect is the result of contrast discrimination; the closer the letters the less dip in the illumination between them. If the letter is an I, one has a simple grating, and the resolution of the two are identical; if the letters are equated for contrast, the result parallels that obtained for gratings. This also explains why, with central fixation of a line of

Table 10-1. Monoyer arithmetic acuity chart

Line on chart	*Decimal notation*	*Snellen notation*	*Visual angle (minutes)*
11	0.1	6/60	10
10	0.2	6/30	5
9	0.3	6/19.8	3.3
8	0.4	6/15	2.5
7	0.5	6/12	2
6	0.6	6/10	1.67
5	0.7	6/8.6	1.43
4	0.8	6/7.5	1.25
3	0.9	6/6.7	1.11
2	1.0	6/6	1
1	1.1	6/5.5	0.91

Table 10-2. Green geometric acuity chart

Line on chart	*Decimal notation*	*Snellen notation*	*Visual angle (minutes)*
12	0.1	6/60	10
11	0.12	6/48	8
10	0.16	6/38	6.3
9	0.20	6/30	5
8	0.25	6/24	4
7	0.32	6/20	3.2
6	0.4	6/15	2.5
5	0.5	6/12	2
4	0.63	6/10	1.6
3	0.8	6/7.5	1.25
2	1.0	6/6	1
1	1.5	6/4	0.65

*Unacceptable, however, are attempts to vary letter size by degree of ametropia or dioptric blur. To blur the subsequent line by exactly 0.25 D is not only impossible but in effect places acuity on an optical instead of a physiologic basis.

letters, one can recognize the end letters but not the ones in between. The contrast of the end letters is not interfered with.

Visual acuity and visual efficiency

Visual efficiency would be a purely ophthalmologic problem were it not that the other side of the coin represents visual loss. And here the solution involves economic, legislative, juridical, and moral considerations. There is considerable diversity in federal, state, and local standards of compensation for an industrially injured eye, sometimes by as much as five times.

A certain degree of ambiguity surrounds such terms as "visual loss," "loss of efficient vision," "loss of an eye," or the "loss of useful vision in one eye." Sometimes it is difficult to decide what vision existed prior to an injury, as in the case of an unsuspected amblyopia (hence the importance for everyone to have an acuity on file). Nor are there any simple rules to govern compensation for loss of earning capacity, the risk of further damage, or the grave consequences of losing the only remaining eye. Also to be individually evaluated are such factors as loss of color vision, cosmetic defects, scotomas, metamorphopsia, lagophthalmos, muscle anomalies that do not lead to diplopia (such as a palsy of the lateral rectus in an only seeing eye), disturbances in accommodation and adaptation, epiphora, and the cost of treatment, not to speak of the mental and physical suffering.

To estimate the decrease of percentage visual efficiency it is necessary to define efficient vision. We cannot rely on the engineer's energy input/output formula, and thresholds do not include pain, discomfort, itching, burning, and other symptoms to which the human machine is subject. The economic and juridical definition tries to take some of these factors into account by considering peripheral vision, motility, and binocular vision, as well as the central visual acuity. But there is no standard definition of efficient vision and no absolute number that can be assigned to visual loss. The diagnosis of visual loss is a clinical and professional judgment, based not only on acuities but also on a history, and evaluation of symptoms and signs, consideration of normal ranges and errors of measurement, differences between acceptable and optimum vision, and the actual and potential capacities of the patient.

It is clear that the Snellen fraction does not express visual efficiency. Indeed the courts take a dim view of such claims, pointing out that they have no support in testimony or reason. If 20/40 implies 50% visual loss, it is manifestly absurd that 20/60 implies 33% loss. A variety of suggestions have therefore been advanced to convert the central acuity measurement into a more realistic expression of vision. The most useful of these is based on the experiments of Snell and Sterling in 1924. They used meshed glass planes (etched by cross-lines) of such density that the vision of a series of observers with 20/20 vision was reduced to 20/40 through one plate. By using a series of such plates, it was found that six destroyed all useful vision (20/400), three resulted in 20/100 vision, and two resulted in 20/65 vision. It follows that one plate reduced vision by 16.6%, accepting 20/400 as no useful vision.* By further assuming that 20/20 is 100% vision and 20/200 is 20% vision, they computed the constant for a geometric progression of visual efficiency: $VE = 0.83625^{(n-1)}$, where n is the visual angle. This was accepted as an expression of the efficiency of central vision by the 1925 Committee on Compensation for Eye Injuries of the American Medical Association and is still used.

In 1955 a committee appointed by the Council on Industrial Health of the American Medical Association issued a revised report that incorporates separate tables for distance and near acuities, in which near acuities are given more weight (for example, whereas 20/100 means 50% loss, 14/140 corresponds to 98% loss; conversely 14/28 represents only 10% loss because the patient is still able to read ordinary newsprint). This is in accord

*There is some difficulty here; the plates duplicate the effect of a cataract or extensive corneal opacification but not the acuity loss of localized macular disease.

with the concept that (a) distance and near acuities may differ significantly, and (b) distance and near acuities do not represent equal degrees of visual disability. The Committee further recommended that aphakic vision be evaluated as 50% of the best visual efficiency obtainable with a correcting spectacle lens; if vision is 20/80 (60%), the central visual efficiency is considered to be 0.50 of 60% or 30%. Moreover, if the aphakic eye is the only useful eye, an additional 25% is allowed in estimating binocular efficiency. Contact lens correction was not recommended, since it might not be tolerated at a subsequent time, and spectacle correction might provide less improvement. (Additional considerations of motility, peripheral acuity, and binocular efficiency are considered in Chapter 15.)

Acuity and refractive error

The theory and practice of refraction is based on acuity as the criterion of the optimum lens correction. If for some nonoptical reason, acuity does not come up to standard, the usual subjective methods of refraction may become unreliable and must be modified. Conversely, to utilize the patient's discrimination of lens changes, the conditions should allow for the best resolution possible. The ametropic eye imposes limits on acuity because of retinal image blur. The image is out of focus, the light is spread out across the retinal mosaic, and the contour gradients are less sharp. In elementary optical theory, a blur circle can be defined with mathematic precision; in reality it is at best a diffraction pattern, usually with aberrations because of wide apertures. The light distribution is a caustic, with some parts clearer and some parts more luminous and, in case of polychromatic light, spread out by wavelength. Unless accommodation is totally inactive, the portion of the caustic on the retina will vary, and involuntary eye movements shift the image over a collection of foveal cones. At the borders the illumination gradient is less steep; hence the dip between mounds of light representing critical detail fuses more readily.

A considerable amount of effort has been wasted in attempts to correlate the degree of ametropia with decrease in acuity. The acuity measurements fail to consider that the patient may squint and squirm and squeeze his eyes and tilt, turn, and twist his head. Ametropia measurements ignore pupil diameter, adaptation, contrast, and illumination. Most eyes have a different ametropia with and without cycloplegia by objective or subjective criteria, and the effect on acuity varies with the amount and axis of astigmatism. Both acuity and ametropia are gross measurements taken with many factors uncontrolled. To express the relation by a formula is to try to turn two clinical sows into a mensurational silk purse.

Refractive amblyopia

It has been known for some time that patients with high ametropia and without strabismus may have subnormal vision despite adequate lens correction. After wearing the lens for a period of weeks or months, however, the vision improves without further treatment. It is more likely in myopia with or without astigmatism and generally in the more ametropic eye, although it may also be bilateral. It can present itself in adults as well as children. The mechanism is not known, but its functional character is indicated by the reversability.

REFERENCES

Abbott, W. A.: Records of visual acuity, Am. J. Ophthal. **1**:71-72, 1918.

Abraham, S. J.: Bilateral ametropic amblyopia, J. Pediatr. Ophthal. **1**:57-61, 1964.

Adjutant General's Office: Studies in visual acuity, Department of the Army, Personnel Research report No. 742, 1948, United States Government Printing Office.

Adler, F. H., and Fliegelman, M.: Influence of fixation on visual acuity, Arch. Ophthal. **12**:475-483, 1934.

Andersen, E. E., and Weymouth, F. W.: Visual perception and the retinal mosaic, Am. J. Physiol. **64**: 561-594, 1923.

Appelle, S.: Perception and discrimination as a function of stimulus orientation: the "oblique effect" in man and animals, Psychol. Bull. **78**:266-278, 1972.

Arnulf, A., Dupuy, O., and Flamant, F.: Les microfluctations d'accommodation de l'oeil et l'acuité visuelle pour les diamètres pupillaires naturels, Compte. Rend. Acad. Sci. **232**:349-351, 1951.

Aubert, H., and Forster, R.: Beiträge zur Kenntniss des indirecten Sehens, Graefe's Arch. Ophthal **3**:1, 1857.

Avant, L. L.: Vision in the Ganzfeld, Psychol. Bull. **64**:246-258, 1965.

Ball, R. J., and Bartley, S. H.: Effects of intermittent monochromatic illumination or visual acuity, Am. J. Optom. **47**:519-525, 1970.

Blackwell, H. R.: Luminance difference thresholds, Handbook Sensory Physiol. **7**(4):78-101, 1972.

Blakemore, C., and Campbell, F. W.: On the existence of neurones in the human visual system selectively sensitive to the orientation and size of retinal images, J. Physiol. (London) **203**:237-260, 1969.

Bock, R. H.: Amblyopia detection in the practice of pediatrics, Arch. Pediatr. **77**:335-339, 1960.

Campbell, F. W., and Green, O. G.: Optical and retinal factors affecting visual resolution, J. Physiol. (London) **176**:576-593, 1965.

Campbell, F. W., and Gubisch, R. W.: Optical quality of the human eye, J. Physiol. (London) **186**:558-578, 1966.

Campbell, F. W., and Kulikowski, J. J.: Orientational selectivity of the human visual system, J. Physiol. (London) **187**:437-445, 1966.

Campbell, F., Kulikowski, J., and Levinson, J.: The effect of orientation on the visual resolution of gratings, J. Physiol. (London) **187**:427-436, 1966.

Campbell, F. W., and Robson, J. G.: Application of Fourier analysis to the visibility of gratings, J. Physiol. (London) **197**:551-566, 1968.

Carmichael, L., and Dearborn, W. F.: Reading and visual fatigue, New York, 1947, Houghton Mifflin Co.

Chall, J. S.: Learning to read: the great debate, New York, 1967, McGraw-Hill Book Co.

Cohn, T. E., Thibos, L. N., and Kleinstein, R. N.: Detectability of luminance increment, J. Opt. Soc. Am. **64**:1321-1327, 1974.

Crawford, J. S., Shagass, C., and Pashby, T. J.: Relationship between visual acuity and refractive error in myopia, Am. J. Ophthal. **28**:1220-1225, 1945.

Dayton, G. O., Jensen, G., and Jones, M. H.: Visual acuity of infants measured by means of optokinetic nystagmus and oculogram, Invest. Ophthal. **1**:414-420, 1962.

Ditchburn, R. W., and Ginsborg, B. L.: Involuntary eye movements during fixation, J. Physiol (London) **119**:1-17, 1953.

Drever, J. D.: Perceptual learning, Ann. Rev. Psychol. **11**:131-160, 1960.

Dreyer, V.: On the exactness of visual acuity determination charts with decimal, Snellen, and logarithmic notation, Acta Ophthal. **42**:295-306, 1964.

Enoch, J. M.: Physical properties of the retinal receptors and response of retinal receptors, Psychol. Bull. **61**:242-251, 1964.

Enoch, J. M.: The need for standards in tests of vision, Am. J. Ophthal. **72**:836-837, 1971.

Fantz, R. L.: The origin of form perception, Sci. Am. **204**:66-72, 1961.

Ferree, C. E., and Rand, G.: The testing of visual acuity, the comparative merits of test objects and a new type of broken circle as test object, Am. J. Ophthal. **17**:610-618, 1934.

Ferree, C. E., and Rand, G.: New method of rating visual acuity, J. Gen. Psychol. **25**:143-176, 1941.

Fink, W. H.: An evaluation of visual acuity symbols, Am. J. Ophthal. **28**:701-711, 1945.

Flamant, F.: Étude de la repartition de lumière dans l'image retinienne d'une fente, Rev. Opt. **34**:433-459, 1955.

Flom, M. C., Heath, G. G., and Takahashi, E.: Contour interaction and visual resolution: contralateral effects, Science **142**:979-980, 1963.

Foxell, C. A. P., and Stevens, W. R.: Measurements of visual acuity, Br. J. Ophthal. **39**:513-533, 1955.

Fry, G. A.: The significance of visual acuity measurements without glasses, Am. J. Optom. **25**:199-210, 1948.

Fry, G. A.: Blur of the retinal image, Columbus, 1955, Ohio State University Press.

Gibson, E. J., and Walk, R. D.: The "visual cliff," Sci. Am. **202**:64-71, 1960.

Giraud-Teulon, M.: Congrès International d'Ophthalmologie, Paris, 1862, J. B. Bailliere.

Glezer, V. D., Levshina, L. I., Nevskaya, A. A., and Prazdnikova, N. V.: Studies on visual pattern recognition in man and animals, Vision Res. **14**:555-583, 1974.

Graham, C. H.: Sensation and perception in an objective psychology, Psychol. Rev. **65**:65-76, 1958.

Graham, C. H., editor: Vision and visual perception, New York, 1965, John Wiley & Sons, Inc.

Green, J.: On a new series of test letters, Trans. Am. Ophthal. Soc. **4**:68, 1869.

Hecht, S., and Mintz, E. V.: The visibility of single lines at various illuminations and the retinal basis of visual resolution, J. Gen. Physiol. **22**:593-612, 1939.

Hecht, S., Shlaer, S., and Pirenne, M.: Energy, quanta, and vision, J. Gen. Physiol. **25**:819-840, 1942.

Hering, E.: Outlines of a theory of the light sense, translated by L. M. Hurvich and D. Jameson, Cambridge, 1964, Harvard University Press.

Hofstetter, H. W.: From 20/20 to 6/6 or 4/4? Am. J. Optom. **50**:212-220, 1973.

Hulsman, H. L.: Visual factors in reading; with implications for teaching, Am. J. Ophthal. **36**:1577-1586, 1953.

Hurvich, L. M., and Jameson, D.: The perception of brightness and darkness, Boston, 1966, Allyn & Bacon, Inc.

Jameson, D., and Hurvich, L. M.: Theory of brightness and color contrast in human vision, Vision Res. **4**:135-154, 1964.

Kepler, J.: Ad vitellionem astronomiae pars optica, 1604.

Kepler, J.: Dioptrice, 1611.

Kimura, J.: The effect of letter position on recognition, Can. J. Psychol. **13**:1-10, 1959.

Koenderink, J. J., and van Doorn, A. J.: Method of stabilizing the retinal image, Appl. Optics **13**:955-961, 1974.

Lebensohn, J.: Snellen on visual acuity, Am. J. Ophthal. **53**:152-155, 1962.
Linksz, A.: On writing, reading, and dyslexia, New York, 1973, Grune & Stratton, Inc.
Lit, A.: Visual acuity, Ann. Rev. Psychol. **19**:27-54, 1968.
Low, F. N.: Peripheral visual acuity, Arch. Ophthal. **45**:80-99, 1951.
Luckiesh, M.: Test charts representing a variety of visual tasks, Am. J. Ophthal. **27**:270-275, 1944.
Luckiesh, M., and Taylor, A. H.: Visual acuity at low brightness levels, Am. J. Ophthal. **27**:53-56, 1944.
Ludvigh, E.: Effect of reduced contrast in visual acuity as measured with Snellen test letters, Arch. Ophthal. **25**:469-474, 1941.
Ludvigh, E.: Direction sense of the eye, Am. J. Ophthal. **36**:139-143, 1953.
Lythgoe, R. J.: Measurement of visual acuity, special report No. 173, Medical Research Council, London, 1932, H. M. Stationery Office.
Marshall, W. H., and Talbot, S. A.: Recent evidence for neural mechanisms in vision leading to a general theory of sensory acuity. In Kluver, H., editor: Visual mechanisms, Biological Symposia, vol. 7, Lancaster, 1942, Jacques Cattell Press, pp. 117-164.
Matthews, M. L.: Appearance of Mach bands for short durations and at sharply focused contours, J. Opt. Soc. Am. **56**:1401-1402, 1966.
Meyer-Arendt, J.: Transfer functions in optical and visual testing. 1. Contrast transfer, Am. J. Optom. **45**:507-511, 1968.
Morris, A., Katz, M. S., and Bowen, J. D.: Refinement of checkerboard targets for measurement of visual acuity limens, J. Opt. Soc. Am. **45**:834-838, 1955.
National Center for Health Statistics: Binocular visual acuity of children: demographic and socioeconomic characteristics, United States, Department of Health, Education and Welfare publication 72-1031, series 11, No. 112, 1973, U. S. Government Printing Office.
Ogilvie, J. C., and Taylor, M. M.: Effect of length on the visibility of a fine line, J. Opt. Soc. Am. **49**:898, 1959.
Ogle, K. N.: On the problem of an international nomenclature for designating visual acuity, Am. J. Ophthal. **36**:509-521, 1953.
Ogle, K. N.: Visual acuity. In Straatsma, B. R., and others, editors: The retina, Berkeley, 1969, University of California Press, pp. 443-483.
Overington, I.: Interaction of vision with optical aids, J. Opt. Soc. Am. **63**:1043-1049, 1973.
Pieron, H.: The sensations, New Haven, 1952, Yale University Press.
Prince, J. H., and Fry, G, A.: Correction for the guessing bias in the Landolt ring test, Am. J. Ophthal. **46**:77-80, 1958.
Raab, E. L. Refractive amblyopia, Intern. Ophthal. Clin. **11**:155-168, 1971.
Randall, H. G., Brown, D. J., and Sloan, L. L.: Peripheral visual acuity, Arch. Ophthal. **75**:500-504, 1966.
Ratliff, F.: Mach bands: quantitative studies on neural networks in the retina, San Francisco, 1965, Holden-Day, Inc.
Riggs, L. A.: Visual acuity. In Graham, C. H., editor: Vision and visual perception, New York, 1965, John Wiley & Sons, Inc., pp. 321-349.
Riggs, L. A.: Responses of the visual system to fluctuating patterns, Am. J. Optom. **51**:725-735, 1974.
Robinson, H. M.: Why pupils fail in reading, Chicago, 1946, University of Chicago Press.
Segall, M. H., Campbell, D. T., and Herskowitz, M. J.: The influence of culture on visual perception, New York, 1966, Bobbs-Merrill Co.
Senders, V. L.: Physiological basis of visual acuity, Psychol. Bull. **45**:465-490, 1948.
Shlaer, S.: Relation between visual acuity and illumination, J. Gen. Physiol. **21**:165-188, 1937.
Sloan, L. L.: Measurement of visual acuity, Arch. Ophthal. **45**:704-725, 1951.
Sloan, L. L., and Brown, D. J.: Area and luminance of test object as variables in projection perimetry, Vision Res. **2**:527-541, 1962.
Sloane, A. E., and Gallagher, J. R.: Changes in vision during adolescence, Am. J. Ophthal. **33**:1538-1542, 1950.
Snell, A. C., and Sterling, S.: Percentage evaluation of macular vision, Arch. Ophthal. **54**:443-461, 1925.
Snellen, H.: Test-types for the determination of the acuteness of vision, ed. 4, London, 1868, Williams & Norgate.
Snellen, H.: Optotypi and visum determinatum, Berlin, 1896, Hermann Peters.
Stevens, S. S.: On the psychophysical law, Psychol. Rev. **64**:153-181, 1957.
Stevens, S. S.: To honor Fechner and repeal his law, Science **133**:80-86, 1961.
Stuart, J. A., and Burian, H. M.: A study of separation difficulties, Am. J. Ophthal. **53**:471-477, 1962.
Teller, D., Morse, R., Borton, R., and Regal, D.: Visual acuity for vertical and diagnoal gratings in human infants, Vision Res. **14**:1433-1439, 1974.
Thorn, F., and Boynton, R. M.: Human binocular summation at absolute threshold, Vision Res. **14**:445-458, 1974.
Thurstone, L. L.: A factorial study of perception, Psychometric Monograph No. 4, Chicago, 1944, University of Chicago Press.
Tinker, M. S.: Illumination standards for effective and easy seeing, Psychol. Bull. **44**:435-450, 1947.
Tolhurst, D. J.: Separate channels for the analysis of the shape and the movement of a moving visual stimulus, J. Physiol. (London) **231**:385-402, 1973.
Troland, L. T.: Analysis of literature concerning dependency of visual acuity upon illumination intensity, Trans. Illum. Eng. Soc. **26**:107-196, 1931.
van den Brink, G., and Bouman, M. A.: Visual acuity depending on spherical correction, Ophthalmologica **138**:222-224, 1959.
Voipio, H., and Hyvarinen, L.: Objective measurement of visual acuity by arrestovisography, Arch. Ophthal. **75**:799-802, 1966.

Walls, G. L.: The vertebrate eye and its adaptive radiation, Bloomfield Hills, 1942, Cranbrook Institute of Science.

Walls, G. L.: Factors in human visual resolution, J. Opt. Soc. Am. **33**:487-505, 1943.

Weale, R. A.: Problems of peripheral vision, Br. J. Ophthal. **40**:392-415, 1956.

Wertheim, T.: Ueber die indirekte Sehscharfe, Z. Psychol. Physiol. Sinnersorg. **7**:172-189, 1894.

Westheimer, G.: Visual acuity, Ann. Rev. Psychol. **16**:359-380, 1965.

Westheimer, G.: Visual acuity, Handbook Sensory Physiol. **7**(4):170-187, 1972.

Weymouth, F. W.: Visual sensory units and the minimal angle of resolution, Am. J. Ophthal. **46**:102-113, 1958.

Wildman, K. N.: Visual sensitivity at an edge, Vision Res. **14**:749-755, 1974.

Wolin, L. R., and Dillman, A.: Objective measurement of visual acuity, Arch. Ophthal. **71**:822-826, 1964.

Yamada, E.: Some structural features of the fovea centralis in the human retina, Arch. Ophthal. **82**: 151-159, 1969.

Yonemura, G. T.: Luminance threshold as function of angular distance from an inducing source, J. Opt. Soc. Am. **52**:1030-1034, 1962.

11

Cycloplegics and cycloplegia

In the garden of refractive problems, the use of cycloplegics remains a hardy perennial; an orchid when the patient cannot cooperate, a thorn if he is predisposed to glaucoma, and a forget-me-not when he leaves the office. The purpose of cycloplegia is not so much to obtain a more "scientific" refraction but to permit examination of infants and children or disinclined, dissimulating, and disorganized adults. It allows differentiation between spastic and psychic causes of asthenopia, between true myopia and pseudomyopia, and between the accommodative and nonaccommodative components of esotropia in eyes with small pupils and hazy media. It is applicable to both objective and subjective methods of refraction, retinoscopy, astigmatic dial, or cross cylinder. But it is not a guarantee of accuracy or an excuse for sloppy technique.

Cycloplegia is not always complete, not aways equal, and not always properly timed. The increased aberrations, reduced depth of focus, altered accommodation-convergence, spurious off-axis measurements, occasional toxic and allergic effects, and the additional time and inconvenience to the patient are the price paid for immobilizing the ciliary muscle and dilating the pupil. Moreover, few people would wear a lens prescribed only from cycloplegic measurement very long or very comfortably.

Cycloplegics are indicated because it is occasionally difficult to refract wihout them and, more importantly, because the information provided adds to our diagnostic data. The question is not whether to use cycloplegics but when.

CLINICAL PHYSIOLOGY

The innervation of the ciliary muscle like other smooth muscle is from the autonomic nervous system, which is anatomically classified into thoracolumbar or sympathetic and craniosacral or parasympathetic divisions. The parasympathetic division is the chief supply to the ciliary muscle; stimulation causes accommodation, and inhibition causes relaxation. The sympathetic division provides for some minor relaxation but is pharmacologically impotent and need not be considered.

A characteristic of autonomic innervation (adrenal gland excepted) is that it always involves two nerve fibers, a preganglionic nerve fiber and a postganglionic nerve fiber, which synapse, in the case of the parasympathetic, in a ganglion close to the organ innervated. The ciliary ganglion is the relay station for the eye.

The motor innervation of the ciliary muscle originates in the midbrain in a portion of the oculomotor complex termed the "Edinger-Westphal nucleus." The preganglionic parasympathetic fibers accompany the somatic motor fibers of the third nerve, leaving its inferior branch by a short motor root to the ciliary ganglion. The ganglion, a small 1 × 2 mm rectangular body in the apex of the orbit, gives rise to the postganglionic parasympa-

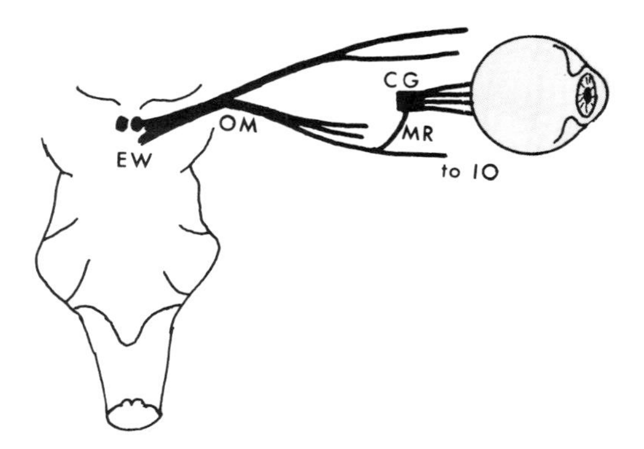

Fig. 11-1. Parasympathetic innervation to the ciliary muscle. The preganglionic fiber originates in the Edinger-Westphal nucleus *(EW)*, travels with the oculomotor nerve *(OM)*, and enters the ciliary ganglion *(CG)* through a short motor root *(MR)* that goes to the inferior oblique muscle *(IO)*. The postganglionic fibers originate in the ciliary ganglion and enter the eye through twenty or thirty short ciliary nerves.

thetic fibers that enter the eyeball by way of twenty to thirty short ciliary nerves (Fig. 11-1).

At a ganglionic synapse and at the neuroeffector parasympathetic junction, nerve stimulation results in the liberation of an active chemical mediator, acetylcholine, and such nerves are termed "cholinergic." The brief duration of action of acetylcholine is due to a group of inhibitory tissue enzymes, the cholinesterases, present in high concentration at the nerve terminals. There are specific and nonspecific cholinesterases, the first most effective in terminating the action of acetylcholine and the second acting less rapidly and of unknown physiologic importance. (One is involved in limiting the action of succinylcholine.)

Cholinergic drugs may directly stimulate the effector (e.g., pilocarpine), inactivate cholinesterase, prevent the destruction of acetylcholine (either reversibly like physostigmine or irreversibly like isoflurophate), and some have a combination of actions (neostigmine). Acetylcholine is a poor cholinergic drug because of its rapid hydrolysis.

Although acetylcholine is the only chemical cholinergic mediator, there are at least three different cholinergic receptors: within the ganglion (blocked by such antiganglionic blocking agents as tetraethylammonium); at the neuromuscular junction with skeletal muscle (blocked by curare); and at postganglionic parasympathetic neuromuscular junctions (blocked by atropine). The transmission at the neuroeffector site of the ciliary muscle belongs to the last, sometimes termed "muscarinic," type.

Cycloplegics belong to the group of anticholinergic agents whose basic action is blocking the postganglionic neuroeffector response. They combine without intrinsic action with the receptors of the smooth muscle cells. Since the innervation of the iris is similar to the ciliary body, all cycloplegics also dilate the pupil but at lower drug concentration.

The blocking action of cycloplegics does not prevent the release of acetylcholine but its combination with the receptor. Such competitive blocks may be overcome by supplying more transmitter and an appropriate time interval.

Cholinergic blocking agents increase heart rate; inhibit the smooth muscle contraction of bronchioles, gastrointestinal tract, bladder, and uterus; decrease the secretions of salivary, sweat, mucous, and lacrimal glands; and dilate cutaneous blood vessels. They also produce a fall in blood pressure and depress or stimulate the central nervous system.

PHARMACOLOGY

The desired effect of a drug is termed its "therapeutic effect"; undesired effects are "side effects." The mydriasis accompanying atropine is a side effect, when the drug is used as a cycloplegic. An "idiosyncrasy" is a drug effect which differs from that obtained in the majority of patients treated with an equivalent dosage (such as the hallucinations occasionally seen with scopolamine or cyclopentolate). An "intolerance" refers to an exaggeration of a typical drug response, such as the prolonged duration of cycloplegia in some patients. "Hypersensitivity" is a reaction that differs from either the expected pharmacologic response or from known toxic reactions. Frequently, the terms "drug allergy" and "drug hypersensitivity" are used interchangeably. Hypersensitive reactions include blood dyscrasias, angioneurotic edema, asthma, drug fever, conjunctivitis, dermatitis, and anaphylaxis, the last being potentially the most serious. These unwanted and unpredictable effects, although less common with topically administered drugs, are an expression of individual differences, with occasional adverse consequences on vision, general health, or life itself.

In case of a severe drug reaction, the patient is made to lie down, and the airway and blood pressure is checked. Hypotension is controlled with subcutaneous epinephrine (if there is no history of cardiac disease); antihistamines may be given intravenously; oxygen may be given for cyanosis; and sedatives may be given for convulsions.

Any cycloplegic or mydriatic can induce an acute attack of glaucoma in predisposed eyes. The factors involved are largely anatomic. In such eyes, pupil dilation causes the peripheral iris folds to block a narrow chamber angle, and the increased pressure in the posterior chamber pushes the relaxed folds further, resulting in a sharp, rapid rise in intraocular pressure. Cycloplegics should therefore be used with caution in eyes with shallow chambers, small anterior segments associated with hyperopia, and in patients with a history of intermittent blurred vision or pain at night. The absence of a previous rise in pressure after cycloplegia is no guarantee that it may not occur, and a chamber of normal depth may be blocked if the iris is close to the trabeculum (plateau iris).

Most ophthalmic drugs are administered topically in liquid or ointment vehicles. Drug penetration depends on the integrity and metabolic characteristics of the corneal barrier, the main entrance port to the interior of the eye. Drugs in liquid vehicles are rapidly mixed with the tears and eliminated through the lacrimal drainage system, a process enhanced by blinking and ocular movements. Conjunctival and nasolacrimal absorption occurs to a limited extent and is responsible for systemic reactions. Ointments provide longer contact with the eye but interfere with corneal wound healing; liquids are better absorbed but more rapidly inactivated.

Ointment bases are generally petrolatum, lanolin, or peanut and castor oil, with an emulsifier in compound bases. Problems with ointments are both chemical (drug absorption) and physical (drug dispersion). Thus a twenty times greater atropine concentration is required in water in oil emulsions than in oil in water emulsions.

Methylcellulose is a synthetic, water-soluble compound frequently used as a lubricant and drug vehicle. It produces a sticky effect and is often replaced with hydroxypropyl methylcellulose. Methylcellulose is stable and does not support the growth of microorganisms, and its viscosity permits longer drug contact with the eye. Polyvinyl alcohol is another commonly used drug vehicle and is less viscous than methylcellulose. Surface-acting agents such as benzalkonium chloride, polysorbate 80, and polyoxyl 40 stearate may be added to increase drug penetration.

Both liquids and ointments may become contaminated by bacteria, fungi, or viruses, and only drugs obtained from reputable manufacturers should be used. These products conform to requirements for sterility of the *United States Pharmacopoeia*. All ophthalmic medications are required to be sterile and to provide for maintenance of reasonable sterility once the container is opened. To this end, preservatives are used in multidose containers such as chlorobutanol, benzalkonium, mercurials, and esters of benzoic acid. The effectiveness with which they combat secondary contamination depends on the time interval the preservative is allowed to act (i.e., the dosage frequency) the total volume, the type of container and dropper, the care with which the instillation is performed, and exposure to cross infection.

Tonicity and pH of ophthalmic medications generally approximate that of tears; unfortunately some drugs are not stable at this pH, and a compromise must be reached. Most compounds instilled into the conjunctival sac are quickly changed to the pH of tears but will produce varying degree of discomfort or burning if initially different. Tonicity and pH are controlled by buffering solutions and sodium chloride concentration. Over a period of time, potency decreases, and solutions may become discolored, contain sediment, decomposition products, or become contaminated. Such drugs should be discarded.

CYCLOPLEGIC AGENTS

Following is a brief description of the most commonly used cycloplegic agents, their dosage, duration of action, and side effects.

Atropine

Drugs of the atropine family have been known for about 2,000 years. During the sixteenth century they were fancied for cosmetic purposes, hence the term "belladonna." The generic name comes from Atropos, one of the three fates whose office it was to cut the thread of life, no doubt a reflection of its potential toxicity.

Atropine produces both cycloplegia and mydriasis. The mydriasis can be potentiated by the simultaneous use of sympathomimetics, but the duration is too persistent for fundus examination or for refraction of adults. Although atropine exerts no direct effect on intraocular pressure of a normal eye, it can induce an acute attack of glaucoma in eyes with narrow chamber angles. In addition to its cycloplegic effect, atropine is also used in the treatment of anterior uveitis and keratitis, to break posterior synechiae, to inhibit accommodative spasm, and as a substitute for occlusion in hyperopic eyes.

Atropine is absorbed through the cornea, and qualitative differences in effectiveness depend on the rate of penetration. Onset of cycloplegia occurs within 1 hour, reaches a maximum in 12 to 24 hours, and lasts up to 2 weeks. Atropine acts directly on the postganglionic neuroeffector receptors and can be overcome to some extent by sufficient dosage of parasympathomimetics.

Side effects are due to systemic absorption, chiefly as the result of transconjunctival penetration and absorption from the nasolacrimal system. Atropine is hydrolyzed in the liver and excreted in the urine (a drop of the latter in a cat's eye is an old test for atropine poisoning). Like all cholinergic blocking agents, atropine inhibits salivation, perspiration, and lacrimal secretions and produces vasodilation, tachycardia, fall in blood pressure, depression of gastrointestinal motility, and varying degrees of central nervous system stimulation (although it is used in the management of parkinsonism). The frequent use of systemic belladonna preparations as gastrointestinal antispasmodics should be kept in mind in evaluating patients with accommodative insufficiency of unknown cause.* Symptoms of atropine poisoning (hot as a hare, dry as a bone, red as a beet, blind as a bat, mad as a hatter) may be misdiagnosed as an acute childhood febrile disease. The fatal dose is estimated to be from 1 to 2 gm (as contrasted to the usual preanesthetic dosage of 0.5 mg). A 10 ml bottle of 1% atropine thus contains enough drug to kill several children.

Allergic reactions to atropine are fairly common, resulting in an eczematous dermatoconjunctivitis. The conjunctivitis is generally of the papillary type with eosinophilia. There is marked itching, and the skin of the eyelids becomes red, swollen, and weepy. Occasionally the preservatives used are an unsuspected cause of allergy. Allergic reactions should be differentiated from the effects of drug irritation, which produce a nonspecific watery inflammation of the conjunctiva without dermatitis or eosinophilia (frequently seen with miotics such as eserine, pilocarpine, and carbachol). Ophthalmic tests for drug allergy are potentially dangerous and are not recommended; it is more prudent to substitute another agent in the presence of positive history.

*Among commonly used systemic medications that have a cycloplegic effect are the following: antispasmodics such as amprotropine (Syntropan), methantheline (Banthine), and oxyphenonium (Antrenyl); antihistamines such as promethazine (Phenergan); tranquilizers (phenothiazines); antiparkinsonian group such as trihexyphenidyl (Artane); and the ganglionic blocking agents used to combat hypertension. Conversely, ciliary spasm may be induced by any of the parasympathomimetics, as in patients beginning antiglaucoma therapy, with drugs such as morphine and the digitalis group.

Atropine is available in solutions of methylcellulose, hydroxypropyl methylcellulose, buffered aqueous, or caster oil and in ointments of petrolatum and mineral oil. Concentrations are 0.12%, 0.5%, 1%, 2%, 3%, or 4% in solutions and 0.5% and 1% in ointments. The lowest concentration compatible with the clinical problem should be used, preferably in small individual dose containers. Drug concentrations over 1% are seldom needed for refraction and that only for strabismic children. The probability of systemic absorption is decreased by the ointment form, and there is less danger of accidental swallowing, but the applied dosage may be erratic. Solutions for home use are best given in minimal amounts, such as 1 ml "Dropperettes" (the 0.25% in oil is a good compromise). Careful instructions, preferably written, are given to the parent as to the time, amount, and technique of instillation and for disposal of the remaining drug.

The conventional atropine dosage for refraction of strabismic children is 1 drop of the 1% solution, or one ointment application three times a day for 3 days preceding the refraction. Where this schedule orginated is not known, but it is probably excessive, since maximum cycloplegia can be achieved in 12 to 24 hours. The 3-day routine may have been an allowance for inadequate home instillation. A 1-day routine is adequate if the parent is judicious in instilling the drug. This still avoids an unhappy child in the office and disruption of routine.

Scopolamine

Scopolamine (hyoscine), is an occasional substitute for patients sensitive to atropine. It is available in 0.2% to 0.25% solutions in hydroxypropyl methylcellulose or ointment. In these dosages, its cycloplegic effect and duration of action is similar to 0.5% atropine. Unlike atropine, its action is somewhat less uniform, and it has a depressant effect on the central nervous system.

Oxyphenonium

Oxyphenonium bromide (Antrenyl) is a long-acting cycloplegic similar in effect to atropine and scopolamine. It is usually used as a gastrointestinal antispasmodic. A solution for ophthalmic use may be prepared by 1% concentration in 1:5000 solution of benzalkonium chloride (Havener, 1974). No cross sensitivity seems to occur with atropine. Side effects include a depressant effect on the central nervous system and a ganglionic blocking action in high dosage.

Homatropine

Homatropine is a synthetic ester similar but less potent than atropine. The side effects are similar to those of atropine, and the same precautions apply. The drug is used in 2% to 5% solutions in methylcellulose. Dosage is 1 drop repeated in 5 minutes for routine refraction. Onset of cycloplegia is from 10 to 30 minutes, and duration of action is from 36 to 48 hours. Maximum cycloplegia occurs in about 30 minutes to 2 hours. Homatropine is frequently used in conjunction with 1% hydroxyamphetamine (Paredrine), or 2% cocaine, whose sympathomimetic action potentiates the mydriasis. Cocaine, however, has a deleterious effect on corneal epithelium, and its possible toxic effects are not to be overlooked. Although at one time these cycloplegic combinations were popular for routine refraction, they have been replaced by newer short-acting agents.

Cyclopentolate

Cyclopentolate (Cyclogyl) is a potent short-acting cycloplegic and mydriatic introduced about two decades ago. Its cycloplegic effect is superior to homatropine, its onset more rapid, and its duration shorter. The mydriatic effect is also excellent. There appears to be no cross sensitivity with atropine. As with all cycloplegics, the drug is less effective, or less rapid, in patients with deeply pigmented irises. Ocular instillation results in some transient burning and discomfort, which requires reassurance. Systemic toxic effects are rare, including ataxia, visual hallucinations, or psychotic reactions. The reactions are temporary, and only one case of permanent change has been reported and involved an older patient with borderline senile dementia.

Cyclopentolate is available in 0.5% to 2%

solutions in Gifford's buffer. The 1% solution is adequate for most children and adults. One drop is instilled in each eye and repeated in 5 minutes. Cycloplegia occurs in 20 to 30 minutes and last about 6 to 8 hours. Eyes with darkly pigmented irises may take longer (45 minutes), and mydriasis can be potentiated with phenylephrine. Because of the rapid wearing off, the refraction must be reasonably timed to coincide with the maximum cycloplegic effect, and the accommodative inhibition checked by one of the tests for residual accommodation.

Tropicamide

Tropicamide (Mydriacyl) is a rapidly acting cycloplegic of short duration. The mydriatic effect is comparable to phenylephrine (although the latter is more effective in older patients). Tropicamide is available in 0.5% to 2% solutions, but the latter is seldom used. The vehicle is distilled water with phenylmercuric nitrate as a preservative. The usual dosage for refraction is 1%, 1 drop repeated at least three times in 5-minute intervals. (The 0.5% solution is not effective for cycloplegia.) Maximum cycloplegia occurs in 20 to 25 minutes but wears off so rapidly (effectiveness drops to 79% after 45 minutes) that the timing of the refraction is crucial, a schedule which is not always easy to achieve in a busy office practice. Even at maximum cycloplegia, the residual accommodation may be 3 D or more by the time the second eye is refracted.

EVALUATING THE CYCLOPLEGIC EFFECT

The intelligent use of cycloplegics depends on the clinical problem, the age of the patient, the pigmentation of the iris, the concentration, frequency and reliability of the instillations, and the time interval before the patient is examined. Moreover there are individual variations in response. Published data on the precise effect of a particular cycloplegic after 30 minutes are of little help if the refraction must be postponed because of an intervening office emergency. The only safe and reasonable method of evaluating the cycloplegic effect is to measure the residual accommodation—at least approximately, since no precise clinical method is available. The main problem is that unless the patient's refractive error is already known, the ability to read a near target may be the result not of active accommodation but of high myopia. Of course absent pupillary light constriction is no criterion of cycloplegia and is entirely unreliable.

The simplest method is to place a +3 D lens in front of the eye. If the eye is assumed to be emmetropic, the far point will be at 33 cm, and the patient should not be able to read small print brought any nearer if there is zero accommodation. The extent to which the near point (converted to diopters) differs from 3 D is a measure of the residual accommodation. The technique can be varied by placing the near point target at the nose and moving it further from the eye until the letters are recognized. Using this technique, residual accommodation with the same drug (e.g., homatropine) ranges from 0.87 D to over 2 D, based on differences in the target, depth of focus, and inherent resolution.

In Duane's method, a +3 D sphere is added to the distance correction. With one eye occluded, the patient fixes a Jaeger 2 test at 33 cm. The target is moved nearer until the line blurs and then moved away until it blurs again. The dioptric difference in the two blur points is the residual accommodation. The method presupposes that the eye is emmetropic; hyperopia will cause an apparent decrease, and myopia will cause an apparent increase in the range. Prangen (1931) in a classic paper on what constitutes satisfactory cycloplegia stated that it was his practice to estimate the refractive error carefully before adding the +3 D lens but gave no hint as to how this estimate was made in advance.

Another method is to place a +3 D sphere before the corrected eye and a reading target at some near distance and reduce the plus until the print can no longer be made out. The added minus represents the residual accommodation.

In these blur point methods, depth of

focus, inherent variability of acuity, type of target, ability to interpret blurred images, motivation to do so, or the criteria by which ametropia is estimated in advance are rarely stated. So it comes as no surprise that when these factors were specifically controlled in a laboratory situation (Brickley and Ogle, 1953), residual accommodation with a homatropine-cocaine combination (1/50 grain of each) was found to be 0.04 D, a negligible amount. It follows that blur points after cycloplegia measure more than residual accommodation, factors which cannot be easily controlled in clinical practice.

Dynamic retinoscopy has been used as an objective method of measuring residual accommodation. The technique is not as objective as one might suppose because the patient is required to fixate a near target which happens to be attached to the retinoscope instead of a reading card.

The patient is instructed to fixate at the plane of the retinoscope with the distance correction in place. Even without cycloplegia, a *with* motion will be noted, indicating that accommodation lags behind the dioptric stimulus distance; that is, the eye is hyperopic for the test distance by 0.25 D to 1.50 D (depending on the detail of the target, pupil size, spherical aberration, and amplitude). A dioptric reversal of +2.25 D at 40 cm over the distance cycloplegic finding would represent a fair estimate of adequate cycloplegia. Alternatively the examiner moves nearer the eye to obtain "neutral," which represents the near point of accommodation that is to be compared to the near point obtained without cycloplegia.

All factors considered, any of these approximate methods are better than no method at all. Dynamic retinoscopy is particularly practical, since it can be used directly after conventional neutralization and requires the least time. The patient simply fixates the instrument (or a near point card with a hole in it), and the conventional retinoscopy is continued to the new end point. If residual accommodation is found to exceed 1 or 2 D (depending on one's criterion),* more drug is instilled.

*The value of 2 D of residual accommodation, frequently quoted as representing adequate cycloplegia, includes at least 1 D of depth of focus.

CLINICAL PROCEDURE

Cycloplegics are indicated for the refraction of any patient in whom accommodative relaxation is deemed necessary, such as those with ciliary hypertonicity, spasm, and accommodative esotropia, and in children when subjective cooperation cannot be reliably obtained. Cycloplegia is not synonymous with validity and can be used with both subjective and objective refractive techniques. Cycloplegics should not be used as a substitute for mydriatics in fundus examination and are seldom required in patients over 50 years of age, since effective mydriasis can be achieved with sympathomimetics in this age group.

A satisfactory routine is to follow the cycloplegic with a postcycloplegic refraction on a subsequent visit. Most new patients are dilated for fundus study before the refraction anyhow. To avoid the inconvenience of two visits, it is feasible to reverse the procedure and perform both examinations on the same day, but separate tests of biologic functions with all their attendant fluctuations are more desirable. Patients (and clinicians) may be tired, distracted, or disinterested, all of which can be rectified on a subsequent visit. Moreover patients are not lost to follow-up after having their eyes dilated. Jackson wrote:

> I find the most useful rule is never to prescribe a spherocylindric correction on one determination, but on another day to make a second determination, starting with what I found before and correct the errors I made in my first examination. I have seen one pair of eyes five times before satisfying myself that I have the correction for the contracted pupils that would be of greatest service.*

And this from the Dean of American refractionists.

Refraction under cycloplegia can follow the usual fogging technique, although fogging is not necessary, and the steps can be expedited to ensure binocular inhibition of accommodation. Retinoscopy should concentrate on the

*From Jackson, E.: Norms of refraction, J.A.M.A. **98**:132-133, 1932.

central pupil to avoid peripheral aberrations. Artificial pupils (metal discs of varying apertures) are seldom necessary. Contrary to common opinion, near point phorias and fusion tests can be performed under cycloplegia, although their interpretation is, of course, different than when accommodation is active. In fact, by comparing distance and near phorias, one can obtain a measure of residual accommodation if the AC/A ratio is known a priori.

Generally, the differences between cycloplegic and noncycloplegic results are not great. In 70% to 80% of cases, findings agree within ±0.50 D. The agreement on cylinder axis is even higher (85%); the agreement on cylinder power is more variable (80%) because of the effectivity of the additional sphere uncovered under cycloplegia. As might be expected, differences are more likely to occur in younger patients than in older ones, in hyperopes than in myopes, and between different examiners or different techniques. One must not conclude that cycloplegic and noncycloplegic refraction always gives the same information. The purpose of cycloplegia as a diagnostic technique is to uncover the exception—not the rule, and diagnostically important differences will be found in accommodative esotropia and pseudomyopia.

After the use of cycloplegics, the patient should be warned of the reduced near vision (or distance vision in hyperopia), especially with regard to driving, climbing, or the operation of fast-moving machinery. The instillation of a cholinergic agent is sometimes required after short-acting cycloplegics but should not exceed the duration of action of the cycloplegic (Table 11-1).

Table 11-1. Cycloplegic agents and their effects

Drug	*Time for mydriasis*	*Time for maximum cycloplegia*	*Duration of adequate cycloplegia*	*Duration*	*Approximate residual accommodation*
Atropine 1%	30 min to 1 hr	12 to 24 hr	24 hr	10 to 18 days	+
Scopolamine 0.25%	30 min to 1 hr	1 hr	2 hr	4 to 6 days	+
Homatropine 5%	30 min	1 hr	1 to 2 hr	36 to 48 hr	++
Cyclopentolate 1%	20 min	20 to 45 min	30 min	6 to 8 hr	++
Tropicamide 1%	20 min	20 to 35 min	15 min	2 to 6 hr	+++

Cycloplegics and myopia

In 1811 William Wells, a presbyopic London oculist, persuaded his young assistant, Dr. Cutting, to use belladonna on one of his eyes and measure its effect on accommodation (using the reflection of a candle flame in a thermometer bulb as test object). The effect was not only dramatic but also continued for 9 days. Cutting was apparently myopic, and Wells soon tried it on some of his myopic patients, since "Being in the possession of a new instrument, I next attempted to gain, by means of it, some illustration of the changes which the vision of shortsighted persons undergo from age."* Wells not only discovered the cycloplegic effect of belladonna but also became the forerunner of a distinguished group of followers who have tried to prevent or at least alleviate the progression of myopia by cycloplegics ever since.

The use of cycloplegics in myopia assumes that the condition is, at least in part, the result of ciliary spasm or hypertonicity induced by near work. A treatment schedule may range from weekly use of atropine to nightly use of a short-acting cycloplegic (presumably to break up the accommodative spasm of the day). A realistic evaluation of such treatment cannot be based on simple acuity measurements because of the inherent subjective variations of practice, blur interpretation, depth of focus, contrast, illumination, and type of target. Moreover myopia progression is unpredictable even in the two eyes of the same patient, controls are not

*From Wells, W. C.: Observations and experiments on vision, Philosoph. Trans. R. Soc. London **101**:378-391, 1811.

feasible, and timing is a matter of years, not months or weeks. Ciliary theory does not account well for the (now unquestioned) axial elongation of higher degrees of myopia, the type which leads to secondary pathologic degeneration of the retina and vitreous and whose prevention is of prime concern. At best the lens flattening induced by eliminating ciliary tonus cannot exceed 2 D, a rather insignificant proportion of a 15 to 20 D myopia. And if the treatment is aimed toward the reduction of small degrees of myopia, the use of cycloplegia over the required period of years would seem to be a high price in lost education and recreation just to wear a lesser powered minus lens. Some might consider the treatment worse than the disease. Moreover the ciliary theory does not account for the facts that the crystalline lens of myopic eyes is flatter on the average; that hyperopes who must accommodate more than myopes at any distance do not all become myopic; that myopia stabilizes around age 20 years despite continued or increased near work; that peasants who do not write, do not read, and do no near work become myopic just about as frequently as industrial workers; and that despite a century of attempts to prevent myopia by inhibiting accommodation* (with cycloplegics, undercorrection, bifocals, base-in prism, or taking the minus lenses off for reading), the results have been a dismal failure.

*We see later that when the vision is very blurred, accommodation may in fact become active—an undesirable effect of undercorrecting myopia for those who wish to inhibit accommodation. On the other hand, some writers advocate parasympathomimetics in myopia to stimulate ciliary tonus and thus prevent the choroid from stretching!

REFERENCES

Abbot, W. O., and Henry, C.: Paredrine, a clinical investigation of a sympathomimetic drug, Am. J. Med. Sci. **193:**661, 1937.

Abraham, S. V.: The use of several new drugs as substitutes for homatropine, Am. J. Ophthal. **36**(2):69-74, 1953.

Abraham, S. V.: A preliminary report on the use of "bis-tropamide" in the control of myopia, J. Pediatr. Ophthal. **1:**39-48, 1964.

Abraham, S. V.: Control of myopia with tropicamide, J. Pediatr. Ophthal. **3:**10-22, 1966.

Adcock, E. W.: Cyclopentolate (Cyclogyl) toxicity in pediatric patients, J. Pediatr. **79:**127-129, 1971.

Bannon, R. E.: The use of cycloplegies in refraction, Am. J. Optom. **24:**513-568, 1947.

Barbee, R. F., and Smith, W. O.: A comparative study of mydriatic and cycloplegic agents, Am. J. Ophthal. **44:**617-622, 1957.

Beckman, H.: Pharmacology: the nature, action, and use of drugs, Philadelphia, 1961, W. B. Saunders Co.

Bedrossian, R. H.: The effect of atropine on myopia, Ann. Ophthal. **3:**891-897, 1971.

Borthne, A., and Davanger, M.: Mydriatics and age, Acta Ophthal. **49:**380-387, 1971.

Bothman, L.: Homatropine and atropine cycloplegia, Arch. Ophthal. **7:**389-398, 1932.

Bothman, L.: Refractive errors in the same eyes while under the influence of homatropine, scopolamine, and atropine, Trans. Am. Acad. Ophthal. **42:**822-826, 1937.

Brickley, P. M., and Ogle, K. N.: Residual accommodation under homatropine—cocaine cycloplegia, Am. J. Ophthal. **36:**649-659, 1953.

Carpenter, W. T.: Precipitous mental deterioration following cycloplegia with 0.2% cyclopentolate HCl, Arch. Ophthal. **78:**445-447, 1967.

Chance, J. C., Ogden, E., and Stoddard, K. B.: The effect of undercorrection and base in prism upon the myopic refractive state, Am. J. Ophthal. **25:**1471-1474, 1942.

Cher, I.: Experience with Cyclogyl, Trans. Ophthal. Soc. U. K. **79:**665-670, 1959.

Cruise, R. R.: The abuse of atropine in refraction work, Trans. Ophthal. Soc. U. K. **29:**245, 1909.

Davanger, M.: The pupillary dilation curve after mydriatics, Acta Ophthal. **49:**565-571, 1971.

Duke-Elder, S.: System of ophthalmology: The foundations of ophthalmology, vol. 7, St. Louis, 1962, The C. V. Mosby Co.

Fraunfelder, F. T., Hanna, C., Cable, M., and Hardberger, R. E.: Entrapment of ophthalmic ointment in the cornea, Am. J. Ophthal. **76:**475-484, 1973.

Friedman, B.: Comments on the teaching of refraction, Arch. Ophthal. **23:**1175, 1940.

Gettes, B. C.: Three new cycloplegic drugs, clinical report, Arch. Ophthal. **51:**467-472, 1954.

Gettes, B. C.: Drugs in refraction, Int. Ophthal. Clin. **1:**237-248, 1961.

Gettes, B. C.: Tropicamide, a new cycloplegic mydriatic, Arch. Ophthal. **65:**632-635, 1961.

Gettes, B. C.: Choice of mydriatics and cycloplegics for diagnostic examinations in children, Int. Ophthal. Clin. **3:**879-884, 1963.

Gettes, B. C., and Belmont, O.: Tropicamide, comparative cycloplegic effects, Arch. Ophthal. **66:**336-340, 1961.

Gettes, B. C., and Leopold, I. H.: Evaluation of five cycloplegic drugs, Arch. Ophthal. **49:**24-27, 1953.

Goodman, L. S., and Gilman, A.: The pharmacological basis of therapeutics, New York, 1965, The Macmillan Co.

Grambill, H. D., Ogle, K. N., and Kearns, T. P.: Mydriatic effect of four drugs determined with pupillograph, Arch. Ophthal. **77**:740-746, 1967.

Grant, M. W.: Physiological and pharmacological influences upon intraocular tension, Pharmacol. Rev. 7:143-182, 1955.

Harris, L. S.: Cycloplegic-induced intraocular pressure elevations, in normal and open-angle glaucomatous eyes, Arch. Ophthal. **79**:242-246, 1968.

Harris, L. S., Galin, M. A., and Mittag, T. W.: Cycloplegic provocative testing after topical administration of steroids, Arch. Ophthal. **86**:12-14, 1971.

Havener, W. H.: Ocular pharmacology, ed. 3, St. Louis, 1974, The C. V. Mosby Co.

Hoefnagel, D.: Toxic effects of atropine and homatropine eyedrops in children, N. Engl. J. Med. **264**:168-171, 1961.

Holland, M. G.: Autonomic drugs in ophthalmology. III. Anticholinergic drugs, Ann. Ophthal. **6**:661-665, 1974.

Hosaka, A., and Ohashi, T.: Treatment of pseudomyopia with topical Cyclogyl, Jap. J. Clin. Ophthal. **23**:907-911, 1969.

Kennerdell, J. S., and Wucher, F. P.: Cyclopentolate associated with two cases of grand mal seizure, Arch. Ophthal. **87**:634-635, 1972.

Lancaster, W. B.: Refraction correlated with optics and physiological optics, Springfield, Ill., 1952, Charles C Thomas, Publisher.

Lazenby, G. W., Reed, J. W., and Grant, W. M.: Anticholinergic medication in open angle glaucoma, Arch. Ophthal. **84**:719-723, 1970.

Leydhecker, W.: The effect of homatropine upon the tension of the normal and glaucomatous eye, Am. J. Ophthal. **39**:459, 1955.

Lowe, R. F.: The natural history and principles of treatment of primary angle-closure glaucoma, Am. J. Ophthal. **61**:643-650, 1966.

Ludlam, W. M., Weinberg, S. S., Twarowski, C. J., and Ludlam, D.: Comparison of cycloplegic and noncycloplegic ocular component measurement in children, Am. J. Optom. **49**:805-818, 1972.

Mayer, L. L.: Dilatation of the pupil for ophthalmoscopic evaluation, J.A.M.A. **133**:38-39, 1939.

Miranda, M. N.: Residual accommodation, a comparison between cyclopentolate 1% and a combination of cyclopentolate 1% and tropicamide 1%, Arch. Ophthal. 87:515-517, 1972.

Moncreiff, W. F., and Scheribel, K. J.: Further studies concerning homatropine cycloplegia and paredrine with special reference to rate of accommodative recovery, Am. J. Ophthal. **25**:839-843, 1942.

Obianwu, H. O., and Rand, M. J.: The relationship between the mydriatic action of ephedrine and the colour of the iris, Br. J. Ophthal. **49**: 264-270, 1965.

Prangen, A. deH.: What constitutes satisfactory cycloplegia? Am. J. Ophthal. **14**:665-671, 1931.

Priestley, B. S., and Medine, M. M.: A new mydriatic and cycloplegic drug, Am. J. Ophthal. **34**:572-575, 1951.

Ragorshek, R. H.: Residual accommodation, Am. J. Ophthal. **36**:1086-1092, 1953.

Reber, W.: Comparative potency of hyoscin and scopolamine hydrobromide in refraction work, J.A.M.A. **50**:1323, 1908.

Risley, S. D.: The value of homatropine hydrobromate in ophthalmic practice, Am. J. Med. Sci. **82**:113, 1881.

Robb, R. M., and Petersen, R. A.: Cycloplegic refraction in children, J. Pediatr. Ophthal. **5**:110-114, 1968.

Sato, T.: The causes and prevention of acquired myopia, Tokyo, 1957, Kanehara Shuppan Co., Ltd.

Schimek, R. A., and Lieberman, W. J.: The influence of Cyclogyl and neosynephrine on tonographic studies of miotic control in open-angle glaucoma, Am. J. Ophthal. **51**:781-784, 1961.

Sheard, C.: An objective method of determining the monocular amplitude and range of accommodation by dynamic means. In American encyclopedia of ophthalmology, Chicago, 1920, Cleveland Press.

Smith, S. E.: Mydriatic drugs for routine fundal inspection; a reappraisal, Lancet **2**:837-839, 1971.

Smolen, V. F., and Schoenwald, R. G.: Drug absorption analysis from pharmacologic data, J. Pharm. Sci. **60**:96-103, 1971.

Tammisto, T., Castren, J. A., and Marttila, L.: Intramuscular administered atropine and the eye, Acta Ophthal. **42**:408-417, 1964.

Wahl, J. W.: Systemic reaction to tropicamide, Arch. Ophthal. **82**:320-321, 1969.

Wang, E. S. N., and Hammarlund, E. R.: Corneal absorption reinforcement of certain mydriatics, J. Pharm. Sci. **59**:1559-1563, 1970.

Wells, D. W.: Cycloplegics in refraction, Am. J. Ophthal. **11**:120-124, 1928.

Wells, W. C.: Observations and experiments in vision, Philosoph. Trans. R. Soc. London **101**:378-391, 1811.

Windsor, C. E., Burian H. M., and Milojevic, B.: Modification of latent nystagmus, Arch. Ophthal. **80**:657-663, 1968.

Wolf, A. V., and Hodge, H. C.: Effects of atropine sulfate, methylatropine nitrate (metropine) and homatropine hydrobromide on adult human eyes, Arch. Ophthal. **36**:293-301, 1946.

Young, F. A.: The effect of atropine on the development of myopia in monkeys, Am. J. Optom. **42**:439-449, 1965.

12

Objective methods of refraction

The instrument, wrote S. Weir Mitchell, trains the individual by exacting accuracy and teaching care, creating a wholesome appetite for precision, which eventually becomes habitual. The laser, automated refractor, ultrasound, and supercold have made diagnosis more penetrating, surgery more daring, and prognosis more optimistic. Technology feeds our hunger for mechanical explanations and mathematic solutions. As a transient disease, it views man as a predictable bundle of reflexes or the eye as an aggregate of precise optical components that must only be measured more often and more accurately. Its immediate consequence is an ocean of data beyond our capacity to classify, manipulate, or even retrieve. We find ourselves in danger of becoming technicians floundering in a sea of trivial measurements. Fortunately most devices perish in the using, and memory serves by holding fast the best. Let us therefore appraise what is new, use what is good, discard what is transient, and put aside the fear of inevitable error.

KERATOMETRY

The keratometer is said to measure the power of the cornea objectively. The statement is incorrect on two counts. It measures curvature and not power, and its objectivity is tainted by an assumed refractive index. It is, however, the instrument par excellence to evaluate corneal topography for fitting contact lenses. Its diagnostic use in refraction is limited not only by the assumed index but also because thickness and posterior corneal curvature are not measured; readings are confined to a small central optical zone, which may or may not be the center of toricity or coincide with the fixation axis. Since the keratometer deals only with the cornea, whereas retinoscopy or subjective tests measure the total refractive status, including lenticular astigmatism, at the spectacle plane, one would not expect the results to coincide. Nevertheless, various attempts have been made to predict one from the other.

The keratometer was invented by Helmholtz (1854) as a laboratory instrument to measure the optical constants of the eye (hence the alternative name "ophthalmometer"). The clinical instrument we owe to Javal and Schiotz. Opinion on its diagnostic value in clinical refraction has tended to run to extremes. Stirling (1941), for example, considered it no blessing at all and believed it should be abolished in every clinic, whereas Tait (1951) wrote that it gave the most precise of all refractive measurements.

Procedure

The eyepiece should be adjusted individually to the examiner's vision and the calibration of the instrument periodically checked against a steel sphere of known radius. Current instruments give the reading in both focal power and radius of curvature; in older models the radius must be computed from the index (which is usually but not always 1.3375) as follows:

$$\text{Radius} = \frac{337.5}{\text{Power reading}}$$

The actual clinical procedure varies from one instrument to the next, since each has its own peculiarities of mechanical and optical construction. Most modern keratometers are one-position instruments; the power in each principal meridian is measured without shifting the target mires. Focusing is by the coincidence method; a blur appears doubled. The readings are in millimeters of radius or in diopters or both. The range varies from 3 to 52 D or from 35 to 60 D, depending on the instrument. Some models provide for direct viewing of the cornea under 15× magnification for measuring corneal diameter and contact lens inspection, and some provide for controlled eccentric fixation to evaluate peripheral corneal topography (topogometry).

The patient is seated in front of the instrument, and the unexamined is eye occluded. Chin and forehead must be securely fixed, and centering is maintained by having the patient fixate the reflection of his own eye. The examiner focuses the mires by eliminating doubling. This adjustment is sensitive, and one hand is kept on the focusing wheel (or joystick) throughout the procedure. A central reticle target such as a cross hair is centered on the cornea to bring the optic axis of the instrument in line with the patient's fixation axis. The primary meridians are found by aligning the mire limbs. An off-axis condition is rectified by rotating the entire instrument head. The axis scale is calibrated degrees. There are two power drums, one for the horizontal meridian and one for the vertical meridian (or meridians nearest to them). It is conventional to read the horizontal meridian first (sometimes called the "primary meridian"), by superimposing the plus signs at the side of the mire image; the vertical meridian is read by aligning the minus sign. (Some instruments use plus signs for both meridians.) The mires are kept focused in the meantime by eliminating doubling. (It is generally impossible to measure both principal meridians with the same focus adjustment.) Several readings are taken to ensure accuracy. The patient's head and eyes should remain rigid throughout the test, and blinking is avoided as much as possible. Limits of tolerance are ±0.25 D and 3°. Particular care is required in maintaining the fixation of an aphakic eye, and this may require an auxiliary fixation target for the opposite eye. It is conventional to record the keratometer findings in minus cylinder form. For example, if the reading in the horizontal is 45.00 D, and the reading in the vertical is 42.00 D, −3.00 D × 90 is recorded as the correcting minus cylinder.

Interpretation

In the diagnosis of refractive error, as contrasted to the topographic survey required for contact lens fitting, the keratometer may give us a hint whether a given ametropia is axial or refractive. Generally, a large flat cornea will be associated with an axially elongated myopic eye; a small highly curved cornea in a nearsighted eye would point toward refractive myopia. Unfortunately there is no way of clinically confirming when these estimates are ill proportioned. This is not the fault of the instrument but of the unpredictable combinations of ocular optical components.

A variety of empirical rules have been advanced over the years to make keratometer results more nearly comparable to those which measure total ametropia such as retinoscopy. The best known is Javal's rule, frequently rediscovered and recriticized despite the author's specific limitations. The rule attempts to predict the total astigmatism of the eye (i.e., cornea and lens) from the keratometer reading as follows:

Total astigmatism = 1.25 (Corneal astigmatism) + Lenticular astigmatism

The correction factor of 1.25 is multiplied by the corneal astigmatism, which must be written in minus cylinder form (oblique astigmatism is ignored). The correction is for the change in effectivity in computing the spectacle from the corneal refraction. The lenticular astigmatism varies with age: 0.25 D up to age 30 years, 0.50 D up to age 60 years, and 0.75 D over 60 years. In Javal's rule it is added as a minus cylinder axis 90. For example if the keratometer reads 46.50 D in the 180 axis and 44.25 D in the 90 axis, -2.25×90 is recorded as the corneal astigmatism. If the patient is 35 years old, Javal's rule predicts the total astigmatism to be -2.75×90.

One problem is that Javal's rule does not consider the effect of the spherical component on the cylinder effectivity. Since the keratometer does not measure the spherical error except indirectly (for example, one could assume that 44.00 D arbitrarily represents emmetropia), the sphere is determined by retinoscopy or subjective refraction. In the range of 8 D hyperopia, for example, a keratometric cylinder reading of −3 D has an effective power at the spectacle plane of only −2.31. If the spherical error is in the range of 8 D myopia, the spectacle cylinder is −3.87 D. (These effectivity computations are identical to those based on a vertex distance of 13.75 mm in Chapter 7.) Javal's rule predicts a cylinder of −3.75 whatever the spherical ametropia.

A more useful empirical rule (Nadbath, 1958) for estimating the correcting cylinder (at a vertex distance of 13.75 mm) is as follows: Modify the sphere (as found by retinoscopy or subjective tests) by dividing by 7 if a minus power or by 14 if a plus power. Modify the keratometric minus cylinder by dividing by 3. The product of the two fractions gives the correction factor to be added algebraically to the keratometric cylinder. For example, if the spherical correction is +8 D and the cylinder is -3×90, then $\frac{8}{14} \times \frac{3}{3} = +0.57$. The estimated spectacle cylinder is therefore −2.43 D. The sign of the correction factor is the same as the spherical error, and any lenticular astigmatism must be added to the final answer. If the keratometer cylinder is -4×90 and the patient is 35 years old, the estimated spectacle cylinder is -3.74×90.

It will be noted that in strong myopic errors the spectacle cylinder (written in minus cylinder form) is less than the keratometer predicts, and in strong plus corrections such as in aphakia, the spectacle cylinder is larger than the corneal cylinder. The lenticular astigmatism, which is small in younger people, is generally ignored in fitting cosmetic contact lenses. There is no point in estimating the spectacle cylinder from the keratometer in refraction, since it can be measured more easily, accurately, and directly by retinoscopy and subjective tests. With contact lenses, however, only corneal astigmatism is corrected, and the residual cylinder may need to be incorporated in the prescription. Residual astigmatism can be estimated either by comparing the keratometer readings at the spectacle plane to the retinoscopy or by repeating the retinsoscopy or subjective refraction through a trial contact lens.

The keratometer has diagnostic value in following the course of progressive myopia, pterygium, keratoconus, and healing of corneal scars and in preliminary estimates of refractive error in eyes with hazy media. In keratoconus the mires appear distorted or irregular; a similar change may be demonstrable in the corneal edema after the wearing of an improperly fitted contact lense. Keratometric distortions also occur in trachoma and interstitial or herpetic keratitis, after ocular trauma, and in some corneal dystrophies.

In aphakia the lens component is missing; hence keratometry is hailed as an invaluable and time-saving aid. This is a misconception. The refraction of the aphakic eye like any other eye depends on all optical components, including a presurgical axial error. Furthermore, no lens will help the aphakic patient with macular degeneration, optic atrophy, or retinal detachment; the correction to prescribe is the one with which the patient can see.

Keratoscopy

In keratoscopy, corneal curvature is evaluated by the appearance of reflected images. The Placido disc, which is the earliest instrument, consists of concentric black and white rings, usually increasing in width from the center out so that the reflections are equally spaced. The greater the corneal curvature the smaller the image. In astigmatism, images are elliptic (the long axis in the meridian of least curvature), and in irregular astigmatism, the symmetry is disturbed. Modern instruments incorporate photographic recording or photoelectronic scanning. The optical principle is the same as keratometry, but since the target consists of many parts, it extends the topography of the cornea measured. The main difficulty is in analyzing the results, which is generally done by comparison to photographs of known spheres (the method used by Gullstrand). Not only is the cornea somewhat elliptic and does not follow a simple mathematic function, but the angle from which the photograph is taken changes the quantitative interpretation.

When two grids of similar spacing are superimposed, a new pattern is formed by the intersecting lines. Aside from its op-art flavor, these moiré patterns permit the detection of minute variations in surface regularity. The pattern is rectilinear if the surface is spherical, deformed if it is not, and indistinct at the periphery in either case. It is analyzed by superimposing the negative on one obtained from a standard sphere.

OBJECTIVE OPTOMETERS

Optometers locate the far point without patient responses, although they require the patient's cooperation in fixation and accommodative relaxation. Some models are glorified ophthalmoscopes, calibrated to read in diopters of ametropia; recent instruments are automated and computerized, incorporate an array of sophisticated gadgetry, and require an operator with "only a few hours training." This, according to its enthusiasts, will replace the "complex procedure of retinoscopy followed by tedious subjective judgments by the patient." Of course this charming balderdash is invariably written by engineers and biophysicists, never by clinicians. No practical refractionist would want to prescribe a lens, no matter how scrupulously measured, for a blind eye. Readers will have to make up their own minds about the complexity of the retinoscope vis-à-vis these instruments (considering the price). Retinoscopy of course is the only procedure that could conceivably be replaced in the refractive routine. More serious, however, is the notion that optometers provide a "prescription" rather than an optical measurement and that refraction may consequently be delegated to a technician. Such recommendations only reveal a basic misunderstanding of what refraction is all about.

The *Rodenstock refractometer* consists of an illuminated test plate placed at the anterior focus of a convex lens. Light is reflected into the eye by a mirror through a prism whose movement is equivalent to altering the test distance. By adjusting the vergence of the entering beam, the retinal image, viewed by the observer through a telescope, is brought into focus. The test object is rotated to various meridians for the evaluation of the astigmatism. The *Hartinger optometer* incorporates a double pinhole; the endpoint is a coincidence alignment of images in the primary meridians. The *Fincham optometer* also uses the coincidence principle. Light from a slit source is reflected into the patient's eye through an eccentric part of the pupil. If the eye is ametropic, it will intersect the retina above the optic axis in hyperopia and below the axis in myopia. By adjusting the optical distance of the target, the entering vergence is altered to focus on the axis as seen through a doubling apparatus. The observer simply aligns one side of a split field with the image of the slit seen in the other. The vergence of the entering rays is a measure of the ametropia. Provisions are made for measuring the refractive error in various meridians and for different degrees of eccentricity. The instrument may also be adapted for the objective measurement of accommodation.

In the *infrared optometer* (Campbell, Allen, Roth, and others), two narrow pencils of light pass through the pupil on either side of the optic axis to form a single image in emmetropia and double images in ametropia (Scheiner principle). Doubling is abolished by altering the vergence of the incident beam. The emerging rays form an aerial image, which activates two photocells. The entering beam is interrupted by a sectored disc so that doubling results in a to and fro activation of the photocells, which are connected to give zero output when the image is single. By using infrared rays, the eye is not dazzled, the pupil does not constrict, and accommodation is not activated. A somewhat similar principle is used in a clinical instrument recently marketed, the *Dioptron.* The image is projected on the retina as a series of moving light and dark bars, which the optical scanner views through a series of stationary bars. Depending on the focus, the light and dark bars of the moving and stationary systems coincide, and the photocell reads the illumination level of the emerging beam. Using infrared rays, the instrument scans four meridians, and a computer evaluates the result in terms of sphere, cylinder, and axis by fitting the data points to a sine curve. The *Ophthalmetron* is a self-recording retinoscope that provides its own end point and gives the result in graphic form. Utilizing infrared illumination and photoelectric detectors, it scans the eye at the rate of 720 sweeps per second. The optical principle resembles retinoscopy in that the far point is located by an advance-recede method. The detector, on a carriage, moves nearer or farther, depending on the vergence of the emerging beam, and plots the results when it observes "no motion." The system then rotates about its own axis to scan the other meridians, with the whole process taking about 3 seconds!

Electrophysiologic measurement of ametropia seems to be the ultimate in objective methods at least for lower animals. Indeed the response characteristics (in cats) of individual ganglion cells to various lens changes have been measured by microelectrodes. More practical is the *visually evoked response* (VER) in which minute electrical discharges from the visual cortex on brief stimulation of the eye are monitored by topical scalp electrodes and their waveforms extracted by computer summation techniques. The VER has been used to differentiate between unilateral amblyopia and macular disease and for objective perimetry, color vision, and acuity measurements. It is adapted to objective optometry by noting the responses to variations in target size and defocus.

Automated refractors will in due course play a significant role in investigating incidence and prevalence of ametropia in large population samples previously inaccessible, in cases in which communication and cooperation is minimal, in screening, and in providing preliminary optical data for refinement by conventional refractive techniques. Automated refractors will not save the clinician time or effort, they will not replace professional judgment (how often does one prescribe unmodified retinoscopic findings), and they do not "represent the beginning of a change of the patient/one doctor philosophy which has served as the traditional basis for eye care"—at least not until the instrument can take a history, analyze the patient's visual needs, evaluate the symptoms, and anticipate the therapeutic effect.

RETINOSCOPY

Retinoscopy is the simplest and most informative method of objective refraction, practical with infants, illiterates, and unreliable or uncooperative patients. Although many refractionists cherish the ambition to become so expert in retinoscopy that no credence need be placed on other methods, the experienced clinician will seldom prescribe from it if he has the option of confirming his results by subjective tests. And therein lies its chief value because it provides a starting point for subjective testing and independent confirmation of subjective answers.

The discovery of retinoscopy dates back to a chance observation by Sir William Bowman in 1859 that a peculiar reflex movement occurs in astigmatic eyes examined with the mirror ophthalmoscope. Writing his friend Donders, Bowman commented on the importance of axial observation, the advantage of a large pupil, and the linear "shadow" seen in astigmatism. The clinical implications were not appreciated until Cuignet (1873) used the plane mirror to measure astigmatic errors, and the optical theory was explained by

Landolt (1878) and Priestley-Smith (1884), who suggested the name "shadow test." Various names have been proposed but are only of historic interest, and "retinoscopy," introduced by Parent in 1881, appears firmly established. Of course the instrument does not really measure the retina, which is no more visible than it is with the ophthalmoscope.

Basic principles

If a practical clinician wants to frighten himself, he has only to look at some textbook diagrams and equations used to "explain" the theory of retinoscopy—so he settles for a cookbook description of its mechanics and never gains an insight into this important technique, which is no more complicated than lens neutralization.

Hold a +10 D trial lens (representing the optical system of the patient's eye) at arm's length, and move a pencil up and down at various distances behind the stationary lens. As you view the pencil through the lens, note the "with," "against," and "neutral" motion. The target is out of focus, and the movement appears to take place in the plane of the lens. The speed of the movement and the size of the "reflex" increase near the neutral position; thus the reflex is not peculiar to the eye, no real shadows are involved, and there is no black box magic about the instrument. This experiment differs from lens neutralization in that the "target" rather than the lens is moved. In retinoscopy the target is an illuminated patch on the patient's fundus, moved by tilting the instrument.

The practical criterion of ametropia is the far point, and that is what the retinoscope measures. With or against motion occurs according to whether the instrument is inside or outside the eye's far point—and remains neutral when the far point coincides with the peephole. Rather than hunt for it, the far point is brought to the instrument with trial lenses. Since the far point represents the static refraction, the patient must relax accommodation by fixing a distant target.

Optical principles

The optical principles of retinoscopy are best understood in two stages: the illumination system and the observation system. The illumination system begins at the instrument bulb and ends on the patient's retina. The observation system begins on the patient's fundus and ends in the examiner's eye. The two systems are independent; the first deals only with the light entering the patient's eye, and the second deals with the emerging rays.

Illumination system. Early retinoscopes were simply plane or concave perforated mirrors that reflected light into the patient's eye from a source beside the head. Some teachers still recommend them just as some farmers prefer mules to tractors, but it will save time to proceed directly to the best quality, self-luminous instruments.

Modern retinoscopes duplicate the concave or plane mirror by an adjustment of the condensing lens relative to the light sources (*S*), both in the handle of the instrument (Fig. 12-1). A knob or sleeve on the handle provides the adjustment. In some instruments it moves the lens and in others the bulb; consequently one must try it out (or read the instructions) to see which does what. When the source is inside the focal point of the condensing lens, the emerging rays diverge; when the source is outside the focal point, the rays converge. In either case, the rays are redirected by the angulated mirror (always plane) into the patient's eye as if coming from an "apparent source" (*S'*). The position of the apparent source above or below the optic axis determines which part of the fundus is illuminated.

Parallel or divergent rays leaving the instrument correspond to the "plane mirror effect" (not to be confused with "plano" vergence). The plane mirror adjustment produces not only plano vergence but all degrees of divergence as well. The apparent source sits on the end of an imaginary bar projecting backward from the instrument; the less divergent the beam the further the apparent source. When the retinoscope is tilted down (the top of it forward), the apparent source moves up, and its image on the patient's retina moves down. Thus a plane mirror adjustment moves the fundus image in the *same* direction as the rotation of the instrument (Fig. 12-2, *A*).

A "concave mirror effect" results in converging rays. With maximum convergence,

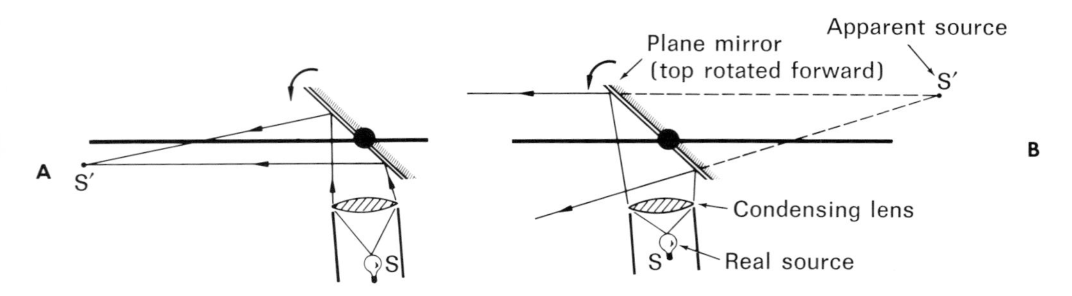

Fig. 12-1. By changing the position of the light source relative to the condensing lens in the handle of the retinoscope (sleeve up or down), the vergence of the beam leaving the instrument is altered. **A,** Light source outside anterior focal point of condensing lens: converging rays. By pulling the bulb away from the lens, the beam is made convergent, and the apparent source *(S′)* can be placed between the patient's eye and the instrument. **B,** Light source inside anterior focal point of condensing lens: diverging rays. The closer the bulb (real source) is to the lens the more divergent the beam; the more divergent the beam the nearer the apparent source to the mirror.

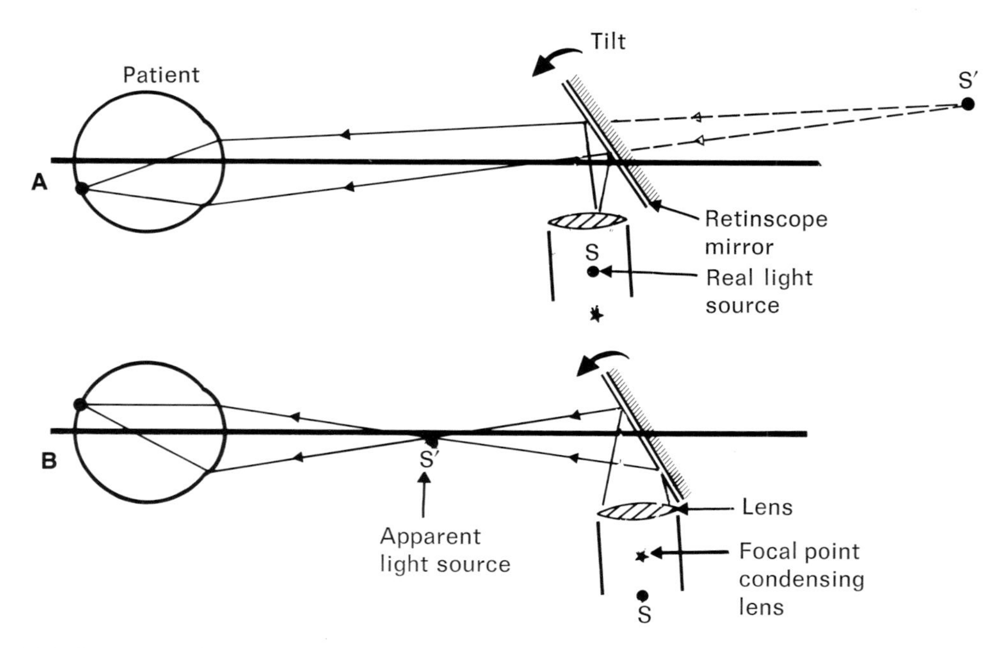

Fig. 12-2. By rotating the retinoscope, the top forward in this case, different parts of the patient's fundus are illuminated. **A,** If the instrument has a plane mirror adjustment with the beam divergent, the apparent source *S′* is above the axis, and the lower part of the patient's fundus is illuminated. **B,** If the retinoscope has a short focus concave adjustment (apparent source between the eye and the instrument), the same movement illuminates the upper part of the patient's eye. The adjustment of the instrument therefore determines the kind of reflex motion we see because the reflex takes its origin from the illuminated fundus patch and begins on opposite sides of the axis.

the apparent source falls *between* the eye and the examiner. Lesser degrees of convergence place the apparent source behind the patient's eye. These adjustments are termed "short" and "long focus concave mirror effects." In the long focus adjustment, the fundus image is displaced in the *same* direction as the rotation of the retinoscope (like the plane mirror); in the short focus adjustment, the fundus image is displaced *opposite* to the rotation of the retinoscope (Fig. 12-2, *B*). If the term "concave mirror" is used without qualification, the short adjustment is always assumed.

In summary the illumination system produces an illuminated area in the patient's fundus. The retinoscope can be adjusted to emit parallel, divergent, or convergent rays. The rays are reflected toward the patient's eye by a plane angulated mirror and appear to come from an apparent source whose location determines the position of the illuminated fundus patch. The patch is displaced in the same direction as the rotation of the retinoscope for all adjustments *except* the short focus concave. Only when the apparent source is between the eye and the instrument is the fundus image opposite to its rotation. None of these fundus displacements are influenced by the patient's ametropia.

Observation system. In the observation system, the rays originate at the patient's fundus. They are reflected from the pigment epithelium and underlying choroid, hence the reddish-orange color. These rays are headed for the far point, and their vergence

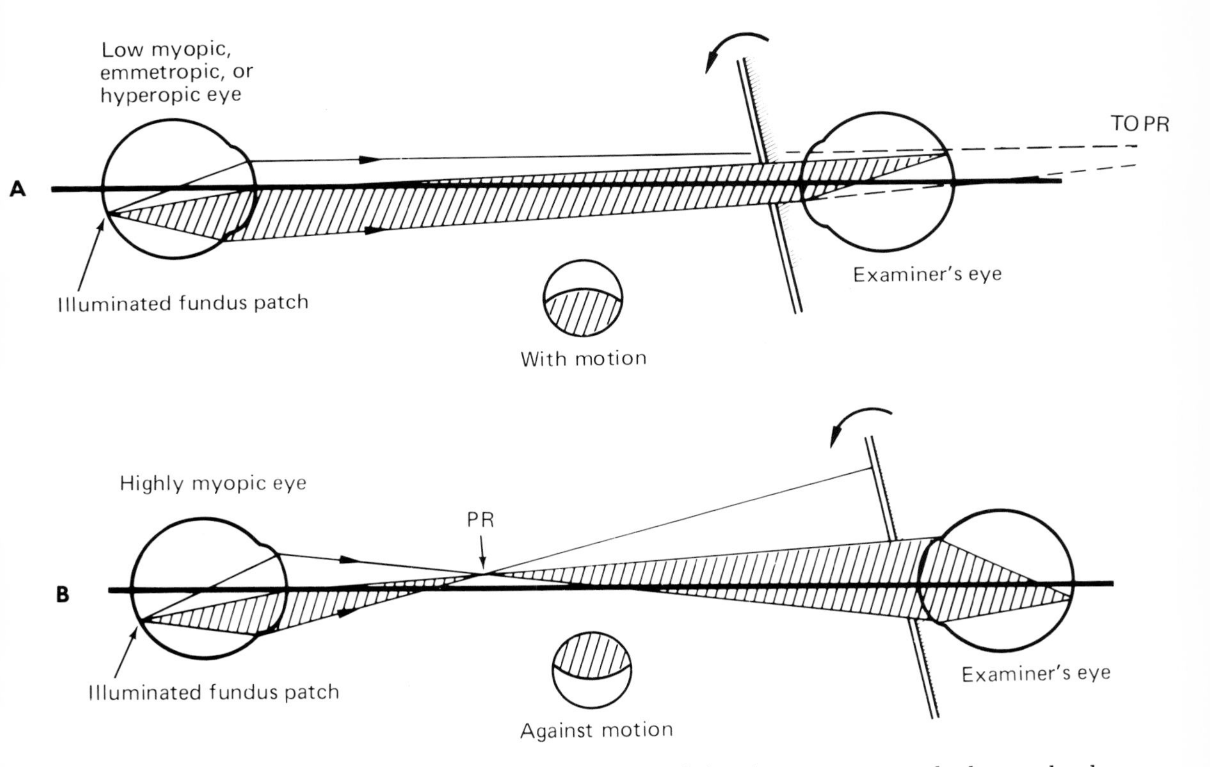

Fig. 12-3. Observation system in retinoscopy. It is assumed that the instrument in both cases has been adjusted to a plane mirror, the top rotated forward and the lower part of the fundus illuminated. The observation system begins with this fundus patch, and rays emerging from the eye are headed for its far point. **A,** In the case of low myopia, the PR is behind the examiner's eye, and he sees the reflex occupying the lower half of the pupil. Since the instrument was rotated down and the reflex is down, this is called a "with motion." **B,** If the eye is highly myopic (greater than the working distance), the far point falls between the patient's eye and the instrument and is recorded as an "against motion." Only such a highly myopic eye exhibits against motion when the instrument has a plane mirror adjustment; all other degrees of ametropia give a with motion. All motions are reversed if the instrument is adjusted to a short focus, concave effect because the emerging rays will then originate from a fundus patch on the contrary side of the axis.

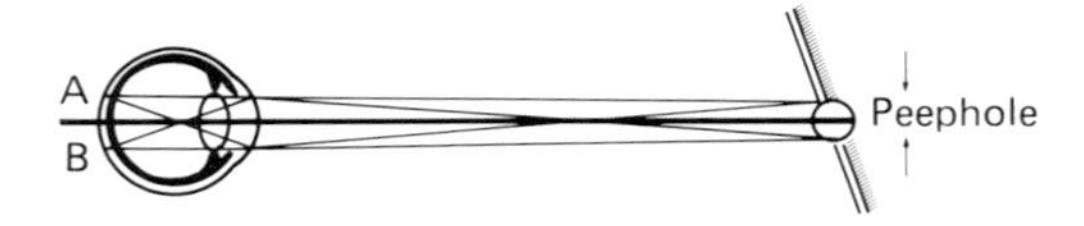

Fig. 12-4. The end point in retinoscopy is termed "neutral." The diagram illustrates that at neutral the far point of the eye coincides with the peephole of the instrument, and all the emerging rays enter the examiner's eye. If the incident rays fall outside the area *AB* (by tilting the instrument), none of the rays can enter the peephole. The examiner sees the pupil completely dark or completely illuminated. Note that neutral does not characterize emmetropia but an eye made so myopic that its far point coincides with the retinoscope, whose position depends on the working distance.

characterizes the ametropia. Convergence indicates myopia, divergence indicates hyperopia, and plano vergence indicates emmetropia.

Some of the emerging rays enter the peephole of the retinoscope to form an image on the examiner's retina. The examiner "projects" this image to the plane of the patient's pupil, which is the object of his attention and for which his eye is focused. He does not see the illuminated fundus patch directly but views it through the optical system of the patient's eye, which acts as a strong magnifier. This is the "reflex."

Let us assume the retinoscope is adjusted to a plane mirror effect, tilted forward so that the illuminated fundus patch is *below* the optic axis of the patient's eye. Fig. 12-3, *A*, shows a hyperopic eye with the rays starting in the lower fundus, emerging from the pupil like a diverging cone whose apex is the virtual far point. The rays that enter the peephole come from the lower part of the pupil, those from above are cut off by the mirror. The dark area has been called a "shadow," but since it is only an area of nonillumination, we restrict our attention to the lighted part or "reflex." The reflex is confined to the lower pupil, and since the retinoscope was rotated down (forward), its displacement is in the same direction as the rotation of the instrument, that is, a with motion.

If the eye were emmetropic, the reflex would again be below, which is identical to that in hyperopia. Indeed with motion is seen in all ametropias in which the far point does *not* lie between the eye and the instrument.

The far point falls between the eye and the instrument in high myopia. The emerging rays now converge to a real focus, that is, they cross before entering the examiner's eye (Fig. 12-3, *B*). The rays entering the peephole come from this image and therefore strike the examiner's fundus below *his* optic axis. He sees it in the upper part of the patient's pupil as an against motion.

When the far point reaches the peephole, no motion is seen, and the reflex either fills the pupil or is totally absent. This is the end point or neutral; the peephole is now conjugate to the patient's retina. In Fig. 12-4, *AB* is the fundus area conjugate to the peephole. As long as the incident rays fall within it, all the emerging rays must enter the peephole. The pupil fills with light irrespective of its size. If the incident rays fall outside the area *AB*, none of the emerging rays can strike the peephole, and the pupil is dark. Note that neutral does not characterize emmetropia but an eye made myopic by the dioptric equivalent of the working distance. Only if one held the instrument at 6 meters would neutral obtain in the emmetropic eye. Hence to obtain the net result the dioptric equivalent of the working distance must be subtracted from the neutralizing lens power.

In the previous description of reflex motion, the retinoscope was assumed to have a plane mirror adjustment. If the knob is moved to the short focus concave mirror effect, all motions are reversed because the fundus image is initially displaced to the opposite side of the optic axis. Neutral, of course, stays the same.

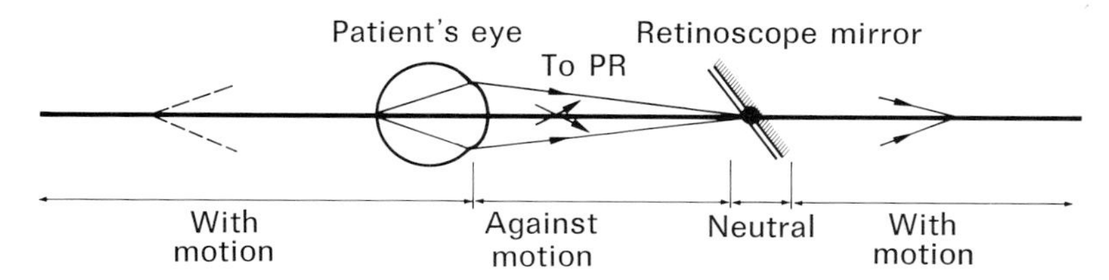

Fig. 12-5. The presence of with or against motion for a plane mirror–adjusted retinoscope depends on the position of the far point. Against motion is seen under these circumstances only when the far point falls between the patient's eye and the instrument. All other refractive errors exhibit with motion. If the far point falls in the area of the instrument, the result is neutral. Because of spherical aberration, it is not a precise point but a range of uncertainty.

Table 12-1. Retinoscopic reflex

	Plane mirror or long focus concave mirror		*Short focus concave mirror*	
	Hyperopes, low myopes, and emmetropes	*High myopes*	*Hyperopes, low myopes, and emmetropes*	*High myopes*
Rotation of retinoscope	Down	Down	Down	Down
Displacement of apparent source	Up	Up	Down	Down
Displacement of fundus image	Down	Down	Up	Up
Displacement of pupil reflex	Down	Up	Up	Down
Direction of motion	With	Against	Against	With

If the retinoscope beam is focused on the plane of the patient's pupil, an interesting effect appears; the pupil fills with light, and no motion is visible whatever the refractive error. It looks like neutral, but it is due to the incident rays and not the emerging rays—hence it is called "incident neutral." Incident neutral is a neat method of retroilluminating corneal and lenticular opacities, which appear black against the uniform red fundus glow.

In summary, the retinoscopic reflex may be with, against or neutral. The motion depends on the vergence of the emerging rays, which in turn depends on the ametropia. With the plane mirror adjustment, an against motion is seen only when the far point lies between the eye and the instrument, and a neutral reflex is seen when the far point coincides with the peephole; all other refractive errors yield with motion (Fig. 12-5). The long focus concave mirror adjustment gives results identical to the plane mirror. The short focus concave mirror adjustment reverses all reflex motions except neutral, which remains the same. These results are summarized in Table 12-1.

An apparent neutral, termed "incident neutral," can be created by focusing the retinoscope beam in the plane of the patient's pupil.

Instrumental factors

A bright, sharp reflex is easier to evaluate than a dim, blurred one and can be achieved by decreasing the size of the fundus image, increasing its brightness, and sharpening its focus.

The fundus acts as a diffusing screen and tends to blur the reflex borders. By decreasing the size of the fundus image, diffusion is reduced. To obtain a small image, the retinoscope bulb can be capped with a pinhole, but this reduces intensity, hence the "hairline" filament bulb. The filament's thickness serves as a pinpoint source if the beam is driven across the eye in a direction perpendicular to its length. This is the principle behind streak retinoscopy.

The brighter the fundus image the brighter the reflex. Fundus illumination can be increased by a higher voltage, a smaller working distance, a dilated pupil, or a larger peephole.*

A sharply focused fundus image results in a crisper reflex. Since the far point is conjugate to the fundus, it follows that placing the apparent source at the far point will give the sharpest reflex. The position of the apparent source, that is, the vergence of the retinoscopic beam, can be adjusted by the knob on the instrument. If the far point is at 200 cm and the beam has a −0.50 D vergence, the reflex appears clearest and brightest, since all the incident rays now participate in forming the fundus image. This then is the purpose of the plane and concave mirror adjustments provided by the self-luminous streak instrument.

It is true that the location of the far point is initially unknown, but whatever the ametropia, the far point must fall at the peephole at neutral. The apparent source therefore should be as close to the peephole as possible at the conclusion of the test. This means that the plane mirror effect is the optimum instrument adjustment for the critical evaluation of neutral.

*Too large a peephole reduces the reflecting surface, and less light enters the patient's eye; a partially silvered peephole reflects some emerging rays but induces internal reflections; thus the last word has yet to be said on retinoscope design.

This reasoning can be carried a step further. Whatever the initial ametropia and wherever the far point, the instrument beam can always be adjusted to obtain the sharpest reflex. Since we are looking at the reflex it is only a matter of adjusting the knob (or sleeve) of the retinoscope to "enhance" its appearance. We know that when the reflex appears sharp and bright, the apparent source will coincide with the far point wherever it may be. For example, the beam would be slightly divergent for low myopes, somewhat convergent for hyperopes, and parallel for emmetropes (Fig. 12-6). As the ametropia is neutralized, the beam is made progressively more divergent so as to end up with maximum divergence at neutral. The instrument should always have maximum plane mirror adjustment to evaluate neutral. Note that whenever the reflex is enhanced, there is always a with motion, irrespective of the ametropia. For example, in high myopia the far point lies between the patient's eye and the instrument, and the crispest reflex is seen when the retinoscope has a concave mirror adjustment, that is, a with motion.

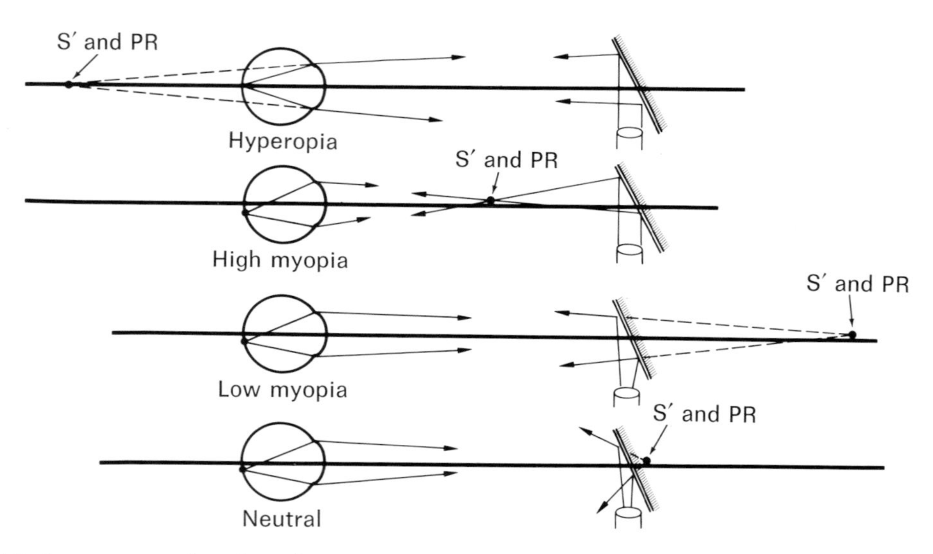

Fig. 12-6. Sequence to show how the position of the apparent source can be adjusted to coincide with the far point of the eye in various types of ametropia. Clinically the procedure is termed "enhancing the reflex."

With motion therefore always gives the crispest reflex for any kind of refractive error. It is not an arbitrary preference by beginners but an optical fact, and it applies to experts as well as to beginners.

In theory the vergence of the beam entering the patient's eye could be matched to the vergence emerging from it and the knob of the instrument calibrated in diopters. It is not practical because working distances vary, and neutral is not a precise point but a zone of uncertainty. The principle, however, can be used to estimate the ametropia. For example, a myopic eye requires beam divergence to enhance the reflex. Now part of the beam falls across the face (or refractor), and its width is proportional to this particular divergence; the greater the divergence the greater the width. A convergent beam is needed to enhance the reflex in high hyperopia; therefore the visible band across the face is narrow. Since maximum beam divergence places the apparent source at the peephole, the facial band is always widest at neutral, and the closer one comes to neutral, the wider it becomes (Fig. 12-7). Estimating refractive error by observing the facial band actually works better in low ametropia, since the difference in width is readily noticed. Copeland could by this method accurately estimate refractive error by one sweep of the retinoscope. But proficiency can only be gained by observing the width of the facial band in every patient, with every

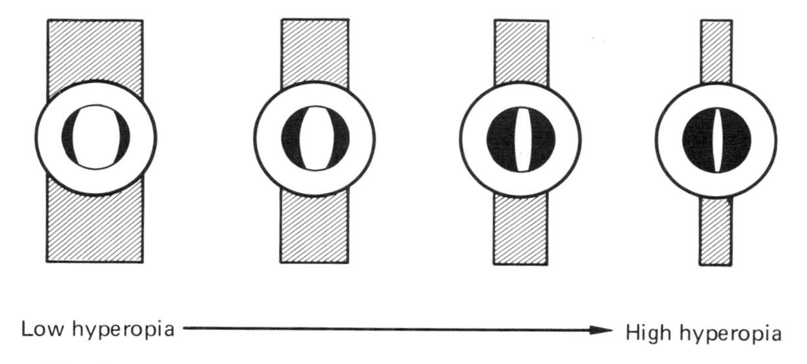

Fig. 12-7. The width of the reflex depends on the degree of ametropia—the nearer the refractive error to the neutral point (which depends on the working distance) the larger the part of the pupil occupied by the reflex. The width of the intercept with the periorbital area is an indication of the vergence of the beam leaving the instrument, more divergent on the left than on the right. By noting this width when the reflex is clearest and brightest, the ametropia can be estimated.

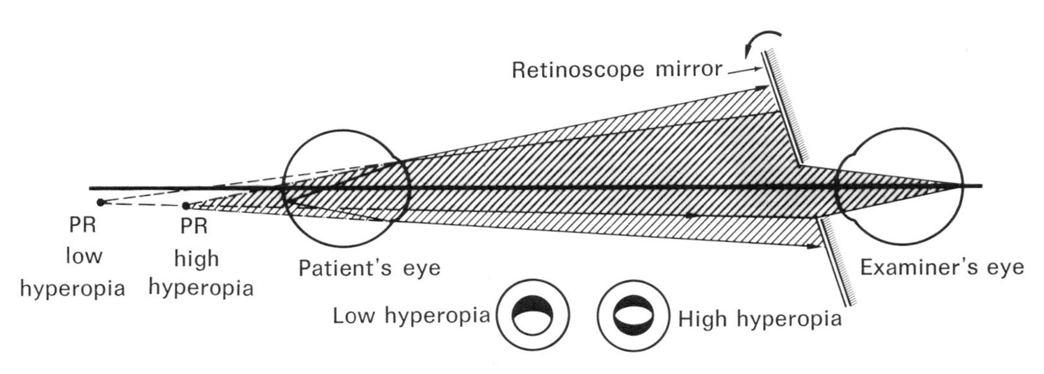

Fig. 12-8. Diagram showing why an eye with greater hyperopia gives a smaller reflex. Since the far point of the more hyperopic eye is closer to it, the emerging rays are more divergent and occupy a smaller proportion of the patient's pupil.

kind of refractive error, always maintaining the same exact working distance. Only with sufficient experience should one attempt this estimating technique in the absence of confirming data.

Characteristics of the reflex

Ametropia is the most important factor in the appearance of the reflex; the closer one approaches neutral, the larger, brighter, and faster the reflex. Its size is explained by Fig. 12-8, which shows a high and a low hyperopic eye superimposed. The far point of the less hyperopic eye is further, and consequently the rays emerging through its pupil are more "concentrated" and occupy a greater proportion of the pupil area. In the more hyperopic eye, many rays are cut off without emerging, and those that do occupy a smaller portion of the pupil. Comparing these two eyes, we see that the less hyperopic eye has the larger reflex, and the principle holds for all refractive errors. The nearer the eye approaches neutral the larger the reflex; the greater the ametropia the smaller the reflex.

Since more rays enter the peephole and therefore the examiner's eye in low ametropia, the reflex also becomes brighter.* Indeed the illumination at the peephole plane is increased in proportion to the square of the reflex size. In high ametropia (e.g., aphakia), the reflex is very dim. To brighten it, one need not increase lamp voltage but bring the far point nearer neutral with trial lenses, which immediately brighten it. The increased reflex brightness also produces a color change from red to pink, but the vagaries of fundus pigmentation do not make this a reliable end point.

In high ametropia the reflex is small, and therefore a greater instrument tilt is needed to drive it from one end of the pupil to the other; its speed is slower. In low ametropias the reflex is large, and the slightest retinoscope tilt displaces it. Thus speed decreases with higher ametropia and is defined as the ratio of reflex to retinoscope motion. The ratio approaches infinity at neutral, which is therefore an infinitely fast motion.

*At neutral the reflex dims again because the peephole is now conjugate with the fundus. The only reason a reflex is seen at all is because the diffusion circles from the out-of-focus light source overlap.

Clinical experience demonstrates that neutral is not a precise end point but a range of uncertainty between the last with motion and first against motion. The zone manifests itself in one of two ways: either several lens changes are required to produce a reversal at a fixed working distance, or one must advance or recede toward the patient's eye between the point where one type of motion disappears and the other appears for a fixed trial lens. The second is the linear reciprocal of the first. To locate the neutral point one must bracket this neutral zone; the larger the zone the more difficult it is to find its midpoint.

Several optical factors contribute to the neutral zone, and the most important is spherical aberration. Fig. 12-9 shows ocular spherical aberration as a range of far points all starting from the retina but at different pupil eccentricities. If the retinoscope is at position *B*, the examiner sees against motion peripherally and with motion in the central pupil. The greater the width of the pupil the greater this scissors motion. The neutral range may approach 1.50 D, and judging which motion predominates in the central pupil is part of the art of retinoscopy. Less art is needed if the pupil is not widely dilated.

Other factors influencing the neutral zone include the depth of focus (the range for which the examiner cannot distinguish one motion from the other) and chromatic aberration (since the reversal point must be chosen from a potpourri of red, orange, and yellow rays).

Reflex in astigmatism

In the cover test for strabismus the deviation is judged not by the static position of the eye but by its refixation movement. Similarly, in retinoscopy, with and against motion are judged not by the position of the reflex in the pupil but by the movement itself. The paradox is that at neutral where judgment is most critical, the reflex motion

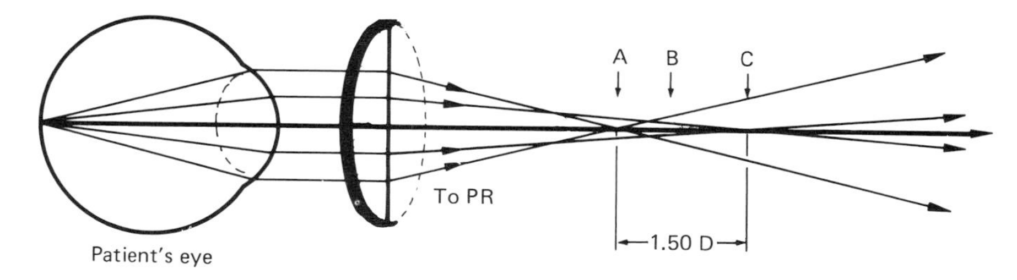

Fig. 12-9. Spherical aberration in terms of a range of far points for a fixed retinal focus. An examiner viewing the eye with the retinoscope held at position *B* would see against motion in the periphery and with motion in the center of the same pupil.

is most rapid and therefore most difficult to classify. The purpose of the advance-recede technique is to obtain more definite end points by which to bracket the neutral zone.

Astigmatic eyes exhibit reflex movements that vary in speed, direction, or both. The greatest difference, naturally, is seen when scanning the primary meridians. A primary meridian is best identified by more or less eliminating one motion first with spherical trail lenses, by the approach-recede method until one meridian neutralizes out, or by focusing the instrument beam on the plane of the pupil (incident neutral). Any of these techniques will abolish the movement in one primary meridian. The streak is now rotated 90°, and the remaining motion is easily seen and neutralized at leisure. Since we prefer with motion, the meridian of slowest against (or fastest with) motion is neutralized first and will represent the spherical meridian. The second meridian may be neutralized with spheres or, preferably, with plus cylinders. Note that a commitment to with motion requires the use of plus cylinders.

A characteristic of astigmatism is "oblique" motion, that is, a motion whose direction differs from that of the streak outside the pupil. For example, if we scan the horizontal meridian and the reflex moves in the forty-fifth meridian, we have an oblique motion. Conversely, a reflex motion that exactly parallels the movement of the streak is referred to as "clean." Oblique motion should not be confused. One may have clean motions in the primary meridians of oblique astigmatism and oblique motions when sweeping "off axis" in vertical or horizontal astigmatism. The characteristic of spherical ametropia is clean motions in *all* meridians.

Oblique motion always means astigmatism, which is evident whenever we scan a nonprimary meridian, that is, when we are off axis. It is most apparent when one meridian is already neutralized and we are scanning to find the other. To eliminate the oblique motion we must scan precisely along a primary meridian, and this is the best way to find the primary meridian.

The astigmatic beam emerging from the eye produces a banded, or elongated, reflex. This is best understood in terms of a spot retinoscope. In spherical ametropia the reflex is a circular disc, but in astigmatism over 1.50 D, the disc shows an obvious elliptic distortion. The distortion follows from Sturm's conoid, but the orientation of the band depends on which meridional far point is nearest to the patient's pupil. The clinician's eye is focused on the patient's pupil, and the line image closest to the examiner's retina is also the one nearest the patient's pupil.

The elongation of the reflex with the spot retinoscope is apparent only if astigmatism is high. But with the streak retinoscope, the banded appearance is obvious even in small astigmatic errors. Of course, the reflex with the streak instrument is banded even in spherical ametropia, but it is more banded in astigmatism. Since the band is enhanced by rotating the streak, the difference in "bandedness" in the primary meridians is easily recognized. For example, in an eye corrected by $+1 + 1 \times 180$, the reflex band is enhanced when the streak is horizontal.

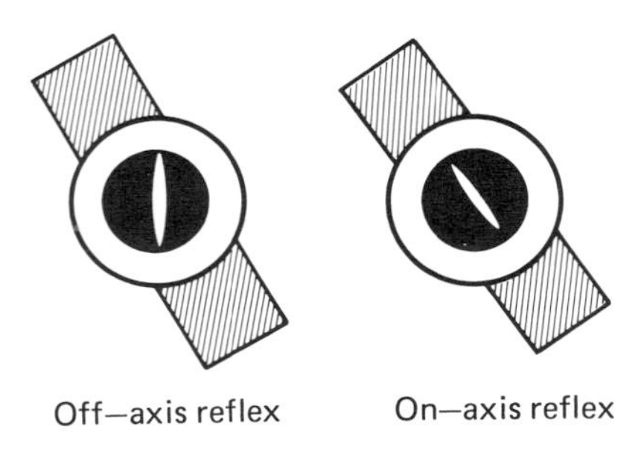

Fig. 12-10. Technique of locating astigmatic axis by aligning reflex and streak.

After neutralizing the less hyperopic meridian, a horizontal band remains, and it is easy to position the correcting plus cylinder axis parallel to it.

The banded reflex in streak retinoscopy produces a dramatic oblique motion when scanning off axis; it persists in moving in a different direction from the streak outside the eye as if the reflex had a mind of its own. The reason is unusual. Suppose the ocular astigmatism is 45°, and we scan the horizontal meridian; the reflex moves in the forty-fifth meridian. Now observe a pencil held at 45° through a 1-inch hole in a paper sheet held about 1 foot away. Move the pencil horizontally but maintain its oblique position; as seen through the hole, it appears to move obliquely (there must be no identifying marks on the pencil). Oblique motion is therefore a kind of optical illusion, which is not to imply that there is anything psychic about it.

In addition, when the streak is off axis, there is a "break" between the bands inside and outside the pupil (Fig. 12-10). This break is the stationary equivalent of oblique motion and can be observed without sweeping by simply rotating the streak. When the streak lies in either primary meridian, the break disappears, and the two bands appear continuous.

Aligning the bands inside and outside the pupil, that is, eliminating the break, is the easiest method of locating the primary meridians with a streak retinoscope. It is particularly satisfying to the beginner because of its simplicity and accuracy. The visibility of the break depends on a fair amount of astigmatism, and as the eye is neutralized, it becomes progressively more difficult to see. But oblique motion is always apparent, even when the remaining astigmatic error is small. Some retinoscopists never progress beyond the "break" technique, like the microscopist without a fine focus. To become truly expert, one should learn to use the break to identify the primary meridians and refine the axis by eliminating oblique motions.

Notes on terminology

Teaching streak retinoscopy invariably involves a certain confusion of terminology. The term "meridian" refers to the axis of the eye as if it were an optic cross. Thus an eye 1 D hyperopic in the ninetieth meridian requires +1 × 180 cylinder to correct. To evaluate the ametropia in the ninetieth meridian, we scan it with the streak horizontal moved up and down. The movement of the streak is thus identical to that used in hand neutralization, and the position of the streak is that of the line whose movement one neutralizes. The "streak" refers to the elongated band of light emitted by the instrument whose width and position can be adjusted by regulating and rotating the sleeve of the retinoscope. The "facial band" refers to the intercept of the streak with the orbital area of the face or the refractor opening. The "reflex" refers to the lighted portion of the pupil and represents the vergence of the rays reflected from the patient's fundus. The direction of the streak and reflex coincide in clean motions (spherical ametropia or on-axis scanning in astigmatism). A vertical streak lies in the ninetieth meridian but is used to scan the power of the one hundred eightieth meridian of the eye.

REFERENCES

Allen, M. J., and Carter, J. H.: An infrared optometer to study the accommodative mechanism, Am. J. Optom. **37**:403-407, 1960.

Baldwin, W. R., and Stover, W. B.: Observation of laser standing wave patterns to determine refractive status, Am. J. Optom. **45**:143-151, 1968.

Bennett, A. G., and Francis, J. L.: Retinoscopy and ophthalmoscopy. In Davson, H., editor: The eye, vol. 4, New York, 1962, Academic Press, Inc., pp. 181-193.

Ben-Tovim, N.: Cylinder dioptometry, Arch. Ophthal. **84**:260-271, 1970.

Bestor, H. M.: The interpretation of dynamic skiametric findings, Am. J. Physiol. Opt. **1**:223-233, 1920.

Bizzell, J. W., Hendricks, J. C., Goldbert, M. F., Patel, M. K., and Robbins, G. F.: Clinical evaluation of an infrared refracting instrument, Arch. Ophthal. **92**:103-108, 1974.

Borish, I. M.: Clinical refraction, ed. 3, Chicago, 1970, Professional Press, Inc.

Bowman, W.: Contribution to the general history of conical cornea, Ophthal. Hosp. Rep. 1859-1860, p. 157.

Campbell, F. W., and Robson, J. G.: High-speed infrared optometer, J. Opt. Soc. Am. **49**:268-272, 1959.

Cuignet, F.: Keratoscopie, Rec. Ophthal., 1873, p. 14.

Duffy, F. H., and Rengstroff, R. H.: Ametropia measurements from the visual evoked response, Am. J. Optom. **48**:717-728, 1971.

Dwyer, W. O., Granata, D., Bossin, R., and Andreas, S.: Validity of the laser refraction technique for determining spherical error in different refractive groups, Am. J. Optom. **50**:222-225, 1973.

Dwyer, W. O., Kent, P., Powell, J., McElvain, R., and Redmond, J.: Reliability of the laser refraction technique for different refractive groups, Am. J. Optom. **49**:929-931, 1972.

Floyd, R. P., and Garcia, G.: The Ophthalmetron, a clinical trial of accuracy, Arch. Ophthal. **92**:10-14, 1974.

Gullstrand, A.: Photographic-ophthalmometric and clinical investigations of corneal refraction, translated by W. M. Ludlam with an appendix by S. Wittenberg, Am. J. Optom. **43**:143-214, 1966.

Harter, R. M., and White, C. T.: Effect of contour sharpness and check-size on visually evoked cortical potentials, Vision Res. **8**:701-711, 1968.

Harter, M. R., and White, C. T.: Evoked cortical responses to checkerboard patterns; effect of check size as a function of visual acuity, Electroencephalogr. Clin. Neurophysiol. **28**:48-54, 1970.

Hill, R. M., and Ikeda, H.: "Refracting" a single retinal ganglion cell, Arch. Ophthal. **85**:592-596, 1971.

Hodd, F. A. B.: Retinoscopy. In Transactions of the London Refraction Hospital, London, 1947, Hatton Press, Ltd.

Jackson, E.: The measurement of refraction by the shadow test, or retinoscopy, Am. J. Med. Sci. **89**:404-412, 1885.

Jackson, E.: Skiascopy, Philadelphia, 1895, Edwards & Docker Co.

Javal, E., and Schiotz, H.: Un ophthalmometre pratique, Ann. Ocul. (Brussels) **86**:5-21, 1881.

Knoll, H. A.: Measuring ametropia with a gas laser; preliminary report, Am. J. Optom. **43**:415-418, 1966.

Knoll, H. A., and Mohrman, R.: The Ophthalmetron; principles and operation, Am. J. Optom. **49**:122-128, 1972.

Knoll, H. A., Mohrman, R., and Maier, W. L.: Automatic objective refraction in an office practice, Am. J. Optom. **47**:644-649, 1970.

Landolt, E.: Refraction and accommodation of the eye, Edinburgh, 1886, Y. J. Pentland.

Lebensohn, J. E.: A simplified astigmometer, Am. J. Ophthal. **32**:1128-1130, 1949.

Lindner, K.: Die Bestimmung des Astigmatismus, Basel, 1927, S. Karger.

Lundlam, W. M., Wittenberg, S., and Rosenthal, J.: Measurement of the ocular dioptric elements utilizing photographic methods, Am. J. Optom. **42**:394-416, 1965.

Millodot, M., and Riggs, L. A.: Refraction determined electrophysiologically, Arch. Ophthal. **84**:272-278, 1970.

Nadbath, R. P.: The rule of seven or fourteen thirds, Arch. Ophthal. **60**:534-536, 1958.

Parent, H.: De la keratoscopie, practique et theorie, Rec. Ophthal. Feb., 1880, p. 65.

Parent, H.: Diagnostique et détermination objective de l'astigmatisme, Rec. Ophthal. **3**:229-252, 1881.

Pascal, J. I.: Modern retinoscopy, London, 1930, Hatten Press, Ltd.

Placido, A.: Novo instrumento de esploracao da cornea, Period Ophthal. Prat. **2**:27, 1880.

Priestley-Smith: A simple ophthalmoscope for the shadow test, Ophthal. Rev. 1884.

Rigden, J. D., and Gordon, E. I.: The granularity of scattered optical laser light, Proc. Inst. Radio Eng. **50**:2367-2368, 1962.

Roth, N.: Recording infra-red coincidence optometer, Am. J. Optom. **39**:356-361, 1962.

Rundles, W. Z.: Streak retinoscopy, Arch. Ophthal. **21**:833-843, 1939.

Safir, A., Knoll, H., and Mohrman, R.: Automatic objective refraction, Trans. Am. Acad. Ophthal. Otolaryngol. **74**:1266-1275, 1970.

Sheard, C.: Ophthalmometry and its application to ocular refraction and eye examination, Am. J. Physiol. Opt. **1**:357-397, 1920.

Sloan, P., and Polse, K. A.: Preliminary clinical evaluation of the Dioptron, Am. J. Optom. **51**:189-197, 1974.

Steiger, A.: Ueber Beziehungen zwischen Myopie und Astigmatismus, Z. Augenheilkd. **20**:97-118, 1908.

Stirling, A. W.: Random reminiscences of last century European ophthalmologists, Am. J. Ophthal. **26**:727-748, 1941.

Swaine, W.: Retinoscopy, Optician **109**:171-172, 335-337, 1945.

Swaine, W.: Retinoscopy, Optician **110**:35-37, 71-74, 1945.

Tait, E. F.: Textbook of refraction, Philadelphia, 1951, W. B. Saunders Co.

Volk, D. L.: Objective methods of refraction, Am. J. Ophthal. **39**:719-727, 1955.

von Sallman, L.: Untersuchungen ueber den Linsenastigmatismus, Arch. Ophthal. **131**:492-504, 1934.

Warshawsky, J.: High-resolution optometer for the continuous measurement of accommodation, J. Opt. Soc. Am. **54**:375-379, 1964.

Westheimer, G.: A method of photoelectric keratoscopy, Am. J. Optom. **42**:315-320, 1965.

Wilson, D. C.: Dynamic optometer, J. Opt. Soc. Am. **64**:235-239, 1974.

13

Clinical retinoscopy

Retinoscopy has been described as the art that best characterizes the skillful refractionist. Like chess it is easy to learn but difficult to master, and clinical proficiency is not obtained en passant, in a week, or overnight. An understanding of the principles will allow variations in method, refinements in technique, and an appreciation of its limitations.

Although good work can be done with any retinoscope, a self-luminous instrument seems an obvious choice. Greater simplicity is claimed for the spot retinoscope, but the streak is more versatile and permits refinements not otherwise possible. Whatever the instrument, it should be kept in good adjustment with several spare bulbs on hand.

Most beginners prefer a dilated pupil, helpful if the media are hazy or when accommodative and fixation control are not feasible, but generally it should be avoided. A large pupil widens the neutral zone and increases the uncertainty of the end point. Fixation is maintained by a target distant enough to relax accommodation and interesting enough to hold the attention. A large target or toy with a fogging lens before the opposite eye under reduced illumination helps inhibit accommodation. If not, it will soon be evident by wide reflex fluctuations and by poor correlation with the subjective.

Since the patient fixates a target behind the examiner's head, retinoscopy is inevitably off center of the patient's visual axis, introducing errors due to coma and oblique astigmatism. The smaller the eccentricity the less the induced error; the greater the working distance the easier one can scan close to the visual axis. Although axial retinoscopy can be done with cycloplegia, corneal reflections make it difficult, and the dilated pupil makes it inaccurate. Switching to the alternate eye helps but is easier said than done; most clinicians find that one eye serves them more reliably for the critical judgment of reflex movements.

To enhance contrast, retinoscopy is best performed in a semidarkened room within the limits of frightening the patient and stumbling over furniture. A refractor expedites the examination, but with children, individual lenses must be held up, preferably before the eye being examined. Most patients need reminding to keep their eyes open, or one soon scans a setting sun. A working distance within comfortable reach of the refractor conserves time and effort. A refractor expedites the examination, since axis and power can be modified without taking one's eye off the reflex. Usual working distances are 50 cm, 66 cm, and 100 cm; the closer the distance the brighter the reflex, but the less the leeway for error. The further the distance the larger the neutral zone; hence linear errors have smaller dioptric consequences. A "working lens" to compensate for the distance saves us the trouble of a subtraction, but the increased reflections, absorptions, and aberrations are not worth it. A tape measure helps until the working distance becomes automatic by the use of visual and kinesthetic clues. A certain flexibility is advisable in cases of high ametropia, hazy media, or productive cough. Finally, retinoscopy should be a brief procedure—the longer one stares at an unrecognizable reflex the more confusing it becomes.

PROCEDURE OF STREAK RETINOSCOPY

Adjust the retinoscope to maximum plane mirror effect, which should result in the widest streak.* (In the Copeland retinoscope, this corresponds to the sleeve in the highest position.) Practically all refractive errors, including emmetropia, will now show with motion. The only exception is high myopia (far point between eye and instrument), which shows against motion. The motion of the reflex is studied by sweeping the pupil. Wide excursions are not necessary; the movement is at the wrist, not the shoulder. The instrument is held firmly in *both* hands so that beam vergence (sleeve up and down) and axis (sleeve rotation) can be manipulated simultaneously. (Practice scanning ocular meridians smoothly and quickly by driving the beam across successive lines of a clock dial held at the working distance.)

In streak retinoscopy the bulb filament acts as a quasi-pinpoint source. The streak is therefore driven perpendicular to the meridian studied (Fig. 13-1). Thus to study the vertical meridian of the eye, the streak is horizontal and moved up and down.

*See notes on terminology at the end of the previous chapter.

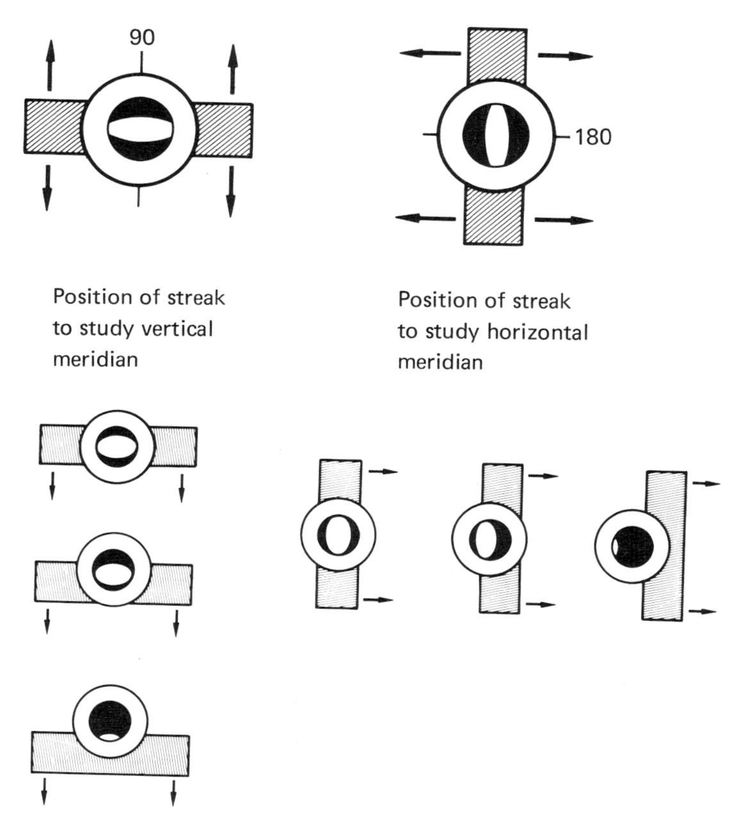

Fig. 13-1. Method of scanning ocular meridians; the streak is horizontal to evaluate the power in the vertical ocular meridian and vertical to evaluate the 180° meridian. The streak is therefore driven perpendicular to the axis under investigation (just as in hand neutralization, we look at a horizontal line when moving the lens up and down). The lower figure shows a with motion in the vertical and an against motion in the horizontal meridians of the eye.

In the previous chapter we saw that with motion is always brighter, clearer, and easier to see; hence retinoscopy should be based on the analysis of with motion. To neutralize with motion, add plus power; to neutralize against motion, add minus power. If an initial against motion is seen, add enough minus sphere to convert to a with motion before proceeding.

The best technique is the one with the most reproducible end point. By comparing retinoscopic and subjective results, practicing on model and animal eyes, and analyzing discrepancies, one eventually comes to rely on the instrument even when subjective confirmation is not feasible.

Spherical ametropia

Consider an eye with 1 D spherical hyperopia. Scanning we see with motion at the same speed in all meridians. The reflex moves exactly like the streak; that is, the motion is "clean," and if the streak is rotated, no break occurs in the band inside and outside the pupil. If the working distance is 1 meter, the eye is 2 D from neutral; hence the reflex is fairly small. As neutralization proceeds by adding plus, the motion becomes faster, and the reflex gets larger. With a +2 D lens, the motion becomes infinitely fast, and the pupil fills with light. The eye is now 1 D myopic, with its far point at the peephole of the instrument held at 1 meter. The "net" retinoscopy is obtained by subtracting 1 D from the total trial lens power (+2 D) and is recorded as +1.00 D sphere.

If the working distance had been 50 cm, the same eye would have been initially 3 D from neutral, the motion slower, and the reflex narrower. Neutral would occur with a +3 D trial lens, and 2 D is subtracted to obtain the net result. The appearance of the reflex therefore depends on both ametropia and working distance.

Observing the reflex of the 1 D hyperopic eye with the instrument held at 50 cm, adjust the sleeve (or knob) from its maximum plane mirror position slightly downward, and note that the reflex becomes brighter and sharper. When it appears brightest and clearest, the beam leaving the retinoscope converges toward the far point 100 cm behind the eye. If the retinoscope had been calibrated for this working distance, this sleeve position would be labeled +1 D. Now note the width of the facial band (outside the eye) at this enhanced position; it corresponds to 1 D of instrument convergence and is a kind of calibration. If we always work at this distance and make a mental note of it in various degrees of ametropia, the facial band provides a good estimate of refractive error.

Continue to move the sleeve of the retinoscope until the beam focuses in the plane of the pupil. The streak is now very narrow, the pupil fills with light, and no reflex motion is visible regardless of the ametropia. This incident neutral should not be confused with the true neutral of emerging rays heading toward the far point.

Return the sleeve to the "enhanced" reflex position, and neutralize the eye with trial lenses. Note that as these are added, the sleeve must be moved farther up to maintain a bright, clear reflex. At neutral the rays coming out of the patient's eye focus at the peephole and the optimum reflex is seen with the instrument at maximum plane mirror adjustment. The sleeve should always be at the top position when evaluating neutral.

At neutral, the reflex is so large and moves so fast that it may be difficult to analyze; hence we move nearer the eye for a definite with motion, then recede for a definite against motion and bracket the interval. The clinical neutral point is at the midpoint of this zone of uncertainty, which is chiefly due to spherical aberration of the eye. By working closer to the eye, the neutral range is smaller and easier to bracket; smaller pupils also give narrower zones. Under cycloplegia, scanning and analysis should be confined to the central pupillary area.

The neutral zone can also be identified (with somewhat less precision) by trial lenses. For example, if a definite with motion is seen with +0.75 D and a definite against motion with +1.25 D, the neutral range is 0.50 D. This corresponds to a linear range of 50 cm at 1 meter but only 13 cm at a 0.5 meter working distance; hence the closer working distance gives less room for error.

In eyes with hazy media, it may be necessary to approach within ophthalmoscopic distances, and in some eyes the reflex divides like the blades of a scissors, swirls about, or assumes other bizarre shapes difficult to classify. These irregularities are due to asymmetry, tilting, eccentricities, opacities, or distortions of refracting surfaces. A tilted trial lens also causes reflex irregularity, always more evident with dilated pupils. Some irregularities occur in all eyes as a kind of background noise one learns to ignore with experience. Pronounced irregularities seen in keratoconus, pterygium, corneal scars, or cataracts may make retinoscopy impossible but does not preclude improvement with optical aids. A subjective examination, if feasible, should always be performed.

After completing the retinoscopy, advance and recede to confirm that all meridians neutralize out simultaneously.

Astigmatism

Astigmatic eyes show differing motions in different ocular meridians, but there is always a maximum and a minimum representing the primary meridians. In regular astigmatism, these are perpendicular to each other; hence we begin by more or less neutralizing one meridian. Since we intend to use with motion, we neutralize that meridian which will leave a with motion in the other. Therefore neutralize the meridian that shows the fastest with motion or the slowest against motion first.

The spherical component can be neutralized in one of three ways: by trial lenses, by approaching or receding from the eye, or by incident neutral. Trial spheres is the usual way; the others are for rapid estimation. Whatever the method, the streak is then rotated 90° to neutralize the cylindrical meridian.

Suppose the eye has 1 D hyperopia in the vertical meridian and 2 D hyperopia in the horizontal. The spherical component is the one having less hyperopia and is neutralized first (streak horizontal, moved up and down). The streak is now rotated 90° and driven across the horizontal meridian (streak vertical, moved left to right). We see 1 D of with movement, and its size, brightness, and speed is commensurate to this degree of refractive error.

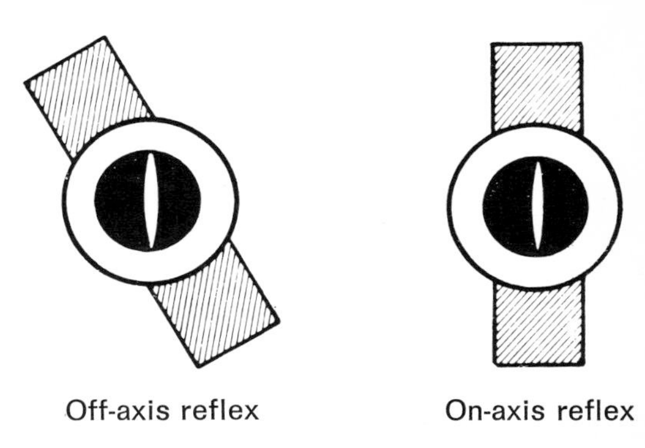

Fig. 13-2. Method of locating the astigmatic axis by the "break" technique.

We are now ready to locate the precise axis of the correcting cylinder. Starting with the streak vertical, rotate it 5° to either side and note a "break" between reflex and streak. Continuity occurs only if the streak is exactly on axis (Fig. 13-2). Place the axis of the plus cylinder parallel to the reflex; the cylinder axis, the streak, and the reflex should now all be in line. It is not necessary to read the axis at this stage; our attention remains riveted on the reflex in the pupil.

The axis may be further refined by observing that all scans give clean motion only when the streak is exactly on axis (vertical in this example). If the streak is rotated to 45° or 135° and we scan these oblique meridians in the uncorrected eye, the reflex continues to move in the horizontal, a "skew" or oblique motion characteristic of an off-axis condition. The skew disappears only when scanning along a true axis. If the astigmatic error is fairly high (over 0.75 D), the "break" method works well, but for smaller cylinders the elimination of oblique or skew motion is more accurate. More important, whatever the initial astigmatism, it gets smaller as the eye is neutralized, and at the end point, only oblique motion remains as a reliable criterion of the correct axis.

Once the cylinder axis is correct, proceed to neutralize for power by increasing plus and bracketing the neutral zone. Since the sphere cannot be determined exactly until after the cylinder is found, alternately scan both primary meridians by "balancing." This achieves the correct end point in which all meridians neutralize simultaneously. Three or four alternating sweeps, modifying the sphere and cylinder as necessary, are usually required.

An older method of neutralizing astigmatism employs only spherical trial lenses. This technique, still popular with those who use a spot retinoscope, gives the correction in the form of an optic cross, and the spherical results are recorded for each primary meridian. The technique has the advantage of only a single trial lens before the eye at any one time, thus reducing reflections and aberrations. But the axis must be identified by the trial lens handle (or an axonometer). Astigmatism is neutralized by first noting the spherical error for one meridian, and then increasing plus (based on with motion) to neutralize the other primary meridian. Spherical neutralization does not allow the refinements of retinoscopy, in which the correcting cylinder itself is used as a tool to locate the precise axis. It is not recommended for streak retinoscopy; indeed few clinicians use it routinely. It has an occasional role in eyes with hazy media or high refractive errors in which excess reflections obscure the visibility of reflex movements.

Some are taught the use of plus cylinders, some select it from experience, and some have it thrust on them because it is the only kind in the available refractor. Plus cylinders are the logical consequence of using with motion; minus cylinders are the logical choice in the subjective fogging technique. If the refractor has only minus cylinders, we can still use plus cylinder technique by transposing. For example, if we require +50 × 90, the same result is obtained by a +0.50 sphere combined with −0.50 × 180. The minus cylinder axis is, of course, placed perpendicular to the reflex.

For every increase in −0.25 D cylinder, one increases the sphere by +0.25 D, which will transpose to the equivalent plus cylinder. Since both the sphere and minus cylinder refractor wheels are rotated for each cylinder change, this transposition has become known as the "double click" method. Only with motion is neutralized as usual.

Binocular retinoscopy

One source of error in retinoscopy is not keeping the unexamined eye fogged. Suppose each eye is 1 D hyperopic, and we examine the right eye while the left fixates a distant target. If the left eye is dominant, 1 D accommodation is activated to maintain clear distance vision, and since accommodation is binocular, the right eye ends up with +1 D undercorrection. If the right eye is dominant, it relaxes accommodation as plus is added until the vision becomes blurred, at which point the patient may switch to the left eye and accommodate for distance, resulting in a change in reflex motion. Indeed the opposite eye is likely to fixate during retinoscopy whatever the dominance, since you are shining a light in the eye being examined. The error can be avoided by keeping a fogging lens on the unexamined eye. How much fog is required can be determined by a preliminary binocular scan.

After neutralizing each eye, compare one to the other along each primary meridian to ensure that there has been no intermittent change in fixation and accommodation. Avoid the tendency to take less time with the second eye on the assumption that the ametropia is symmetrical. In strabismus it may be necessary to resort to alternate cover to maintain proper fixation of the eye being measured. The net retinoscopy is recorded for later comparison with the subjective and also helps in self-analysis.

Summary of technique

Retinoscopy logically follows a series of steps to answer the following questions.

Is the eye spherical or astigmatic? If all meridians appear identical and the motion is clean, the error is spherical. If the reflex differs in some meridians and a break or oblique motion is detected, the eye is astigmatic. Whatever the initial reflex movement, add enough plus to convert all meridians to a definite with motion. Neutralize the spherical error by bracketing the neutral point (advance-recede method), and confirm that all meridians neutralize simultaneously. If the eye is astigmatic, residual motion remains in one primary meridian.

Where are the primary meridians? The primary meridians are those demonstrating no break, that is, reflex and streak alignment. Confirm by eliminating oblique motion. In regular astigmatism the primary meridians are 90° apart, with the wider band belonging to the one nearer neutral.

What is the spherical component? With the instrument set for maximum "plane mirror adjustment" (i.e., sleeve at the topmost position), identify the primary meridians. In the plus cylinder technique, the meridian showing the fastest with motion (or slowest against) is the spherical component and is neutralized first. This leaves a with motion to be neutralized by a plus cylinder.

What is the cylindrical component? Having neutralized the spherical meridian, the streak is rotated 90°. Place the plus cylinder axis coincident or parallel to the streak. Increase plus cylinder power until the last with motion disappears or the first against motion appears. Bracket the zone of reversal for correct power. Refine axis by rotating streak several degrees to either side and noting break or oblique motion. Return to spherical meridian, and alternately compare both primary meridians while moving closer or further from the eye. All meridians should now neutralize simultaneously.

TECHNIQUE OF CYLINDER RETINOSCOPY

Cylinder retinoscopy is a refinement, introduced by Jackson (1895) and popularized by Pascal, in which the trial cylinder is used as a tool to find the correct axis. The principle is best understood by pretending that the ocular astigmatism results from a "built-in" eye cylinder to be neutralized by an opposite-powered trial cylinder. For example, an eye corrected by +1 × 90 has a built-in eye cylinder of −1 × 90. Since the eye cylinder and the correcting glass cylinder are always of opposite sign, their proper alignment, axis to axis, corrects the astigmatism. If the axis is misplaced, however, the two cylinders are crossed obliquely, creating a new "false" astigmatism whose primary meridians differ from either the eye or the trial cylinder axis. The new astigmatism may be illustrated by the following experiment:

Hold a +1 D and a −1 D trial cylinder axis vertical superimposed about 1 foot from your eye, and observe a vertical line about 3 feet away. Moving the combination horizontally as in hand neutralization, no motion is seen because the resultant power is plano. Now hold the two cylinders with axes crossed at right angles; the result is a cross cylinder, and with motion can be seen in one meridian, against motion in the other.

Now rotate the minus cylinder axis to 80°, and let the plus cylinder axis remain at 90°. The target line appears tilted clockwise. Holding the two cylinders firmly together, the target rights itself if the combined cylinders are rotated to two positions: 40° and 130°. These are the newly created primary meridians of the oblique cylinder combination. Moving the two cylinders together, note that the motion in the 40° meridian is against, the motion in the 130° meridian is with. The motions are always opposite because the combination acts as a weak cross cylinder.

By misaligning the axes of the two cylinders 10°, we have created a new astigmatism whose primary axes are 40° and 130°. The shift to 40° for example, represents a fourfold displacement for the 10° misalignment. Since the plus cylinder represents the correcting cylinder, note that it must be moved toward the against-moving, 40° (or myopic) meridian to match the (true) axis of the minus (in this case) eye cylinder.

Technique for axis

The results of the previous experiment can be applied directly to clinical retinoscopy. Assume that the vertical (spherical) meridian of an eye has been neutralized, leaving a with motion in the horizontal (cylindrical) meridian. Proceeding in the usual fashion, that is, streak vertical moved left to right, the correcting +1.00 D cylinder axis is vertical coincident with the streak. Suppose, however, that the true axis of ocular astigmatism is not 90° but 80°. Scanning with the trial cylinder axis at 90°, we should note an oblique motion. The very existence of this oblique motion tells us we are off axis; that is, the inaccurately placed trial cylinder has created a new astigmatism (Fig. 13-3).

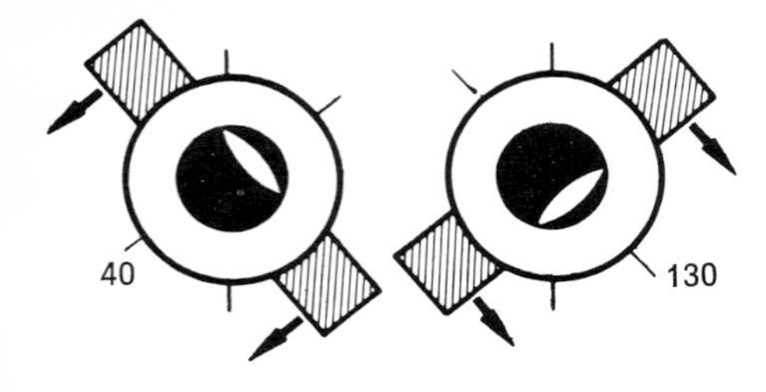

Fig. 13-3. The principle of cylinder retinoscopy is illustrated in this schematic diagram. The +1 D trial cylinder axis has been placed at 90°, but the true ocular astigmatism is at 80°. A new false astigmatism is created by this malposition whose primary meridians are 40° and 130°. Scanning the 40° meridian, we see an against motion; scanning the 130° meridian we see a with motion. The 40° meridian is therefore the guide meridian, and the plus cylinder axis is turned toward it (in 5° steps).

If we scan along the new primary meridians, the reflex motion would be clean only in the 40° and 130° meridians, and we see an against motion in the first and a with motion in the second.* These opposite movements in the primary meridians of the newly created astigmatism are entirely analogous to the movements observed by hand neutralizing the two displaced cylinders in our trial cylinder experiment. The 10° axis misplacement results in a fourfold shift, the extent of which should be recalled whenever one is tempted to rotate a patient's cylinder to fit some arbitrary notion of symmetry.

We found in the lens experiment that the correcting plus cylinder axis must be rotated toward the against-moving or myopic meridian of the newly created (false) astigmatism for correct alignment. Similarly, in retinoscopy the plus cylinder axis must be rotated toward the against meridian, called the "guide meridian." The guide meridian in our example is 40°, and the plus cylinder axis is accordingly rotated from 90° to 40° (i.e., clockwise). The axis need only be turned slightly at first, perhaps 5° to 85° (remember that factor of 4). Scanning again and still observing oblique motion, we rotate the axis to 75°. We now find the guide (against) meridian has shifted to the opposite side (around meridian 123°), which means we have rotated it too far; consequently we move it back toward 80°. Since this is now the correct axis, clean motions will be seen in all meridians. Of course, the motions need not be equal, but the directions always correspond to that of the sweep.

*Recall that the streak must lie in the 130° meridian moved from lower left to upper right to scan the 40°. The streak is always perpendicular to the meridian whose power is being studied. All axes values, of course, refer to the standard notation.

The entire procedure takes longer to describe than to perform. It is not necessary to compute the new axes of the false astigmatism but only to detect their presence and identify which one shows against movement, since that is the guide. Since we are looking at the reflex, it is not even necessary to remember that but only juggle the trial cylinder axis until clean motion is observed, which can only occur if the cylinder axis is correct.

This technique of axis refinement can be used in every case. Suppose we located the axis by the usual break and oblique motion technique and now want to check its accuracy. We deliberately misplace the axis perhaps 10° and study the resultant oblique motions. The cylinder is then realigned by the technique outlined, and if all is well, one arrives at the previous result. Cylinder retinoscopy works best when the spherical error is neutralized first, and only simple astigmatism remains to be evaluated.

For minus correcting cylinders, the guide meridian is the with-moving (or hyperopic) meridian, and the axis of the correcting minus cylinder is turned toward it.

Technique for power

The deliberate misalignment of the correcting cylinder axis to reconfirm its position may be used also to evaluate cylinder power. Thus far it was assumed that the trial cylinder was of correct power but at the wrong axis; however, our trial lens cylinder may also be too strong or too weak, with the two following consequences: The primary meridians of the newly created, false astigmatism will not lie on equal sides of the bisector between the eye cylinder and the correcting cylinder, and the speed of the reflex differs in the two meridians, with either the against or the with motion predominating.

When the correcting plus cylinder axis is rotated away from the true axis perhaps 10° and its power is too weak, one meridian lies closer and the with movement predominates; if too strong, one meridian lies further from the axis, and the against movement is more marked.

The cylinder power can be refined by the predominant motion, which is difficult, or by the displaced meridians, which is easy. This time the meridians of the newly created, false astigmatism are shifted only by changing cylinder power, (a phenomenon that prompted Lindner to call it "mobile astigmia"). Assume the cylinder axis remains positioned 10° from the true axis throughout. The principle is another application of the behavior of obliquely crossed cylinders. When one cylinder of correct power but opposite sign is displaced from another, a new false astigmatism is created whose new primary axes are found at (45°) + (half the angle of cylinder displacement). For example, if the plus cylinder is at 100° and the minus (or eye) cylinder is at 90°, the new axes will be 45° on either side of the 95° meridian (50° and 140°)—providing, however, that the two cylinders are of equal opposite power. If the displacement from the true axis is small (within 20°), we expect to find the new astigmatic meridians approximately 45° on either side. For example, it is at 47½° for 5°, 50° for 10°, and 55° for 20° off-axis displacement. Mere inspection will show that the new meridians are about half a right angle on either side of the cylinder axis as positioned. If the cylinder power is correct, this 45° shift is expected or "normal." If the cylinder power is incorrect, the shift will not be 45° but something else, that is, "abnormal." The new meridians are no longer symmetrical about the correcting cylinder axis; one is closer, the other further, and neither is 45° from the cylinder axis as positioned.

We must now agree from which meridian to measure. Since the against meridian was the guide for aligning the axis, we retain it for the power. The "guide angle" is therefore the one between the correcting cylinder axis and the against (myopic) meridian. This angle is smaller than normal (less than 45°) if the plus correcting cylinder is too weak; larger than normal (greater than 45°) when too strong. The refinement can be made, however, even if one forgets the rules, by arbitrarily increasing and decreasing cylinder power until the new meridians of false astigmatism are symmetrically (45°) disposed about its positioned axis.

Summary

Cylinder retinoscopy is a method of refining the cylinder by identifying oblique motions when the cylinder axis is misplaced. If the cylinder power is correct or nearly so, these motions will be opposite and lie in meridians roughly 45° on either side of the positioned cylinder axis. The correcting plus cylinder axis is to be rotated toward the against-moving meridian, which acts as guide. (The guide meridian for minus cylinders is the with-moving meridian.) The technique works best when the spherical error is neutralized first and when the ocular astigmatism is at least 1.00 D.

If the power is incorrect, it may be refined by deliberately misplacing the cylinder axis 10°. The cylinder power is increased or decreased until the axes of the newly formed astigmatism lie symmetrically about the bisector between the true and the trial cylinder axes.

STRADDLING TECHNIQUE OF CYLINDER RETINOSCOPY

The straddling technique (Copeland) is similar to cylinder retinoscopy, but a streak retinoscope is essential (adjusted for plane mirror effect). Before proceeding, recall that in streak retinoscopy, the beam is driven perpendicular to the meridian under investigation; for example, to scan the 90° meridian the streak is horizontal, moved up and down. Now the streak may be rotated to any meridian; for example, it may be placed coincident (parallel) with the 90° meridian. Here our

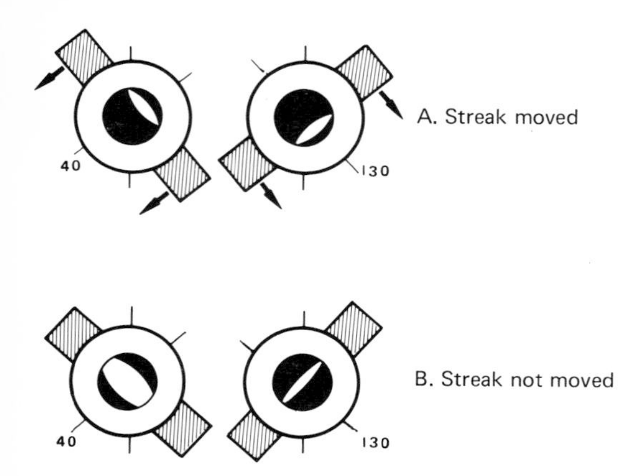

Fig. 13-4. Comparison of cylinder (**A**) and "straddling" technique (**B**) in retinoscopy. In **A**, the meridians are scanned (streak perpendicular to the meridian studied); in **B**, the streak is not moved, but rather the examiner moves nearer or farther from the eye, and the streak is parallel to the meridian investigated.

intention is not to scan the 90° meridian but only to evaluate the appearance of a stationary reflex band for break, width, and brightness. The vertical reflex band depends as always on the power of the horizontal meridian of the eye. Although this can be called an "analysis" of the vertical meridian, it differs from the usual scanning method in that the streak is parallel rather than perpendicular to the meridian being studied. If this sounds confusing, be of good cheer, since even the experts have mixed it up (Fig. 13-4).

Suppose we have an eye that is 1 D hyperopic in the vertical meridian and 2 D hyperopic in the horizontal meridian. The vertical is the spherical meridian (less hyperopia), and the horizontal is the cylindrical meridian (since we intend to use plus cylinders). To analyze the cylindrical meridian the streak in the conventional technique is vertical, moved left to right, and the axis of the trial cylinder is placed coincident with the streak. Suppose, however, the true axis of ocular astigmatism is not 90° but 80° and we discover an oblique motion. We are now ready to apply the straddling technique to refine the axis.

Technique for axis

The streak is placed *parallel* (i.e., coincident), with the meridians about 45° on either side of the positioned trial cylinder axis. We find the exact location by rotating the streak (not scanning) until the break disappears in the reflex band. In our example, this will occur when the streak lies *in* (parallel to) the 130° or 40° meridians. The reflex bands may not be equally clear in the two meridians; so we move closer to the eye until both are easily seen. (If the bands looked alike in width and brightness, the cylinder axis would have been correct.) In this example the band seen when the streak is in the 130° meridian will be wider and brighter (closer to neutral) than the band in the 40° meridian. Slowly receding from the patient but holding the streak steadily in the 130°, the band will widen until it fills the pupil (neutral). Rotating the streak to the 40° meridian, the reflex here remains "banded." The expansion of the band in the 130° meridian as it approaches neutral is no different from any other neutralization. But the difference in the two bands shows that the correcting cylinder axis is misaligned. The rule is to turn the plus cylinder axis in the direction of the band that *does not* fill the pupil (or neutralize) first (that is, the 40° meridian is the guide meridian). Note that there has been no scanning at all, however, only observation of a stationary reflex as we approach or recede from the eye and a rotation of the streak to observe its behavior in the two meridians.

Both the straddling and cylinder retinoscopy techniques naturally lead to the same result; in our example the 40° meridian is the guide meridian in either method. In cylinder retinoscopy the guide meridian is identified by the against motion when scanned (streak perpendicular to it). In the straddling method, it is identified by the residual stationary band (streak *parallel* to it) as we recede from the eye (Fig. 13-5).

With minus trial cylinders in the straddling technique, the axis is turned toward the meridian in which the band *does* neutralize first.

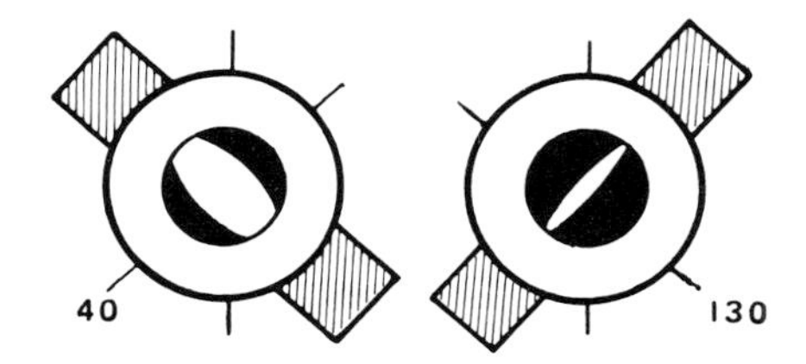

Fig. 13-5. Straddling technique of retinoscopy. The examiner moves toward the patient's eye until a with motion is obtained in both primary meridians of the false astigmatism (assumed to be 40° and 130° in this illustration). Receding from the eye, the reflex that lies *in* the 130° meridian will be seen to neutralize first (since it is wider), whereas the reflex in the 40° meridian still appears "bandlike." The guide meridian is the one that does *not* neutralize first; that is, the 40° meridian and the plus cylinder axis is turned toward it. The two techniques of retinoscopy (cylinder and straddling) obviously lead to the same result.

Technique for power

To refine cylinder power by the straddling technique, assume that the axis is correctly positioned and the streak is parallel to the true principal meridians. In our example, the spherical meridian is tested when the streak is horizontal; the cylindrical meridian is tested when the streak is vertical. The streak position here is the same as in conventional retinoscopy. Without scanning, simply observe the width of the two bands as you approach and recede from the eye. If the bands fill the pupil equally and simultaneously, the cylinder power is correct. If, on receding from the eye, the band in the spherical meridian fills the pupil first, the plus cylinder is too *weak;* if the band in the cylindrical meridian fills the pupil first, the plus cylinder is too *strong.* (The rule is reversed for minus cylinders.)

Summary

The straddling technique is a modification of cylinder retinoscopy in which observation is directed to the appearance of a stationary reflex band. The cylinder axis is refined by rotating the streak 45° to either side of the assumed true axis. The two bands are observed by moving toward the eye and then receding and alternately comparing them; one band is seen to fill the pupil first if the axis is incorrect. The guide meridian (for plus cylinders) is the band that does *not* neutralize first. (The guide meridian for minus cylinders is the meridian containing the band that *does* neutralize first.) The correcting cylinder axis is turned toward the guide meridian.

To refine the power, compare the band in the spherical and cylindrical meridians. If the spherical band fills the pupil first, the plus cylinder power is too weak; if the cylindrical band fills the pupil first, the plus cylinder is too strong. The converse is true for minus cylinders.

ESTIMATING TECHNIQUES

Estimating ametropia by retinoscopy without lenses minimizes reflections and aberrations and is useful when lenses are unavailable or impractical or as a preliminary to conventional neutralization. The validity of these estimates depends greatly on dexterity and experience, but in low ametropias, estimating may lead to gross errors. Since retinoscopy is such a brief procedure, even with exquisite attention to detail, methods that yield potentially inaccurate results should not be used unless there is no alternative. It is generally better to repeat familiar techniques, even if they take longer. This is particularly true when retinoscopy is the only available refractive technique, as in infants and children for whom subjective confirmation is not feasible.

All estimating techniques depend on long practice and careful observation of previously identified criteria. No definite rules can be given because they vary with individual technique and working distance. Generally, they are based on one or more of the following principles: linear distance, appearance of the band, spiraling, and retinoscopic focus.

Linear distance

After the reflex is bracketed by the approach-recede method, a mental note is made of the linear distance at which neutral

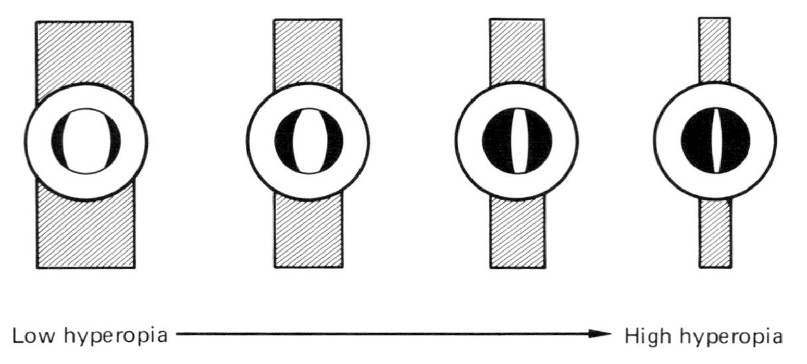

Fig. 13-6. Technique of estimating the refractive error by the width of the facial intercept—the width of the intersection of the streak with the periorbital area of the face or the refractor.

is obtained. The difference in dioptric equivalent at which each principal meridian neutralizes is the cylinder power.

Appearance of band

The reflex is enhanced to its optimum appearance by adjusting the retinoscopic beam. The width of the facial band outside the eye is determined by the vergence of the instrument beam, which coincides with the far point, and hence is proportional to the ametropia. The wider the beam the closer the eye is to neutral. The exact width of the facial band varies with the examiner's working distance. Experience with its appearance in different degrees of ametropia at this distance allows one to estimate the ametropia (Fig. 13-6).

Spiraling (Copeland)

In the spiraling technique, motion is eliminated in one meridian of the astigmatic eye by the approach-recede method. Locate the axes and without changing the sleeve position, rotate the streak 90° and scan the opposite meridian. The direction, speed, and size of the reflex indicates cylinder power.

Retinoscopic focus

In this technique the reflex is enhanced to its optimum appearance, and the real image of the instrument's beam is identified by catching it on a card or screen. The far point is its distance from the eye. Obviously this method works only for high degrees of ametropia.

SOURCES OF ERROR IN RETINOSCOPY

The "validity" of a test merely means that the test measures what it is supposed to. Both retinoscopy and subjective refraction measure ametropia; so logically we compare the two (what else is there?). Which should be the standard depends on the book one reads. In 1928 Percival wrote: "It is well not to waste too much time on subjective tests, as a more accurate determination can be made in far shorter time by objective tests." About twenty-five years later Lancaster (1952) uses practically the same words to say the opposite: "The wisest use of time and effort devotes very brief time to retinoscopy and concentrates on the far more exact and rewarding subjective tests." Most clinicians take a middle view, accepting the inevitable need for objectivity in infants or illiterates but confirming instrumental results with subjective testing whenever feasible.

Certain differences between retinoscopy and subjective methods must be expected. A major cause is accommodative fluctuation, controlled by fogging in subjective testing but only indirectly in retinoscopy. In retinoscopy the light rays are reflected from the epithelium; in the subjective test the focus is on the photoreceptors. In the subjective test the patient selects the end point and the portion of the image caustic that gives him the best resolution. Retinal physiology and cortical psychology play no role in retinoscopy but are an integral part of subjective interpretation and discrimination. Finally, and least important, are the factors under the examiner's control: the care in bracketing the end point, the eccentricity of scanning, and the attention to details of technique, which may lead to errors rather than true differences. Nevertheless, numerous studies have shown that both methods agree best for axis and least for spherical power, a result that should surprise no one.

Most of the differences between retinoscopy and subjective refraction are small enough or predictable enough for considerable confidence to be placed in both, and this allows us to rely on the retinoscope exclusively in children or illiterates without subjective confirmation. But occasionally, even with good technique, the results fall short of achieving the visual improvement obtained with subjective tests. Such unexplained instances emphasize the need to correlate the two techniques whenever possible.

The reliability of the retinoscope has been questioned because different clinicians do not always arrive at the same result. We are told that the conclusion is "statistically significant," but this is only a technical phrase, which does not mean it is important, scientific, or even useful. Better that the same clinician miss the answer consistently by 0.50 D than vary by ±1.00 with an average error of zero. One can drown in an average of 2 feet of water. There is no point in comparing one retinoscopist to another simply by how much each misses the subjective on the average. It is consistency that counts.

Reliability is a matter of experience, and optimum performance may not be gained for several years. Attention to detail, variations in technique, and the use of refinements will achieve respectable results. In the final analysis, the diagnostic value of the retinoscope depends on the clinician who knows how to use it.

Practice on model eyes

A model eye (sometimes called "schematic") is of great help in gaining proficiency in clinical retinoscopy. Commercial models consist of two telescoping portions to adjust axial length with variable pupils. A homemade model can be made from a spool and an old camera lens. It is best to set the model for "emmetropia" at the selected working distance and change "refractive error" by lenses. After experience is gained in observing spherical and astigmatic reflexes, obtain some discarded spectacle lenses and neutralize these with trial lenses to within ±0.25 D and 5° (depending on power). The greater the cylinder the more accuracy should be achieved on axis. Start with large pupils and progress to smaller ones. Practice various techniques, studying their advantages and limitation.

REFERENCES

Allen, M. J.: How to select a retinoscope, J. Am. Optom. Assoc. **40**:920-923, 1969.

Bader, D. A., and Levi, D. M.: Television retinoscopy—a new teaching and research tool, Am. J. Optom. **49**: 119-122, 1972.

Bradford, R. T., and Lawson, L. J.: Clinical evaluation of the Rodenstock refractometer, Arch. Ophthal. **51**:695-700, 1954.

Brubaker, R. F., Reinecke, R. D., and Copeland, J. C.: Meridional refractometry. I. Derivation of equations, Arch. Ophthal. **81**:849-852, 1969.

Brubaker, R. F., Reinecke, R. D., and Newman, J. S.: Meridional refractometry. II. Data reduction, Arch. Ophthal. **83**:570-573, 1970.

Copeland, J. C.: The refraction of children with special reference to retinoscopy, Int. Ophthal. Clin. **3**:959-970, 1963.

Copeland, J. C.: Streak retinoscopy. In Sloane, A., editor: Manual of refraction, Boston, 1970, Little, Brown & Co.

Copeland, J. C.: Copeland streak retinoscope, Rochester, N. D., Bausch & Lomb Optical Co.

Fantl, E.: The cylinder rotation test, Am. J. Ophthal. **34**:1730-1734, 1951.

Freeman, H., and Hodd, F. A. B.: Comparative analysis of retinoscopic and subjective refraction, Br. J. Physiol. Opt. **12**:8-19, 1955.

Hirsch, M. J.: A skiascopic procedure for visual screening surveys, Am. J. Optom. **27**:587-591, 1950.

Hirsch, M. J.: The variability of retinoscopic measurements when applied to large groups of children under visual screening conditions, Am. J. Optom. **33**:1-7, 1956.

Hodd, F. A. B.: The measurement of spherical refraction by retinoscopy, Transactions of the International Optical Congress, London, 1951, British Optical Association.

Jackson, E.: Skiascopy and its practical applications, Philadelphia, 1895, Edwards & Crocker Co.

Klein, M.: Principles of retinoscopy, Br. J. Ophthal. 28:157-176, 1944.

Klein, M.: Retinoscopy in astigmatism, Br. J. Ophthal. **28**:205-220, 1944.

Kuether, C. L.: Photographic recording of the retinoscopic reflex, Am. J. Optom. **49**:113-118, 1972.

Lindner, K.: Die Bestimmung des Astigmatismus durch Schattenprobe mit Cylindergläsern, Berlin, 1927.

Obstfeld, H.: Retinoscopy with the ophthalmoscope, Ophthal. Opt., pp. 59-72, Jan. 23, 1971.

Pascal, J. I.: Modern retinoscopy, London, 1930, Hatton Press, Ltd.

Reimers, P. L., Cohn, T. E., and Freeman, R. D.: The influence of bias upon retinoscopy, Am. J. Optom. **50**:647-652, 1973.

Rosengren, B.: A method of skiascopy with the electric ophthalmoscope, Acta Ophthal. **26**:215-221, 1948.

Safir, A., Hyams, L., and Philpot, J.: Studies in refraction. I. The precision of retinoscopy, Arch. Ophthal. **84**:49-61, 1970.

Srampelli, B.: La schiascopia statica, Bull. Ocul. **21**: 414-430, 1942.

Volk, D.: Objective methods of refraction; a comparison of the Rodenstock eye-refractometer and the Reid streak retinoscope in determining the refractive status of the eye, Am. J. Ophthal. **39**:719-727, 1955.

Weinstock, S. M., and Wirtschafter, J. D.: A decision-oriented flow chart for teaching and performing retinoscopy, Trans. Am. Acad. Ophthal. Otolaryngol. **77**:732-738, 1973.

Zugsmith, G. S.: Use of coated lenses in retinoscopy, Arch. Ophthal. **39**:383-385, 1948.

14

Subjective methods of refraction

Some distinguished authorities hold the point of view that subjective refraction is somehow unscientific because it depends on the patient's intelligence and cooperation and that it should therefore be considered only to verify the refractive error as estimated objectively. Despite this charitable view of the refractionist's intelligence, this sentiment seriously underestimates the value of subjective responses. To say that a refractive technique is imprecise because it is subjective is to discard a century of painfully accumulated psychophysical measurements. It means throwing out the psychologist's favorite baby along with the tears of frustration bathing subjective refractive techniques. Only subjective responses can guide a "scientifically based professional practice." We must not confuse scientific with instrumental or professional with modern, and practice is already too impersonal.

The history guides the direction of the examination. Symptoms determine which tests are to be included; the patient's discrimination determines how often the tests need be repeated; and the intelligence of the patient determines the reliability to be attached to the answers. Clinical experience demonstrates that patients who are uncomfortable, hurried, distracted, or frightened are unlikely to give valid answers. It is up to us to put patients at ease, to proceed at their speed, to communicate in their language, and to select the most appropriate tests and the most efficient apparatus for the problem at hand. Most patients appreciate occasional encouragement and some reassurance that their responses are helpful, especially toward the end point where small changes lead to indecision and natural timidity to avoid mistakes.

Subjective refraction depends on visual acuity. The end point is not maximum resolution but optimum lens correction under the clinical circumstances. The circumstances are average illumination, physiologic pupil size and adaptation, practical target distances, and unlimited discrimination time. The targets are usually letters that evaluate recognition and not resolution. They differ in typography, legibility, contrast, and shape, and they are influenced by familiarity, motivation, and expectation. Acuity is influenced by factors other than refractive error. A good optical image, no matter how sharp, will not help the eye with amblyopia, cataract, or macular disease. And subjective methods may lead to gross errors when the patient cannot or will not cooperate.

The end result of subjective refraction is the strongest plus or weakest minus lens that gives acceptable acuity. It does not necessarily follow that this is also the lens that will be prescribed. Subjective tests like objective tests provide measurements, not prescriptions; the prescription depends on a diagnosis. Entering into the final decision are the patient's symptoms, habits, requirements, previous prescription, the effectiveness of the previous prescription, and the binocular cooperation of the eyes. We may deliberately overcorrect the accommodative esotrope, undercorrect the exotrope, or patch one eye of the amblyope.

The vision after refraction should be at least as good as the patient might have obtained himself by trial and error. Tests so complicated that they confuse both patient and examiner lead only to results that are out of joint and compound fracas. The optical and physiologic effects and limitations of each test or technique must be constantly kept in mind. A novel or unusual answer does not always mean a new law of optics has been uncovered; the patient may have fallen asleep behind the refractor.

Fronmüller in 1843 first urged physicians to use trial lenses instead of having patients select their own from a box. Jaeger in 1854 proposed the near point targets still in use, and in 1862 Snellen published his optotypes. Cylinders were ordered as "down" or "down and slightly out" until H. Knapp suggested the standard notation in 1866. The father of scientific refraction is unquestionably Frans Cornelius Donders (1818-1889), who placed the diagnosis and treatment of ametropia and accommodative anomalies on a rational basis. Precise measurement of small refractive errors belongs to this century. Tscherning, for example, could see little use for steps smaller than 1 D. But it is not sufficient to say that ophthalmology has made enormous progress, since scientific refraction is still in its infancy, and the treatment and prevention of ametropia leaves much to be desired.

FOGGING TECHNIQUE

The keystone of subjective refraction is the control of accommodation—the end point is the refractive status of the eye in repose.

A practical technique of inhibiting accommodation based on the optical effect of plus lenses is called "fogging." The name describes the principle. Vision is artificially blurred to such an extent that ciliary contraction would only serve to make it worse. The fogging lens converts all ametropias to a state of optical myopia. But this optical inhibition is not always complete or successful. If the fog is too great and all contrast washed out, accommodation may actually become active as in empty space myopia. The adequacy of the fog is therefore periodically checked by the ability to read the letter chart.

Optical principles

The optical principles of the fogging technique are best illustrated by an example. An eye with compound hyperopic astigmatism viewing a distant chart accommodates to place the circle of least confusion on the retina. The astigmatic dial appears symmetrically blurred. Minus lenses only inhibit accommodation proportionately and could be

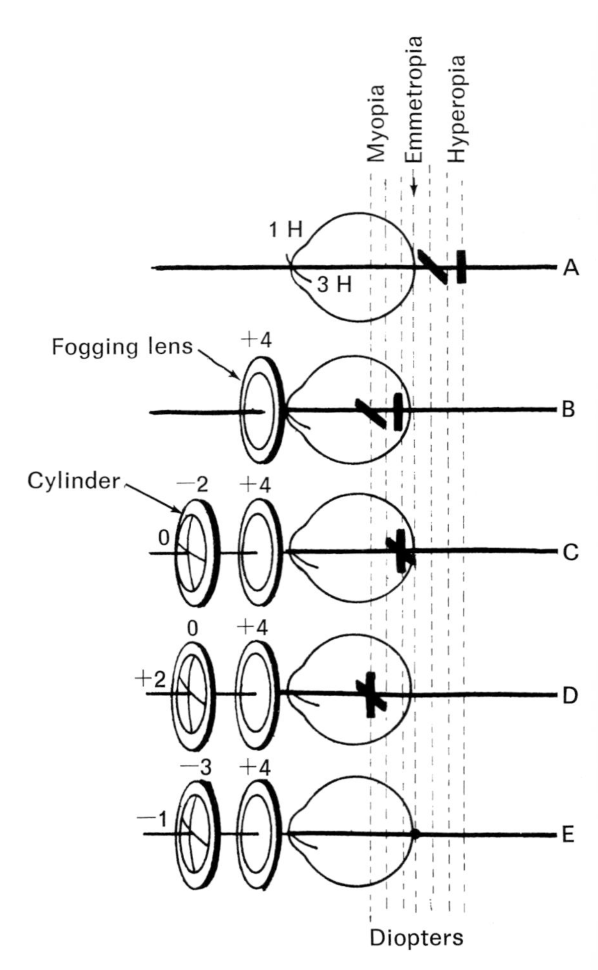

Fig. 14-1. Schematic diagram of the optical sequence of placing a +4 fogging lens in front of an eye with compound hyperopic astigmatism, then applying plus or minus cylinders, and including the final correction. This or a similar picture should flash through the examiner's mind as he proceeds with his subjective techniques of refraction.

added almost to the limit of the amplitude without much change in acuity. The results are meaningless. Consider an eye hyperopic 1 D in the vertical meridian and 3 D in the horizontal meridian (Fig. 14-1, *A*). Since the vertical ocular meridian has more plus power, both focal lines lie behind the retina with the horizontal image nearer. Viewing an astigmatic dial without accommodating, the 3-9 line appears clearest. By accommodating 2 D, the circle of confusion is placed on the retina, equal-mixed astigmatism results, and all dial lines are now equally blurred. Suppose, however, we added a +4 D fogging lens (Fig. 14-1, *B*). The vertical ocular meridian is now 3 D myopic, the horizontal meridian is 1 D myopic, and both focal lines are anterior to the retina. This time the vertical line is closer, and the 12-6 line appears clearest on the dial. Additional accommodation only makes all lines more blurred.

A fogging lens therefore converts all forms of astigmatism to compound myopic. As the fog is reduced from +4 D to +3 D as in our example, the vertical image line rests on the retina and the horizontal image is 2 D in front. On the astigmatic dial, the 12-6 line now has maximum sharpness compared to the horizontal. This is the point of optimum contrast—the ideal unfogging point. For reasons that will be apparent, we generally do not wish to unfog this far. A more practical end point is when one line on the dial appears just noticeably sharper and darker. This is the point where the darkest spoke of the clock dial can just be discriminated as three distinct lines (about 20/40 Snellen acuity). With other dials it depends on the angular subtense of their spokes, and the Snellen equivalent varies with the illumination, contrast, pupil size, degree of fog, and resolution.

The principle of correcting astigmatism is to collapse the interval of Sturm by cylinders. The cylinder moves one focal line relative to the other. A plus cylinder moves the clearer focal line to the level of the more blurred one; a minus cylinder moves the more blurred focal line to the level of the clearer one (Figs. 14-1, *C* and *D*). With minus cylinders, the patient judges when the lines are equally clear, not equally blurred; hence minus cylinders are preferred with fogging.

As the more blurred focal line approaches the clearer one, the interval of Sturm becomes narrower, and the circle of confusion approaches the retina (Fig. 14-1, *E*). If we unfog too much, the cylinder stimulates accommodation by pushing the confusion circle behind the retina. Thus, although cylinders move only one focal line, they can activate accommodation by shifting the circle of least confusion.

With minus cylinders, the blurred focal line approaches the clearer one until all lines appear equal but blurred. The degree of blur depends on how far the clearest focal line was anterior to the retina to begin with. Hence the fogging lens should be strong enough to prevent the minus cylinder from overshooting but not so strong that the resultant blur interferes with the discrimination of equality on the dial. If the fogging lens is adequate, it is not necessary to change it as the minus cylinder is increased.

Fogging is obviously binocular and must be maintained throughout the test. In reducing the fog, the lower plus lens is placed in the trial frame before the higher one is removed; with the refractor, it is only necessary to turn the lens wheel. A convenient starting point is the gross retinoscopic finding. The working lens (+1 D for 1 meter and +2 D for ½ meter) serves as the fogging lens. If the retinoscopy is reasonably reliable, unfogging can proceed directly on the astigmatic dial. If vision with the working lens is unequal, this suggests either that the patient accommodated or that the working distance was altered during the test. Note that fogging is *not* reduced as long as the vision improves but only to the point at which the patient can just discriminate one line clearest on the astigmatic dial. This is the reason the dial lines are made larger than the minimal 1-minute angle. Unfogging to find the astigmatic error should not be confused with unfogging to the optimum sphere (to best Snellen acuity).

PROCEDURE WITH ASTIGMATIC DIALS

Astigmatic dials have their advantages, but not every patient is a good candidate for them. The secret of success can be summarized in one word—demonstrate. Place a 1 D cylinder before the uncovered, slightly fogged eye. Ask if one line on the dial appears darker and clearer than the others. Now rotate the cylinder 30° and repeat the question. If no difference is noted or no line stands out, one might as well proceed to another technique. This simple preliminary test will save much time and grief. By now most patients will understand what you mean by clearer and darker and, by knowing what to look for, give more valid answers.

Several precautions apply to all forms of astigmatic dial testing. The fog must be adequate but the vision not so poor as to prevent discrimination. The patient's head should not be tilted, and his forehead should be firmly apposed to the forehead rest of the refractor. If the pupil is very small, as with miotics, the increased depth of focus reduces contrast, and larger lens changes are needed. The illumination of the chart must be evenly distributed, and the spokes should subtend at least a 2-minute visual angle (fly specks are permissible only if symmetrical). Try the test on yourself to understand what is expected.

Optical principles

The optical principles of the dial are based on astigmatic imagery. Since each object point forms a conoid of Sturm, the image depends on which part of the interval is intersected by the retina. For example, if the horizontal focal line lies on the retina, the image of any object point is a horizontal line. If the object is a cross, all points of the cross, whether from its horizontal or vertical limb, form horizontal image lines. The horizontal limb gives overlapping horizontal lines and thus appears clear, thin, and dark. Simultaneously the vertical limb also gives horizontal lines and consequently appears wider, blurred, and gray (because of the spacing between the lines). In questioning the patient we ask which line appears clearer *and* darker. To exaggerate this contrast, some astigmatic dials consist of dashed rather than continuous spokes.

The fog is reduced until one line appears clearer and darker than the others. If the test follows retinoscopy, some preliminary knowledge of the axis and amount of astigmatism will already be available and serves as a guide. If all the lines appear equal, no astigmatism is present. This should be confirmed by an induced astigmatic error with a trial cylinder. If one line appears clearer and darker, instruct the patient on identification. On the clock dial this is easy; on the sunburst dials a ruler is handy (to be held parallel to the line). Better yet is a projected dial with a rotating pointer that can be manipulated by the examiner.

The minus cylinder axis is placed perpendicular to the clearest line. For example, if the patient reports the vertical line darkest, the minus cylinder axis is placed horizontal. With the clock dial, the rule of 30 is useful: Multiply the smallest clock number designation by 30. Thus if the 12-6 line is clearest, $6 \times 30 = 180$ gives the minus cylinder axis. Of course, not all dials are clock dials; hence the sunburst types generally have attached calibrations. (Be sure they are for minus cylinders.) If the refractor has only plus cylinders, it is necessary to transpose by the "double click" method as described in Chapter 13. Another alternative is to use a plus cylinder refractor for retinoscopy and a trial frame and lenses for the subjective test.

If the patient reports that a group of lines on the clock dial appears clearer, it is necessary to interpolate. The patient is instructed to compare the relative clarity and select the best, second best, and third best line. For example, suppose the 10-4 line is clearest, while the 11-5 and 3-9 lines are judged equally blurred but clearer than the remaining. The correct minus cylinder axis is 120. If the 10-4 and 3-9 lines are equally sharp, the correct axis is midway between 120 and 90, that is, at 105. If the 10-4 line is best and the 3-9 line better than the 11-5 line, the minus cylinder axis is taken as 10° from the

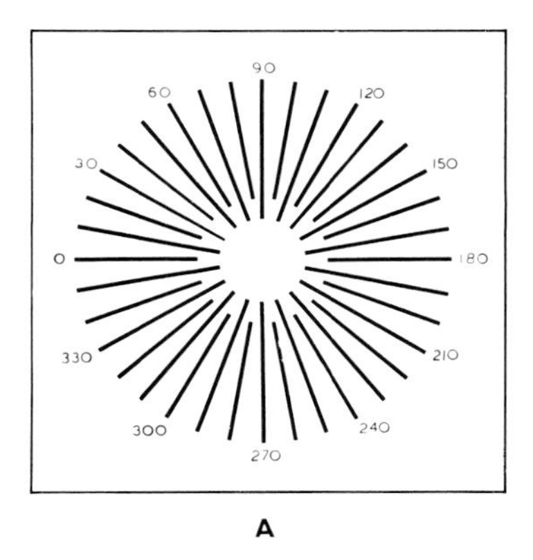

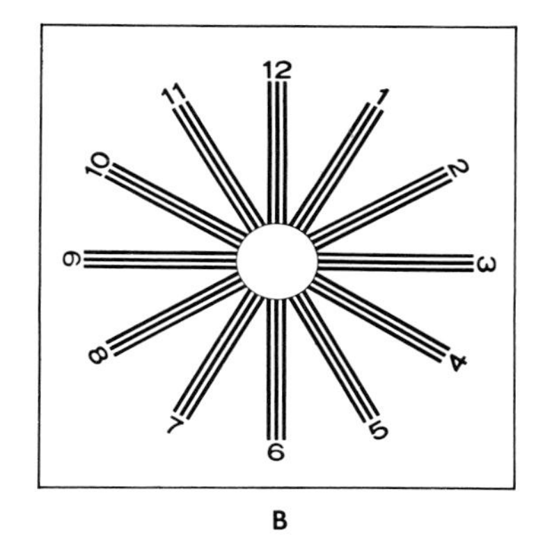

Fig. 14-2. Lancaster-Regan dial (**A**) and clock dial (**B**).

best line, that is, at 100. It will be evident that for all its theoretical simplicity, the clock dial leaves something to be desired. The problem of interpolation is seldom mentioned in the refraction manuals, and since it is unlikely that all astigmatic errors conveniently fall at 30° intervals, an entire afternoon could be spent with one patient.

The patient's attention is now directed to the clearest line and the line perpendicular to it. The power is increased in 0.25 D steps until the lines reverse; that is, the line that was originally the clearest becomes the poorest. (If the cylinder axis was improperly chosen, a new meridian will appear clearer, and no reversal will occur if the astigmatism is irregular.) On reversal, the cylinder is reduced, and the midpoint is bracketed. The eye should remain slightly fogged throughout.

Fixed dials

The clock dial, illustrated in all elementary books, is useful only to explain the principles of astigmatic testing and is soon abandoned in practice. More serviceable is a sunburst dial such as the Lancaster-Regan dial, which is calibrated at 15° intervals. The numbers correspond to those on the refractor or trial frame and give the minus cylinder axis directly in standard notation. Identification is simplified if combined with a rotating arrow by which to locate the selected line. The axis of the minus correcting cylinder lies perpendicular to this line, and the chart is so calibrated (Fig. 14-2, *A* and *B*).

Rotating dials

Presenting target lines only in the primary meridians causes less distraction and confusion. The trick is to find the correct meridians. This is the purpose of rotating dials (Fig. 14-3).

The *rotating cross* (Fig. 14-3, *A*) or T (Fig. 14-3, *C*) is coupled with a sunburst dial by a synchronized gear mechanism; the primary meridians on the sunburst are indicated by a rotating arrow. Once the correct axis is identified, the cross or T is presented, and its limbs automatically lie in the indicated meridians. The patient can now concentrate on equalizing the two limbs of the cross. One advantage of this combination, which comes together on a slide that fits the acuity projector, is that after the cylinder power is found, the axis can be rechecked on the sunburst. In high degrees of astigmatism, it is not unusual to find a subsequent modification in axis required after finding the power.

The axis, of course, can be found by rotating the cross itself. This works best as a refinement of the axis already determined with the retinoscope. The *Robinson-Cohen cross* (Fig. 14-3, *E*) consists of broken black lines on a red background. One of the lines is identified by two small black dots, and the red background serves to inhibit accommodation. The line with the black dots is placed

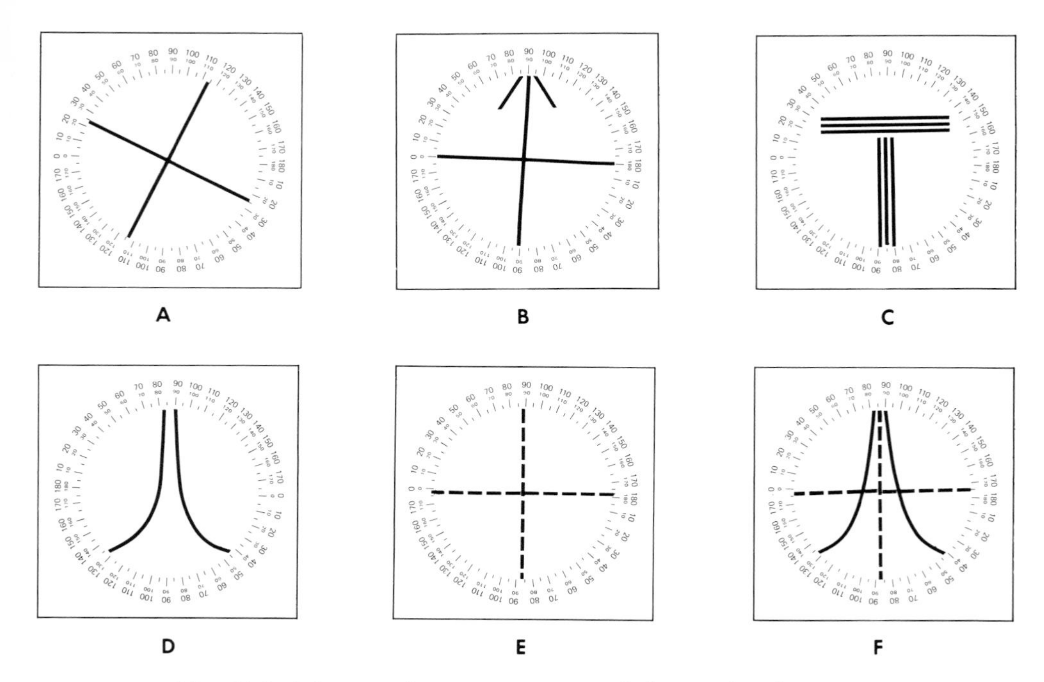

Fig. 14-3. Selection of rotating astigmatic dials. For details see text.

vertically, and, as the fog is reduced, the patient is asked if one line appears clearer. If both are alike, the cross is rotated 45°, and the procedure is repeated. If one line appears clearer, an attempt is made to straddle the axis. The cross is rotated until the lines appear equally clear on each side of the indicated meridian. Then the dial is rotated 45° until the line marked with the two dots lies in the clearest meridian; the indicated minus cylinder axis is perpendicular to this line, and the dial is so calibrated.

The *Marano dial* adds the principle of color mixing to the end point. The target is a rotating cross made of alternating red and green dashes. When one limb lies in the astigmatic meridian, it appears not only clearer but also yellow, because of the overlapping red and green colors. The procedure is similar to other forms of rotating dials. When the cross is equalized by cylinders, the limbs revert to alternate red-green dashes. An additional feature is a single red-green line geared to the cross, which may be exposed alone to locate one principal astigmatic meridian. The color fusion works best when the target is projected on an aluminized screen in a darkened room. The Marano dial comes on a standard slide, including Snellen letters, that fits the acuity projector.

The *Lebensohn astigmometer* (Fig. 14-3, *B*) is representative of "arrow" dials. It consists of a cross, with an attached arrowhead. The arrow points outward, and the two short lines are separated 60°. The dial is rotated on a wall chart calibrated for either direct or mirror viewing. After reducing the fog, the target is rotated until the "arrow line" appears darkest and clearest. The wings of the arrow should now appear equally sharp. If one wing is darker, the cross is rotated in the direction *away* from the darker wing in 5° steps until they appear equal. The arrow now points to the correct axis, and the patient is directed back to the primary limbs of the cross, which are equalized by minus cylinders (axis perpendicular to the arrowhead limb) in 0.25 D steps until reversal occurs. The Lebensohn dial is easy to use, simple to understand, and may be further refined by the Crisp-Stine test described on p. 237.

One modification of this principle is a rotating arrowhead to locate the darkest line on a standard sunburst dial. When the arrow reaches the indicated axis, the patient compares the wings. Since the arrow points inward here, the axis is turned *toward* the darker wing.

The *Raubitschek dial* (Fig. 14-3, *D*) has a unique arrow consisting of two curved paraboloid lines. The lines start parallel and end in opposite directions at the base. The portion making up the tip is perpendicular to that making up the base; hence all meridians are represented by each arc. When the arrow is rotated, even a slight deviation from the true axis of astigmatism results in the appearance of a darker "shadow" portion on one limb. The section of the curve corresponding to the dark area is parallel to the meridian of greatest ocular refractive power. As the arrow is rotated, the dark area jumps from one side to the other. The axis is correct when both limbs are equally sharp and symmetrically dark near the tip. (The examiner says to the patient: "Here is a road becoming narrower at one end. Tell me which side of the road looks clearer and darker at the narrow end.") The eye is slightly fogged, and the arrow is turned in the direction away from the darker side to obtain equality; if turned too far, the shadow jumps to the opposite side. The axis of the minus-correcting cylinder is placed perpendicular to the arrow.

The amount of astigmatism was originally determined by inserting the trial cylinder 20° from the true axis and rotating the Raubitschek to reestablish equality. The power was computed (or read from a scale), based on unequal, opposite crossed, oblique cylinders. Bannon (1958) greatly simplified the procedure by adding a dashed cross that bisects the paraboline arrow (Fig. 14-3, *F*). Once the axis is found, the patient's attention is directed to the cross, which is then equalized (minus cylinder axis perpendicular to the arrow) as with the standard rotating cross. This combines the sensitivity of the paraboline arrow for axis with the simplicity of the conventional cross for power. The rotating paraboline arrow is unique and holds the attention while uninfluenced by previous experience.

The *Bannon "paraboline" test* is available as a slide to fit the acuity projector combined with a sunburst dial with rotating arrow. The dial is presented first (patient slightly fogged) and the approximate axis identified. The paraboline arrow then automatically lies in the indicated meridian for refining the axis, and the cylinder power is determined on the dashed cross.

Summary

All astigmatic dials depend on careful control of accommodation. If the fog is inadequate, the test is bound to be a failure. As the fog is reduced, one line or group of lines will appear clearer and darker than others to the astigmatic eye. This criterion is first demonstrated to the patient by inserting a trial cylinder. Most patients should be able to detect a 0.50 cylinder. The axis of the cylinder is identified by straddling and bracketing. With the fogging technique, minus rather than plus cylinders provide the more easily discriminated end point. The minus cylinder is placed with its axis perpendicular to the clearest line. The axis is recorded in standard notation, the way it is written in the final prescription. Fixed dials allow simultaneous comparison of an axis change as the power is modified. Rotating dials allow easier identification of the axis because attention is confined to the primary meridians.

All dials require careful instructions and the ability to discriminate differences. Some targets are more ingenious than others, but they take longer. Diagnosis can be expedited by starting with the retinoscopic finding and selecting the most suitable target for each patient. Common mistakes are improper fogging, unclear instructions, and transposition errors. Performed correctly, the reliability approaches 0.90. Many clinicians prefer to refine and confirm the results further with additional techniques such as the cross cylinder.

CROSS CYLINDER TECHNIQUE

The cross cylinder is a compound lens having a net minus power in one principal meridian and a net plus power in the other. For example, a minus 1 D sphere combined with a 2 D plus cylinder is a 1 D − 1 D cross cylinder. The most commonly used cross cylinders are 0.25 − 0.25; 0.50 − 0.50, and 1.00 − 1.00, with the 0.50 D having the widest application. Unlike what one might expect, it is the axes that are calibrated; a red dot indicates the minus cylinder axis; a white dot, the plus cylinder axis. The axes are indicated because that is how we write prescriptions.

In 1887 at a meeting of the American Ophthalmological Society, Jackson first described the cross cylinder technique for power: "The astigmatic lens as described by Stokes (can be modified) as a spherical of one kind with a cylindrical of the other kind of twice the strength. For years I have used such a lens to hold in front of the appropriate correction to determine if a cylindrical lens, or a modification of a cylindrical lens already chosen, will improve it; and it is far more useful, and far more used, than any other one lens in my trial set." Twenty years later, its value in refining the axis was realized. The technique has changed little, a firm testimonial to the soundness of Jackson's conception.

Procedure

The cross cylinder comes in a metal frame with a handle that straddles the two principal axes. The lens can be hand-held in front of the trial frame or manipulated in a self-centering mounting attached to the refractor. The second alternative is less likely to lead to inadvertent rotation. The lens is flipped (or twirled) so that the patient is presented with two views of the target, generally the 20/20 or 20/30 line of Snellen letters. The patient is told that both "pictures" will be somewhat blurred and to choose the one that is clearer or less blurred of the two. The examiner should demonstrate by flipping, or the patient invariably answers that both are more blurred than the previous lens.

Unlike the astigmatic dial, the cross cylinder presents two different views in quick succession. The change therefore is more vivid than the gradual blurring of spokes on a dial seen simultaneously. Moreover, it emphasizes that the cross cylinder should be flipped rather quickly, since it is in comparing the two positions rather than the blur of each that counts. The examiner must impress this point on the patient so as to anticipate the common complaint that he is flipping the lens too rapidly. The examiner says: "Do not study each picture or try to read the letters, but give me your first impression as to which is clearer." The cross cylinder test has its greatest utility in *refining* the axis and power of the cylinder already determined (by retinoscopy or astigmatic dial).

By changing the sequence of the two twirled positions, guesswork is eliminated and the answers verified—an important advantage adding to the "objectivity" of the test. Additional verification is obtained by periodically comparing the acuity with the indicated cylinder correction (cross cylinder removed). The cross cylinder test applies equally well to refining plus or minus trial cylinders, and its popularity is due in no small measure to the fact that one can proceed directly to it from the retinoscopy (always plus cylinders).

For reasons explained under optical principles, the cross cylinder is not used with the eye under fog but with 20/20 or 20/30 acuity. If the eye is fogged, the results are inaccurate. Accuracy also depends on maintaining the same spherical equivalent, which means changing the sphere proportionately with each cylinder change—one reason for a refractor rather than a trial frame with loose lenses.

The mechanics of the test take longer to explain than to perform. It is divided into two phases: the refinement for axis and the refinement for power. The test is performed monocularly; the axis is generally determined first, followed by the power, and concluded with a reexamination of axis.

Technique for axis. The cross cylinder is held straddling the correcting cylinder axis (Fig. 14-4, *A*). Its handle therefore parallels (or coincides with) the axis of the trial cylinder. The cross cylinder is flipped, and the patient is asked which gives the sharper picture. If both positions appear the same, the trial cylinder is at the correct axis. This may be confirmed by rotating it to a wrong axis, for example, 10° away, and proceeding with the test. If one flip position is better

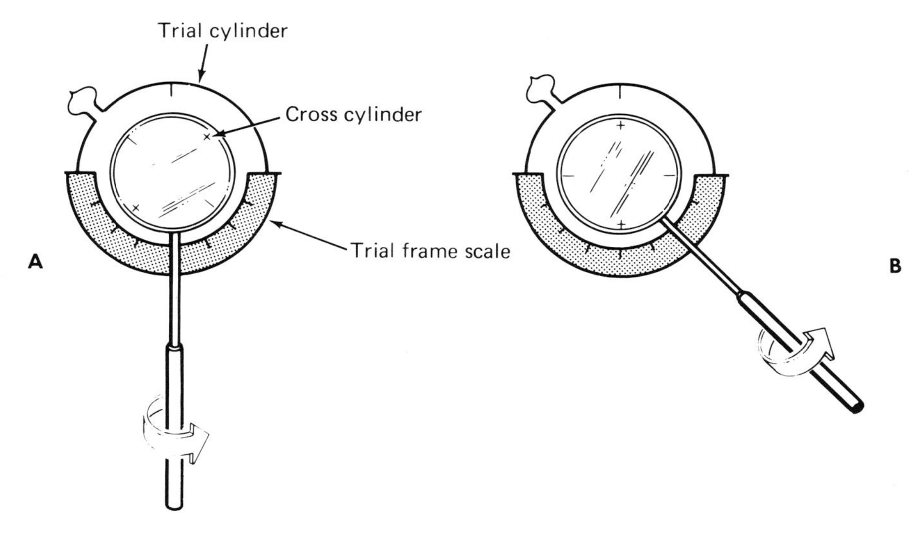

Fig. 14-4. The cross cylinder is used as shown in **A** to refine the trial cylinder axis and in **B**, to refine the trial cylinder power.

than the other, the trial cylinder axis is to be modified. The direction in which it is rotated depends on the sign.

The axis of a plus trial cylinder is rotated toward the white dot flip position if that is reported clearer and away from the white dot position if that is worse. For example, if the trial cylinder axis is at 90°, and if the patient prefers the flip position in which the white dots are at 135° to the one with the white dots at 45°, the trial cylinder axis will be rotated to 95°. The modification of the trial cylinder is never more than 5° or 10° and bears no relation to the power of the cross cylinder. It does depend on the patient's ability to detect the differences in the two flip positions. With plus trial cylinders, therefore, all axis refinements are made on the basis of the position of the white dots—they are analogous to the "guide" meridian in cylinder retinoscopy.

With minus trial cylinders, the red dots serve as a guide. The minus cylinder axis is rotated toward the minus (red) dot flip position if it is better and away from the red dot position if it is more blurred.

The trial cylinder is refined by rotating its axis until a reversal is obtained. Rotations become smaller as one brackets the midpoint. If one direction is consistently reported as better than the other, then the patient did not understand the test, the conditions are inappropriate, or the trial correction is grossly incorrect. To avoid being led around the clock, the examiner should check the acuity periodically and note if vision improves as the trial cylinder more nearly approaches the supposedly correct axis. A go-around is almost inevitable if the astigmatism is small, for example, 0.25 or 0.12 D, and indicates that the patient cannot discriminate the small differences but responds because he thinks he must. Occasionally the reason is different. Some patients have a distinct preference for vertical lines, which helps differentiate Snellen letters (where vertical strokes do predominate). If the eye is overfogged, a vertical line on the retina will allow better recognition of the letters, and one tends to overcorrect against-the-rule astigmatism. Again the solution is to check the acuity periodically during the test. Alternatively one can select a symmetrical target such as a Maltese cross, Verhoeff rings, or checkerboard patterns.

The end point in all cross cylinder refinements is not maximum clarity but equal blurredness. This should be emphasized in asking the patient to make final discriminations. To ease the patient's concern that the vision at the end seems worse than when the procedure was started, remove the cross cylinder to demonstrate the improved acuity.

Technique for power. Cylinder power is refined by placing the cross cylinder axes coincident with the primary axes of ocular astigmatism (Fig. 14-4, *B*). This time the handle of the instrument straddles the primary ocular meridians. In one flip position the red dot axis coincides with the trial cylinder axis; in the other, the white dots coincide with it. The patient viewing the 20/20 or 20/30 Snellen letter indicates which is clearer or less blurred. Again, the examiner should explain that both pictures are more blurred, but the patient is only to compare the two and not estimate absolute blurredness. If the trial cylinder power is correct, both flip positions are equally blurred (or equally clear). This is confirmed by arbitrarily changing the trial cylinder and repeating the test.

The cylinder power is modified as follows: If the trial cylinder is plus, increase power if the preferred position is the one with the white dots coincident with its axis, and reduce power if the red dots coincide with its axis. The converse applies if the trial cylinder is minus. Dioptric changes are made in the smallest increments that are discriminated—usually in 0.25 D steps. Changes are unrelated to the amount of astigmatism or the power of the cross cylinder.

Since the smallest practical change on the spherical wheel of the refractor is 0.25 D, the sphere is increased by +0.25 D when the plus cylinder is reduced by 0.50 D and conversely. Only in this way can the size of the confusion circles by compared.

The end point of the cross cylinder test is not maximum acuity but equal blurredness; the final blur depends on the power of the cross cylinder. When the ametropia is fully corrected, one focal line lies as much anterior as the other is posterior to the retina. When the cross cylinder is flipped, their position is reversed. The collapse of the astigmatic interval is reached only when the cross cylinder is removed.

Optical principles

The optics of the cross cylinder for power is simply an illustration that an ametropic eye 0.50 D myopic, for example, prefers −0.50 to +0.50 D.

Let us assume the eye has 0.50 D of simple myopic astigmatic error with the rule and is fogged 0.50 D to compound myopic astigmatism. Both focal lines are now anterior to the retina, one by 1 D, and the other by 0.50 D (Fig. 14-5). The retinal image consists of oval blurs 0.50 D × 1.00 D. If a ±0.25 D cross cylinder is added minus axis vertical, the first focal line is displaced another 0.25 D farther, whereas the other is brought 0.25 D nearer to the retina. The retinal image consists of ovals 0.25 D × 1.25 D. When the cross cylinder is flipped, the astigmatic interval is collapsed to 0.75 D spherical myopia, and the retinal image is a blur circle 0.75 D × 0.75 D. The patient, asked to discriminate between "b" and "c," must compare two blurred images not much different from each other, one of which is oval and the other circular. If the target consists of Snellen letters, the discrimination will be rather difficult.

If the initial condition is equal-mixed astigmatism, however, one focal line is 0.25 D anterior, and the other is 0.25 D posterior to the retina. Applying the ±0.25 D cross cylinder axes parallel to the trial cylinder, in one position the interfocal interval is increased by 0.50 D, and the retinal image consists of blur circles 0.50 D × 0.50 D. When the cross cylinder is flipped, the interfocal interval is eliminated, and the patient easily sees the difference (Fig. 14-6). The choice in equal-mixed astigmatism is easier than under fog in which both images are not only blurred but blurred in different ways as well. In all cases therefore the cross cylinder test for power is used with the eye in equal-mixed astigmatism. Thus only confusion circles of different size are compared.

To maintain the circle of least confusion on the retina as the trial cylinder power is changed, the spherical power must also be changed. Recall that a cylinder has a "spherical effect." For example, with an interfocal

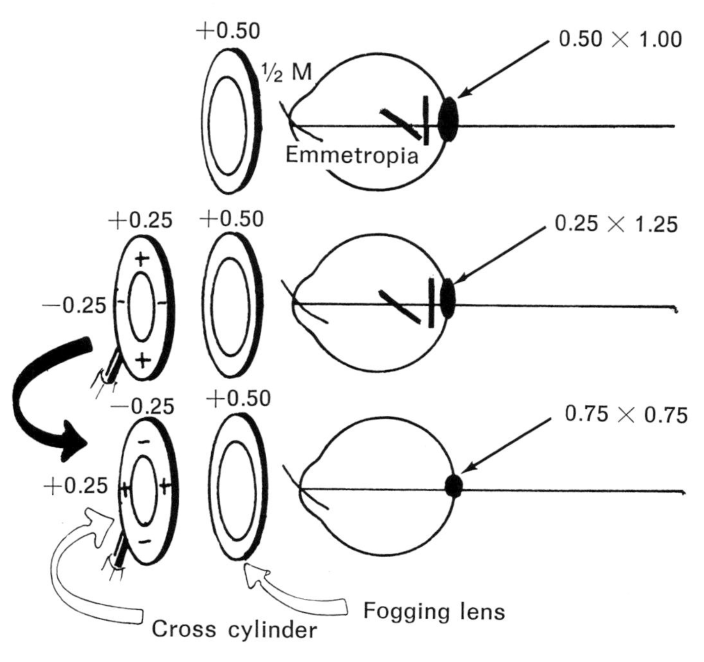

Fig. 14-5. Optical sequence in using a cross cylinder with the eye fogged. The patient must compare blur circles of different shape as well as different size.

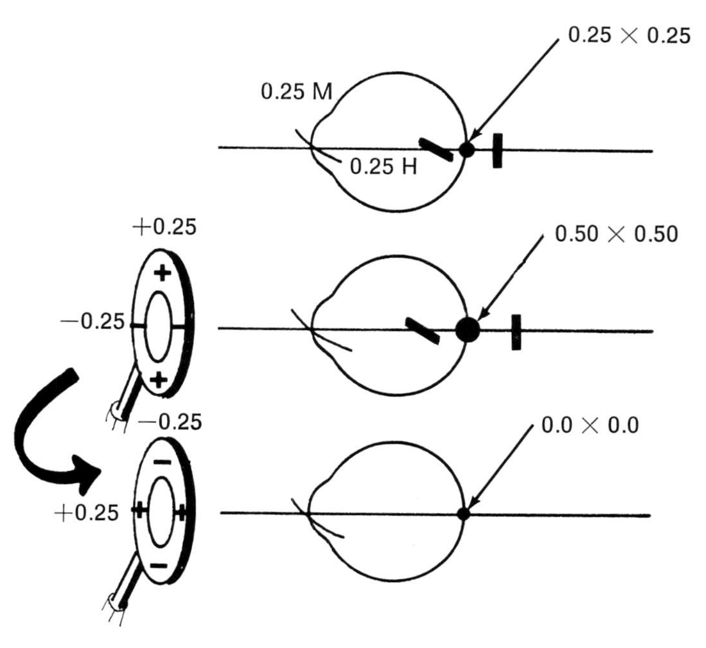

Fig. 14-6. Optical sequence in using the cross cylinder without fogging. The patient compares only blur circles of different size. This diagram emphasizes the importance of maintaining the eye in a condition of equal-mixed astigmatism throughout the cross cylinder test.

interval of 2 D, the circle of least confusion is at 1 D. If a 0.50 D cylinder is added, one focal line approaches the other, but the interfocal interval is narrowed to 1.50 D, and the circle of least confusion lies at 0.75 D. A 0.25 D shift in the dioptric midpoint results, which is equal to half the added cylinder. Therefore to maintain the circle of confusion on the retina the sphere must be changed by half of every cylinder change. With the refractor, it means one click of the sphere wheel for every double click of the cylinder wheel. (The cross cylinder as contrasted to a simple cylinder has a spherical equivalence of zero and does not itself change the position of the circle of least confusion. The power of a 50-50 cross cylinder can be written as +0.50 = 1.00 × red or −0.50 + 1.00 × white.) The cross cylinder only expands or contracts the interval of Sturm.

The optical principle of the cross cylinder test for axis depends on the effects of opposite, obliquely crossed cylinders. Suppose an eye whose ocular astigmatism is represented by −1 D × 80° (simple hyperopic astigmatism) is corrected by a +1 D cylinder mistakenly placed at 90°. We have already seen that the result is a new astigmatism whose primary meridians are 130° and 40°. When the cross cylinder is now applied, with its axes 45° from the trial cylinder axis, the plus (white dots) axis falls nearer 30°, and this position will appear clearer. The plus trial cylinder is therefore turned in that direction, perhaps 10° to 80°. This is now the correct axis, and thus flipping the cross cylinder will result in equal blur for the two flip positions.

Summary

The Jackson cross cylinder is a simple and quick method of refining the power and axis of the trial cylinder. Unlike the astigmatic dial, it allows discrimination between two different images in quick succession, a comparison that is simpler for most patients than judging blurredness of astigmatic spokes. The cross cylinder test is easier to understand and quicker to perform. It should not be used with the eye fogged, and the spherical equivalent must be maintained for each cylinder change (easy with the refractor but clumsy with loose trial lenses). The cross cylinder can be used to refine either plus or minus trial cylinders, and it is especially suitable for confirming the retinoscopy. Random presentation is an important advantage in checking patient answers. The reliability is similar to that of the astigmatic dial. The most common failures in technique include too much fog, not maintaining the spherical equivalent, and using a cross cylinder power too small for the patient to discriminate. Results should be confirmed by comparing Snellen acuity with the original and modified astigmatic corrections.

Choosing between the cross cylinder and the astigmatic dial is like asking a child to select a favorite parent. Every clinician becomes attached to the technique with which he is familiar, but no choice need be made. The methods are complementary, not exclusive, and there is every reason, if the time and the patient's discrimination permits, to use one to confirm the other.

OTHER TECHNIQUES FOR EVALUATING ASTIGMATISM

There are probably as many variations for testing astigmatism as there are refractionists. Most clinicians eventually settle on a routine they find easy to communicate to their patients and reliable in results. A sobering array of modifications will soon convince the reader that no single test is best for every patient.

Rotating the trial cylinder

Rotating or "rocking" the trial cylinder is perhaps the oldest but least accurate method of finding the poper astigmatic correction. The cylinder is arbitrarily rotated hither and yonder and the acuity noted. It works fine for a 6 D cylinder, but unless one plans to spend the weekend at it, it is difficult to believe the errors possible for a cylinder less than 1 D. Recall that when the correcting cylinder is displaced slightly from the true axis of ocular astigmatism, a new astigmatism is created whose primary meridians are approximately 45° away—if the power is correct. The two distorted views of the Snellen letters depend both on power and axis. Not only is it difficult to choose between letters that are distorted in different ways, but also the change for low errors is small and, when approached gradually, almost imperceptible.

A more useful modification is one in which the Snellen letters are replaced by an astigmatic sunburst dial. The patient is slightly fogged (0.25 to 0.50 D) and when the cylinder is rotated away from the true axis, there is a sudden dramatic change in the blackest line from 45° on one side to 45° on the other. The stronger the cylinder the smaller the rotation required to elicit it. The change in lines is easier to recognize than blurred Snellen letters, but it works only if the patient is familiar with astigmatic dials and can discriminate them.

Crisp-Stine test

The Crisp-Stine technique is basically an application of cross cylinders to the rotating cross astigmatic dial. The patient is slightly fogged, and the rotating cross straddles the axis of the correcting cylinder. For example, if the axis found by retinoscopy is 180°, the cross limbs are presented at 45-135. The cross cylinder is now placed with axes parallel to the axes of the cross (45-135), and the patient is to compare the sharpness of the limbs in the two flip positions. If the limbs appear equal, the axes is correct; if different, the correcting plus cylinder axis is moved toward the white dot flip position if that is better or away from it if there is worse (the converse is true for minus trial cylinders). After adjusting the trial cylinder axis, the rotating cross is repositioned so as to again straddle the new axis, and the test is repeated. The end point is reached when the cross limbs appear equally clear or reverse for the two flip positions.

In the test for power the rotating cross limbs are placed coincident or parallel to the axes of ocular astigmatism. The cross cylinder axes coincide with the rotating cross limbs, and the patient compares the clarity of the lines for the two flip positions. The trial cylinder power is modified as in the standard cross cylinder power refinement. The end point is when the dial lines are equally blurred or reversed for the two flip positions. The chief difficulty with the Crisp-Stine technique is the time it takes to explain it. One of the desirable features of the cross cylinder is that it involves the comparison of more or less blurred Snellen letters, a criterion with which most people are familiar. Introducing the rotating dial adds a new difficulty unless the patient is already familiar with the astigmatic chart.

Stenopeic slit

The stenopeic slit is 2 mm wide in a black disc to fit the trial frame. It is rotated to various meridians and the position that gives the best acuity noted. Lenses are tried to improve the vision further. The slit is next rotated 90°, and this meridian is similarly refracted with appropriate spheres. The result is the spherocylindrical corrections in terms of the optic cross. For example, if the best acuity with the slit vertical is −3.00 D and the best acuity with the slit horizontal is +1.00 D, the spherocylindrical prescription is +1.00 − 4.00 × 180.

In irregular astigmatism the slit positions that give best vision may not be perpendicular, and the glass prescription is compounded from the obliquely crossed cylinders. The actual lens is not generally as effective as the slit because the pinholing effect is absent, but it may provide the best vision under the circumstances.

Contact lenses

In severe or irregular astigmatism a subjective refraction may be attempted through a plano, trial contact lens. A preliminary pinhole test will indicate if the vision can be optically improved despite the failure of spectacle lenses. Dramatic improvement can be obtained with contact lenses in keratoconus or in other forms of irregular corneal astigmatism and should be given a trial. For example, in keratoconus the fluid (or tear) lens may neutralize up to 90% of the corneal toricity, and an acuity increase from 20/400 to 20/20 is not unusual, depending on how much secondary corneal opacification has developed. The technique is based on trial and error fitting, since keratometer readings are unreliable.

EVALUATING THE SPHERE

After the cylindrical component has been correctly identified with respect to axis and power, the Snellen acuity chart is presented and the eye is fogged to 20/60. The examiner then unfogs to best vision with maximum plus or least minus sphere. With each lens change, there should be a demonstrable improvement in acuity by the number of lines or letters correctly identified. Some writers recommend unfogging to "standard" (20/20) rather than to maximum acuity especially in myopia. There appears to be no valid physiologic reason for this.

The final sphere will depend on the vergence of the incident light, that is, the distance of the chart. At 20 feet, this means the eye will be 0.17 D undercorrected; at 4 meters, 0.25 D of accommodation is required for optical infinity. Aside from the exigencies of office space, the clinical history suggests the visual requirement. For example, a prepresbyopic engineer whose distance vision is confined to 2 or 3 meters might be given an overcorrection sufficient to postpone the need for separate reading glasses, an expedient not possible in a bus driver for example.

The unfogging end point is reached when a further plus reduction does not improve acuity. The patient may report that an additional −0.25 to −0.50 D makes the letters smaller and darker; this effect is due to the smaller blur circles induced by accommodation, and miosis and should be avoided.

The spherical component of the subjective refraction can be checked or refined by one or more of the following techniques.

Duochrome test

The duochrome (or bichrome) test is a refinement based on the chromatic aberration of the eye. It replaces the earlier cobalt blue, filter test of Helmholtz. The cobalt transmission is asymmetrical about the spectral midpoint. The duochrome principle is that optimum acuity is assumed for polychromatic light when the midpoint of the spectrum—yellow—focuses on the retina. Since red and green focus equidistant from yellow, the end point is achieved when targets of these two colors appear equally blurred. The myope sees the red target clearer, the hyperope sees the green clearer, and the emmetrope sees them equally blurred.

The test is performed monocularly and does not require normal color vision. A red-green filter on a slide inserted in the acuity projector converts any portion of the chart into a background half red and half green (self-illuminated split charts are also available). It does not matter that the letters themselves remain black; the colored ground suffices to make it work. One row of letters is presented, usually one line above maximum acuity to allow for the chromatic blur. The letters on both colors are now slightly more blurred than they were with the white ground, but the *letters* on the two colors are to be compared. Minus sphere is increased if red is better and plus sphere is increased if green is better, until the two sides appear equal. Despite differences in brightness and variations in the filters, the duochrome test is sensitive to ±0.25 D lens changes. Its appeal is due in no small measure to the ease and simplicity of the end point, and it can be explained to a 3-year-old child. It is the nearest clinical method to refracting with monochromatic light. The test may fail if astigmatism is not properly corrected since a separate conoid of Sturm results for each color. Some patients respond peculiarly because of a psychologic color preference, lenticular changes, or an unusual degree of chromatic aberration. Myopes may show continued preference for red and hyperopes for green, but these inconsistencies can be checked by acuity with Snellen letters. The red-green test has no inhibitory action on ciliary tonus, and consequently the usual fogging precautions apply.

Cross cylinder

The spherical component may be refined further by the cross cylinder. The patient views an astigmatic cross or T chart at 20 feet monocularly. A 0.50 − 0.50 cross cylinder, red axis vertical, is placed in the trial frame, and the patient judges the equality of the lines. If the vertical lines are darker, minus sphere is increased; if the horizontal lines are darker, plus sphere is increased. It is assumed that astigmatism has been adequately corrected, and the usual fogging precautions apply. The principle of the test is that the cross cylinder creates an optical, equal-mixed astigmatism in which +0.50 D is added to the vertical ocular meridian and −0.50 D is added to the horizontal. The horizontal focal line thus images in front of the retina, and the vertical line images behind. The end point of the test is to place the circle of confusion of this artificially induced astigmatism on the retina.

Refraction through old glasses

Refining the refraction by adding plus and minus spheres to the patient's old glasses has been recommended as a rapid and useful technique. It is unquestionably rapid, but few clinicians would care to stake their professional reputation on someone else's prescription. An obvious source of error is a lens that deliberately overcorrects for accommodative esotropia or intermittent exotropia.

BINOCULAR EQUALIZATION

After the subjective examination has been completed, an identical procedure is followed for the other eye. Since each test is performed monocularly and the accommodative conditions could conceivably differ, or the final vision in each eye may vary for other reasons, equalization is attempted. Both eyes are fogged and unfogged simultaneously. The vision should blur and clear symmetrically on alternate cover, prism dissociation, or polarized targets. Attempts are made to equalize the vision in the two eyes by appropriate lens changes. Obviously the technique will not work in ambylopia or monocular retinal disease. The purpose of equalization is to confirm the adequacy of the patient's responses on the refractive test and the symmetry of accommodation. A better optical end point is often reached when the patient can compare the vision simultaneously in the two eyes instead of depending on monocular clarity. The difference should not exceed 0.25 to 0.50 D. If 0.25 D reverses the better eye, it is conventional to leave the dominant eye with the better vision.

SUMMARY OF SUBJECTIVE ROUTINE

The following is a ten-step summary of subjective refraction based on a reasonably accurate preliminary retinoscopy:

1. Occlude one eye and fog monocularly, using the retinoscopic finding as a baseline. A retinoscopic working lens may serve as initial fog (to at least 20/60 Snellen vision).
2. Present the astigmatic dial. Reduce fog slowly until one line appears clearer and darker than the others. If all lines are equally clear, astigmatism is absent. If one line is clearer, add minus cylinder axis perpendicular to the clearest line.
3. Equalize lines on the astigmatic dial. Refine the axis with the rotating cross, rotating cylinder, or Crisp-Stine test.
4. Present a single 20/30 Snellen line and equalize with duochrome test. Add minus sphere if the red side is clearer, plus sphere if the green side is clearer.
5. Present the 20/25 Snellen line and refine the cylinder axis with the Jackson cross-cylinder. Modify the cylinder according to whether it is plus or minus, and the red dot or white dot position is preferred.
6. After the axis is correctly identified, refine cylinder power with the cross cylinder. Maintain the spherical equivalent for each cylinder change by increasing plus sphere 0.25 D for every 0.50 D minus cylinder change (or conversely for plus cylinders).

7. Fog and unfog monocularly to best vision on the Snellen chart.
8. Repeat the procedure for the opposite eye.
9. Equalize vision in the two eyes by alternate occlusion or polarized targets.
10. Fog and unfog binocularly to best vision. Record the result and compare to cycloplegic findings.

REFERENCES

Bannon, R. E.: Recent developments in techniques for measuring astigmatism, Am. J. Optom. **35:**352-359, 1958.

Bannon, R. E., and Walsh, R.: On astigmatism, Am. J. Optom. Monograph No. 7, 1945.

Bennett, A. G.: Trial cases ancient and modern, Ophthal. Opt. **6:**964-967, 1011-1014, 1061-1066, 1966.

Berens, C.: The eye and its diseases, ed. 2, Philadelphia, 1949, W. B. Saunders Co.

Berens, C., Brackett, V., and Taylor, B. E.: Duochrome television eye exerciser, Am. J. Ophthal. **43:**771-772, 1957.

Biegel, A. C.: The cylinder rotation test with the astigmatic dial, Am. J. Ophthal. **59:**277-290, 1965.

Borish, I. M.: Clinical refraction, ed. 3, Chicago, 1970, Professional Press, Inc.

Burnett, S. M.: Astigmia or astigmatism—which? Am. J. Ophthal. **20:**374-379, 1903.

Cowan, A.: Refraction of the eye, ed. 3, Philadelphia, 1948, Lea & Febiger.

Crisp, W. H.: A plea for the more general use of the cross cylinder, Am. J. Ophthal. **6:**209-214, 1923.

Crisp, W. H.: The cross cylinder tests especially in relation to the astigmatic axis, Trans. Ophthal. Soc. U. K. **51:**495-514, 1931.

Crisp, W. H., and Stine, G. H.: A further very delicate test for astigmatic axis, using the cross cylinder with an astigmatic dial and without use of letter charts, Am. J. Ophthal. **32:**1065-1068, 1949.

Donders, F. C.: Accommodation and refraction of the eye, London, 1864, The New Sydenham Society.

Egan, J. A.: A resume of cross cylinder application and theory, Survey Ophthal. **1:**513-529, 1956.

Emerson, E.: A fogging lens, Am. J. Ophthal. **59:**1048-1051, 1965.

Frey, R. G.: Chromasia of the eye employed as an aid in the determination of refraction, Klin. Monatsbl. Augenheilkd. **129:**534-543, 1956.

Friedenwald, H.: Binocular metamorphopsia produced by correcting glasses, Arch. Ophthal. **21:**204, 1892.

Friedenwald, J.: A new astigmatic chart, Am. J. Ophthal. **7:**1, 1924.

Friedman, B.: Acceptance of weak cylinders at paradoxic axes, Arch. Ophthal. **24:**720-726, 1940.

Friedman, B.: The Jackson crossed cylinder: a critique, Arch. Ophthal. **24:**490-499, 1940.

Fronmüller, G. T.: Ueber die Auswahl der Brillengläser, J. Chir. Augenheilkd. **32:**174-187, 1843.

Giles, G. H.: The principles and practice of refraction, London, 1965, Hammond, Hammond & Co., Ltd.

Green, J.: On a new system of tests for the detection and measurement of astigmatism, Trans. Am. Ophthal. Soc. **5:**131-143, 1868.

Hirsch, M. J.: Changes in astigmatism after age forty, Am. J. Optom. **36:**395-405, 1959.

Hofstetter, H. M.: The correction of astigmatism for near work, Am. J. Optom. **22:**121-134, 1945.

Hughes, W. L.: Change of axis of astigmatism on accommodation, Arch. Ophthal. **26:**742-749, 1941.

Jackson, E.: The astigmic lens (crossed cylinder) to determine the amount and principal meridians of astigmia, Ophthal. Rec. **17:**378-383, 1907.

Jackson, E.: Accuracy in the measurement of refraction, Ann. Ophthal. **18:**703-712, 1909.

Jackson, E.: How to use the cross cylinder, Am. J. Ophthal. **3:**321-323, 1920.

Javal, E.: Sur le choix des verres cylindriques, Ann. Ocul. (Brussels) **53:**50-60, 1866.

Lancaster, W. B.: Subjective test for astigmatism, especially astigmatic charts, Trans. Am. Acad. Ophthal. Otolaryngol. **20:**167-191, 1915.

Landolt, E.: The refraction and accommodation of the eye, Philadelphia, 1886, J. B. Lippincott Co.

Linksz, A.: Determination of the axis and amount of an astigmatic error by rotation of the trial cylinder, Arch. Ophthal. **28:**532-651, 1942.

Linksz, A., and Triller, W.: Biastigmatism; evaluation and criticism of the refractive technique advocated by Marquez, Am. J. Ophthal. **27:**992-1002, 1944.

Lloyd, R. I.: The first trial case, Am. J. Ophthal. **18:**753-754, 1935.

Mackenzie, W.: Diseases of the eye, London, 1866, Longmans, Green & Co.

Marano, J. A.: Subjective astigmatism determined in minutes, Optom. Weekly, April 15, 1962.

Marquez, M.: The great usefulness of bicylindric combinations in the exploration of astigmatism, Am. J. Ophthal. **25:**1458-1463, 1962.

Neumueller, J.: The effect of the ametropic distance correction upon the accommodation and reading addition, Am. J. Optom. **11:**20-28, 1937.

O'Brian, J. M., and Bannon, R. E.: The fogging method of refraction, Am. J. Ophthal. **31:**1453-1459, 1948.

Pascal, J. I.: The ioskiascopy test, Arch. Ophthal. **7:**378-382, 1932.

Pascal, J. I.: Fundamental differences between crossed cylinders and line chart astigmatic tests, Arch. Ophthal. **24:**722-730, 1940.

Pascal, J. I.: A self-setting cross cylinder, Am. J. Ophthal. **23:**1039-1041, 1940.

Pascal, J. I.: The Jackson crossed cylinder, Arch. Ophthal. **25:**355-356, 1941.

Pascal, J. I.: Intrinsic variability of astigmatic errors, Arch. Ophthal. **32:**123-124, 1944.

Pascal, J. I.: A new application of the Raubitschek arrow test, Eye Ear Nose Throat Mon. **33:**473-477, 1954.

Prangen, A. deH.: Some problems and procedures in refraction, Arch. Ophthal. **18**:432-447, 1937.

Raubitschek, E.: The Raubitschek arrow test for astigmatism, Am. J. Ophthal. **35**:1334-1339, 1952.

Regan, J. J.: Astigmatic dials in refined refraction, Arch. Ophthal. **17**:788-796, 1937.

Schwarting, B. H.: Cross cylinder scan test for astigmatism, Arch. Ophthal. **82**:330-331, 1969.

Sloane, A. E.: Manual of refraction, Boston, 1961, Little, Brown & Co.

Smart, F. P.: Some observations on crossed cylinders, Arch. Ophthal. **24**:999-1000, 1940.

Smith, D.: Refraction under cyclodamia, Am. J. Ophthal. **9**:896-903, 1926.

Souter, W. N.: The refractive and motor mechanism of the eye, Philadelphia, 1910, Keystone Publishing Co.

Stine, G. H.: The Crisp-Stine test for astigmatism and the Lebonsohn astigmometer, Am. J. Ophthal. **33**:1587-1590, 1950.

Tait, E. F.: Textbook of refraction, Philadelphia, 1951, W. B. Saunders Co.

Verhoeff, F. H.: The "V" test for astigmatism and astigmatic charts in general, Am. J. Ophthal. **6**:618-620, 1923.

Williamson-Noble, F. A.: A possible fallacy in the use of the cross cylinder, Br. J. Ophthal. **27**:1-12, 1943.

Wunsh, S. E.: The cross cylinder, Int. Ophthal. Clin. **11**:131-153, 1971.

15

Special techniques in refraction

Every good refractionist achieves similar results; every poor one makes his own mistakes. The flexibility to select the most appropriate test for the problem at hand is the hallmark of the experienced clinician. Nearly everyone requires spectacles at some time in life, but methods applicable to adults will not work for children, and further modifications are necessary for infants and patients with subnormal vision, aphakia, or strabismus. Prescribing a new lens for a previously examined myopic teenager may only take a few minutes, but evaluating a child with amblyopia may take hours over several visits. In this chapter we consider some modifications in refractive procedure that will expedite the examination in routine cases and help in the diagnosis and management of more difficult problems.

In 1886 Landolt considered the characteristics of a good refractionist:

> The choice of glasses is a delicate operation. He alone is successful in it who, to a perfect theoretical acquaintance with the subject, adds the intelligent observation of each patient. It does not suffice to know the action of lenses, from every point of view, and the workings of the visual organ. We must know, more than this, how to individualize cases—that is to say, we must take into account the state of refraction and accommodation, that of the muscles of the patient's eyes, the particular purpose of his wearing glasses, his peculiar habits, and many other circumstances to which attentive observations in practice teaches us to attach importance.*

*From Landolt, E.: The refraction and accommodation of the eye, Philadelphia, 1886, J. B. Lippincott Co.

BINOCULAR REFRACTION

In conventional techniques of refraction, one eye is examined while the other is occluded. Binocular refraction attempts to carry out the examination under circumstances more nearly approaching those that exist in the daily use of the eyes.

With both eyes open, binocular reflexes and normal disparity operate, the visual axes remain parallel, and the accepted lens is less likely to overcorrect an exophoria. Binocular examination permits balancing the accommodative stimulus in anisometropia, compensating for cyclofusional movements, detecting suppression, and measuring stereoscopic vision, aniseikonia, heterophorias, and fixation disparity. In latent nystagmus, acuity may improve dramatically if neither eye is occluded.

Alternate occlusion

Alternate occlusion is frequently used to compare the vision in the two eyes. By rapid timing, the total disruption of binocular reflexes is partially prevented. In a technique called "phase difference haploscopy," the alternation is carried out mechanically above flicker-fusion frequency. The optical path of two projectors, producing dissimilar views for each eye, is interrupted by sectored discs rotating at a frequency of 100 cycles per second. This rotation is too rapid to be perceived by the patient. In front of each eye, an additional sectored disc rotates at the same frequency. All four discs are driven by synchronized motors so geared that the on-phase of one eye corresponds with the off-phase of the other. Monocular images are projected on a screen seen by both eyes, which serves to maintain peripheral fusion. Targets are acuity symbols, astigmatic dials, and arrows for evaluating muscle imbalances or size differences. The arrows are aligned by turning the projectors.

Fogging

The eye not being tested is fogged, for example, by the retinoscopic working lens. Although the fog prevents sharp vision, it is sufficient to maintain binocular fixation. In the technique of "cyclodamia," a +1.50 fogging lens is added to both eyes to relax accommodation, and the blurred vision (about 20/60) is compared and equalized. Equality is checked by moving the target nearer or by reducing the fogging lenses. One problem with this method is that a blurred dominant eye may interfere with the acuity of the other.

Anaglyphs

When a red-green field is viewed through appropriately matched filters, the green field is eliminated from the eye wearing the red filter and conversely. Black targets are visible to both eyes and maintain binocular fixation. Selected monocular and binocular targets thus can be introduced, but allowance is made for chromatic aberration by reversing the targets and repeating the measurement. Anaglyphs are particularly suitable for testing children, mapping suppression scotomas (Fig. 15-1), and maintaining fixation of a poorly seeing eye in perimetry.

Stereoscopic methods

A stereoscopic method of binocular refraction, previously used by Landolt, consists of a series of test types viewed by one eye and an empty field seen by the other. The two circles, superimposed when the eyes are parallel, induce the eyes to relax accommodation. The ideal refraction stereoscope would be 20 feet long; in practice, the instrument is short, with lenses to collimate the light, and it induces proximal accommodation. The stereo principle is incorporated in many commercial vision screeners.

An ingenious binocular method of refraction (Turville, 1946) is based on the principle

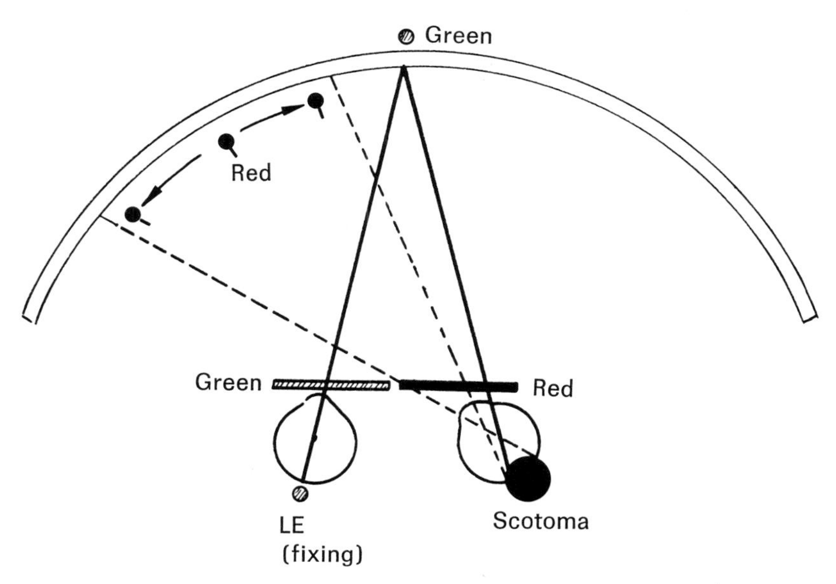

Fig. 15-1. Anaglyph technique to measure suppression scotoma on a perimeter. The red-green filters prevent one eye from seeing the other's target. In this illustration the left eye fixates and maintains the alignment of the esotropic right eye. Unfortunately this method is not readily applicable to young children.

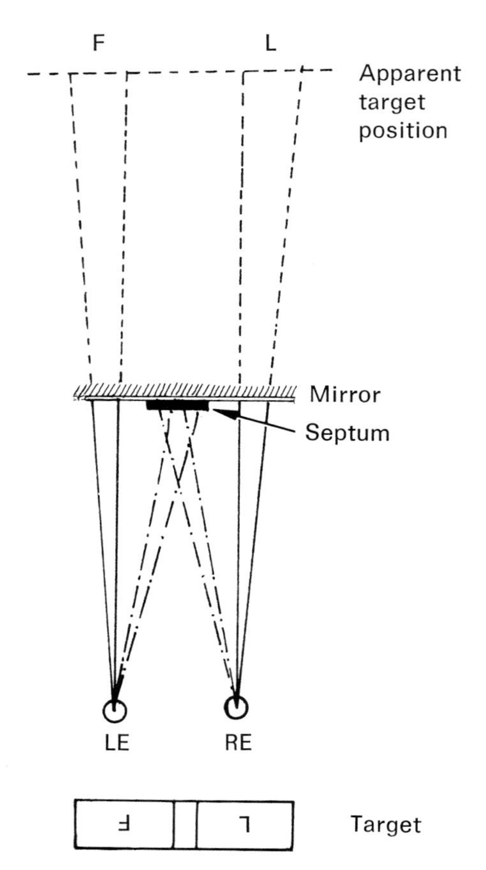

Fig. 15-2. Optical principles of the septum technique of binocular refraction (Turville). The real targets are presented to the mirror, which is partially obscured by a central septum so that each eye sees its own target.

of bar reading applied to distant vision (Fig. 15-2). When a bar or septum is placed midway between the eyes and the page of a book, parts of the page are visible only to each eye. Similarly, a septum (such as a strip of cardboard 25 mm wide hung from the ceiling) positioned at the intersection of the diagonals running from each fovea to a distant, doubled chart (for example, an F for one eye and an L for the other) allows each eye to see its own target if properly aligned; the latter is the chief difficulty. A recent modification involves a split chart for mirror viewing, with an unsilvered vertical strip on the mirror as a septum (or the mirror itself is split and the two parts slightly angulated).

Polarization

Polarized targets viewed through appropriate filters permit evaluating monocular vision under binocular conditions. Vectographic slides* are now available for distance and near and for adults or children.

The vectographic process is a controlled deposition of iodine crystals on a stretched polyvinyl alcohol film. A target printed in this manner appears opaque through the appropriate analyzer. The targets are in turn placed between two stretched transparent films with their axes at right angles. The orientation of the target polarization determines whether it is seen by one eye or the other or by both. Unlike previous methods of polarization, the illumination is not reduced twice (by polarizer and analyzer) but involves only one attenuation at the analyzer, which allows considerably brighter targets. Targets are mounted on a slide for transillumination for projection in a standard acuity projector or are printed on a surface sprayed with nondepolarizing aluminum paint to be viewed by ordinary illumination.

Vectographic slides allow testing each eye independently while both remain open. The analyzing filters can be worn as separate glasses, inserted into the trial frame, incorporated into the accessory lens disc of the refractor, or suspended in front of the refractor lens cells (Fig. 15-3).†

REFRACTION AT THE READING DISTANCE

A separate refraction at the reading distance is seldom made except in presbyopia. There are, however, a variety of circumstances in which near vision is decreased, such as in high ametropia, nystagmus, nuclear sclerosis, high astigmatism, aphakia, lens subluxation, macular disease, accommodative spasm or insufficiency.

Most refractors have a calibrated steel rod for near targets. The rod fits midway between the eyes, and the target holder can be displaced toward one side or the other. Of the many near point targets, the most convenient

*American Optical Co., Southbridge, Mass.

†The filters should be checked to see that their axes correspond to the targets (45-135 for the vectograph slides but 90-180 for the British polarized units). Head tilting is not a problem, since more than 30° is required to alter visibility.

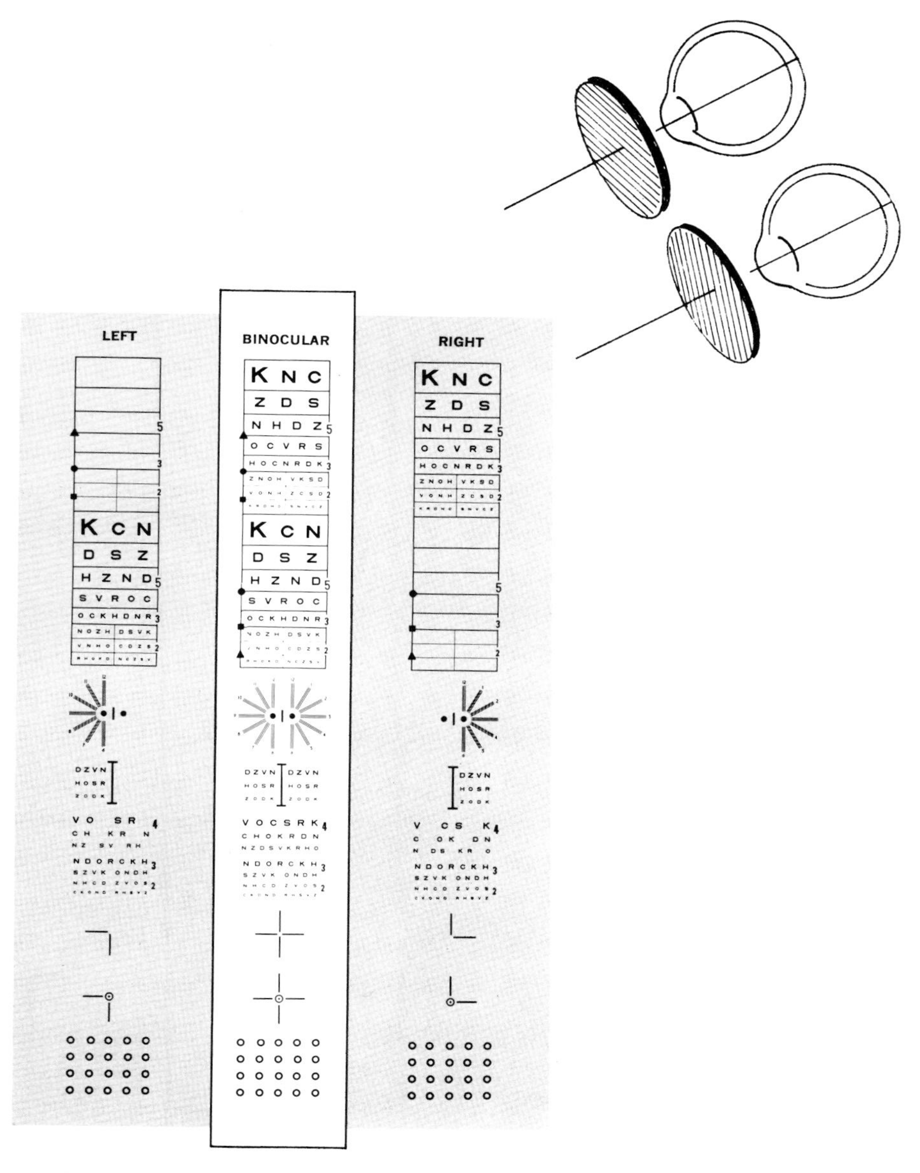

Fig. 15-3. Polarization technique of binocular refraction. The patient views the vectographic slide (American Optical Co.) through polarizing filters. The targets permit evaluation of suppression, phorias, fixation disparity, and stereoacuity.

allow rapid change by a sliding or rotating card and include "reduced" Snellen letters, pictures, rings, grids, or print. The target distance is selected according to occupation, the length of the patient's arms, and the required visual range.

Binocular techniques also are applicable to near testing by means of stereoscopes, anaglyphs, haploscopes, or polarization. Self-illuminated near targets also are available (Freeman, Osterberg, Rodenstock). The Freeman unit* is an internally illuminated box that the patient holds at his reading distance. Different targets are illuminated at

*Keeler Optical Products, Inc., Philadelphia, Pa.

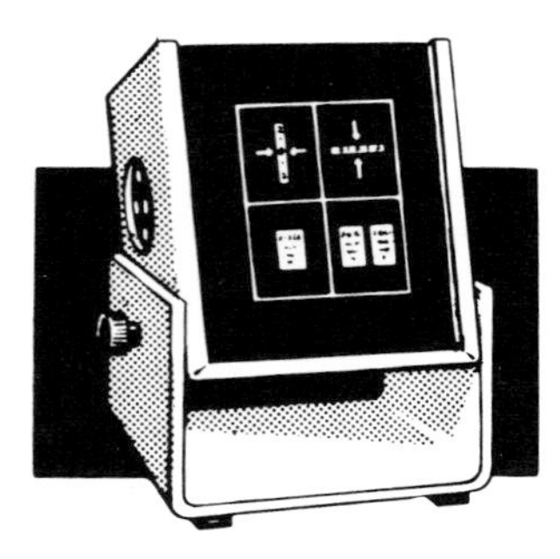

Fig. 15-4. The small box lantern (Bernell Corp.) is held by the patient at his reading distance. Appropriate slides measure near acuity, phorias, fixation disparity, and Worth fusion. A slide for detecting scotomas (Amsler grid) is also available.

the push of a button, including letters, Landolt rings, polarized and red-green rings, or an oblique tangent scale for measuring lateral and vertical phorias. The letters function as an ordinary reading card, the red green rings work on the duochrome principle, the polarized rings allow binocular equalization, and the phoria is measured with a hand-held Maddox rod, with the patient indicating the number intercepted by the line. The test allows the patient to hold the unit in the normal reading position, but the prescription must be transferred to a trial frame. Another self-illuminated lantern from the Bernell Corporation provides targets for near acuity, red green equalization, fixation disparity, Amsler grid, heterophoria, and Worth four dots (Fig. 15-4). The unit is made of light plastic which is easily held or suspended, and the slides are interchangeable.* A series of near point vectographic slides are available from American Optical Co. The targets follow the same pattern as the distance test but may be viewed by ordinary room illumination through appropriate analyzers. In addition, there is a slide for aniseikonia, stereoacuity (from 10 minutes to 12 seconds of arc), and astigmatism (astigmatic dial, Maltese cross or Verhoeff rings for cross cylinder) and a screening test for stereopsis in children. Because of reduced contrast on the acuity slides, an apparent reduction of half a line may be anticipated.

*An expedient is a luminescent plate, similar to that used for night lights, which is attached to the near point rod of the refractor. Various targets may be clipped to the plate for transilluminated viewing.

In evaluating near point refraction or other near clinical tests, recall that the refractor or trial frame should be readjusted for the near interpupillary distance. The illumination should be moderate, or the accompanying miosis will invalidate the end points.

Since a spectacle lens only corrects the eye for distant vision, it effectively alters the accommodative requirement for near objects more or less than that of a naturally emmetropic eye (Chapter 7). Moreover the effectivity operates for each primary meridian, resulting in a different cylinder. Since near vision means more plus sphere, the required cylinder is generally increased (one reason for prescribing the full cylinder for distance vision). This difference is purely optical and not the result of "sectional" or astigmatic accommodation.

REFRACTION METHODS WITHOUT LETTERS

An occasional patient with language problems, subnormal vision, or irregular astigmatism may require a technique that does not depend on reading letters. These methods are frequently used in laboratory investigations and in testing malingerers.

Chromatic techniques

Techniques based on chromatic aberration are monocular, and the usual fogging precautions apply. The cobalt blue filter mentioned previously transmits the extreme ends of the visible spectrum; the blue is refracted most and the red least. A small distant light source is viewed through the filter. The myope sees a red center with a blue surround; the hyperope, the opposite (Fig. 15-5). In astigmatism the colored images are distorted into lines or ellipses. For example, in equal-mixed astigmatism the blue and red ellipses form a cross in their long axes. Discrimination and description of what is seen must be of a high order, which makes the test impractical.

The duochrome technique, generally used

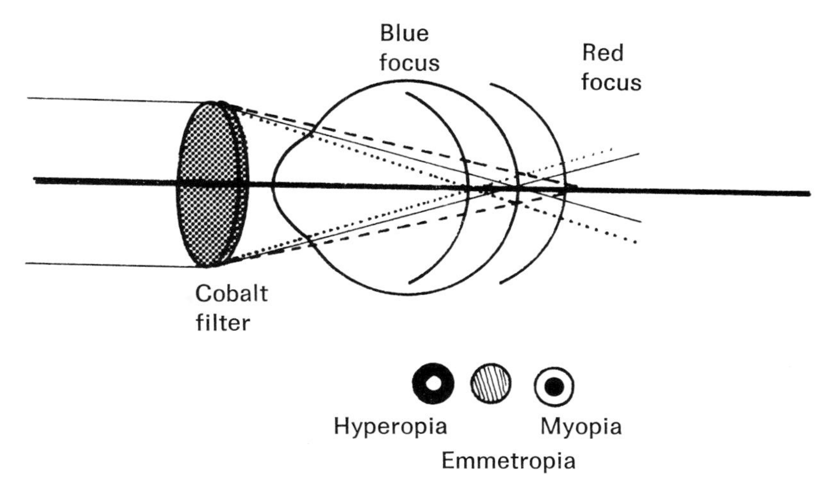

Fig. 15-5. Bichrome technique of refining refraction. The cobalt filter transmits only the long and short wavelengths. The myope sees a red center with a blue halo; the hyperope, the opposite. In astigmatism the circles are ellipses of differing shape. The target is a small light source at 20 feet.

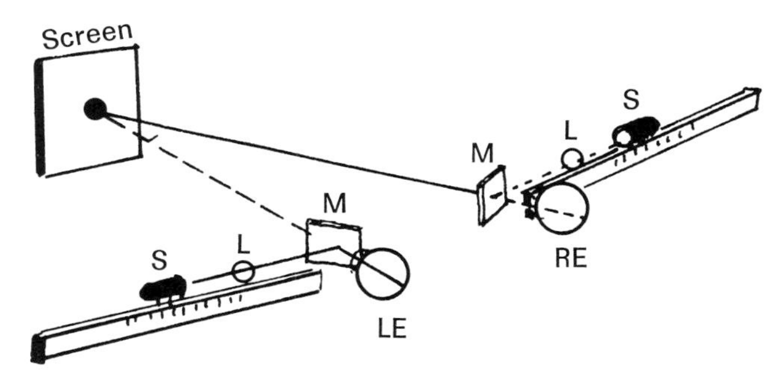

Fig. 15-6. Stigmatoscopy incorporated into a haploscope. The small light sources *S* slide along a graduate scale (to vary accommodation), and the arms may be angulated (to vary convergence). Binocular fixation is maintained by viewing the distant target through the half-silvered mirrors *M*. The light from each source is collimated by the lenses *L*, and additional lens cells are provided in front of the eyes to correct the ametropia.

with letters, can also be applied to Verhoeff circles, which consist of two concentric rings. Vision is equalized on the larger ring and refined with the smaller. The rings on the red field appear clearer to the myope, those on the green field appear sharper to the hyperope, and appropriate trial lenses are added for equalization. The cylinder is determined by the usual cross cylinder technique. Bichromatic methods have been hailed as making retinoscopy unnecessary because they "only" show an average difference of 0.90 D!

Stigmatoscopy

One of the oldest methods of refraction is based on retinal diffusion circles formed by a small light source (Fig. 15-6). The more the eye is out of focus the larger the diffusion circles. The image is spread out by diffraction, aberrations, and imperfections in the ocular media, and the circles vary with pupil diameter; hence the end point is the best subjective focus. In astigmatism the image is a line or ellipse, depending on where the retina intersects the interval of Sturm. The principle is easily visualized by holding a small pinhole in front of a diffused light and viewing it through a +20 D sphere combined with a +3.00 D cylinder (the sphere to bring the far point within arm's reach). As the card is moved closer or further from the eye, characteristic parts of the interval of Sturm are demonstrated; indeed this was the method by which Airy discovered his own astigmatism in 1825. Hyperopes accommodate; hence the technique involves fogging and then reducing plus, until the diffusion circles have the smallest diameter. Although Gullstrand considered this to be the most accurate subjective method of refraction, it requires more than a modicum of discrimination. In astigmatism, the distortion is not seen unless the cylinder is fairly high (about 1 D).

A number of off-shoots of stigmatoscopy belong more to the category of parlor tricks than to legitimate clinical methods, but their exotic names assure them at least passing mention in textbooks. The "Cinescopic" method is essentially a modification of the pinhole or stenopeic slit used for self-neutralization. The pinhole or slit, held close to the eye, is passed across the pupil; the target appears motionless to the emmetrope, a with motion is seen by the myope, and an against motion by the hyperope. These movements are neutralized by appropriate spheres. In velonoskiascopy a thin wire replaces the slit and acts, optically, like the opaque area between two Scheiner pinholes. The wire casts a shadow of a distant light source, which remains fixed for the emmetrope, moves "with" for the myope, and "against" for the hyperope. A target stripe in differing orientations measures astigmatism.

Scheiner disc

The double pinhole, or Scheiner disc, gives a highly accurate end point in measuring the diameter of retinal diffusion circles by converting a blur into diplopia, which disappears when the image is in focus. The double pinholes (whose separation must not exceed the diameter of the pupil) are carefully centered in front of the eye, and a target, such as a small light source or the head of a pin, is brought within arm's reach by a strong convex fogging lens (Fig. 15-7). The farthest position at which diplopia disappears is the far point. In astigmatism, there are two far points, one for each primary meridian. The end point can be further enchanced by fitting each pinhole with a different color or by replacing one with a miniature Maddox rod. Ametropia is diagnosed by alternate cover of each pinhole, noting if the diplopia is crossed (hyperopia) or uncrossed (myopia). The larger the ametropia the greater the separation of the images. Theoretically this method provides an elegant end point (and is incorporated in many subjective optometers), but exact centration makes it impractical except for research purposes. If one pinhole slips off the pupil, the diplopia disappears even though the eye is not corrected.

Laser refraction

Laser refraction is something new and of considerable potential value in screening refractive errors. When a surface is illuminated with coherent light, it assumes a granular appearance, called the "speckle effect." An emmetrope viewing the pattern and moving his head sees a stationary pattern; the myope sees the patterns move in a direction opposite to the head movement; the hyperope, in the same direction. When a rotating drum is substituted for head motion, the motion is seen easily even by children, and the accuracy is said to be ±0.25 D. The beam is rotated to various meridians to measure astigmatism. The laser technique is applicable to patients of limited intelligence and to the evaluation of eyes with hazy media and cataracts. Since an Argon beam is used, allowance is made for the red color to compensate for aberration (an inexpensive green laser is not yet available).

Sensitometry

This technique is based on brightness contrast thresholds. A target on a uniform gray ground is viewed through a graduated filter, and its visibility plotted for various lenses

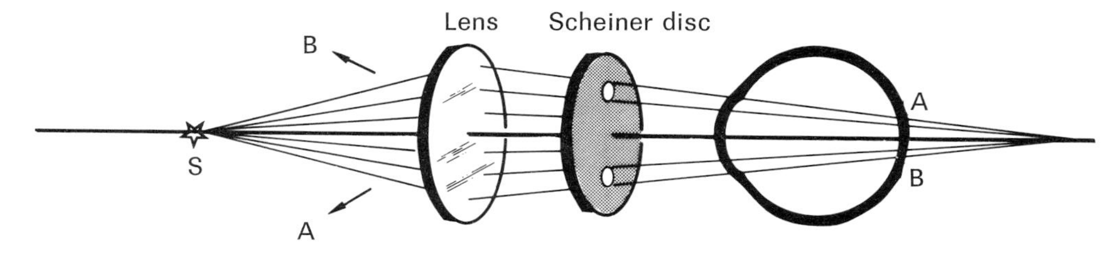

Fig. 15-7. The Scheiner double pinhole disc converts the small light source S into monocular diplopia. The lens compensates for the small fixation distance. In this example of a hyperopic eye, alternate occlusion of the pinhole gives a "crossed" diplopia. Quantitative measurements are made by changing either lens power or target distance.

at each threshold. This is the only method of refraction that considers target contrast, but it is too time-consuming for clinical work.

MALINGERING

Malingering refers to a deliberate attempt to mislead the examiner as to the cause or presence of ocular disease or disability. In functional illness, the patient deceives himself as well as the examiner. This distinction is not always clear, and an individual must be considerably disturbed to feign blindness to gain one's ends. Legal and compensation considerations make differentiation, no matter how distasteful, necessary. The price for being wrong may range from an indignant patient to a lawsuit. Most writers advocate a careful ocular examination (as distinguished from a careless examination, which takes less time and is less likely to arouse belligerence). The diagnosis is suggested by the patient's general behavior, the symptoms and their lack of correlation with objective findings, and the history and purpose of the examination. A number of ingenious tests are helpful to evaluate simulated visual loss, but they must be administered with finesse. The patient is instructed to keep both eyes open, responses are evaluated for consistency, and facial expression and speech mannerisms noted. Hysterical patients generally wish to be helped and take an interest in the examination; malingerers are often passive, resent both test and examiner, and may become agitated if confronted by exposure. Psychiatric consultation may be required to establish a definite cause for functional illness; the refraction is negative evidence and of no help in providing a cure. Malingering may also be positive, for example, by children who wish to please a parent or the clinician.

Visual behavior

Blind patients lean backward when walking to avoid obstacles to their heads; malingerers lean forward and seem to seek out things to bump against. Conversely, supposedly blind patients who walk without stumbling over obstacles can probably see them. Patients feigning blindness may refuse to look at an object presented to them. Blind patients can generally bring their two index fingers together, while malingerers may claim great difficulty.

Objective tests

Objective tests for total blindness include pupil reflexes, aversion and lid closure reaction to a menacing movement, and optokinetic nystagmus. Elaborate electrophysiologic tests are usually not practical. A useful test is the fusional vergence response to a base-out prism placed over the supposedly blind eye.

Subjective tests

Most helpful are those subjective tests that compare consistency of visual loss at various test distances (e.g., 10, 20, and 40 feet). The test is indicated if, in the absence of nystagmus, cataract, or fundus lesion, there is considerable difference between distance and near acuity. A similar idea is used in comparing visual fields at different distances or for various target sizes.

When a high plus sphere is placed over the good eye and a low sphere over the simulated bad one, the acuity should be decreased, provided the patient keeps both eyes open and does not know which eye is being tested. A strong plus and minus cylinder may be placed before the eye with simulated poor vision. The patient is instructed to rotate one cylinder to obtain best acuity, and the test is repeated several times for consistency. The cylinder axes should agree within 15°.

Anaglyph or polarized targets, in which the patient sees some letters monocularly and some binocularly, are helpful in comparing acuity. There are numerous designs in which one target is seen with monocular vision and another with binocular vision (an elaboration of the F-L combination to yield an E). More elaborate are the pseudoscopes in which by means of mirrors the right eye sees the left target, and conversely. By interposing a bar (e.g., a 1-inch wide ruler) between the nose and the printed page, the patient with binocular vision continues to read without interruption, whereas some letters are obscured in monocular vision.

Diplopia tests

Diplopia tests are based on the fact that only those with binocular vision will experience double vision with a dissociating prism. The simulated blind eye is occluded, and a single prism base down with its edge straddling the pupil is placed before the good eye; the patient should now have monocular diplopia. The poor eye is uncovered, and simultaneously, the prism is moved up to cover the pupil; if the patient continues to see double, it proves vision is present in the poor eye.

The most difficult malingering problems are those in which poor vision is claimed in both eyes. Concrete evidence is difficult to establish and may require prolonged and repeated observation.

PEDIATRIC REFRACTION

The history obtained from the parents is frequently cloaked by guilt feelings that the difficulties might have been avoided. Parents should be reassured that poor reading habits, poor posture, too much television, or too little light do not inevitably lead to ametropia; an adequate examination can be performed even though the child cannot read; crosschecks are incorporated for responses of doubtful reliability; and spectacles are not always necessary. Details of obstetric, neonatal, early developmental, and familial anomalies are obtained in cases of congenital diseases and strabismus. Some insight is gained by questions regarding school progress and performance. During the history, the child is observed with respect to general behavior and the ability to follow directions and to communicate answers. The presence of a squint, ptosis, exophthalmos, head tilt, nystagmus, eccentric fixation, skeletal deformities, or muscular spasms can be established at a glance.

Meanwhile the little patient, confronted by a strange examiner, a darkened room, and overpowering machinery, floating in a cycloplegic haze, and pressed for answers, may hesitate or give the first reply that comes to mind. With the right key to his personality, the curiosity comes out, and answers will be direct and candid. "Let us play" is as good as "obey," and the results are more assured. However, children are also more easily bored, more easily distracted, and their discrimination may leave something to be desired. They seldom complain of eyestrain or even poor vision and they never grumble about improper glasses—they just refuse to wear them. All this makes refracting children more challenging and effective results more rewarding.

Visual acuity

Acuity targets are selected appropriate to the patient's age. Infants are tested by pupil reaction and lid closure to bright light, fixation and following movements, optokinetic nystagmus, and menace response. A Marcus Gunn pupil sign helps differentiate functional from organic visual loss. The sign is best elicited by alternately directing a light from one eye to the other. In the involved eye,

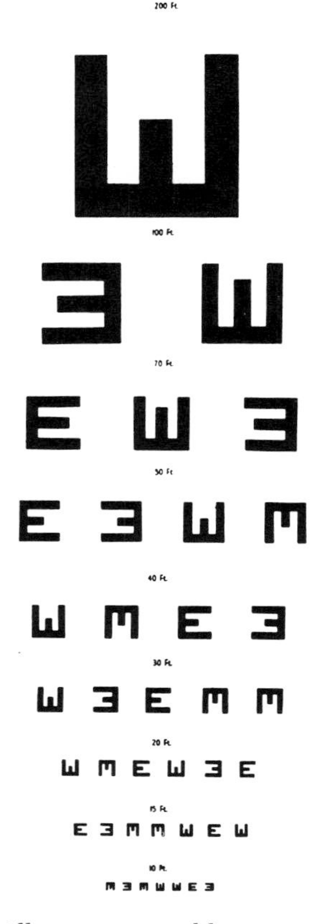

Fig. 15-8. Illiterate or tumbling E test is the simplest, most effective technique of evaluating acuity in children 3 to 6 years old. The child is asked to point in the direction of the "legs of the table."

some of the pupillary fibers are knocked out, causing a reduced or poor constriction followed by a consensual dilation because the other eye is now unstimulated. The pupil apparently dilates in response to light—a paradoxical reaction. Optokinetic nystagmus can be tested by drawing a homemade tape (a dozen alternate 1-inch squares of black and white cloth sewn together) between the fingers. If the infant's attention can be gained, there will be a slow following movement and a rapid return to the primary position. The direction is classified according to the quick component, and the response is compared to the other eye for various directions. The test is calibrated by different squares, different distances, or the size of an interposed object that will stop the nystagmus by engaging the fixation.

Fixation is evaluated with different-sized objects, such as bright-colored toys, or lights (the child is told to "blow out the light"). Various-sized marbles thrown on the floor are sometimes recommended, but they only make noise and seldom measure acuity. A large colorful lollypop is a secret weapon when all else fails. Inability to fixate indicates loss of central vision until proved otherwise. The acuity should be approximately 20/200 at age 1 year; 20/30 by age 3 years; and 20/20 by age 5 years.

The illiterate Snellen test (Fig. 15-8) is probably the best and most reliable of all acuity tests for children old enough to cooperate (or those who insist they know their letters from *Sesame Street*). Even a stupid parent can teach the "E game" to a clever 3-year-old child. Unlike toys or pictures of toys or matching toys, it is easily explained and checked, and it does not depend on previous familiarity. The E becomes a table, comb, fork, three fingers, or whatever analogy is suitable. A small box with an E of varying size printed on each face is readily handled by the examiner or an assistant and is better randomized than a wall chart. The test is usually carried out at a nearer distance, such as 10 feet, while the parent covers one eye. If peeking is suspected, the unexamined eye is covered by an adhesive occluder. Evidence of poor acuity such as rubbing, blinking, squinting, or stumbling may be noted in younger children.

Cycloplegia

The choice of a cycloplegic depends on the clinical problem, age, iris pigmentation, reliability of the parent in instilling the drug, and the history of allergic reactions. Most children can be dilated in the office with short-acting cycloplegics; for those with strabismus, atropine is given for home use. Instruct the parents on the amount and manner of instillation (emphasize *both* eyes, especially in strabismus) and caution them with respect to toxic signs. Explain that dilation occurs an hour after the first instillation, but the full-indicated dosage is still required; otherwise medication is frequently discontinued. Give instructions regarding disposal of leftover medication. Sunglasses may be given, which the children usually refuse to wear. Reassure the parents that they are for comfort and not to protect the eye from "harmful" radiations. A few minutes spent on these details will save subsequent telephone calls.

Retinoscopy

Objective refraction usually can be carried out without sedation, and general anesthesia is seldom required for refraction, although it can be attempted (under cycloplegia) if the fundus must be evaluated for other reasons. Keratometry has been recommended for the examination of children, presumably because it is objective, but I have never found it of much use, and the time is better spent on a diligent retinoscopy. The results obtained depend on the clinician's experience; hence children should not be examined by beginners learning technique; mistakes cannot be subjectively rectified. Retinoscopy is performed with hand-held trial lenses while the child is held in the parent's lap. A moving toy, blinking light, or projected cartoon serves for fixation. A preliminary scan, using one of several estimating techniques almost always can be done without objection and then confirmed with trial lenses. For some reason, children resist the lenses more than the light.

The break and oblique motion technique is used to identify the axis, and the primary meridians are neutralized with spheres. Care must be taken to hold the lenses perpendicular and reasonably near the spectacle plane; a +6 D sphere 2 cm from the spectacle plane will result in over 1 D effectivity error. The retinoscope also provides a check on the adequacy of cycloplegia (recall that pupil dilation is not a reliable sign). If the child is old enough to wear a trial frame, a refractor is better. It greatly expedites the examination, and timing is the key to success in refracting children. The parent helps to keep the head perpendicular and apposed to the forehead rest. Cylinder retinoscopy thus can be used for refinement. The less we can rely on subjective answers, the more effort must be expended on obtaining valid neutral points, and there is no time to improvise or to depend on unfamiliar estimating techniques. The most important thing to remember about retinoscopy in children is to repeat it.

Subjective methods

Subjective methods are confined necessarily to older children able to respond to instructions and to report differences. Begin with interesting targets, short distances, and large lens changes. Children generally do better with the cross cylinder than the astigmatic dial, and most respond reliably on the duochrome, Maddox rod, Worth four dot, and stereo fly tests. All can be readily checked and their reliability confirmed.

In evaluating the discrimination of children, a different technique is used as compared to adults. Rather than relying on "better" or "worse" responses, present the entire Snellen chart, and for each lens change, ask the child to read down as far as he can; thus an actual improvement in the number of lines (or letters per line) can be demonstrated. It may not be brief, but it is safe. Consistency, correlation with the retinoscopic finding, and demonstrable improvement in acuity are the key to validity. If in doubt, repeat the examination at another time on another day with different tests; children are frequently more relaxed on the second or third visit when rapport has been established and fear of the examination dispelled.

After the examination, the results should be explained to the parents and the child. If glasses are prescribed, written instructions are useful, since children cannot be expected to remember when they are supposed to wear them. A reward system sometimes helps in getting them to wear a patch, a prism, or a lens.

Fitting spectacles to an infant or child can be an adventure, not edifying but necessary. The youngster's motivation is astigmatic, and the mother has an apocalyptic image of a perpetual optical cripple. Less than half a dozen frames are available in a 32 eye size ("Teddy Baer" from the American Optical Co., "Nipper" from the

Bausch & Lomb Optical Co., Marine "Tech" from Marine Optical Co., and "Twinkle" from Titmus Optical Co.), which must be fitted to a small nose and round face and somehow kept in proper position in front of the eyes. The frame preferably should be plastic, the lenses as small as possible for appearance and weight, the temples of the riding bow type, and with a headband (Glass-strap, Pro-Spec) if necessary. All lenses must be impact-resistant glass or plastic. Fresnel lenses and prisms are more convenient than clip-ons. If bifocals are ordered, the straight across (e.g., American Optical "Executive") is best, with the dividing line fairly high (check if the child does not hold near objects closer than without the bifocals, thus vitiating their effect). Patching is best accomplished with adhesive occluders and not clip-ons, tie-ons, or lens masking. Partial occlusion can be obtained with nail polish or masking tape. Binocular aphakic children can be treated satisfactorily with spectacles; the optical challenge is in keeping them properly positioned. Contact lenses may be fitted in selected monocular cases, and intraocular lenses may be attempted by the more daring surgeon of wide experience.

TECHNIQUES IN AMBLYOPIA

"Amblyopia" (literally "blunted vision") is a term used to denote reduced vision in the absence of any discoverable organic cause. Somewhere along the line the phrase "ophthalmoscopically invisible cause" became attached to the definition with the result that neuropathies like those due to alcohol or malnutrition and such congenital disorders as achromatopsia are still classified under amblyopia. By this reasoning, a severed optic nerve produces amblyopia until the disc becomes white. To assign poor vision to organic amblyopia when no cause can be discovered is bad enough, but to call it "amblyopia" simply because there is no visible lesion in the fundus is nonsense.

When the reduced vision of one eye of a child cannot be improved beyond a certain point or at all by patching, orthoptics, pleoptics, corrective spectacles, filters, or hypnosis, we say the amblyopia is organic. It may be due to obstetric fundus hemorrhages now absorbed, atrophy of geniculate cells, cortical circuits that failed to develop, metabolic insults, or those mysterious toxic disturbances that always complete the list of unknown causes; but until we can clinically examine the visual pathway beyond the fundus, there is no good way to make the diagnosis by exclusion. Besides, the histochemist and biophysicist would not accept our clinical differentiation between functional and organic, and the amblyopia that starts as a functional disorder in early life ends nonfunctionally fixed in later years, and no one knows what organic changes, if any, occur in the visual system that now make the vision irrecoverable.* Only one positive way to diagnose functional amblyopia is by a demonstrable improvement in vision after treatment. On the other hand, the absence of fixation, a high anisometropia, or a unilateral squint in a young child is enough presumptive evidence that amblyopia already exists or will develop to institute preventive treatment.

Characteristics of functional amblyopia

Amblyopia occurs in 1% to 3% of the population, depending on the investigator, the sample, and the criterion of defective acuity. More amblyopes are without obvious strabismus than with it, and among strabismics it is slightly more common in esotropes than exotropes. Of course, there are more esotropes to begin with, and some of the nonstrabismic amblyopes may turn out to have "subclinical" deviations (microtropia).

Characteristically the acuity is reduced and cannot be improved by such optical means as a pinhole or lenses. An exception is the child with moderately high ametropia (usually myopia) whose vision improves to normal after wearing correcting lenses for 1 to 6 weeks (refractive amblyopia). The vision is better for single symbols than for a series in a row (crowding phenomenon). This is not a characteristic of amblyopia but of reduced resolution; it parallels decreased acuity even

*When light is excluded from the eye of a kitten or chimpanzee, vision fails to develop because of cellular degeneration. There is never any recovery, no known histologic evidence exists that this occurs in humans (even in patients with congenital cataract or ptosis), and it in no way resembles what most clinicians understand as amblyopia.

in normal eyes and has no prognostic significance. The amblyopic eye may have relatively less impairment in acuity with dark adaptation compared to the normal eye, but there is no real evidence that amblyopia represents a quasiscotopic state, or a take-over by the peripheral retina. Electrical responses show no characteristic differences between amblyopic and normal eyes. The fixation of the amblyopic eye tends to be unsteady or eccentric, and micronystagmoid movements are frequently demonstrable with the ophthalmoscope, especially if there is an accompanying vertical imbalance. An ocular deviation is more likely to occur in anisometropia when both eyes are hyperopic. It is generally impossible to decide whether the amblyopia is primary or secondary to the strabismus or to the anisometropia. In children old enough to cooperate, a central scotoma may be demonstrable under binocular testing.

Pathophysiology

The mechanism of amblyopia is unknown. Implicit in most views are the notions of disuse or inhibition. In the theory of disuse, the loss of vision is believed to be the result of lack of function like muscle atrophy of lower motor neuron paralysis. Obviously the matter is not that simple, since the amblyopic eye continues to be exposed to lights and patterns; the disuse is of central fixation rather than of light stimulation. Off-shoots of this theory include congenital organic defects—a notion that does not explain the remarkable plasticity of the defect such as when amblyopia is transferred to the previously better eye after occlusion. The theory of inhibition, most logically advanced by Chavasse, considers that the visual loss follows a central nervous system process, is active only in binocular vision, and is analogous to the suppression of physiologic diplopia, which eventually becomes fixed or obligatory. The earlier the age of inhibition the more profound the amblyopia and the poorer the prognosis for recovery.

Current views are more in accord with active inhibition, although the manner of its operation and fixity are still in dispute. A kind of disuse or rather a lack of incentive for the development of fusion probably occurs in high anisometropia in which the retinal image of one eye is grossly blurred. Strabismus does not usually accompany the amblyopia, because there is no need to further suppress an already out-of-focus image. If the better eye is hyperopic, however, deviation may be induced by the accommodative convergence. When the retinal images are of approximately equal clarity, but an innervational disturbance has brought about misalignment, the result may be alternating fixation without amblyopia or unilateral squint with suppression, depending on oculomotor dominance. The mechanism of suppression, similar to "retinal" rivalry of normal binocular vision, is unique in amblyopia in that it involves the fovea, an area not normally subject to suppression. It prevents seeing two different objects in the same direction, which tends to fix the pattern by conditioning. In addition, there is suppression of the area in the squinting eye that comes in line with the object of fixation, an area which has been termed the "false macula." This eliminates diplopia, and in time this "physiologic hole" is used for fixation even when the good eye is covered. The stability of fixation depends on the rigidity of the new motor relationship; the looser the hookup the more the drift. The shifting pattern of eccentric fixation is evaluated by ophthalmoscopy and observation. When the eccentric area is directly stimulated by an ophthalmoscopically projected star, patients old enough to cooperate frequently report that they do not see it "straight ahead" (eccentric viewing). Patching the amblyopic eye has little effect on eccentricity, since it does not alter the motor relationship between the eyes. Conversely, a visual improvement sometimes follows surgical correction if occlusion of the better eye is reinstituted.

Evaluating amblyopia

The patient's history should establish the age of onset of amblyopia, duration, previous treatment, strabismus (initially alternating or intermittent), and evidence of birth injuries, ocular trauma, and congenital and neurologic diseases. The prognosis is often worse in amblyopia of early onset, long duration, and without previous treatment.

Acuity is evaluated with single and multiple letters, great care being taken to prevent peeking. A clue to the degree of visual loss in infants can be obtained by the vigor with which they object to having the good eye patched. In older children, the response to pinholes and lenses can be measured. The answers, characteristically, become more hesitant as the limits are approached, with frequent switching of letters; if mature enough, the child may report that letters adjacent to the ones fixated appear clearer. In organic visual loss a neutral density filter is more deleterious than in amblyopia. A suppression area may be demonstrated by binocular viewing of polarized or anaglyph targets.

Fixation in infants is more easily evaluated with a penlight by observing the corneal reflection as the light is moved about. A large angle kappa should not be confused with eccentric fixation. In older children, fixation is evaluated by projecting a small star onto the fundus with the ophthalmoscope. The patient views the star, and the examiner compares the fixation area relative to the foveal reflex (Fig. 15-9). The foveal reflex is sometimes described as dull or lacking luster, but this is mostly wishful (mechanical) thinking.

Haidinger brushes are an entoptic phenomenon, characteristic of the macula (either dichroic pigment or Henle's layer), seen when viewing a plane polarized field; a blue filter enhances its appearance, and a rotating field prevents adaptation (Fig. 15-10). The entoptic rotating brushes resemble an airplane propeller, especially when a picture of an airplane is presented simultaneously. The patient reports the location of the propeller with respect to the nose of the plane. Haidinger brushes depend on the anatomic integrity of the macula, and their presence is presumptive evidence of functional visual loss. If the brushes are not superimposed on the airplane in monocular fixation, it indicates eccentric fixation; the separation denotes the angle between the fixation and foveal axes. To verify this, reverse the rota-

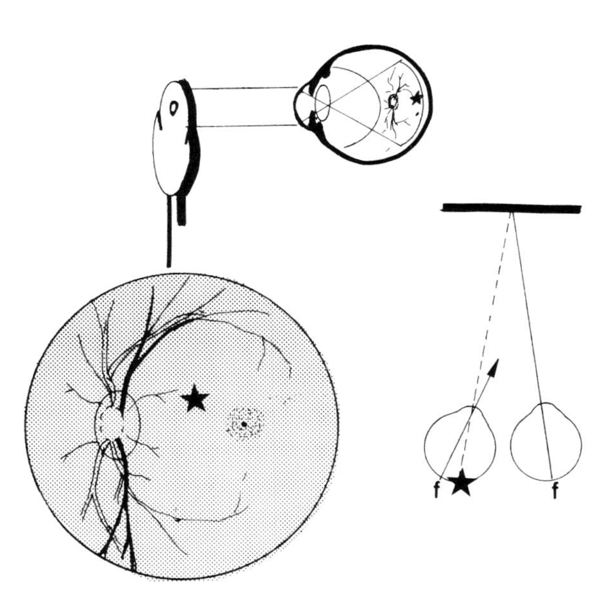

Fig. 15-9. Visuscope technique for evaluating eccentric fixation. The patient with one eye occluded is asked to fixate the small star projected into the fundus, and its position is compared to that of the fovea. In this diagram of a case of esotropia, the fundus area used for fixation lies between the disc and the fovea.

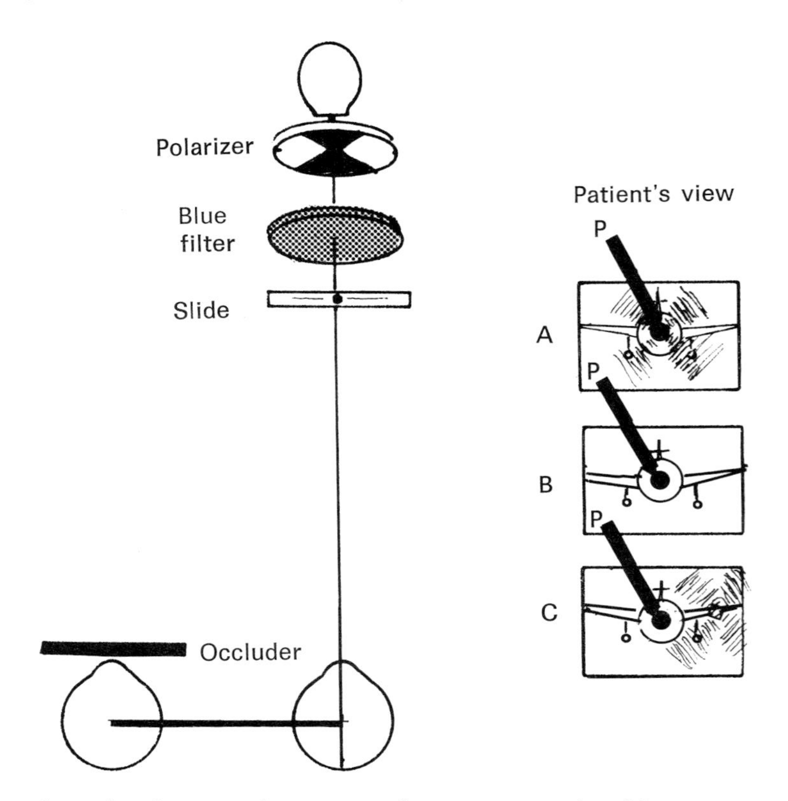

Fig. 15-10. Haidinger brushes test. The entoptic phenomenon is induced by a rotating polarizer viewed through a blue filter. A slide with a small aeroplane enhances the impression of a rotating propeller. The patient indicates fixation on the hub of the aeroplane by means of the pointer *P* and reports whether the propeller coincides with it *(A)*, is absent *(B)*, or displaced *(C)*.

tion of the brushes (by reversing the motor or by a cellophane filter), and ask if the brushes are seen "directly." With a pointer, the patient indicates which target he is fixating. The pointer may be moved to the nose of the plane, to one side, or to the other, indicating the stability of eccentric fixation. The test should be conducted first on the normal eye, since for reasons unknown, some normals fail to perceive the brushes.

In amblyopia, a central relative scotoma of variable size may be demonstrable in older children with binocular viewing. A simple method is red-green glasses and complementary colored targets arranged according to which eye is to be used for fixation. The Goldmann perimeter or projection tangent screen serves for measuring. The conditions should approximate normal seeing, since dramatic targets tend to break down suppression even though it is operative in everyday vision. Polarized targets or the disappearance of a Maddox rod line as it approaches the light spot seen by the fixing eye also help indicate the size and degree of the suppression scotoma. Stereograms (amblyoscope, synoptophore) are used to evaluate suppression at the reading distance.

Ideally the complete examination of an amblyopic child should include an accurate refraction with and without cycloplegia, fundus examination, color vision tests, suppression fields, and a search for latent nystagmus. Techniques for evaluating sensory and motor cooperation of the eyes are considered in Chapters 17 and 25.

Treatment of amblyopia

The best treatment of amblyopia is prevention, which means early screening by pediatricians, general practitioners, and parents. The presence of intermittent squint, habits of closing one eye, rubbing the eye, or head tilt should arouse suspicion. The notion that "they will grow out of it" is an old superstition, difficult to erase, and can only be combated by a wider knowledge of the dangers of delay. Many parents procrastinate because they think an eye examination is impossible if the child cannot read the chart. Although amblyopia cannot be diagnosed with certainty in infants and young children, the fixation and cover test, the behavior on occlusion, and the retinoscopic evidence of anisometropia generally are sufficient to make a decision regarding treatment.

The treatment of amblyopia is occlusion of the better eye: early occlusion and total occlusion of sufficient duration to evaluate the effect. In children of school age in whom vision is very poor, it may be necessary to postpone treatment until a vacation period; in infants or young children, there should be no delay. The most effective occluder is of the adhesive type, which prevents peeking (with appropriate instruction to the parent regarding skin cleansing and hygiene). Cycloplegia of the better eye instead of occlusion generally is to be avoided and is useless of course if the eye is only slightly hyperopic or myopic. The occluder is worn throughout the waking hours and is continued until no further improvement is demonstrable. The vision and fixation pattern are checked at 2-week intervals to prevent occlusion amblyopia. After a satisfactory vision (about 20/40) or an alternating fixation pattern is established, treatment may be continued by alternate or partial occlusion. In latent nystagmus, occlusion should be partial, reducing the vision of the fixing eye to a level just below that of the amblyopic one. Rarely occlusion precipitates or exaggerates a deviation. Several explanations are usually required before the parents understand why patching does not simultaneously improve strabismus or surgery is still required. It is generally conceded that occlusion may be discontinued if no improvement is obtained after 2 to 3 months of adequate treatment.

In the presence of amblyopia, the surgical treatment of strabismus is embarrassingly unpredictable, and attempts are always made first to improve vision by occlusion. If cosmetic surgery is undertaken, a therapeutic trial of occlusion may be tried afterward, since the altered mechanics sometimes changes the position of the functional scotoma and creates a more favorable opportunity.

Occlusion therapy may be supplemented or, when unsuccessful, replaced by more active treatment. Active treatment includes orthoptics, pleoptics, filters, or optical aids and is usually reserved for older age groups. It should be attempted in any patient who has lost the vision in the better eye. The early promise of these treatments has not been born out, and most clinicians favor occlusion of the better eye as the primary treatment. Whatever the treatment, the cooperation of patient and parents is essential in carrying it to a successful conclusion.

VISION SCREENING TESTS

At the beginning of every new school term, the waiting room is filled with children who have failed a vision screening test. Inevitably a certain number of referrals are found to be unnecessary, and some children who require care are overlooked. But clinicians and parents should not become annoyed, since the perfect screening test has yet to be invented. If the test is inefficient, we must help design better ones; if improperly administered, we should offer school administrators and nurses better training.

It is in the nature of things that most ocular difficulties reflect themselves in reduced visual acuity. The acuity test therefore is the simplest, the most easily administered, and the most efficient screening test, outweighing all other ancillary tests combined. Under proper supervision, with appropriate targets at the proper distance and illumination, acuity tests are applicable to any school grade and some preschoolers. If educators decry, as they frequently do, the lack of funds for specialized testing equipment and the lack of trained personnel to administer the tests, let them consider the simple acuity chart hung on a wall. If *all* children could be tested by this means alone, a considerable step toward eliminating amblyopia would have been taken.

True, acuity will not detect hyperopia, moderate astigmatism, color defectives, poor stereoacuity, or reading problems; but even extensive screening tests do not diagnose intermittent squint, congenital cataracts, or retinoblastoma. Moreover there is no agreement as to how much hyperopia or astigmatism inevitably requires correction, and an over-referral to one is an under-referral to another. The observation of a red eye by a school nurse, poor reading performance by a teacher, ocular incoordination by a parent, a white reflex by a pediatrician is more valuable than a battery of phoria or color vision tests. The question of what to test must be balanced by the time, cost, equipment, and special training required to evaluate the results; the more tests are hung on the tail of the acuity chart the greater the number of failures and over-referrals. Such quasi-visual activities as perceptual span, Gestalt completion, reading rate and comprehension, hand-eye coordination, puzzle solving, vocabulary, and truncated intelligence tests resemble a dossier and are of doubtful clinical ophthalmologic value. The more diagnostic screening tests become, the greater the chance of procrastinating with subsequent visual difficulties. Finally, statisticians play a game called "factor analysis"; no matter how many different tests are included, they will be sure to point out that the number is excessive, since all measure the same factor "x."

Preschool vision screening

An ideal screening program would consist of regular, periodic, and complete ophthalmologic examination beginning at birth. Since this is impractical, an attempt is made to recognize those visual problems particularly likely to occur in the early years, especially those amenable to effective treatment. The discovery and correction of amblyopia ranks high in this aim. Early treatment greatly increases the chance of success, whereas irreparable damage is the consequence of delay. Preschool vision screening is directed basically at the measurement of acuity as early as feasible.

A variety of vision tests are available, using different symbols, pictures, test distances, and test symbol spacing. Generally, young children do better at closer (equivalent) test distances, such as 10 feet; in indicating vertical rather than horizontal directions with single rather than multiple targets; and in viewing "real" targets instead of pictures inside a box. With pictures, children frequently identify only their favorites, with symbols only those they are familiar with, and with toys it becomes a problem getting them back. Many clinicians prefer the illiterate E test as a reliable, easily checked, and readily taught test. It gives the most information for the time and effort spent.

School vision screening

Several vision screening tests have been widely used on school populations. Currently available are the Eames Eye Test (World Book), Massachusetts Vision Test (Welch-Allyn Inc.), Ortho-Rater (Bausch & Lomb Optical Co.), Sight Screener (American Optical Co.), School Vision Tester (Titmus Optical Co.) and Telebinocular (Keystone View Co.). All the tests include a distance acuity; in addition, some have tests for near acuity, lateral and vertical phorias, fusion, stereopsis, color vision, hyperopia, astigmatism, visual fields, and suppression. The Eames Test has a clock dial for checking astigmatism and, along with the Massachusetts Vision Test, a +1.50 D lens for evaluating uncorrected hyperopia. The Sight Screener incorporates polarized vectographic targets, the Ortho-Rater and School Vision Tester have an attachment for screening visual fields, and the Telebinocular is of open construction to permit monitoring eye movements.* Machine tests are more

*Other vision screening instruments are manufactured by Rodenstock, Zeiss, and Goodlite; a model from the American Automobile Association is designed specifically for evaluating drivers, and one by Tracor is semiautomated.

compact, are easy to transport, and require some training to administer, but some children are afraid of the box. The pass-fail criteria are similar for all tests, and their intercorrelation is in the 0.80 to 0.90 range. In general, a repeat examination is recommended before referral is advised. There is no consensus as to which tests are most helpful in adding to the selectivity of the acuity test without undue over-referral. Undoubtedly over-referral is the lesser of two evils, but whether a high phoria or low stereoacuity is related to school performance and indicates the need for a complete eye examination has yet to be established. In many cases the test battery is chosen by the school authorities, beyond the clinician's control, to be accepted with a tolerant attitude that generally rubs off on the parents.

Visual skills

Resolution acuities, the three degrees of fusion, color vision, and lateral and vertical phorias are sometimes grouped together as "visual skills." Skills are considered capacities that are "acquired and can be modified, changing with age, and which may be given to an employee by professional eye care." Although acuity and fusion fit vaguely into this category, color vision and phorias are more questionable. One either has normal color vision, or one has not. Any skill attached to it involves hue discrimination, color naming, or perhaps color aptitude (like the painter or designer), none of which is tested by color screening tests. Does a high phoria mean greater or less skill? The "skill" is in fusional vergences, not the phoria. It would have been more logical to measure the fusional vergences, as in relating hyperopia to the accommodation amplitude.

Industrial workers sometimes are placed in certain jobs according to their visual skills "profile." Large numbers of employees, evaluated as efficient or inefficient by other criteria, are tested and their visual skills measured. Then those likely to do well on certain jobs are preselected. Although the idea is logical and statistically reasonable, the physiologic basis of such a classification is questionable. Undoubtedly color blindness is a handicap to a worker assembling color-coded wires, but that a particular phoria score makes the worker more successful, less

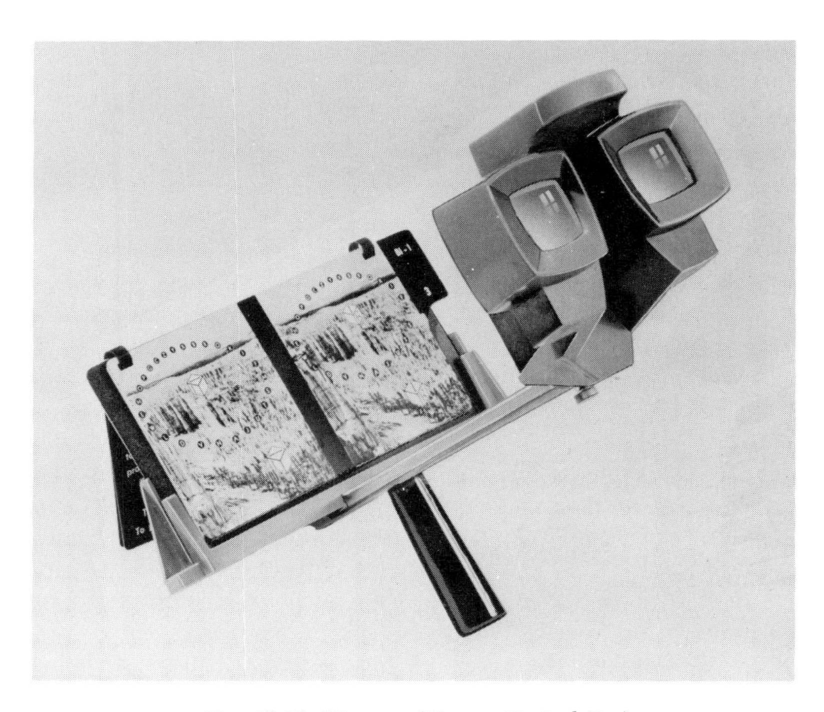

Fig. 15-11. Biopter. (Titmus Optical Co.)

accident-prone, and more likely to stay on the job with fewer days absent are predictions only a statistician could pull out of his hat. Efficiency is based on so many variables that to deny an individual a job or transfer him to another merely because his stereoacuity does not come up to a designated figure may lead to unnecessary hardships and occasional tragedies. We cannot even agree on the visual requirements for driving a car, an activity in which there is no shortage of statistics.

A useful application of the skills is to compare the results with and without lenses or with the old lenses if worn. Tests for size differences, distortions, and horizontal and vertical phorias in the reading field are helpful in predicting the patient's adjustment to new lenses.

A simplified, open Brewster stereoscope (Bioptor) is a reasonably priced instrument for measuring visual skills. The targets are stereograms, ring-bound, and flipped in sequence (Fig. 15-11). The holder is set at the appropriate far and near distances. Targets are also available (Titmus) for home or office orthoptics, such as cheiroscopic drawings. A record of the findings may be given the parent, referring physician, or school nurse.

TECHNIQUES IN INDUSTRIAL OPHTHALMOLOGY

More patients than ever are referred to the ophthalmologist after industrial surveys and driver examinations, for the treatment of injuries, and for assessing the extent of ocular damage and visual loss after industrial trauma.

B. Ramazini (1635-1714) generally is considered the "Father of Occupational Medicine." His interest was aroused when he witnessed the distress of workmen cleaning cesspits: "Of the diseases of those who work upon minute things which strain the eyes, I should advise workmen not only to use spectacles, but to intermit their work now and then and refresh their eyes with a diversity of objects." Sad to say, Ramazini developed eye trouble himself which terminated in blindness.

Every job requires some visual ability, some more critically than others. The watchmaker needs clear vision at 6 inches, the accountant at 16 inches, the musician and housewife at arm's length, the bus driver at 20 feet. The near refractive tests and the reading prescription depend on each patient's requirements. The accommodative flexibility needed by the typist-receptionist makes separate reading glasses a nuisance, whereas standard bifocals hinder reading above eye level. Symptoms arise in workers engaged in prolonged near work not only because of such intrinsic factors as heterophoria, anisometropia, or presbyopia but also because of poor illumination, glare, inadequate contrast, or monotony of the visual task. The range of vision deteriorates in a dimly lit work area, threading a needle with black thread against a dark background will be difficult despite high stereoacuity, and matching colors under sodium or mercury lamps may be impossible. Some visual capacities are necessary on some jobs and not on others. (An uncorrected presbyope could not work a lathe but might make a good watchman.) Sharp distance vision has been said to be a detriment to hosiery loop workers (although I cannot see the harm in it). Good color discrimination is a sine qua non to the painter and designer—or at least used to be.

The industrial worker is subject to both major and minor eye injuries deserving special attention. Safety lenses, helmets, face masks, or shields are required by cutters, chippers, grinders, and caustic and chemical handlers.

Ocular radiation hazards

Until the advent of the laser, the principal optical hazard was ultraviolet, such as in the case of the amateur astronomer who insists on direct viewing of solar eclipses. Harmful rays are those below 320 nm, but high attenuation by glass and plastic only require enclosing the source with these materials. Unshielded ultraviolet rays are absorbed by the cornea and produce keratoconjunctivitis (e.g., welder's flash). Infrared radiations (e.g., from open furnaces) cause heat cataracts and corneal opacities, now only of historic interest. Ultraviolet is more hazardous because the patient is not aware of the exposure. It is also easier to eliminate infrared sources. The laser hazard depends on the type of exposure and is of increasing consequence because of wider use in therapeutics, communications, and the military. Atomic radiation not only burns the beholder but blasts the blastocyst. Finally, even ordinary (visible) light at high intensities and prolonged exposure may produce retinal damage (at least in rodents). Types of absorptive lenses and their indications were considered in Chapter 5.

Estimating visual loss

"Visual efficiency" as defined by a subcommittee of the Council of Industrial Health and accepted by the American Medical Association at the 1955 general meeting is the degree or percentage of competence of the eyes to accomplish their physiologic functions, including (1) corrected visual acuity for distance and near, (2) visual fields, (3) ocular motility with absence of diplopia, and (4) binocular vision. Although these functions are not of equal importance, vision is inef-

Table 15-1. Percentage of central visual efficiency corresponding to central visual acuity notations for distance

Snellen		*Visual angle in minutes*	*Percent central visual efficiency*	*Percent loss of central vision*
English	*Metric*			
20/16	6/5	0.8	100	0
20/20	6/6	1.0	100	0
20/25	6/7.5	1.25	95	5
20/32	6/10	1.6	90	10
20/40	6/12	2.0	85	15
20/50	6/15	2.5	75	25
20/64	6/20	3.2	65	35
20/80	6/24	4.0	60	40
20/100	6/30	5.0	50	50
20/125	6/38	6.3	40	60
20/160	6/48	8.0	30	70
20/200	6/60	10.0	20	80
20/300	6/90	15.0	15	85
20/400	6/120	20.0	10	90
20/800	6/240	40.0	5	95

Table 15-2. Percentage of central visual efficiency corresponding to central visual acuity notations for near

Snellen	*Jaeger*	*Point*	*Visual angle in minutes*	*Visual efficiency*	*Percent loss of central vision*
14/14	1−	3	1.0	100	0
14/18	2−	4	1.25	100	0
14/22	—	5	1.6	95	5
14/28	3	6	2.0	90	10
14/35	6	8	2.5	50	50
14/45	7−	9+	3.2	40	60
14/56	8	12	4.0	20	80
14/70	11	14	5.0	15	85
14/87	—	—	6.3	10	90
14/112	14−	22	8.0	5	95
14/140	—	—	10.0	2	98

ficient without the coordinated action of all factors. Color vision, adaptation, and accommodation are considered dependent functions and are not included in calculating percentage loss of visual efficiency.

A visual acuity of 20/20 or better (measured on a standard Snellen chart) is considered 100% acuity for distance vision, and Snellen 14/14, Jaeger 1, or 3-point type (with presbyopic corrections, if necessary) is considered 100% near acuity (measured at 14 inches or 36 cm). Revised percentage loss based on statistical studies and experience in disabilities is given in Tables 15-1 and 15-2.

The visual field, determined with a white target subtending 0.5° (3 mm at 330 cm) under an illumination not less than 7 footcandles, is plotted on standard field charts in eight (45°) principal meridians. The minimum normal extent for the 3/330 white isopter is as follows:

Temporally	85°
Down and temporally	85°
Down	65°
Down and nasally	50°
Nasally	60°
Up and nasally	55°
Up	45°
Up and temporally	55°

These figures are less than average to allow for poor or delayed subjective responses or for undue prominence of nose or brow. Visual field efficiency of each eye is obtained by adding the number of degrees of the eight principal radii given above (i.e., 500) and dividing by 5. Thus a normal field in 500/5 = 100%. A field contracted 10° in all eight meridians would have an efficiency of 420/5 = 84%. Central scotomas affect central acuity and are not taken into consideration in computing field efficiency.

Visual efficiency as related to motility is evaluated in terms of the absence of diplopia and binocular vision. Except for the reading position, diplopia is not considered a significant visual disability unless present within 30° of central fixation. The extent of the diplopia in the various directions of gaze (on the perimeter at 33 cm or on the tangent screen at 1 or 2 meters) is determined for each 45° meridian. The diplopia is measured by the separation of a test light (without filters or prisms) and plotted. Diplopia within 20° of central fixation represents 100% loss of motility efficiency in one eye. In peripheral diplopia the loss is proportionately less.

If the loss of binocular vision is the result of disease or injury or diplopia is absent because of suppression, the degree of disability depends on the importance of depth perception to the patient's work. Loss of binocular vision without other ocular abnormalities probably does not represent more than a 50% loss. Binocular vision is tested with any standard stereoscope or amblyoscope. (The cosmetic factor of strabismus is considered separately.)

In evaluating the visual efficiency loss of one eye, the distance and near acuity, ocular motility, and diplopia and binocular vision is averaged. Thus if the average distance and near acuity is 50%, field efficiency is 40%, and motility efficiency is 60%, the total efficiency of this eye is $0.2 \times 0.4 \times 0.6 = 0.048$ or 4.8% of the total visual efficiency. An efficiency less than 10% is considered to be a total visual loss.

If one eye is lost and the other eye is normal, the visual efficiency of the remaining eye is multiplied by 3 to compute binocular efficiency (BE):

$$\text{BE} = \frac{(\%\ \text{efficiency of better eye} \times 3) + (\text{efficiency of worse eye})}{4}$$

Thus the binocular efficiency of a patient blind in one eye is 75%, or a binocular loss of 25%. When disease or injury involves both eyes, a loss of efficiency in ocular motility is used only in computing the efficiency of the less efficient of the two eyes. The estimation of the visual efficiency of the better eye therefore is based only on the central visual acuity and on the visual field of this eye. If the binocular visual efficiency is less than 10%, the patient is industrially blind.

Loss of color vision, adaptation, cosmetic deformity, accommodation disturbances, metamorphopsia, entropion or ectropion, lagophthalmos, and epiphora may result in disability, the value of which cannot be calculated.

It is recommended that final estimation of visual efficiency be deferred for 3 months after inflammation, 6 months after surgery, and at least 12 months (but not more than 16 months) after disturbances of ocular motility, sympathetic ophthalmia, traumatic cataract, or optic nerve atrophy.

These recommendations are not meant to change the legal opinions expressed by the courts or by compensating insurance commissions.

REFRACTION OF THE AGED EYE

The wisdom of age comes from seeing every part in its relation to the whole; unfortunately vision is frequently blunted by the cumulative effects of wear and tear, reduced nutrition, and metabolic exhaustion. The corneal sensitivity decreases; the lens acquires vacuoles and discoloration; the anterior chamber becomes shallower, the iris more rigid, and the pupil miotic; the vitreous degenerates; acuity and color vision become less keen; hyperopia increases; critical fusion frequency decreases; and the retina receives up to one third less light. To add to this gloomy picture, reaction time lengthens, perceptual discrimination deteriorates, and with progressive loss of visual resources, a feeling of helplessness and anxiety arises that leads to dependency and insecurity.

The history establishes the visual needs of older patients the duration of existing visual disability, their tolerance to previous lenses or subnormal vision aids, occupational and recreational habits, general mobility, availability of help from a mate or companion, and the presence of concurrent ocular or systemic disease. Motivation with respect to visual improvement is a key factor in planning treatment, ranging from cataract extraction to a magnifying loupe.

Refraction is carried out at a more leisurely pace, with more tolerance for instability and indecision. Objective tests may have to be depended on if answers are unreliable. Retinoscopy, often difficult because of small pupils and hazy media, is best done under sympathomimetic dilation (e.g., phenylephrine). All objective tests naturally are confirmed by subjective responses, if possible, because of the frequently associated macular and optic nerve diseases. Subjective tests usually require closer testing distances, stronger illumination, larger targets, better contrast, and greater trial lens changes. A trial frame often is more suitable than a refractor because of the strong powers involved, the closer distances with subnormal vision aids, the adjustments to distortion, magnification, and bifocals, which may seriously interfere with mobility. A Halberg trial cell clips readily on to the patient's spectacle frame and provides three lens cells to which may be added modifying spheres, cylinders, or prisms. The entire combination is placed in the lensometer to obtain the correct back vertex power.

The astigmatic axis can be better evaluated by arbitrarily increasing cylinder power, then refining with astigmatic dial or cross cylinder. A stenopeic slit may help in eyes with cataracts, corneal degenerations, vitreous opacities, and repaired retinal detachments. Pinholes on the other hand, frequently reduce retinal illumination so much that the vision is made worse instead of better. The distance and reading acuity often differ, a characteristic finding in nuclear sclerosis and macular disease. Projected acuity charts are generally of little use if the vision is less than 20/100. Assorted reading material, clipped from newspapers or magazines, should be available for testing practical reading vision.

In the presence of a dense cataract, the visual prognosis may be evaluated by large targets, two-point light discrimination (specify distance and separation), color vision, and the accuracy of light localization. The history is crucial in evaluating whether the visual loss is consistent with the progression of the lens opacity; it may be checked with other members of the family or in driver's examination records. An estimate of the effect of the cataract can be obtained by comparing the patient's acuity to the clarity with which fundus details can be seen with the ophthalmoscope.

Strong reading adds or magnifiers may significantly improve near vision despite hazy media or retinal pathology. Success sometimes depends less on the magnifier than on the steadiness with which it can be held. The best magnifier is the weakest one that fulfills the patient's needs. Telescopic distance aids are seldom well tolerated; the combination of contact lens and spectacle is occasionally recommended as a useful telescope but only by those who have never tried it.

Strong visual aids are practical only for monocular vision, and the nearer the reading distance the more interference with illumination. If the motivation is adequate, however, many distortions that would be intolerable to a normal-seeing eye are acceptable to one with poor resolution. Additional help can be obtained with nonoptical visual aids such as large-type books, reading stands, writing guides, needle threaders, and high-intensity lamps.

Despite the fact that we are all subject to planned obsolescence, a refraction should be attempted in every patient, and no one should be dismissed as beyond help until optical and nonoptical techniques have been given a fair trial. A significant proportion of poorly seeing eyes can be "cured" by simply providing the correct spectacles.

Dispensing glasses to older people requires attention to fitting details and optical tolerance because of the strong powers involved. Lenses should be lightweight, carefully centered, aspheric if possible, and fitted as near to the eye as feasible. A separate reading glass may be a better choice than bifocals for patients with mobility problems. Frames should be light, sturdy, and with a nose pad designed to distribute weight and to permit minor adjustments.

REFERENCES

Allen, H. F.: Testing of visual acuity in preschool children, Pediatrics **19**:1093-1100, 1957.

Allen, H. F.: Incidence of amblyopia (editorial), Arch. Ophthal. **77**:1, 1967.

Amigo, G.: Binocular balancing techniques, Am. J. Optom. **45**:511-522, 1968.

Apt, L., editor: Diagnostic procedures in pediatric ophthalmology, vol. 3, International Ophthalmology Clinics, Boston, 1963, Little, Brown & Co.

Arruga, A., editor: International strabismus symposium; an evaluation of the present status of orthoptics, pleoptics, and related diagnostic and treatment regimens, Basel, 1968, S. Karger.

Arruga, A.: The use of space diagnostic methods and of prismotherapy in the treatment of sensory alterations of convergent squint. In The First International Congress of Orthoptics, St. Louis, 1968, The C. V. Mosby Co., pp. 62-76.

Bangerter, A.: Amblyopie Behandlung, ed. 2, Basel, 1955, S. Karger.

Bannon, R. E.: A study of astigmatism at the near point, with special reference to astigmatic accommodation, Am. J. Optom. Monograph No. 12, 1946.

Bannon, R. E., Cooley, F. H., Fisher, H. M., and Textor, R. T.: The stigmatoscopic method of determining the binocular refractive status, Am. J. Optom. **27**:371-384, 1950.

Barnard, W.: Treatment of amblyopia by inverse occlusion and pleoptics, Br. Orthopt. J. **19**:19-30, 1962.

Benton, C. D.: Analysis of results with the Massachusetts Vision Test with recommendations for improving its accuracy, Am. J. Ophthal. **36**:363-364, 1953.

Binder, H. F., Engel, D., Ede, M. L., and Loon, L.: The red filter treatment of eccentric fixation, Am. Orthopt. J. **13**:64-69, 1963.

Birren, F.: Safety on the highway, Am. J. Ophthal. **43**:265-270, 1957.

Blum, H. F.: Carcinogenesis by ultraviolet light, Princeton, 1959, Princeton University Press.

Blum, H. L., Peters, H. B., and Bettman, J. W.: Vision screening for elementary schools, the Orinda study, Berkeley, 1959, University of California Press.

Borg, G., and Sundmark, E.: A comparative study of visual acuity tests for children, Acta Ophthal. **45**:105-113, 1967.

Burian, H. M.: Pathophysiologic basis of amblyopia and of its treatment, Am. J. Ophthal. **67**:1-12, 1969.

Burian, H. M., and Luke, N. E.: Sensory relationships in 100 consecutive cases of heterotropia, Arch. Ophthal. **84**:16-20, 1970.

Capobianco, N. M.: Pleoptic treatment of amblyopes with central and eccentric fixation, Am. Orthopt. J. **10**:33-53, 1960.

Charman, W. N.: On the position of the plane of stationarity in laser refraction, Am. J. Optom. **51**:832-838, 1974.

Cogan, D. G., and Kinsey, V. E.: Action spectrum of keratitis produced by ultraviolet radiation, Arch. Ophthal. **35**:670-677, 1946.

Committee on Medical Aspects of Automotive Safety: Visual factors in automobile driving and provisional standards, Arch. Ophthal. **81**:865-871, 1969.

Cornwell, A. C.: Electroretinographic responses following monocular visual deprivation in kittens, Vision Res. **14**:1223-1227, 1974.

Cuppers, C.: Moderne Schielbehandlung, Klin. Monatsbl. Augenheilkd. **129:**579-604, 1956.

Davis, J. C.: Correlation between scores on Ortho-Rater tests and clinical tests, J. Appl. Psychol. **30:** 596-603, 1946.

Dayton, G. O., Jensen, G., and Jones, M. H.: Visual acuity of infants measured by means of optokinetic nystagmus and oculogram, Invest. Ophthal. **1:**414, 1962.

Dews, P. B., and Wiesel, T. N.: Consequences of monocular deprivation on visual behavior in kittens, J. Physiol. (London) **206:**437-455, 1970.

Diskan, S. M.: A new visual screening test for school children, Am. J. Ophthal. **39:**369-374, 1955.

Donnelly, W. L.: The congenital structural anomalies of the ocular muscles, Am. Orthopt. J. **20:**87-90, 1970.

Dunipace, D. W., Strong, J., and Huizinga, M.: Prediction of nighttime driving visibility from laboratory data, Appl. Optics **13:**2723-2734, 1974.

Eames, T. H.: New approach to testing eyes of school children, Am. J. Ophthal. **24:**1170-1173, 1941.

Enoch, J.: Receptor amblyopia, Am. J. Ophthal. **48**(2): 262-274, 1959.

Enos, M. V.: Suppression versus amblyopia, Am. J. Ophthal. **27:**1266-1271, 1944.

Eskridge, J. B.: A binocular refraction procedure, Am. J. Optom. **50:**499-505, 1973.

Fantl, E. W., and Perlstein, M. A.: Refraction in cerebral palsy, Am. J. Ophthal. **63:**857-863, 1967.

Fletcher, M. C., Abbott, W., Girard, L. J., Guber, D., Silverman, S. J., Tomlinson, E., and Boyd, J.: Results of biostatistical study of the management of suppression amblyopia by intensive pleoptics versus conventional patching, Am. Orthopt. J. **19:**8-30, 1969.

Flom, M. C., Heath, G. C., and Takahashi, E.: Contour interaction and visual resolution: contralateral effects, Science **142:**979-980, 1963.

Flynn, J. T., McKenney, S. G., and Dannheim, E.: Brightness matching in amblyopia, Am. Orthopt. J. **21:**38-49, 1971.

Fonda, G. E.: Method of testing visual acuity in the amblyopic eye, Am. Orthopt. J. **8:**149-150, 1958.

Fowler, F.: The treatment of amblyopia, Am. Orthopt. J. **6:**19-28, 1956.

Fox, S. L.: Industrial and occupational ophthalmology, Springfield, Ill., 1973, Charles C Thomas, Publisher.

Franceschetti, A. T., and Burian, H. M.: Visually evoked responses in alternating strabismus, Am. J. Ophthal. **71:**1292-1297, 1971.

Friedman, B.: Entoptic methods in clinical investigations, Am. J. Ophthal. **26:**235-244, 1943.

Friesen, H., and Mann, W. A.: Follow-up study of hysterical amblyopia, Am. J. Ophthal. **62:**1106-1115, 1966.

Goldschmidt, M.: A new test for function of the macula lutea, Arch. Ophthal. **44:**129-135, 1950.

Gordon, D. M., Zeidner, J., Zagorski, H. J., and Uhlaner, J. E.: A psychometric evaluation of Ortho-Rater and wall chart test, Am. J. Ophthal. **37:**699-705, 1954.

Gorman, J. J., Cogan, D. G., and Gillis, S. S.: An apparatus for grading visual acuity of infants on the basis of opticokinetic nystagmus, Pediatrics **19:**1088-1092, 1957.

Gould, A., Fishkoff, D., and Galin, M. A.: Active visual stimulation: a method of treatment of amblyopia in the older patient, Am. Orthopt. J. **20:**39-45, 1970.

Gradle, H. S.: Another test for malingering, Am. J. Ophthal. **20:**300, 1937.

Hallden, U.: An explanation of Haidinger's brushes, Arch. Ophthal. **57:**393-399, 1957.

Haseltine, S. L.: Tests for vision malingering, Eye Ear Nose Throat Mon. **17:**49, 1938.

Helveston, E. M., and von Noorden, G.: The appearance of the fovea in strabismic amblyopia, Am. J. Ophthal. **64:**687-688, 1967.

Hirsch, M. J., and Wick, R. E.: Vision of children, Philadelphia, 1963, Chilton Books.

Holt, L. C., editor: Pediatric ophthalmology, Philadelphia, 1964, Lea & Febiger.

Ikeda, H., and Wright, M. J.: Is amblyopia due to inappropriate stimulation of the "sustained" pathway during development? Br. J. Ophthal. **58:**165-175, 1974.

Imus, H. A.: Industrial vision techniques, Am. J. Ophthal. **32:**145-152, 1949.

Irvine, S. R.: An ocular policy for public schools, Am. J. Ophthal. **24:**779-787, 1941.

Jagerman, L. S.: Visual acuity measured with easy and difficult optotypes in normal and amblyopic eyes, J. Pediatr. Ophthal. **7:**49-54, 1970.

Jobe, F. W.: An anlysis of visual performance in relation to safety, Am. J. Optom. **25:**107-116, 1948.

Johnson, D. S., and Antuna, J.: Atropine and miotics for treatment of amblyopia, Am. J. Ophthal. **60:**889-891, 1965.

Jonkers, G. H.: The examination of visual acuity of children, Ophthalmologica **136:**140-144, 1958.

Keeney, A. H.: Ophthalmic pathology in driver limitation, Trans. Am. Acad. Ophthal. Otolaryngol. **72:** 737-740, 1968.

Keiner, E. C. J. F.: Pathogenesis of eccentric fixation, Am. J. Ophthal. **63:**20-22, 1967.

Krimsky, E.: Children's eye problems, New York, 1956, Grune & Stratton, Inc.

Kuhn, H. S.: Eyes and industry, ed. 2, St. Louis, 1950, The C. V. Mosby Co.

Kuhn, H. S.: Professional participation inside industry. In Symposium on Industrial and Traumatic Ophthalmology, St. Louis, 1964, The C. V. Mosby Co., p. 13-17.

Kupfer, C.: Treatment of amblyopia ex anopsia in adults, Am. J. Ophthal. **43:**918-922, 1957.

Kuwabara, T.: Retinal recovery from exposure to light, Am. J. Ophthal. **70:**187-198, 1970.

Lancaster, W. B.: Stigmatoscopy, Trans. Am. Ophthal. Soc. **32:**120, 1934.

Lasky, M. A.: Simulated blindness, Arch. Ophthal. **27:**1038-1049, 1941.

Lazich, B. M.: Amblyopia ex anopsia, new concept of its mechanism and treatment, Arch. Ophthal. **39:** 183-192, 1948.

Levi, L.: Vision in communication. In Wolf, E., editor: Progress in optics, **8**:345-372, 1970.
Limpaecher, E.: Amblyopia therapy: methods and results, Am. Orthopt. J. **19**:97-103, 1969.
Linksz, A.: Pathophysiology of amblyopia, J. Pediatr. Ophthal. **1**:19-25, 1964.
Lippmann, O.: Vision of young children, Arch. Ophthal. **81**:763-775, 1969.
Luckiesh, M., and Moss, F. K.: Characteristics of sensitometric refraction, Arch. Ophthal. **27**:576-581, 1941.
Luckiesh, M., and Moss, F. K.: Comparison of a new sensitometric method with usual techniques of refraction, Arch. Ophthal. **30**:489-493, 1943.
Maxwell, J. C.: On the unequal sensibility of the fovea centrale to lights of different colours, Rep. Br. Assoc. Adv. Sci. **2**:12, 1856.
Mayer, L. L.: Eyesight in industry, Arch. Ophthal. **28**:375-405, 1942.
Miles, P. W.: Binocular refraction, Am. J. Ophthal. **31**:1460-1466, 1948.
Miller, E. F.: Investigation of the nature and cause of impaired visual acuity in amblyopia, Am. J. Optom. **32**:10-28, 1955.
Morgan, M. W.: The Turville infinity binocular balance test, Am. J. Optom. **26**:231-250, 1949.
Nawratzki, I., and Jampolsky, A.: A regional hemiretinal difference in amblyopia, Am. J. Ophthal. **46**:339-344, 1958.
Nordenson, J. W.: Ophthalmic international standards, Br. J. Ophthal. **35**:496-501, 1951.
Novak, J. F.: Are the eyes right for the job? In Symposium on Industrial and Traumatic Ophthalmology, St. Louis, 1964, The C. V. Mosby Co., pp. 18-34.
Novak, J. F.: How disabled is the injured eye? In Symposium on Industrial and Traumatic Ophthalmology, St. Louis, 1964, The C. V. Mosby Co., pp. 193-207.
Osterberg, G.: A Danish pictorial sight-test chart for children, Am. J. Ophthal. **59**:1120-1123, 1965.
Parks, M. M.: Technique for evaluating vision in infants and children, Int. Ophthal. Clin. **3**:721-733, 1963.
Pitts, D. G.: A comparative study of the effects of ultraviolet radiation on the eye, Am. J. Optom. **47**:535-546, 1970.
Press, E., and Austin, C.: Screening of preschool children for amblyopia; administration of tests by parents, J.A.M.A. **204**:767-770, 1968.
Pugh, M.: Foveal vision in amblyopia, Br. J. Ophthal. **38**:321-331, 1954.
Reinecke, R. D.: Malingering in children, Int. Ophthal. Clin. **2**:837-842, 1962.
Reinecke, R. D., and Cogan, D.: Standardization of objective visual acuity measurements: optokinetic nystagmus vs Snellen acuity, Arch. Ophthal. **60**:418-421, 1958.
Rosenthal, A. R., and von Noorden, G. K.: Clinical findings and therapy in unilateral high myopia associated with amblyopia, Am. J. Ophthal. **71**:873-879, 1971.
Rubin, W. R.: Reverse prism in the treatment of strabismus and amblyopia, Am. Orthopt. J. **16**:62-64, 1966.
Ryan, V.: A critical study of visual screening, Am. J. Optom. **33**:227-257, 1956.
Savitz, R. A., Valadian, I., and Reed, R. B.: Vision screening of preschool children at home, Am. J. Public Health **55**:1555-1562, 1965.
Schapero, M.: Amblyopia, Philadelphia, 1971, Chilton Book Co.
Sheridan, M. D.: Vision screening of very young or handicapped children, Br. Med. J. **5196**:453-456, 1960.
Shipley, T.: Haidinger's brushes in the clinical haploscope, Arch. Ophthal. **70**:176-177, 1963.
Shirley, S. Y., and Gauthier, R. J.: Recognition of color lights by color defective individuals, Can. J. Ophthal. **3**:224-253, 1968.
Sjogren, H.: New series of test cards for determining visual acuity in children, Acta Ophthal. **17**:67-68, 1939.
Sliney, D. H., and Fraesier, B. C.: Evaluation of optical radiation hazards, Appl. Optics **12**:1-24, 1973.
Sloan, L., and Naquin, H.: A quantitative test for determining the visibility of the Haidinger brushes; clinical applications, Am. J. Ophthal. **40**:393-406, 1955.
Sloan, L., and Rowland, W. M.: Comparison of Ortho-Rater and Sight-Screener tests of heterophoria with standard clinical tests, Am. J. Ophthal. **34**:1363-1375, 1951.
Sloane, A. E., editor: Refraction in children, vol. 2, International Ophthalmology Clinics, Boston, 1962, Little, Brown & Co.
Sloane, A. E., and Gallagher, J. R.: A comparison of vision screening tests with clinical examination results, Am. J. Ophthal. **35**:819-830, 1952.
Sloane, A. E., and Rosenthal, P.: School vision testing, Arch. Ophthal. **64**:763-780, 1960.
Snell, A. C.: A treatise on medicolegal ophthalmology, St. Louis, 1940, The C. V. Mosby Co.
Stuart, J. A., and Burian, H. M.: A study of separation difficulties; its relationship to visual acuity and in normal and amblyopic eyes, Am. J. Ophthal. **53**:471-477, 1962.
Sugar, H. S.: Binocular refraction with cross cylinder technique, Arch. Ophthal. **31**:34-42, 1944.
Swan, K. C.: Esotropia following occlusion, Arch. Ophthal. **37**:444-451, 1947.
Tiffin, J.: Industrial psychology, New York, 1942, Prentice Hall, Inc.
Turville, A. E.: Outline of infinity balance, London, 1946, Raphael's Ltd.
Unsworth, A. C.: A discussion of ocular malingering in the armed services, Am. J. Ophthal. **28**:148-160, 1945.
Urist, M.: Fixation anomalies in amblyopia ex anopsia, Am. J. Ophthal. **52**:19-28, 1961.
van Wien, S.: The Leland refractor, Arch. Ophthal. **24**:104-111, 1940.

Verhoeff, F. H.: Description of a reflecting phorometer, Am. J. Physiol. Opt. **7**:39-41, 1926.

Victor, M.: Tobacco-alcohol amblyopia: a critique of current concepts of this disorder, Arch. Ophthal. **70**:313-318, 1963.

von Noorden, G. K.: Classification of amblyopia, Am. J. Ophthal. **63**:238-244, 1967.

von Noorden, G. K.: Some aspects of functional versus organic amblyopia. In The First International Congress of Orthoptics, St. Louis, 1968, The C. V. Mosby Co., pp. 115-118.

von Noorden, G. K.: Current concepts of amblyopia. In Fells, P., editor: First Congress of the International Strabismological Association, St. Louis, 1971, The C. V. Mosby Co., pp. 197-214.

von Noorden, G. K.: Factors involved in the production of amblyopia, Br. J. Ophthal. **58**:158-164, 1974.

von Noorden, G. K., and Helveston, E. M.: The influence of ocular position on fixation behaviour and visual acuity. In Fells, P., editor: First Congress of the International Strabismological Association, St. Louis, 1971, The C. V. Mosby Co., pp. 137-148.

von Noorden, G. K., Springer, F., Romano, P., and Parks, M.: Home therapy for amblyopia, Am. Orthopt. J. **20**:46-50, 1970.

Wald, G., and Burian, H. M.: Dissociation of form vision and light perception in strabismus amblyopia, Am. J. Ophthal. **27**:950-963, 1944.

Weale, R. A.: The aging eye, New York, 1963, Harper & Row, Publishers.

Woo, G.: A report on the refractive error distribution of the nonamblyopic eyes in unilateral amblyopes. Can. J. Optom. **32**:101-104, 1970.

Yasuna, E. R., and Green, L. S.: An evaluation of the Massachusetts Vision Test for visual screening of school children, Am. J. Ophthal. **35**:235-240, 1952.

16

Accommodation

Accommodation is one of those subjects about which much that is supposed to be known has yet to be discovered. The anatomy is controversial, the mechanics theoretical, the innervation doubtful, the stimulus debated, the resting state in flux, the pharmacology uncertain, the clinical syndromes ineffectual, and the treatment empirical. Aside from these limitations, there is enough material to fill a chapter in most textbooks on refraction, including this one.

The optical function of the eye is to form clear images on the retina. Since only one object plane can be conjugate to the retina at any time, accommodation is required whenever the target distance changes. In addition, the eyes converge and the pupil constricts in a triad of responses. In refraction, lenses are used to alter the stimulus to accommodation, prisms are used to alter convergence, and the effect of one on the other is evaluated. The association is manifestly disturbed in such conditions as accommodative esotropia or pseudomyopia. Less evident and more subtle are milder motility disturbances that may still result in discomfort, blurred vision, or intermittent diplopia. Accommodation is thus of clinical interest not only for its own sake but also for changes it can induce in other ocular functions.

PRACTICAL ANATOMY

The accommodative apparatus consists of the crystalline lens, suspensory ligaments, and ciliary body. The lens, a grossly gelatinous mass, is microscopically composed of fibers arranged in onionlike laminae. Their juncture form suture lines and the zones of discontinuity seen with the slit lamp microscope. New fibers of lower refractive index are added throughout life; the lens periphery never stops growing, and its center dies before we are born. The new fibers are responsible for the increased hyperopia of senility.

Surrounding the lens substance is an elastic capsule, thicker at the equator, accounting for the parabolic shape of the lens. The capsule is the force behind accommodation, the motor that alters lens curvature. Left to its own devices, the lens sags like the proverbial fat boy into a spherical lump; the suspensory ligaments keep it from sagging. The ligaments are firmly attached to the capsule, sometimes obstinately so to the consternation of the cataract surgeon. The ciliary ring, which gives rise to the suspensory ligaments, dilates or constricts by the contraction of the ciliary muscle. The capsule is like a balloon suspended by stringed spokes in the center of a bicycle wheel. When the rim (ciliary ring) expands, tension is exerted on the capsule, and the lens flattens. Near vision results by decreased tension on the capsule, a theory attributed to Helmholtz. Considering the evidence, one may have difficulty understanding why it is still called a "theory." Its only serious rival, proposed by Tscherning at the turn of the century, just survives by textbook repetition.

In Porterfield's *Treatise on the Eye* (1759), one finds the following interesting passage: "Some maintain, that according as objects are at different distances, this (crystalline) humor becomes more or less convex, which does indeed very well account for distinct vision at all distances . . . (they) say that the ligamentum ciliare, which arises all around from the inside of the circle of the choroid where it joins the uvea does by its contraction draw the edge of the crystalline, to which it is attached all around, towards that circle; and by that means makes it broader and flatter than before, when objects are at a distance from the eye; and that when relaxed, the crystalline recovers its convexity by the elasticity of its parts."*

Porterfield does not elaborate on who the "some" were who held this view (a hundred years before Helmholtz), but one was undoubtedly Henry Pemberton, a London physician, who wrote a treatise on the subject in 1719 and coined the term "Se accommodat." Donders dismissed all pre-Young theories as "little more than loose assertions," and Pemberton was forgotten. Even Young's 1801 demonstrations made no splash because we find MacKenzie in his textbook on diseases of the eye four decades later attributing accommodation to a back and forth motion of the lens.

The ciliary muscle in mammals is a smooth muscle composed of fibers classically described in three groups: the longitudinal fibers of Brücke, the circular fibers of Müller,† and the radial fibers. Currently this division is out of fashion in favor of a more syncytial arrangement. The stable origin of the ciliary muscle is at the scleral spur. With accommodation, there is an increase in aqueous drainage, as evidenced by higher C values (facility of aqueous outflow in tonography), a contributing factor to the higher incidence of glaucoma in presbyopes. The ciliary muscle fibers are inserted into the choroid, which is pulled forward, and this resistance contributes to the precision of accommodation.

Of the various possible focusing mechanisms, nature seems to have tried them all. In birds and reptiles, the ciliary body squeezes the lens equator to deform its anterior surface. The lamprey flattens its cornea, and the teleost moves its lens but accommodates for distance. The fruit bat has its visual cells in levels ready for any distance, and the horse's retina is tilted to the same purpose. The cat has a stenopeic pupil and the chambered nautilus, a pinhole. Most remarkable is the four-eyed kingfisher who comes complete with bifocals, one system for aquatic vision and another for aerial vision.

Parabolic lenses are seen only in primates with central capsular thinning. The capsule of lower mammals is uniformly thick, and their lenses are spherical; the primate lens loses its parabolic shape when the capsule is removed.

The ciliary muscle shares with other smooth muscles a double autonomic innervation. When stimulated by a minus lens or near fixation, there is a latent period of about 0.3 second, characteristic of smooth muscle and much longer than for extraocular muscle contraction. Ciliary tonus is influenced also by general autonomic outflow such as emotion and exercise, and it varies in different planes of general anesthesia.

The parasympathetic innervation is ten times more effective than the sympathetic. Sympathetic innervation thus accounts for only one tenth of the range of accommodation. The parasympathetic fibers have their origin in the Edinger-Westphal nucleus, relay in the ciliary ganglion, and reach the eyeball through short ciliary branches. The nerve fibers branch just before entering the muscle into a rich plexus. Participating in this plexus are the sympathetic fibers. Sensory fibers travel with the first division of the trigeminal nerve and originate in the intermuscular connective tissue. The pain of "asthenopia" presumably follows stimulation of these fibers either by muscle stretch or the accumulation of metabolic waste products.

Parasympathetic stimulation results in accommodation and when excessive, as with pilocarpine or eserine, causes accommodative spasm. The pain associated with this spasm is localized within and above the eye, particularly in the region of the eyebrows. It bears little resemblance to the pain of prolonged reading or hyperopia. Parasympathetic block results in cycloplegia as with atropine instillation.

The role of the sympathetic nervous system in accommodation remains controversial,

*From Porterfield, W.: Treatise on the eye, Edinburgh, 1759, G. Hamilton & J. Balfour.

†The Müller is Heinrich the anatomist, not Johannes the physiologist. The palpebral and orbital muscles are also named after Heinrich. Johannes was responsible for the doctrine of specific nerve energies. Brücke worked in Johannes' laboratory. All the names are variously spelled.

ranging from the usual antagonistic autonomic function to no function at all. Subconjunctival or retrobulbar injection (but not topical administration) of epinephrine causes ciliary relaxation, and stimulation of the cervical sympathetic nerves causes a small decrease in the dioptric power of the eye, but these are not likely to have much physiologic significance. It is not known whether the sympathetic fibers innervate the ciliary muscle directly or reduce ciliary mass by arteriolar constriction, thus increasing zonular tension.

Little is known of the supranuclear connections controlling accommodation. An increase in the refraction of both eyes has been obtained by faradic stimulation of area 19 of the occipital cortex. These responses were invariably associated with pupillary constriction and convergence.

The projection of area 19 to the Edinger-Westphal nucleus is by way of the internal corticotectal tract, but the exact fiber connections have not been traced.

When the ciliary ganglion is stimulated electrically and the refractive change measured objectively, the state of accommodation is directly proportional to the frequency for a constant electromotive force. The gradation of muscle contraction thus depends on the number of neurons and the frequency of their discharge.

CLINICAL PHYSIOLOGY

Accommodation includes a peripheral effector (lens, suspensory ligaments, and ciliary body), a sensory discriminator (retina), and a central receiving station (cortex). The retina discriminates image blur and sends this information to the central receiver where, coupled with data from the oculorotary muscles, it is evaluated, and modifying impulses are sent back to the effector. The entire process resembles a servomechanism but, unlike simpler reflexes, requires attention, memory, abstraction, and motivation and thus includes the highest functional levels of the cortex.

Objective measurements of accommodation (e.g., by high-speed, infrared photography of lenticular reflections) show that the response may lag behind the stimulus. The extent of the lag depends on the detail and contrast of the target, the depth of focus, the available amplitude, and the resolution of the eye. The depth of focus is a nonreactive zone (about 0.75 D for average pupil size), which must be exceeded before a response occurs. Inside this zone, no blur is recognized; outside it, the least degree of blur that can be discriminated is about 0.2 D but falls off sharply with progressive degrees of ametropia.

With steady fixation, accommodation fluctuates (about 0.2 D with a frequency of two cycles per second), a fluctuation that increases with dim illumination and in empty fields. It is tempting to view such oscillations as a self-correcting, feedback mechanism, for example, to account for the poor accommodative responses in eyes with amblyopia or subnormal vision, except that the fluctuations decrease rather than increase with small pupil diameters.

Ciliary tonus

Ciliary tonus varies with age, and younger eyes require a greater cycloplegic concentration. Usual clinical methods of evaluating ciliary tonus include comparing the cycloplegic findings, gauging the influence of a fogging lens, contrasting positive and negative relative accommodation, and noting any fluctuations of the retinoscopic neutral point.

After prolonged contraction, as in uncorrected hyperopia, or inactivity, as in uncorrected myopia, the muscle may become hypertonic or hypotonic. That this is rare is illustrated by the fact that hyperopes are not inevitably esophoric, or are myopes inevitably exophoric, and there is no histologic evidence of atrophy or hypertrophy of the muscle. Nevertheless, some hyperopic patients show a high amount of latent hyperopia or even pseudomyopia. Accommodative esotropia responds to accommodative inhibition, and the accommodative theory of myopia is implicit in such treatments as undercorrection, bifocals, cycloplegics, base-in prism, or interdicting near work.

When fixation changes from far to near, the time required is longer than from near to far, although all responses are completed in less than 1 second. The accommodative inertia increases with age and is an early symptom of presbyopia.

The notion that accommodation is a clue to distance was first advanced by Berkeley. He thought that the visual apparatus gauges distance by a kind of introspective strain felt as a result of ciliary contraction. The same problem became Wundt's maiden effort in experimental psychology, and he concluded that, in fact, convergence was more important because the binocular threshold was so much lower than the monocular in distance discrimination. As is well known, the ability to judge absolute distances, such as the location of a briefly illuminated light in a dark room viewed monocularly, is extremely poor (for a 22-meter distance, the range can be from 12 to 30 meters). Moreover there is little evidence for proprioception from the ciliary muscle. There may be a monitoring of the innervation, but more likely the feedback is visual by means of retinal blur.

Stimulus to accommodation

Under normal-seeing conditions all the monocular cues such as overlap, geometric and aerial perspective, texture and density gradients, illumination, retinal blur, perceived distance, as well as such binocular factors as convergence and disparity, operate to stimulate accommodation. It requires a laboratory situation to study them individually.

When an isolated target is brought nearer one eye, not only does the retinal image blur, but its angular size increases as well. This increased magnification contaminates all "push-up" amplitude measurements unless taken through a Badal setup, which is seldom clinically feasible. By keeping target distance constant and interposing lenses, only retinal blur is altered. Accommodation is practically always correct; positive to minus lenses or negative to plus lenses, providing illumination is good, amplitude is adequate, and the blur is not excessive. Yet by some inherent wisdom, accommodation remains passive when the target itself is blurred, for example, by defocusing the projector. Fincham (1951) investigated this ability of the eye to distinguish between plus blur, minus blur, and target blur. He found that one criterion by which the eye makes the correct discrimination is chromatic aberration. Since most targets are polychromatic, a plus lens brings long wavelengths into sharper focus, and a minus lens brings short wavelengths into better focus, whereas target blur does not alter either. When the experiment was repeated under monochromatic light, the accommodative response of 60% of his subjects was indeed in error. How does one account for the correct responses in the remaining 40%? Fincham postulated that scanning eye movements help by activating the Stiles-Crawford effect. Since the blur induced by a plus lens causes the rays to cross anterior to the retina, whereas minus lenses make them cross behind, the photoreceptors are stimulated at a different obliquity in each case, and this guides the response in the correct direction.

Retinal blur is not an inevitable stimulus to accommodation, however, as illustrated by comparing the effect of a fogging lens on atropinized and unatropinized eyes. The acuity under cycloplegia is found to be better on the average, indicating that some eyes accommodate in response to a plus lens, others relax, and still others remain passive. Correlated with these three kinds of accommodative responses are three types of accommodative vergence changes, demonstrated by a decrease, increase, or no change in the lateral phoria. Prolonged wearing of the fogging lens does not seem to alter the characteristic response; hence the phoria provides a measure of the effectiveness of fogging. These results throw some doubt on the validity of equalizing accommodation because failure to obtain identical acuity may not be due to improper refraction but unequal fogging.

With excessive fogging, the accommodative response becomes progressively more inaccurate, ending in the positive contraction of "empty-space myopia". It follows that to obtain a predictable accommodative response in clinical refraction, targets should be adequately illuminated and provide appropriate detail, with enough contrast for the available ocular resolving power.

Although it has been known for some time that accommodation stimulates convergence, the converse has been denied until recently. The reason it is not easily demonstrable is that whenever accommodation is caused to change by convergence, a blur results, which in turn alters the accommodation. This self-correcting mechanism requires special experimental designs to circumvent. One way is to photograph the lenticular change at high speed using infrared light or a small artificial pupil that so greatly enchances depth of focus as to make any blur due to ciliary change invisible. Such experiments reveal that the change in accommodation is about 0.2 D per prism diopter. Unlike the AC/A ratio, this ratio decreases with age. Convergence accommodation has been proposed as the primary stimulus to accommodation, with retinal blur as a secondary "fine-focus" adjustment.

When a target approaches the eye, there are changes in visual angle, brightness, overlap, elevation, detail, and perspective. The totality creates a proximal effect, which can be reproduced to some extent by merely "thinking" of nearness or farness. We are taught to do this in ophthalmoscopy so as to avoid proximal accommodation and convergence, but even experienced observers fail when using a stereoscope set for infinity (in fact, stereogram designers seldom set their targets for infinity). The same "mental effort" is used in orthoptic treatment of esotropia and exotropia.

Resting state of accommodation

The resting state of accommodation is currently a puzzle. Most clinicians assume that accommodation falls to zero when an emmetropic eye or an adequately corrected ametropic eye fixates a distant object. Although some additional accommodation may be inhibited by cycloplegia, this is seldom corrected. If we think of accommodation as a reflex, a true zero input must mean more than simply distance fixation, which still involves active vision. Zero input requires reduced illumination or reduced contrast; however, under these circumstances, accommodation, instead of relaxing, becomes positive to the extent of about 0.65 D measured objectively. This is the night myopia or empty-space myopia described previously. It means that the clinician who pushes plus in his refraction may actually stimulate accommodation, not relax it as he perhaps intended.

Physical and physiologic accommodation

A distinction between physical and physiologic accommodation was proposed by E. Fuchs. Flieringa and van der Hoeve (1924) suggested the term "myodiopter" to represent the ciliary contraction producing 1 D of physical accommodation under normal conditions. A half-paralyzed ciliary muscle thus exerts 2 myodiopters for 1 D of lenticular change. The distinction is the basis of the two views of presbyopia; either fewer myodiopters are used to change the less elastic lens and more ciliary muscle power becomes latent, or a greater number of myodiopters are required to produce unit change in accommodation with increasing age.

MEASUREMENT OF ACCOMMODATION

Clinical measurements of accommodation, whether objective or subjective, have limited accuracy because of many variables not easily controlled in practice. Illumination, depth of focus, target size, contrast, visual angle, lens effectivity, monocular and binocular cues, kinesthetic sense, rate of change, and individual variations in inertia and amplitude have their effects. Since most clinical measurements are subjective, the motivation to keep the target clear and the interpretation of partially blurred targets add additional weight to the scale of individual differences. For example, the decrease in amplitude with age according to Donders and Duane is given in most textbooks. It is an optical infirmity common to both men and women, emmetropes and ametropes, diagnosed and treated by every refractionist worthy of the name.

Yet when depth of focus is controlled, the debility is found to occur at least twenty years earlier than Donders taught, and ten years earlier than Duane measured.

Amplitude of accommodation

The amplitude of accommodation is the dioptric difference between the far point and near point of the eye. The linear distance between these two points is the range of accommodation. Clinically the far point is initially unknown; hence it is conventional to correct the refractive error first. Despite the correction, more accommodation is required by the hyperope than the myope because of the effectivity of the spectacle lens (Chapter 7). Moreover the distance correction itself varies with the criterion the clinician has set for himself in defining emmetropia. If he pushes plus to 20/20, he will measure a greater amplitude than if he prescribes for 20/15 acuity. Since the far point and the near point represent the two extreme limits of the depth of focus, the true amplitude is generally from 0.50 to 1.50 D lower, depending on illumination and pupil size. Thus a presbyopic add measured under good light turns out to be inadequate at a poorly lit workbench. And the amplitude is higher if the patient holds the target himself because of reinforcing kinesthetic clues.

A variety of targets are used to measure accommodation. To prevent the lag commonly found with gross targets or those that permit blur interpretation, Duane recommended a single, fine thread, and Landolt used a slit in an internally illuminated cylinder. Two closely spaced lines also work well because when the target blurs, a third line seems to appear between them. Because such targets require high discrimination, they may actually be too demanding; hence most clinicians use letters as a compromise between controlled stimulation and everyday seeing. Because this compromise is not always achieved, overcorrecting presbyopia is the most common error made by beginning refractionists.

It is conventional to measure amplitude by (1) moving a target toward the eye until it blurs or (2) keeping the target distance constant and interposing minus lenses. Each has its limitations. The first, or "push-up," method does not allow for the increased visual angle, which makes the target recognizable although blurred. In young patients, illumination decreases drastically as the target approaches the eye (it is not possible to measure more than 10 to 12 D adequately with a refractor). If the reciprocal is measured as with the Prince ruler, a small linear distance error (easily due to depth of focus or reaction delay) leads to large errors. For example, a near point of 10 cm means 10 D amplitude, but a near point of 8 cm means 12.5 D amplitude—a 2 cm difference corresponds to 2.5 D accommodation. Finally, almost every patient thinks this test mildly silly and fails to maintain maximum focus—which can be easily checked by repeating it several times. Children generally do better when the target is moved from blur to clear, whereas with adults both approaches should be averaged. The clinician must decide whether to use an end point of first blur or blur out. It does not matter which one he chooses as long as he is consistent.

In the minus lens to blur technique, the test distance is 40 cm or whatever is suitable, and the reciprocal is added to the measured amplitude. Thus if the target blurs with −4.00 D at 40 cm, the amplitude is recorded as $4 + 2.50 = 6.50$ D. In presbyopia, an auxiliary plus lens may be required to bring the target within range, and this is deducted from the measured amplitude. In all cases, the fixation is tied to the target distance so that convergence cannot be mustered to bring in additional accommodation. In addition, there is some minification from the lenses, and if the test is done with the refractor, the instrument cannot be easily altered to the appropriate PD or depressed to the usual reading position. Thus the minus to blur technique is not comparable to the push-up method.

Relative accommodation

The amplitude measures the quantitative aspect of accommodation; there is also a qualitative aspect representing how much and how long accommodation can be comfortably sustained. This qualitative aspect is implicit in the notion of leaving a certain proportion of the amplitude in reserve (values range from one third to two thirds). One method by which it is clinically evaluated is the relative freedom of accommodation and convergence.

In normal binocular vision, it is possible to place plus or minus lenses before the eyes and maintain single vision. Similarly, base-out and base-in prisms of limited strength can be placed before both eyes without blurring the target. The lenses measure relative accommodation, the prisms measure relative convergence. Donders first made experimental measurements of these limits and plotted the results graphically to define the zone of clear single binocular vision. It can be readily demonstrated, however, that there is always a certain amount of convergence induced by accommodation. This is manifest by an increased esophoria or decreased exophoria under dissociation. The absence of diplopia in relative accommodation measurements does not mean that the innervation to convergence remains constant but rather that the increasing convergence is neutralized by negative fusional convergence (or fusional divergence). When negative fusional convergence is exhausted, no further increase in accommodation is possible without diplopia. Rather than accept double vision, most patients stop accommodating, and the target blurs. Positive relative accommodation is therefore an indirect measure of negative fusional convergence, and negative relative accommodation is an indirect measure of positive fusional convergence. Indeed the correlation between them is linear and fairly high. The interval between first blur and blur out in response to plus or minus lenses with fixed convergence represents the effects of depth of focus and blur interpretation—factors that contaminate all accommodative measurements.

Negative relative accommodation (NRA) is measured by adding plus spheres binocularly at the reading distance (usually 40 cm). Lenses are added in 0.25 D steps until the target (a fine line of print) first begins to blur. A time allowance is made for "clearing" the target. Accommodative convergence is relaxed in proportion to the positive fusional vergence. Average value for 40 cm is +2.00 D ± 0.50 D. If the distance correction has been adequately measured, the NRA cannot exceed +2.50 D at a 40 cm testing distance.

Positive relative accommodation (PRA) is more difficult to measure because it is influenced by amplitude, accommodative insufficiency, and the motivation to keep the target clear. Minus lenses are added binocularly in 0.25 D steps until the target blurs. The limit is set by negative fusional convergence. Average values at 40 cm are −2.37 D ± 0.50 D for nonpresbyopes. In both the PRA and NRA tests, the illumination should not be so excessive as to artificially improve the results. The patient must not confuse the change in perceived size with target blur, and presbyopes may require additional plus help, which is subtracted from the PRA and added to the NRA. The sum of the PRA and NRA is not a measure of the amplitude of accommodation but an indirect measure of the amplitude of fusional vergences at that test distance.

Accommodative flexibility

An indication of the accommodative flexibility may be obtained by inserting a −1 D and +1 D lens binocularly and noting the time it takes to clear the target. An increased accommodative inertia is characteristic of early presbyopia.

Dynamic retinoscopy

Dynamic retinoscopy is an objective method of evaluating the accommodated eye. First suggested by Cross (1911), the technique was further elaborated by Sheard (1920), Tait (1929), Pascal (1941), and others. The attractive simplicity of the method is somewhat offset by the fact that no one is quite sure what to do with the results once they have been obtained.

The test is performed binocularly through the distance correcting lens, with the patient instructed to fix the retinoscope or an attached target. Each eye is scanned briefly, the examiner working at his usual distance. One would now expect to see neutral, since the patient's retina is conjugate to the instrument, but instead there is a with motion requiring from +0.25 to +1.50 D to neutralize. This is the "low neutral point." The with motion indicates a "lag" of accommodation behind convergence, that is, an initial hyperopia for the near fixation distance despite well-illuminated targets and adequate amplitude. Its exact significance is still undecided, since some is accommodative insufficiency, some is spherical aberration, and some is the result of depth of focus.

Once the low neutral point is identified, continue to add plus sphere (binocularly) to obtain reversal. Generally, the neutral point does not reverse for a number of lens changes, and an additional +0.50 to +3.00 D may be needed to produce an against motion. This is the "high neutral point." It is interpreted as negative relative accommodation proportionate to the addition of plus lenses. (It is assumed that there is no "latent" refractive error.) Unfortunately the high neutral point does not correlate well with other measures of NRA, and much of the controversy about dynamic retinoscopy centers around whether it really measures latent ametropia or the accommodation-convergence relationship.

Early proponents of the technique in their enthusiasm for an objective measure of accommodation probably overstated the case. Some give as many as twelve different applications of dynamic retinoscopy on the assumption, one suspects, of hitting a small target with a shotgun scatter. Among these are methods of determining the static refraction from the dynamic findings, the uncovering of latent refractive error, the detection of astigmatic and unequal accommodation, and the evaluation of cycloplegia. To measure accommodative amplitude the examiner moves closer to the patient until an against motion is obtained. The distance is the near point. An initial against motion, with fixation at the plane of the retinoscope, is supposed to indicate accommodative spasm or a myopic overcorrection. Since accommodative spasm is rare, the other is more likely. Dynamic retinoscopy has also been used to determine the presbyopic add but has no advantage over other methods because it depends on the patient accurately focusing the target, which happens to be attached to the instrument. One application of dynamic retinoscopy that suprisingly is seldom used is in strabismus. In measuring tropia at distance and near, it is useful to determine whether (even with cycloplegia) there is an accommodative element that contributes to the angle of squint. Adding a +3.00 D sphere (the usual technique) still leaves one in the dark concerning whether the patient really relaxes accommodation, and retinoscopy is a helpful objective measure of residual accommodation.

ANOMALIES OF ACCOMMODATION

Clinical measurements of accommodation may reveal responses that are normal, insufficient, excessive, absent, sluggish, or unequal. These findings may be accompanied by symptoms of blurring, sleepiness, discomfort, and failure of performance or comprehension, aggravated by continued use of the eyes for near tasks. Such symptoms are classically grouped under the heading of "asthenopia." If the correction of ametropia is the prerequisite of clear vision, the treatment of accommodative anomalies is generally considered indispensable for comfortable vision.

Accommodative fatigue

Explicit in the term "asthenopia" and implicit in keeping a certain proportion of the accommodative amplitude in reserve is the notion of accommodative fatigue. Although there is no constellation of refractive findings characteristic of asthenopia, the symptoms are fairly clear. The eye has a normal appearance, the ocular movements are not restricted, the convergence of the visual lines present no difficulty, and the acuity is good; but in reading, writing, and other close work, especially by artificial light or in a dim place, the object after a short time becomes indistinct and confused, and a feeling of fatigue and tension sets in, especially above the eyes, necessitating a suspension of work. The person so affected may involuntarily close his eyes, rubs his hand over the forehead and eyelids, and after some moments' rest, once more sees distinctly. But the same phenomena soon reappear more rapidly than before.

Asthenopics, wrote Donders, have sometimes a sad past and live in a gloomy future. The gloom was due to the opposition to convex lenses by most oculists. Initially confused with amblyopia (e.g. Taylor's debilitus visus), it was still attributed to atonic retinas and congested choroids by Lawrence and von Graefe, and Mackenzie considered the disease incurable. The notion that the cause was a spasmodic contraction of the extraocular muscles led to various forms of tenotomies. Bohm (1845) first unconditionally recommended (blue) convex lenses, and Stellwag von Carion referred asthenopia exclusively to presbyopia. Donders believed he had discovered the cause in the hyperopic structure of the eye, and the supposed anomaly of accommodation became an anomaly of refraction. Fifty years later, there is a subtle shift of emphasis from muscular to nervous energy. Gould in his *Biographic Clinics* pointed to the possibility that such men as Darwin and Huxley died before their time because they had wasted their nerve energy through the eyes. This truncated history illustrates the various trends that still permeate our thinking about refractive errors in general, and eyestrain in particular.

If eyestrain is the symptomatic manifestation of accommodative overactivity, it is useful to separate fatigue from impairment. Fatigue is a subjective experience, dependent on boredom, frustration, and poor motivation. Impairment is work decrement, an objective manifestation subject to measurement. The fatigue may predate, coexist, or follow the impairment.

Various techniques have been used to induce accommodative impairment: prolonged reading, fine print, poor contrast, decreased illumination, close working distances, glare sources, flicker, lowered oxygen tension, noise, and other distractions. Ergographic techniques of repeated "push-up" amplitude to fatigue accommodation were a disappointment, taken objectively or subjectively. Objective criteria such as recession of the near point, speed, accuracy, blink rate, critical fusion frequency, heart rate, and galvanic skin response, showed both an increase and a decrease. Some subjects showed a more remote near point, attributed to fatigue, others a closer near point, attributed to practice. By any subjective criterion, a significant proportion of subjects are "tired" before they start or after a few trials. The symptoms depend more on motivation than motor decrement and as much on inclination as on impairment. Thus clinical symptoms occur with low as well as high hyperopia presumably because of the disturbed association between accommodation and convergence. Presbyopes, who should provide the classic example of accommodative fatigue, are in fact generally asymptomatic except for the frustration in attempting to read. The reason given is that since the lens is sclerotic, no additional muscular effort is used to see clearly (there is both evidence for and against this view as we see later).

A diagnosis of accommodative asthenopia therefore must be made with circumspection. Fatigue may arise from diverse causes, some only remotely connected with ciliary activity and slight errors of refraction may or may not bear a proportional relationship to symptoms. Unquestionably a considerable number of patients obtain relief from minimal refractive corrections impossible to explain on an energy basis. Whether the effect is caused by eliminating a trigger mechanism, muscle imbalance, aniseikonia, fixation disparity, restoring the equilibrium of an unstable autonomic nervous system—or an unstable personality—or simply a fortuitous placebo effect, which encourages better ocular hygiene, remains to be determined. A neurotic element is no doubt often involved, which may only mean an anxious, tense, or frightened patient.

Accommodative spasm

The term "accomodative spasm" refers to an undesirable increase in accommodation beyond that required for clear vision. The condition may occur as an isolated phenomenon, including miosis and convergent strabismus, which responds to cycloplegics. In one case of a 12-year-old girl, the spasm followed a viral upper respiratory infection. She was treated with atropine, which straightened the eyes (from 40Δ eso). Medication was continued for 2 months, and the condition resolved spontaneously with no sequelae.

Chronic, uncorrected hyperopia may lead to accommodative spasm. The ciliary hypertonicity has been classified as tonic if constant and clonic if intermittent. The pupil is often normal, the condition is binocular, and treatment consists of correcting the associated refractive error.

Accommodative spasm also occurs in iritis, cyclitis, corneal disease, irritative lesions of the trigeminal nerve, cyclic oculomotor spasm, and after the instillation of parasympathomimetics and anticholinesterases. The diagnosis is based on signs and symptoms of the associated disease or the history of drug instillation.

Symptoms of accommodative spasm include intermittent blurred vision, ocular discomfort, headache, brow ache, photophobia, and diplopia. Clinical findings usually show hyperopia, reduced amplitude of accommodation, miosis, esophoria or esotropia, a variable neutral point in retinoscopy, and unexpected acuity responses to fogging lenses. The diagnosis is established by comparing cycloplegic and postcycloplegic findings.

Pseudomyopia is a rare condition attributed to accommodative spasm, in which the patient appears to be myopic. Distance vision is decreased and is improved by minus lenses. Unlike axial myopia, there are symptoms of discomfort after the use of the eyes. The diagnosis is easily made with cycloplegia. Although pseudomyopia is rare, it has been postulated as a stage in the development of myopia.

Accommodative insufficiency

In accommodative insufficiency the amplitude is less than anticipated for the patient's age. It may be intermittent or constant, with or without associated ocular disease. A subdivision is a condition in which the accommodation is adequate but cannot be sustained except for short periods.

Functional accommodative insufficiency, or premature presbyopia, is attributed to a variety of causes, mostly unknown.* The symptoms include discomfort, intermittent exotropia, and inability to read for any prolonged period. The signs are a decreased positive relative accommodation, exophoria, and decreased amplitude. A frequent cause is uncorrected or inadequately corrected hyperopia in which a large latent component is missed because of inadequate cycloplegia. The diagnosis is made by evaluating the amplitude, the positive relative accommodation, and the ciliary tonus (over 0.50 D is considered an excess). A qualitative test of accommodative flexibility helps in estimating how long comfortable near vision can be sustained. It should be recalled that presbyopia may have an early onset in certain races and latitudes, probably a reflection of differences in nutrition and life-style.

Paralysis of accommodation

Ciliary muscle paralysis occurs in inflammatory diseases of the uveal tract, acute glaucoma, fascicular and peripheral oculomotor palsies, diabetic and toxic neuropathies, and nuclear and supranuclear lesions of the brain stem. The most common cause used to be diphtheria; today it is blunt trauma with associated paralytic mydriasis and iris tear or inadvertent cycloplegia. Contrary to what one might expect, it is rarely found in systemic muscle disease.

Accommodative strabismus

The connection between esotropia and accommodation has been known for over a century but it was Donders who clearly established the relation to hyperopia. Convergent squint, he emphasized, almost always depends on hyperopia, and divergent squint is usually the result of myopia. The problem is, of course, that although many squinters are hyperopes, many more hyperopes are nonsquinters.

*Partial cycloplegia can be produced by a variety of systemic medications as described in Chapter 11.

To explain the discrepancy, a host of theories have been proposed by both major and minor ophthalmologic prophets, ranging from the plausible to the ludicrous. They range from de la Hire's malplaced maculas and Buffon's ocular inequality to Dieffenbach's disproportionate extraocular muscle function, Donders' hyperopia, Alfred Graefe's mechanicoelastic predominance of one extra ocular muscle group, Priestley-Smith's disturbance of cortical nervous centers, Claud Worth's congenital fusion faculty deficiency, Chavasse's obstacles to fusion reflexes, with many minor variants in which are implicated vertical imbalances, unrecognized palsies, aniseikonia, anisometropia, amblyopia, abnormal AC/A ratio, suppression, spasmogenic tendencies, loss of nervous energy, and anomalous coupling of corresponding points. Some surgeons have come full circle, back to the quasimechanical defects of individual extraocular muscles. Compounding the difficulties, a terminology has grown up, surrounding the subject with a deadening effect on any original thought.

Whatever the something plus that induces some hyperopes to become squinters, it is clear that it cannot be the hyperopia per se. The amount of deviation is not proportional to hyperopia; many esotropes are emmetropes or even myopes. The angle of deviation is often greater than that usually associated with the accommodation needed to overcome the hyperopia. One should expect the deviation to fluctuate with each fixation distance, to disappear when accommodation is inhibited, to be greater at near than at distance, and to vary with amplitude. To the extent that these effects hold true, the case is classified as typical; to the extent that they do not, it is an atypical esotropia. When one cannot specify, one can always classify; if no cow is available, one must make do with a goat.

Surgical experience demonstrates anatomic variations in size, shape, and insertion of extraocular muscles, but whether these are causes or results of the deviation remains unclear. It is difficult to explain how a mechanical defect preserves the symmetry of deviated movement.* A mechanical cause would affect versions and alter the position of rest. Indeed this concept is invoked to explain why older patients who lose the vision in one eye develop exotropia, whereas younger ones (under 15 years) usually develop esotropia. An unrecognized muscle palsy (for example, after some febrile disease) that is subsequently resolved but leaving a secondary contracture would account for the residual deviation under cycloplegia, with underaction or overaction of certain muscles. The onset would be sudden, associated with diplopia, with some initial evidence of commitance—all presumably masked by the early age at which it occurs. Mechanical factors do not explain intermittent squint, variations in the angle of deviation with cycloplegia or lenses or in sleep and under anesthesia. Moreover the idea does not tie in well with the high, familial incidence of strabismus. A central (innervational) etiology would account for intermittency and the changes in distant and near fixation, correcting hyperopia, the influence of anesthetics, or the effects of such central-acting drugs as diphenylhydantoin (Dilantin). A nervous hyperirritability is indeed present in some but not all squinters. Yet innervational changes can be induced with lenses and prisms without inducing squint, so an additional factor must therefore be present, generally referred to as the "fusion sense," or "compulsion to fusion." It may be a sensorial faculty or a motor reflex (probably the latter). Thus even excessive accommodation and excess convergence do not lead to strabismus if fusional vergences are available to compensate. We do not know why these reflexes are not mobilized in squinters. On the other hand, the absence of fusion does not lead inevitably to a deviation (as in cases of unilateral blindness without squint) if the position of rest is reasonably orthophoric. To establish which mechanism or which combination of mechanisms is operative in a particular situation taxes the ingenuity of the clinician who knows from experience the time and trouble necessary, especially in young children, to satisfactorily investigate the function of the eyes. Superimposed on

*There is no case of concomitant squint in which the angle of deviation does not show slight variations with different directions of gaze due to mechanical kinematics favoring depression or elevation.

the deviation itself are so-called adaptive mechanisms, including amblyopia, suppression, and anomalous correspondence. Anomalous correspondence is a peculiar phenomenon in which a new physiologic coupling is supposed to occur between the fovea of the fixing eye and some extrafoveal fixation point in the deviated eye. The peculiarity is that no squinter ever actually makes use of it to regain a new kind of binocular vision.

The "typical" case of accommodative esotropia develops around age 2 years; the deviation is greater at near than at distance; there is moderate hyperopia (usually with anisometropia); the AC/A ratio (measured by comparing the distance and near deviations) is "normal"; and the squint is intermittent (at least in the early stages), with eventual regional suppression and amblyopia. The diagnosis is made by the improvement in the angle of squint with cycloplegia or plus lenses. The differential diagnosis includes pseudostrabismus (large angle "kappa" and prominent epicanthal folds), Duane's syndrome, and nonaccommodative or "essential" esotropia in which the angle is usually larger, of earlier onset, and uninfluenced by cycloplegia.

Accommodative esotropia frequently deteriorates, for reasons unknown, into a situation in which accommodative inhibition no longer modifies the deviation. This means the diagnosis must be established early, preferably in the stage of intermittency (by a reliable refraction) and before amblyopia has become established. The treatment in early stages is medical rather than surgical, including cycloplegia, convex lenses (with or without clip-ons or bifocals), miotics, and occasionally central-acting drugs. Amblyopia is diagnosed by eccentric fixation and reduced vision and is treated by occlusion.

Miotics

Both direct-acting parasympathomimetics (pilocarpine) or indirect-acting anticholinesterases (isoflurophate, echothiophate iodide, and demecarium bromide) may be used in accommodative esotropia. Pilocarpine has a direct effect on the ciliary muscle, reducing the required innervation. Anticholinesterases are more effective and enhance the effect of normal acetylcholine or accommodation. Since the convergence is tied to accommodation by a central synkinesis, less convergence is stimulated, and the response AC/A ratio is altered. The anticholinesterases have, however, some undesirable side effects, including burning, hyperemia, myopia, formation of iris cysts, lens opacities, and even retinal detachment; hence the dosage must be carefully regulated and the patient checked periodically.

REFERENCES

Abraham, S. V.: The use of miotics in the treatment of convergent strabismus and anisometropia, Am. J. Ophthal. **32**:233-240, 1949.

Adamson, J., and Fincham, E. F.: Effect of lenses and convergence upon state of accommodation, Trans. Ophthal. Soc. U. K. **59**:163-179, 1939.

Allen, M. J.: An investigation of the time characteristics of the eyes—historical review, Am. J. Optom. **30**:78-83, 1953.

Allen, M. J.: The stimulus to accommodation, Am. J. Optom. **32**:422-431, 1955.

Alpern, M.: Variability of accommodation during steady fixation at various levels of illuminance, J. Opt. Soc. Am. **48**:193-197, 1958.

Alpern, M.: Accommodation. In Davson, H., editor: The eye, vol. 3, ed. 2, New York, 1969, Academic Press, Inc., pp. 217-254.

Alpern, M., Ellen, P., and Goldsmith, R. I.: The electrical response of the human eye in far-to-near accommodation, Arch. Ophthal. **60**:592-602, 1958.

Arnulf, A., Dupuy, O., and Flamant, F.: Les microfluctations d'accommodation de l'oeil, Ann. Opt. Ocul. **3**:109-118, 1955.

Asher, H.: Stimulus to accommodation in normal and asthenopic subjects, Br. J. Ophthal. **36**:666-675, 1952.

Baldwin, W. R.: Accommodative characteristics of a group of myopic adults, Am. J. Optom. **42**:237-247, 1965.

Berens, C., and Sells, S. B.: Experimental studies of fatigue of accommodation, Am. J. Ophthal. **33**:47-58, 1950.

Biersdorf, W. R.: The utility of the anterior lens Purkinje image as a measure of accommodation, Am. J. Optom. **37**:352-362, 1960.

Biggs, R. D., Alpern, M., and Bennett, D. R.: The effect of sympathomimetic drugs upon the amplitude of accommodation, Am. J. Ophthal. **48**:169-172, 1959.

Breinin, G. M., Chin, N. B., and Ripps, H.: A rationale for therapy of accommodative strabismus, Am. J. Ophthal. **61**:1030-1037, 1966.

Brown, R. H.: "Empty-field" myopia and visibility of distant objects at high altitudes, Am. J. Psychol. **70**:376-385, 1957.

de Buffon, G. L.: Sur le cause due strabisme ou des yeux louches, Memoire de l'Academie des Sciences, Paris, 1743.

Burch, P. G.: Accommodation during general anesthesia, Arch. Ophthal. **81**:202-206, 1969.

Campbell, F. W., and Primrose, J. A. E.: The state of

accommodation of the human eye in darkness, Trans. Ophthal. Soc. U. K. **73**:353-361, 1953.
Campbell, F. W., Robson, J. G., and Westheimer, G.: Fluctuations of accommodation under steady viewing conditions, J. Physiol. (London) **150**:579-594, 1959.
Campbell, F. W., and Westheimer, G.: Dynamics of accommodation responses of the human eye, J. Physiol. (London) **151**:285-295, 1960.
Cogan, D. G.: Accommodation and the autonomic nervous system, Arch. Ophthal. **18**:739-766, 1937.
Cornsweet, T., and Hewitt, C.: Servo-controlled infrared optometry, J. Opt. Soc. Am. **60**:548-554, 1970.
Cross, A. J.: Dynamic skiametry in theory and practice, New York, 1911, A. J. Cross Publishing Co.
Donders, F. C.: Accommodation and refraction of the eye, London, 1864, The New Sydenham Society.
Duane, A.: Normal values of the accommodation at all ages, Transactions of the Section on Ophthalmology of the American Medical Association, pp. 383-391, 1912.
Duane, A.: Anomalies of accommodation clinically considered, Trans. Am. Ophthal. Soc. **14**:386-398, 1915.
Duane, A.: Studies in monocular and binocular accommodation with their clinical applications, Am. J. Ophthal. **5**:865, 1922.
Duane, A.: Are the current theories of accommmodation correct. Am. J. Ophthal. **8**:196, 1925.
Duane, A.: Subnormal accommodation, Arch. Ophthal. **54**:566-587, 1925.
Fincham, E. F.: The functions of the lens capsule in the accommodation of the eye, Trans. Opt. Soc. London **30**:101-117, 1928.
Fincham, E. F.: The accommodation reflex and its stimulus, Br. J. Ophthal. **35**:381-393, 1951.
Fincham, E. F.: The reflex reaction of accommodation. In Transactions of the International Optical Congress, London, 1951, British Optical Association, pp. 105-114.
Fincham, E. F.: The proportion of ciliary muscle force required for accommodation, J. Physiol. (London) **128**:99-112, 1955.
Fisher, R. F.: Elastic constants of the human lens capsule, J. Physiol. (London) **201**:1-19, 1969.
Fleming, D. G.: The role of the sympathetics in visual accommodation, Am. J. Ophthal. **43**:789-794, 1957.
Fleming, D. G.: A mechanism for the sympathetic control of visual accommodation, Am. J. Ophthal. **47**:585-586, 1959.
Flieringa, H. J., and van der Hoeve, J.: Arbeiten aus dem Gebiete der Akkommodation, Graefe's Arch. Ophthal. **114**:1-46, 1924.
Flom, M. C., and Goodwin, H. E.: Fogging lenses: differential acuity response in the two eyes, Am. J. Optom. **41**:388-392, 1964.
Fry, G. A.: Indirect skiametry for determining static refraction, Optom. Weekly, pp. 499-500, June 3, pp. 527-529, June 10, 1943.
Goldmann, H., and Aschmann, A.: Studien über Akkommodation, Ophthalmologica **111**:182-186, 1946.
Grant, V. W.: Accommodation and convergence in visual space perception, J. Exp. Psychol. **31**:89-104, 1942.
Heath, G. G.: Accommodative responses of totally color blind observers, Am. J. Optom. **33**:457-465, 1956.
Heath, G. G.: Components of accommodation, Am. J. Optom. **33**:569-579, 1956.
Heath, G.: The influence of visual acuity on accommodative responses of the eye, Am. J. Optom. **33**:513-524, 1956.
Heine, L.: Die Anatomie des akkommodirten Auges, Graefe's Arch. Ophthal. **49**:1-7, 1899.
Henderson, T.: Anatomy and physiology of accommodation in mammalia, Trans. Ophthal. Soc. U. K. **46**:280-314, 1926.
Hensen, V., and Voelckers, C.: Experimental Untersuchungen über den Mechanismus der Akkommodation, Kiel, 1868.
Hermann, J. S., and Johnson, R.: The accommodation requirement in myopia, Arch. Ophthal. **76**:47-51, 1966.
Hess, C.: Die Refraktion and Akkommodation des menschlichen Auges und ihre Anomalien. In Graefe-Saemisch Handbuch Gesamte Augenheilkunde, ed. 3, Leipzig, 1910, Engelmann.
Hill, R. V.: The accommodative effort syndrome: pathologic physiology, Am. J. Ophthal. **34**:423-429, 1951.
Howe, L.: Measurement of fatigue of the ocular muscles. In Transactions of the Section on Ophthalmology of the American Medical Association, 1912.
Hughes, W. L.: Change of axis of astigmatism on accommodation, Arch. Ophthal. **26**:742-749, 1941.
Ivanoff, M. A.: On the influence of accommodation on spherical aberration in the human eye, J. Opt. Soc. Am. **37**:730-731, 1947.
Jacobsen, J. H., Romaine, H. H., and Halberg, G. P.: The electrical activity of the eye during accommodation, Am. J. Ophthal. **46**:231-238, 1958.
Jampel, R. S.: Representation of the near-response on the cerebral cortex of the macaque, Am. J. Ophthal. **48**:573-382, 1959.
Kirchhof, H.: A method for the objective measurement of accommodation speed of the human eye, translated by H. Knoll and M. J. Allen, Am. J. Optom. **27**:163-178, 1950.
Knoll, H. A.: A brief history of nocturnal myopia and related phenomena, Am. J. Optom. **29**:69-81, 1952.
Knoll, H. A.: An infra-red skiascope and other infra-red ophthalmic research instruments, Am. J. Optom. **30**:346-350, 1953.
Koomen, M., Scolnik, R., and Tousey, R.: A study of night myopia, J. Opt. Soc. Am. **41**:80-90, 1951.
Landolt, E.: The refraction and accommodation of the eye, Philadelphia, 1886, J. B. Lippincott Co.
Luckiesh, M., and Moss, F. K.: Functions of relative accommodation, Am. J. Ophthal. **24**:423-428, 1941.
Ludlam, W. M., Wittenberg, S., Giglio, E. J., and Rosenberg, R.: Accommodative responses to small changes in dioptric stimulus, Am. J. Optom. **45**:483-506, 1968.
Mackenzie, W.: A practical treatise on the diseases of the eye, Boston, 1833, Carter Hendee & Co.

Marg, E.: An investigation of voluntary as distinguished from reflex accommodation, Am. J. Optom. **28**:347-356, 1951.
Mark, H. H.: On the accuracy of accommodation, Br. J. Ophthal. **46**:742-744, 1962.
Martin, H. G.: Practical measurement of accommodation and convergence, Am. J. Ophthal. **22**:406-412, 1939.
Mellerio, J.: Ocular refraction at low illumination levels, Vision Res. **6**:217-237, 1966.
Morgan, M. W.: The nervous control of accommodation, Am. J. Optom. **21**:87-93, 1944.
Morgan, M. W.: The resting state of accommodation, Am. J. Optom. **34**:347-353, 1957.
Morgan, M. W., Mahoney, J., and Olmstead, J. M. D.: Astigmatic accommodatin, Arch. Ophthal. **30**:247-249, 1943.
Muller, J.: Elements of physiology, translated by W. Baly, 2 vol. London, 1842, Taylor & Walton.
Nadell, M. C., and Knoll, H. A.: The effect of luminance, target configuration and lenses upon the refractive state of the eye, Am. J. Optom. **33**:86-95, 1956.
Naylor, E. J., Shannon, T. E., and Stanworth, A.: Stereopsis and depth perception after treatment for convergent squint, Br. J. Ophthal. **40**:641-651, 1956.
Olmstead, J. M. D.: The role of the autonomic nervous system in accommodation for far and near vision, J. Nerv. Ment. Dis. **99**:794-798, 1944.
Otero, J. M.: Influence of the state of accommodation on the visual performance of the human eye, J. Opt. Soc. Am. **41**:942-948, 1951.
Pascal, J. I.: Practical applications of dynamic retinoscopy, Arch. Ophthal. **27**:859-862, 1941.
Perkins, R. B.: Clinical measurement of the physiologic position of rest of the crystalline lens, Am. J. Optom. **48**:343-350, 1971.
Phillips, S., Shirachi, D., and Stark, L.: Analysis of accommodative response times using histogram information, Am. J. Optom. **49**:389-400, 1972.
Porterfield, W.: Treatise on the eye, Edinburgh, 1759, G. Hamilton & J. Balfour.
Reese, E. E., and Fry, G. A.: The effect of fogging lenses on accommodation, Am. J. Optom. **18**:9-16, 1941.
Ripple, P.: Variation of accommodation in vertical directions of gaze, Am. J. Ophthal. **35**:1630-1634, 1952.
Ripps, H., Chin, N. B., Siegel, I. M., and Breinin, G. M.: The effect of pupil size on accommodation, convergence, and the AC/A ratio, Invest. Ophthal. **1**: 127-135, 1962.
Roelofs, C. O.: Apparent refraction and accommodation in their dependence upon the distance between eye and glasses, Ophthalmologica **37**:37-47, 1957.
Saladin, J. J., Usui, S., and Stark, L.: Impedance cyclography as an indicator of ciliary muscle contraction, Am. J. Optom. **51**:613-625, 1974.
Sheard, C.: Dynamic skiametry. In American Encyclopedia of Ophthalmology, Chicago, 1920, Cleveland Press.
Siebeck, R.: Accommodation and binocular vision, Halle, 1957, Carl Marhold Verlag.
Smithline, L. M.: Accommodative response to blur, J. Opt. Soc. Am. **64**:1512-1516, 1974.
Swegmark, G.: Studies with impedance cyclography on human ocular accommodation at different ages, Acta Ophthal. **47**:1186-1206, 1969.
Tait, E. F.: A quantitative system of dynamic skiametry, Am. J. Optom. **3**:131-155, 1929.
Urist, M. J.: Atypical accomodative esotropia, Am. J. Ophthal. **41**:955-964, 1956.
van der Hoeve, J., and Flieringa, H. J.: Accommodation, Br. J. Ophthal. **8**:97, 1924.
Wald, G., and Griffiin, D. R.: The change in refractive power of the human eye in dim and bright light. J. Opt. Soc. Am. **37**:321-336, 1947.
Warwick, R.: The ocular parasympathetic nerve supply and its mesencephalic sources, J. Anat. **88**:71-93, 1954.
Westheimer, G.: The effect of spectacle lenses and accommodation on the depth of focus of the eye, Am. J. Optom. **30**:513-519, 1953.
Westheimer, G.: Accommodation measurements in empty visual fields, J. Opt. Soc. Am. **47**:714-718, 1957.
Wundt, W.: Principles of physiological psychology, New York, 1902, The Macmillan Co.
Young, F. A.: Refraction of the monkey eye under general anesthesia, Vision Res. **3**:331-339, 1963.
Young, T.: On the mechanism of the eye, Philosoph. Trans. R. Soc. London **91**:23-88, 1801.

17

Elements of ocular motility

Vision depends not only on resolution but also on precise foveal alignment. Oculomotor disturbances prevent the most meticulous refraction from achieving efficient, comfortable seeing. Clinicians have therefore come to look beyond optics, which is static, to the dynamic rules governing ocular movements. Because the rules are incomplete, the treatment of motility imbalances and strabismus remains empirical, but they help in selecting the best that can be realized if not the most that can be imagined.

PRACTICAL ANATOMY

All ocular movements result from the coordinated action of six extraocular muscles. With the exception of the inferior oblique, the muscles originate at the apex of the orbit in intimate relation to the optic nerve. Indeed the origin of the superior and medial recti includes the dural sheath, hence the painful movements of optic neuritis. The four recti muscles insert in front of the equator and pull backward; the two obliques insert behind and pull forward. On balance, there is normally no translatory movement of the globe. The muscle tendons insert and merge with the sclera at various distances from the limbus, which serves as a surgical landmark (Fig. 17-1).

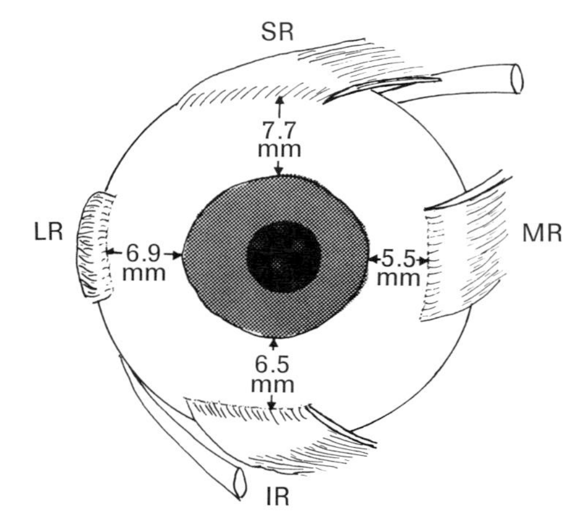

Fig. 17-1. The limbus serves as a surgical landmark for the tendinous insertions of the anteriorly attaching extraocular muscles. The medial rectus *(MR)* is nearest, the superior rectus *(SR)* farthest, and a line connecting the midpoints of each insertion forms a diverging spiral (of Tillaux).

Although their insertions are not parallel, it is convenient to think of the muscles as acting in pairs: lateral-medial recti, superior-inferior recti, and superior-inferior obliques. Each pair has a common axis of rotation, and all have a common hypothetical center of rotation fixed 13.5 mm behind the cornea. Each muscle rotates the eye by exerting a force tangent to the first point of contact with the globe—the physiologic insertion. As the eye turns, the arc of contact between physiologic and anatomic insertion changes, but the direction of pull remains constant (Fig. 17-2). The direction of pull or line of force in conjunction with the center of rotation forms the muscle plane. The paired muscles have a common muscle plane, and their action is antagonistic. A simultaneous recession or resection of a muscle pair has 25% greater effect than consecutive surgery on each separately.

The direction of pull of the superior-inferior rectus is about 27° from the anteroposterior axis; the direction of pull of the superior-inferior oblique is 51°; and the axis of the lateral-medial rectus is horizontal (Fig. 17-3). It follows that the contraction of vertical rotators produces not only elevation-depression but also adduction-abduction and inrolling-outrolling. These secondary actions change with the position of the globe.

Under normal circumstances, secondary actions are balanced by the coordination of all extraocular muscles to produce smooth movements, usually without rolling. The primary and secondary actions of the individual muscles, acting as if each were present without the others, are given in Table 17-1.

Individual muscles differ not only in their direction of pull but also in length, weight, and cross section. Since maximum shortening

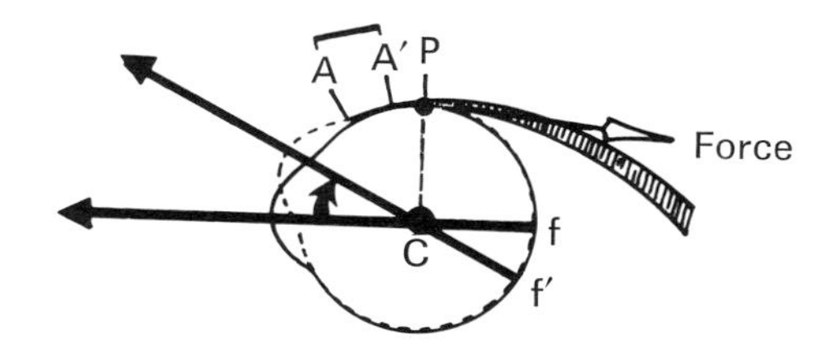

Fig. 17-2. Schematic diagram showing the anatomic insertion *(A)* and physiologic insertion *(P)* of an extraocular muscle. The anatomic insertion is fixed relative to the eye; the physiologic insertion is fixed relative to the orbit. When the globe rotates, the arc of contact changes from *AP* to *A′P*, but the direction of pull or line of force remains the same. The axis of rotation is therefore independent of eye position.

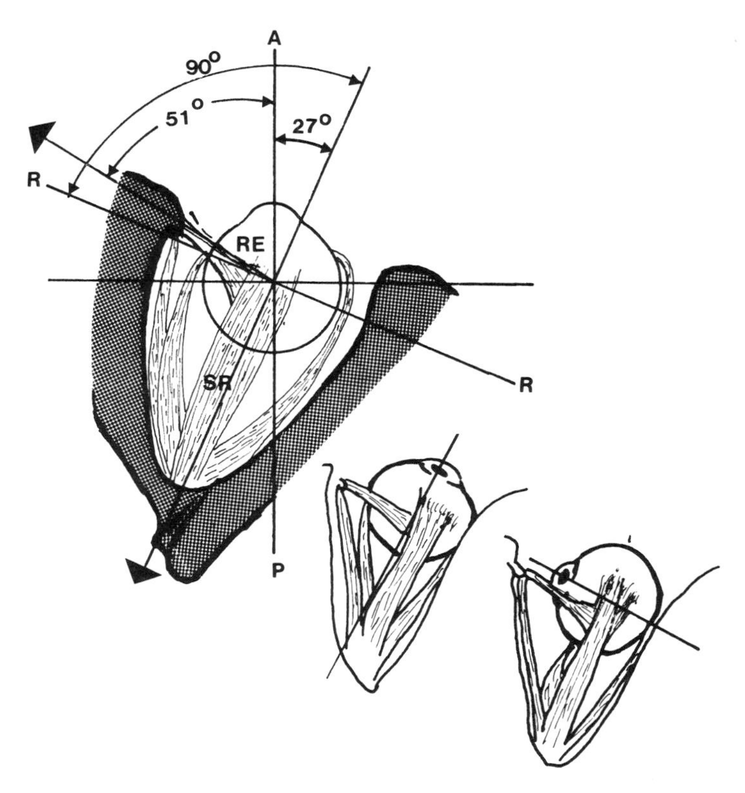

Fig. 17-3. Semischematic diagram illustrating the lines of force and axes of rotation of the vertical muscles. The superior and inferior rectus have a common axis *(RR)*, and so do the superior and inferior obliques. The change of action of these unvarying axes of rotation when the eye is abducted or adducted is illustrated in the smaller sketches. All diagrams refer to the right eye.

Table 17-1. Primary and subsidiary functions of extraocular muscles

Muscle	*Primary function*	*Subsidiary action(s)*
Medial rectus	Medial rotation (in)	Nil
Lateral rectus	Lateral rotation (out)	Nil
Superior rectus	Upward rotation or elevation	Medial rotation Incyclorotation
Inferior rectus	Downward rotation or depression	Medial rotation Excyclorotation
Superior oblique	Incyclorotation or depression	Outward rotation
Inferior oblique	Excyclorotation or elevation	Outward rotation

is about one third the length and since the average muscle is 40 mm, the anticipated shortening is about 13 mm in theory, somewhat less in practice because of restricting fascia and check ligaments. The more a muscle is stretched the greater the shortening, hence the increased surgical result obtained in large angle strabismus for the same recession or resection.

The purpose of a muscle operation is to alter the contractive ability or direction of pull or both and thus modify its capacity to rotate the eye. In a recession, for example, the muscle effect is weakened because it starts from a shortened position of rest, there is less opposition to its antagonist, and the arc of contact is decreased. A variety of intangibles makes it impossible to mathematically predict the result from so many millimeters of surgery; indeed some surgeons perform the same operation whatever the deviation, and others prefer a free tenotomy to allow the globe to come to rest where it may. So if the procedures are equivocal, the results are never dull, and only well-documented experience and attentive study will prevent unpleasant surprises.

The amplitude of eye movements never measures the strength of a muscle because only about 5% of the fibers are active at a time. Taking the globe diameter as 24 mm and its weight as 8 gm, the estimated pull required to turn the eye at its observed velocity is 5 gm. The contractive power of each muscle is 500 to 1000 gm; hence the available power exceeds the requirement by over a hundred times. Motility disturbances are therefore seldom due to weakness or fatigue. For example, one half of the superior rectus muscle fibers can be mobilized to elevate the lid (in a Motais repair for levator palsy) without affecting elevation.

Extraocular muscle fibers are thinner, with a larger nerve to muscle ratio than other striated muscles, making possible finely graded movements. The muscle fibers are divided into slow or tonic fibers mediating postural and vergence reflexes, and fast or twitch fibers mediating versions. The fast fibers have a fibrillar structure and are innervated by thick nerve fibers; slow fibers have a more uniform ("field") structure and are innervated by small efferents. There is some evidence that the latter are derived from the autonomic nervous system.

Surrounding the eyeball is the fibroelastic connective tissue capsule of Tenon whose sleevelike extensions receive the muscles. The capsule, a kind of universal joint, minimizes friction, serves as a bed for the bulbar conjunctiva, and the subcapsular space receives the aqueous in a surgical iridencleisis. The capsular sleeves are interconnected by a connective tissue intermuscular membrane, which must be divided in muscle surgery to obtain the full effect. Other extensions form secondary attachments to the globe and check ligaments to the orbit.

The anterior ciliary arteries supply blood to the muscles, and their branches participate in the nutrition of the iris and cornea, which may be compromised in extensive muscle surgery. The veins drain into the superior and inferior ophthalmic veins.

The superior oblique is innervated by the fourth cranial nerve, the lateral rectus by the sixth cranial nerve, and the other muscles by the third cranial nerve. Sensory innervation is provided by the trigeminal complex, which has reflex connections to the vagus.

Undue muscle traction during a surgical procedure causes bradycardia and may lead to cardiac arrest, hence the wisdom of preanesthetic atropine. Autonomic innervation maintains the tone of the muscles and their arteries.

Adhesions of the superior rectus muscle to the levator of the lid may lead to lid droop if interfered with. The medial rectus is the strongest; the lateral rectus the only muscle with two heads, and the superior oblique is the longest (about 60 mm) because it has two bellies, one before and one after passing through the trochlear pulley. The inferior rectus is the shortest of the recti muscles, and the inferior oblique is the shortest of all muscles and the only one to take origin from the anterior orbit. Both the lateral rectus and superior oblique are frequently palsied by a blow on the head; the inferior oblique may be entrapped in a blowout fracture of the orbit; anomalies of the trochlea or superior oblique tendon cause Brown's syndrome; fibrosis of the lateral rectus is at least one cause (and used to be the only one) of Duane's syndrome; and a fibrotic replacement of the medial rectus may result in strabismus fixus.

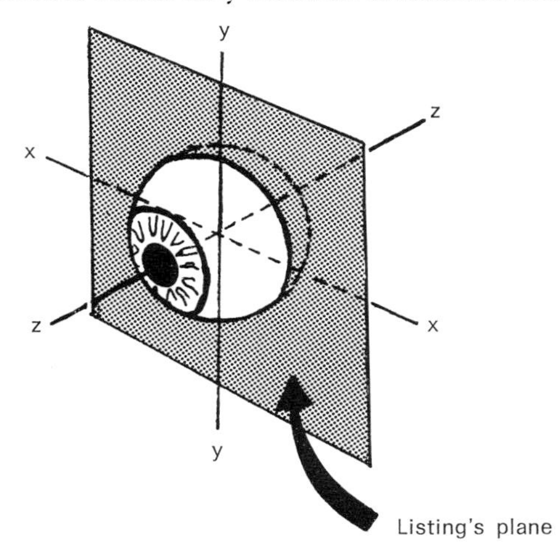

Fig. 17-4. Listing's plane is an equatorial plane fixed in the head; it coincides with the equatorial plane of the eye only when the eye is in the primary position. All ocular movements can be imagined to occur about an axis that lies in Listing's plane and that is perpendicular to a second plane containing the initial and final points of fixation and the center of rotation of the eye.

KINEMATICS

A line from the object of fixation to the center of rotation is the "fixation axis"; a plane containing the fixation axes of both eyes is the "fixation plane"; and the line connecting the two centers of rotation is the "interocular baseline" (or, in practice, the interpupillary distance). The extent to which the fixation point may move is the "motor field"—roughly 50°. Normally head movements extend this field greatly. For reasons unknown, up-gaze becomes more limited in older people. The range of ocular movements is tested by simply asking the patient to follow an interesting target. The directions of greatest clinical importance coincide with the primary fields of the six extraocular muscles, hence six "cardinal" directions. These best isolate the muscle actions, although they are not necessarily the positions in which the muscle is most powerful. If both eyes show limited movements, the result is a gaze palsy; if one eye lags progressively, it is a muscle palsy.

The primary position is the arbitrary zero point of ocular rotations; the eyes fixate a distant target in the horizontal plane, visual axes are parallel, vertical meridians of both corneas are vertical and parallel, with the head erect and immobile. Starting from the primary position, all movements may be resolved about a system of coordinates, such as vertical, horizontal, and anteroposterior in the Fick (or Helmholtz) system, but more simply into a polar system (Listing). The horizontal and vertical axes fixed in the head defines Listing's plane (Fig. 17-4). All movements can be imagined as occurring about an axis in this plane; the specific axis is perpendicular to a second plane that contains the center of rotation and the initial and final points of fixation. This rule, named after Listing although he never published it, does not mean the eye must actually turn about the indicated axis but only that the result is as if it did. *Listing's law* can be verified subjectively by afterimages or plotting the blind spot and objectively by observing a conjunctival vessel, corneal marker, or photography. Although the rule holds true for versions, it breaks down completely for vergences; the globes extort relative to each other about 1° for every 10° of convergence.

A model eye or "ophthalmotrope" can be constructed from a rubber ball. The lines of muscular insertions are drawn in relation to the equator and the three primary axes. By means of pins and strings, the action of individual muscles and muscle combinations can be demonstrated. The effect of rotation to secondary and tertiary position on the vertical meridian (so-called false torsion) is readily visualized. Note that the inferior and superior obliques are named anatomically, but functionally the first elevates, and the second depresses the eye.

Any movement about the horizontal or vertical axis in Listing's plane is termed a "cardinal movement," and the eye reaches a "secondary" position. An oblique movement places the eye in a "tertiary" position, which is always accompanied by a certain degree of rolling or torsion—the inevitable consequence of the movement of spherical bodies. Since this rolling is not caused by any specific muscle, it has been termed "false torsion." This is not to imply that it is illusory because according to *Donders' rule*, the angle of false torsion is always the same for a given tertiary position regardless of how the eye reaches it. The requirements of spatial orientation would not tolerate anything else, since the same retinal points must be stimulated in both eyes when I look up and to the right, regardless of where I looked before. The presence of false torsion permits a retrospective definition of the primary position; it is that position from which a cardinal movement does not result in false torsion.

A controversial seesaw in the literature about false torsion and merry papers on kinematics with roller coaster analytic geometry and spherical trigonometry totters the understanding but only teeters the brink of the clinical problem. After exhausting every yawn, one finally comes to agree with Goethe's Faust:

> He that would study and portray
> A living creature, thinks it fit
> To start with finding out the way
> To drive the spirit out of it.
> This done, he holds within his hand
> The pieces to be named and stated,
> But finds the spirit tie that spanned
> And knit them, has evaporated.

Objective measurements of eye movements and position leave much to be desired; so we rely on simple direct observation of corneal reflections. When the first Purkinje image is centered in the pupil, however, it delineates the pupillary and not the visual axis. The angle the pupillary axis makes with the visual, optic, or fixation axis constitutes a veritable greek alphabet soup. The only one which can be measured clinically is correctly called "lambda" but is almost universally called "kappa" and is the angle between the pupillary axis and line of sight at the entrance pupil (not the nodal point). The line of sight is usually nasal to the pupillary axis and, if so, is referred to as "positive" (Fig. 17-5). A large positive angle kappa may simulate a divergent squint. A positive angle kappa is added to esotropia (considered positive) or subtracted from exotropia (considered negative) to arrive at the true angle of deviation; the converse is true when angle kappa is negative. The angle is most easily measured clinically on the perimeter or a modification thereof.

The terminology of eye movements is bedeviled by an apathy of imprecision. Binocular conjugate rotations used to be called "versions," are currently called "saccades," and are referred to as "ductions" in the optometric literature. Monocular rotations should be called "ductions," although they are mensurational artefacts, since no eye normally ever moves alone. Disjunctive binocular movements are "vergences," frequently confused with one subvariety—fusional vergences—but are in fact also initiated by monocular accommodation, awareness of nearness, or even in anticipation of stimulation. A rolling or torsion is a "cyclorotation," but whether true or false, not a cyclophoria. A "heterophoria" is a tendency to deviation, never manifest under binocular conditions by definition. A heterotropia is a "squint" in technical jargon, which confuses the patient who identifies the term with blepharospasm. A "comitant squint" does not mean the visual axes are always parallel, since the angle of deviation changes with the vergence. An "incomitant squint" may or may not be paralytic; it only implies that the deviation varies during versions. A failure of the visual axes to intersect exactly on the plane of fixation during fusion is a "fixation disparity," also called "retinal slip" or "associated phoria," but it is not a tropia, micro or otherwise. The "visual axis" lies in objective space and is the hypothetical construction line connecting the object of fixation with

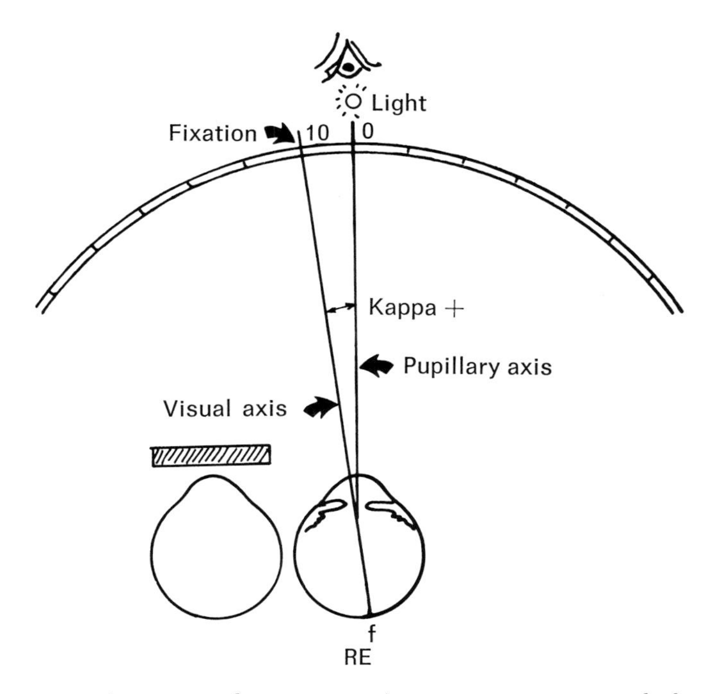

Fig. 17-5. The perimeter (or any similar instrument) serves to measure angle kappa. The examiner sighting above the stationary light centers the Purkinje reflex in the pupil of the sighting eye by having the patient follow a second target along the arc of the perimeter. If the visual axis is nasal to the pupillary axis as in the sketch, angle kappa is said to be positive.

Table 17-2. Synergists

Muscle	*Primary action*	*Secondary action*	*Synergists*
Lateral rectus (LR)	Abduction	None	SO and IO
Medial rectus (MR)	Adduction	None	SR and IR
Superior rectus (SR)	Elevation	Adduction Incycloduction	IO MR and IR SO
Inferior rectus (IR)	Depression	Adduction Excycloduction	SO MR and SR IO
Superior oblique (SO)	Incycloduction	Depression Abduction	SR IR LR and IO
Inferior oblique (IO)	Excycloduction	Elevation Abduction	IR SR LR and SO

Table 17-3. Antagonists

Muscle	Action	Antagonists	
		Primary	*Secondary*
Lateral rectus (LR)	Abduction	MR	SR and IR
Medial rectus (MR)	Adduction	LR	SO and IO
Superior rectus (SR)	Elevation Adduction Incycloduction	IR	SO SO, LR, and IO IO and IR
Inferior rectus (IR)	Depression Adduction Excycloduction	SR	IO SO, LR, and IO SO, SR
Superior oblique (SO)	Incycloduction Depression Abduction	IO	IR SR and IO SR, MR, and IR
Inferior oblique (IO)	Excycloduction Elevation Abduction	SO	SR IR and SO SR, MR, and IR

the fovea. It should not be confused with the visual direction of the fovea (sometimes called the "principal visual direction"). A more appropriate term for "visual axis" would be "line of sight" (in analogy with Hering's "lines of directions") or, better yet, "foveal fixation line", but the terminology is probably too ingrained for change. The "fixation axis" connects the object of regard to the center of rotation of the eye.

CLINICAL PHYSIOLOGY

In every ocular movement, each muscle is aided by another whose action is similar, that is, which acts synergistically (Table 17-2). The superior rectus and inferior oblique muscles cooperate in elevating the globe; conversely, the lateral and medial recti oppose each other and are antagonists when the eye is in the primary position (Table 17-3).

Because each muscle has subsidiary actions, "synergism" or "antagonism" refers only to the primary actions; subsidiary functions may or may not cooperate (as the rolling components of the superior rectus and inferior oblique muscles).

A simple graphic illustration (Fig. 17-6) shows that in a pure elevation the superior rectus must be about 1.7 times as effective as the inferior oblique whose chief effect is outrolling. The inferior oblique muscle's contribution is limited to the amount of rolling that can be neutralized by the superior rectus. Such considerations, due to the angled axes of rotation, emphasize that eye movements always require the cooperative action of more than one muscle. If a muscle is paralyzed, however, the secondary actions become manifest and help in the diagnosis. It remains a puzzle why so many clinically diagnosed muscle palsies do not result in more obvious cyclorotations.

Secondary actions of the muscles change with eye position; so in restoring a large esotropia to parallelism, the superior and inferior recti, for example, become weaker adductors, and this may result in an overcorrection. Conversely, a 63° esotropia would effectively eliminate the superior and inferior recti as elevators and depressors.

The economic coordination of the extraocular muscles is assured by *Sherrington's law* of reciprocal innervation; a contraction does not proceed simultaneously in opposing muscles. The same economy applies to binocular movements, expressed by *Hering's law:* the innervation to each eye is equally distributed. These two rules form the cornerstone of motility diagnosis. The cooperating muscles in each eye that turn them toward the same direction are termed "yoke" muscles and are shown in Fig. 17-7 for the diagnostic (or cardinal) gaze directions.

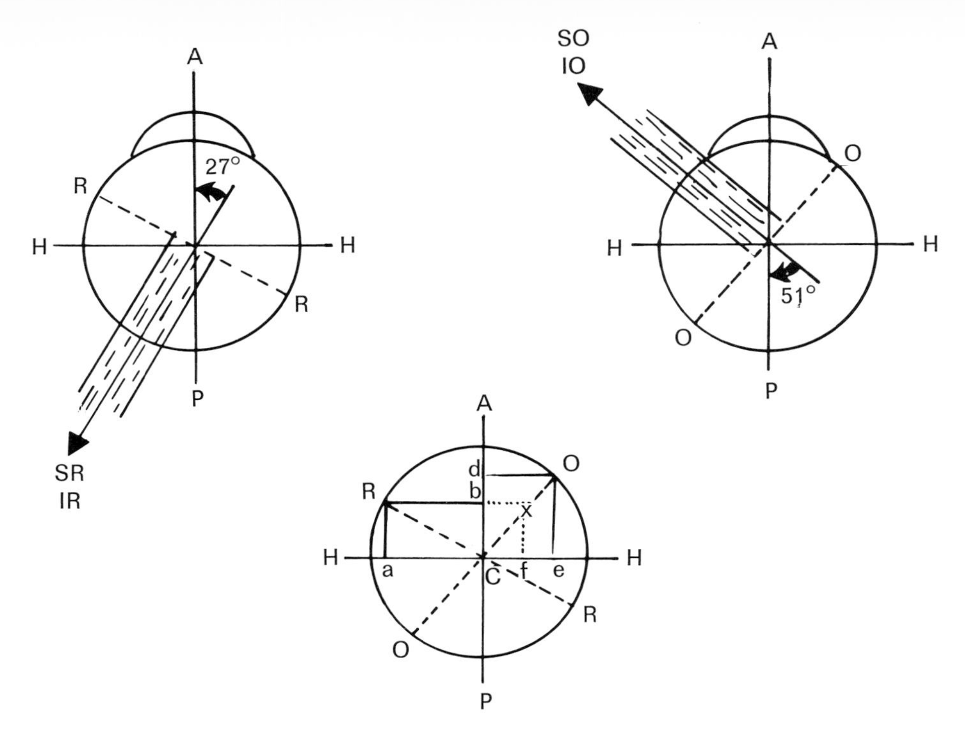

Fig. 17-6. Schematic diagrams illustrating the cooperation of the extraocular muscles. *Top,* The lines of force and common axes of rotation of the superior-inferior recti and superior-inferior obliques. *Bottom,* A composite showing only the axes of rotation. All diagrams refer to the right eye. The lower diagram shows, for example, that in a simple elevation, a maximum contraction of the superior rectus *(CR)* produces *Ca* elevation and *Cb* inrolling. A maximal contraction of the inferior oblique *(CO)* produces *Ce* elevation and *Cd* outrolling. Since no rolling occurs in pure elevation, the inferior oblique can only assist the superior rectus by contracting *CX*, so that the inrolling-outrolling components *(Cb)* are neutralized. The elevating contribution of the inferior oblique is therefore *Cf*. The proportions are as the sines of the angles: $\frac{\sin 51°}{\sin 27°} = 1.7$.

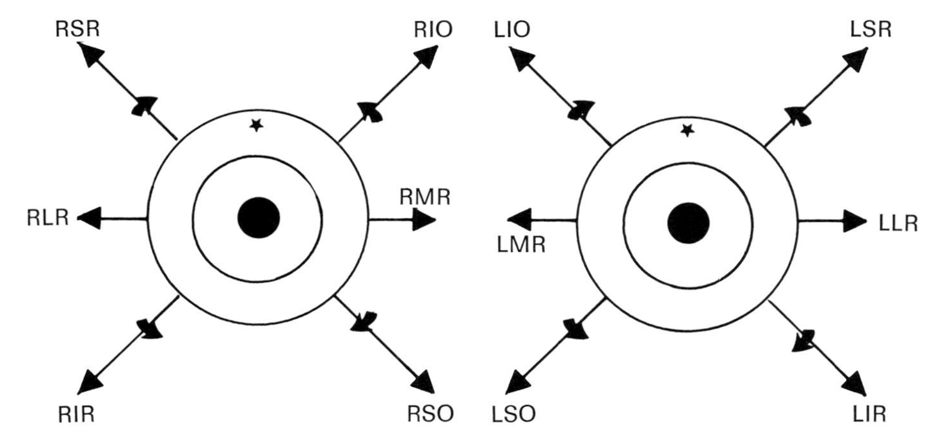

Fig. 17-7. Schema illustrating the yoke muscles of both eyes, with patient facing examiner, in the six diagnostic directions of gaze. Note that these are *not* the primary actions of the muscles acting as if each were present alone. The diagnosis of motility disorders is based on diagnostic gaze directions, that is, the directions which best indicate a particular muscle action, not necessarily its maximum force. The rotational component of the vertical muscles is illustrated by a small arrow (important in interpreting Bielschowsky's sign). There are six diagnostic directions because there are six extraocular muscles.

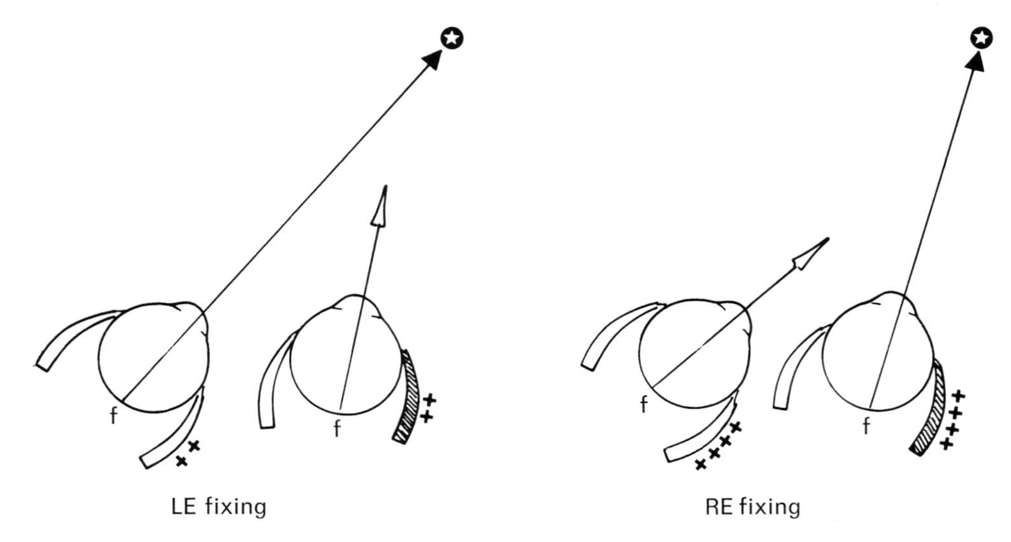

Fig. 17-8. Application of Hering's law is illustrated in a recent palsy of the right lateral rectus. When the left eye fixates, the right eye lags because the normal innervation (arbitrary two units) is insufficient to achieve full contraction of the palsied muscle. When the right eye fixates, more than normal innervation (arbitrary four units) is required to prod the right lateral rectus. The same innervation now sent to the normal left eye causes an excessive inward rotation; the secondary angle of deviation is larger than the primary.

Hering's law is best illustrated by a recent partial paralysis. For example, if the right lateral rectus is palsied, and it is the fixing eye, more innervation must be sent to it, hence to the yoked left medial rectus. The result is that the secondary deviation is larger than the primary deviation when the normal eye fixates (Fig. 17-8). This difference must be clinically elicited and is important in diagnosis. Fixing with the palsied eye is not unusual, either because it is dominant or because the double images are more separated and hence easier to suppress.

In addition to the incomitance of the deviation when the palsied eye fixates, its antagonist in the same eye has less opposition, and the action of the yoke muscle of the antagonist (in the other eye) is also reduced. This gives rise to an apparent palsy of the normal eye. Chavasse called it "inhibitional palsy of the contralateral antagonist," which describes it accurately, if not euphoniously.

In long-standing paralysis there may be readjustment of innervation and anatomic changes due to contracture. Although fibrosis is seldom demonstrable, one theory of congenital strabismus is that it begins as an unrecognized palsy, which later becomes comitant.

A voluntary movement across the field of vision causes retinal displacement, although objects remain stationary. If the eyeball is displaced artificially by pushing it up with the finger, objects appear to move. These results are reversed when viewing afterimages. Some allowance is therefore made in ordinary seeing for the contraction of the muscles. Helmholtz deduced that we know we have moved our eyes in the first instance because we "consciously judge the effort of will" required to move them, or, in modern terms, the brain keeps a record of the efferent innervation. William James, supported by Sherrington, concluded that muscle proprioceptors subserve visual position sense. Until muscle spindles were discovered, it was easy to agree with Helmholtz; now that the receptors are known to be abundant in extraocular muscles, no one is quite sure. The general consensus is that Helmholtz's innervation theory is correct, which leaves the function of the spindles in the air. They probably play a role in recording gross changes and in regulating tonus, perhaps metabolic recovery after prolonged contraction.

Von Graefe first noted that after recent palsy the visual feedback and innervation record are in disaccord. Since more impulses are required to prod the weakened muscle, the brain judges the eye to have turned further than it actually has. Asked to point at the target (without seeing his hand), the patient points past it. This past-pointing tends to support the Helmholtz innervation theory of eye position sense.

Individual extraocular muscle fibers obey the *all-or-none law* and are extremely fast-acting both with respect to latent period and contraction time. The electrical activity of the muscle can be picked up by small electrodes placed within the muscle substance (electromyography). Technical precautions require proper electrode positioning, the chief difficulty being that the activity of untapped motor units remains unknown. In saccadic eye movements there is a short burst of activity, with reciprocal inhibition of the antagonist as predicted by Sherrington's law. The initial burst is believed necessary to overcome the viscoelastic inertia of the eyeball. Following or pursuit movements show a gradation of firing or recruitment, with simultaneous gradual inhibition of the antagonist. During the waking state, a continuous discharge is recorded even when the eye is still, suggesting activation of alternating motor units to maintain tonus. Fortunately other electrophysiologic characteristics are outside the province of this book because some seriously conflict with established clinical rules of ocular motility. But electromyography is useful in differentiating muscular from nervous paralysis.

Neurologic control of eye movements is organized in evolutionary levels, ascending from the brain stem to the cerebral cortex. Each higher level utilizes the lower one with little change in the primitive pattern. Voluntary gaze movements originate in the frontal cortex. Stimulation of Brodman's area 8 results in conjugate ocular deviations to the opposite side, whereas bilateral stimulation near the same area causes vertical conjugate movements and lid opening and closing. These results confirm Jackson's dictum that movement patterns and not individual muscles are represented in the cortex.

Involuntary (e.g., pursuit) movements originate in the occipital cortex (Brodman's area 19) and are retinally elicited but require attentive observation, hence "psycho-optical" reflexes. These movements are not as strong or as constant as those obtained by frontal stimulation, suggesting the dominance of voluntary over automatic responses. The frontal and occipital projections discharge by separate paths to the midbrain and are distributed by collicular and para-abducens gaze centers to the nuclei of the third, fourth, and sixth cranial nerves. Collectively the nuclei are interconnected to each other and to the vestibular system by the oldest tract of the central nervous system—the median longitudinal fasciculus.

Exploring all the stops along the median longitudinal fasciculus is an adventure best delegated to the neuro-ophthalmologist. The vestibular mechanism is the chief integrater of spatial orientation relative to posture and acceleration. Although no vertical or horizontal gaze centers have ever been anatomically demonstrated, they are postulated on the basis of such clinical conditions as Foville's and Parinaud's syndromes and probably represent localized circuits in the reticular formation (until recently a neurologic dark continent). Little is known of cerebellar control, but it is probably integration as for other body movements. Even less is known of vergence control, probably located in the occipitoparietal cortex.

TYPES OF OCULAR MOVEMENTS

Although the same muscles are used for all eye movements, they may be grouped into fairly specific mechanisms according to function. Under normal circumstances these movement systems work simultaneously unless dissociated by diseases that attack one or the other preferentially.

Postural system

Oddly enough, the first function of the eye muscles was not so much to move the eyes but to keep them still, that is, to maintain the constancy of the retinal image despite changes in the position of the head and body. Only in a motionless panorama is orientation possible and detection of movements feasible. Magnus first divided these nonoptic reflexes into "static" reflexes, which deal with gravity, and "statokinetic" reflexes, which deal with angular acceleration. Both systems have an active discharge pattern at rest; changes are brought about by an increase or decrease in the rate of discharge. Complete cessation, as in eighth nerve section, causes severe vertigo.

If my head is tilted to the right, the pattern of discharge of the otolith end organs changes, and the eyes roll in a counterbalancing counterclockwise direction. Also affected are proprioceptive receptors in the neck and visual feedback from the retina, which serve to modify muscle tone, in a continuous monitoring process to prevent spatial disorientation.

Angular acceleration or deceleration (as in spinning the head) stimulates the semicircular canals, which are interconnected with the oculomotor nuclei according to the plane of rotation. The system is tested by caloric irrigation (several milliliters of warm or cold water in the external auditory canal for 40 seconds with the patient recumbent and head elevated 30°). The normal response is a jerk nystagmus, with the fast component away from the side stimulated by cold and toward the side stimulated with warm (the mnemonic "cows" helps recall which: cold, opposite, warm, same). Bilateral stimulation causes vertical nystagmus. Because the illusion of movement is created, the patient may complain of nausea and spinning sensation or even tend to fall toward the direction of the slow component. In infants the system may be tested by actual rotation.

True vertigo is associated with pallor, sweating, nausea, or vomiting. Onset and duration, hyperventilation, allergy, hearing deficits and tinnitus, trauma, systemic diseases, and other associated neurologic signs such as ataxia, paralysis, nystagmus, and psychologic orientation are important in the evaluation of the dizzy patient. Drugs like barbiturates, diphenylhydantoin (Dilantin), and alcohol produce toxic vestibular disturbances, often with an associated vertical nystagmus.

Nystagmus is a rhythmic involuntary oscillatory movement of one or both eyes in one or more directions of gaze, which may be symmetrical, asymmetrical, horizontal, vertical, or rotary, with or without dizziness. Ocular nystagmus, often with poor vision, may be congenital or acquired. Spasmus nutans occurs in early life, is associated with head tilt and head nodding, and resolves without treatment. Congenital nystagmus persists through life and is usually horizontal, with typically better near than distance vision. Latent nystagmus is important as a possible cause of poor vision when one eye is occluded. Vestibular nystagmus is frequently associated with vertigo and hearing difficulties. Gaze nystagmus probably occurs most frequently in neurologic diseases. All forms of nystagmus are classified according to the fast component.

Saccadic system

When I turn my eyes to fixate one object or another equally distant from me, the foveas are aligned. These movements are extremely rapid (up to 700° per second), vision is suspended during the movement, and there is limited capacity to change once initiated. Because they originate in the frontal lobe, saccades are also designated as "rapid" or "voluntary" eye movements, but the distinction between volition and reflex is often blurred (e.g., in the case of a sudden sound) and, when related to "free will," borders on one of the more narcotic areas of philosophy. The character of a saccade is not essentially different if one eye is covered but becomes more complex if the change in fixation also involves a change in distance.

Saccadic fixation and refixation depends on the integrity of the fovea. At birth, saccades are not used, and only after about 2 weeks does the infant look at a light. Combined oculographic recording and observation of optokinetic nystagmus by Dayton and associates (1964) indicates acuity of 20/150 and perhaps better; many newborns were able to fixate with the general macular area, suggesting more rapid differentiation than previously suspected. In contrast, vergences do not develop until later; accommodative vergence depends on acuity, and fusional vergences depend on binocular acuity (e.g., unequal in anisometropia) and, for want of a better term, "compulsion to fusion." Congenitally blind patients never develop refixation saccades, and their pendular nystagmus is a kind of visual ataxia due to loss of retinal position-sense feedback.

In reading the eyes make a series of short, jerky saccadic movements, interrupted by pauses lasting some 0.2 to 0.3 seconds during which words are perceived. Vision during the movement is too poor for discrimination if it occurs at all. Reading speed therefore depends on the duration and number of pauses in sweeping the page. The number of pauses varies with practice and the nature of the reading task. Most reading problems are seldom due to oculomotor difficulties, based on objective measurements (e.g., photography of corneal reflections as with the "ophthalmograph" or variations in potential as with electroculography).

During early stages of sleep or general anesthesia, the eyes undergo rhythmic or discontinuous, dissociated movements. They disappear in deep sleep, and the eyes assume an eccentric position. Rapid eye movements are said to occur during dreams, an analyst's nightmare.

Pursuit system

The distinction between shifting the eye from one target to another and following a continuously moving target was recognized by Dodge (1903). Unlike saccades, pursuit movements are slow, tracking responses (about 30° per second), and vision is present throughout their course. They are essentially the same whether carried out with both eyes or one, providing there is no change in target distance. (Pursuit vergences have only been studied in conjunction with and therefore confused with fusional vergences.) Pursuit movements depend on target velocity; if the target moves too rapidly, the response is interrupted by saccades, and the eyes cogwheel in a jerky manner (hence the distinction between smooth and jerky pursuit movements). The neurologic origin of lateral pursuit movements is in the occipital cortex of the contralateral side (bilateral for vertical pursuit) projecting to the vertical midbrain and horizontal pontine gaze centers, where they are intimately related to the saccadic projections from the frontal lobes.

Nature has wisely provided a means for shifting fixation and for maintaining fixation on a moving target. In a comatose patient, fixation is impossible, and if the head is moved, the more primitive postural reflexes cause the eyes to deviate opposite to the head turn like the eyes of a doll (hence "doll's head" phenomenon).

Optokinetic nystagmus (OKN) elicits both smooth pursuit (slow phase) and saccadic (fast refixation) movements and can be used to evaluate both systems. Thus if the OKN drum is moved from left to right, the slow pursuit phase depends on the integrity of the left occipital lobe and the fast refixation saccade, on the integrity of the right frontal lobe. Fibers connecting the occipital and frontal areas travel through the parietal lobe, and deep lesions there cause loss of the fast component of OKN when moving the drum toward the side of the lesions.

OKN responses depend only on the integrity of motor pathways as illustrated by defects independent of hemianopia, although usually they are associated. The reflex does depend on foveal integrity, providing an objective if indirect measure of visual acuity.

Vergence system

Vergences regulate the visual axes of the two eyes relative to each other, such as in convergence and divergence. The control is attributed to the occipitoparietal cortex with projection to a vergence center near the third nerve nucleus, although the details are not known. Characteristically these are slow movements (about 20° per second) with a relatively long latent period. Lateral vergences, although binocular, may be initiated by monocular stimulation (tonic, accommodative, and proximal stimuli), as well as retinal disparity. (They are more fully discussed in Chapter 18.) Vertical and cyclorotational vergences are independent of accommodation and probably occur mainly under artificial test conditions.

Everyone with normal binocular vision has had an experience similar to that described by Helmholtz: "When I begin to get sleepy in the evening from reading, or when out of courtesy to the company I try to keep my eyes open after a long dinner, I am apt to see double" and "as soon as my attention is aroused and I begin to recover myself, the double images generally fuse rapidly together again." In practice, a weak, dissociating prism substitutes for gaseous guests, and if the disparity is not too large, there is a brief period of double vision followed by an automatic, involuntary "fusional movement" to regain unification. Diplopia is not necessary for fusional

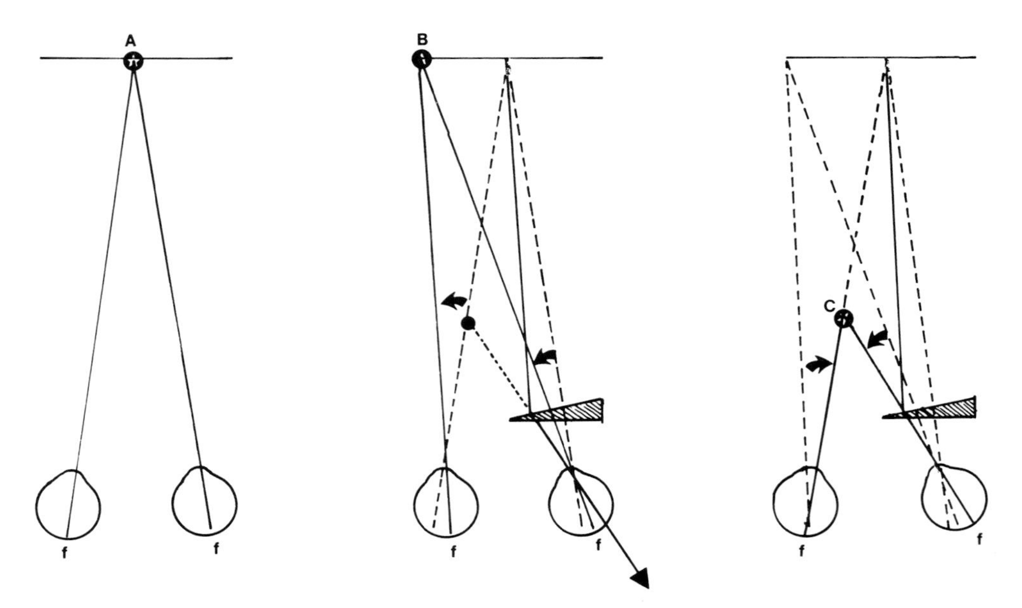

Fig. 17-9. Diagram illustrating that the movement following interposition of a base-out prism before one eye still follows Hering's law. There is a symmetrical saccade to the left followed by a symmetrical convergence. The bilateral distribution of innervation remains constant.

movements, however, since the prism may be so weak that no double vision is ever noted. Although fusional responses to a weak monocular prism seem to defy Hering's law, the innervation to the two eyes actually remains equal through consecutive versions and vergences (Fig. 17-9).

Fusional movements may be horizontal, vertical, or cyclorotational, each with a characteristic amplitude, and each induced by a different kind of disparity. The disparate stimuli are usually similar enough to permit fusion, but the movement can also be *initiated* by unlike stimuli (e.g., horizontal line for one eye, vertical line for the other), although the response is not completed, and the eyes then drift to the phoria position. Conjugate and disjunctive eye movements may respond independently; each can react regardless of whether the other is being stimulated or is responding. Usually the two are associated; the separate components can be differentiated by their individual time courses as in Fig. 17-9. The stimulus for fusional movements is disparity, whether the two ocular stimuli differ in shape, contrast, or luminance. Split-brain preparations further indicate that the detection of disparity occurs in the visual cortex and not the lower centers. There are individual differences in preferential responses to crossed, near-zero, and uncrossed disparities. In some "stereoblind" individuals, one or more of these responses may be absent.

Although there is no question that disparity is the adequate stimulus for fusional movements, there is some confusion about the role of peripheral and central retinal stimulation. For example, by means of polarized projectors, it is possible to create a vertical disparity of the background (e.g., a page of printed text) and induce a foveal diplopia while the foveal targets remain unchanged. This does not mean that peripheral disparity is more effective than bifoveal fixation, since diplopia is induced only if the area of peripheral stimulation is large and the foveal stimulation weak (e.g., two central dots or lines). Vertical disparity is also more effective than horizontal, whatever the foveal stimulation. If the foveal stimuli have strong contour, peripheral disparity is less effective. These results indicate that figure-ground organization has its consequences, not that peripheral and central fusion are independent neurophysiologic processes. A peripheral diplopia may also be induced by central disparity (as in the direct-comparison eikonometer target described in Chapter 19); the result simply depends on the number of retinomotor units activated. But the peripheral bulk effect of induced fusional movements give us an orthoptic lever in training ocular alignment despite central suppression.

Horizontal fusional movements are most easily demonstrated in a patient with a large heterophoria. If one eye is occluded and then uncovered, the visual axes return to the target of fixation as if by magnetic attraction and independent of the will. The patient may experience a momentary diplopia (crossed or uncrossed according to the phoria), and the two images quickly flow together.

"Diplopia" is a subjective phenomenon; hence only "I" can see double. The doubling refers to egocentric space (not oculocentric directions, which deal with the position of objects relative to each other and to the fovea). If my eyes are dissociated by a strong vertical prism, I see the same target in two egocentric directions. In fact, it is difficult for me to say which one is "straight-ahead" unless I have a strongly dominant eye, in which case one looks more substantial. This is in contrast to two different targets seen in the same egocentric direction, for example, the bird and cage in a haploscope. To distinguish these often confused concepts, Walls (1948) at one time introduced the terms "heterolocal diplopia" (one target in two directions), and "homolocal diplopia" (e.g., the "lustre" when a black and white square in a stereoscope are superimposed). Normal binocular vision would be "homolocal haplopia," but unfortunately the terms never caught on.

The amplitude of horizontal fusional movements is clinically evaluated by interposing a progressively stronger base-out or base-in prism for a constant fixation distance until fusion is abolished.

Vertical fusional movements are measured by interposing a base-up or base-down prism until dissociation occurs. The vertical amplitude is almost always much smaller than the horizontal amplitude (hence the vertical imbalance of anisometropia causes the most difficulty).

Cyclofusional movements are disjunctive rotations about the anteroposterior axis in response to a torsional disparity. A pair of vertical or horizontal lines presented haploscopically are tilted toward or away from each other. The amplitude may range from 10° to 33° depending on test conditions. There is some evidence that torsional disparity up to ±5° may result in fusion despite the absence of ocular movements (measured objectively). Apparently fusion is maintained under certain circumstances within Panum's areas without cyclorotations. Cyclofusional problems induced by oblique cylinders, or off-axis cylinders, present no problems at all; since the vertical and horizontal meridians are inclined toward each other in the same eye, no cyclotorsion can compensate for them, and it is doubtful that any effort to fusion is made or that such effort gives rise to symptoms. Unlike cyclofusional movements, torsional movements following the inclination of the head are conjugate reflex rotations in the interest of maintaining binocular vision. Recent evidence suggests that they are less extensive than previously thought.

Vergence testing constitutes an important phase of routine refraction and is frequently considered as a criterion of visual efficiency. It is therefore disappointing that results vary considerably in patients with comfortable binocular vision and in the same patient with different techniques of measurement. Thus the size of the eye movement is usually smaller than the prismatic increases and acaounts in part for the perceptual distortions during prism vergences. Moreover fusional amplitude depends on the immediately preceding stimulus, the rate at which the disparity is introduced, whether the disparity is continuous or in steps, the location and extent of the retina stimulated, and the amount of detail included in the target.

A conventional technique is to introduce progressive horizontal or vertical disparity (e.g., by rotary prisms) until the patient recognizes a blur or diplopia, then gradually reduce the prism until single binocular vision is regained (recovery). The test is done through the emmetropic correction, and the results are modified by an existing heterophoria. Rotary prisms give larger and more reproducible results than a prism bar and permit smoother, symmetrical increases. Vergences are measured at distance and near (13 or 16 inches). Blur points are obtained only with horizontal vergences (not with divergence at 20 feet if the emmetropic correction is worn). The recovery point varies according to the instructions given the patient, and all vergence tests should follow a regular routine in the sequence of the refractive examination for subsequent comparison.

There is some evidence for a specific, visually evoked response wave for binocular fusion, although the same wave may also be induced by monocular stimulation and binocular stimulation with patternless targets. Further investigation is needed to substantiate a specific fusion component of the human visually evoked response (Kawasaki and associates, 1970)

EVALUATING OCULAR MOTILITY DISORDERS

In the diagnosis of motility disorders, wrote Chavasse, not only are all of us free to form and express an opinion, but also, and this is particularly democratic, the various opinions have a most exhilarating diversity; if one authority says the right superior rectus is paralyzed, we can be sure that another equally eminent will blame the left superior oblique. The difficulty, of which Chavasse was well aware, is not that the tests are unreliable but that the little patients are so unconstrained and uncooperative. What squeamish clinician, faced by a squirming, squawking, squinting infant, has not dreamed of a simple, rapid, objective, reproducible method of evaluating ocular motility. The numerous techniques and gadgets in the literature suggest that it has not yet been found.

History

The symptoms of oculomotor disturbances include discomfort, headaches, pain, limitation of movement, diplopia, nausea, dizziness, amblyopia, faulty depth perception, visuotactual incoordination associated with abnormal movements, deviations, nystagmus, and perhaps other neurologic signs.

In children the history is obtained from the parents and referring physician and includes the nature of the problem (constant or intermittent); age of onset; variability of the deviation; progression; ocular mannerisms such as closing one eye or tilting the head; change in deviation with anger or fatigue; associated illness or injury, including prematurity, previous treatment, and results; visual development, performance at home or school, and familial history of oculomotor disorders.

During the history, unobtrusive observation will reveal the presence and type of deviation, epicanthal folds, ptosis, abnormal movement, nystagmus, head tilt, accuracy of fixation, preferred eye, anisocoria, proptosis, enophthalmos, and associated neurologic signs such as paresis, ataxia, spasms, hydrocephalus, and psychologic affect. The more that can be learned during these preliminary stages the greater the facility and specificity of the examination proper.

Ocular examination

Examination should be appropriate to the level of physical and mental development of the child and explanations appropriate to the educational background of the parent. As with other pediatric procedures, it takes patience and perserverance. Following movements, pupillary reflexes, vestibular reflexes (by rotating the infant), optokinetic responses (to a tape carried in the clinician's pocket), retinoscopy, and ophthalmoscopy can be carried out quickly, while the infant is in the mother's lap, having his bottle, or even asleep. Fundus examination may reveal an optic atrophy, coloboma, central toxoplasma scar, or even a retinoblastoma as a cause of the deviation, while a preliminary retinoscopy serves to establish the presence and type of ametropia or anisometropia and incidentally retroilluminate corneal or lenticular opacities.

Motility examination

Diagnostic motility tests may be objective or subjective but must always be simple and rapid enough to be applicable to children because of the trend toward earlier recognition and diagnosis. For children under age 3 years, one is limited to objective methods; in older children and adults a correlation of objective and subjective results detects anomalous sensorimotor adaptations. Direct observation of ocular movements requires no special equipment and, if carried out systematically with an understanding for physiologic nuances, will yield much useful information. We wish to determine whether a deviation is present under conditions of nat-

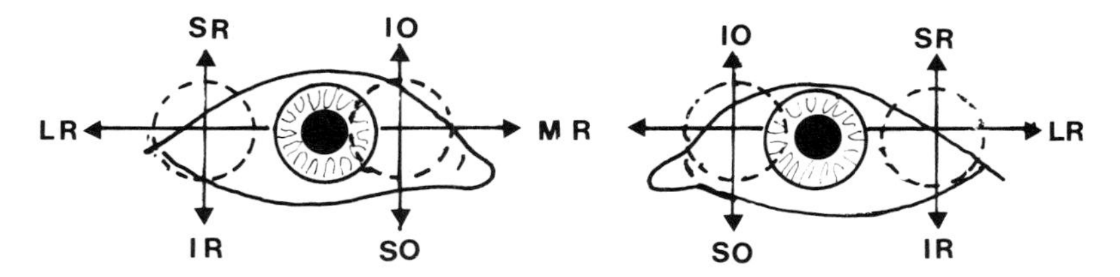

Fig. 17-10. Another diagram of the six diagnostic directions of gaze. A quick sketch of the ocular position in various directions, quantified by symbols or prism diopters, is useful for later analysis and comparison. The fixing eye must be indicated.

ural seeing if possible; if it is a phoria or a tropia; the amount of deviation at distance and near; whether the deviation is constant or intermittent, comitant or incomitant, and accommodative or nonaccommodative; the accuracy of fixation; quality of fusion; visual acuity; and whether there is an associated head turn or tilt.

With each eye covered in turn, the extent of the motor field and the accuracy of fixation is determined. The child may sit in his mother's lap, and the examiner steadies the head with his left hand while holding the fixation target with his right. Gaze directions are tested by turning the head or swiveling the chair. The distance fixation target may be an animated toy for smaller children or a Snellen letter for older ones. Near fixation targets consist of pictures, letters, or small toys that control accommodation. A "muscle light" is rarely adequate for near testing unless under cycloplegia; indeed a bright penlight may effectively dissociate the eyes by breaking fusion. Some conclusion can often be reached about acuity and the dominant eye by observing the fixation pattern. The dominant eye is the fixing eye in unilateral squint or amblyopia. The neurologic integrity (if not the quantity) of convergence and divergence is checked by the "push-up" test.

Binocular movements are tested in the diagnostic positions of gaze (Fig. 17-10). The vertical yoke muscles can be recalled by noting that if one is an elevator, the other is a depressor; if one belongs to the right eye, the other belongs to the left eye; if one is a rectus muscle, the other is an oblique (thus the right superior rectus (RSR) and the left interior

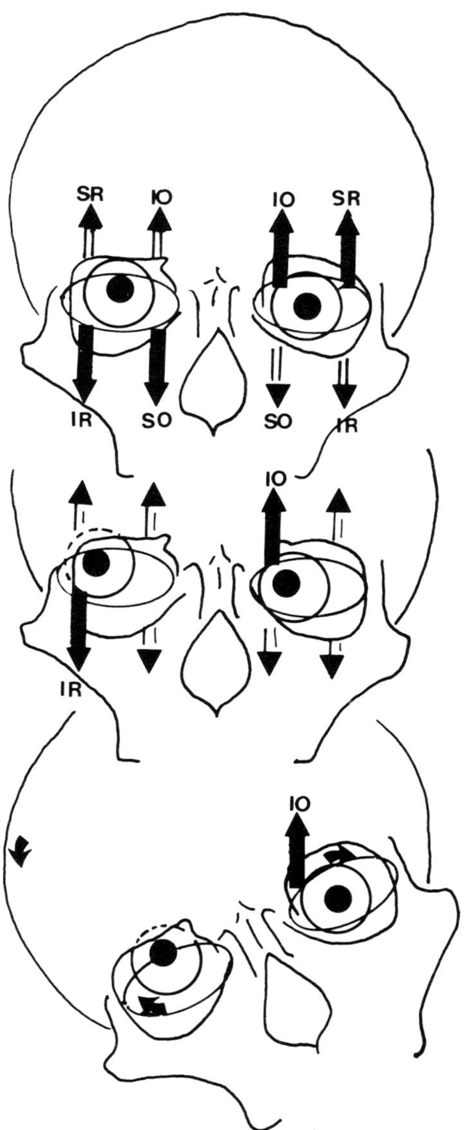

Fig. 17-11. Three-step diagnostic procedure for identifying an isolated vertical muscle palsy. For details see text.

oblique (LIO) are yoke muscles in gaze up and to the right). A dissociation of binocular movements that varies with gaze suggests a muscle palsy; the absolute degree of incomitance is less important than the relative difference in the rotation of the two eyes.

For example, in right hypertropia (Fig. 17-11), the four possible palsied muscles are the right depressors, the right inferior rectus (RIR) and the right superior oblique (RSO) muscles, or the left elevators, the left inferior oblique (LIO) and the left superior rectus (LSR). Determine whether the hypertropia is greater in right or left gaze; if it is greater in right gaze, it implicates the muscles whose prime function is to turn the eyes up and to the right (LIO and RIR). To differentiate between them, one can evaluate whether the right hypertropia is increased by looking down and to the right (RIR weakness) or increased by looking up and to the right (LIO weakness), but this evidence may be lacking in long-standing palsies (over 6 months) and is also confused by inhibitional palsy of the contralateral antagonist when the palsied eye fixates. A better differentiation is provided by the Bielschowsky head tilt test. If the right hypertropia increases by forcibly tilting the head toward the right shoulder, the LIO is implicated; if the hypertropia increases by tilting the head toward the left shoulder, the RIR is implicated.

The "Bielschowsky sign" is due to the fact that the eyes roll in compensation to a head tilt (vestibular reflex). If the head is tilted toward the right shoulder, the extorting muscles in the left eye are the LIR and the LIO. If the LIO is palsied, the depressing action of the LIR is unopposed; hence the left eye is depressed, exaggerating the right hypertropia. The patient is instructed to look straight ahead throughout the test, and since the eyes remain in the primary position, there is no overaction of the yoke of the palsied muscles.

Associated with an isolated vertical muscle palsy, particularly when an oblique is involved, is a head tilt. The patient may choose to turn the head so that the paretic muscle interferes least with binocular single vision. To confound any hard-and-fast rules, the head may be turned either to avoid diplopia or to exaggerate it (thus making suppression easier).

The *cover test* is the most commonly used technique of objectively evaluating motility. In younger children, the examiner's left thumb serves as a convenient occluder while the palm is used to steady and rotate the head. One eye is covered and the other eye observed; if the uncovered eye makes no movement when either eye is covered, there is no deviation (unless there is severe amplyopia with eccentric fixation, in which case the child objects strongly to having the good eye occluded).

If no deviation is found, the examiner should proceed to the *alternate cover test*. Each eye is covered in turn, but instead of removing the cover, it is quickly switched to the opposite eye (so as not to permit fusion). If no shift of fixation occurs, there is orthophoria, assuming both eyes are fixing centrally. If the uncovered eye turns in, there is exophoria; if out, esophoria; if up, hypophoria; and if down, hyperphoria. In a vertical phoria, the relative separation of the eyes is maintained; that is, the right eye turns up under cover by as much as the left eye turns down under cover. In the rare case of "alternating hyperphoria" (or "occlusion hypertropia"), the covered eye always turns up. It should be clear that in the cover test, the refixation movement is observed in practice; the covered eye deviates and takes up fixation when the cover is removed. In the presence of a phoria, the speed and facility of bifixation when both eyes are uncovered suggest the quality of the fusional vergences. Although rudimentary fusional movements may be observed in strabismus, they are never complete; indeed this is the outstanding motor defect.

If one eye is markedly amblyopic, its fixation may be so poor, that the deviation is larger than when the good eye fixates. The result is an apparent incomitance, not to be confused with a muscle palsy. Another interesting phenomenon is a small eye movement or flick (from 1Δ to 6Δ) when the preferred eye is covered in the unilateral cover test (tropia); but with the alternate cover test, a considerably greater movement is observed (phoria). The flick is more common in esodeviations than exodeviations and may be found in persons with comfortable binocular vision. It may represent an usually large fixation disparity or a very small residual tropia (microtropia) following surgical correction or

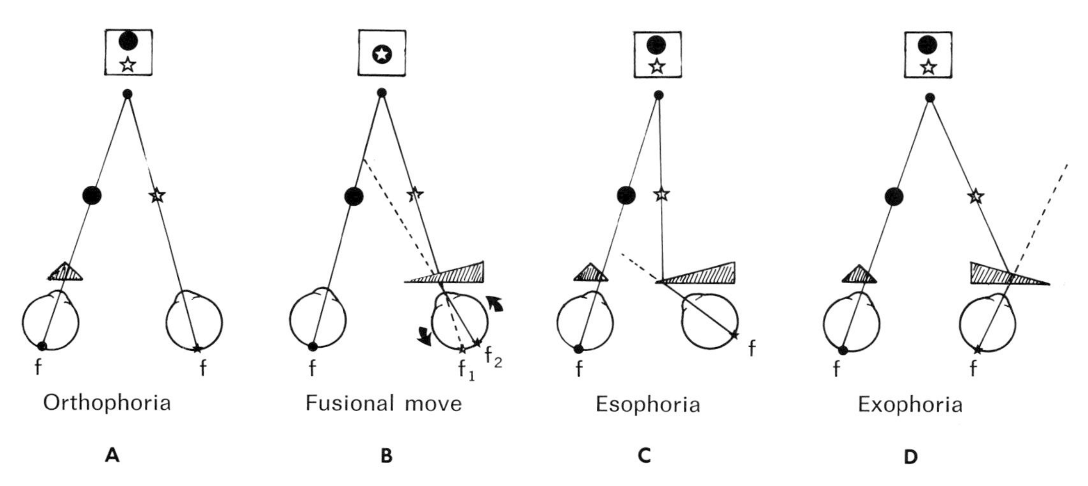

Fig. 17-12. Series of diagrams illustrating the effect of a lateral prism on the eyes. The eye turns toward the apex of the prism only in **B**, a fusional movement in the interest of maintaining binocular vision. In **A**, **C**, and **D**, the eyes are dissociated by an additional vertical prism, and the lateral prism only deviates the light ray on to the fovea. No prism is required in orthophoria, base out measures esophoria, and base in measures exophoria.

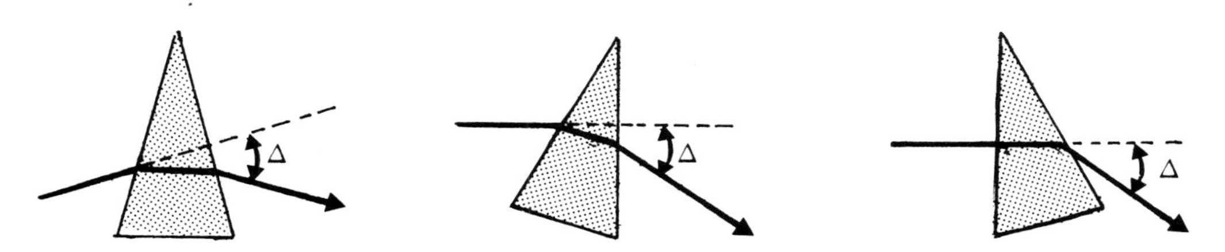

Fig. 17-13. Diagrams illustrating that the change in deviation produced by a prism depends on its orientation with reference to the incident ray. Recall that minimum deviation occurs only with symmetrical passage, but ophthalmic trial prisms are calibrated according to the deviation produced when they are held perpendicular to the line of sight, that is, normal to the visual axis. If the prisms are used to measure a near deviation, its distance from the eye becomes critical; the further the prism the greater the decrease in effective deviation; that is, the eye turns less than the prism power suggests.

orthoptics. Diplopia is apparently prevented by a small suppression scotoma within the deviated eye, and the unusual feature is the simultaneous presence of a phoria and tropia.

The qualitative cover test can be quantified by prisms; the refixation movement is neutralized until no further movement is observed. Individual square prisms, rotary prisms, or a prism bar may be used; individual prisms are preferred when the deviation has both horizontal and vertical components. The prism power measures the deviation (Fig. 17-12). The direction of the prism base is out to compensate for esodeviations (phoria or tropia), base in for exodeviations, base down in front of the right eye for a right hyperdeviation, and base up in front of the right eye for a left hyperdeviation. For large horizontal deviations, several prisms may need to be combined, preferably split between the two eyes. To leave the examiner's hands free for the prisms and occluder, an assistant or the patient holds the fixation target. The prism should be held with one face perpendicular to the line of sight* and as close to the patient's eye as possible, or the measurement

*Prisms with an apical angle larger than 42° cannot be held perpendicular to the incident light without total internal reflection and hence must be split between the two eyes.

will overestimate the true deviation (Chapter 7) (Fig. 17-13).

In rare instances in which the cover test is not feasible, the more approximate *Hirschberg test* may be used. The corneal reflections are compared in the deviating and nondeviating eye, and the centration of the Purkinje image in the deviated eye is judged relative to its distance from the pupil center (Fig. 17-14). The usual figure of 7° per millimeter displacement is too low (the cover test gives about 11° per millimeter in comparable cases as pointed out by Wheeler in 1943 and more recently by Jones and Eskridge in 1970). Image displacement is a complex function of the rotation angle at the entrance pupil (rather than the center of rotation); the curvature of the cornea and sclera differ; there may be some translatory movements of the globe; and the examiner views the cornea from an oblique angle.

The *Krimsky prism test* is a modification of the Hirschberg technique in which prisms are placed before the fixing eye until the light reflex is observed to be centered in the deviating eye. A rotary prism is satisfactory, and the test is done at 13 inches. The light reflex is always shifted in the direction of the apex of the prism. In oblique deviations, the vertical misalignment is neutralized first. The Krimsky test is a simple, rapid method that may be usefully refined by a subsequent, quantitative cover test. Care should be taken that the light is not so intense as to dissociate the eyes and that the patient looks at the light and not at the prism.

All corneal reflection tests require modification by angle kappa, which must be determined separately (but not easily in children). An exception is the *Priestley-Smith tape* (Fig. 17-15, *A*) because the corneal reflection in the squinting eye occupies the same position as it previously did in the fixing eye. The standard method of objective strabismometry is the *perimeter* or *Maddox tangent scale* (Fig. 17-15, *B* and *C*), neither particularly suitable for young children.

The cover test was popularized by Duane, White, and Brown. The perimeter was introduced by Javal. The first comprehensive description of motility disturbances is to be found in a text by Alfred Graefe (1858), a victim of confusion with his more illustrious contemporary cousin, Albrecht von Graefe. Albrecht devised the vertical prism dissociation test for heterophoria, introduced iridectomy and linear cataract extraction, and invented the "knife," each sufficient for immortality. Alfred was the editor of the Graefe-Saemich *Handbuch*, the ophthalmologic bible of his day. Julius Hirschberg deserves to be remembered for more than a mildly inaccurate motility test. He wrote the classic history of ophthalmology (nine volumes), applied the electromagnet for extracting intraocular foreign bodies, and found time to edit a practical clinical ophthalmic journal. Bielschowsky wrote the motility revision of Graefe's text.

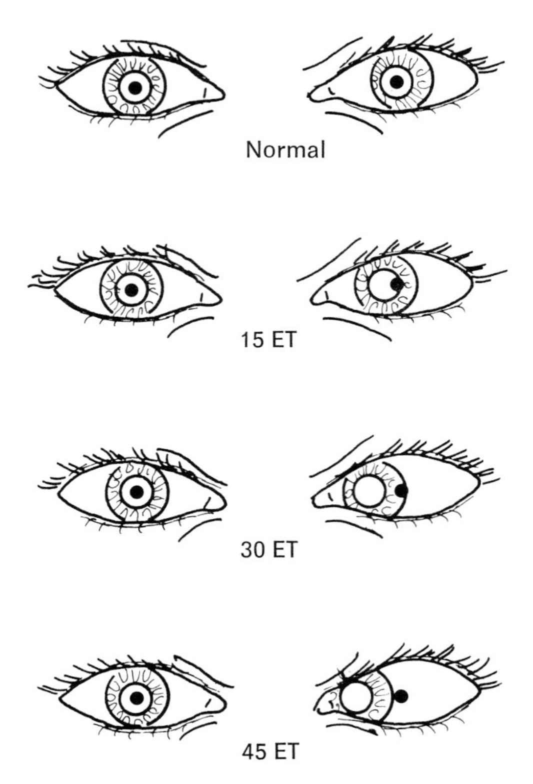

Fig. 17-14. The Hirschberg test attempts to quantify the deviation according to the position of the light reflex from a penlight held at the reading distance. The figures shown, in degrees, are approximate, however, and are probably too low by a factor of 2.

The *"forced duction" test* is useful in identifying mechanical causes for deviation. With the patient under local anesthesia, the eye is grasped at the limbus with a forceps, rotating the globe away from the line of force to be tested (e.g., toward the nose with the forceps at the temporal limbus to test the

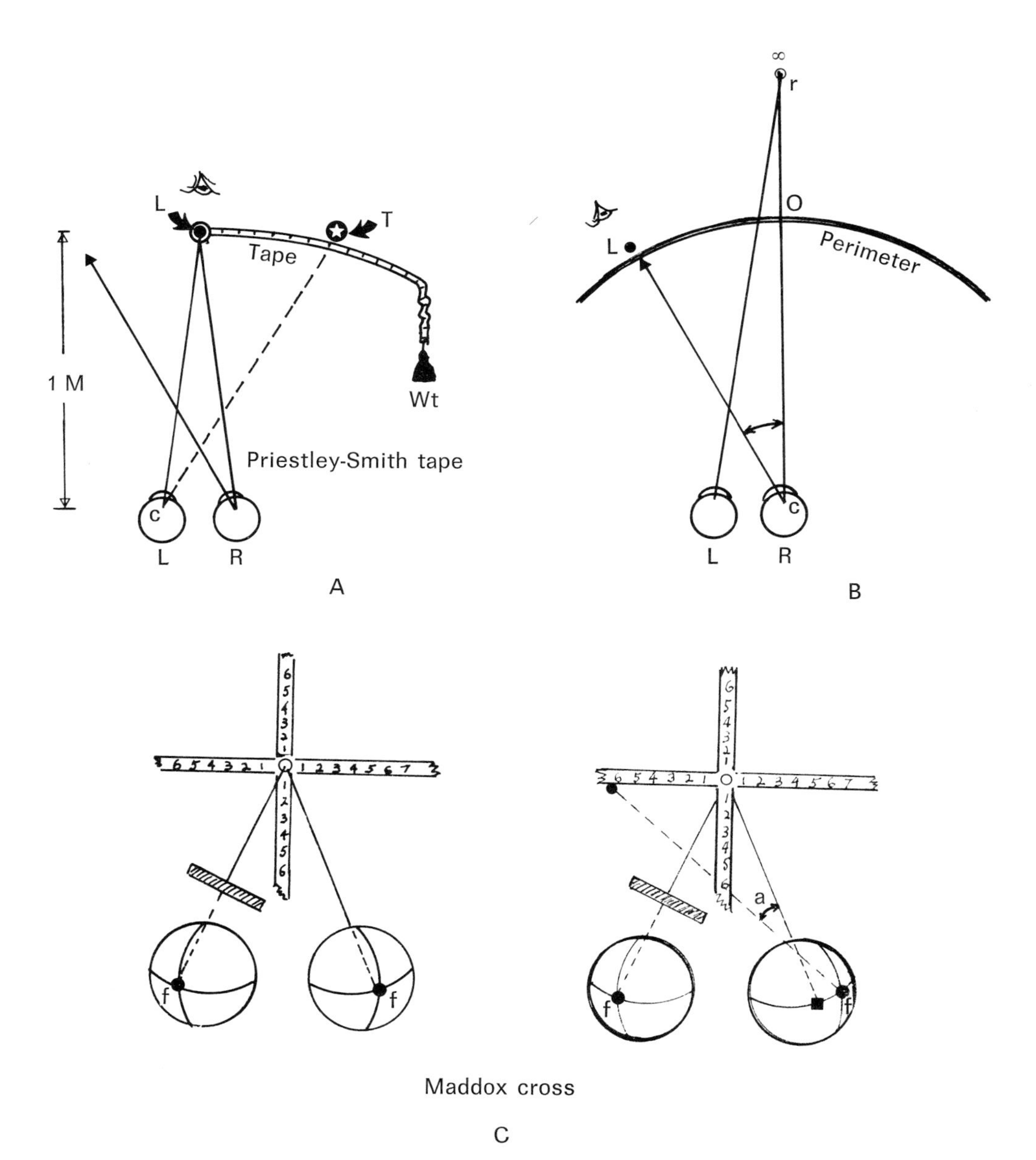

Fig. 17-15. Some representative methods of objective strabismometry. **A,** The Priestley-Smith tape consists of a ring to which is attached a 1-meter string (to define the distance from the eyes) and a graduated tape (to measure the deviation). The examiner sights above a light (e.g., a Finoff transilluminator) held inside the ring and centers the Purkinje reflex in the fixing eye. The patient is then instructed to follow a second target (e.g., the examiner's finger) moved along the graduated tape until the reflex in the squinting eye occupies the same position it previously did in the fixing eye. The angle represents the primary deviation. **B,** In the perimeter method, according to Javal, the patient fixates a distant target aligned with the zen marker of the arc. The examiner sighting above a light moved along the arc centers the reflex in the pupil of the deviated eye. **C,** In the Maddox cross, the principle is the same as the perimeter. The cross is also applicable for measuring the angle subjectively by noting the number the right eye sees superimposed on the red spot (red glass LE). Both eyes must have normal retinal correspondence and central fixation.

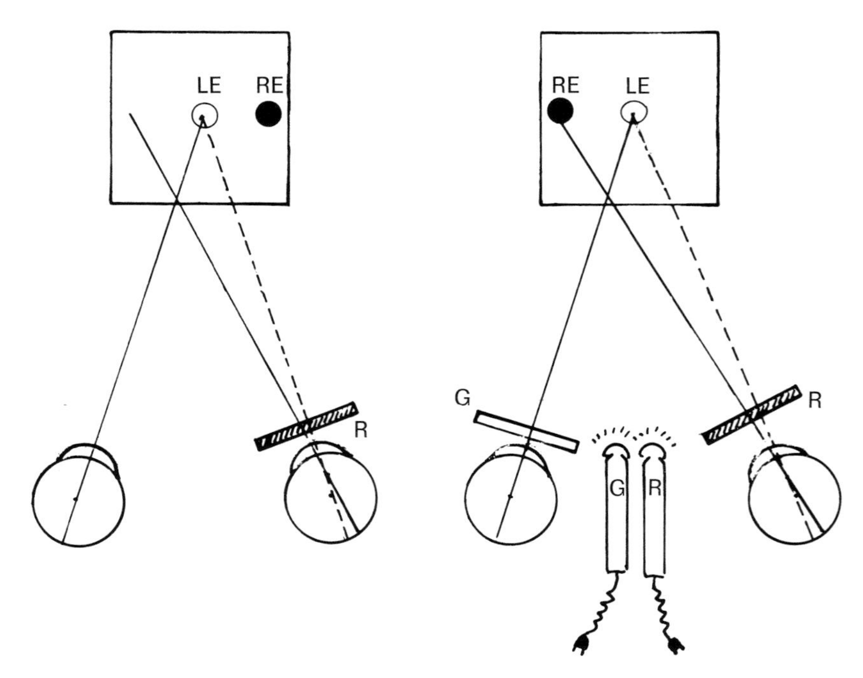

Fig. 17-16. Two methods of analyzing diplopia. *Left*, The red glass test in which both eyes see the same target. *Right*, Represented by the Lancaster or Hess red-green test, each eye sees its own target. Note that for the illustrated esotropia, the diplopia is uncrossed in the first and crossed in the second.

restriction of the lateral rectus). The test may be done as an office procedure in adults and in children under sedation. The restriction is characterized both by the feeling of resistance and the limitation of the amplitude of movement.

Subjective testing is necessarily restricted to older children who can describe the nature and extent of diplopia and absent correspondence and suppression does not obscure the response.

A subjective variant of the cover test is the *Duane parallax test* or the *Verhoeff phi phenomenon*. In the first, the patient describes the relative position of the diplopic targets as the cover is alternated; in the second, he notes the "jump" with rapid alternation. Eliminating the "phi" with prisms is a rather precise end point, but it does not always work because of suppression, ambiguous point of reference, and the need for accurate observation and description.

Subjective diplopia tests may be divided into two types; those in which the same target stimulates both eyes and those in which each eye is stimulated by a different target (Fig. 17-16).

Representative of the first group is the *red glass test*. A deep red filter (dense enough to obscure all but the target, which is a vertical bar light or circle with vertical markers to identify torsion) is placed before one eye. The patient describes the relation and separation of the double targets in the cardinal directions, which form a diagnostic picture (Fig. 17-17). The filter is placed before the opposite eye to detect secondary deviations. The test is qualitative but can be quantified by a calibrated scale or screen.

The vigorous debate of the last century whether strabismus should be measured in angular or linear units has deflated but has never been settled; we speak of the "angle" of squint but measure it in prism diopters despite the fact that there is no linear relation between the two, and the discrepancy is considerable for large angles. Thus when the orthoptist, using her synoptophore, reports a deviation of 40°, the deviation is not $40 \times 1.74 = 69.6\Delta$ but $\tan 40 = 84\Delta$ (although it would be the same number of centrads).

Diplopia, assuming normal correspondence,

depends on the localization of retinal stimuli. These are most easily recalled in terms of an imaginary binoculus (or "cyclopean eye") representing the two superimposed real eyes (Fig. 17-18).

> The rule of diplopia is a paradox quite,
> To differentiate pictures incongruous,
> The one on the left
> Goes with the eye on the right
> And vice versa, if the response is anomalous.

In physiologic diplopia, the object of fixation is always seen single; only more remote or nearer (nonfixated) targets are doubled but shadowy and insubstantial so that they practically never intrude into consciousness. This doubling promptly disappears when the involved target is fixated.

In pathologic diplopia, one object is distinct, and one is blurred. The diplopia is constant (unless there is suppression) despite changes in fixation and causes confusion and disorientation. The diplopia may be lateral, vertical, or more rarely, cyclorotational, usually a combination of these.

Representative of the second group is the *Lancaster "projection" test.* The patient wears red-green goggles and views monocular red and green targets projected by two flashlights. The examiner, holding one flashlight, projects a red "doughnut"; the patient, holding the other light, is instructed to "shoot"

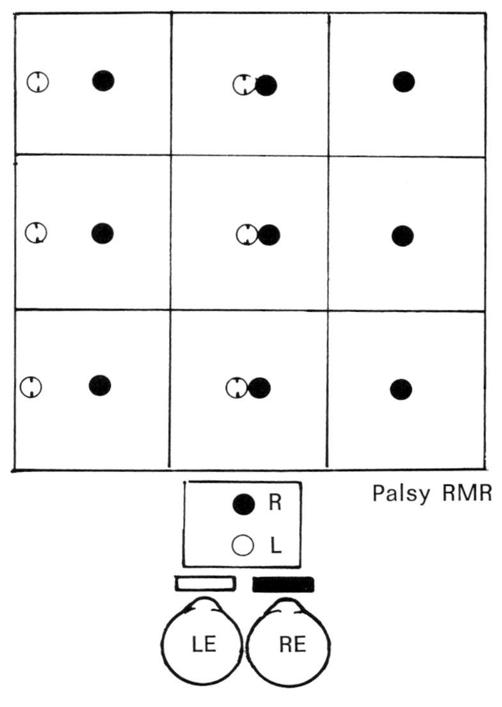

Fig. 17-17. Illustration of the application of the red glass test to a recent palsy of the right medial rectus. The deviation increases in the direction of action of the palsied muscle.

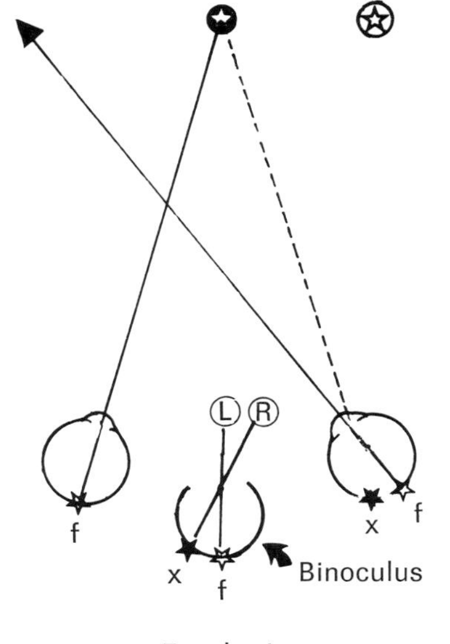

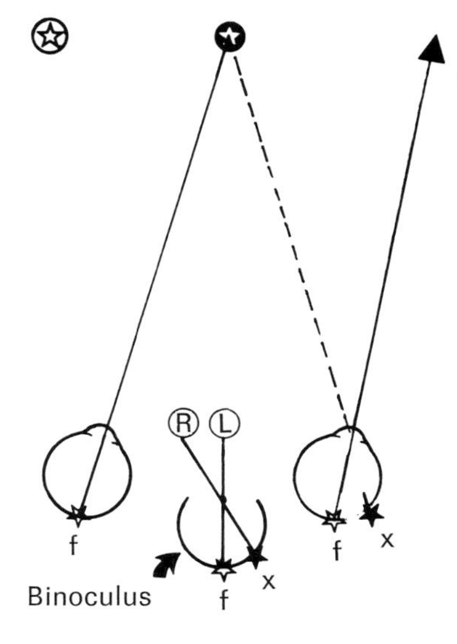

Exophoria

Fig. 17-18. The basis of the diplopic images in the red glass test (both eyes see the same target) is best illustrated by means of an imaginary binoculus, representing the superimposition of both retinas. In esodeviation, the eccentrically stimulated point *x* is to the left of the binoculus fovea, hence localized to the right. The converse is true for exodeviations.

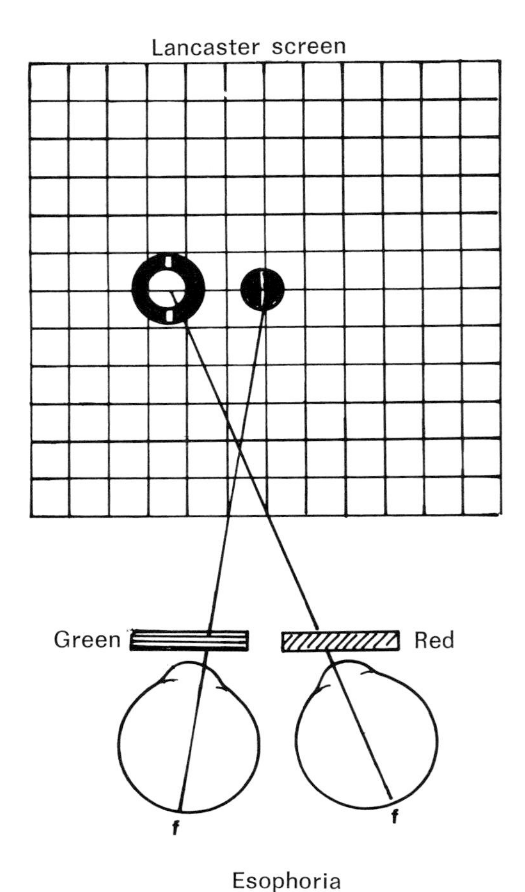

Fig. 17-19. Application of the red-green test in esophoria. Since each eye sees its own target, the images are localized according to the visual axes directly, and the diplopia is crossed. The calibrated screen quantifies the deviation.

the green ball inside the doughnut. The separation of the targets is the angle of deviation (phoria or tropia). Unlike the red glass test, the diplopia is not reversed but matches the visual axes of the eyes (Fig. 17-19). Because the targets fall on corresponding points, the foveas, they appear superimposed when separated by the amount and direction of the deviation. Torsion is indicated by vertical markers on the projected targets. If the red-green goggles are reversed, secondary deviations can be demonstrated. In normal retinal correspondence the targets appear superimposed when separated by the objective angle of deviation. Calibration is provided by a graduated screen (e.g., 2Δ at 4 meters). The test works in some children as young as 3 years of age and also provides an instructive demonstration to the parent.

HETEROPHORIA

A phoria is a tendency for the eyes to deviate relative to each other under dissociation; it represents the magnitude of the task that fusion has to perform to maintain binocular vision. Fusion can be disrupted by covering one eye, dissimilar targets, interposing displacing prisms, filters, polarizers, distorting one image, or consecutive stimulation and afterimages. Each of these methods has been used in the measurement of heterophoria. The terminology was introduced by Stevens (1887). Orthophoria was initially defined as visual parallelism but has come to apply to any fixation distance providing the visual axes intersect at the object of regard when fusion is broken. If there is an outward deviation after dissociation, it is an exophoria; if inward, esophoria; if upward, hyperphoria; and if downward, hypophoria (Fig. 17-20). Cyclorotation consequent to dissociation is a cyclophoria; if the tops of the vertical meridians of the cornea incline toward each other, it is an incyclophoria; and if they incline away from each other, it is an excyclophoria. "Anisophoria" means the phoria varies with the direction of the gaze. Since horizontal phorias change with accommodation, hence ametropia, a "basic" phoria means that the distance measurement was taken through the emmetropic correction. In vertical phorias the affected eye must be specified.

Heterophoria of some degree is probably universal; orthophoria is an oculomotor ideal as infrequently realized as emmetropia, its optical counterpart. The exact incidence depends on the precision of measurement and how long fusion is abolished. An incidence of 84% to 98% is not unusual for lateral phorias, with about 20% for vertical phorias and less than 5% for cyclophorias. Incidence figures mean little unless the fixation distance is specified.

A brief example will clarify the factors involved. Consider a patient with distance orthophoria, an interpupillary distance of 60 mm, an AC/A ratio of

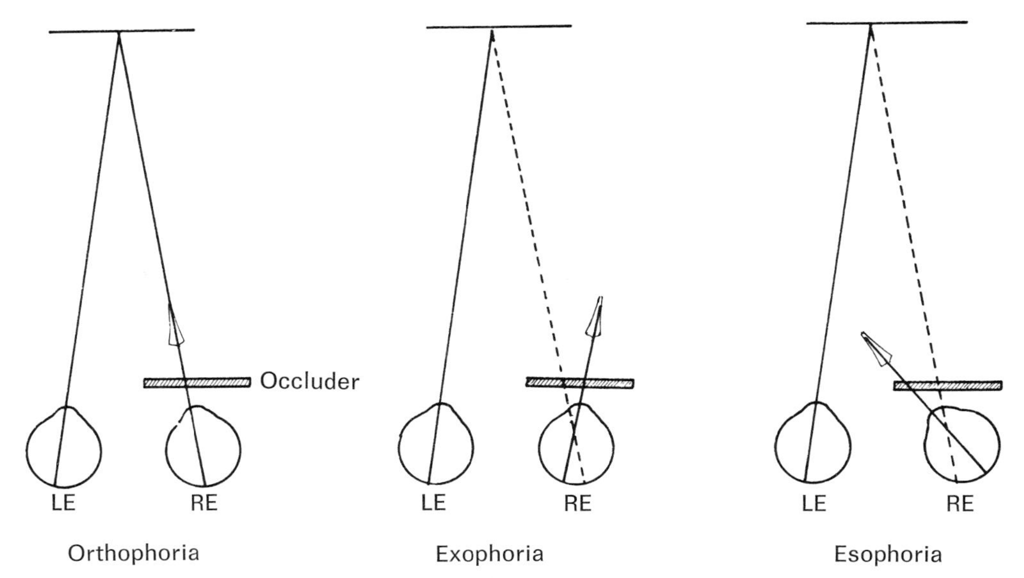

Fig. 17-20. The definition of heterophoria is based on the position the eyes assume relative to each other when dissociated (an occluder in this example).

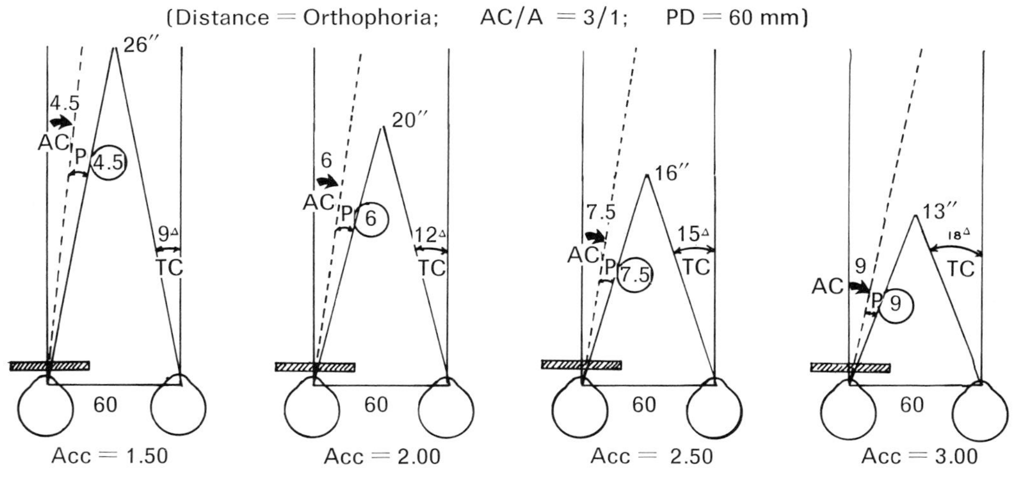

Fig. 17-21. Series of diagrams illustrating that the near lateral phoria depends on the testing distance. The closer the target the larger the phoria for the same AC/A ratio and interpupillary distance. The distance phoria is assumed to be ortho, and proximal convergence is ignored. *AC* = accommodative convergence; *P* = phoria; *TC* = total convergence.

3/1, and negligible proximal convergence. If the lateral phoria is measured at 26 inches, accommodation is 1.50 D, the total required convergence is $1.50(6) = 9\Delta$, the accommodative convergence is $1.50(3) = 4.5\Delta$, and the patient manifests $9 - 4.5 = 4.5\Delta$ exophoria. (The meaning of these computations is discussed more fully in Chapter 18.) If this calculation is repeated for a fixation distance of 20 inches, the accommodative requirement is 2 D, and since the AC/A ratio and PD do not vary, the phoria will be $12 - 6 = 6\Delta$ exophoria. At a fixation distance of 16 inches, the phoria is 7.5Δ exophoria; and at 13 inches, 9Δ exophoria (Fig. 17-21).

The example emphasizes that it is not possible to attach "clinical significance" to a particular phoria value, although this is frequently specified on driver's and flyer's examinations. Even when the test distance is stipulated, it is not the absolute phoria value but its relation to the compensating fusional vergence that determines visual efficiency.

Etiology

The cause of heterophoria is not known. Among the fuzzy causes, based on fuzzier evidence, are orbital and ocular asymmetries, maldevelopment and degeneration of the extraocular muscles, uncorrected oblique astigmatism, irregular excitation of the lower centers, disturbances of proprioception, incoordination of higher centers, reflex influences of the trigeminal nerve, senility and chronic debility, insufficient rest, and overindulgence in tobacco, alcohol, and sex. Moreover it has become traditional to emphasize the close relationship between heterophoria and heterotropia; for example, esophoria is a "tendency" and esotropia is a "manifest" inward deviation. This resemblance, although genuine, is superficial and to exaggerate it is to do less than justice to the pathophysiology of strabismus. There is more than a quantitative difference between heterophoria, in which fusion may not be used, and heterotropia, in which fusion cannot be used. An occasional blood pressure flick is not a coronary curtain call, one night cough does not portent a tuberculous finale, and a stanza with a starlet is not a passion play. Intermittent fusional lapses in normal binocular vision are fortunately rare, always temporary, and practically never result in deviations of the magnitude seen in squinters. Adults accidentally blinded in one eye develop exotropia whatever the preexisting phoria. Even granting that many forms of strabismus go through a stage of (mostly undiscovered) intermittance, the dissociated phoria during the binocular stage is smaller than the deviation in the tropia stage, and the associated phoria may be in the opposite direction. Moreover normal sensory cooperation of the eyes (stereoacuity, fusion, rivalry) is always present with heterophoria but is seldom achieved by squinters. Suppression and "anomalous correspondence" does not characterize heterophoria but is common in strabismus. An anatomic shift in the position of rest or an unrecognized muscle palsy may explain an unusually large distance phoria but ties in poorly with the symmetry of the condition and the constancy of the AC/A ratio over the greater range of reading distances.

Although phorias change with anoxia, ethanol, general anesthesia, ametropia, or fixation distance, these results can be readily explained in terms of variations in tonic, accommodative, proximal, and postural reflexes without static, kinetic, or neurogenic scenarios. Rather than a "forme fruste" of strabismus, heterophoria is a symptom of the innervational balance of well-known reflexes. When the eyes are artificially dissociated, the phoria becomes manifest and provides a clinical opportunity to study not the deviation but the neural imbalance that causes it. This imbalance differs qualitatively from strabismus in which some reflexes have either never developed or cannot be mobilized.

Measurement

The same principles underlie all methods of measuring phorias. Fusion is interrupted in some way, and the eyes drift to the phoria position. The deviation is then measured on a calibrated screen or with prisms. The method of disrupting fusion is said to influence the deviaton, but this is doubtful, since various techniques give similar results, and there is no ready way to calibrate the degree of dissociation. If illumination to one eye is progressively diminished by a photometric wedge, the break occurs at a level that is constant for the same individual but varies from one to the next. In comparing the facility of fusion through a monocular red glass or other filter, it is the transmission, not the color that influences the result. The phoria position, however, is influenced by previous tonicity, as when fusional vergences are measured immediately prior to determining the phoria. Tonicity can also be built up slowly by wearing an anisometropic correction, therapeutic prisms, orthoptic exercises, or other measures. This is the mechanism for the observed changes in vertical phorias after the prescription of compensating prisms. Because the effect is prolonged, the return to normal tonus may take days or weeks, so the patient wearing a prism gets symptoms when it is removed and is reluctant to give it up. Conversely, prolonged occlusion yields different results than a phoria measured immediately after dissociation. It is still debated which is the more useful finding, since the second more nearly represents the actual use of the eyes. The velocity of the relaxation movement is itself a complex function of previous eye movements, fusional stress, and the magnitude of the heterophoria. The clinical implication is that phorias should be determined consistently in the refraction routine, preferably before the measurement of fusional vergences. Obviously the emmetropic correction is worn with due consideration for any uncorrected latent hyperopia or lens-induced prismatic imbalance.

Unlike vertical phorias, in lateral phorias the tonicity factor is usually overshadowed by accommodation. Lateral phorias vary predictably with change in the accomodative stimulus (not necessarily the accommodative response, which is the only one measured clinically). This constancy is expressed by the AC/A ratio. It follows that accommodation must be controlled in all phoria measurements. As an extreme example, if a near phoria is being measured and the patient through lack of attention or instruction abandons looking at the near target, the clinician simply measures the far phoria over again and erroneously assumes the two are equal. This is one reason near phorias under cycloplegia have little diagnostic value. Moreover, if the data are obtained in early stages of cycloplegia or while the effect of the drug is wearing off, the ciliary effort may exceed the response, with unpredictable consequences. The conventional clinical technique of accommodative control is to have the patient read out random letters while the phoria is being measured. The problem is simplified but not eliminated for distance phorias. Accommodation may become active (and therefore stimulate accommodative convergence) if the blur is too great or the contrast too small. If the depth of focus is excessive, such as in miosis due to antiglaucoma drugs or strong light or in cases of subnormal vision in which blur discrimination is blunted, the accommodative stimulus may differ significantly from the simple optics of the fixation distance.

The awareness of the nearness of a target results in proximal accommodation and convergence and hence changes the phoria. The amount is generally small but unpredictable (it may be large in squinters who can modify the angle of deviation by "throwing the eyes out of focus" consequent to thinking of a distant or near object). In measuring the AC/A ratio, for example, it is best to keep this factor constant by repeating the near phoria at the same test distance.

Postural reflexes may influence phorias by changes in head or body position; hence the head should be in the natural inclination and the patient seated comfortably without undue strain or anxiety. Ideal test conditions seldom obtain in the refracting room, but the refractive error should be adequately corrected, accommodation controlled, attentive cooperation gained, and the most appropriate test selected. Unlike the examination of motility disorders in children, most phoria measurements are made on older children and adults as part of a routine refraction, and more precise subjective techniques can be used.

The cover test and its subjective variants (the Duane parallax, Verhoeff phi phenomenon, and cover and Maddox rod) follow the same principles outlined in previously. Accommodative control is not readily feasible with a muscle light; hence these techniques should be modified to include a detailed reading target. Quantification is achieved by rotary or individual prisms or a prism bar.

The action of a prism is to deviate light rays toward its base, but the effect on the eyes depends on whether or not they are dissociated. If a small base-out prism is placed before one eye while binocular vision is maintained, the eye rotates toward its apex in the interest of preserving fusion. If the eyes are then dissociated, there is an outward rotation to regain the physiologic rest position—the base-out prism has induced an exophoria. On the other hand, if the same prism is placed before dissociated eyes, there is no ocular rotation; the prism merely displaces the chief ray onto the fovea if the eye is esophoric. Hence a base-out prism *measures* esophoria. Conversely, a base-in prism *induces* an esophoria under binocular conditions but *measures* a manifest or existing exophoria. Once the chief ray coincides with the deviated fovea, the patient reports that the targets are aligned. In some cases, as the targets approach each other, one of them disappears and may not reappear until it is on the opposite side of that seen by the other eye. The

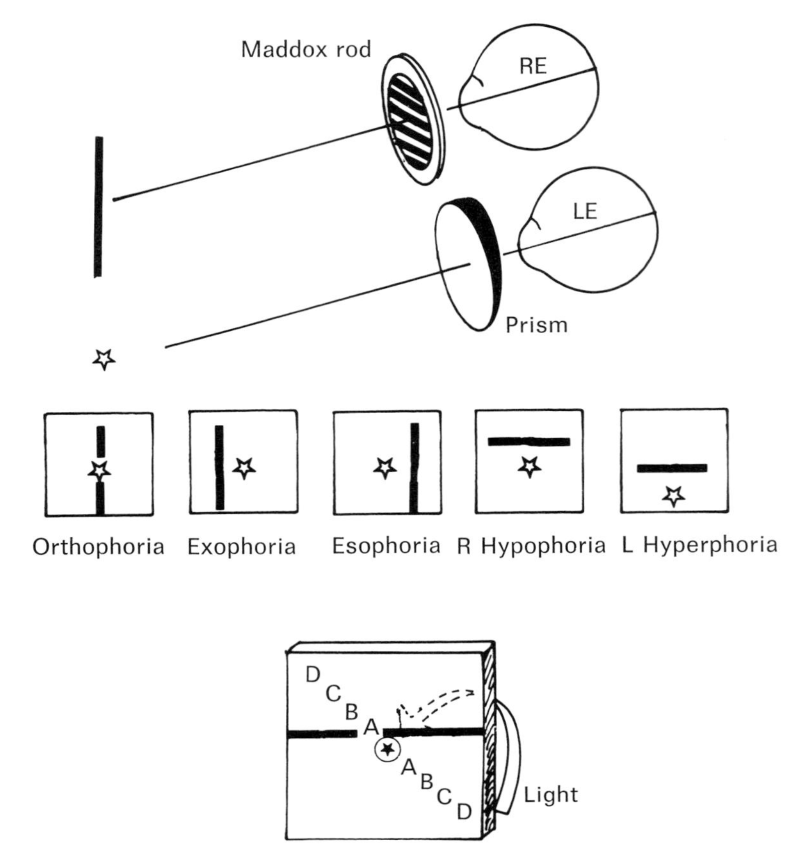

Fig. 17-22. Maddox rod test for heterophoria. For near testing, the small, hand-held Bernell or Freeman unit may be used, since accommodative targets are provided.

prismatic equivalent is one measure of the extent of the suppression scotoma.

All prismatic measurements depend on the position of the prism face with respect to the visual axis, the vertex distance, the fixation distance, and whether there are contributing prismatic effects from the emmetropic correction, especially in higher powers, since the eye looks through the lens eccentrically.

The *Maddox rod* (Fig. 17-22) is novel and easy to understand, hence a popular method of measuring phorias. It makes no apparent difference whether the rod is placed in front of the dominant or nondominant eye or whether it is red or white, the room is light or dark, or the streak long or short. The target is a distant light source, and the measuring prism is placed before the same or opposite eye. To measure horizontal phorias, the rod is horizontal, producing a vertical streak; base out or base in are added until the streak is seen superimposed on the light. To measure vertical phorias, the rod is vertical, producing a horizontal streak; base up or down is added for alignment. To measure cyclophorias, parallel Maddox rods are placed before both eyes, and the patient reports whether the streaks are inclined toward each other. The rotation of one rod (in degrees) required to restore parallelism is a measure of the cyclophoria. The Maddox rod may be combined with letters or numbers for near testing (e.g., in the Freeman near vision unit or Bernell lantern) to ensure accommodative control. A muscle light by itself should not be used for near phoria testing as it provides no precise accommodative stimulus.

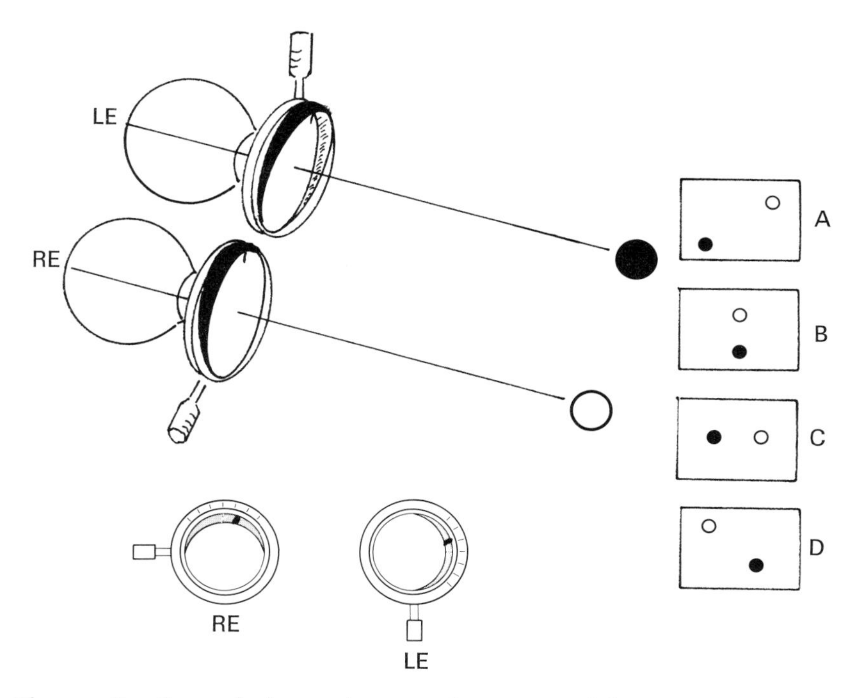

Fig. 17-23. The von Graefe test for heterophoria employs a vertical dissociating prism and a horizontal measuring prism. The initial appearance is shown in **A.** The final position for measuring lateral phorias is shown in **B** and for vertical phorias, in **C.** In **D,** the lateral prism is excessive.

The *von Graefe technique* employs two rotary prisms, conventionally arranged about 6Δ base up before the left eye and about 12Δ base in before the right eye. The patient observes a distant or near target, which may be a single letter or a line of letters. He now sees two targets one above and to the right of the other. The horizontal prism is altered to align the two targets one above the other for lateral phorias; the vertical prism is altered to align the targets side by side for vertical phorias (Fig. 17-23). The chief advantage of this method is that the same conditions apply to distance and near phorias, accommodation is controlled, and the instrumentation is conveniently at hand on the refractor. Moreover additional spheres are easily added and the phoria repeated to measure the AC/A ratio.

A variety of *stereoscopic methods* may be applied to measuring phorias, ranging from the Lancaster red-green projection test to the Thorington arrow. The principle is that each eye sees its own target, separated by a septum as in the Maddox wing test, by colored or polarized filters, or by intermittent alternate stimulation. An arrow may be presented to one eye and a line of calibrated numbers, letters, or pictures to the other. The patient indicates the position of the arrow, or the arrow may be aligned with the zero marker by prisms (Fig. 17-24). More fluctuation generally occurs on this test as the patient changes fixation toward the various symbols.

Management

The management of heterophorias depends on whether symptoms are present and is directed at the pertinent mechanism. The treatment may include altering accommodative vergence by modifying the lens prescription, relieving prisms, drugs, orthoptics, and, in rare instances, muscle surgery. Vertical phorias are more likely to cause symptoms than horizontal phorias because the compensating fusional vergence amplitudes are smaller. The most frequent cause of vertical

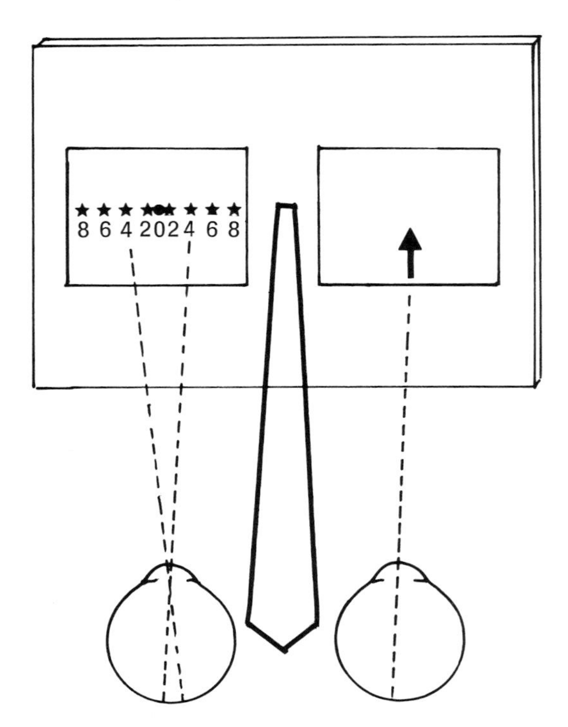

Fig. 17-24. In the stereoscopic variant of measuring phorias (e.g., Maddox Wing), each eye sees its own target. The number the arrow points to is calibrated to give the lateral phoria directly. A similar target is used for vertical phorias. The targets should be used at the appropriate distance for which they were designed.

phorias are probably anisometropia and anisometropic prescriptions and are considered in Chapter 19. Lateral phorias are discussed in Chapter 18. Cyclophoria is seldom measured routinely because it is rare, no treatment is indicated, and no satisfactory treatment is available. Alternating hyperphoria (occlusion hypertropia, dissociated vertical divergence) is a rare condition in which the covered eye deviates upward, whichever eye is covered. The cause is not known, and the treatment is directed at associated anomalies (latent nystagmus, hypertropia, abnormal head posture). In many cases, binocular vision is normal, and no treatment is required.

REFERENCES

Adler, F. H.: Reciprocal innervation of extra-ocular muscles, Arch. Ophthal. **3**:318-324, 1930.

Allen, M. J.: The dependence of cyclophoria on convergence, elevation and the system of axes, Am. J. Optom. **31**:297-307, 1954.

Alpern, M.: Muscular mechanisms. In Davson, H., editor: The eye, vol. 3, ed. 2, New York, 1969, Academic Press, Inc., pp. 1-252.

Alpern, M.: Eye movements, Handbook Sensory Physiol. **7**:303-330, 1972.

Ames, A.: Cyclophoria, Am. J. Physiol. Opt. **7**:3-38, 1926.

Apt, L., Isenberg, S., and Gaffney, W. L.: The oculocardiac reflex in strabismus surgery, Am. J. Ophthal. **76**:533-536, 1973.

Aserinsky, E., and Kleitman, N.: Two types of ocular motility occurring in sleep, J. Appl. Physiol. **8**:1-10, 1955.

Bender, M. B., editor: The oculomotor system, New York, 1964, Harper & Row, Publishers.

Bender, M. B., and Weinstein, E. A.: Functional organization of the oculomotor and trochlear nuclei, Arch. Neurol. Psychiatr. **49**:98-106, 1943.

Berens, C.: A prism bar of thermoplastic material for measuring high degrees of squint, Am. J. Ophthal. **23**:805, 1940.

Bielschowsky, A.: Congenital and acquired deficiencies of fusion, Am. J. Ophthal. **18**:925-937, 1935.

Bielschowsky, A.: Disturbances of the vertical motor muscles of the eye, Arch. Ophthal. **20**:175-200, 1938.

Bielschowsky, A.: Lectures on motor anomalies, Hanover, 1940, Dartmouth College Publications.

Blodi, F. C., and van Allen, M. W.: Electromyography of extraocular muscles in fusional movements, Am. J. Ophthal. **44**:136-144, 1957.

Boeder, P.: The cooperation of the extraocular muscles, Am. J. Ophthal. **51**:469-481, 1961.

Boghen, D., Troost, B. T., Daroff, R. B., Dell'Orso, L. F., and Birkett, J. E.: Velocity characteristics of normal human saccades, Invest. Ophthal. **13**:619-623, 1974.

Bourdon, B.: La perception visuelle de l'espace, Paris, 1902, C. Reinwald.

Breinin, G. M.: Analytic studies of the electromyogram of human extraocular muscle, Am. J. Ophthal. **46**:123-142, 1958.

Breinin, G. M.: The electrophysiology of extraocular muscles, Toronto, 1962, University of Toronto Press.

Breinin, G. M.: The structure and function of extraocular muscle—an appraisal of the duality concept, Am. J. Ophthal. **72**:1-9, 1972.

Brindley, G. S., and Merton, P. A.: The absence of position sense in the human eye, J. Physiol. (London) **153**:127-130, 1960.

Burian, H. M.: Fusional movements—role of peripheral retinal stimuli, Arch. Ophthal. **21**:486-691, 1939.

Burian, H. M.: The place of peripheral fusion in orthoptics, Am. J. Ophthal. **30**:1005-1012, 1947.

Burian, H. M.: Etiology of heterophoria and heterotropia. In Allen, J. H., editor: Strabismus Ophthalmic Symposium II, St. Louis, 1958, The C. V. Mosby Co., pp. 174-183.

Cantonnet, A., Filliozat, J., and Fombeure, G.: Le strabisme, Paris, 1932, N. Maloine.

Chamberlain, W.: Restriction in upward gaze with advancing age, Am. J. Ophthal. **71**:341-346, 1971.
Chavasse, F. B.: Worth's squint, Philadelphia, 1949, Blakiston Co.
Cogan, D.: Neurology of the ocular muscles, Springfield, Ill., 1956, Charles C Thomas, Publisher.
Cooper, S., and Daniel, P. M.: Muscle spindles in human extrinsic eye muscles, Brain **72**:1-24, 1949.
Cooper, S., Daniel, P. M., and Whitteridge, D.: Afferent impulses in the oculomotor nerve from the extrinsic eye muscles, J. Physiol. (London) **113**:463-474, 1951.
Cridland, N.: Measurement of heterophoria, Br. J. Ophthal. **25**:145-167, 189-225, 1941.
Crosby, E. C.: Relations of brain centers to normal and abnormal eye movements in the horizontal plane, J. Comp. Neurol. **99**:437, 1953.
Daroff, R. B.: Control of ocular movement, Br. J. Ophthal. **58**:217-223, 1974.
Davies, T., and Merton, P. A.: Recording compensatory rolling of the eyes, J. Physiol. (London) **140**:27-28, 1958.
Dayton, G. O., Jones, M. H., Aiu, P., Rawson, R. A., Steele, B., and Rose, M.: Developmental study of coordinated eye movements in the human infant, Arch. Ophthal. **71**:865-875, 1964.
Dodge, R.: Five types of eye movements in the horizontal plane of the field of regard, Am. J. Physiol. **8**:307, 1903.
Duane, A.: Some new tests for insufficiencies of the ocular muscles, together with a system of abbreviation suitable for note taking, N. Y. Med. J. **50**:113-118, 1889.
Duane, A.: The value of the screen test as a precise means of measuring squint, Ann. Ophthal. **12**:620-630, 1903.
Duane, A.: Congenital deficiency of abduction in association with impairment of adduction, retraction movements, contraction of the palpebral fissure, and oblique movements of the eyes. Arch. Ophthal. **34**:133-159, 1905.
Duke-Elder, S., and Wybar, K.: System of ophthalmology: ocular motility and strabismus, vol. 6, St. Louis, 1973, The C. V. Mosby Co.
Eakins, K. E., and Katz, R. L.: The role of the autonomic nervous system in extraocular muscle function, Invest. Ophthal. **6**:253-360, 1967.
Ellerbrock, V. J.: Experimental investigation of vertical fusional movements, Am. J. Optom. **26**:327-337, 388-399, 1949.
Ellerbrock, V. J.: Inducement of cyclofusional movements, Am. J. Optom. **31**:553-566, 1954.
Fender, D.: Control mechanisms of the eye, Sci. Am. **211**:24-33, 1964.
Fink, W. H.: Vergence test—evaluation of various techniques, Am. J. Ophthal. **31**:49-57, 1948.
Fink, W. H.: Surgery of the vertical muscles of the eye, Springfield, Ill., 1962, Charles C Thomas, Publisher.
Flom, M. C.: Some interesting eye movements obtained during the cover test, Am. J. Optom. **35**:67-71, 1958.
Ford, F. R., and Walsh, F. B.: Tonic deviation produced by movements of head, Arch. Ophthal. **23**:1274-1284, 1940.
Fulton, J. F.: Physiology of the nervous system, ed. 3, New York, 1949, Oxford University Press.
Gay, A. J., Newman, N. M., Keltner, J. L., and Stroud, M. H.: Eye movement disorders, St. Louis, 1974, The C. V. Mosby Co.
Goldstein, J. H., Clahane, A. C., and Sanfilippo, S.: The role of the periphery in binocular vision, Am. J. Ophthal **62**:702-706, 1966.
Graefe, A.: Klinische Analyse der Motilitatsstorungen des Auges, Berlin, 1858, Hermann Peters.
Graybiel, A., and Woellner, R. C.: A new and objective method for measuring ocular torsion, Am. J. Ophthal. **47**:349-352, 1959.
Green, E. L., and Scobee, R. G.: Position of the Risley prism in the Maddox-rod test, Am. J. Ophthal. **34**:211-217, 1951.
Grupposo, S. S., and Pomerantzeff, O.: A new hand deviometer, Am. J. Ophthal. **58**:1063-1064, 1964.
Guyton, J. S., and Kirkman, N.: Ocular movement, Am. J. Ophthal. **41**:438-476, 1956.
Haddad, H. M., and Zulouf, M.: Deviometry in noncomitant deviations, Ann. Ophthal., pp. 55-69, Jan., 1971.
Healy, E.: The treatment of phorias in adults, Am. J. Ophthal. **31**:703-708, 1948.
Hebbard, F. W.: A new method of recording small eye movements using corneal reflections, Am. J. Optom. **41**:241-247, 1964.
Henderson, J. W.: The neuroanatomy of ocular motility and strabismus. In New Orleans Academy of Ophthalmology: Strabismus, St. Louis, 1962, The C. V. Mosby Co., pp. 56-99.
Hering, E.: Spatial sense and movements of the eye, Baltimore, 1942, Williams & Wilkins Co.
Hermans, T. G.: Torsion in persons with no known eye defects, J. Exp. Psychol. **32**:307-324, 1943.
Hirschberg, J.: Ueber Messung des Schielgrades und Dosirung der Schieloperation, Centralbl. Prakt. Augenheilkd. **9**:325-327, 1885.
Hyde, J. E.: Some characteristics of voluntary human ocular movement in the horizontal plane, Am. J. Ophthal. **48**:85-94, 1959.
Irvine, S. R.: A simple test for binocular fixation, Am. J. Ophthal. **27**:740-746, 1944.
Irvine, S. R.: Measuring scotomas with the prism displacement test, Am. J. Ophthal. **61**:1177-1187, 1966.
Irvine, S. R., and Ludvigh, E.: Is ocular proprioceptive sense concerned in vision? Arch. Ophthal. **15**:1037-1049, 1936.
Jackson, J. H.: Remarks on the diagnosis and treatment of diseases of the brain. In Selected writings, London, 1932, Hodder & Stoughton.
James, W.: The principles of psychology, New York, 1890, Henry Holt & Co.
Jampel, R. S.: Multiple motor systems in the extraocular

muscles of man, Invest. Ophthal. **6**:288-293, 1967.

Jampel, R. S.: The fundamental principle of the action of the oblique ocular muscles, Am. J. Ophthal. **69:** 623-638, 1970.

Jampolsky, A., editor: Ocular deviations, vol. 4, International Ophthalmology Clinics Boston, 1964, Little, Brown & Co.

Jampolsky, A.: A simplified approach to strabismus diagnosis. In the New Orleans Academy of Ophthalmology: Symposium on strabismus, St. Louis, 1971, The C. V. Mosby Co., pp. 34-92.

Javal, E.: Manuel theoretique et practique du strabisme, Paris, 1896, G. Masson.

Jokl, A.: Julius Hirschberg, Am. J. Ophthal. **48**:329-339, 1959.

Jones, R., and Eskridge, J. B.: The Hirschberg test—a reevaluation, Am. J. Optom. **47**:105-114, 1970.

Kaivonen, M.: Diagnosis of paretic vertically—acting muscle from compensatory head posture, Acta Ophthal. **39**:370-376, 1961.

Kawasaki, K., Hirose, T., Jacobson, J. H., and Cordella, M.: Binocular fusion, Arch. Ophthal. **84**:25-28, 1970.

Keiner, G. B. J.: New viewpoints on the origin of squint, The Hague, 1951, Martinus Nijhoff.

Keiner, G. B. J.: Physiology and pathology of the optomotor reflexes, Am. J. Ophthal. **42**:233-251, 1956.

Kreiger, A. E.: Orbital blow-out fractures, Am. Orthop. J. **21**:64-76, 1971.

Krewson, W. R.: Comparison of oblique extra-ocular muscles, Arch. Ophthal. **32**:204-207, 1944.

Krimsky, E.: The stereoscope in theory and practice, Br. J. Ophthal. **21**:161-197, 1937.

Krimsky, E.: The cardinal anglometer, Arch. Ophthal. **26**:670-674, 1941.

Krimsky, E.: Fixational corneal light reflexes as an aid in binocular investigation, Arch. Ophthal. **30**:505-520, 1943.

Lancaster, W. B.: Detecting, measuring, plotting, and interpreting ocular deviations, Arch. Ophthal. **22**:867-880, 1939.

Lancaster, W. B.: Fifty years experience in ocular motility, Am. J. Ophthal. **24**:485-496, 619-624, 741-748, 1941.

Lancaster, W. B.: Terminology in ocular motility and allied subjects, Am. J. Ophthal. **26**:122-132, 1943.

Lang, J.: Management of microtropia, Br. J. Ophthal. **58**:281-292, 1974.

Laurence, J. Z.: The optical defects of the eye and their consequences, asthenopia and strabismus, London, 1865, R. Hardwicke.

Lawrence, W.: Treatise on the eye, Philadelphia, 1843, Lea & Blanchard.

Levine, M. H.: Evaluation of the Bielschowsky head-tilt test, Arch. Ophthal. **82**:433-439, 1969.

Listing, J. B. Cited in Ruete, C. G. Th.: Lehrbuch Ophthalmologie, Braunschweig, 1855, Friedrich Vieweg & Sohn.

Ludvigh, E.: Extra foveal visual acuity as measured with Snellen test letters, Am. J. Ophthal. **24**:303-310, 1941.

Ludvigh, E.: Control of ocular movement and visual interpretation of environment, Ach. Ophthal. **48:** 442-448, 1952.

Ludvigh, E.: Temporal course of relaxation of binocular duction (fusion) movements, Arch. Ophthal. **71**:389-399, 1964.

Ludvigh, E., and McKinnon, P.: Dependence of the amplitude of fusional convergence movements on the velocity of the eliciting stimulus, Invest. Ophthal. **7**:347-352, 1968.

Ludvigh, E., McKinnon, P., and Zaitzeff, L.: Relative effectivity of foveal and parafoveal stimuli in eliciting fusion movements, Arch. Ophthal. **73**:115-121, 1965.

Lyle, T. K.: Torsional diplopia due to cyclotropia and its surgical treatment, Trans. Am. Acad. Ophthal. Otolaryngol. **68**:387-411, 1964.

Lyle, T. K., and Foley, J.: Subnormal binocular vision with special reference to peripheral fusion, Br. J. Ophthal. **39**:474-487, 1955.

Mackworth, J. F., and Mackworth, N. H.: Eye fixations recorded on changing visual scenes by the television eye-marker, J. Opt. Soc. Am. **48**:439-445, 1958.

Marlow, F. W.: Recent observations on prolonged occlusion test, Am. J. Ophthal. **16**:519-527, 1933.

Marlow, F. W.: Muscle balance in myopia, Arch. Ophthal. **13**:584-597, 1935.

Marquez, M.: Supposed torsion of eye around visual axis in oblique directions of gaze, Arch. Ophthal. **41**:704-717, 1949.

Martens, T. G.: Use of prisms in the treatment of phorias, Am. Orthopt. J. **7**:71-75, 1957.

Maxwell, J. T.: Muscle balance determinations at reading distance, Arch. Ophthal. **26**:98-101, 1941.

Mein, J.: Newer methods of investigating strabismus, Br. J. Ophthal. **58**:232-239, 1974.

Metz, H. S., and Rice, L. S.: Human eye movements following horizontal rectus muscle disinsertion, Arch. Ophthal. **90**:265-267, 1973.

Miller, J. E.: Electromyographic pattern of saccadic eye movements, Am. J. Ophthal. **46**:183-186, 1958.

Miller, J. E.: Recent histologic and electron microscopic findings in extraocular muscle, Trans. Am. Acad. Ophthal. Otolaryngol. **75**:1175-1185, 1971.

Morgan, M. W.: The reliability of clinical measurements with special reference to distance heterophoria, Am. J. Optom. Monograph No. 177, 1955.

Nauheim, J. S.: A preliminary investigation of retinal locus as a factor in fusion, Arch. Ophthal. **58**:122-125, 1957.

Niederecker, O., Mash, A. J., and Spivey, B. E.: Horizontal fusional amplitudes and versions, Arch. Ophthal. **87**:283-285, 1972.

Parks, M. M.: Isolated cyclovertical muscle palsy, Arch. Ophthal. **60**:1027-1035, 1958.

Parks, M. M.: Recent developments in sensory testing, Am. Orthopt. J. **15**:85-91, 1965.

Pascal, J. I.: Effect of version and vergence movements on ocular torsion, Am. J. Ophthal. **40**:735-737, 1955.

Pasik, T., Pasik, P., and Bender, M. B.: The superior

colliculi and eye movements, Arch. Neurol. **15**:420-436, 1966.

Peter, L. C.: The extraocular muscles, Philadelphia, 1927, Lea & Febiger.

Posner, A.: Nocomitant hyperphorias considered as aberrations of postural tonus of musculature apparatus, Am. J. Ophthal. **27**:1275-1279, 1944.

Priestley-Smith, D.: A tape measure for strabismus, Ophthal. Rev. **7**:349-352, 1888.

Putnam, O. A., and Quereau, J. V.: Precisional errors in measurement of squint and phoria, Arch. Ophthal. **34**:7, 1945.

Raab, E. L.: Dissociative vertical deviations, J. Pediatr. Ophthal. **7**:146-151, 1970.

Rashbass, C.: The relationship between saccadic and smooth tracking eye movements, J. Physiol. (London) **159**:326-338, 1961.

Rashbass, C., and Westheimer, G.: Disjunctive eye movements, J. Physiol. (London) **159**:339-360, 1961.

Rashbass, C., and Westheimer, G.: Independence of conjugate and disjunctive eye movements, J. Physiol. (London) **159**:361-364, 1961.

Robinson, D. A.: The oculomotor control system: a review, Proc. Inst. Elec. Eng. **56**:1032-1049, 1968.

Robinson, D. A., and Fuchs, A. F.: Eye movements evoked by stimulation of frontal eye fields, J. Neurophysiol. **32**:637-648, 1969.

Romano, P. E., and von Noorden, G. K.: Limitations of cover test in detecting strabismus, Am. J. Ophthal. **72**:10-12, 1972.

Rosenbaum, A., and Metz, H. S.: Saccadic velocity measurmenet, Am. J. Ophthal. **77**:215-222, 1974.

Savage, G. C.: Ophthalmic myology, Nashville, 1911, McQuiddy Printing Co.

Scobee, R. G.: Anatomic factors in the etiology of heterophoria, Am. J. Ophthal. **31**:781-795, 1948.

Scobee, R. G.: The oculorotary muscles, ed. 2, St. Louis, 1952, The C. V. Mosby Co.

Scobee, R. G., and Green, E. L.: Tests for heterophoria, Am. J. Ophthal. **30**:436-451, 1947.

Scott, A. B.: Extraocular muscles and head tilting, Arch. Ophthal. **78**:397-399, 1967.

Scott, A. B.: Active force tests in lateral rectus paralysis, Arch. Ophthal. **85**:397-404, 1971.

Scott, A. B., and Collins, C. C.: Division of labor in human extraocular muscles, Arch. Ophthal. **90**:319-322, 1973.

Sewell, J. J., Knobloch, W. H., and Eifrig, D. E.: Extraocular muscle imbalance after surgical treatment for retinal detachment, Am. J. Ophthal. **78**:321-323, 1974.

Sherrington, C. S.: The integrative action of the nervous system, New York, 1906, Charles Scribner's Sons.

Sloane, A. E.: Analysis of methods for measuring diplopia fields, Arch. Ophthal. **46**:277-310, 1951.

Spaeth, E. B.: The vertical element in the causation of so-called horizontal concomitant strabismus, Am. J. Ophthal. **31**:1553-1566, 1948.

Stevens, G. T.: Motor apparatus of the eyes, Philadelphia, 1906, F. A. Davis Co.

Stuart, J. A., and Burian, H. M.: Changes in horizontal heterophoria with elevation and depression of gaze, Am. J. Ophthal. **53**:274-279, 1962.

Sugar, H. S.: An evaluation of results in the use of measured recessions and resections in the correction of horizontal concomitant strabismus, Am. J. Ophthal. **35**:959-967, 1952.

Thompson, D. A.: Measurements with cover test vs. troposcope, Am. Orthopt. J. **2**:47, 1952.

Urist, M.: Vertical muscle paresis, Am. J. Ophthal. **57**:1007-1037, 1964.

Urist, M.: Pseudostrabismus caused by abnormal configuration of the upper eyelid margin, Am. J. Ophthal. **75**:455-457, 1973.

Urrets-Zavalia, A.: Significance of congenital cyclovertical motor defects of the eyes, Br. J. Ophthal. **39**:11, 1955.

van Wien, S.: The influence of torsional movements on the axis of astigmatism, Am. J. Ophthal. **31**:1251-1260, 1948.

Verhoeff, F. W.: Occlusion hypertropia, Arch. Ophthal. **26**:780-795, 1941.

von Graefe, A.: Beiträge zur Lehre vom Schielen, Arch. Ophthal. **3**:177, 1857.

von Noorden, G. K., and Helveston, E. M.: Influence of eye position on fixation behavior and visual acuity, Am. J. Ophthal. **70**:199-204, 1970.

von Noorden, G. K., and Mackensen, G.: Pursuit movements of normal and amblyopic eyes, Am. J. Ophthal. **53**:477-487, 1962.

von Noorden, G. K., and Maumenee, A. E.: Atlas of strabismus, ed. 2, St. Louis, 1973, The C. V. Mosby Co.

Walls, G. L.: The evolutionary history of eye movements, Vision Res. **2**:69-80, 1962.

Walls, G. L.: Is vision ever binocular, Opt. J. Rev. Optom. **85**:33-43, 1948.

Warwick, R.: Representation of the extraocular muscles in the oculomotor nuclei in the monkey, J. Comp. Neurol. **98**:449-504, 1953.

Westheimer, G.: Kinematics of the eye, J. Opt. Soc. Am. **47**:967-974, 1957.

Wheeler, M. C.: Objective strabismometry in young children, Arch. Ophthal. **30**:720-736, 1943.

White, J. W.: Hyperphoria, diagnosis and treatment, Arch. Ophthal. **7**:739-747, 1932.

White, J. W.: The screen test and its application, Am. J. Ophthal. **19**:653-659, 1936.

Wiesel, T. N.: Effects of monocular deprivation on the cat's visual cortex, Trans. Am. Acad. Ophthal. Otolaryngol. **75**:1186-1192, 1971.

Winkelman, J. E.: Central and peripheral fusion, Arch. Ophthal. **50**:179-183, 1953.

Wolff, E.: The anatomy of the eye and orbit, ed. 6, revised by Last, R. J., Philadelphia, 1968, H. K. Lewis & Co., Ltd.

Wolter, J. R.: The relation of horizontal saccadic and vergence movements, Arch. Ophthal. **56**:685-690, 1956.

Wong, G. Y., and Jampolsky, A.: Agenesis of three

horizontal rectus muscles, Ann. Ophthal. pp. 909-915, Sept., 1974.

Yarbus, A. L.: Eye movements and vision, New York, 1967, Plenum Press.

Zeeman, W. P. C.: Conservative treatment of strabismus, Doc. Ophthal. **7-8**:527, 1954.

Zuber, B. L., and Stark, L.: Saccadic suppression, elevation of visual threshold associated with saccadic eye movements, Exp. Neurol. **16**:65, 1966.

Zweifach, P. H., Walton, D. S., and Brown, R. H.: Isolated congenital horizontal gaze paralysis, Arch. Ophthal. **81**:345-350, 1969.

18

Accommodation and convergence

Whenever fixation changes from a distant to a near object, each eye accommodates, and both eyes converge in the interest of clear, single binocular vision. A failure of these two precise adjustments to coordinate results in motility imbalance, inevitable in most forms of ametropia and intractable in most forms of strabismus. Symptoms include discomfort, blurred vision, vertigo, and intermittent diplopia, especially in reading or other close work. A disparity between accommodation and convergence that exceeds the patient's capacity of adaptation may give rise to suppression, amblyopia, or strabismus. The etiology is complex and the mechanism poorly understood. Refinements of neurophysiologic testing are not always in accord, are not always practical, and sometimes contradict psychophysical results. Treatment, based on limited anatomic facts and unlimited physiologic speculation, has tended to remain empirical. If the surgeon's motto is "be bold," the orthoptist whispers "but not too bold." Ranging from poetic admonitions to conserve nervous energy to mundane injunctions to cut the plus and get a deposit, most refractionists hasten to qualify their prescription before being obliged to retract the prismatic consequences. As Percival wrote at the turn of the century:

> This seems to explain the difference of opinion of ophthalmologists on the subject of faulty muscular tendencies. Some, laying stress on their observation that such faulty tendencies in some cases do not give rise to asthenopia, are inclined to disregard them in all cases, and to confine their attention to the exact correction of refractive errors. Others urge the correction of every muscle anomaly as soon as it is discovered, attributing to it almost any ache or pain, or indeed anything else the patient may suffer from. The position I take is that if the relation between the two functions is unfitted to present requirements, and if there is not sufficient faculty of adaptation, we should make the glasses suit the patient, instead of vainly attempting to make the patient suit the glasses.*

This has a cheerful ring for clinical ears. It emphasizes that the evaluation of motility imbalances is not merely a technical ritual but a diagnostic tool to be individually applied when indicated by specific signs and symptoms.

CLINICAL PHYSIOLOGY

Vergence movements compounded with faster versions occur continuously in normal vision. Vergences are generally associated with accommodation and miosis, although some components function independently and may be abnormal to a different degree. For example, a large accommodative vergence coupled with hyperopia and inadequate fusional vergence may result in accommodative estropia; a low accommodative vergence with a large tonic divergence may result in intermittent exotropia and end as a constant squint if the associated hyperopia is overcorrected.

*From Percival, A. C.: The relation of convergence to accommodation and its practical bearing, Ophthal. Rev. **11**:313-328, 1892.

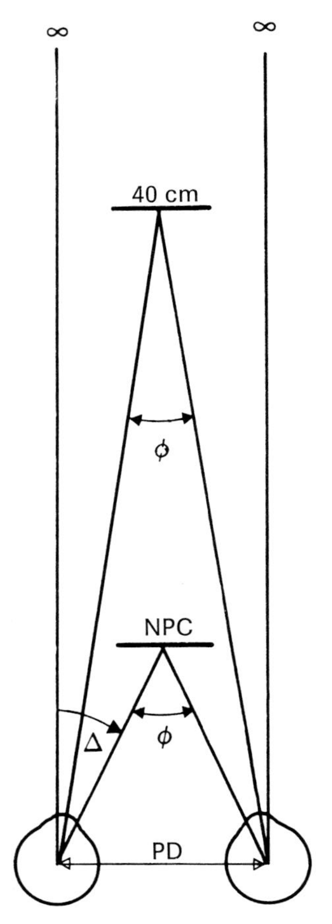

Fig. 18-1. Convergence may be measured in angular (ϕ) or linear (Δ) units.

In symmetrical convergence, each eye turns equally about its fixed hypothetical center of rotation 13.5 mm from the cornea. In asymmetrical vergence it becomes more complex. When a target is brought nearer along the visual axis of one eye, all the rotation is in the other in flagrant violation of Hering's law, (fortunately the only known exception). A line connecting the two centers of rotation is the interocular distance—the base of a triangle whose apex is the fixation point (Fig. 18-1). For practical purposes, the smaller interpupillary distance is substituted.

Convergence may be measured in any angular units, but linear ones are more convenient, and of these, the simplest is the meter angle (Nagel, 1861). The meter angle (like the diopter) is the reciprocal of the fixation distance (in meters) and gives a unit numerically identical to the accommodation required for the same distance. Although this makes it easy to remember, one person's meter angle differs from another according to the PD. It will be evident from Fig. 18-2 that a wider PD requires more convergence to fixate the same distance. Since the meter angle ignores PD, a more representative unit is the prism diopter (or centrad for larger rotations). The prism diopter also relates directly to the prismatic effects of spectacle lenses. The number of prism diopters required to bifixate a target at a given distance is obtained by multiplying the equivalent meter angle (MA) by the PD (in centimeters):

$$\Delta = (\text{MA})\ (\text{PD})$$

Thus an individual with a PD of 60 mm requires $3 \times 6 = 18\Delta$ of convergence to fixate at ⅓ meter. Note that neither meter angle nor prism diopter gives any information as to the actual convergence mobilized but only the rotation required for a given distance.

A minor source of confusion results from Nagel's conception of the meter angle as a monocular unit; that is, at ⅓ meter, *each* eye rotates 3 meter angles. Hence both eyes converge 6 meter angles. Since convergence is always binocular, most authors have dropped this "small" unit in favor of the large one, which applies to *both* eyes. The large meter angle is used throughout this book, and it means both eyes converge 3 meter angles to fixate at ⅓ meter. When converted to prism diopters, the resultant also applies to both eyes.

Near point of convergence

The closest point that can be bifixated is the near point of convergence, the furthest point is the far point, and the difference is the range or amplitude of convergence. The far point of convergence is practically at infinity, although some divergence from parallelism can be forced by base-in prism. The near point like its accommodative counterpart is measured with a small sphere (e.g., 2 mm perimetry) brought slowly toward the root of the nose until the target doubles. Some patients may confuse blur with doubling, and some fail to recognize either. A red glass placed before one eye dramatizes the

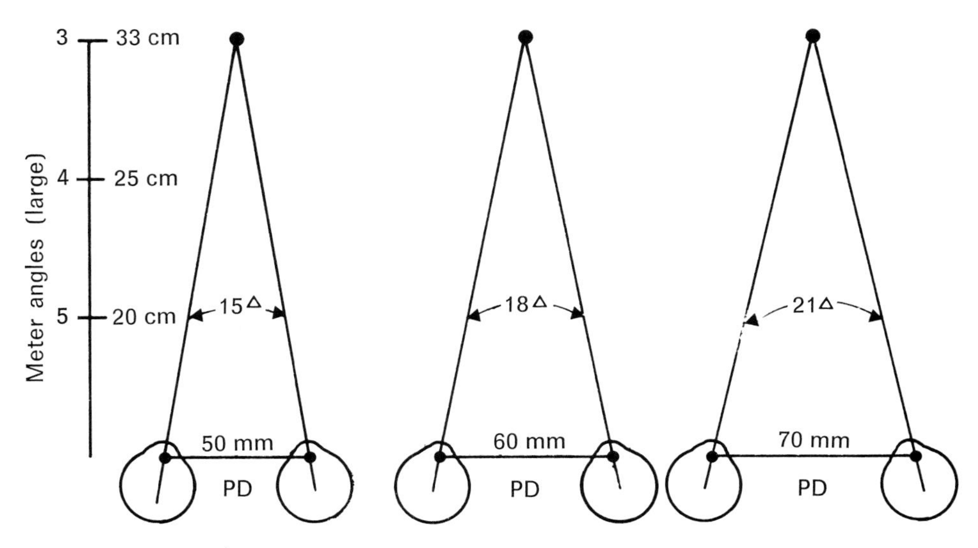

Fig. 18-2. Convergence in prism diopters varies with the interpupillary distance *(PD)*. Convergence in meter angles depends only on the fixation distance.

diplopia but also breaks fusion earlier. More reliable is objective observation of the deviation of the nondominant eye at the near point. Comparing objective and subjective near points has little diagnostic value except as a clue to the patient's attention span. The distance of the break point from the root of the nose (in millimeters or centimeters) defines the near point. An additional 25 mm (to the center of rotation of the eye) may be added, although such accuracy is seldom required in clinical work. But several readings provide more valid data, especially at these close distances where small linear errors lead to large variations. For example, a near point of 6 cm means 100Δ of convergence amplitude (with a 60 mm PD); a near point of 8 cm reduces this to 75Δ. So a diagnosis of convergence spasm or insufficiency should seldom be made on the basis of the near point alone but correlated with other findings, such as a large exophoria or reduced positive fusional amplitude, and the symptoms.

Unlike accommodation, the convergence near point is not influenced by age or refractive error. But is is modified by motivation and kinesthesis; hence encouragement is necessary, and the amplitude is larger if the patient holds the target. Even so, the same motivation may not apply at home or on the job, resulting in a clinical overestimate of effective, that is, sustainable convergence.

Although all components are represented in the convergence amplitude, not all are equally representative; accommodation and fusion supply more innervation than tonus and proximity. First proposed by Maddox, the classification of convergence into four components—or at least into a response activated by four types of stimuli—is the basis of the modern view of the accommodative convergence relationship:

> Strictly speaking, there are four elements of convergence though the first and the third are perhaps closely related. The four are: (1) Tonic; (2) Accommodative; (3) Convergence due to knowledge of nearness, or in other words "voluntary convergence," for we cannot without special practice converge the eyes voluntarily under ordinary conditions without thinking of a near object; (4) Fusional convergence. Of these four elements I have included the second and third under the one name of "accommodative convergence" to simplify practical work. On looking at a near object, the voluntary and accommodative elements bear very definite proportions to each other in different individuals.*

*From Maddox, E.: The clinical use of prisms and the decentering of lenses ed. 2, Bristol, 1893, John Wright & Co.

Tonic convergence

Most measurements have a starting point, but with convergence the "zero" hangs in the air at the anatomic position. Devoid of innervation, this position of dead rest is not easily defined without invoking a corpse and not easily measured except in total ophthalmoplegia. Electromyography, with subjects under general anesthesia, confirms that in most people the anatomic position of rest is slightly divergent. Mechanical and elastic orbital contents are sometimes implicated in exotropia, but normal eyes never reach the divergence of the orbital axes. The anatomic position, of course, is seldom known clinically.

A physiologic zero point is more easily defined if not more easily measured. It is the position of tonic equilibrium in which all the muscles are ready to spring into action. Tonic innervation brings the eyes from the anatomic to the physiologic position of rest, but its operation is not clear. Definitions of tonus range from low-grade innervation of inactive muscle to maintenance of the eye in a particular static position, including the binocular fusion position. Current usage requires ocular dissociation and the elimination of proximal effects, which makes the test for tonic vergence identical to the distance phoria. Of course, tonic convergence is measured from the anatomic position, whereas the distance phoria is measured from the primary position (Fig. 18-3). Nevertheless, it is reasonable to assume that excessive tonus brings the eyes to a position of relative distance esophoria; a tonus deficiency leads to a distance exophoria. In this context, exophoria is a minus value because it represents the amount by which tonic vergence fails to bring the eyes to parallelism.

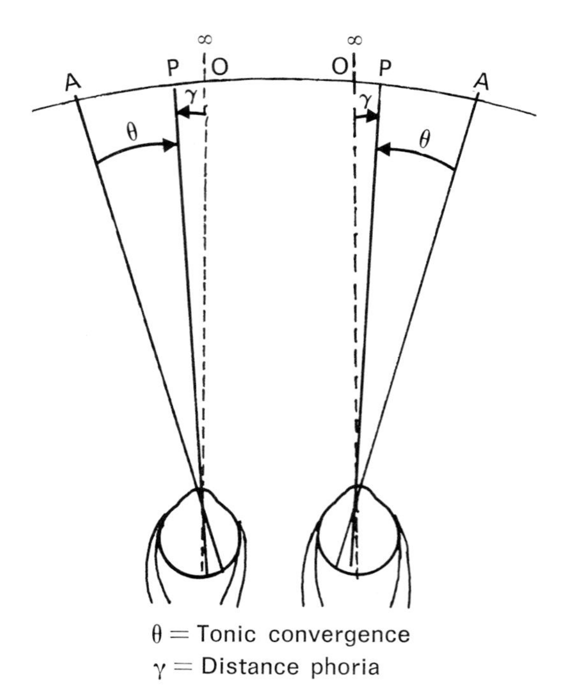

Fig. 18-3. Tonic vergence (θ) is measured from the anatomic position of rest *(A)* to the physiologic position of rest *(P)*. The distance phoria (exophoria in this example) is measured from parallelism *(O)* to the dissociated position.

The sources of tonic impulses are only poorly understood. Hence the "physiology" of the passive position may be questioned in a patient accustomed to binocular vision, and there is little "rest" to a position maintained by active postural reflexes. There is indirect evidence that tonic contraction is maintained by muscle fibers of different size, innervated by small-sized (possibly autonomic) nerve fibers. The prolonged response of extraocular muscles to drugs such as succinylcholine suggests a slow motor system, which may be the substrate of tonus. Electromyography shows a stream of impulses in alternating motor units, perhaps substituting for each other to reduce fatigue. Contributing to tonic innervation are postural reflexes activated by head position and acceleration. The tonic system is fairly stable; changes in posture, gaze direction, or relation of the target to the horizon seem to have a negligible effect on the distance phoria.

The cerebral cortex inhibits tonic vergence, and drugs such as ethanol and barbiturates inhibit this inhibition, causing a larger distance esophoria or smaller exophoria. This inhibition varies with fatigue and general health, a factor implicated in precipitating strabismus. A muscle under stretch also exhibits greater tonus; hence an advancement has greater effect than a recession in surgical

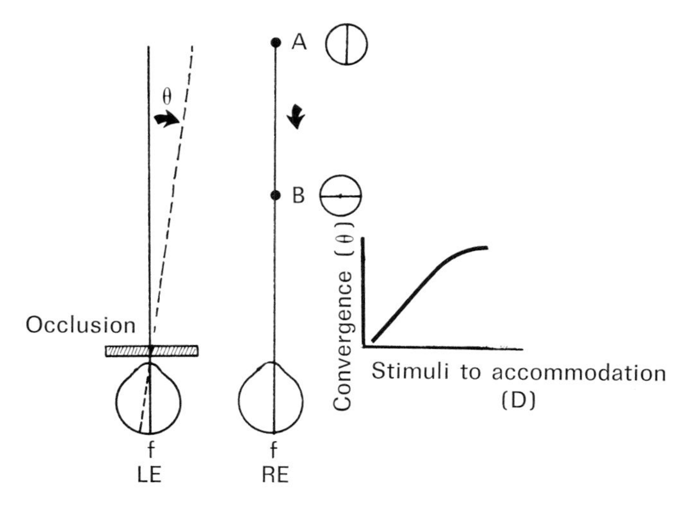

Fig. 18-4. Muller's experiment demonstrating the inward turning of the covered eye when the fixing eye changes accommodation from target *(A)* to a nearer target *B*. The angle of convergence thus induced and plotted against the stimulus of accommodation gives a line where slope is the AC/A ratio (inset). The linearity holds except at the limits of the accommodative amplitudes.

treatment of squint. A change in tonus with age is believed to be responsible for the inward deviation of a blinded eye in the young and an outward deviation in adults.

The position of physiologic rest is no longer considered merely the result of negative convergence, since active divergence can be demonstrated by electromyography. In intermittent convergent strabismus, for example, the firing of the lateral rectus muscle can be recorded directly when fusion is restored, indicating active contraction. The position of rest represents a balance between tonic convergence and tonic divergence; an imbalance occurs in congenital esotropia or exotropia. Although this is an oversimplification, it distinguishes these forms of strabismus from those in which accommodation (or hyperopia) is a chief factor.

Accommodative convergence

The convergence directly associated with accommodation is most easily demonstrated by the cover test, first used by the physiologist J. Muller in 1826. Muller noted that the occluded eye turns inward whenever the other changes fixation from far to near (Fig. 18-4). The rotation represents a change in the horizontal phoria and varies directly with accommodation, except at the limits of the accommodative amplitude.

The association between accommodation and convergence was not only described as early as 1759 by Porterfield, but in addition, he commented on how the association becomes established: "This change in our eyes whereby they are fitted for seeing distinctly at different distances, does always follow a similar motion of the axes of vision with which it has been connected by use and custom." From this he concluded that a fixed and necessary connection becomes established between them, a view that is surprisingly modern. Donders and later Landolt, faced by the obvious disparity between accommodation and convergence in ametropia, supposed the relation to be learned and elastic. It took another fifty years to establish that the flexibility is between accommodation, and total convergence not accommodative convergence.

The change in accommodative vergence per unit change in accommodation is the AC/A ratio, conventionally expressed as a fraction; accommodative convergence in prism diopters is the numerator, and accommodation in diopters is the denominator. Of the general population 65% to 90% have an AC/A ratio of from 3Δ to 5Δ per diopter (Fig. 18-5). A similar distribution is found in most population samples regardless of age,

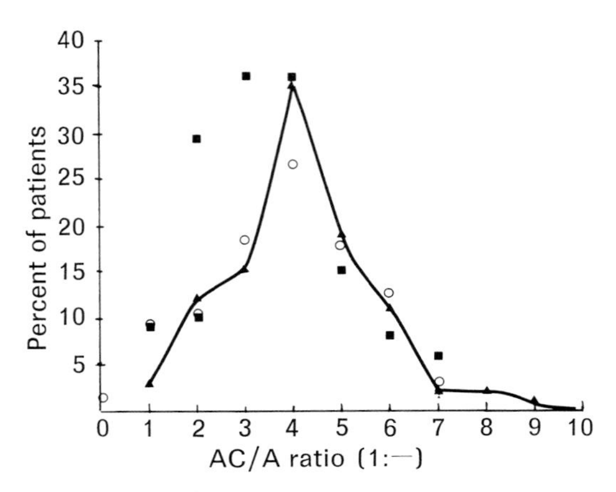

Fig. 18-5. The distribution of the AC/A ratio in 150 patients under age 45 years follows approximately a bell-shaped curve. Circles show the results of Morgan and Peters (1951) for 200 cases over 45 years; and the squares, the data for 104 cases of Ogle and Martens (1957).

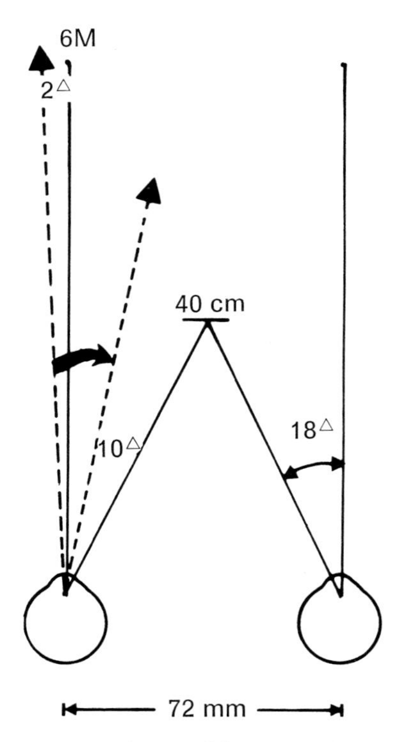

Fig. 18-6. Interrelation of distance and near phoria and accommodative and total convergence. In this example the interpupillary distance (PD) is 72 mm; hence the total convergence required for 40 cm is 2.5 × 7.2 = 18Δ.

refractive error, or symptoms of discomfort.

The AC/A ratio varies between individuals but is fairly constant for each, and has come to be accepted as a unique physiologic characteristic. Repeated measurements show little day-to-day variability (less than 1Δ). This constancy is maintained despite refractive correction, orthoptic training, drugs, or surgery. Since Donders, there has been speculation regarding the manner in which the AC/A relation is built up, that is, whether it is learned or innate. A learned association is suggested by the apparent adaptation to ametropia and to correcting spectacles. An innate basis is suggested by the demonstration of a fixed AC/A ratio in strabismics in whom the two functions could not have come to be associated by use. Families with an esotropic propositus also reveal a significant difference from the average AC/A ratio, and a study of identical twins shows less variance between them than other siblings.

The AC/A ratio in most people is about half their interpupillary separation; hence a fairly large exophoria results at the reading distance. If the ratio equaled the PD, no fusional vergence would be required at any fixation distance. On the other hand, an individual with a 72 mm PD and 2Δ of exophoria at far requires 7.2 × 2.5 = 18Δ of total convergence to fixate a target at 40 cm (Fig. 18-6). If his AC/A ratio is 4/1, the accommodative vergence is 4 × 2.5 = 10Δ, leaving a residual of 10Δ to be supplied by the fusional component. The 10Δ is the near exophoria at 40 cm. A moderate near phoria is so common that it is sometimes called "physiologic exophoria" because it causes no discomfort. Whether a motility imbalance is symptomatic depends not only on the size of the phoria but also on the compensating fusional vergence. Even orthophoria is no help when fusion is absent; the least lag of accommodation, the slightest deviation from emmetropia, and the smallest change in posture converts an orthophoria without fusion into a deviation without single vision. The term "orthotropia" is sometimes applied to this condition; obviously it can only be true at one fixation distance.

Accommodative vergence may be demonstrated under binocular conditions by a change in fixation disparity. If minus lenses are added binocularly (over the distance prescription) while the subject observes a distant disparity target (maintaining peripheral fusion), an eso shift results, that is, the visual axes intersect in front of the plane of the target. When the prism required to produce the same disparity is plotted against this lens power, the slope is the AC/A ratio. This calculated ratio is remarkably similar to that obtained by conventional phoria methods, and the similarity suggests that factors influencing the phoria operate even when the eyes are not dissociated. (Fixation disparity is considered in more detail in Chapter 25.) The disparity results confirm the assumption that the phoria-induced fusional stress operates under conditions of binocular vision even though it is seldom measured under those conditions.

Results such as those just discussed are the basis of the conclusion reached by Ames and Gliddon (1928) that relative accommodation does not exist, since accommodative vergence invariably changes with accommodation even when the eyes are not dissociated. But total convergence retains enough constancy (by proportionate changes in fusional vergence) to prevent overt diplopia. Fixation disparity thus buffers the vergence mechanism, permitting slight inaccuracies and absorbing stressing forces.

Accommodative vergence is directly related to the effort to accommodate, which is in turn proportional to the accommodative stimulus at least over the usual clinical measurement range. At the limits of the accommodative amplitude, the relation between the ciliary stimulus and response breaks down; the response tends to lag behind the stimulus (called the "accommodative lag"), partly the result of increased depth of focus.

An additional discrepancy occurs because of the optics of the situation. An AC/A ratio of 4/1 does not have the same meaning when referred to the spectacle and principal plane of the reduced eye. On the basis of Pascal's accommodative unit (Chapter 7), a 3 D myope exerts 2.7 D of accommodation, and a 3 D hyperope exerts 3.36 D of ocular accommodation through their spectacle correction compared to a natural emmetrope (at ⅓ meter). For a 60 mm PD, the myope's actual AC/A ratio is thus larger (4.4/1) and the hyperope's AC/A ratio is smaller (3.3/1) than the assumed value of 4/1. This may be one reason why an accommodative esotrope does better when given miotics (or cycloplegics) in addition to spectacles.

It follows that it is not always possible to achieve the same accommodative convergence balance in spectacle-corrected ametropes as in a natural emmetrope. Fortunately plus spectacle lenses also have a base-out prismatic effect when reading, which tends to compensate for the induced esophoria. The larger accommodative effort required of the spectacle-corrected hyperope could be eliminated by contact lenses. Most hyperopes, however, cannot be convinced to wear contact lenses by motility considerations alone, preferring an overplus or near add. The same considerations apply to accommodative esotropes who are usually too young for contact lenses but adapt nicely to executive bifocals. But in anisometropia, contact lenses may be the only means to achieve an optical and motility balance.

Proximal convergence

In using a stereoscope or similar binocular instrument, the eyes generally converge more than required by the optical conditions. This overconvergence is so common that stereogram designers allow for it even when the instrument is set for distant viewing ("instrument convergence"). The cause is an awareness of the nearness of the target. The effect operates directly through retinal image size or indirectly through secondary cues when size is held constant or ambiguously reversed as in Fig. 18-7. To a greater extent

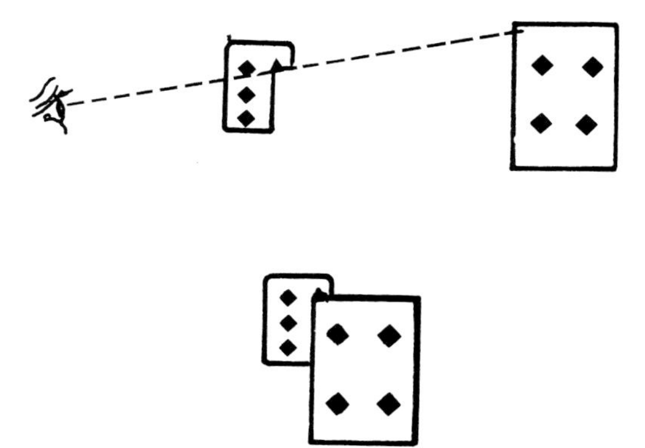

Fig. 18-7. Experimental set-up to demonstrate the effect of size and distance on proximal convergence.

than other components, proximal convergence is influenced by past experience and motivation (hence the synonym "psychic convergence"). Retinal image size affects vergence, which in turn modifies accommodation. Since the latter is self-correcting, it is hard to decide which response is primary and which is secondary.

In investigating the accommodative convergence relation, one variable is manipulated, and the effect on the other is measured. But the induced change modifies the stimulus conditions by feedback mechanisms. Thus a vergence change modifies the accommodation. Closed-loop conditions predominate in clinical measurements; hence the induced effect is not truly independent, one reason for the difficulty in demonstrating convergent accommodation.

Proximal vergence correlates directly with fusional convergence and inversely with accommodative vergence. Presumably those patients requiring more positive fusional vergence (e.g., in reading) develop a larger proximal effect by conditioning. Conversely, if the fusional requirement is small or negative, a proximal stimulus may cause divergence. Generally, the nearer the target the greater the proximal effect. It is not known whether proximal vergence is a distinct entity or a modified form of tonic vergence. Patients with strabismus often show a large proximal effect. Indeed in Cantonnet's orthoptic technique, the patient is instructed to "think" of a distant object in esotropia or a near object in exotropia, thus modifying the deviation by "mental effort."

Fusional convergence

Accommodative vergence usually provides most of the rotational requirements for near vision, but it does not provide all that may be demanded, and the deficit is supplied by fusion. If normal binocular vision is interrupted, the eyes assume a fusion-free position. When the dissociating element is removed, bifoveal fixation is regained without any apparent effort and independent of volition. The double "images" irresistibly flow together as if by magnetic attraction, a process vividly termed "compulsion to fusion." The stimulus for fusional movements is egocentric disparity and goes to completion if the two targets are sufficiently similar in size, shape, color, and brightness. The velocity of the horizontal movement is a function of the disparity, taking about 0.8 seconds for completion. The disparity is continuously monitored; so changes can occur if the disparity is changed before the movement is completed. Less is known about the time characteristics of vertical fusional movements. Human fusional responses appear to be made up of two distinct components: a central process limited by Panum's areas and a compensatory motor movement. For example, small torsionally disparate retinal images are fused by the central mechanism without any evidence of cyclofusional rotations. There is some evidence that Panum's limits for uncrossed disparity differ from crossed disparity, each limit behaving independently as convergence is altered. The extent to which fusional movements can follow alternating patterns is still controversial. The cortical mechanisms for disparity detection are considered in Chapter 25.

Clinically fusional movements are evaluated with distance and near fixation by interposing vertical or horizontal prisms. The maximum prism that permits fusion is a measure of the amplitude under these test conditions. Larger fusional amplitudes are obtained for detailed foveal targets of large area, good contrast, and sharp resolution. Horizontal disparities induce larger fusional amplitudes than vertical disparities.

In addition to a gradual change in disparity, the eyes also respond to sudden or steplike changes ("jump vergences"). All clinical measurements are technically jump vergences, since a manual measuring prism cannot be rotated with perfect smoothness. Jump vergences tend to anticipate the stimulus pattern with rapid flicks in the correct direction. When the prism exceeds the fusional capacity, the eyes drift slowly back to the phoria position. A sudden change breaks fusion earlier, suggesting that prisms be introduced slowly, smoothly, and symmetrically. A Risley rotary prism obviously serves this purpose better than individual loose prisms or a prism bar.

An interesting phenomenon is the aftereffects of fusional movements on phorias. After measuring fusional amplitude, dissociation does not result in the same phoria that would have been obtained if fusion had not been previously stressed. The phoria changes in the same direction as the fusional movement; base out has a greater effect than base in; and the time required to return to the initial state is longer the greater and more rapid the stress. This not only has practical implications in terms of the sequence of measuring phorias but also in evaluating phorias per se. When the basic phoria is high and large fusional vergences are required to maintain binocularity, a different result is obtained when fusion is abolished briefly, compared to prolonged monocular occlusion. Two opposite inferences have been drawn: (1) that prolonged occlusion is necessary to obtain a valid phoria measurement, and (2) that brief dissociation is more likely to approximate normal-seeing conditions. A similar problem occurs in evaluating strabismic deviation, that is, whether the conditions should be natural (or "free space") or whether the test should be a stressing probe. Most clinicians adopt the attitude that the test most likely to be found reliable in practice is probably most valid in theory.

If phorias are the demand, the fusional amplitudes are the reserves to maintain clear, single vision. When the demand equals or exceeds the reserve, the result is uncomfortable binocular vision or comfortable monocular vision with suppression. The fusional reflexes thus buffer instabilities between accommodation and convergence. The failure of this buffer is a characteristic motor defect of strabismus.

Neurophysiology

In normal binocular vision, convergence, accommodation, and miosis (and perhaps elevation of the lid when the target is above the horizon) characterize the "near complex." Miosis is practically always associated with accommodative vergence; only 50% of the population show a pupil response with changes in fusional vergence.

Of subjects tested by either fixation disparity or phorias but not necessarily both, 8% to 10% show a nonlinear relationship between accommodation and convergence. What is surprising is that two such different effector mechanisms should be associated at all. Consider that accommodation and miosis are controlled by smooth muscle and convergence, by skeletal muscle; the first are controlled by autonomic nerve fibers and the second, by somatic nerve fibers. Accommodation is potentially monocular; convergence (with one exception) is invariably binocular. Accommodation may be selectively inactivated by cycloplegics; convergence is selectively paralyzed in certain midbrain syndromes. Whereas convergence is the direct consequence of muscular contraction, accommodation is a refractive change one step removed from ciliary action. Yet the linear relation between accommodation and convergence is not due to fortuitous cancellation of clinical inaccuracies for the same result if obtained by more precise experimental methods. Over intermediate fixation distances (those in the practical reading range), the response AC/A ratio measured objectively (by retinoscopy, electromyography, photography, or impedance cyclography) is also linear. Although the response AC/A ratio is not a practical clinical measurement, the stimulus AC/A ratio is related to it by a constant factor, providing one uses a detailed test target, ensures an adequate accommodative amplitude and emmetropic correction, and avoids excessive depth of focus. Alpern and co-workers (1959) have shown that a fair estimate of the response AC/A ratio can be obtained by multiplying the stimulus AC/A ratio by the empirical constant 1.08.

The neurophysiologic basis of the accommodation-convergence synkinesis is probably a central excitation that innervates ciliary muscle, medial recti, and the iris simultaneously. Jampel (1958, 1959) showed that weak faradic currents to the preoccipital cortex of monkeys result in the triad near response. The stability of the system is most likely maintained by a servomechanism that makes the AC/A ratio resistant to changes induced by optical, pharmacologic, orthoptic, or surgical means.

A central linkage also explains the action of drugs on the accommodation-convergence synkinesis. In incomplete cycloplegia, the AC/A ratio may increase by as much as eight to ten times its normal value. This is not a direct effect on convergence but a partial block at the ciliary myoneural junction requiring greater innervational effort. It follows that to obtain accommodative relaxation by cycloplegia (e.g., in accommodative esotropia), drug concentration must be high enough and allowed to act long enough to demonstrate the futility of increased ciliary effort. Thus partial cycloplegia may actually increase convergence and is sometimes used in the management of intermittent exotropia. When giving atropine for esotropia, the parents should be warned that the deviation may increase in the initial stages of drug action. An analogous effect is seen in ametropic patients first receiving correcting lenses; the accommodation may be initially more or less than required, apparently changing the AC/A ratio. This change, however, returns to the precorrection value after a few days or weeks.

Drugs that enhance acetylcholine paradoxically act the same as cycloplegics, although the mechanism differs. In the presence of powerful anticholinesterases the innervation required for ciliary contraction greatly diminishes and so does the associated accommodative vergence. The drug effect is not due to the associated miosis, despite the increased depth of focus, because the result is the same even when pupil constriction is prevented by simultaneous sympathomimetics. Again the effect is evident only if the drug is given in sufficient dosage and over some time. The beneficial effect in esotropia comes from the improved cosmetic appearance. It is doubtful that fusion is affected, although it may bring the deviation within range of the available amplitude. In some cases, for reasons unknown, the effect remains even after the drug is discontinued.

Ethanol and barbiturates in mild doses inhibit fusion by reducing fusional amplitudes. They also decrease accommodative vergence as evidenced by a larger *near* exophoria; the AC/A ratio changes even though saccades and pursuit movements are not affected. Unfortunately the effect is temporary and without the prismatic hangover necessary for a permanent cure of strabismus.

When both eyes turn in the same direction (version), no cyclotorsion is induced; in convergence there is an excyclorotation whose magnitude is a function of the target distance. The effect is too small for ordinary clinical measurements, but the amount is similar whether the vergence is induced by accommodative or fusional stimuli, suggesting a common final path. The nucleus of Perlia was traditionally believed to play this role, a belief that still persists as a sentiment if not as a neurologic command. Based on the classic Nissl technique, Warwick (1955) concluded that the medial recti are innervated by cells in the vertical columns of the lateral oculomotor nuclei. Isolated extirpation of these muscles in monkeys *(Macaca mulatta)* were never observed to have an effect on cells in central or paramedian groups, whereas they regularly demonstrated retrograde degeneration after extirpation of the superior rectus. It is doubtful therefore that the nucleus of Perlia is a "center" of convergence in humans.

Relative convergence

A disparity between accommodation and convergence is inevitable in ametropia; the uncorrected hyperope must accommodate more and the myope less than the emmetrope for the same convergence. A rigid synkinesis would preclude binocular vision, whereas an acquired association would require relearning a new synkinesis with every lens change.

Donders coined the terms "relative accommodation" and "relative convergence" to describe the degree of independence between the two functions. That we continue to see clearly and singly despite the interposition of lenses or prisms is now explained in terms of convergence components. Since accommodation always induces an associated vergence, it is total convergence that remains constant (by a proportionate change in the fusional component). Thus the myope must muster more fusional vergence to supplement his deficient accommodative vergence as compared to the emmetrope. When the myopia is corrected, less fusional vergence is needed in proportion to the increased accommodation. When the ametropia is large, the AC/A ratio high, or the fusional vergences low, adaptation to new lenses may not be possible, resulting in blurred or double vision. Since blurred vision is more acceptable than diplopia, the patient usually complains that his new glasses are "too strong."

Relative accommodation is determined by adding binocular plus or minus spheres over the emmetropic prescription. Minus spheres measure positive relative accommodation; plus spheres, negative relative accommodation. The end point is target blur, and the total range is the amplitude of relative accommodation. Relative convergence is measured by prisms through the distance correction (carefully centered). The end point may be blur or diplopia. Base-out prism measures positive realtive convergence; base-in, negative relative convergence. Since the innervation is shared by both eyes, a slight prismatic difference does not matter, but an attempt is made to change the prism over each eye symmetrically. The eye that ultimately deviates is the nondominant one; the diplopia is crossed in exophoria and uncrossed in esophoria.

Patients also report an apparent size and distance difference as prisms are introduced. The target appears smaller with an inward motion (base-out prism) and larger with an outward motion (base-in-prism), mnemonically called the "SILO" effect. The explanation is that a base-out prism causes the target to appear nearer than its true position, since the estimation of distance varies with innervational effort to convergence. Because the retinal image does not get larger as it would if the target had really come closer, it is judged smaller. Base-in prism has the converse effect.

PRACTICAL MEASUREMENTS

Clinical as contrasted to experimental measurements leave many factors uncontrolled and uncontrollable: lenses may or may not activate accommodation, prisms may or may not induce ocular rotations, the patient may or may not attend to the target, and the examiner may or may not interpret the responses correctly. Accommodative tests rarely allow for pupil or target size, illumination, or apparent distance; and convergence tests do not allow for tonic fluctuations of frustration or fatigue. Spherical and chromatic ocular aberrations are ignored; retinal adaptation and ciliary tonus are assumed to remain constant; and the effects of practice are neglected. Since the speed and variety of ocular movements are impossible to measure, diagnostic judgment must make up for lack of clinical test precision.

Heterophorias are the basis of most clinical methods of evaluating the accommodation convergence relation. To obtain valid results, ametropia is adequately corrected, acuity balanced, and accommodation controlled. For example, a bright penlight may evoke no accommodative effort, and the examiner, instead of measuring a near phoria, simply measures the far phoria again. The Graefe prism test is thus more applicable than the Maddox rod for near testing, particularly since rotary prisms permit the use of print

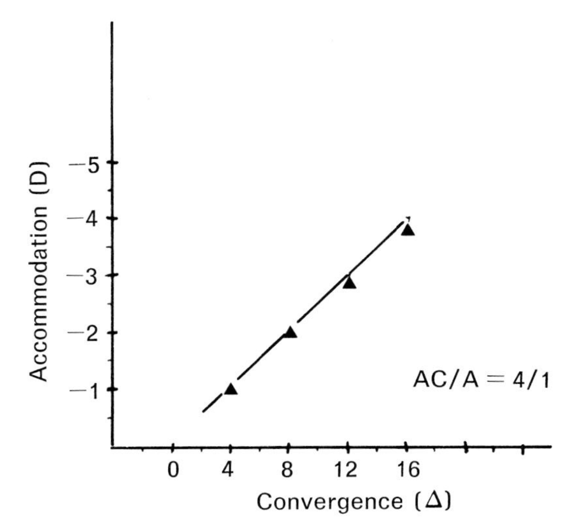

Fig. 18-8. The plot of accommodative convergence for intermediate levels of accommodation is linear. The slope of the line is the AC/A ratio.

and allow the examiner to monitor the accommodative effort. With children, the objective cover test is a necessary alternative, but even here the near target should contain enough detail to evoke a predictable accommodative response. With these precautions, phorias can be measured with a high degree of accuracy.

Although the linear extent of the AC/A ratio is not known in advance, it usually holds for the clinical test range. For example, if an initial near phoria of 6 exo is changed to 2 exo by a −1.00 D sphere, one can anticipate 2 eso through a −2.00 D sphere or 10 exo through a +1.00 D sphere, since all are consistent with a 4/1 AC/A ratio (Fig. 18-8). This provides a means to check reliability by comparing the effects of progressive stimuli. The results may also be portrayed graphically but since any two points plot a line, several readings are required if reliability is in question.

Tonic convergence

Test conditions for tonic convergence are the same as for distance heterophoria. The patient wears his distance correction, and fusion may be dissociated by alternate cover, Maddox rod, or dissociating prism. Accommodation control hinges on a careful refraction, and the result is probably influenced by the size, detail, and context of the target (even though it should not be from purely optical considerations).

Tonic convergence per se is not actually measured because the anatomic position of rest is unknown; a distance exophoria is interpreted as a low tonic vergence and esophoria as a high tonic vergence.

Accommodative convergence

Accommodative vergence brings the eyes from the distance to the near dissociated position (assuming tonic convergence to remain constant and proximal convergence to be negligible). It is expressed by the AC/A ratio, which gives the vergence change per unit change of accommodation. The accommodative stimulus may be a change in ametropic correction or added lenses at a fixed reading distance or by altering the fixation distance proper.

The *gradient method* is the most direct, the most reliable, and the simplest technique of evaluating the AC/A ratio. No computation is necessary, the PD need not be considered, and proximal convergence is held constant. Minus spheres are preferred because they give more reliable responses than accommodative inhibition. Since the stimulus is small, an error in measuring the phoria critically alters the results; hence suspicious findings are checked by repeating the test through additional spheres.

The *distance-to-near phoria method* requires less time but more computation. No additional test is made; one simply compares the distance and near phoria. Accommodative vergence is computed by substracting the phoria from the total convergence (which depends on fixation distance and PD). Computation may be made mentally or by means of a nomogram, sketch, graph, or formula. For example, if the distance phoria is 1 exo, the near phoria 4 exo, and the PD = 70 mm, the total convergence required at ⅓ meter is $3 \times 7 = 21\Delta$. The accommodative convergence is the change from 1 exo at far to 4 exo at near or 18Δ. Since the accommodative stimulus is 3 D, the AC/A ratio is 18/3 or 6/1. In these computations the distance phoria is taken to represent tonic vergence (exophoria is minus, and esophoria is plus):

$$\text{Accommodative vergence} = \text{Total convergence} - \left(\text{Tonic vergence} + \text{Fusional vergence}\right)$$

This method ignores proximal convergence.

In most methods of measuring the AC/A ratio, fusion is disrupted. In the *fixation disparity method,* binocular vision with peripheral fusion is maintained. The test target consists of identical (e.g., polarized) targets seen by each eye through polarizing filters for peripheral fusion and a small central area containing uniocularly seen vernier lines. When accommodation is stimulated, the central lines become displaced relative to each other, although peripheral fusion is maintained. The visual axes may cross anterior (esodisparity) or posterior (exodisparity) to the plane of the target. The disparity may be measured by a screw and knob mechanical attachment to the polaroid projectors or the lines themselves. The results are read in minutes of arc (visual angle).

Objective methods include dynamic retinoscopy, photography, and electromyography. In a typical laboratory setup, the stimulus to accommodation is altered by lenses, and vergence is measured by rotating the haploscope arms, which pivot about each eye's center of rotation. Pupil size is controlled, and a Badal arrangement maintains constant retinal image size. Retinoscopy, stigmatoscopy, photography of Purkinje images, or other methods monitor accommodation objectively through half-silvered mirrors. A recent refinement is electro-oculography to measure the vergence. Objective techniques are not necessarily more valid than subjective ones and occasionally suffer from the gross fluctuations peculiar to sophisticated instrumentation.

Proximal convergence

There is no reliable method of measuring proximal convergences under clinical conditions, but an estimate is obtained by comparing the AC/A ratio as determined by the gradient and the distance-to-near phoria technique. Since the gradient is based on measurements at the same distance, proximal convergence does not enter into the result; in the distance-to-near phoria method, there is an actual change in distance so that proximal convergence is stimulated. One therefore expects the latter to give a higher AC/A ratio (about 1Δ per meter of fixation distance). Closer distances give greater proximal effects but with considerable individual variation.

Fusional convergence

When weak horizontal prisms are placed before the eyes of an individual maintaining normal binocular vision, and their power is slowly increased, fusion is eventually broken, and the target doubles. With moderate attention, however, it will be noted that the target blurs before it doubles. This blur represents the accommodative vergence mobilized after fusional vergence has been exhausted. The accommodative vergence is positive in response to a base-out prism and negative in response to a base-in prism. Changes continue at the same rate after the blur, eventually ending in diplopia.

Since the first blur can be measured but the accommodation active at the break point is not known, the blur is a more valid end point of fusional amplitudes. The target should contain enough detail to make the blur recognizable, and the patient's attention must be called to it, or none will be reported. If no blur is noted, the break is taken as the end point. Of course, no blur is expected in emmetropia when measuring negative fusional vergence at 6 meters. If there is suppression, the target will appear to move left or right.

After the break, the prism is slowly reduced until binocular vision is regained, and this recovery is taken to indicate the facility of fusion. Like other fusional measurements it varies with rate and symmetry of prism change, the test detail, the extent of the visible field, as well as intrinsic amplitudes. The sequence of testing also influences the result; it is conventional to measure base-out amplitudes before base in and distance before near fusion. Data may be recorded sequentially as blur, break, and recovery. Approximate averages and standard deviations are as follows:

	Blur	*Break*	*Recovery*
Distance			
Base out (Δ)	16 ± 3	21 ± 3	10 ± 3
Base in (Δ)	None	7 ± 3	4 ± 3
Near (40 cm)			
Base out (Δ)	17 ± 3	22 ± 3	11 ± 3
Base in (Δ)	13 ± 3	20 ± 3	13 ± 3

Since phorias are compensated in binocular vision, it is necessary to distinguish between true and measured fusional amplitude. The measured amplitude is the reading shown on the rotary prism (or the distance from zero on the synoptophore arm); the true amplitude depends on the phoria. Thus a base-out reading of 20Δ is interpreted as a true positive amplitude of 20Δ in orthopho-

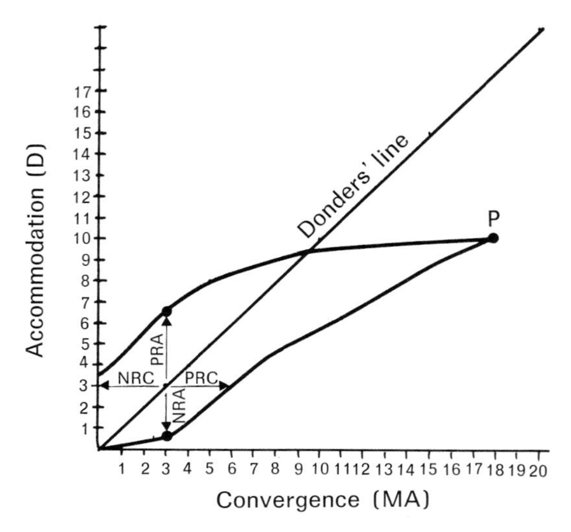

Fig. 18-9. Donders graph as modified by Landolt to show the zone of binocular vision. *PRA* and *NRA* are the positive and negative relative accommodation; *PRC* and *NRC* are the positive and negative relative convergence. The Donders line represents points of equal accommodation and convergence when measured in diopters and meter angles, respectively.

ria, of 30Δ with 10 exo, and 10Δ with 10 eso at that distance. In comparing fusional demand (the phoria) to fusional reserves, the patient uses his true amplitude in binocular vision. Unfortunately the terminology is confusing, since "absolute," "relative," "total," "partial," and "reserve fusional amplitudes" are all used by different writers for the same concepts.

Graphic Methods

A graphic method to represent the relation between accommodation and convergence was first used by Donders. He determined, for a series of points situated in the median plane, the strongest minus and plus spheres that could be overcome while maintaining single binocular vision. These relative ranges of accommodation were plotted, and to them Landolt subsequently added the ranges of relative convergence (Fig. 18-9). (Diopters were not used in Donders' time; Landolt first applied modern units.)

The line connecting points of equal accommodation and convergence has come to

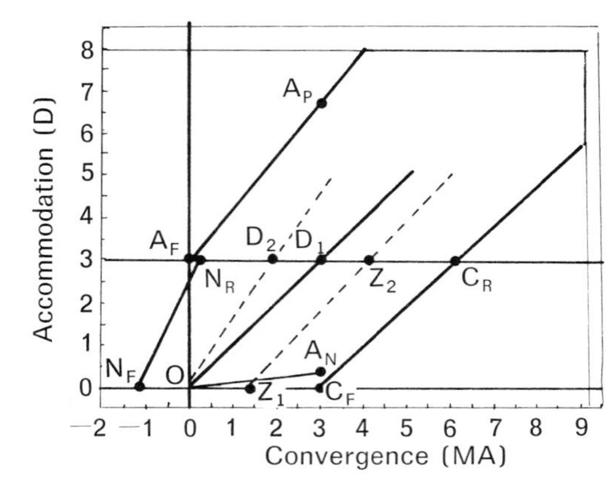

Fig. 18-10. Type of graph used by Sheard to outline the zone of comfort for fixation at 6 meters and 33 cm. The symbols are as follows:

O—Center of coordinates; also negative relative accommodation at 6 meters

A_F—Positive accommodation (minus lens) at 6 meters

A_P—Positive accommodation (minus lens) at 33 cm

A_N—Negative accommodation (plus lens) at 33 cm

N_F—Divergence or abduction (prisms, base in) at 6 meters

N_R—Divergence or abduction (prisms, base in) at 33 cm

C_F—Convergence or adduction (prisms, base out) at 6 meters

C_R—Convergence or adduction (prisms, base out) at 33 cm

OD_1—Line of equal accommodation and convergence

OD_2—Line of accommodation and convergence with the balance at the reading point indicative of 1-meter angle (6 prism diopters) of accommodative exophoria

$OD_2Z_2Z_1$—Area of comfort as proposed by Percival

be known as the "Donders," or "demand" line. Donders supposed that accommodation would be comfortably sustained only for a fixation distance for which the positive relative accommodation (minus sphere to blur) was proportionately large compared to the negative relative accommodation (plus sphere to blur). Landolt added the qualification that not more than one third of the absolute range of convergence could be maintained without asthenopia. Percival used the same measurements to map the "zone of comfort," which he defined as the middle third of the range

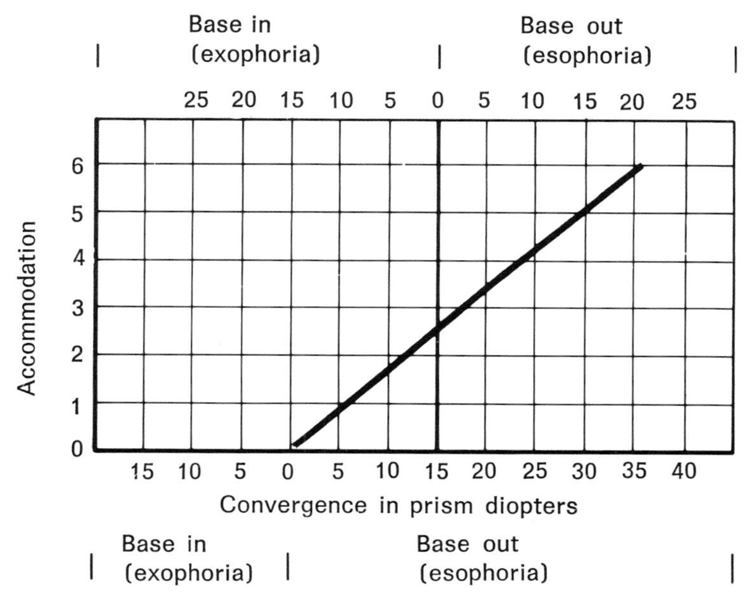

Fig. 18-11. Modern clinical graph used to plot the zone of binocular vision.

between positive and negative relative convergence. According to Percival, the Donders line should fall within this zone. Sheard later added consideration of the phoria line because it represents the real starting point of ocular rotations (Fig. 18-10).

Although graphic methods need not be used routinely, they allow checking consistency and reliability, extrapolating to stimuli and distances not actually tested, and summarize much information at a glance. A type of graph currently used for this purpose is shown in Fig. 18-11. It will be evident that relative accommodation is zero at the far point and near point of convergence. At the far point, accommodation cannot be relaxed because it is already inhibited to zero; at the near point, accommodation cannot be stimulated, since convergence is at its maximum.

SOME CLINICAL APPLICATIONS

The linkage of accommodation and convergence cannot be voluntarily dissociated; yet a disparity occurs in ametropia, presbyopia, heterophoria, and strabismus that may or may not be symptomatic. Conversely, when a refractive error is corrected, the previous adjustment is altered, which may or may not be desirable. The great economic problem in ocular motility is to balance the fusional demand with the supply.

Ametropia

In ametropia, more or less accommodative effort is required than in emmetropia for the same convergence. This imbalance may not cause symptoms if fusional reserves are adequate, but if correcting lenses throw the fusional demand outside the binocular range, they may not be tolerated.

Hyperopic corrections tend to increase exophoria and reduce esophoria; myopic corrections tend to increase esophoria and reduce exophoria. Their tolerance depends on the available fusional amplitudes. By comparing the demand to the reserve, a decision on lens modification can be made. Thus if the AC/A ratio is high, considerable vergence change can be brought about by a relatively minor optical modification, but if the ratio is low, this may not be feasible. For example, a 3 D hyperope with an AC/A ratio of 4/1 receiving his first spectacles relaxes his accommodative vergence 12Δ. A precorrection orthophoria becomes 12 exo. An available fusional amplitude of 10Δ might have compensated for the orthophoria but not the exophoria, and the lenses are not comfortable unless the plus is reduced. Conversely, if the initial phoria had been 12 eso, the full hyperopic correction gives the desirable orthophoria with no fusional strain. In this example, the AC/A ratio was 4/1; if the ratio had been 2/1, the lens change required to eliminate the esophoria would need to be +6 D and would not be tolerated for constant wear. Here a better alternative is reducing the phoria by base-in prisms or building up the positive fusional vergence by base-out exercises. The final ametropic prescription therefore depends not only on retinoscopy and astigmatic dial but also on the motility changes induced by the new lenses—that is why we measure the motility changes.

Presbyopia

Since accommodation decreases with advancing age, one might expect an increased ciliary effort associated with a tendency to esophoria. In fact, the results are generally opposite. There is less innervation to accommodation and an increased incidence of exophoria. This occurs despite the fact that the AC/A ratio does not change; the average is about the same as for nonpresbyopes. These findings tend to support the Helmholtz-Hess-Gullstrand view of the pathophysiology of presbyopia.

The presbyope receiving his first pair of reading glasses now makes less accommodative effort, and the resultant exophoria leads to intermittent diplopia if the fusional vergences are inadequate. The diplopia may be erroneously attributed to an improperly centered bifocal seg. Alternatively the patient rejects the reading glass as "too strong," although the cause is oculomotor rather than optical. A large basic exophoria, high AC/A ratio, and low positive fusional amplitudes are particularly likely to cause intolerance to large, sudden increases in the reading prescription. In such cases, the reading add is increased slowly, thus allowing a gradual mobilization of fusional reflexes. A strong light helps to tide the patient over the period of undercorrection. Base-in prism, either by decentration, clip-ons, or membrane prisms are often useful and well tolerated. Home orthoptics such as base-out stereo training cards and finger-to-nose exercises through a base-out prism are helpful.

Heterophoria

Lateral heterophorias represent the degree to which accommodation and convergence are out of step in achieving clear, single binocular vision. The disproportion is innervational; the significance is not in muscular mechanics but in the physiology of the AC/A ratio. The clinical evaluation of heterophoria is based therefore on comparing the initial phoria without correction (or with the old correction if worn) to the phoria through the anticipated prescription.

Large phorias may be asymptomatic if the fusional amplitudes are adequate, but a change in prescription may throw the demand outside the comfortably sustainable range of compensation. The principle of supply and demand is the basis for modifying the lens prescription, ordering prisms, or orthoptics. In these decisions, symptoms and signs must be considered and allowances made for the variance of psychophysical measurements, which are biologic ranges determined by rather coarse clinical tests. As in all refractive disorders, treatment depends on symptoms and signs, not numbers.

If accommodation is not properly controlled (or the ametropia corrected), a patient with heterophoria may demonstrate an instability of fixation under monocular or binocular conditions (fixation disparity). This does not mean that heterophoria leads to sensory adaptations (eccentric fixation, suppression, "anomalous correspondence," etc.) as seen in strabismus. Nor is the visualization of Haidinger brushes a necessary criterion of foveal suppression, since some normal observers cannot perceive them. Finally, the sweeping generalization that patients with heterophoria have poor depth perception is not warranted. Practically everyone has a phoria at some fixation distance; the correlation between phorias and stereoacuity depends on the tests used to measure each.

The use of relieving prisms for lateral phorias is more controversial than for vertical imbalances. Not only are horizontal fusional amplitudes larger, but also accommodation adjustments are available to modify the fusional demand. When optical modifications are not feasible because of a low AC/A ratio or minimal ametropia and the history is suggestive, relieving prisms may be considered (base out for esophoria and base in for exophoria). Monocular occlusion may also be tried as a diagnostic measure. In some instances, a horizontal phoria is combined with a vertical phoria, and treatment of the second provides relief. Alternative solutions may include orthoptics or surgery.

The most suggestive symptoms and signs are intermittent diplopia, head tilt or rotations, headaches, nausea, and a collection of unsuccessful spectacles.

According to Percival, only the middle third of the total positive and negative fusional range can be used with comfort for fixation distances up to ⅓ meter. If the convergence demand is not located in the area of comfort, prisms are indicated. Sheard maintained that the heterophoria should also be considered; specifically, that the fusional reserve amplitude be twice as large as the phoria. For example, a 10Δ exophoria at 40 cm would require a minimum of 20Δ positive fusional convergence for continuous, comfortable vision. If the measurements do not meet this criterion, the appropriate horizontal prism is prescribed. Sheard's formula for the indicated prism (in prism diopters) equals two thirds of the heterophoria minus one third of the compensating fusional vergence. For example, if the exophoria is 10Δ and the vergence amplitude 8Δ at 40 cm, the indicated prism is $\frac{2}{3}(10) - \frac{1}{3}(8) = 4\Delta$ base in. Worrell and associates (1971) recently analyzed the results obtained in forty-three cases when prisms were prescribed by Sheard's criterion. Each patient received two identical sets of spectacles, one with the prism and the other without. Neither the patient nor the researchers knew which pair contained the prism. The subjective responses were then analyzed. They found eleven of thirteen patients preferred prism base in for distance esophoria. Prisms were not preferred more than would be expected by chance for distance exophoria or near esophoria. Presbyopes preferred base-in prisms for near exophoria. This experimental double-blind procedure has obvious promise for testing other forms of conventional clinical wisdoms. Although this technique is not generally possible in private practice, a therapeutic trial with membrane (Fresnel) prisms is a valuable alternative.

It is generally agreed that prism base in is better tolerated than prism base out because the latter places an additional burden on convergence. Vertical prisms are indicated more often than horizontal prisms because vertical imbalances cannot be compensated by changes in accommodation and convergence. Other factors being equal, orthoptics is more useful in young patients and prisms more useful in older ones, since visual reflexes are more elastic in the young and the compensating fusional potential more expansive. The general health, mental stability, and relief experienced by previous optical or orthoptic treatment are considered, orbital and facial asymmetries evaluated, the fit and the centering of spectacles appraised, and the placebo therapeutic effect estimated. Such theoretical considerations are good advice for thought but sometimes poor proposals for practice. Prisms are static, whereas fixation sprints and scampers. Prismatic relief is limited to a particular fixation distance, a specific reading point, a characteristic meridian, and a limited direction in that meridian. They cause spatial distortions, stereoscopic misjudgments, and sensory adaptations. Fortunately, the vast majority of heterophorias requires no special treatment; most can be managed by proper refraction and appropriate modification of the optical prescription.

The orthoptic management of heterophoria is directed at the fusional amplitudes. It is doubtful that fusional vergences can be created by training, but when present, their breadth can be increased. Treatment depends on the mature and motivated cooperation of patient and parents. Simple home exercises require nothing more complicated than a loose prism (base out to increase positive fusional vergence; base in to increase negative fusional vergence). The television screen is an ideal accommodative target, and the exercise can be graded by varying the distance from the screen. Accommodation must of course be controlled, which is why simple finger-to-nose exercises are useless to enhance fusional amplitudes. On the synoptophore, the arms are rotated to create a base-out or base-in effect. This exercise can be supplemented by lenses (plus lenses for a fixed convergence require increased positive fusional vergence; minus lenses require increased negative fusional vergence). Polarized projection permits the dissociation of central and peripheral fusion in cases of foveal suppression. Control markings indicate the accommodative and vergence response. In younger children, an objective indication of fusional responses is gained by observing pupil constriction, and the targets should provide interesting, colorful detail for those who cannot read. Orthoptic treatment may be supplemented by pharmacologic and optical measures (cycloplegia, miotics, Fresnel prisms, bifocals, etc.). Periodic reexamination is advisable, since the results of instrumental training do not always carry over to everyday seeing.

In summary, symptomatic heterophorias may be treated by reducing the phoria or increasing the fusional amplitude or a combination of both. A near phoria may be altered by changing the accommodative stimulus, providing the AC/A ratio is high enough to yield proportionately large changes for small optical modifications. Distance phorias are less easily modified but can be influenced by overcorrecting or undercorrecting the refractive error, tracing all latent accommodation, and pharmacologically shifting the required innervation. Alternatively a relieving prism can be ordered. Fusional vergences may be amplified by prisms for a fixed accommodation or lenses for a fixed convergence, both by formal and home exercises, with and without specialized instrumentation. The use of neostigmine bromide (to facilitate neuromuscular transmission), antiparkinson drugs (for their effect on the reticular formation), and cerebral depressants or stimulants have been reported in the literature, but these treatments are experimental, the results have not been proved, and side effects are not without risk. Adequate fusional amplitudes also develop spontaneously in response to ordinary use of the eyes. This is Nature's orthoptics, and it often works well if not interfered with by unnecessary and incorrect spectacles.

APPENDIX

The clinical use of statistics

Uncertainty is a certain impediment to interpretation; fractured measurements lead to stumbling diagnoses and treatment that merely cripples along. Although "for instance" is not proof, given enough instances one can handicap the odds. Every clinical measurement is an instance that deviates from the "true" value, which we can never know. Since people differ with respect to astigmatism and accommodation as they do in weight or political preference, we must know the average anticipated values. But how much of a deviation from this average must there be before we suspect a dysfunction or poor technique or inaccurate instrumentation? To repeat a phoria is one thing; to recommend prolonged, expensive, perhaps dangerous treatment is another. And having recommended it, what are the chances of success or failure? To answer these and similar questions, statistics can help. I do not mean statistics as an inductive branch of mathematics, for classifying research data, or for interpreting conclusions in the literature (although it does that) but as a necessary clinical tool for evaluating individual measurements. To the practitioner, published statistical data substitute for controlled experiment—controls that it would be impractical for him to obtain for himself. For every refractive examination is, in a sense, an experiment that tests the relevant body of optical and physiologic knowledge based on existing information of what is normal.

As an illustrative example, suppose we find a near phoria of 10 exo, measured under

Table 18-1. Distribution of near induced phorias in hypothetical sample consisting of ten cases selected at random and assumed to be normally distributed

	Scores X	*Deviation (d)*	d^2	*z*	%
Patient 1	2 exo	−4Δ	16	−1.60	5.5
Patient 2	4 exo	−2Δ	4	−0.82	20.0
Patient 3	6 exo	0Δ	0	0	50.0
Patient 4	6 exo	0Δ	0	0	50.0
Patient 5	8 exo	+2Δ	4	+0.82	20.0
Patient 6	3 exo	−3Δ	9	−1.20	11.0
Patient 7	7 exo	+1Δ	1	+0.40	34.0
Patient 8	5 exo	−1Δ	1	−0.40	34.0
Patient 9	9 exo	+3Δ	9	+1.20	11.0
Patient 10	10 exo	+4Δ	16	+1.60	5.5
(N = 10)	(ΣX = 60)	(Σd = 20)	(Σd^2 = 60)		
	Mean = 6 exo		Average deviation = 2		
	Median = 6 exo		Standard deviation = 2.45		
	Mode = 6 exo		Variance = 6		
	Standard error of the mean = .82				

clinical conditions. What are the chances of obtaining the same result if we repeated the measurement once, twice, or a hundred times? Obviously there is neither time nor opportunity, so we rely on the results obtained by someone who actually did repeat the findings. But how comparable are these results to those obtained from our individual patients? To analyze the problem, imagine a hypothetical researcher had assembled ten cases (N = 10) and obtained the data shown in Table 18-1.

The second column of Table 18-1 shows the raw scores for each patient in column 1. The arithmetic mean $\left(\frac{\text{sum of scores}}{\text{number of scores}}\right)$ is 60/10 = exo. The median of this sample is also 6 exo—as many patients have a larger than a smaller phoria.* Column 3 shows the deviation of each score from the mean. The algebraic sum of these deviations is always zero, but if we disregard sign, and add the scores and divide by the number of cases (20/10) = 2, we have the average deviation. The average deviation is useful if the scores are not normally distributed. Most biologic data, however, arrange themselves along a bell-shaped curve. If we make the assumption that phorias are normally distributed, we can make use of the known properties of the normal (or Gaussian) curve. To do so we need to compute the standard deviation.

The standard deviation (σ) is obtained by squaring each deviation in column 3 to give the results in column 4 of Table 18-1. Add these and divide by the number of cases: (60/10) = 6. The value 6 is known as the variance. The standard deviation is the square root of the variance ($\sqrt{6} = 2.45$). To make use of the normal curve, convert each deviation to standard deviation units (or "z" score). Thus the first deviation in Table 18-1 is −4; hence −4/2.45 gives a z score of −1.6. Proceeding similarly for each deviation (a small computer helps), we obtain the results recorded in column 5. This completes all the necessary computations.

Normal curve values are found in all handbooks of physics and chemistry or statistics texts. A simplified form is given in Table 18-2, and its significance is illustrated in Fig. 18-12. The figure shows the normal distribution

*Three statistic averages, the mean, median, and mode, are ways of summarizing frequency distributions. They represent the central tendency but give no clue to the dispersion or variance of the distribution. The mean is the most common average and is most closely related to the mathematic theory of probability. The median is less influenced by extreme fluctuations (for example, it would be nonsense to speak of the mean IQ of a group of children, since intelligence is a property of individuals). The mode is the item that occurs most frequently, or that which may be regarded as "typical," (e.g., the most frequent annual income in a community). The popular reference to the "average man" refers to the mode, but frequently no single well-defined type exists; moreover the mode of a combined group cannot be computed from the modes of its components.

Table 18-2. Selected values from tabulated normal probability curve

z score	*Area from mean to z score**	*Area in larger portion†*	*Area in smaller portion‡*	*Ordinate at z*
0.00	.0000	.5000	.5000	.3989
0.10	.0398	.5398	.4602	.3970
0.20	.0793	.5793	.4207	.3910
0.30	.1179	.6179	.3821	.3814
0.40	.1554	.6554	.3446	.3683
0.50	.1915	.6915	.3085	.3521
1.00	.3413	.8413	.1587	.2420
1.50	.4332	.9332	.0668	.1295
1.60	.4452	.9452	.0548	.1109
1.96	.4750	.9750	.0250	.0584
2.00	.4772	.9772	.0228	.0540
2.33	.4901	.9901	.0099	.0264
2.58	.4951	.9951	.0049	.0143
2.81	.4975	.9975	.0025	.0077
3.00	.4987	.9987	.0013	.0044
3.10	.4990	.9990	.0010	.0033
3.30	.4995	.9995	.0005	.0017
3.70	.4999	.9999	.0001	.0004

*Computed by subtracting 0.5000 from values in column 3.
†Computed by subtracting values in column 4 from 1.000.
‡Computed by subtracting values in column 3 from 1.000.

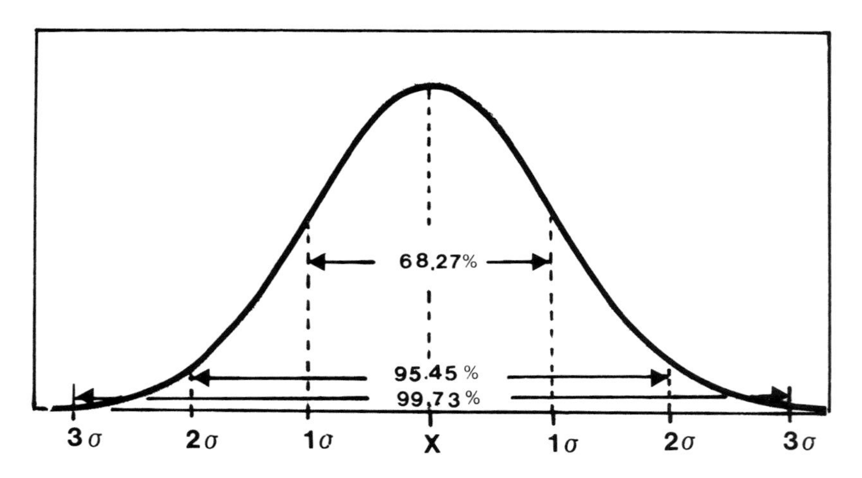

Fig. 18-12. Normal curve.

(N = 1), and the abscissa gives the z score. The z score is applicable to any normally distributed sample as compared to the actual values of a particular sample.

Suppose we wish to determine in this sample of ten cases how often one can expect a deviation of 4Δ (the first recorded deviation in column 3 of the first table) to occur *by chance?* Converted to z scores, we have 4/2.45 = 1.6 standard deviation units. Now the area above this z score is given in column 4 of Table 18-2 as 0.0548 or 5.5%. Thus we can expect to find a deviation of −4Δ (or+4Δ since the curve is symmetrical) to occur by chance only 5.5% of the time. If we were willing to be wrong about five times in a hundred, we might say the deviation was *not* due to chance. Perhaps accommodation fluctuated, attention wandered, the instructions were not clear, the prism scale was inaccurate, or the patient had a convergence spasm. Whatever the reason, a 4Δ deviation would be suspicious at the 5% level of confidence. The columns of Table 18-2 thus answer the following questions:

Column 1—the z score
Column 2—the proportion of cases expected to fall between the mean and any given z score
Column 3—the proportion of cases above a certain z score
Column 4—the porportion of cases below a certain z score
Column 5—the proportion of cases expected to have the z score in question (i.e., the ordinate of the curve)

Some other useful findings can be obtained from the normal curve. Note that the probability of obtaining a z score of +1.96 or higher is 2.5% (from column 4 of Table 18-2). Therefore the area between ±1.96 standard deviation units contains 95% of cases. In our sample of ten cases, 95% of the population fall between 6 ± 4.8, that is, between 1.2 exo and 10.8 exo. Similarly, the mean ±2.58 standard deviation units contains about 99% of cases. (The mean $\pm 1\sigma$ contains 68%; the mean $\pm 2\sigma$ contains 95.45%; and the mean $\pm 3\sigma$ contains 99.73% of cases as shown in Fig. 18-12.) Thus we could predict that 12.33Δ exo (equivalent to a z score of 2.58) would only occur 0.5% of the time by chance. In psychophysical measurements, the 5% prediction is considered fairly safe; the 1% prediction very safe. (In constructing an atomic reactor, the confidence limits would, of course, have to be much better).

It is true that the absolute value of a phoria is less important than its relative size compared to the compensating fusional vergences. But that is the point. If the exophoria is high, we might cut the plus or prescribe a base-in prism. If the phoria is not excessive and the fusional amplitude is low, we might prescribe orthoptic exercises. It is by knowledge of the averages that we make the differential diagnosis. The key question therefore is how representative is the sample to which we compare our results? Did the sample population have the same AC/A ratio; was the ametropia fully corrected; were pupil diameter, general illumination, target size, and fixation distance similar to the conditions under which we perform the test in the refracting room? To return to our hypothetical sample, it is evident that if we analyzed another ten cases, the results would probably differ; and if we plotted the means of ten samples of ten cases each, the results would probably fall on a normal curve similar to the individual scores. But the chances are that the mean of the means is more representative and closer to the "true" value of the universe of possible measurements. Since we can never measure everyone, the true value is never known. Can we make an estimate of it from a single sample? That is, if we had only one sample of ten cases can we predict the universe mean at a certain level of confidence? The statistic required is known as the "standard error of the mean." It is computed by dividing the standard deviation of the sample by the square root of the number of cases in the sample minus one. In our sample, the standard error of the mean is $\frac{2.45}{\sqrt{9}} = 0.82$, and represents the variance of the sample mean from the universe mean. The greater the variance the less representative our sample. Since a z score of ±2.58 occurs by chance 1% of the time (from Table 18-2), the sample variance is (0.82) (2.58) = 2.1, or the estimated universe mean may range from 6 ± 2.1 or from 3.9 exo to 8.1 exo at the 1% level of confidence.

A glance at the formula for the standard error of the mean shows that the sample might become more representative by reducing the standard deviation or increasing the number of people sampled. The variance can be reduced by more rigid testing conditions and by more careful selection. But a more homogeneous sample may be self-limiting (for example, in comparing the refractive findings of children seen in a clinic with their higher proportion of strabismus to the general population; conversely, general statistics on trachoma would be of little use in studying an American Indian population). Obviously case selection must be tempered by common sense; at least for our purposes, the sampling must be similar to the type of patient and test conditions used in practice. Disraeli was wrong. It is not statistics that lie; it is their inappropriate interpretation that leads to fallacies, thus proving practically anything.

Suppose we wanted to investigate the effect of a particular orthoptic treatment on near phorias. We might compare the mean phoria of ten cases before and after treatment and assume the first is the universe mean, and the mean after treatment is a chance fluctuation from this universal average. Such a formulation is called the "null hypothesis"; it is always stated in the form that observed differences are *not* significant but are due to chance. The null hypothesis may then be rejected at a given level of confidence. The application of the standard error of the mean difference is termed the "critical ratio." It enables us to determine (in standard deviation units) how far below the average difference a difference of zero falls.

In evaluating the significance of a particular clinical measurement we need to know whether the test is consistent (gives the same result on repeated application) and whether it measures what it is supposed to. The first is called "reliability" and the second, "validity." Reliability may be determined by repeating the test and comparing the paired measurements. One then computes the standard error of the difference and tests the null hypothesis that the observed differences are due to chance. For good reliability it should be possible to accept the null hypothesis at a high level of confidence. A more useful technique, however, is to determine the coefficient of correlation of paired measurements, which allows us to predict the second from the first. For example, to attach significance to an orthoptic procedure that we believe can change the near phoria, the mean change must be larger than the change that would have been observed if no training had been given. The principles underlying measures of reliability constitute sampling theory. In clinical practice, reliability depends on such factors as an unambiguous end point, clear instructions, standardized test conditions, repeated observations, and calibrated instrumentation, not to speak of gross errors in reading the end point or recording the result. By far the largest variable error is the subjective or personal error—the fact that our patients are not trained observers. Obviously there is little point to reducing an instrumental error from 10 to 0 if the personal error is 100 since the combined variable error would be changed little (assuming the two variable errors are unrelated). Indeterminate errors due to uncontrollable variables increase the variability of the sample and enter into all measurements. They may be assumed to follow a normal distribution and tend to cancel each other with repeated measurements.

Whether a test measures what it is supposed to depends first on a decision as to what we want to measure. This decision is theoretical, not statistical; we do not care about the phoria in a patient with a steamy cornea complaining of acute pain in the eye. Here we are more interested in the validity and reliability of the tonometer.

A valid test presupposes reliability, although the converse does not necessarily hold; a highly consistent central acuity tells us little of the damage to the visual system in uncontrolled chronic glaucoma, at least not until the disease has already reached catastrophic proportions. There are no clear-cut methods to confirm the validity of a test. We assume a test is valid if it correlates with other tests that presumably measure the same function. When Alfred Binet invented the IQ test, he assumed that intelligence ties in with school achievement; hence he included those measures that correlated positively with school performance. To this day, no one knows what intelligence *is*, only that results obtained by such tests enable us to predict certain aspects of behavior. Tests that measure the same function need not necessarily give identical results, since conditions differ, but a battery of tests suggest a trend. So we have more than one test for astigmatism, for relative accommodation, or for anemia. We have more confidence in a diagnosis of convergence insufficiency if the near point, phorias, fusional vergences, and characteristic symptoms all point in the same direction.

A test is more likely to be valid if changes in the test conditions lead to predictable consequences established by general physiologic knowledge of the function tested. For example, when a near phoria is repeated through a series of plus or minus lenses, we anticipate predictable changes according to the AC/A ratio. Phoria techniques that fail to evidence these changes are not likely to be valid.

Statistics are numbers that easily give one a false sense of precision, particularly when they exceed the accuracy of the measurements or the logic of what is measured. The birthrate of a particular community may be 152.5 per month, but this does not mean that half a fetus appears at regular intervals. We can correlate accommodative amplitude with age, but it would be ridiculous to prescribe a reading add by age alone. The clinician who looks at his patients through statistical glasses is in danger of forgetting that averages are completely useless as a substitute for individual diagnosis. Even insurance companies, whose computations of general death rate are accurate to the last decimal, cannot predict the demise of a particular individual. Statistics are no substitute for clinical thinking. Correlations imply relationships that are not necessarily causal. The connection between retrolental fibroplasia and prematurity turned out to be through a third factor, ambient oxygen. The fact that the progression of myopia decreases among army recruits does not mean enlistment will prevent myopia. Improper case selection may lead to the conclusion that diabetic retinopathy is decreasing, whereas it is in fact increasing in older age groups; the decrease is a spurious result of the faster general population growth. The incidence of ametropia in different populations cannot be compared unless the basis of case selection and the diagnostic criteria are similar. Moreover methods of collecting data change; is the higher incidence of reported cancer due to an actual increase in the disease or better methods of diagnosis? Thus the distinction between fact and hypothesis is not always clear; many so-called facts are only mummified theories. Priestley summarized the problem best:

> Very lame and imperfect theories are sufficient to suggest useful experiments which serve to correct those theories, and give birth to others more perfect. These, then, occasion further experiments, which bring us still nearer to the truth; and in this method of approximation, we must be content to proceed, and we ought to think ourselves happy, if, in this slow method we make any real progress.*

*From Priestley, J.: The history of discoveries relating to vision, light, and colours, 1772.

REFERENCES

Abraham, S. V.: Nonparalytic strabismus, amblyopia and heterophoria; a clinical presentation (privately published).

Allen, M. J.: An investigation of the time characteristics of accommodation and convergence of the eyes, Am. J. Optom. **30**:393-402, 1953.

Alpern, M.: The after effect of lateral duction testing on subsequent phoria measurements, Am. J. Optom. **23**: 442-447, 1946.

Alpern, M.: Accommodation and convergence with contact lenses, Am. J. Optom. **26**:379-387, 1949.

Alpern, M.: The zone of clear single vision at the upper levels of accommodation and convergence, Am. J. Optom. **27**:491-513, 1950.

Alpern, M.: Testing distance effect on phoria measurement at various accommodation levels, Arch. Ophthal. **54**:906-915, 1955.

Alpern, M.: The position of the eyes during prism vergence, Arch. Opthal. **57**:345-353, 1957.

Alpern, M.: Vergence and accommodation. II. Is accommodation vergence related merely to the accommodative stimulus, Arch. Ophthal. **60**:358-360, 1958.

Alpern, M.: Muscular mechanisms. In Davson, H., editor: The eye, vol. 3, New York, 1962, Academic Press, Inc.

Alpern, M., and Ellen, P.: A quantitative analysis of the horizontal movements of the eyes in the experiment of Johannes Mueller, Am. J. Ophthal. **42**:289-296, 1956.

Alpern, M., and Hofstetter, H. W.: The effect of prism on esotropia—a case report, Am. J. Optom. **25**:80-90, 1948.

Alpern, M., Kincaid, W. M., and Lubeck, M. J.: Vergence and accommodation. III. Proposed definitions of the AC/A ratios, Am. J. Ophthal. **48**:141-148, 1959.

Ames, A., and Gliddon, G. H.: Ocular measurements, reprinted from Transactions of the American Medical Association Section on Ophthalmology, Chicago, 1928.

Anderson, E. C.: Treatment of convergence insufficiency: a review, Am. Orthopt. J. **19**:72-77, 1969.

Armaly, M. F., and Burian, H. M.: Changes in the Tonogram during accommodation, Arch. Ophthal. **60**:60-69, 1958.

Backer, W. D., and Ogle, K. N.: Pupillary response to fusional eye movements, Am. J. Ophthal. **58**:743-756, 1964.

Balsam, M. H., and Fry, G. A.: Convergence accommodation, Am. J. Optom. **36**:567, 575, 1959.

Bannon, R. E.: Diagnostic and therapeutic use of monocular occlusion, Am. J. Optom. **20**:345-358, 1943.

Bielschowsky, A.: Divergence excess, Arch. Ophthal. **12**:157, 1934.

Biggs, R. D., Alpern, M., and Bennett, D. R.: The effect of sympathomimetic drugs upon the amplitude of accommodation, Am. J. Ophthal. **48**:169-172, 1959.

Breinin, G. M.: The nature of vergence revealed by electromyography, Arch. Ophthal. **54**:407-409, 1955.

Breinin, G. M.: The position of rest during anesthesia and deep sleep, Arch. Ophthal. **58**:323-326, 1957.

Breinin, G. M.: Accommodative strabismus and the AC/A ratio, Am. J. Ophthal. **71**:303-311, 1971.

Burian, H. M.: Anomalies of convergence and divergence functions and their treatment. In The New Orleans Academy of Ophthalmology: Symposium on strabismus, St. Louis, 1971, The C. V. Mosby Co., pp. 223-232.

Capobianco, N. M.: The subjective measurement of the near point of convergence and its significance in the diagnosis of convergence insufficiency, Am. Orthopt. J. **2**:40-42, 1952.

Childress, D. S., and Jones, R. W.: Mechanics of horizontal movement of the human eye, J. Physiol. (London) **188**:273-284, 1967.

Chin, N. B., and Breinin, G. M.: Ratio of accommodative convergence to accommodation, Arch. Ophthal. **77**:752-756, 1967.

Christoferson, K. W., and Ogle, K. N.: The effect of homatropine on the accommodation–convergence association, Arch. Ophthal. **55**:779-791, 1956.

Cogan, D. G.: Accommodation and the autonomic nervous system, Arch. Ophthal. **18**:739-766, 1937.

Cohen, M. M., and Alpern, M.: Vergence and accommodation. VI. The influence of ethanol on the AC/A ratio, Arch. Ophthal. **81**:518-525, 1969.

Costenbader, F. D.: Diagnosis and clinical significance of the fusional vergences, Am. Orthop. J. **15**:14-20, 1965.

Cridland, N.: The relation between accommodation and non-fusional convergence, Trans. Ophthal. Soc. **69**: 567-674, 1949.

Cushman, B., and Burri, C.: Convergence insufficiency, Am. J. Ophthal. **24**:1044-1052, 1941.

Davis, J. C., and Jobe, F. W.: Further studies on the AC/A ratio as measured on the orthorater, Am. J. Optom. **34**:16-25, 1957.

Duane, A.: A new classification of the motor anomalies of the eye, based upon physiological principles, Ann. Ophthal. **6**:84-122, 1897.

Dyer, J. A., and Martens, T. G.: Surgical treatment of divergence excess, Am. J. Ophthal. **50**:297-302, 1960.

Emmes, A. B.: A statistical analysis of the accommodation–convergence gradient, Am. J. Optom. **26**:474-482, 1949.

Eskridge, J. B.: Age and the AC/A ratio, Am. J. Optom. **50**:105-107, 1973.

Fender, D. H., and Nye, P. W.: An investigation of the mechanism of eye movement control, Kybernetik **1**:81-88, 1961.

Fincham, E. F., and Walton, J.: The reciprocal actions of accommodation and convergence, J. Physiol. (London) **137**:488-508, 1957.

Flom, M. C.: Variations in convergence and accommodation induced by successive spherical lens additions with distance fixation—an investigation, Am. J. Optom. **32**:111-136, 1955.

Flom, M. C.: On the relationship between accommodation and accommodative convergence, Am. J. Optom. **37**:474-482, 517-523, 619-632, 1960.

Fry, G. A.: An analysis of the relationship between phoria, blur, break, and recovery findings at the near point, Am. J. Optom. **18**:393-402, 1941.

Fry, G. A.: Research in accommodation and convergence, Am. J. Optom. **30**:169-176, 1953.

Fry, G. A.: The effect of age on the AC/A ratio, Am. J. Optom. **36**:299-303, 1959.

Hebbard, F. W.: Foveal fixation disparity measurements and their use in determining the relationship between accommodative convergence and accommodation, Am. J. Optom. **37**:3-26, 1960.

Hermann, J. S., and Samson, C. R.: Critical detection of the accommodative convergence to accommodation ration by photosensor-oculography, Arch. Ophthal. **78**:424-430, 1967.

Hofmann, F. B., and Bielschowsky, A.: Ueber die Willkur entzogenen Fusions bewehgungen der Augen, Pflügers Arch. Gesamte Physiol. **80**:1-40, 1900.

Hofstetter, H. W.: The proximal factor in accommodation and convergence, Am. J. Optom. **19**:67-76, 1942.

Hofstetter, H. W.: The zone of clear single binocular vision, Am. J. Optom. **22**:301-333, 361-384, 1945.

Hofstetter, H. W.: Accommodative convergence in squinters, Am. J. Optom. **23**:417-437, 1946.

Hofstetter, H. W.: Accommodative convergence in identical twins, Am. J. Optom. **25**:480-491, 1948.

Hyde, J. E.: Some characteristics of voluntary human ocular movements in the horizontal plane, Am. J. Ophthal. **48**:85-94, 1959.

Ittelson, W. H.: Visual space perception, New York, 1960, Springer Publishing Co.

Ittelson, W. H., and Ames, A.: Accommodation, convergence, and their relation to apparent distance, J. Psychol. **30**:43-62, 1950.

Jampel, R. S.: Representation of the near response in the cerebral cortex of the macaque, Am. J. Ophthal. **48**:573-582, 1959.

Johnson, D. S.: Some observations on divergence excess, Arch. Ophthal. **60**:7-11, 1958.

Jones, B. A.: Orthoptic handling of fusional vergences, Am. Orthopt. J. **15**:21-19, 1965.

Kent, P.: Convergence accommodation, Am. J. Optom. **35**:393-406, 1958.
Knapp, P.: The clinical management of accommodative esotropia, Am. Orthopt. J. **17**:8-13, 1967.
Kramer, L. W.: Miscellany in orthoptic procedures, Am. Orthopt. J. **17**:34-38, 1967.
Lancaster, W. B.: Physiology of disturbances of ocular motility, Arch. Ophthal. **17**:983-993, 1937.
Maddox, E. E.: The clinical use of prisms and the decentering of lenses, ed. 2, Bristol, 1893, John Wright & Co.
Manley, D. R.: Classification of esodeviations. In Manley, D. R., editor: Symposium on horizontal ocular deviations, St. Louis, 1971, The C. V. Mosby Co., pp. 3-48.
Mann, I.: Convergence insufficiency, Br. J. Ophthal. **24**:373-390, 1940.
Manson, N.: Anemia as a factor in convergence insufficiency, Br. J. Ophthal. **46**:674-677, 1962.
Marg, E., and Morgan, M. W.: The pupillary near reflex; the relation of pupillary diameter to accommodation and the various components of convergence, Am. J. Optom. **26**:183-198, 1949.
Marg, E., and Morgan, M. W.: The pupillary fusion reflex, Arch. Ophthal. **43**:871-878, 1950.
Martens, T. G., and Ogle, K. N.: Observations on accommodative convergence, Am. J. Ophthal. **47**:455-463, 1959.
Michaels, D. D.: A clinical study of convergence insufficiency, Am. J. Optom. **30**:65-71, 1953.
Morgan, M. W.: A comparison of clinical methods of measuring accommodation-convergence, Am. J. Optom. **27**:385-396, 1950.
Morgan, M. W.: Relationship between accommodation and convergence, Arch. Ophthal. **47**:745-759, 1952.
Morgan, M. W.: Accommodation and vergence, Am. J. Optom. **45**:417-454, 1968.
Morgan, M. W., and Olmstead, J. M. D.: Quantitative measurements of relative accommodation and relative convergence, Proc. R. Soc. Exp. Biol. Med. **41**:303-307, 1939.
Morgan, M. W., and Peters, H. B.: Accommodation-convergence in presbyopia, Am. J. Optom. **28**:3-10, 1951.
Mould, W. L.: Recognition and management of atypical convergence insufficiency, J. Pediatr. Ophthal. **7**:212-214, 1970.
Nagel, W. A.: Das Sehen mit zwei Augen und die Lehre von der identischen Netzhautpunkten, Leipzig, 1861, C. F. Winter.
Norn, M. S.: Convergence insufficiency, Acta Ophthal. **44**:132-138, 1966.
Ogle, K. N.: Researches in binocular vision, Philadelphia, 1950, W. B. Saunders Co.
Ogle, K. N., and Martens, T. G.: On the accommodative convergence and the proximal convergence, Arch. Ophthal. **57**:702-715, 1957.
Parks, M. M.: Abnormal accommodative convergence in squint, Arch. Ophthal. **59**:364-380, 1958.
Pascal, J. I.: On convergence tests, Am. J. Ophthal. **26**:967-969, 1943.
Passmore, J. W., and MacLean, F.: Convergence insufficiency and its management, Am. J. Ophthal. **43**:448-456, 1957.
Payne, J. W., and von Noorden, G. K.: The ratio of accommodative convergence to accommodation in strabismus with A- and V-patterns, Arch. Ophthal. **77**:26-28, 1967.
Percival, A. S.: The relation of convergence to accommodation and its practical bearing, Ophthal. Rev. **11**:313-328, 1892.
Pereles, H.: Ueber die relative Akkommodationsbreite, Graefe's Arch. Ophthal. **35**:84-115, 1889.
Porterfield, W.: Treatise on the eye, Edinburgh, 1759, G. Hamilton & J. Balfour.
Posner, A.: Divergence excess considered as an anomaly of the postural tonus of the muscular apparatus, Am. J. Ophthal. **27**:1136-1142, 1944.
Reinecke, R. D.: AC:A nomogram, Arch. Ophthal. **77**:788, 1967.
Reinecke, R. D.: Accommodative convergence/accommodation ratio and exotropia. In Manley, D. R., editor: Symposium on horizontal ocular deviations, St. Louis, 1971, The C. V. Mosby Co., pp. 141-148.
Riggs, L. A., and Niehl, E. W.: Eye movements recorded during convergence and divergence, J. Opt. Soc. Am. **50**:913-930, 1960.
Ripps, H., Chin, N. B., Siegel, I. M., and Breinin, G. M.: The effect of pupil size on accommodation, convergence, and the AC/A ratio, Invest. Ophthal. **1**:127-135, 1962.
Robinson, D. A.: The mechanics of human vergence eye movements, J. Pediatr. Ophthal. **3**:31-37, 1966.
Sabin, F. C., and Ogle, K. N.: Accommodation-convergence association, Arch. Ophthal. **59**:324-332, 1958.
Schapero, M., and Levy, M.: The variation of proximal convergence with change in distance, Am. J. Optom. **30**:403-416, 1953.
Scobee, R. G., and Green, E. L.: A center for ocular divergence; does it exist? Am. J. Ophthal. **29**:422-434, 1946.
Scobee, R. G., and Green, E. L.: Relationships between lateral heterophoria, prism vergence, and the near point of convergence, Am. J. Ophthal. **31**:427-441, 1948.
Sears, M., and Guyer, D.: The change in the stimulus AC/A ratio after surgery, Am. J. Ophthal. **64**:872-876, 1967.
Sheard, C.: Zones of ocular comfort, Am. J. Optom. **7**:9-25, 1930.
Sloan, L. L., Sears, M. L., and Jablonski, M. D.: Convergence-accommodation relationships, Arch. Ophthal. **63**:283-306, 1960.
Tait, E. F.: Report on results of experimental variations of stimulus conditions in responses of accommodation convergence reflex, Am. J. Optom. **7**:199-206, 1932.
Tait, E. F.: Fusional vergence, Am. J. Ophthal. **32**:1223-1230, 1949.
Tait, E. F.: Accommodative-convergence, Am. J. Ophthal. **34**:1093-1107, 1951.

Tamler, E., Jampolsky, A., and Marg, E.: An electromyographic study of asymmetric convergence, Am. J. Ophthal. **46**:174-182, 1958.

Townes, C. D.: Surgical treatment of hetrophoria, Trans. Am. Acad. Ophthal. Otolaryngol. **49**:338, 1945.

Urist, M. J.: Divergence excess combined with convergence excess in the V syndrome, Am. J. Ophthal. **50**:765-783, 1960.

van Hofen, R. C.: Partial cycloplegia and the accommodation-convergence relationship, Am. J. Optom. **36**: 22-39, 1959.

von Noorden, G. K.: Divergence excess and simulated divergence excess: diagnosis and surgical management, Doc. Ophthal. **26**:719, 1969.

Warwick, R.: The so-called nucleus of convergence, Brain **78**:92-114, 1955.

Westheimer, G.: Amphetamine, barbiturates, and accommodation-convergence, Arch. Ophthal. **70**:830-836, 1963.

Westheimer, G., and Mitchell, A. M.: Eye movement responses to convergence stimuli, Arch. Ophthal. **55**:848-856, 1956.

Worrell, B. E., Hirsh, M. J., and Morgan, M. W.: An evaluation of prism prescribed by Sheard's criterion, Am. J. Optom. **48**:373-376, 1971.

Zuber, B., and Stark, L.: Dynamical characteristics of the fusional vergence eye-movement system, IEEE Trans. Sys. Sci. Cybernet. **SSC-4**:72-79, 1968.

PART III

EVALUATION

19

Anisometropia, anisophoria, and aniseikonia

To prescribe poorly because the refraction was done ineptly is one thing; to prescribe accurately and create more problems than the patient started with is a disenchantment of anisometropia. The term "anisometropia" (a = not; iso = equal) means unequal refractive error. Since no two eyes are exactly alike if measured with sufficient precision, the definition is arbitrary, and the prevalence depends on the importance attached to symptomatic discomfort, oculomotor disbalance, spatial distortion, or image deformity. The cause is unknown; except for traumatic, surgical, disease, or drug-induced changes, it probably represents a stage of ametropic genetics coupled with local structural idiosyncracies.

The diagnosis of anisometropia presents little difficulty. When, how, or even whether it should be treated remains something of an enigma, optically tantalizing, therapeutically baffling, frequently disillusioning. It is not enough to arrive at a diagnosis; we must know at what stage. No lens will help the child already amblyopic; no equal images benefit the adult who cannot put them together; no binocularity is useful if it distorts the environment. At some time every refractionist discovers the therapeutic value of prisms only to find that exchanging one nuisance for another is not necessarily progress. And although aniseikonia is no longer in dispute, its management remains subject to debate. Restraint is better than recriminations not only because size lenses are so expensive but also because the complaints are so expressive.

SIGNS AND SYMPTOMS

The history establishes the onset, nature, frequency, and duration of complaints; their relation to the use of the eyes; whether symptoms were present with the old glasses or precipitated by the new ones; and the means used to obtain relief. For example, a patient whose previous spectacles become intolerable after a lens replacement makes us reach for our ever handy lens clock and caliper, confident that someone inadvertently changed the base curve, the index, the thickness, or the surface on which the cylinder is gound. A sudden, fluctuating anisometropia suggests looking into the blood sugar; a gradual but progressive onset may mean nuclear sclerosis.

Symptoms may include headaches, discomfort, burning, pulling, drawing, itching, photophobia, or, more commonly and more vaguely, "eyestrain." They are not characteristic of anisometropia; more suggestive are inaccurate distance judgments, spatial distortions, nausea, dizziness, and perhaps intermittent double vision possibly relieved by closing one eye. The patient may describe car sickness or stumbling over steps or that walls appear shrunken, floors tilted, and columns twisted. Frequently, the complaints express what he feels rather than what he sees, and this is likely to be less than enthusiastic if the difficulties are precipitated by new spectacles. Responses differ greatly, and one soon views the optical rules with skepticism and learns a healthy respect for human adaptability. Although what's what is more important than who's who, prismatic indiscretions are less likely if the patient is asymptomatic. Some adaptive mechanisms can be diagnosed from across the room: the twisted glasses, the tilted head, and the turning eye.

Acuity differences are expected in anisometropia but may go unnoticed unless the patient happens to close one eye, and even this may be ignored by a child. Acuity differences are not evident if anisometropic hyperopia is tested without cycloplegia, and in myopia the differences depend on the associated astigmatism, blur interpretation, and whether the lids are squeezed together.

Anisometropic images are not only out of focus but also differ in size and shape, causing flat objects to appear three-dimensional and objects in three dimensions to appear perverted. These spatial distortions disappear in ordinary environments, but there are no physiologic mechanisms to compensate for them; so they reappear when least expected with an annoying insistence and sometimes dangerous persistence in walking, climbing, and driving.

Signs and symptoms of anisometropia may thus be induced by the condition itself, by the patient's attempts to compensate for it, or even by the prescription designed to eliminate it.

PATHOLOGIC PHYSIOLOGY

Clinical problems in anisometropia can be related to three major pathophysiologic mechanisms, distinct in theory but intertwined in practice: unequal acuity, unequal motility, and unequal ocular images. Without equal acuity, binocular vision may fail to develop; without equal motility, it may not be comfortable; without equal images, it may be distorted.

Unequal acuity

In children, unequal acuity is a potential obstacle to the development of normal binocular vision. Most children are hyperopic, and since the eyes cannot muster unequal accommodation, one image remains blurred. The greater the blur the less the incentive to fusion. Since the less hyperopic eye is likely to be used for all visual activities, the other is suppressed and may become amblyopic. Here it is not a matter of anomalous correspondence, eccentric fixation, or ocular deviation; the oculomotor apparatus is normal, but the blur is different, and the more hyperopic eye is out of focus whatever the fixation distance. Its blurred image interferes with the sharply seeing eye, and vision may fail to develop. Amblyopia is less likely in anisometropic myopia not only because myopia appears later in life but also because fair acuity is achieved at least at some near distance.

The depth of amblyopia is generally out of proportion to the degree of anisometropia. It is less severe than in strabismic amblyopia and more easily treated, probably because the need to eliminate confusion and diplopia in strabismus requires a more active, hence more permanent, suppressive adaptation. Anisometropic hyperopia may precipitate strabismus when the dominant eye has a high AC/A ratio because the accommodation mustered to obtain clear vision induces convergence. This esodeviation, normally compensated by the fusional reflexes, becomes manifest if binocular vision has failed to develop. Fortunately strabismus occurs only in a small proportion of anisometropic amblyopes; not so fortunately, without an obvious deviation the diagnosis is more easily missed, and treatment is delayed.

Even should binocular vision develop in small degrees of anisometropia by the skin of the fusional teeth, decompensation may occur whenever the patient is tired, ill, or otherwise stressed. Intermittent double vision, periodic confusion, more suppression, and possibly a permanent squint follows. Although the pathophysiology of strabismus is not fully understood, acuity, suppression, tonic, and accommodative and fusional reflexes play a role, and all are affected to some degree in anisometropia. If these obstacles are to be avoided, anisometropia must be corrected early to give each eye the best possible focus and both eyes the best potential for binocularity. However little we know about strabismus, this much we do know, and this much we can do positively, unequivocally, and without harm.

Unequal motility

When viewing an eccentric object through lenses, the angle the eye turns depends on the induced prismatic effect. Rotation is increased for a plus lens and decreased for a minus lens, according to Prentice's rule: d = cF, where d is the deviation in centimeters, c the distance of the visual axis from the optical center, and F the power in the meridian of rotation. Since the induced prismatic effect is unequal, each eye must undergo a different ocular rotation. For example, in the anisometropic prescription RE: +2 D, LE: −2 D, when both eyes turn 10° down (or about 5 mm for a 28 mm distance of the lens from the center of rotation of the eye), the right eye encounters 0.5 (+2) = 1Δ base up, and the left eye encounters 0.5 (−2) = 1Δ base down, a differential prismatic effect of 2Δ. The more eccentric the fixation the larger these induced effects. Because vertical fusional vergences are small, vertical imbalances are more likely to cause trouble. Of course, the eyes only execute compensatory rotations if binocular vision is present; a prism before a suppressing eye has no effect, although it may throw the image outside the accustomed suppression area and create diplopia where none existed before.

In anisometropia there is a differential prismatic effect not only at a particular point but at every eccentric point as well, and by Hering's law of equal innervation, a readjustment to maintain binocular vision must be made for each change in fixation. The readjustment is between the habitual innervation to the fixing eye and the now conflicting feedback from retinal corresponding points. When anisometropia develops slowly, there is time for adaptation, but a correction may suddenly create a new condition; so the timing as well as the magnitude of the induced imbalances must be considered.

The differential prismatic effect can be expressed either by Prentice's rule or by spectacle magnification. Spectacle magnification compares retinal image size before and after the lens is placed before the eye and is increased by plus lenses and decreased by minus lenses. Assuming that the center for rotation coincides with the nodal point for simplicity, a +3.25 D lens 28 mm from the nodal point produces spectacle magnification of $m = \frac{1}{1 - dF}; \frac{1}{1 - 0.028\,(3.25)} = 1.10$ or 10%.

When the eye views an object 30° from the optic axis, it must therefore make a 10% (30°) = 3° additional rotation corresponding to 3 (1.74) = 5.2Δ, which is the induced prismatic effect of the lens.

Similarly, when the visual axis passes 30° eccentrically in a lens placed 28 mm from the center of rotation of the eye, tan 30° = 0.5774; 0.5774 (2.8) = 1.62 cm from the optical center. From Prentice's rule, 1.62 (3.25) = 5.2Δ, the same result.

The term "anisophoria" was coined by Friedenwald (1936) to describe the changing heterophoria as the eyes fixate through different portions of anisometropic spectacles. (There are also minor phoria variations with oblique rotations even in isometropic individuals, which are ignored here.) In fact even emmetropic eyes undergo unequal vertical rotations in reading across a page held in the frontal plane, since the print is closer to one eye than the other; at a 33 cm distance and a 15 cm by 15 cm page, the amount can exceed 3Δ. But these seem to be immediately compensated for and are never symptomatic.

Friedenwald believed that the chief cause of symptoms in aniseikonia was not unequal retinal images but fluctuating phorias and that the benefit of aniseikonic lenses was due to eliminating anisophoria. One may or may not agree with this view, since neither mechanism is fully understood, but aniseikonia also occurs in the primary position; hence spatial distortions (and presumably discomfort) may appear without ocular rotations. Moreover in axial anisometropia there is anisophoria but no aniseikonia when correcting lenses are placed at the anterior focal point of the eye (or approximately so). Of course, anisophoria is not manifest when binocular vision is maintained (to demonstrate it the eyes must be dissociated), but this does not abolish the fusional strain in the meantime. In the sense that the phoria is artificially induced by lenses, it is "false"; the true phoria has to be computed by subtracting the induced prism. Yet it is the "false" phorias the patient must learn to put up with when he wears his spectacles. So it does not follow that a more accurate measurement is obtained by taking the glasses off because it is the induced phorias that are the potential source of difficulty. As Lebensohn (1953) has emphasized, by comparing intrinsic and induced phorias we obtain a measure of how much readaptation is needed.

It is a simple matter to compute the induced prismatic effect for spheres and spherocylinders whose primary meridians are vertical and horizontal. With oblique cylinders, however, a pure lateral rotation causes a vertical imbalance, and a pure vertical rotation causes a horizontal imbalance.

An insight into these effects can be gained by a simple graphic method of resolving an oblique prism into its horizontal and vertical components. Consider first the spherocylinder +2 + 4 × 120. We wish to compute the prismatic effect at a point *R*, 6 mm down and 4 mm in (Fig. 19-1). Taking the sphere separately, one has from Prentice's rule: 0.6 (+2) = 1.2Δ base up and 0.4 (+2) = 0.8Δ base out. To compute the prismatic effect of the cylinder, draw a circle as indicated, and locate the axis at 120° with a protractor. The line drawn from *R* perpendicular to the axis represents the oblique prism. Note that the base is toward the axis (since this is a plus cylinder), and it lies in the 30°

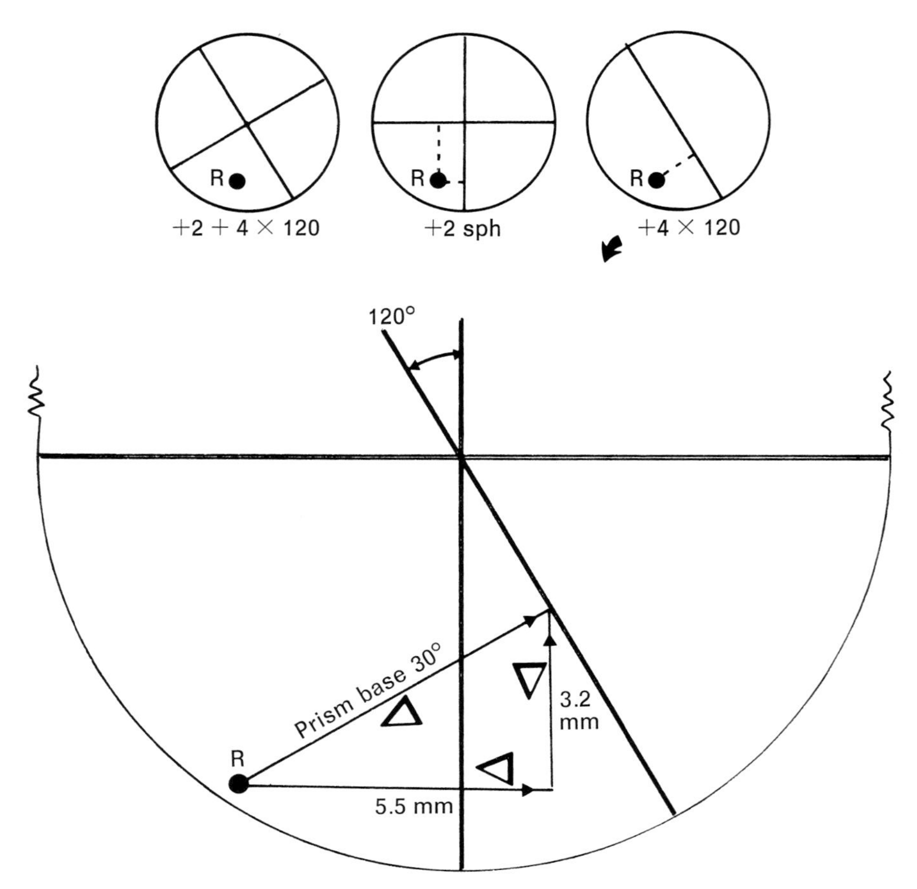

Fig. 19-1. Graphic method for computing prismatic effect of oblique cylinders. In spherocylindrical prescriptions, the sphere and cylinder are treated separately. This diagram applies only to the latter. Draw a circle and indicate axis by measuring with a protractor. Drop a perpendicular from the oblique prism, which may be resolved into its horizontal and vertical components by completing the right triangle. Using an appropriate scale (for example, 1 mm = 1 cm), the prismatic power can then be computed from Prentice's rule. The small triangles indicate the direction of the prism.

meridian. To resolve it into vertical and horizontal components, draw a perpendicular line and complete the right triangle. Choosing suitable units (e.g., 1 mm = 1 cm), the length of the vertical line represents the vertical prism: 0.32 (+4) = 1.28Δ base up in this example. The length of the horizontal line is the horizontal component: 0.55 (+4) = 2.2Δ base out. The total induced prismatic effect at *R* is the sum of the effects of the sphere and cylinder: 2.48Δ base up and 3Δ base out.

Based on the same construction, Fig. 19-2 shows the effect of a pure downward rotation for a series of cylinder axes from 90° to 180°. In each case, the induced oblique prism is represented by the dotted line perpendicular to the axis in question. Taking each intersection and projecting them on the X-X and Y-Y axes, we get the horizontal and vertical components. Thus if the downward rotation is 6 mm from the optical center of a simple cylinder axis 150°, the horizontal component is *OC* (measured on X-X), and the vertical component is *RC* (measured on Y-Y). Note that as the cylinder axis approaches 180°, the vertical prismatic component increases, but the horizontal component reaches a maximum when the cylinder axis is 135° (or 45°) and then decreases.

Fig. 19-3 shows the effect of horizontal rotation through oblique cylinders. As the eye turns laterally, there is an oblique prismatic effect indicated by the line perpendicular from *R* to the cylinder axis. Axes are shown from 90° to 180°. The nearer the axis approaches 90° the greater the horizontal prism (increasing from *RA* to *RO*), whereas the vertical prismatic component (indicated by *OA*, *OB*, *OC*, etc.) increases to a maximum at 135° (or 45°) and then decreases again.

Whereas the horizontal prismatic effect induced by vertical rotations through oblique cylinders is negligible (due to larger fusional amplitudes), the vertical prismatic effect induced by horizontal rotations is not. Convergence alone may thus change the near vertical phorias, especially when the cylinder axis is at 45° or 135°, the meridians causing the most difficulty. Not only is the vertical

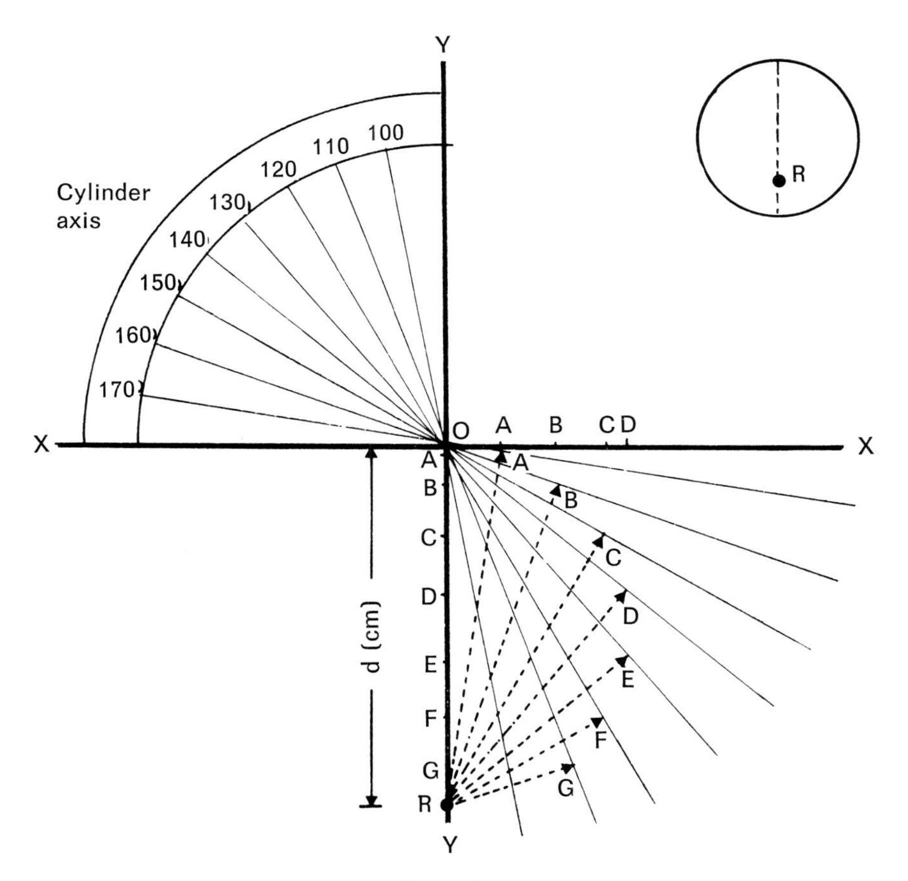

Fig. 19-2. Prismatic effect of oblique cylinder for vertical rotations. For details see text.

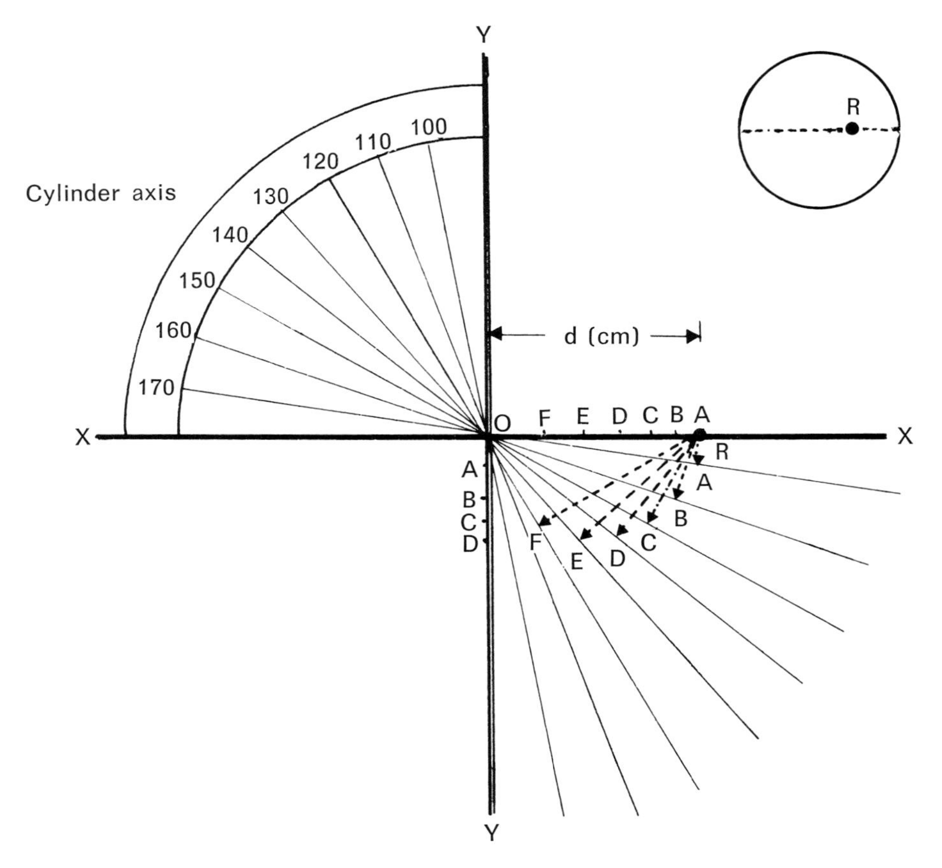

Fig. 19-3. Prismatic effect of oblique cylinder for horizontal rotations. For details see text.

prism a maximum here, but also the anisophoria fluctuates most in all directions.

As with all sensorimotor adaptations, some patients adjust better than others, and the amount of induced prismatic effects that by all odds should prove intolerable can sometimes be overcome. Although 1Δ or 2Δ of vertical imbalance are frequently cited as the limit, two or three times as much can be tolerated after an initial adjustment period. Minutes to hours later the orthophoric patient wearing a vertical prism returns to orthophoria despite the prism; a patient with a phoria regains the same value despite an induced prism (providing it is within the limits of binocular vision). Failure to take these adjustments into account can lead to prescribing increasingly stronger prisms for an apparently persistent vertical phoria.

Adaptations may also be demonstrated by the fixation disparity technique (Ogle, 1953), in which a disparity with forced vergence tends to return to the original value despite the prism and regains that value when the prism is removed. Some patients can overcome as much as 6Δ vertical prism, although recovery (when the prism is removed) is prolonged. In some cases, patients adapt to half the prism, whereas others adapt to the full amount after a day or two. Why some patients compensate fully and others do not is unknown, but this is no doubt responsible for the different views on therapeutic prisms in the clinical literature. Significantly, Ogle reported that five of his seven subjects whose eyes were prismatically forced into vertical divergence reported no discomfort. So induced prismatic effects of lenses should be measured, not computed as if the patient had no capacity for adaptation.

Unequal ocular images

No two eyes are ever identical; there are always differences in dioptric power, axial length, aberrations, and resolving power. Under certain circumstances, ocular images differ in size or shape, and abnormal symptomatic incongruities are termed "aniseikonia." The incongruities are not absolute but relative to each other; hence aniseikonia is always an anomaly of binocular vision.

Lancaster (1942) differentiated between "normal" and "abnormal aniseikonia." The first refers to physiologic retinal disparity; the second to measurable spatial distortions. Normal aniseikonia or physiologic disparity is the inevitable result when an object lies partly outside the horopter, that is, when it extends in three dimensions. An arrow extending away from us has one retinal image larger than the other, and this inequality is the stimulus for stereopsis. An arrow in the frontal plane viewed by asymmetrical convergence also results in unequal images, but because a record of the innervation pattern is kept, it appears in the frontal plane. Because binocular vision has a built-in elasticity, called "Panum's fusional areas," two ocular images of different size may fall on noncorresponding points but still fuse (Fig. 19-4). Since fusional areas will not stretch, there is a limit outside which there is physiologic diplopia and suppression. All this is expected, normal, and physiologic—the basis of stereoscopic vision.

Retinal disparity was almost discovered by Leonardo da Vinci. He recorded in his notebook that when the two eyes view a sphere, the space behind the ball is visible because each eye sees partly around it. Had he selected any other shape, he might have noted that the object itself appears differently to each eye. So the discovery was left to Wheatstone: "It will now be obvious why it is impossible for the artist to give a faithful representation of any near solid object, that is, to produce a painting which shall not be distinguished in the mind from the object itself. When the painting and the object are seen with both eyes, in the case of the painting two *similar* pictures are projected on the retina, in the case of the solid object the pictures are *dissimilar;* there is therefore an essential difference between the impressions on the organs of sensation in the two cases, and consequently between the perceptions formed in the mind; the painting therefore cannot be confounded with the solid object."*

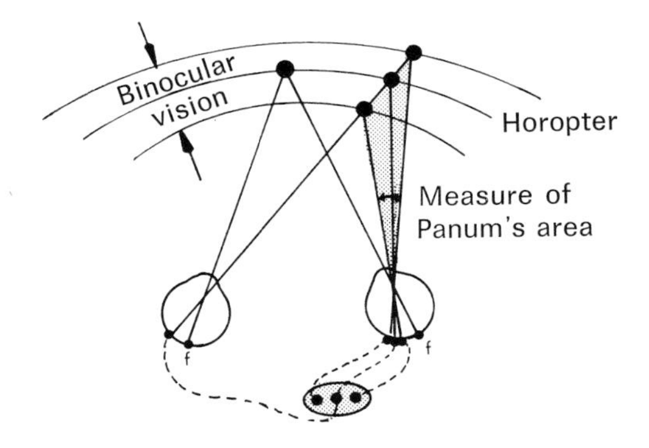

Fig. 19-4. Diagram illustrating projection of Panum's area on the horopter. Note that the presence of these corresponding areas, within which there is single binocular vision, implies that the horopter has "thickness."

Abnormal, or clinical aniseikonia (we shall use the term only in this sense), refers to those situations in which ocular images are unequal when they should be equal or equal when they should be unequal. If we view a horizontal arrow in a frontal plane through a "size" lens placed before one eye, one image is magnified as it might have been if the arrow actually extended away from us, and the brain interprets the flat figure as three-dimensional, (Fig. 19-5). The principle was clearly stated by Helmholtz in his treatise on physiologic optics:

> The general rule determining the ideas of vision that are formed whenever an impression is made on the eye, with or without the aid of optical instruments, is that such objects are always imagined as being present in the field of vision as would have to be there in order to produce the same impression on the nervous mechanism, the eyes being used under ordinary normal conditions.

Perceptions must make sense, must have meaning, and must fit with experience; hence

*From Wheatstone, C.: Contributions to the physiology of vision. I. On some remarkable and hitherto unobserved phenomena of binocular vision, Philosoph. Trans. R. Soc. London **128**:371-394, 1838.

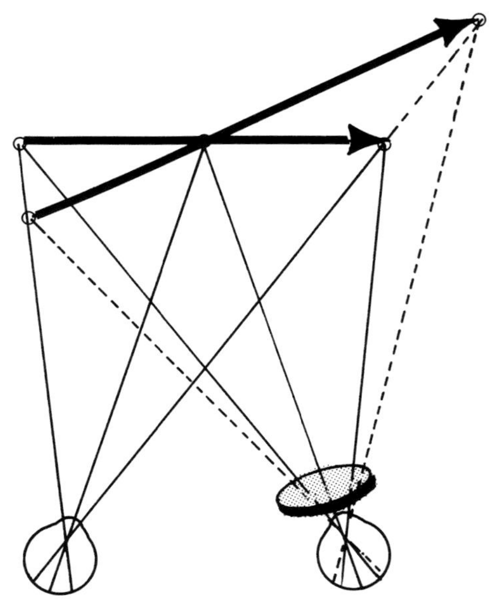

Fig. 19-5. Effect of "size lens" placed before the right eye is to make the frontal arrow appear in three dimensions. The size lens can actually be any spectacle lens that magnifies the retinal image compared to the other eye. The magnification may be produced by a power or a shape difference.

when a size lens is worn a few hours or days, the spatial distortions disappear. A patient with aniseikonia does not see others with asymmetrical arms or legs (though his own face may appear initially asymmetrical in a mirror) because monocular cues and past experience inform him otherwise; hence the reason for Helmholtz's last qualification—under "ordinary normal conditions." The conditions include monocular as well as binocular cues: size, brightness, perspective, overlap, and gradients of density and texture all operating simultaneously. The more natural the environment the easier it is to suppress the erroneous picture both eyes see together in favor of the correct one each eye sees separately.

Erroneous binocular cues are never physiologically compensated and reappear whenever monocular cues are reduced or absent, such as in a leaf room.* We recognize that binocular vision based on normal disparity is superior to monocular vision in the quantitative perception of relative distances. We may fail to recognize that it involves more than a stereothreshold. When one ocular image differs from the other, not only is the stereothreshold lowered, but objects are altered in size, shape, inclination, and position as well. The patient sees objects not only larger but also nearer; so he misses the curb, spills the coffee, hits the garage door, or worse. Moreover, since ocular movements depend on where objects are seen, the innervation is faulty. These perceptual distortions, only manifest with binocular vision, are the essence of the pathophysiology of aniseikonia. Strabismics, who suppress or alternate, and uncorrected monocular aphakes do not have clinical aniseikonia; in the first group, binocular vision did not develop, and in the second group, it is not usually possible (at least with spectacles).

Aniseikonia can be created in two emmetropic eyes by a size lens, which is nothing more than a solid glass Galilean telescope or afocal lens. This type of magnification is induced by all real lenses, which have some thickness and front surface power, and is called the "shape factor." In addition, there is the magnification due to thin lens power as in anisometropia. A "built-in" size lens pro-

*The "leaf room" is a rectangular room whose walls are covered with artificial vines and leaves individually adjusted to stand out uniformly from the surface. The rough surface of the leaves, properly illuminated, provides contours for stereoscopic vision, yet introduces only a minimum number of monocular cues limited mostly to perspective so that the observer can recognize distortions in shape.

duces intrinsic aniseikonia; if magnification is due to the correcting spectacles, as is usually the case, the aniseikonia is induced.

Size differences in aphakia and marked anisometropia were long known, but those less than 5% were thought unimportant because of Panum's fusional areas and poor peripheral acuity. It was a lawyer with an unusual interest in physiologic optics, Adelbert Ames, who investigated the perceptual effects of small size differences around 1919 and became the guiding spirit in the now legendary Dartmouth Eye Institute, with its lengendary interdiciplinary group of ophthalmologists (Burian, Carleton, Lancaster, Linksz), optometrists (Bannon, Ellerbrock), and biophysicists (Boeder, Imus, Madigan, Ogle, Gliddon).

It is convenient to express the ocular size difference in percent. A 3% aniseikonia means one retinal image is 3% larger than the other. Since clinical aniseikonia is measured not by retinal image size but by stereoscopic distortions, it depends on the capacity to discriminate relative depth differences. Some people are more discriminating than others; so there are tolerances to aniseikonia as there are to astigmatism, heterophoria, or blurred images. This was already recognized in the first paper ever published on aniseikonia (Ames and co-workers, 1932), in which some subjects were affected by 0.25% size differences and others were unaffected by a 5.25% size lens. The lower limit is actually not surprising when we recall the normal threshold of stereoacuity.

For a 1° foveal field, a 0.25% size difference represents 0.0025° or 9 seconds of arc, well within the average stereo capacity. In fact, the low threshold is another kind of puzzle: If slightly unequal images fall within Panum's fusional areas (estimated to be about 5 minutes at the fovea), how is the size difference discriminated at all? The reason is in part that Panum's areas are defined by the limits of double vision rather than stereoscopic vision; that Panum's areas are themselves magnified; and that peripherally, acuity decreases and the size of Panum's areas increases. But how a small horizontal disparity makes itself known through a change in depth localization, even though the images are fused, has never been completely solved.

The optics of aniseikonia is based on relative spectacle magnification (Chapter 7), which compares retinal image size in the lens-corrected ametropic eye to the retinal image in the standard emmetropic eye. If ametropia is axial, there is no difference in relative spectacle magnification, providing the correcting lens is placed at the anterior focal point of the eye (Knapp's rule). Some modification in image size can be produced by fitting the stronger lens closer to the eye's anterior focal point. If both lenses are fitted at the same distance (usually 2 mm inside the anterior focal point, which is assumed to be at 16.66 mm from the front surface of the reduced standard eye), there is about 0.25% magnification per diopter of anisometropia.

In refractive anisometropia, assuming axial length remains the same as the standard emmetropic eye, relative and spectacle magnification are the same and depends only on the power of the correcting lens (F) and its distance (d) from the nodal point (or entrance pupil):

$$m = \frac{1}{1 - dF}$$

A difference between the two eyes can therefore be altered by fitting the weaker lens further from the eye than the stronger one. If both lenses are fitted at the same distance, the magnification is about 1.5% per diopter of anisometropia.

It is not possible to clinically determine whether ametropia is axial or refractive, and since spectacles are generally fitted inside the eye's anterior focal point, the magnification in axial and refractive ametropias are in opposite directions. An empirical value of 1% magnification difference per diopter of anisometropia represents a fair average estimate based on the assumption that ametropia is partially axial and partially refractive. It follows that there is no absolute correlation between anisometropia and aniseikonia, and even two emmetropic eyes may suffer aniseikonia if they differ in refractive power (each, of course, with its proportionate axial length).

Anisometropia is by far the most significant cause of aniseikonia. Most ametropias under 4 D are due to abnormal correlation of normal optical components (correlation ametropia).

The axial length is out of proportion, although it may still fall within the range seen in emmetropic eyes. Since the spectacle correction of axial anisometropia induces little relative spectacle magnification, aniseikonia is seldom a problem. When the anisometropia is refractive as in the case of diabetic index myopia, aniseikonia is induced, but the chief annoyance is usually lowered acuity and diplopia, not spatial distortion; indeed binocular vision may be impossible.

In spherical refractive anisometropia, aniseikonia involves all meridians simultaneously—an "overall" size difference. Astigmatism, which is generally refractive, causes "meridional" magnification. Recall that the magnification of a plus cylinder (or the minification of a minus cylinder) is in the power meridian perpendicular to the axis. A plus cylinder axis 90° placed before one eye magnifies its image horizontally compared to the other, and the object appears further on the side corresponding to the larger image. This is the classic geometric disparity of stereoscopic vision. In a stereophotograph the same horizontal disparities are produced not by cylinders but by taking the picture from two horizontally separated points. Horizontal disparity can work both ways: create a depth difference where none exists or eliminate a depth difference actually present.

If a plus cylinder is placed before one eye axis horizontal, a vertical image size disparity is created (within the limits of binocular vision) with a peculiar result called the "induced size effect" (Ogle, 1939). Vertical disparity does not lead to stereopsis, but the perceptual effect turns out as if it did, only opposite. Objects on the vertically magnified side appear *nearer,* as if increasing the vertical dimensions of one ocular image "induces" an apparent horizontal increase in the opposite eye. The term "induced size effect" is based on this analogy, but the reason for it is not known. The visual system seems to respond under these circumstances in the only way it can; since vertical disparity has no stereoscopic effect, it reacts as if a horizontal disparity had been created in the other eye. Significantly, the induced effect only occurs if there are at least two vertically separated contours in the binocular field to stimulate fusional innervation. Moreover, the induced effect increases to a maximum, then decreases again with increasing vertical disparity. This is not due to fatigue because it can be reproduced by starting from larger disparities and progressing to smaller ones.

In oblique astigmatism, to the perceptual effects of horizontal and vertical disparities are added rotary deviations in which vertical lines appear inclined toward or away from the observer. These rotations should not be confused with cyclophoria because the meridians rotate in opposite directions in the same eye. But a large proportion of patients (about 70%) with binocular oblique astigmatism show incorrect spatial orientation (under appropriate test conditions) even though they are not aware and hence seldom complain of them in ordinary surroundings (Burian and Ogle, 1945). Some never completely adjust to these distortions even in ordinary environments and cannot wear their correcting cylinders. It is these so-called grief cases in which the cylinder must either be eliminated or prescribed in graduated doses, depending on the acuity, the previous prescription, and the symptomatic response. Empirical correction of aniseikonia seldom works in astigmatism, and appropriate size lenses must be computed by translation. Most clinicians limit themselves to a Cervantian tilt of cylinder axis or a Quixotic reduction of cylinder power.

Even when the ocular images have been equalized for distance (e.g., by altering vertex distance or base curve), it does not follow that reading targets have also been equalized; in fact some inequality generally remains because of optical effectivity. And in all cases of aniseikonia, there is necessarily a differential prismatic effect on ocular rotations; ocular movements are no longer the same as they were before the lens was placed before each eye (spectacle magnification) and differ in the two corrected eyes (relative spectacle magnification). Finally, in addition to the optical incongruity in the vertical, horizontal, or oblique meridians, there may be asymmetrical differences from corneal scars or keratoconus, which can only be corrected by a contact lens.

CLINICAL MEASUREMENTS

The examination is planned according to the symptoms and, since these are nonspecific, includes evaluation of acuity, refraction, motility, binocular status, and size effects, or aniseikonia.

Acuity

Because anisometropia is frequently associated with amblyopia, check acuity with and without correction. The younger the patient the greater this possibility and the more important early diagnosis and early treatment. The most obvious facts are the ones most often forgotten, and it cannot be too strongly emphasized that amblyopia can exist without an obvious strabismus. Suppression is frequently confined to the foveal area, and peripheral fusion keeps the eyes in line. In older children, acuity can be measured with the tumbling E test; in younger children one must depend on fixation patterns, optokinetic nystagmus, or reaching for small candy spheres. The Visuscope, Haidinger brushes, amblyoscope, and polarized targets are useful in older children. An organic basis for visual loss should be ruled out; conversely, an "ophthalmoscopically visible lesion" does not mean a functional amblyopia cannot also be present.

Refraction

Only by inactivating accommodation can we establish a baseline by which to compare one eye to the other; so cycloplegia is practically mandatory in children and in nonpresbyopic hyperopic anisometropes. The keratometer may provide a clue to the character of the ametropia; equal corneal curvatures in unequal refractive errors suggest axial ametropia; unequal curvatures correlated with total astigmatism suggest refractive ametropia. Retinoscopy is the only guide in children too young for subjective testing and should be repeated several times. Treatment of amblyopia depends as much on providing a sharp optical image for the poor eye as on occluding the better one.

Subjective refraction, either with a refractor or trial frame, requires proper lens centering for far and near measurements, since inaccuracies will lead to diagnosing nonexistent oculomotor imbalances. A tilt of the instrument or the head behind the instrument can introduce significant vertical prism. A military rigidity is not necessary, nor do we want to induce torticollis; hence the patient should hold his head naturally while testing.

Trial frames permit more natural posture, more accurate fitting, and more appropriate near point measurements. We can actually see whether the patient turns his head or his eyes in reading and by how much. The vertex distance can also be measured more precisely, especially in high errors in which fitting the final lenses at a different distance may alter motor imbalance and magnification. An improperly angulated frame induces a new cylindrical component, which may add to an existing cylinder and change the amount and axis of astigmatism. Trial lenses should be accurately placed, and faulty lenses can themselves introduce unknown prismatic effects. They should have front surface cylinders whose magnification is known and can be translated by the laboratory.

The axis of astigmatism must be accurately identified with respect to a possible asymmetry in the same eye and between the two eyes. If the patient has never worn glasses, one may wish to reduce the cylinder or even eliminate it by prescribing the spherical equivalent. The acuity provides a helpful guide to the available options. For example, if the acuity through $-1 + 1 \times 45$ is 20/15 and is 20/25 through a -0.50 D sphere, the spherical equivalent may be a better choice for the initial prescription.

In adult patients with established monocular vision habits, suppression is nine tenths of the law of comfort. Only central suppression may be required so that peripheral vision and perhaps fusion can be achieved with appropriate balancing lenses.

In balancing the prescription, it is often possible to minimize the anisometropic prescription by 0.50 to 1.00 D without a significant loss of acuity, a useful expedient when binocularity is established but the patient has never been corrected or when a change in correction is magnified by several years without an eye examination. In making these adjustments, it is conventional to leave the dominant eye with the better vision.

In high anisometropia and astigmatism, the distance prescription may fail to correct the accommodative balance at the reading distance, thus promoting suppression. In such cases, near point refraction is carried out by the methods described in Chapter 15, and a separate reading lens ordered.

The accuracy of the refractive findings can be checked by comparing objective and subjective measurements and by evaluating the range and relative amplitude of accommodation for each eye in turn.

Motility

The differential prismatic effects of anisometropic corrections are undoubtedly a major cause of symptoms, as emphasized by Jackson seventy five years ago. Unfortunately it is more difficult to measure these effects than one might think. If the patient looks through the optical centers of his lenses, there is no induced heterophoria; so the measurements in the primary position or head-down position in reading will not tell us the difficulties encountered with eccentric fixation. We can visualize them by a useful scheme introduced by Boeder (Fig. 19-6). Suppose the right eye requires +4.50 D and the left, +1.50 D. Combining the two, the anisometropic difference is 3.00 D, right eye. We can picture a series of zones around the optical center with progressively increasing prismatic effect. Dividing the zones into 10° excursions (about 5 mm), one has 0.5 (3) = 1.5Δ in the first zone, 3Δ in the second, 4.5Δ in the third (corresponding to 30° eccentricity), etc. The greater the anisometropia the smaller the central zone of zero imbalance. In astigmatism, the zones are not only restricted but also elliptically distorted. Although anisophoria in eccentric fixation is seldom measured clinically, it is not difficult to visualize the progressive constriction of the patient's useful motor field with increasing anisometropia.

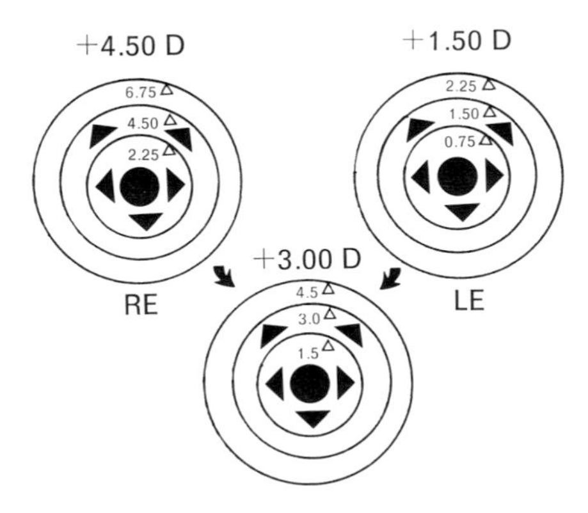

Fig. 19-6. Boeder graph schema to illustrate differential prismatic effects in varying zones of an anisometropic correction. The illustrated zones are at 10° intervals (5 mm). The "binocular" lens represents the dioptric difference directly.

Vertical and lateral phorias are measured at distance and near, with and without correction, or with the old correction if glasses are worn. A change in vertical phorias represents the induced prismatic effect of the new prescription. A phoria is a tendency for the eyes to deviate from their position of binocular fixation (within the limits of fixation disparity) when fusion is disrupted or frustrated.

Phorias represent a kind of innervational strain on the fusional mechanism. By measuring the amplitude of fusional vergences, we obtain a basis for comparing the demand to the compensating reserves. Thus a relatively large vertical phoria may be tolerated if the vertical amplitudes are adequate, whereas a small vertical imbalance may lead to symptoms if the reserves are low. This helps us decide whether to treat the phoria with prisms or build up fusional amplitudes with orthoptics. If the new correction markedly differs from the previous spectacles or if no correction was worn, the size difference reduces the vergence amplitude, and several readings on separate occasions may be required to obtain reliable data.

Binocular status

Suppression and strabismus are complications of anisometropia requiring specific treatment in addition to correcting the refractive error. Evaluation is made by polarization, anaglyphs, and stereoscopic instruments as described in Chapter 15. Vectographic slides are particularly useful and reveal foveal suppression not evident on the Worth four dot test. These slides also have targets for stereoacuity and fixation disparity. The Freeman near vision unit or the Bernell lantern are useful in testing binocular function at the reading distance.

Aniseikonia

Aniseikonia falsifies spatial perception, and its measurement depends on the presence of binocular vision, stereopsis, and at least 20/60 acuity. Although aniseikonia may be an etiologic factor in amblyopia and strabismus, there is no clinical way to measure it under these circumstances or under any circumstances in children. The space eikonometer (Fig. 19-7) or its clinical counterpart (Fig. 19-8) utilizes variable size lenses to measure horizontal, vertical, and inclination errors. The magnification required to restore distortions in the stereoscopic pattern is measured directly. Some qualitative screening tests are also available, mainly stereograms that reproduce specific aniseikonia distortions. Direct comparison methods in which each eye evaluates its own target under binocular conditions may confuse aniseikonia with fixation disparity.

CLINICAL MANAGEMENT

In many cases of anisometropia, the full correction is indicated; in others, a partial correction; in some, no correction. The man-

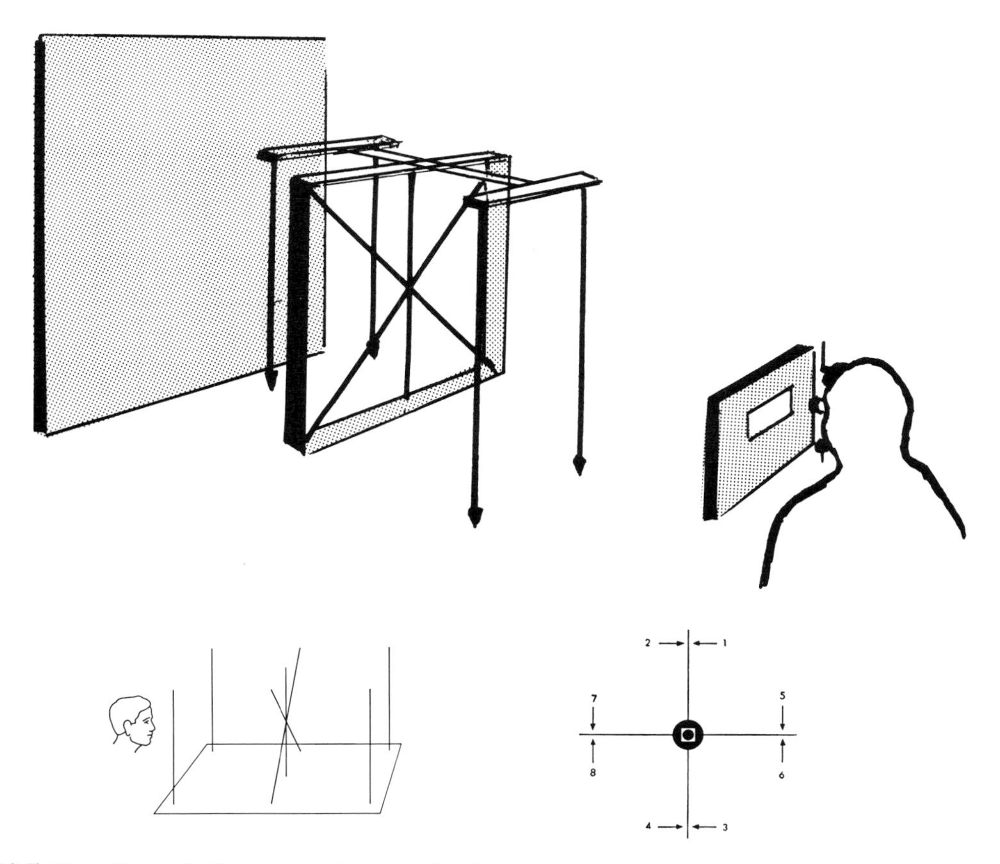

Fig. 19-7. *Top,* Original eikonometer design. The observer evaluates the appearance of the central cross and the two sets of vertical plumb lines through size lenses. *Bottom,* The normal appearance is illustrated next to the direct comparison method in which one half the arrows is seen by one eye, one half by the other. The direct comparison method is obviously influenced by fixation disparity.

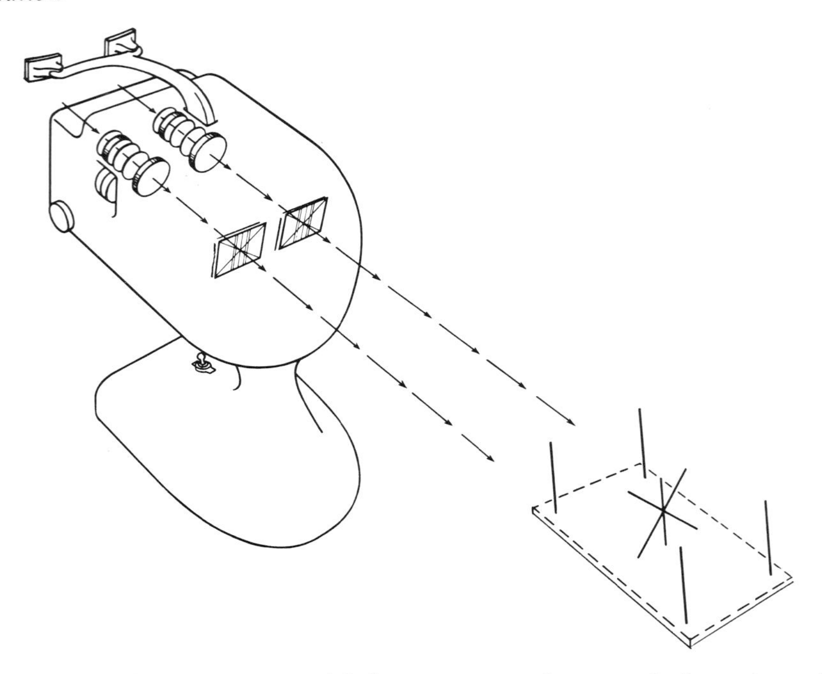

Fig. 19-8. The clinical eikonometer is a modified stereoscope with targets similar to those of the original space eikonometer. It is assumed the patient has binocular vision, effective stereopsis, and acuity. Any perceptual distortions are corrected by means of the incorporated size lenses. Readings for the horizontal, vertical, and inclination errors are a measure of clinical aniseikonia.

agement depends on the symptoms, age, and refraction, not arbitrary optical rules. Here is a child, a little hyperopic on one side, more hyperopic on the other. We order an equal prescription for an unequal refractive error, leaving one image blurred, distorted, and of different size. The result is less incentive to fusion, more tendency to suppression, and a degradation of binocular vision. So much for the notion that anisometropia must never be fully corrected. Here is an adult, myopic on one side, hyperopic on the other. He has never worn glasses and comes to us for presbyopic help. If we decide to help him not only with his reading but also with his anisometropia, the result may be a space odyssey. The patient orbits new worlds of three dimensions soon to return with a verbal fallout of complaints, a blast to our self-confidence. So much for the notion that anisometropia must always be fully corrected. These modest lessons illustrate that in anisometropia, when to prescribe is as important as what to prescribe. There is no point in following therapeutic rules so elastic they fail to hold themselves up.

Anisometropic amblyopia

To little minds all things are great, and in children a little anisometropia can lead to a large amblyopia. The diagnosis of amblyopia in a child too young for subjective testing is not easy, and in the presence of anisometropia, an anomalous fixation pattern of the more ametropic eye is enough to start occlusion therapy. In an older child, particularly in anisometropic myopia, the lenses may not fully correct vision on the first examination, but it frequently improves spontaneously as the glasses are worn. An organic basis for poor vision must of course be ruled out by appropriate ophthalmologic examination.

The treatment of anisometropic amblyopia is full correction of the refractive error and full-time occlusion of the better eye. Some recommend preliminary patching of the amblyopic eye for 2 to 3 weeks to break up eccentric fixation and red filters, neutral density filters, or wafer prisms over the amblyopic eye. Pleoptic treatment has been recommended in selected cases. Of these various methods, total occlusion of the nonamblyopic eye remains the simplest, the most rapid, and the most effective treatment.

Anisometropia like ametropia changes with time, and the refraction should be periodically rechecked. An amblyopic cure also fluctuates if the binocular status is not stable; so vision should be monitored at 1- to 3-month intervals.

Adult anisometropia

In the majority of adult anisometropes, the refractive difference developed slowly, has been apparent, and probably has been corrected. Mostly they have developed binocular vision, and their tolerance may be suprising. It is true that with large lens differences, there are unavoidable prismatic effects when viewing objects eccentrically; the problem is solved by not viewing objects eccentrically, that is, by turning the head instead of the eyes.

This does not mean that the adult anisometrope is immune to symptoms. If the prescription differs greatly from that previously worn, some difficulties can be anticipated. A large change may be necessary because the previous correction was inaccurate or if there has been no examination for several years. The minimum refractive difference necessary to achieve good acuity is ordered either at once or in graduated doses by titrating with temporary spectacles, clip-ons, or Fresnel adds. Even if the required change is small, difficulties can be avoided by not changing base curve, vertex distance, thickness, or the surface on which the cylinder is ground arbitrarily. The previous glasses are checked for prisms that we may have failed to find or for inaccurate decentration to which the patient has become adjusted.

If no glasses are worn, the full correction may still be tolerated and the results rewarding, especially in myopia. Much depends on the age at which the anisometropia developed, and the condition should be explained, the difficulties anticipated, and the rewards demonstrated. Without binocular vision, a sharp image is tepid consolation for insuperable diplopia. If one eye is emmetropic or slightly hyperopic, the patient may have learned to use it for distance vision, with the more myopic eye used for reading. Although this possibility is frequently described and even encouraged, it seldom works in practice. First, the refractive error is rarely conveniently unequal for distance and near; second, there is generally some astigmatism, which reduces the uncorrected acuity; third, most people have a dominant eye that they prefer for fixation whatever the distance; and finally, the inevitable accommodative fluctuation does not encourage stable, comfortable, albeit monocular vision.

A rapidly developing anisometropia in an adult is likely to be refractive, and this condition is considered an ideal indication for aniseikonic correction. Since the cause is usually diabetes, nuclear sclerosis, corneal trauma, or a drug-induced index change, these patients are more disturbed by the loss of acuity and diplopia. Because they cannot compensate for such rapid changes, binocular vision is absent, and spatial distortions do not occur. If they have not learned to suppress, they are better served by occlusion or a frosted lens to eliminate the disconcerting second image—at least until the underlying pathology has been treated or has stabilized.

Anisometropia and presbyopia

Correcting anisometropia with spectacles does not always solve the near acuity problem. In high anisometropia the vision remains unequal at the reading distance because of unequal effectivity. For example, if the right eye is 1 D myopic and the left is 4 D hyperopic, and both are corrected for distance, from Pascal's formula for the accommodative unit: 1 − 3% F, or 0.97 D for the right eye, and 1 + 4% F, or 1.16 D for the left eye. At 40 cm the right eye need only accommodate 2.50 (0.97) = 2.42 D, but the left eye must accommodate 2.50 (1.16) = 2.90 D, leaving 0.50 D of differential accommodation. In astigmatic anisometropia the same considerations apply; that is, the near cylinder differs, which may lead to discomfort, suppression, or intermittent squint. A satisfactory solution may require separate reading glasses.

Most adult anisometropes adjust by lowering the head instead of the eyes in reading. This always works at least until the presbyopic stage when bifocals prevent it without looking above the seg. A separate reading glass or a monocentric bifocal is therefore a better choice. If the patient has never worn glasses and is given bifocals, the prismatic imbalances of the distance and reading add are suddenly and simultaneously thrust on him.

In patients who cannot adapt or whose reading level is fixed by occupation (e.g., a surgeon), the prismatic imbalance can be modified by compensated segs or, in higher degrees (up to 5Δ), by bicentric grinding.

Bicentric grinding or "slabbing off" is an ingenious method of producing two principal axes and two optical centers in the same lens. The purpose is to equalize the prismatic effect of anisometropia at a predetermined reading point. The method is applicable to single vision, bifocal, or trifocal lenses. The principle is best visualized by a shift in the center of curvature of the grinding tool for the front surface. The prism is equalized in concave lenses by grinding down the stronger minus lens; with convex lenses, the weaker lens is ground down to match the stronger. Fig. 19-9 illustrates the optics involved for a planoconcave lens.

The slab-off method is best when prismatic compensation cannot be achieved by decentration. To determine the prism introduced by bicentric grinding we use a lens clock. Measure the curve of the slabbed-off lens on the front surface above the wedge. Next, with the three legs in line with the 90° axis, slide the lens clock

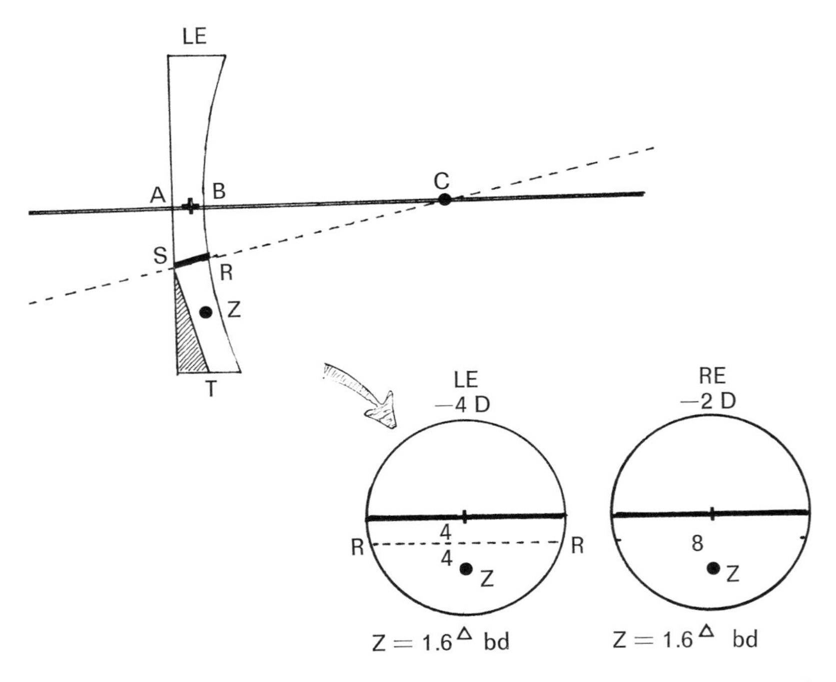

Fig. 19-9. The optical principle of bicentric grinding is illustrated in this diagram. *C* is the center of curvature of the inner concave surface, and *ABC* is the principal axis of the lens. At *R*, *ST* is ground perpendicular to the line *SRC*, creating a second optical center. The point *R* is so chosen that the prismatic effect at the reading point Z matches the other eye. For example, in the prescription RE: −2 D; LE: −4 D, at 8 mm below the optical center, there is a differential prismatic effect of 1.6Δ base down. Slabbing off the left lens, the point *R* is 4 mm below the optical center so that at Z, 8 mm below the optical center, there is 1.6Δ base down, the same as the right eye.

over the surface until the middle leg is exactly on the line of the slab-off. Subtract the first reading to obtain the amount of prism.

Anisometropia and contact lenses

If one eye is hyperopic and the other myopic, and the patient is corrected with spectacles, the myopic eye needs to accommodate less and the hyperopic eye more than a natural emmetrope. If the patient switches to contact lenses, these accommodative requirements are reversed. In axial anisometropia, contact lenses exaggerate the size difference because they are farther from the anterior focal point of the eye. In refractive anisometropia (e.g., unilateral aphakia), on the other hand, a contact lens may enable the patient to obtain some measure of binocular vision.

Prismatic imbalances

Patients with anisometropia are likely to have motility problems; if not, they are sure to have them when they start wearing their new spectacles. Correction may require prismatic relief, which, although it continues to perplex, has not ceased to arouse interest. We cannot expect sound principles of practice if their indication is based on speculation, postulate, or poetry. While foraging in the forest of prismatic infelicities, we must follow the optical arrows without losing sight of the physiologic Indians. Excesses are more likely to be restrained if abuses are recognized. Unquestionably prisms provide relief in certain cases as in intrinsic hyperphoria, which exceeds the compensating vergences, and thus allow comfortable binocular vision not otherwise possible. Similarly, in presbyopia, efficient use of the bifocal seg may require neutralization of the prismatic imbalance at the reading level. But prisms only neutralize the imbalance in one meridian and only in one direction in that meridian. They do not correct anisophoria, and they induce distortions, dispersions, reflections, and aberrations. And for some patients a prism may throw the image outside the suppression area, producing diplopia. Moreover, as Lebensohn (1953) has shown, in two thirds of 200 consecutive cases of anisometropia an intrinsic anisophoria is present or develops to balance the lens-induced, differential prismatic effects. The reluctance of most clinicians to order prisms is thus based on a respectable body of accumulated experience with unhappy results. Indeed most textbooks qualify their indications with a definite maybe.

Prisms may be useful in anisometropia, particularly if fusion cannot be demonstrated without them, but acceptance and the potential for eliminating symptoms must be checked by a trial period. Small vertical prisms (less than 1Δ to 2Δ) are seldom indicated in single vision lenses, and larger imbalances can be reduced by ordering a reduced sphere or cylinder or spherical equivalent. Such modifications depend on acuity and binocularity. Vertical imbalances can be eliminated by head movements, but reduced acuity cannot be compensated for.

Aniseikonia

"Aniseikonia" is a term coined by Lancaster (1938) to describe unequal perceptual images. This is not to imply that we perceive "images" but rather that the resultant spatial distortions only occur in binocular vision, depending on cerebral unification. Of course, all perceptions, monocular or binocular, occur in the brain, and undue emphasis should not be placed on neurophysiologic and psychologic factors of which we know little. In fact the optical incongruities produce effects that are predictable and consistent enough to serve as a major argument that aniseikonia is not a purely psychologic phenomenon.

No scientific discovery is born full grown, full-fledged, and flapping. One is not likely to make much of aniseikonia if in addition to eschewing the word, one ignores the facts. And the facts are that in every ophthalmic practice a certain number of patients fail to obtain relief of their symptoms from refractive treatment. A proportion of these are unquestionably neurotic, a diagnosis that may fail to enkindle the enthusiasm of either patient or refractionist. Surely a psychiatric diagnosis should not be made until all the

Table 19-1. Approximate magnification changes associated with changes in front base curves for various lens powers

Vertex power of lens (diopters)	*Average center thickness (millimeters)*	*Change in front base curve*					
		−4 D	*−2 D*	*+2 D*	*+4 D*	*+6 D*	*+8 D*
+8.00	7.0	−3.5%	−1.7%	+1.7%	+3.5%	+5.2%	+6.9%
+7.00	6.2	−3.0%	−1.5%	+1.5%	+3.0%	+4.5%	+6.0%
+6.00	5.4	−2.6%	−1.3%	+1.3%	+2.6%	+3.9%	+5.2%
+5.00	4.6	−2.2%	−1.1%	+1.1%	+2.2%	+3.3%	+4.4%
+4.00	3.9	−1.8%	−0.9%	+0.9%	+1.8%	+2.8%	+3.7%
+3.00	3.2	−1.5%	−0.7%	+0.7%	+1.5%	+2.2%	+2.9%
+2.00	2.7	−1.1%	−0.6%	+0.6%	+1.1%	+1.7%	+2.2%
+1.00	2.2	−0.8%	−0.4%	−0.4%	+0.8%	+1.2%	+1.6%
0.00	1.8	−0.5%	−0.2%	+0.2%	+0.5%	+0.7%	+1.0%
−1.00	1.4	−0.2%	−0.1%	+0.1%	+0.2%	+0.3%	+0.3%
−2.00	1.0	+0.1%	+0.1%	−0.1%	−0.1%	−0.2%	−0.3%
−3.00	0.9	+0.4%	+0.2%	−0.2%	−0.4%	−0.5%	−0.7%
−4.00	0.9	+0.6%	+0.3%	−0.3%	−0.6%	−0.8%	−1.1%
−5.00	0.9	+0.8%	+0.4%	−0.4%	−0.8%	−1.1%	−1.5%
−6.00	0.9	+1.0%	+0.5%	−0.5%	−1.0%	−1.4%	−1.9%
−7.00	0.9	+1.2%	+0.6%	−0.6%	−1.2%	−1.7%	−2.3%
−8.00	0.9	+1.4%	+0.7%	−0.7%	−1.4%	−2.0%	−2.7%

Table 19-2. Approximate magnification changes associated with changes in eyewire distance for various lens powers

Vertex power of lens (diopters)	*Change in eyewire distance*							
	Lens moved toward the eye				*Lens moved away from eye*			
	1 mm	*2 mm*	*3 mm*	*4 mm*	*1 mm*	*2 mm*	*3 mm*	*4 mm*
+10	−1.0%	−2.0%	−3.0%	−4.0%	+1.0%	+2.0%	+3.0%	+4.0%
+8	−0.8%	−1.6%	−2.4%	−3.2%	+0.8%	+1.6%	+2.4%	+3.2%
+6	−0.6%	−1.2%	−1.8%	−2.4%	+0.6%	+1.2%	+1.8%	+2.4%
+4	−0.4%	−0.8%	−1.2%	−1.6%	+0.4%	+0.8%	+1.2%	+1.6%
+2	−0.2%	−0.4%	−0.6%	−0.8%	+0.2%	+0.4%	+0.6%	+0.8%
−2	+0.2%	+0.4%	+0.6%	+0.8%	−0.2%	−0.4%	−0.6%	−0.8%
−4	+0.4%	+0.8%	+1.2%	+1.6%	−0.4%	−0.8%	−1.2%	−1.6%
−6	+0.6%	+1.2%	+1.8%	+2.4%	−0.6%	−1.2%	−1.8%	−2.4%
−8	+0.8%	+1.6%	+2.4%	+3.2%	−0.8%	−1.6%	−2.4%	−3.2%
−10	+1.0%	+2.0%	+3.0%	+4.0%	−1.0%	−2.0%	−3.0%	−4.0%

optical possibilities have been exhausted. The correction of aniseikonia may bring relief when all other methods have failed and should certainly be tried after all other methods have failed. This advice comes to us from no less an authority than Lancaster:

> Finally there is no doubt that a few causes of eyestrain are curable by correcting aniseikonia. This was greatly exaggerated in its importance and frequency when it was first exploited. All obscure cases should be given the benefit of investigations of that factor as a possible cause of disturbed binocular vision *after* the much more frequent causes have been treated without relief.*

Correction of aniseikonia is seldom indicated in children or in patients who are

*From Lancaster, W. B.: Refraction and motility, Springfield, Ill., 1952, Charles C Thomas, Publisher.

asymptomatic, whose symptoms are not related to the use of the eyes, and whose anisometropia is long-standing or exceeds the limits of binocular fusion, or when there is an obvious fundus lesion. Every anisometropic prescription should preserve the base curve, thickness, index, and vertex distance to which the patient has become adjusted.

The empirical correction of aniseikonia, based on an intelligent estimate of symptoms and refractive findings, may be attempted in spherical anisometropia by modifying the spectacles. Thus magnification may be changed by increasing front curve, thickness, or vertex distance of plus lenses. (Changing the front curve seldom works for minus lenses because the increase in the rear curve also increases the vertex distance.) The amounts can be computed from the formulas given in Chapter 7 or from Tables 19-1 and 19-2.* In cylindrical prescriptions these simple adjustments do not work because they influence each meridian a different amount. Meridional magnification may, however, be modified by bitoric grinding.

Aniseikonia can be avoided, if not eliminated, by not prescribing a lens that gives maximum acuity, that is, by modifying the sphere or cylinder. It is advisable not to create symptoms that were nonexistent by lenses that correct refractive error too completely.

*The table values are based on the general lens magnification formula: $m = \left(\frac{1}{1 - t/n\ F_1}\right)\left(\frac{1}{1 - dF}\right)$, where F_1 is the front surface power, F the back vertex power, t the lens thickness, and d the vertex distance. When magnification is expressed in percent and t and d are expressed in millimeters, for an assumed glass index of 1.5 and with the fractions simplified by expansion, then:

(a) Shape factor: $m_{\%} = \frac{tF_1}{15}$, which gives the result of changing the base curve or lens thickness (with t expressed in millimeters

(b) Power factor: $m_{\%} = \frac{dF}{10}$, which gives the result of changing vertex distance (expressed in millimeters)

A more elaborate derivation is given by Brown and Enoch (1970). The table values were originally given by Berens and Bannon (1963).

REFERENCES

Alcorn, H. W.: Problems of anisometropia, J. Iowa Med. Soc. **49**:88-89, 1959.

Allen, D. C.: Vertical prism adaptation in anisometropes, Am. J. Optom. **51**:252-259, 1974.

Ames, A.: Aniseikonia—a factor in the functioning of vision, Am. J. Ophthal. **18**:1014-1020, 1935.

Ames, A.: The space eikonometer test for aniseikonia, Am. J. Ophthal. **28**:248-262, 1945.

Ames, A.: Binocular vision as affected by relations between uniocular stimulus patterns in commonplace environments, Am. J. Psychol. **59**:333-357, 1946.

Ames, A., and Gliddon, G. H.: Ocular measurements. In Transactions of the Section on Ophthalmology of the American Medical Association, pp. 1-68, 1928.

Ames, A., Gliddon, G. H., and Ogle, K. N.: Size and shape of ocular images. I. Methods of determination and physiologic significance, Arch. Ophthal. **7**:576-597, 1932.

Ames, A., and Ogle, K. N.: Size and shape of ocular images. III. Visual sensitivity to differences in the relative size of the ocular images of the two eyes, Arch. Ophthal. **7**:904-924, 1932.

Bannon, R. E.: Clinical manual on aniseikonia, Buffalo, 1965, American Optical Co.

Bannon, R. E., Neumueller, J., Boeder, P., and Burian, H. M.: Aniseikonia and space perception—after 50 years, Am. J. Optom. **47**:423-441, 1970.

Bannon, R. E., and Textor, R. T.: Analysis of successful and failure aniseikonic cases, Am. J. Optom. **24**:262-276, 1947.

Berens, C., and Bannon, R. E.: Aniseikonia—a present appraisal and some practical considerations, Arch. Ophthal. **70**:181-188, 1963.

Bishop, J. W.: Treatment of amblyopia secondary to anisometropia, Br. Orthopt. J. **13**:64-69, 1963.

Boeder, P.: Analyses of prism effects in bifocal lenses, Southbridge, 1939, American Optical Co.

Brown, R. M., and Enoch, J. M.: Combined rules of thumb in aniseikonic prescriptions, Am. J. Ophthal. **69**:118-126, 1970.

Burian, H. M.: Clinical significance of aniseikonia, Arch. Ophthal. **29**:116-133, 1943.

Burian, H. M.: Influence of prolonged wearing of meridional size lenses on spatial localization, Arch. Ophthal. **30**:645-666, 1943.

Burian, H. M.: History of the Dartmouth Eye Institute, Arch. Ophthal. **40**:163-175, 1948.

Burian, H. M., and Ogle, K. N.: A study of aniseikonia in a case of increasing unilateral index myopia, Am. J. Ophthal. **26**:480-490, 1943.

Burian, H. M., and Ogle, K. N.: Aniseikonia and spatial orientation, Am. J. Ophthal. **28**:735-743, 1945.

Burnside, R. M., and Langley, C.: Anisometropia and the fundus camera, Am. J. Ophthal. **58**:588-594, 1964.

Carleton, E. H., and Madigan, L. F.: Size and shape of ocular images. II. Clinical significance, Arch. Ophthal. **7**:720-738, 1932.

Carleton, E. H., and Madigan, L. F.: Relationship

between aniseikonia and anisometropia, Arch. Ophthal. **18**:237-247, 1937.
Charnwood, Lord: A new test for aniseikonia, Optician, Dec. 26, 1952.
Cibis, P. A., and Haber, H.: Anisopia and perception of space, J. Optic. Soc. Am. **41**:676-683, 1951.
Curtin, B. J., Linksz, A., and Shafer, D. M.: Aniseikonia following retinal detachment, Am. J. Ophthal. **47**: 468-471, 1959.
Cushman, B.: Hyperphoria and some of its problems, Am. J. Ophthal. **40**:332-343, 1955.
Ellerbrock, V. J.: Further study of effects induced by anisometropic corrections, Am. J. Optom. monograph No. 53, 1948.
Ellerbrock, V. J.: The effect of aniseikonia on the amplitude of vertical divergence, Am. J. Optom. **29**:403-415, 1952.
Ellerbrock, V. J., and Fry, G. A.: Effects induced by anisometropic corrections, Am. J. Optom. **19**:444-459, 1942.
Emsley, H. H.: Irregular astigmatism of the eye; effect of correcting lens, Trans. Opt. Soc. London **27**: 28-42, 1925.
Friedenwald, J.: Diagnosis and treatment of anisophoria, Arch. Ophthal. **15**:283-307, 1936.
Gettes, B. C.: Prisms in refraction, Int. Ophthal. Clin. **1**:249-260, 1961.
Guibor, G. P.: Ophthalmic prisms, some uses in ophthalmology, Am. J. Ophthal. **26**:833-845, 1943.
Helveston, E. M.: Relationship between degree of anisometropia and depth of amblyopia, Am. J. Ophthal. **62**:757-759, 1966.
Hermann, J. S.: The specific role of contact lenses in sensorimotor anomalies, Am. Orthopt. J. **16**:30-43, 1966.
Hess, C.: Anisometropie. In Graefe's Handbuch Gesamte Augenheilkunde, Leipzig, 1903.
Horwich, H.: Anisometropic amblyopia, Am. Orthopt. J. **14**:99-104, 1964.
Hughes, W. L.: Aniseikonia, some clinical observations, Am. J. Ophthal. **18**:607-615, 1935.
Hurt, J.: Fusion in anisometropia, Am. Orthopt. J. **21**:101-106, 1971.
Jampolsky, A., Flom, B. C., Weymouth, F. W., and Moses, L. E.: Unequal corrected visual acuity as related to anisometropia, Arch. Ophthal. **54**:893-905, 1955.
Kramer, M. E.: The orthoptic treatment of the vertical motor anomalies, Am. J. Ophthal. **30**:1113-1123, 1947.
Lancaster, W. B.: Nature, scope and significance of aniseikonia, Arch. Ophthal. **28**:767-775, 1942.
Lancaster, W. B.: Refraction correlated with optics and physiological optics, Springfield, Ill., 1952, Charles C Thomas, Publisher.
Lebensohn, J. E.: Anisophoria, anisometropia, and the final prescription, Am. J. Ophthal. **36**:643-649, 1953.
Lebensohn, J.: Nature of innervational hyperphoria, Am. J. Ophthal. **39**:854-859, 1955.
Lebensohn, J.: The management of anisometropia, Eye Ear Nose Throat Mon. **36**:217-222, 1957.
Linksz, A.: Aniseikonia, with notes on the Jackson-Lancaster controversy, Trans. Am. Acad. Ophthal. Otolaryngol. **63**:117-140, 1959.
Linksz, A.: Therapeutic use of prisms, Int. Ophthal. Clin. **11**:269-271, 1971.
Linksz, A., and Bannon, R. E.: Aniseikonia and refractive problems, Int. Ophthal. Clin. **5**:515-534, 1965.
Miles, P. W.: Factors in the diagnosis of aniseikonia and paired Maddox-rod tests, **30**:885-897, 1947.
Ogle, K. N.: Induced size effect, a new phenomenon in binocular vision, Arch. Ophthal. **20**:604-623, 1938.
Ogle, K. N.: Induced size effect, Arch. Ophthal. **21**: 604-625, 1939.
Ogle, K. N.: Induced size effect with the eys in asymmetric convergence, Arch. Ophthal. **23**:1023-1038, 1940.
Ogle, K. M.: Meridional magnifying lens systems in the measurement and correction of aniseikonia, J. Opt. Soc. Am. **34**:302-312, 1944.
Ogle, K. N., and Madigan, L. F.: Astigmatism at oblique axes and binocular stereoscopic spatial localization, Arch. Ophthal. **33**:116-127, 1945.
Panum, P. L.: Physiologische Untersuchungen über das Sehen mit zwei Augen, Kiel, 1858.
Phillipa, C. I.: Strabismus, anisometropia, and amblyopia, Br. J. Ophthal. **43**:449-460, 1959.
Posner, A.: The prescribing of prisms for hyperphoria, Am. J. Ophthal. **34**:197-199, 1951.
Romano, P. E., and Kohn, R.: Aniseikonia due to strabismic amblyopia, Arch. Ophthal. **87**:174-178, 1972.
Roper, K. L., and Bannon, R. E.: Diagnostic value of monocular occlusion, Arch. Ophthal. **31**:316-320, 1944.
Sorsby, A., Leary, G. A., and Richards, M. J.: The optical components in anisometropia, Vision Res. **2**:43-51, 1962.
Stickle, A. W., and Laughlin, D. T.: The use of compensated lenses in ocular motility, Am. Orthopt. J. **19**:104-109, 1969.
Troutman, R. C.: Artiphakia and aniseikonia, Am. J. Ophthal. **56**:602-639, 1963.
Weale, R. A.: Retinal irradiation and aniseikonia. Br. J. Ophthal. **38**:248-249, 1954.
Wheatstone, C.: Contributions to the physiology of vision. I. On some remarkable and hitherto unobserved phenomena of binocular vision, Philosoph. Trans. R. Soc. London **128**:371-394, 1838.
Worrell, B. E., Hirsch, M., and Morgan, M. W.: An evaluation of prism prescribed by Sheard's criterion, Am. J. Optom. **48**:373-376, 1971.

20

Presbyopia

The onset of presbyopia is an irreversible optical failure, an unexplained evolutionary blunder that comes as a psychologic shock. The patient over forty sees his reading vision dim. Soon, he imagines, the muscles will weaken, the skin shrivel, the hand shake, the mind totter, and only an empty shell remain. "Grow old along with me, the best is yet to be." Browning's poetry, I take it, is faultless, and if the physiology is somewhat optimistic, the psychology could not be better. Some are ready to extrapolate from Donders' curve to life expectancy; they think even the shell is doomed in early presbyopia. That is what happens when you play with numbers and do not stop to think. But then the clinician who thinks of his patients only as statistics, on the average, does not think; he only thinks he thinks. He even thinks that what he thinks is practical by confusing the applications of the law of probability, so true in general, so fallacious in particular.

"Presbyopia" means "old eye" or, more properly, the vision of old age, in any case, a poor term for today's life span. It is not a happy term; even Donders had his doubts, and both Landolt and Percival were ready to banish it from the ophthalmologic lexicon. Presbyopia was first applied to anyone who could not see well at a short distance: high hyperopes, absolute hyperopes, and patients with accommodative palsy or insufficiency. Donders restricted it to "the condition in which, as the result of the increase in the years, the range of accommodation is diminished, and the vision of near objects interfered with." But he chose 8 inches for the near distance; so no myope over 5 D could ever become presbyopic by this criterion.

In fact the onset of presbyopia cannot be defined. It varies with age, ametropia, vertex distance, and pupil diameter. It changes with the illumination, size, and contrast of the target and the reading distance. It is modified by the length of the arms, the pliability of the neck, the elasticity of the ego, and the geographic latitude. It fluctuates with ciliary tonus, depth of focus, effort to see, and general and ocular health. And the amplitude decreases not in middle age but almost as soon as we are born. Presbyopia is thus a clinical and not a pathologic entity.

SIGNS AND SYMPTOMS

The history tells the story of progressive accommodative incompetence. It begins with an annoying inertia of focus when switching the gaze to different distances and advances to an inability to carry out prolonged near work without stinging, burning, smarting, or tearing, which eventually leads to disinterest in reading. An incapacity to see fine print or small targets at the accustomed reading distance follows. And when the object is brought nearer, there is the strange effect that the blur increases. Soon ordinary print begins to blur, blotch, smudge, smear, run together, and disappear even when held at arm's length. And all symptoms are intensified under inadequate light, amplified by poor contrast, and exaggerated at the end of the day.

With the waning of the amplitude and the waxing of the near point, the range of accommodation dwindles. The hyperope finds himself enjoying his distance glasses more; the myope enjoys them less. The anisometrope may begin to use one eye for distance vision and the other for near. As accommodation lags, so does the associated accommodative convergence, leading to an increased exophoria. If the load on the positive fusional vergence is excessive, there may be intermittent double vision.

Although fate and nature set a limit, a blurred signal is better than no signal at all. How much blur the patient accepts depends on his occupation, vanity, and accommodative reserve. The watchmaker seeks help before the truckdriver, the short before the tall, the hyperope before the myope, the contact lens–corrected myope before the one wearing spectacles, and the self-effacing before the self-admiring. The character, the severity, and the timing of the symptoms determine not only when to prescribe but also what will be acceptable.

With the first pair of bifocals, the fixation is confusing, the localization confounding, the floor disjointed, and the expectations disenchanted. The segment line is as evident as a flyspeck, and its position disturbing to the neck, discomfiting to the back, and disconcerting to the nerves. And a separate pair of reading glasses is practically guaranteed to be found at the wrong place when needed, if they are found at all. As the add is increased, the newspaper reader finds a new competition in the clarity of the headlines and want ads, the pianist discovers his music is a muddle, the housewife sees her sink smudged, the painter perceives his canvas clouded, and the teacher learns that her intermediate vision is low grade. To recognize the price tag, the desk, the diploma, or the dashboard, it is now necessary to stretch, bend, flex, elasticize, and temporize. It is aggravating, exasperating, and unavoidable. It is the most common complaint of advanced presbyopia.

The amplitude of accommodation diminishes with age in a more or less regular manner. But prescribing a reading add only by age might shock even a statistician, since the numbers disguise a lack of precision. Donders had already cautioned against mistaking the mean amplitude with the range of normal values for that age by plotting his original results as a scattergram. Moreover the near point is always uncertain. One never knows if the patient makes a maximum effort or if the minimum near point was properly measured. The outcome changes when referred to the spectacle, corneal, or equivalent plane of the eye and whether the patient is a natural or lens-corrected emmetrope. And in any event, the results may not be relevant to the visual requirements.

Donders measured the amplitude in a restricted sample of emmetropes (evaluated subjectively) from ages 10 to 80 years; Duane measured 1000 subjects between the ages of 8 and 70 years (ametropia evaluated by cycloplegia). Donders referred his results to the eye's equivalent plane; Duane to the spectacle plane (Tables 20-1 and 20-2). Donders' values are thus lower than Duane's, and Duane's values are more representative of clinical measurements with the correcting spectacles in place.

Hamasaki and associates (1956) evaluated the near point by stigmatoscopy and compared it to the "push-up" method in patients ranging in age from 42 to 60 years. The mean amplitude by stigmatoscopy was found to be significantly lower, a difference attributed to the elimination of depth of focus. If a constant value of 1.50 D was added to the stigmatoscopic amplitude, the result fitted the push-up curve. The results indicate that a significant reduction in amplitude is upon us before either Donders or Duane suspected. But the symptoms are hidden by a providential depth of focus, enhanced by senile and accommodative miosis.

PATHOLOGIC PHYSIOLOGY

The pathogenesis of presbyopia is generally attributed to lenticular sclerosis, although actual hardening of the lens has never been demonstrated, much less measured, and there is no correlation between surgically removed lenses at different ages. (Cataract

Table 20-1. Donders' table

Age in years	10	15	20	25	30	35	40	45	50	55	60	65	70
Amplitude in diopters	14	12	10	8.5	7.0	5.5	4.5	3.5	2.5	1.75	1.00	0.50	0.25

Table 20-2. Duane's table

	Accommodation				*Accommodation*		
Age	*Minimum*	*Mean*	*Maximum*	*Age*	*Minimum*	*Mean*	*Maximum*
8	11.6	13.8	16.1	37	4.5	6.7	8.8
9	11.4	13.6	15.9	38	4.1	6.4	8.5
10	11.1	13.4	15.7	39	3.7	6.1	8.2
11	10.9	13.2	15.5	40	3.4	5.8	7.9
12	10.7	12.9	15.2	41	3.0	5.4	7.5
13	10.5	12.7	15.0	42	2.7	5.0	7.1
14	10.3	12.5	14.8	43	2.3	4.5	6.7
15	10.1	12.3	14.5	44	2.1	4.0	6.3
16	9.8	12.0	14.3	45	1.9	3.6	5.9
17	9.6	11.8	14.1	46	1.7	3.1	5.5
18	9.4	11.6	13.9	47	1.4	2.7	5.0
19	9.2	11.4	13.6	48	1.2	2.3	4.5
20	8.9	11.1	13.4	49	1.1	2.1	4.0
21	8.7	10.9	13.1	50	1.0	1.9	3.2
22	8.5	10.7	12.9	51	0.9	1.7	2.6
23	8.3	10.5	12.6	52	0.9	1.6	2.2
24	8.0	10.2	12.4	53	0.9	1.5	2.1
25	7.8	9.9	12.2	54	0.8	1.4	2.0
26	7.5	9.7	11.9	55	0.8	1.3	1.9
27	7.2	9.5	11.6	56	0.8	1.3	1.8
28	7.0	9.2	11.3	57	0.8	1.3	1.8
29	6.8	9.0	11.0	58	0.7	1.3	1.8
30	6.5	8.7	10.8	59	0.7	1.2	1.7
31	6.2	8.4	10.5	60	0.7	1.2	1.7
32	6.0	8.1	10.2	61	0.6	1.2	1.7
33	5.8	7.9	9.8	62	0.6	1.2	1.6
34	5.5	7.6	9.5	63	0.6	1.1	1.6
35	5.2	7.3	9.3	64 to	—	1.1 to	—
36	4.9	7.0	9.0	72	0.6	1.0	1.6

characteristics are in any case inconclusive, since dehydration and swelling depend on the stage of the opacity rather than age of the patient.) It is now known that the capsule becomes less elastic with age. Fisher (1969) found Young's modulus of elasticity to decrease from 6×10^7 to 1.5×10^7 dynes per square centimeter. But reduced elasticity alone cannot explain presbyopia, only the reduced ability to change lens curvature. To account for the progressive decrease in amplitude, the lens must also become flatter. The lens substance is plastic and molds itself in pliable obedience to the elastic capsule. When the capsule is stripped off (in primates), the lens become flatter, assuming the unaccommodated shape, a change attributed to subtle modification in internal structure. Moreover, as the lens grows and additional fibers are added, not only does volume increase but curvature decreases also; so the young lens is more spherical than the old. The nucleus shrinks relative to the lens as a whole, and the entire lens moves forward, flattening the anterior chamber. In addition, biochemical changes probably alter both interlaminar refractive index and physical fiber malleability. Perhaps most remarkable is that these changes occur with such an exquisite symmetry as to affect all meridians equally and both eyes simultaneously.

The intracapsular mechanism of presbyopia thus takes two forms: a decreased capsular elasticity and an expansive reduction in lens substance pliability. Weale (1962) suggests an antagonism in which the capsular force is more important in young eyes, and the lens substance changes in old ones. This would explain the decreasing amplitude in the prepresbyopic eyes.

If the shape of the accommodated lens is determined by the elasticity of the capsule, the static form is related to the ciliary muscle and suspensory ligaments. As the lens equator approaches the ciliary processes, there is more slack in the suspensory ligaments, and since they do not shrink with age, their mechanical efficiency decreases. Does the muscle contribute in some way, if not to the cause, at least to the signs and symptoms of presbyopia? The role of the ciliary muscle is controversial, with two opposite conceptions.

In the Donders-Duane-Fincham view, the amount of ciliary effort required to produce unit change in accommodation increases with age. More ciliary innervation (a greater number of myodiopters) is required for each diopter of lens change, and a maximum ciliary contraction to reach the near point. In the Helmsholtz-Hess-Gullstrand view, it is assumed that ciliary effort does not change with age. If less lens diopters can be mobilized, less effort (fewer myodiopters) must be expended; hence less ciliary contraction will be required to achieve the near point, and progressively more myodiopters remain unrelated to any refractive change. Each theory leads to experimentally verifiable implications, yet the results are contradictory. Considering only the clinical evidence, the lack of asthenopic symptoms, the invariance of the AC/A ratio with age, and the ciliary response to partial cycloplegia all support the Helmholtz-Hess-Gullstrand theory.

Presbyopia unlike accommodative insufficiency does not result in real discomfort aside from the frustration of not being able to read small print. The absence of symptoms is attributed to lack of ciliary effort. If the decrease in accommodation is lenticular, no additional contraction is mobilized to bring about what cannot be physically achieved. Furthermore, if the effort of the presbyope to accommodate 1 D is the same as in a young individual, the amount of accommodative convergence induced should remain the same. This is what has been generally found: the AC/A ratio indeed remains constant (a small decrease reported is within limits of error). The average AC/A ratio for a presbyopic population is about the same as for a nonpresbyopic sampling (about 4Δ/1 D).

Fuchs first reported that 2% eucatropine (Euphthalmine) caused a 4 D reduction in amplitude in an 11-year-old boy but no change in a 38-year-old man. Duane measured the elapsed time to onset of cycloplegia after homatropine instillation and found the interval shorter in presbyopes than nonpresbyopes. If there was a larger proportion of latent ciliary fibers, Duane reasoned that the cycloplegic effect should take longer. This is not necessarily so, however, since only the active fibers are under tension and might easily be more susceptible to the drug. Other results, such as those by Flieringa and van der Hoeve, using partial cycloplegia (i.e. during the recovery from cycloplegia) have been both confirmed and denied. There is evidence that ciliary effort unrelated to lens form is less than a one to one ratio (myodiopter per lens diopter) as the theory predicts. It is assumed that a proportionate atrophy and sclerosis occurs in the ciliary muscle fibers. Why this is not evident in young eyes where accommodation has already decreased considerably, and how it comes to be binocularly symmetrical is not clear. In an extensive and thorough review of the evidence, Alpern (1962) concludes that in the last analysis, a decision on the theory of presbyopia awaits a decision on the theory of the synkinesis of vergence and accommodation.

CLINICAL EVALUATION

The diagnosis and treatment of presbyopia is probably the most common if not the simplest refractive problem. But it need not be approached with the trepidation suggested by some writers if the pitfalls of treating people as statistics are avoided and the symptoms properly evaluated. By choosing reliable diagnostic tests, it can become one of the most satisfying optical achievements. Southall wrote:

> Nowadays elderly folks in all walks of life have recourse to eye-glasses for reading and writing and for other useful occupations which otherwise could not be conveniently pursued. It is sad to think of the tedium of existence that must have been the lot of many an aged man or woman in the days before spectacles were invented or before they had come into common use.*

There are many tests to evaluate the required presbyopic add; there is only one final prescription. Obviously we do not discard one test or another because it does not correlate with the final result. Each test measures a different aspect under different conditions with different end points, and a rigid adherence to a particular technique is to confuse refractive measurements with refractive prescriptions. The final reading lens is based on a clinical diagnosis, a fabric woven from a web of symptoms, a weft of signs, and a woof of cosmetic fantasies, twisted by ametropia, anisometropia, and binocular motility and textured by the patient's needs, posture, and prosperity.

Plus lens to clear near vision

The most commonly used technique simply adds sufficient plus power to achieve clear vision on a reading card held at the required distance. The patient wears his ametropic corrections, and all measurements are made at the spectacle plane. Some conditions and precautions apply to this and other methods of evaluating presbyopic requirements. The illumination should be adequate but not so intense as to unduly increase depth of focus. The history will suggest what illumination is available at home or on the job, especially if the visual task involves poor contrast. Additional illumination is recommended if there is senile miosis and reduced transmission or crystalline lens discoloration.

The distance of the reading target should duplicate the patient's needs. The widespread use of 13 or 14 inches as the work distance on insurance and compensation forms is an unfortunate, unrealistic, and arbitrary choice and is seldom a practical work distance, which varies with occupation. The distance requirements of a barber differ from those of an accountant, and they even vary within the same occupation; the engineer may read dials at 30 inches, blueprints at 16 inches, and a slide rule at 12 inches. Most people in fact require clear vision at more than one distance, and the more fixation must fluctuate the more useful a bifocal. Many bifocal styles and designs are now available in glass or plastic, standard or corrected curve, one-piece or fused, in powers computed to best supplement the lost accommodative amplitude. Most of them, however, are inflexible and do not provide the range of vision obtainable by the normal ciliary mechanism. This decrease in range becomes particularly evident with the first or second change in add and is the most common complaint of bifocal wearers. Since most people require clear vision at various distances, there is no "all-purpose" bifocal, and the loss of visual range must be made up by approaching or receding from targets located at intermediate distances.

Some average working distances for selected occupations are given in Table 20-3, which shows the large variations. Individual needs are best obtained from the history and by actual measurements, since not all barbers are equally tall, and some engineers paint pictures, collect stamps, or play the piano. Depending on the time spent at these activities, a separate pair of spectacles may be needed for work, play, and street wear.

Acuity requirements also vary, and we can

*From Southall, J. P. C.: Physiological optics, London, 1937, Oxford University Press.

select Snellen equivalent, Jaeger print, or printer's point (Table 20-4), or better, a sample of the actual reading material. It will be seen that the only occasion for reading near 20/20 letters is on a reading card. How many middle-aged housewifes must be walking about with a crick in their neck and a crimp in their back as a result of this foolish fetish for 20/20 at 13 inches. Most patients know their needs best and can indicate a practical target for near acuity. Recall that we are not speaking here of measuring acuity but of using it as a criterion of the optimum reading add. Samples of news print, phone books, magazines, or stock quotations serve as realistic targets, and special requirements such as micrometers or slide rules are best evaluated by actual use.

In those occupations which require a large field of view, separate reading lenses or large segs are indicated. The usual rule is to devote the same percentage area to the seg as the time devoted to the near vision. The required area can be measured with a ruler or with trial bifocal lenses available in sets. A similar procedure is adopted for evaluating intermediate distance vision needs.

Table 20-3. Some average occupational near requirements

	Average (inches)	*Range (inches)*
Housewife	22	16 to 30
Secretary	18	16 to 26
Accountant	16	15 to 28
Dentist	16	12 to 24
Grocery clerk	20	18 to 32
Artist	22	12 to 26
Barber	16	14 to 22
Butcher	20	18 to 28
Carpenter	18	16 to 26
Sales clerk	20	18 to 24
Watchmaker	8	6 to 14
Architect	16	12 to 30
Librarian	16	14 to 32
Surgeon	17	14 to 24

Balancing relative accommodation

In this method, the positive and negative relative accommodative amplitudes (PRA and NRA, respectively) are measured at the selected reading distance. The indicated add is NRA + ½ (relative accommodative amplitude). For example, if the PRA is −1 D and the NRA is +2 D, the relative amplitude is 3 D, and the suggested add is (−1) + ½ (3) = +0.50 D. The principle is that the relative accommodation, being a fraction of the total, is equally divided into amounts in use and in reserve. With the indicated add in the previous example, the PRA and NRA are balanced at 1.50 D each.

Accommodative reserve

As the amplitude decreases, the proportion of accommodation used to that held in reserve increases. Clinical experience suggests that when the reserve drops to less than one half, difficulties are experienced for sustained near work. The indicated add is individually com-

Table 20-4. Near point visual acuity equivalents

Snellen fraction equivalent	*Visual angle (minutes)*	*Jaeger type*	*Printer's point*	*Practical equivalent*
20/20	1.	1+	3	Near point test cards
20/25	1.25	1	4	Bibles
20/30	1.50	2	5	Want ads
20/40	2	3	6	Telephone directory
20/50	2.5	5	8	News print
20/70	3.5	7	10	Books
20/80	4	8	12	Children's books
20/100	5	10	14	Children's books
20/200	10	17	28	Large type books

puted from the measured amplitude and the estimated reading distance. For example, at a reading distance of 40 cm, the indicated add for a 3 D amplitude is 1 D so as to leave 1.5 D in reserve.

There is, of course, no way to be sure that this use/reserve ratio is appropriate for this patient, and its effectiveness will need to be evaluated by trial and error. Indeed the ratio probably varies with ciliary tonus, general health, and the monotony, intermittency, or durability of the seeing task. The chief disadvantage of this technique is that it relies on rather inaccurate amplitude measurements.

Duochrome test

The duochrome test is similar to that used to refine the distance sphere. The target is a self-illuminated Snellen (or other print) chart, half red and half green (e.g., the Freeman near vision unit or the Bernell lantern). This is a monocular test, and the patient does not wear red-green filters. With the approximate reading add over the distance correction, add sufficient plus to make the red side clearer, then reduce until they equalize or reverse. The technique may also be used to confirm an add found by other methods. The patient holds the target and moves it in or out until the two sides appear equally sharp. If the add is too weak, green is better; if the add is too strong, red is better.

Fig. 20-1. Near point target for use with cross cylinder test.

Polarized targets

The polarized target method is similar to that for binocular refraction (Chapter 15). It is used here to confirm the equality of the indicated add for the two eyes by having the patient compare the sharpness of one to the other. Self-illuminated lanterns with a trial frame or vectographic near point cards with a refractor may be used.

Cross cylinder test at near

The principle of the cross cylinder test at near is that the patient is allowed to indicate his accommodative requirement at the reading distance. A cross cylinder red axis vertical is placed before both eyes, and the patient views a cross grid consisting of several vertical and horizontal lines. (The target is on a card attached to the refractor rod.) The overhead illumination must be dimmed (room light on but not overhead refractor lamp), an important part of the test often ignored. The patient wearing the distance correction views the target, and binocular plus is added in 0.25 D steps until the vertical lines are clearer and darker. The plus is then reduced until the horizontal line is darker; this is the end point. The purpose of both cross cylinder and dim illumination is to prevent sharp focus and allow the circle of confusion free choice, so to speak. The technique can be performed with the refractor, and the prescription need not be transferred to a trial frame. The cross cylinder is an excellent test because it gives a clue where the patient "wants his accommodation." Moreover it does not depend on letters, helpful in illiterates.

Range of vision

The range of vision through the reading lens becomes progressively more restricted as the amplitude decreases and the add increases. For example, the range of vision of an eye corrected for distance requiring a +1.00 add to read at 40 cm has an available amplitude of 1.50 D. The range is from 100 cm (no accommodation) to 40 cm (neglecting depth of focus). When the required add is +2.00 D, the available amplitude is reduced to 0.50 D, and the range is from 50 to 40 cm. This represents a reduction from 60 to 10 cm, a considerable decrease in intermediate vision. It means the patient will need to bring near objects within 50 cm to see them.

The test is performed by placing the tentative add in the refractor, or better, the trial frame. The patient is instructed to move the target (an appropriate line of letters) to the furthest and closest distance at which it can be recognized. The add is then modified in 0.25 D steps to bring the midpoint of the range to the optimum distance required. Generally, the range is best left larger on the distaff side for obvious reasons. The effect of arm length and head posture will be especially evident with this technique and is recommended as a final step in all methods of evaluating the add.

Dynamic retinoscopy

Dynamic retinoscopy is objective in the sense that the examiner evaluates the end point; it is subjective in the sense that the patient must fixate on and accommodate for the target held at the reading distance. The target is generally larger (subtending about 1° at the eye) and may be attached to the instrument, or one can scope above it or through a hole in the card. The illumination is reduced just enough to permit evaluating the reflex. Both eyes remain open, and the examiner scans only one meridian, alternating between eyes. With the distance correction in place, a with movement is usually seen. Adding plus spheres in 0.25 D steps, the examiner continues past the neutral point until a definite against motion appears. Suppose it takes a +1 D sphere to obtain neutral and a +1.75 sphere to obtain against motion. The first is the "low neutral"; the second, the "high neutral." This means that at the beginning of the test the eyes were hyperopic for the test distance (hence the with motion), the +1 D brought them into focus (neutral), and they then relaxed accommodation until the PRA was exhausted (against motion). The indicated add is the amount required to focus the target (low neutral) plus a proportion of the positive relative accommodation (generally 0.50 D) to allow for comfortable, sustained activity. Results should be confirmed by one or more subjective techniques.

MANAGEMENT OF PRESBYOPIA

Presbyopia is a relative entity, and a number of conflicting requirements must be met and solved simultaneously. These include the distance refraction, optical jump and displacement, field of view, optical and mechanical fitting problems, accommodation and convergence, anisometropia, and cosmetic appearance. One clinician choosing the proper bifocal proceeds a hero; another, unluckily timing it, is reproached for the same enterprise. No refractionist is a permanent hero to his optician.

Refraction

Early symptoms of presbyopia are more likely to occur in uncorrected hyperopes than myopes and even in corrected hyperopes than emmetropes. Recall from Chapter 7 that ocular accommodation (computed from Pascal's formula) must be mobilized to a greater extent in spectacle-corrected hyperopia because of the effectivity of the vertex distance. A previously uncorrected hyperope can sometimes be managed by a single vision distance correction to be worn for reading. By crowding the plus, the need for a separate reading prescription may be postponed for a year or two depending on the visual demands. In myopes, however, it is seldom possible to cut the distance prescription without an annoying reduction in acuity. The patient who indicates a need for a +2 D add on the near test but who never noted any difficulty until a month before can probably be given half the optical requirement.

Generally, an add of less than 1 D is rarely indicated. The exception is a surgeon or technician who must carry out a precise task at a specified distance. Nevertheless, the rule to give the weakest possible add for the visual requirement is a good one; when in doubt, leave it out. There is no evidence that reading glasses retard or accelerate the further progression of presbyopia, a question frequently asked by patients.

In addition to losing his accommodation, the presbyope is also subject to changes in ametropia, generally toward hyperopia. The amount is variable and can be unpredictable in diabetes or for incipient cataract requiring a new prescription even though the add is unchanged. A greater than average add may be required for patients with subnormal vision. At some point, a decision must be made between bifocals and separate reading and distance glasses. Not every patient is a good candidate for bifocals, and there are difficulties even under ideal conditions. The floor is blurred and may be doubled, near objects are somewhat magnified, the field is decreased and the range of vision reduced, objects jump up and are displaced from their true position and the seg line betrays one's age. A bifocal has limited use when the near target is above the horizon (e.g., a grocery clerk, violinist, television announcer). Separate reading and distance glasses, on the other hand, are often a bigger inconvenience, even to the myope who must take his distance glasses off to read. But it is necessary to demonstrate this inconvenience by actually observing a distance target through the reading add and the near target through the distance glass. It is not necessary to condemn the patient to a year or two of separate spectacles to make him appreciate bifocals later. Moreover, by postponing bifocal prescribing, there is the disadvantage that the add will later be stronger, the optical difficulties greater, and adaptative problems compounded. A compromise is often feasible and appreciated—a single vision distance lens combined with a reading spectacle in the form of a straight-across bifocal (e.g., "Executive" style). This still requires switching glasses but not every time he looks up. Moreover the annoyance of the seg is avoided for such activities as walking or driving in which it is most troublesome. This combination is a useful alternative not only in making a later adjustment to constant bifocal wear but also because the visible field is large and easily used, there is no jump, and in these early stages of presbyopia, the add is not so strong as to make the floor excessively blurred. Indeed the straight-across bifocal is an excellent reading lens for patients who require no distance correction but refuse to slide glasses down their nose when looking up. A half-eye frame also achieves the same effect (if there is no significant ametropia), and is less costly and lighter besides, but not everyone is impressed by their appearance.

If there is considerable astigmatism, the cylindrical correction at near differs for the same reason that the spectacle-corrected myope must accommodate less than the spectacle-corrected hyperope. More accommodation is required in the hyperopic astigmatic meridian; indeed we saw in Chapter 7 that almost 1 D of astigmatism may end up uncorrected at a reading distance of 25 cm for a compound prescription of $+2 + 4 \times 90$ worn 20 mm from the cornea. There is no remedy for this short of a separate pair of reading glasses, since the eye cannot muster sectional accommodation.

Optical jump and displacement

When the visual axis passes across the dividing line from the distance to the reading portion of a bifocal, there is a prismatic displacement giving rise to "image jump." Its cause is evident from Fig. 20-2. As the gaze is lowered, it meets the gradually decreasing base-up effect of the distance convex sphere; at the junction line, however, the

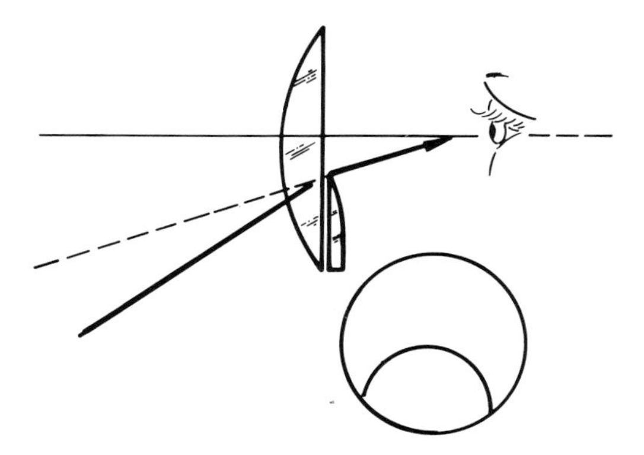

Fig. 20-2. Optical jump induced by bifocal segment.

base-down effect of the segment is introduced. Objects appear suddently to jump up to a new position. Observing a series of book shelves and scanning from the distance to the reading seg, one or more of the shelves disappear as it falls within the lost angular field. The larger the seg and the higher the add the greater the field loss.

The optics of image jump is easily understood by pretending the bifocal is made up of two thin lenses separated by an infinitely thin layer of air. Since jump results from the *difference* between the prismatic effects of the distance and reading portion at the junction line, only the prismatic effect of the add must be considered. Since most adds are plus, the jump is invariably up, and the amount is readily computed from Prentice's rule: $d = cF$, where d is the jump in prism diopters, c is the distance of the optical center of the seg from the dividing line (in centimeters), and F is the power of the add (in diopters). Thus the optical jump of a +2 D round seg of 22 mm diameter is: $(1.1)(2) = 2.2\Delta$ base up.

"Optical displacement" is the sum of the prismatic effects of the distance and reading portions at the reading point. Here we compare the position of the object viewed with and without the lens. For example, the prismatic effect 1 cm below the distance optical center of a +1 D lens is 1Δ base up. With a round seg of +2 D whose optical center is 1 cm, the prismatic effect is 2Δ base down. The total prismatic displacement is the sum or 1Δ base down. If the prescription specifies a cylinder, the power in the vertical meridian must be computed.

It is evident that to minimize jump the optical center of the seg should be as close to the dividing line as possible. Thus the straight-across bifocal has no jump because the optical centers of the distance and near portions coincide. The flat-top bifocal has minimal jump when the optical center of the seg is near the top. Large round seg bifocals have the most jump because the optical center is furthest from the top. None of these effects are influenced by the power of the distance portion; that is, jump depends only on the power of the seg and the location of its optical center.

In the case of optical displacement, however, the round seg (base down) tends to neutralize the base-up effect of a plus distance portion and exaggerate the base-down effect of a minus distance prescription. For the hyperope obtaining his first bifocal, a round top minimizes displacement but gives maximum jump. For the myope, any base-down effect contributed by the add exaggerates the optical displacement, one reason myopes have more difficulty adjusting to bifocals.

The relative importance of jump and displacement has yet to be satisfactorily settled. Some clinicians aim to eliminate jump, others to neutralize displacement. Briefly the arguments are as follows. Jump is not very serious when shifting fixation from far to near because the eye is moving (hence not seeing) as it passes over the dividing line, but it becomes disturbing when looking at the floor or at near objects partly visible above and below the seg requiring a head movement. On the other hand, the patient wearing distance glasses has adapted to a certain visuotactual coordination to localize objects, including their displacement through simple vision lenses. When he obtains bifocals and the optical displacement is neutralized, no new coordination must be learned; he touches objects where he sees them. But since he is unlikely to take his glasses off to compare the "true" and displaced object positions he can easily learn a new association with respect to displacement, whereas jump remains forever. For the myope, either argument leads to the same recommendation, that is, a straight-across or flat-top bifocal. For the hyperope, one view suggests a flat-top bifocal to minimize jump; the other view suggests a round-top bifocal to minimize displacement. My experience is that jump is more annoying to many patients whatever the ametropia.

The problem becomes more complex in anisometropia, in which prismatic displacement differs in each eye and induces a vertical imbalance. With single vision lenses, the anisometrope can avoid the prismatic difference by lowering the head when reading so that the line of sight passes through the optical centers, but with bifocals this is not possible without looking above the segs. The difficulties are solved by compromise. Induced vertical imbalances may be measured (e.g., by a Maddox rod) and corrected when indicated by bicentric grinding, compensated segs, or prisms. Whatever the decision, it should be made by the clinician in cooperation with the optician and not arbitrarily delegated, since some opticians routinely decenter all lenses to eliminate prismatic displacement.

Pascal (1940) arrived at the same conclusion: "A little independent thinking and observation on the part of the physician who comes in actual contact with the patient may disprove some standardized notions of even the leading manufacturers.

The difficulties patients have with modern bifocal lenses is probably largely due to the lenses having two separated optical centers, that is, being bicentric. This causes an image to be formed along two axes. If both sets of light pencils enter the pupil, the object is seen double. If the pupil happens to lie between the two sets of light pencils, neither one enters the eye and there is a gap or blind area. The so-called 'jump' experienced when the eye changes from the upper to the lower portion is most likely due to the doubling and the gap. Similarly, the patient is apt to miss a step and misjudge the position of objects because of the doubling effect and the blind area defects which are inherent in all bicentric bifocal lenses.

The prism value at the reading portion is of little or no significance as long as there is no appreciable difference between the lenses for the two eyes. For example, one can easily get about with, say, a 4 prism diopter, base down or up, for each eye without missing steps or having to fumble for an object. If the step is raised or lowered, so is the foot raised or lowered and to the same extent, so they make perfect contact. The same applies for things to be grasped by the hand. The prisms displace the object and the hand to the same extent, so there is no fumbling. The manufacturers have produced almost ideally invisible bifocal lenses from the cosmetic point of view, but optically it seems they may be on the wrong tract if their ideal is a bicentric bifocal lens."*

*From Pascal, J.: Bifocal lenses, Arch. Ophthal. **20**:553-554, 1940.

Field of view

The size of the bifocal seg determines the field of view. Larger reading fields are required by engineers, architects, or surgeons; smaller fields by golfers, bus drivers, or construction workers. No seg size can be stretched to cover every job; the architect may also like to golf. The factors determining field of view are seg size, pupil diameter, vertex distance, lens power, head movement, and working distance.

The larger the pupil the smaller the field of *clear* vision. Large pupils straddle the seg edges, causing diplopia and confusion. Fortunately the near reflex, increasing age, and good illumination summate to produce miosis so that this effect is negligible, but in aphakes with large iridectomies it must be considered. The closer the lens to the cornea the larger the field of view (limited by base curve, length of lashes, and size of the nose). Strong convex lenses reduce the field, and the patient who turns his eyes more than his head will have more difficulty limiting himself to small seg apertures. New postural habits must be learned such as holding the head stiff and lowering the gaze and turning the head for lateral excursions. Seg size also depends on the reading distance, since the angular field increases in direct proportion to the working distance. Seg size has its limits; for example, a large flat-top seg cannot be ground in a high minus prescription without making the lens excessively thick.

Optical and mechanical factors

Most patients prefer a bifocal that looks like a single vision lens; any bifocal is thus a compromise between appearance and convenience, optics, and mechanics. The round-top seg with its knife edge is least visible; the flat-top seg has a visible ledge. Invisible seamless lenses are available, but they are not invisible to the patient who must put up with a larger transitional confusion zone.

The conventional distance of the top of a bifocal seg is 3 to 4 mm below the optical center of the distance portion. Flat-top bifocals can be fitted lower, often advisable

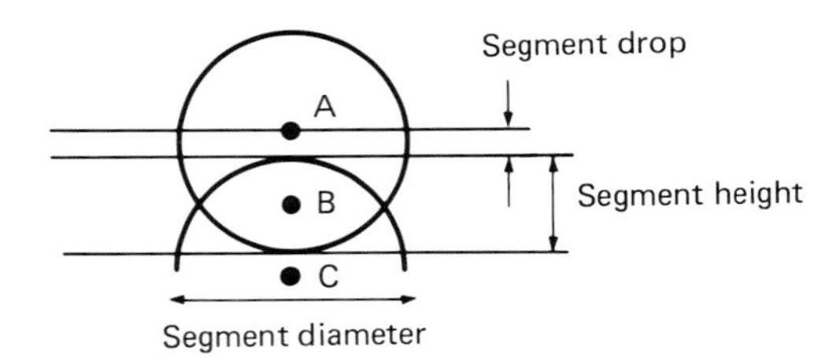

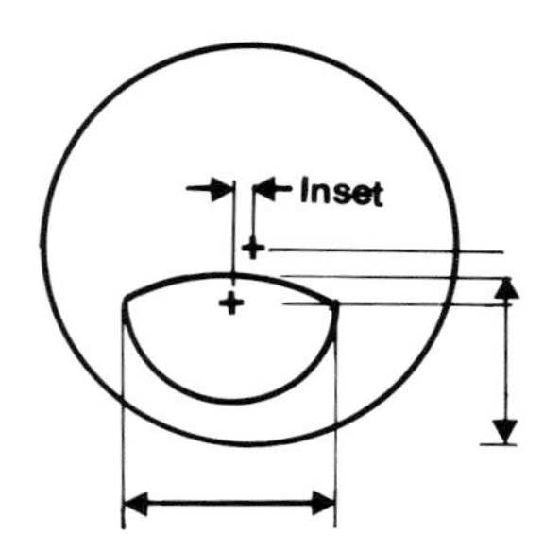

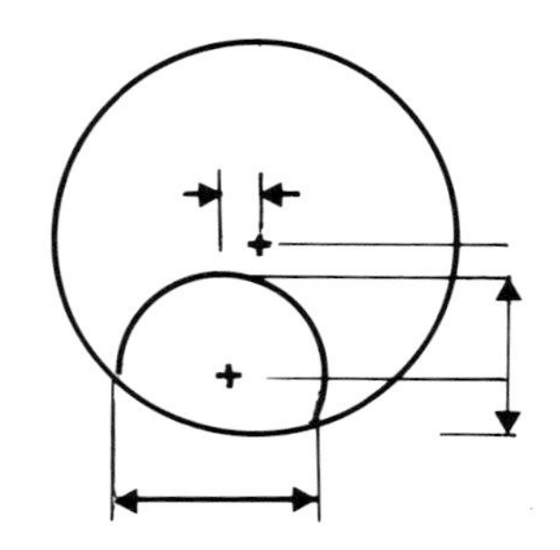

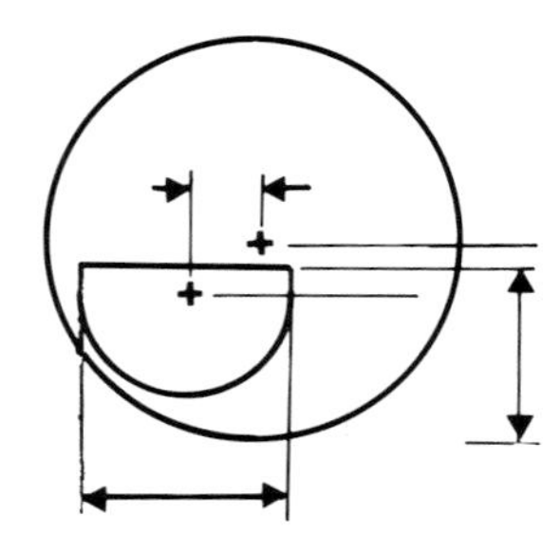

Fig. 20-3. Bifocal segment terminology.

as a first choice. The precise seg position depends on head posture, purpose of the lens (primarily for reading or constant wear), and required field. If he crouches over his work with lowered head, the seg should be higher. A high seg should also be used for children with accommodative esotropia to prevent peeking. In general, the upper edge of the segment is fitted on a level with the edge of the lower lid, the head being in a natural position. In short, the seg should be high enough to be useful and low enough not to interfere with distance vision. A frame that is easily angulated with adjustable nose pads will permit some later repositioning.

Segment terminology

Bifocals are usually dispensed with the top of the seg level with the lower lid margin and the optical center of the distance portion coincident with the pupil when the eyes are in the primary position. The seg height is the vertical distance from the top of the seg to the lowest point of the lens periphery (Fig. 20-3). The seg diameter is the diameter of the circle of which the seg is a part. The seg drop is the vertical distance of the seg top from the distance optical center. Segs are also inset, the spearation between the vertical lines passing through the distance optical center and the midpoint of the seg diameter.

Bifocal segs are decentered nasally to compensate for the convergence of near vision; the amount required is computed or measured directly by ink marking the lens at the intersection of the central pupillary line. The further the lens from the cornea the greater the required decentration. Sterling derived a formula that also takes lens power into consideration

$$\text{Inset} = \frac{1/2 \text{ Interpupillary distance}}{1 + d\,(1/2.7 \text{ cm} - 1/f)}$$

where d is the distance between the lens and the reading target, 2.7 cm is an assumed average distance to the center of rotation of the eye, and f is the focal length of the distance portion of the lens (in centimeters) in the horizontal meridian. Average values range from 2.0 mm for a +4 D to a 1.6 mm for a −5 D (62 mm PD). An approximate formula for seg inset is

$$\text{Inset} = \frac{\text{PD}}{34}$$

Occasionally a facial asymmetry causes one pupil to be closer to the nose than the other, determined by measuring the PD from the root of the nose. In such cases the final seg inset will be asymmetrical.

Image quality varies with bifocal design and construction. The Kryptok has more chromatic aberration for hyperopic corrections but may have less in high minus distance prescriptions. Image quality can be evaluated by comparing fine print with and without the lens through its center and periphery. Corrected curve bifocals give better quality by reducing aberrations, but since the curves of the seg depend on the distance lens, the bifocal design is necessarily a compromise. Aspheric lenticulars are not aspheric in the seg area.

Bifocal appearance and weight should ideally differ little from an equivalent single vision lens, and most bifocal styles are now available in plastic to reduce weight, of particular value in aphakic prescriptions. The seg line can also be coated to reduce reflections. Individualized bifocal dispensing is a major factor in satisfactory adjustment, since all fitting problems involved in single vision lenses are compounded. Standard rules will not apply to the patient who habitually tilts his head, turns his eyes down more than the average 8 mm, or has one eye higher or closer to the nose than the other. With multifocals much more than with single vision lenses, a frame that gets out of adjustment can easily lead to an undesirable prismatic effect or even looking above the seg with one or both eyes. An oversized frame for reading glasses makes looking above it difficult; a small frame in turn limits the size of the seg. An improperly angulated or tilted frame that slips down the nose may cause more dissatisfaction than all the optical theories of prismatic displacement combined.

The rule of thumb that bifocals must never be altered is equivalent to never changing antibiotics. There is every reason to change style if the previous one was ineffective or uncomfortable. A change in size may also be appropriate for different uses, for example, a large seg for work, medium seg for everyday wear, and a small seg for sport. The only rule is choosing a lens that best serves the purpose, with maximum flexibility, optimum appearance, and greatest comfort.

Accommodation and convergence

In prescribing an add for presbyopia, the amount is always plus, and the unused accommodation no longer stimulates its associated accommodative convergence. Since psychic convergence changes little, there is an increased exophoria for a constant AC/A ratio. When the AC/A ratio is high, the add may not be accepted unless the positive fusional convergence reserve can make up the deficit. Inadequate fusion leads to intermittent diplopia or, more likely, attempts to hold the reading farther away, with the patient complaining that his glasses are too strong. Such cases may be helped by reducing the add, increasing the illumination, base-in prism, building up fusion, or trifocals. The trifocal permits the patient to oscillate reading distances, using the intermediate seg when near vision becomes tiresome. It is seldom necessary to abandon binocularity by correcting one eye for distance and the other for near.

Anisometropia

When a patient with anisometropia looks through his new bifocals, he is faced by a vertical prismatic imbalance at the reading point. The symptoms include drawing, pulling, headaches, nausea, or diplopia. The problem is illustrated in Fig. 20-4. The prescription is RE: +2.50, LE: −1.75 with a +2.00 add. The reading point is 6 mm below the distance optical center and 7 mm above the optical center of the seg (assumed to be 22 mm round). For the right eye, the distance lens creates a prismatic effect of (0.6)

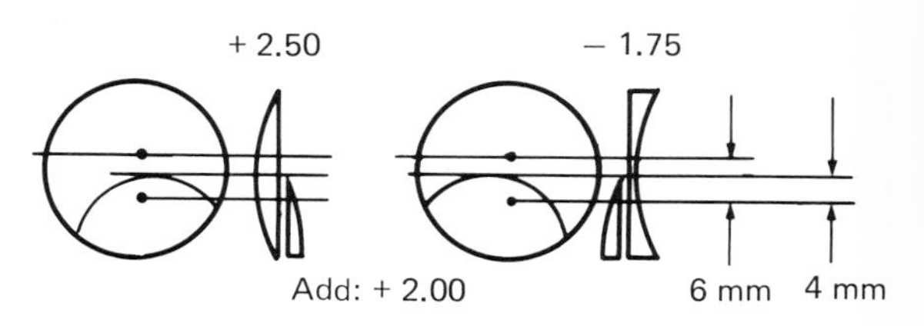

Fig. 20-4. Vertical imbalance at the reading level in anisometropia. For details see text.

(2.50) = 1.5Δ base up and (0.7) (2.00) = 1.4Δ base down in the seg for a total prismatic effect of 0.1Δ base up. A similar computation for the left eye gives (0.6) (1.75) = 1.05Δ base down for distance and (0.7) (2.00) = 1.4Δ base down for the seg, or a total of 2.45Δ base down. The difference between the two eyes is 2.35Δ base down LE.

If the prescription contains a cylinder, the power is the vertical meridian is computed (e.g., from the sine-squared formula). Since the visual axis passes nasal to the distance optical center in reading, there is also an induced horizontal prismatic effect, but the amount is usually small relative to the horizontal fusional vergences.

Vertical imbalances at the reading level are inevitable in anisometropia, whether the patient wears bifocals or not. With single vision lenses, the patient can avoid them by dropping his head instead of the eyes, thus looking through the optical centers of both lenses, and the same applies to single vision reading lenses. So all vertical imbalances can always be eliminated by prescribing two pairs of glasses or a monocentric (straight-across) bifocal.

Rather than arbitrarily correct imbalances by computation, they can be measured. Thus if there is a vertical phoria, it just might be exactly neutralized by the imbalance and require no compensation. The technique suggested by Lebensohn (1949) consists of measuring the intrinsic vertical phoria without any correction, for example, by a handheld Maddox rod prism combination device and fixation light held at the appropriate reading distance. The test can also be repeated with the indicated reading prescription in place. Care must be exercised to hold the device level or by placing the Maddox rod in the trial frame. The patient indicates his vertical phoria subjectively, or the double images are lined up by a measuring prism.

Prismatic compensation is indicated in significant vertical imbalances (over 1 D) and requires measuring the compensating vertical vergences. An estimate of acceptable imbalance is made by noting how the patient holds his head when reading through his previous spectacles.

The prismatic differential at the reading level is computed from Prentice's rule, the key factor being the variance between the eyes. It is conventional to split any prismatic correction if feasible. Prism segs are unfortunately unsightly; not much better are dissimilar segs (e.g., the seg with the large optical radius in front of the eye wearing the stronger plus power in the vertical meridian) because they look poor, and the patient thinks he received the wrong glasses. Compensated segs are feasible but only neutralize small prismatic differences (about 1.2Δ). The best alternative (aside from separate reading glasses or monocentrics) is bicentric grinding or slabbing off as described in the last chapter. Slabbing off compensates for 2Δ to 5Δ of imbalance and is the only method available for trifocals. In convex lenses, the weaker lens is slabbed off; in concave lenses, the stronger lens is slabbed off.

Cosmetic factors

Last and perhaps least important are cosmetic considerations. When the alternatives are explained and demonstrated, it is a rare patient who will insist on two pairs of glasses in the peculiar conviction that switching spectacles is less conspicuous than a segment line. Some patients have convinced themselves that they will never get used to bifocals, but even this dogma has its mislay.

Cosmetic problems are not new, and vanity was with us even before bifocals became popular as evidenced by the rhetorical advice given in 1789 by George Adams, optician to his Royal Highness the Prince of Wales: "Though the effect of time are the certain and inevitable portion of all who live to advanced age, and are neither to be retarded by riches, nor prevented by wisdom; yet such are the weaknesses of the human mind, and such the potentialities of self love, that we all endeavor to conceal from ourselves and others the approaches of age; and no one likes to appear as hastening to that bourne from which none have returned.

These propensities give rise to a variety of artifices, by which each individual endeavors to hide from himself and others, what no artifice can conceal, and which everyone can discover in all but himself; but then their endeavors often contribute to hasten the evils they are meant to conceal. Opticians have daily experience of the truth of these observations, and they are in no instance more fully verified than in the preference given by many to reading glasses (under whatever pretext it may be covered) merely because they think that the decay in their sight, and their advance in age, are less conspicuous by using a reading glass than spectacles."*

Although an invisible bifocal is cosmetically desirable, it should not be bought at the price of optical performance. When segs advertised themselves by accumulated dirt and grime at their edges or down-curving obviousness, their poor public image was probably justified. The popularity of modern larger bifocal segs indicates that visibility is not such a major concern as some practitioners believe. A little tactful encouragement is helpful, and a proper sense of perspective can occasionally be achieved by a reading card with a little poem written by John Gay, about 1727:

I cannot raise my worth too high;
Of what vast consequence am I!
  Not of th' importance you suppose,
Replies a Flea upon his nose:
Be humble, learn thyself to scan;
Know, pride was never made for man.
'Tis vanity that swells thy mind.
What, heav'n and earth for thee design'd!
For thee! made only for our need;
That more important Fleas might feed.

MULTIFOCAL LENSES

"Multifocals," as the name implies, are lenses with more than one focal length, adapted for distance and for one or more near points. Many types and styles are available, each with its own purpose and indication. In no other phase of optical prescribing is attention to detail more rewarding than in selecting the lens most appropriate for the patient's needs.

Bifocals

The first bifocal, as everyone knows, was invented by Benjamin Franklin: "I imagine it will be pretty generally true that the same convexity of glass through which a man sees clearest and best at the same distance proper for reading is not the best for greater distances. I therefore, had formerly two pairs of spectacles which I shifted occasionally, as in traveling I sometimes read and often wanted to regard the prospect. Finding this change troublesome and not always sufficiently ready, I had the glasses cut, and half of each kind associate in the same circle, thus I wear my spectacles constantly, I have only to move my eyes up or down as I want to see distinctly far or near, the proper glasses being always ready."

Bifocals are classified as fused and one-piece (or solid). The first one-piece bifocal, patented in Philadelphia about 1873, was the "Upcurve," combining a planoconvex top with a biconvex lower portion. The first one-piece bifocal to become widely available was introduced in 1910 under the trade name "Ultex" and is still popular. Two distinct curves are ground on the same spherical surface, the cylinder being ground on the other. Its advantages are permanence, achromatism, and the large size available. Two Ultex lenses are split from a single large blank with a central circular segment. If the same blank is used, a seg up to 32 mm can be obtained. All plastic bifocals are of one-piece construction as are the straight-across bifocals. One-piece bifocals can be identified by touch, since all have a ledge or step at the junction of distance and reading portions that distinguishes them from fused designs. This cosmetic annoyance can be minimized by avoiding chipping, keeping the lens clean, and coating the ledge.

The first attempt to gain added power by a change in index rather than curvature resulted in the original "Kryptok," invented (naturally in Philadelphia) by John Borsch, Sr. in 1899. He placed a button of flint glass in a countersink depression in the distance portion. This button was held in place by a cover glass cemented over the entire surface. In 1908 his son produced the first true fused bifocal by heating. He retained the original name (from the Greek "kryptos" meaning hidden, emphasizing its relative invisibility). The Kryptok has a continuous curve, and the powers of the crown and flint areas are in the ratio of their refractive indices. Although

*From Adams, G.: An essay on vision; briefly explaining the fabric of the eye and the nature of vision intended for the service of those whose eyes are weak or impaired, London, 1789.

it is relatively invisible, it is not achromatic, and the color fringes may disturb a fastidious patient. More significant is the reduced depth of focus at the reading distance, which is immediately apparent when comparing the Kryptok to other achromatic bifocals of the same power that utilize barium glass segs. The Kryptok is a round seg of 22 mm diameter; hence the optical center 11 mm below the top edge. The circular seg, of course, may end up as a crescent, depending on the shape of the frame.

In 1926 a series of fused bifocals with shaped segs was introduced under the name "Univis" (a contraction of universal visibility). A portion of the button is in effect removed. The button is a composite of flint and crown or barium and crown. When fused, the crown merges with the crown of the distance portion, leaving a seg of predetermined shape. Precise selection of the physical properties of glass is necessary to give simultaneous expansion. The fused bifocal is one of the most popular designs, and the process has been adopted by all major manufacturers. The knife edge makes a circular, fused seg almost invisible. The flat-top design tends to minimize jump, chromatism is reduced, and the widely used "D" seg has widths ranging from 20 to 28 mm. The seg, as in all fused bifocals, is ground on the outside surface of the lens. In some designs the sharp corners are rounded, with otherwise the same optical properties.

Modern bifocals may have round tops, flat tops, or rounded off flat tops. They can be straight across, invisible, or of variable focus. The most popular seg diameter is 22 mm, with smaller or larger ones as indicated by the required field of view. The flat-top bifocal produces the least jump, the round-top is least visible of the regular bifocals, the straight-across is monocentric, and the variable focus lens has yet to become practical. Most bifocals are available with corrected curves (except the Kryptok) in adds ranging from +0.50 to +4.50 D (higher ones on special order).

The straight-across bifocal (e.g., Executive, Horizon, Dualens, Univis E, and Kurova M) has a seg top at the distance pole. For the flat-top bifocals, the distance is 3 mm for the Panoptik, 4 mm for the Sovereign, 4.5 mm for the Univis B, 5 mm for the Univis D, and 7 mm for the Univis R; and for the round-top bifocals, 11 mm for the Kryptok 22, and 19 mm for the Ultex A. These representative examples do not constitute recommendations. For additional details, manufacturer's literature should be consulted.

Occupational bifocals are specialized designs, for example, for those who require a reading field above eye level or who require only a small distance "hole" in a reading lens. Aspheric bifocals are available for cataract patients and special designs (e.g., myodisc, minus lenticular) for high myopia.

Invisible bifocals, such as the Younger seamless, are one-piece designs (plastic or glass) in which the dividing line is polished out to achieve better cosmetic appearance at the expense of an induced cylindrical effect in the transition area (about 5 mm wide). The lens works reasonably well for low distance and near powers and is indicated for patients who are excessively concerned with appearance.

"Varilux" is a variable focus lens of French design introduced about 1959. A variable lens provides a gradual change of add either over the whole surface or the reading portion. In the "Varilux" the distance and reading portions are constant but between them lies a progression zone in which the power changes as a result of an invisible gradual change in curvature of the front surface (approximating an aspheric). Generating such a curve and the smoothing and polishing processes are complex, requiring special machinery. Optical performance suffers from deformation in the transition zone so that the usable area of the lens is narrow at the top and wider at the bottom of the progression zone. If carefully fitted and precisely centered, definition is adequate. A plastic version is also available. A continuously variable lens, the "Omnifocal," is no longer readily available. A recent variable power combination follows a design by Luis Alvarez (1969 Nobel Laureate in Physics). Two elements, each varying in power along one meridian, are combined perpendicularly for spherical power changes and obliquely for astigmatic corrections. The design holds considerable promise not only for lenses but also for clinical refraction, to which it has already been applied on an experimental basis.

Trifocals

Trifocals provide sharp vision at far, reading, and arm's length. The intermediate power is generally half that of the reading (60% to 70% are also available). Trifocals give greater flexibility but have their disadvantages; the intermediate seg is in the way when not needed, the double line is harder to adjust to, and the distance and reading parts are encroached on to a variable extent. Trifocals are indicated, sometimes as a separate vocational lens, for painters, musicians, bartenders, bakers, or homemakers. They are available in the same basic designs as bifocals or with the segs in special combinations, for example, reading below, intermediate above, and distance in between for those whose arm's length vision involves looking up (e.g., grocers and bartenders). Because more variables are involved, trifocals are more difficult to fit than bifocals.

The major consideration in prescribing trifocals is a specific need for intermediate vision, which the patient should be able to demonstrate by history or actual measurements. Recall that an older pair of reading glasses or bifocals may provide just the right intermediate vision for the musician, occasional cardplayer or weekend painter. Also possible is undercorrecting the distance prescription. Another expedient is a bifocal with intermediate above and reading below and a separate pair of distance glasses. Occupational trifocals, on the other hand, are sometimes the only choice for the required flexibility. The segs can be ordered in any combination or sequence. In the standard trifocal, the top of the intermediate seg is generally placed in line with the lower edge of the pupil, and the reading seg is preferably not displaced more than 2 mm from where it would be for a bifocal. Interference with distance vision by high segs is a common difficulty. Patients with anisometropia, poor acuity, difficulties in adapting to bifocals, high myopia with good intermediate vision, and good depth of focus are poor trifocal candidates. Conversely, those well adjusted to trifocals should not be deprived of their advantages even if current requirements differ from those for which they were initially prescribed. Also available are four-way lenses, with a trifocal lower portion and an additional reading or intermediate seg above.

Dispensing aspects

Although the dispenser generally works independently of the clinician, the proper frame at the proper fit is an integral part of the optical behavior of the lens correction. This is particularly true when prescribing multifocals where a poor dispensing choice may nullify the optics, offset the field of view, undo the flexibility, and frustrate the adaptation. One of the more common complaints of new bifocal wearers is that the segment is too low. Although the first bifocal is best fitted lower, it is easily overdone especially after the frame settles down. Segments do not become invisible by lowering them; they only become useless.

Since seg height depends on the frame, an exact decision cannot be made until these have been selected, properly aligned, and firmly pressed down on the nose for measurement. Rocking arms provide a distinct advantage, since they permit subsequent adjustments, especially as the patient learns a new head posture. But manufacturers habitually deliver these frames with the pads close to permit individual fitting. If the seg height is measured before they are properly set, the seg will later be too low.

The shape of the lower rim of the frame cuts into the reading field. Upswept, asymmetrical shapes are particularly annoying and should be avoided. Some frame shapes are so peculiar that an exact seg position is difficult to specify. Measurements are best made relative to the major reference points of the lens and a horizontal line tangent to the bottom edge. Obviously the frame should have an adequate depth to house the seg. An allowance of no less than 17 mm below the iris is advisable, less for the straight-across or flat-top designs. The entire frame is angulated according to the rules previously outlined to avoid creating a cylindrical effect. Improper angulation is no substitute for a seg that is too high.

Improperly diagnosed bifocal complaints are difficult to correct. The diagnosis is based on an evaluation of symptoms and actual observation of head and eye position. Frequently, the difficulties are related not to improper refraction or cosmetics but improper seg height, size, design, or inset.

REFERENCES

Bannon, R. E.: The presbyopic cripple, Survey Ophthal. **13**:298-302, 1969.

Bennett, A. G.: Variable and progressive power lenses, Optician **160**:421-427, 533-538, 1970.

Berens, C.: New quadrifocal spectacle for the presbyopic ophthalmologist, Arch. Ophthal. **48**:632-633, 1952.

Boeder, P.: Analysis of prismatic effects in bifocals, Southbridge, 1939, American Optical Co.

Cannon, L.: A survey of the reading and working distances of presbyopes, Ill. Med. J. **115**:4-6, 1959.

Crisp, W. H.: Bifocals for juveniles, Am. J. Ophthal. **32**:1005-1006, 1949.

Davis, J. K., and Fernald, H. G.: The one-piece (Franklin type) bifocal, Am. J. Optom. **46**:163-188, 1969.

Duane, A.: Normal values of the accommodation at all ages, J.A.M.A. **59**:1010-1013, 1912.

Fincham, E. F.: The proportion of ciliary muscular force required for accommodation, J. Physiol. (London) **128**:99-112, 1955.

Fisher, R. F.: Elastic constants of the human lens capsule, J. Physiol. (London) **201**:1-19, 1969.

Goldmann, H.: Senile changes of lens and vitreous, Am. J. Ophthal. **57**:1-13, 1964.

Hamasaki, D., Ong, J., and Marg, E.: The amplitude of accommodation in presbyopia, Am. J. Optom. **33**: 3-14, 1956.

Hofstetter, H. W.: A survey of practices in prescribing presbyopic adds, Am. J. Optom. **26**:144-160, 1949.

Hofstetter, H. W.: A longitudinal study of amplitude changes in presbyopia, Am. J. Optom. **42**:3-8, 1965.

Hofstetter, H. W.: Further data on presbyopia in different ethnic groups, Am. J. Optom. **45**:522-527, 1968.

Lebensohn, J. E.: Practical problems pertaining to presbyopia, Am. J. Ophthal. **32**:22-30, 1949.

Lebensohn, J. E.: Bifocal and inter-related issues, Eye Ear Nose Throat Mon. **44**:67-76, 1965.

Leudde, F. W.: What subluxated lenses reveal about the mechanism of accommodation, Am. J. Ophthal. **24**:40-45, 1941.

Mandell, R. B.: Myopia control with bifocal correction, Am. J. Optom. **36**:652-658, 1959.

Neumueller, J.: Mathematical optical viewpoint of presbyopia, Am. J. Optom. **8**:361-376, 1931.

Rambo, V. C., and Sangal, S. P.: A study of India, Am. J. Ophthal. **49**:903-1004, 1960.

Schapero, M., and Nadell, M.: Accommodation and convergence responses in beginning and absolute presbyopes, Am. J. Optom. **34**:606-622, 1957.

Slataper, F. J.: Accommodation of presbyopia and its correction, Arch. Ophthal. **34**:389-397, 1945.

Snell, A. D., and Lueck, I. B.: Presbyopia, Int. Ophthal. Clin. **5**:443-470, 1965.

Teramoto, C.: Studies on presbyopia, Acta Soc. Ophthal. Jap. **63**:1718-1735, 1959.

Volk, D., and Weinberg, J. W.: The omnifocal lens for presbyopia, Arch. Ophthal. **68**:776-784, 1962.

Weale, R. A.: Presbyopia, Br. J. Ophthal. **46**:660-668, 1962.

Westheimer, G.: Accommodation levels during near crossed cylinder test, Am. J. Optom. **35**:599-604, 1958.

Wick, R. E.: The visual examination of the patient past forty, Am. J. Optom. **27**:226-233, 1950.

21

Aphakia

The history of cataract (from the Greek word meaning "waterfall") extends back at least 3000 years to the Code of Hammurabi. Although extraction remains the only effective treatment, the modern ophthalmic surgeon, blessed by better techniques, sharper instruments, delicate sutures, and improved safeguards, achieves ever more successful results. He is also likely to operate earlier, on patients accustomed to medical miracles, who expect them with minimal effort promptly. "No Hungarian peasant," writes Linksz, "planned to drive a car after his cataract had been removed. No Slovak woodcutter had to mount a bus at the curb of a busy street corner. They had no television, and often did not know how to read."* Most of our patients, unfortunately, do not posses le charme slav, that delightful mixture of nonchalance, imperturbability, and indifference. Instead of provincial placidity, we get a suburban bedlam of complaints. Our patients not only want to see, but they want to see as they did before; they resent reaching for objects that are not there, the occasional diplopia, the black and blue shins, and the insecurity of restricted side vision; they are not happy to be at the mercy of the least frame misalignment or lens displacement; they are disturbed by spatial convolutions, distortions, and aberrations; they often reject an unattractive frame or the most desirable lens if it looks like a poached egg. "Aphakia" (the term was coined by Donders) thus hides an overflow of symptoms, a veritable river of difficulties. The surgeon who tingles to extract must sooner or later face this deluge and refract, and the result is not always a gurgle of gratitude. Mixing Gallic and gallows humor, Landolt wrote: "My satisfaction is always mingled with great apprehension when I see come back to me a fine black pupil which I have delivered of its thick veil; the cataract removed from the eye, too often is changed into a torrent of complaints against the operator. And we much oftener see a look of gratitude in the empty orbit of him who has suffered an enucleation than in the eye restored with sight. Enucleation simplifies so many things. Aphakic vision is so complicated."

On the positive side, today's cataract patient has a better chance for good vision because such complications as capsular remnants, vitreous loss, flat chamber, and wound dehiscenses are less frequent. The cause of his symptoms are better understood and more readily alleviated by lighter, better designed lenses, carefully fitted and attractively adjusted. Many difficulties can be eliminated by contact lenses, which can be worn comfortably most of the waking day, often with some measure of binocular vision. Intraocular lenses and other surgical refinements are a future promise of even better optical results; so the patient may indeed walk, in Kirby's phrase, "out of a thick fog into bright and glorious sunshine."

*From Linksz, A.: Optical complications of aphakia, Int. Ophthal. Clin. 5:271-308, 1965.

SIGNS AND SYMPTOMS

Although we learn from experience that we seldom learn from experience, aphakic symptoms are so dramatic and vivid that they cannot be ignored. A keenly sensitive description of the visual experiences of the new aphake was written by A. C. Woods, based on his own reaction to bilateral cataract extraction. It appeared as an anonymous editorial in the *American Journal of Ophthalmology** and should be read by every cataract surgeon.

When the new aphake first receives his correcting spectacles, he finds even his most familiar environment enlarged, distorted, blurred, and doubled. These symptoms are exaggerated when he attempts to move about, leading to overturned chairs, spilled coffee, yelping pets, and other domestic tragedies. Several weeks may be required to accustom himself to this new vision, to grasp objects he reaches for, to manipulate curbs and stairs, to make allowances for the insidious appearance and disappearance of objects in his peripheral field, to accept the generally sinuous movements of the environment, in short to regain his confidence and security through the new lenses. Some limitations, such as the reduced visual field and the loss of binocular vision, cannot be overcome whatever the amount of practice.

Symptomatic difficulties vary from one patient to another, depending on age, prior visual loss, seeing requirements, and personality. Even foreknowledge cannot forestall them, and they are especially abstruse when we try to explain them to a patient, preoccupied, perturbed, and apprehensive about the coming operative ordeal. Such considerations necessarily influence the decision to undertake surgery. "Instances of this kind," concluded Landolt, "are abundant, and I am sure of finding an echo in the hearts of my colleagues when I say that the instructions of aphakic persons often present greater difficulties than the cataract operation itself."

*__35:__118-122, 1952.

PATHOLOGIC PHYSIOLOGY

The absence of the lens, whether surgically removed, congenitally absorbed, or traumatically dislocated, has optic, physiologic, and psychologic consequences. In surgical aphakia the changes are sudden and unexpected, persistent and magnified. During the immediate postoperative period, there is tearing, discomfort, and photophobia. The cornea is edematous, distorted, and smeared with medication. The pupil is dilated, the vitreous hazy, and the eye irritable from traumatic iritis. The patient, preoccupied by the itching and scratching in his eye, the residual facial pain of local anesthesia, the torment in his back, and the chafing of his bladder, is not yet concerned with optical aberrations, gratefully accepting the vision through the +10 D lens the surgeon holds up between dressing changes. These reactions are sufficiently resolved by the end of the third to fifth postoperative day to allow him to return home wearing a temporary correction, such as a +10 D button in an aluminum or plastic shield. If the operated eye is the only seeing eye and especially if the patient is older, some form of vision helps prevent the occasional disorientation, which may have disastrous consequences. One sad experience of a patient getting out of bed, removing his patch, and "washing his eye out" is enough to put a damper on any inclination to bilateral surgery at the same hospitalization and to mobilize the patient's family for intermittent guard duty.

The corneal incision induces an irregular against-the-rule astigmatism. The amount varies with the care with which the wound is closed, whether the sutures are interrupted, symmetrically placed, and at the correct depth. Strangely, not preplacing sutures has made little difference in my experience. As the healing progresses, the astigmatism providentially becomes regular, and the amount decreases, reaching a minimum in 1 to 3 months, one of the criteria in ordering the final lens prescription. Improper wound closure may result not only in a filtering bleb but also in an exaggerated astigmatism and eccentric corneal pole, which will make fitting a subsequent contact lens difficult if not impossible.

If iridectomy be the signature of the cataract surgeon, optical and physiologic considerations recommend a modest endorsement. One or two small peripheral iridectomies prevent pupillary block and preserve a round pupil, which means better depth of focus, less aberrations, and reduced glare. An eccentric pupil compounds the difficulties in fitting a contact lens, and an asymmetrical pupil makes it almost impossible to align a bifocal seg for the two eyes.

Lens removal results in high hyperopia, deepens the anterior chamber, displaces the iris and entrance pupil, and reduces angle kappa. More light now reaches the retina, especially of shorter wavelengths, increasing the vividness of colors previously blocked, and the patient will be grateful for a tinted lens to reduce glare. The cornea becomes the most important refracting component. Its power is 43 D; so the result is a retinal magnification of 58.64/43 = 1.36, or 36%. Since emmetropic corneal powers range from 39 to 47.6 D, the magnification may vary from 50% to 23% compared to the preoperative state and more so in refractive ametropia. Despite this magnification, the induced hyperopia is so large that uncorrected vision is seldom better than 20/200 unless the eye was previously highly myopic. A previously emmetropic eye will require a spectacle correction of about +10.50 D at a vertex distance of 15 mm (or 20 mm from the entrance pupil).

Calculation of the spectacle correction follows the optical principles outlined in Chapter 7. Since anatomic length is not altered from the presurgical 24 mm, the vergence (L′) required to produce a sharp retinal image is $L' = n'/l'$; 1.33/24 mm = +55.5 D. Since the ocular power (F) is now only 43 D, the required incident vergence (L) is $L' - F$; 55.5 − 43 = +12.50 D at the cornea. The spectacle lens (F_s) necessary to produce this vergence at a vertex distance (d = 15 mm) is

$$F_s = \frac{L}{1 - dL}; \quad \frac{+12.50}{1 - 0.015\,(+12.50)} = +10.50 \text{ D}$$

The hyperopia of surgical aphakia is refractive. Hence the relative spectacle magnification (RSM) is the same as the spectacle magnification (SM), that is, $\frac{1}{1 - dF}$, where d is the distance from the entrance pupil and F the spectacle power. The size of the retinal image in the corrected aphakic eye compared to the standard emmetropic eye thus depends only on the power of the correcting lens and its distance from the entrance pupil. Any plus lens therefore inevitably produces some magnification (unless placed inside the eye). A +10.50 D spectacle lens 15 mm from the cornea is about 20 mm from the entrance pupil, giving a magnification of $\frac{1}{1 - 0.02\,(+10.50)} = 1.25$ or 25%.* Even a contact lens is 5 mm from the entrance pupil and results in $\frac{1}{1 - 0.005\,(+12.50)} = 1.06$ or 6% magnification. This is therefore the residual aniseikonia compared to the phakic eye when the phakic eye is assumed to be emmetropic.

There is nothing in these computations that guarantees that the phakic eye *is* emmetropic or, if emmetropic, of "standard" power. Moreover an aphakic eye may also have had some presurgical axial ametropia. For example, if the aphakic eye was previously axially myopic, the RSM for axial ametropia is unity if the correcting lens is placed at its anterior focal point; outside this point (e.g., −16.66 mm) there is minification, and inside there is magnification. If an aphakic eye receives a contact lens correction, the magnification of refractive hyperopia is reduced, but the magnification of axial myopia is increased—two opposite effects. The ideal situation is one in which the aphakic eye is axially hyperopic and the phakic eye is axially hyperopic, with the first corrected by a contact lens and the second by a spectacle lens. Such variations help explain why some monocular aphakes can obtain binocular vision and others cannot, since the aniseikonia can easily range from zero to 15%. Shape factor magnification adds to these effects and depends on front surface power and thickness of the spectacle (or contact) lens.

*Another way to look at this, following Ogle, is to pretend a "thin" minus lens has been placed at the optical center of the crystalline lens, which exactly neutralizes its power. The resultant hyperopia must now be corrected by a plus spectacle lens at some distance (e.g., 20 mm) from it. The hypothetical minus lens and the correcting plus spectacle lens now constitute a Galilean telescope whose afocal magnification is $\frac{1}{1 - dF_1}$ where F_1 is the spectacle lens and d the distance between lenses; hence $\frac{1}{1 - 0.2\,(+10.50)} = 1.25$ or 25%, the same result previously obtained.

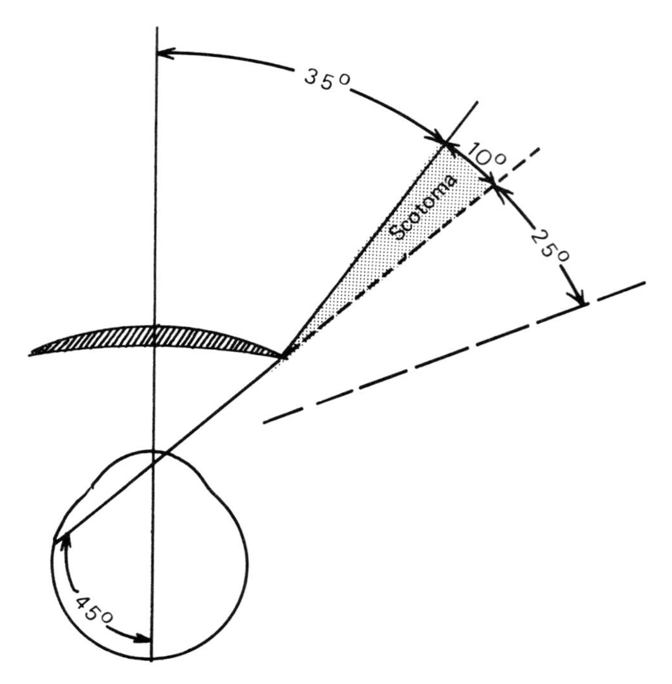

Fig. 21-1. Angular scotoma in aphakia. Only one side is shown, illustrating that a 35° visible field covers 45° of retinal topography and leaves a 10° "scotoma." The scotoma is actually circumferential, and its extent and width varies with the magnification, lens diameter, vertex distance, and direction of view.

Spectacle magnification is a primary source of difficulty for the newly corrected aphake. It causes diplopia, field constriction, scotomas, spatial disorientation, visuotactual incoordination, and aniseikonia.

When the patient looks at the world through his aphakic lens, it is 26% larger than before. When he looks at an object in front of him, he sees it not larger but nearer and reaches short. More important, and more dangerous, a step or curb is further than he thinks, and it is necessary to teach him that in reaching or walking, he must look at his hand or foot simultaneously with the object or curb. Only by relating body and target can he learn the new coordination required for secure mobility.

Magnification means a given retinal area is covered by a smaller visual field. The usual monocular field of about 150° is reduced to $\frac{150}{1.26} = 119°$, yet covers the same retinal topography. (Additional contractions result from peripheral lens aberrations and must be considered in interpreting perimetric fields.) Fig. 21-1 illustrates an aphakic eye corrected by a 40 mm diameter spectacle lens 20 mm from the entrance pupil, covering about 90° of visual field (45° on either side of the fixation point). Corresponding to the nasal 45° retinal area, there is a temporal 45/1.25 = 35° visual field, leaving an angular blind area of 10° in which no object is visible, since it cannot stimulate the retina. As the blind area extends circumferentially, it is called a "ring scotoma," a scotoma not in the pathologic sense of retinal disease but as an optical consequence of spectacle magnification.

Fig. 21-1 illustrates further that there is an area of about 25° *outside* the lens where objects stimulate the retina; that is, the rays may enter the pupil of the eye in the primary position because the normal temporal field of about 70° is not changed by surgery. Acuity for such peripheral objects is very poor and limited to the detection of motion that is just enough, however, for people to seem to appear and disappear as they enter and emerge from the patient's blind area. This

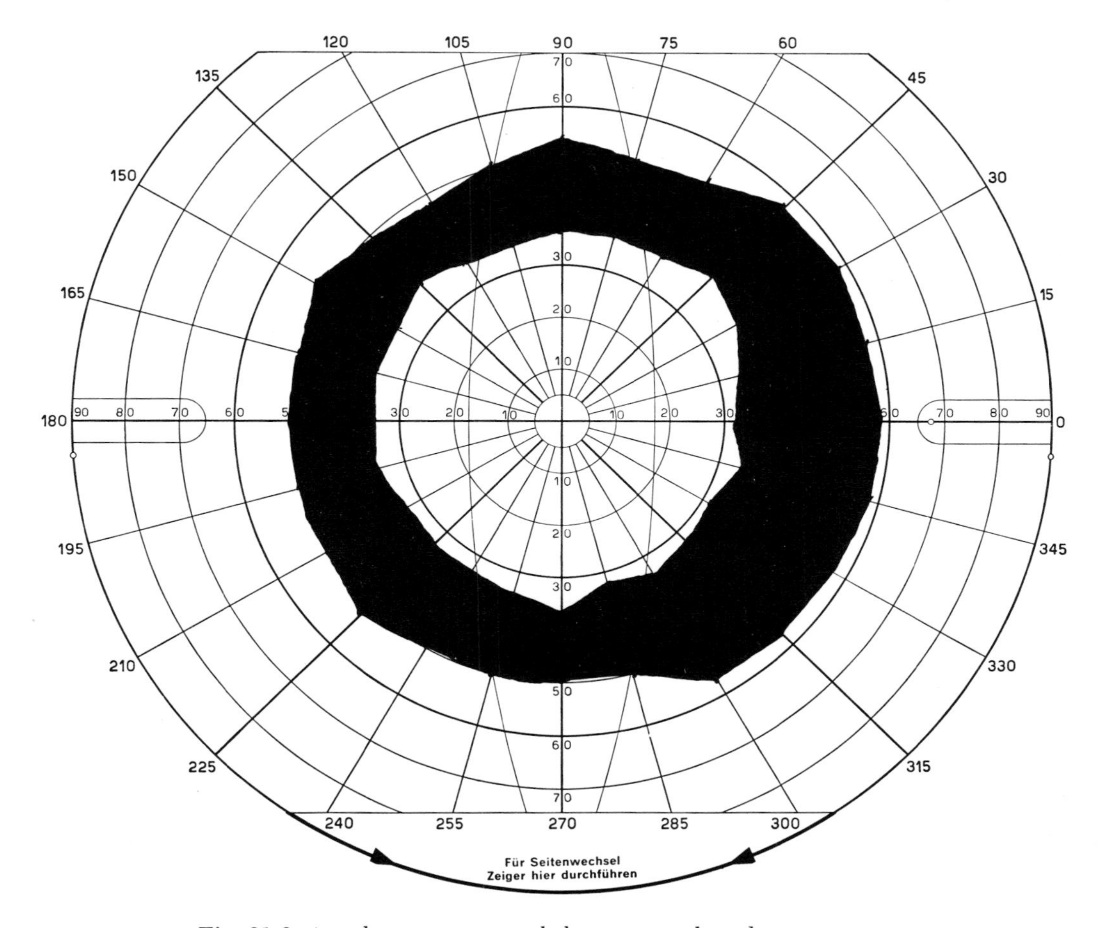

Fig. 21-2. Angular scotoma in aphakic patient plotted on perimeter.

scotoma may be plotted on the perimeter, using large targets (Fig. 21-2).

The extent of the ring scotoma varies with power, diameter, and distance of the spectacle lens and the size of the pupil. Objects within it disappear. When the spectacle-corrected aphake walks into a familiar room, objects at known positions can be avoided by moving the head for exploratory observations. It is when objects are not stationary, as in a crowd of moving people, that they pop in and out of the visual field with the annoying insolence of a jack-in-the-box, resulting in frequent collisions and a string of apologies. Patients are concerned over stumbling and stepping on pets or grandchildren. This insecurity and uncertainty of the peripheral field is a common complaint especially in the intermediate (10 foot) range; in reading the scotoma is eccentric enough so that problems are avoided, and in driving the central visible field extends laterally enough to include the needed side vision.

When the corrected aphakic patient turns his eye, it pivots about the center of rotation, which lies further back than the entrance pupil, so that the macula encounters the lens periphery at a new angle and the ring scotoma moves toward the lens axis. This "roving" ring scotoma and the restricting effect of the lens aperture contributes to objects appearing and disappearing. (It can be duplicated by closing one eye and cupping the thumb and index finger around the open eye. Looking straight ahead, place a small target just within the temporal field and note that as the eye turns to look directly at it, the object disappears behind the finger.) The in-and-out appearance of objects in the field of vision adds to and exaggerates the jack-in-the box effect.

The periphery of strong lenses cause distortions and aberrations in proportion to power, thickness, and profile. Corrected curve lenses and aspherics improve image quality but never completely eliminate distortions and aberrations. The change can be demonstrated by comparing central acuity with that obtained 30° from the optical center, for example. Contact lenses yield better peripheral acuity because they move with the eye. With spectacles, the patient must move his head to continue using the lens center.

One benefit of magnification is improved acuity so that a patient whose real acuity is only 20/25 is able to read 20/20. This may enable him to read large print, such as newspaper headlines, without a reading add and is the reason a weaker add can be ordered for aphakes than comparable phakic presbyopes. Of course, it will not work if resolution is reduced by macular changes or irregular astigmatism, where the reading add may need to be increased to 3 or 4 D or even replaced by a magnifying loupe.

Since aphakic corrections involve strong optics, a small lens displacement gives a large change in effectivity. Recall that the approximate formula for effectivity is $F^2/1000$ per millimeter displacement, which means a +10 D lens changes by $10^2/1000$ or 0.1 D per millimeter displacement. The average aphakic patient can achieve a reading add of 2 D merely by pushing his spectacles 2 cm further down his nose. This permits the use of single vision aphakic corrections in the immediate postoperative period and still allows enough near vision to eat, read a news headline, or even a price tag. The safety features of two pairs of spectacles for an older aphake are not to be ignored. The reading spectacle may be made up as a bifocal, which avoids the seg line when walking or driving. To make optimum use of the displacement effect of single vision aphakic lenses, the spectacle frame should be fitted as near to the eye as possible, which also reduces magnification and increases the field of view. Maximum plus or even overcorrecting (+0.50 to +1.00 D) of distance prescription results in negligible acuity loss, especially when far vision seldom extends beyond the confines of a room, and yet gives some intermediate vision and permits reading large print—all with the same lens.

Diplopia is the reason a monocular aphake cannot use both eyes. The anisometropia is so large that the prismatic effect even within a few millimeters of the optic center cannot be compensated. Monocular aphakes do not suffer from aniseikonia as is frequently stated, since the size difference precludes binocular vision in the first place. As a matter of course, one corrects either the phakic or aphakic eye and prescribes the spherical equivalent for the other. If the phakic eye has poor vision but is dominant, its blurred image through a balancing lens may still intrude on the corrected aphakic eye. Conversely, if the aphakic eye is dominant and the phakic vision good, there is more resistance to accepting a temporary balancing aphakic lens. In deciding which eye to operate on first in bilateral cataracts, the dominant eye (all other factors being equal) should be given preference.

How binocular vision is achieved in unilateral, contact lens–corrected aphakia is not clear, since unless there is a very fortuitous presurgical ametropia, a residual aniseikonia from 5% to 11% remains. Yet many of these unilaterally operated patients are subjectively satisfied, depending, of course, on one's criterion of satisfaction or, for that matter, the criterion of binocular vision. Panum's area easily allows central superimposition with accurate fixation, and within this area one can "fuse" Worth dots and obtain some stereoacuity even with large aniseikonia. Outside central fixation, disparity rapidly exceeds Panum's areas, although they too increase, but the puzzle is why the peripheral diplopia does not induce a central diplopia. Several additional factors probably play a role; for example, most aphakic patients without binocular vision for some time learn to get about by more efficient use of the monocular cues of the seeing eye, and their ability to suppress stands them in good stead after surgery. Moreover the acuity may be imperfect or unequal, and disparity discrimination drops rapidly with decreased resolution.

As if diplopia, distortion, reduced side vision, ring scotomas, and the altered visuotactual coordination were not enough, the new aphake is now given a lens whose cyclopean apparition must be discouraging if not dismaying. The wish to avoid infirmity is overshadowed by the need not to look abnormal, a psychologic fact often glossed over in an older patient but undoubtedly a major factor in contact or spectacle lens acceptance.

APHAKIC REFRACTION

Aphakic refraction is the final stage of cataract surgery, deserving the same meticulous attention to detail as the extraction proper. The patient judges the issue not by the symmetry of the iridectomy, the minuteness of the astigmatism, or the dexterity of the incision but by the visual result. After a poor refraction, the patient can plainly see that he cannot see.

Objective methods

The natural curiosity of the surgeon prompts a preliminary opthalmoscopy and vision check as soon as the initial ocular irritability and photophobia subside. As soon as a visual improvement is noted with a +10 D lens, the first refraction is attempted.

Corneal edema and vitreous haze generally do not permit a reliable retinoscopy unless the pupil is dilated (e.g., with 10% phenylephrine). A refractor expedites this preliminary examination and serves as a baseline for later, more accurate measurements. The plus cylinder technique is used, since with motion is easier to evaluate and is repeated at weekly intervals, refined by subjective techniques. Objective methods are always confirmed by subjective tests, since retinal or optic nerve disease may lower vision despite clear media and perfect optics.

Keratometry is of limited value not only because the instrument gives no clue as to the spherical error, since the cylinder is measured at the corneal instead of the spectacle plane, but also no account is taken of a presurgical axial ametropia. It does help to indicate the progress of the corneal healing, irregular astigmatism, and the centration of the corneal pole relative to the pupil, useful in deciding whether contact lenses are feasible.

Subjective methods

When vision reaches 20/60 or better, standard methods of subjective refraction (astigmatic dial, cross cylinder) are applicable. The refractor is now replaced by a trial frame or trial clip and lenses, since these permit more precise positioning and angulation. Care is taken in adjusting the device to the exact interpupillary distance, to keep the head level, and to prevent excessive lens tilt. The sphere is placed in the rear trial cell, with the cylinder in front using as few lenses as possible to reach the final prescription. The trial frame measures back vertex power only when modern trial lenses are used and properly placed and is then automatically compensated for the air space.

In evaluating vertex distance, do not rely on the zero marker of the trial frame (set for 13.75 mm) because the final frame is often fitted closer. Measure vertex distance with an appropriate instrument or with a stenopeic slit placed in the rear cell and a metal millimeter ruler to the closed lid, allowing 1 mm for lid thickness. Many trial frames do not have an indicator of the pantoscopic tilt (which is probably more important than vertex distance), so this must be estimated. Although exact duplication of the trial prescription may not be possible, modern trial lenses are less likely to lead to error. Since lens shape (or profile) alters back vertex power, the final power is best estimated by placing the total combination in the lensometer. Special cataract trial lenses offer no advantage if more than one lens is required.

Improper frame angulation induces a cylinder whose axis is the axis of rotation. Thus a +10 D sphere rotated 20° about its horizontal axis (pantoscopic tilt) causes 1.25 D plus cylinder axis 180°, a not insignificant amount. This new cylinder adds by oblique combination to any existing astigmatism, producing changes in axis as well as power. The proper lens inclination is perpendicular to the line of sight, with an imaginary line passing through the external canthus. If the final lens is to be a bifocal, the angulation is adjusted for the height of the segment and depends on style and design. An allowance of 0.5 mm downward decentration for every degree of additional pantoscopic tilt is recommended.

An alternative to the trial frame is the so-called provisional lens plan. Because vertex distance and lens angulation is important, the final frame is selected early and properly adjusted and all refractions then made without changing the fit. The Halberg trial cell can be conveniently clipped to a temporary lens, and since overrefraction changes will generally be small, the new combination is read off from the lensometer and the lens modified. Several manufacturers make this service available at nominal cost for a reasonable period until the final power is determined.

Any standard subjective technique can be used for aphakic refraction, but since these patients are older, the magnification strange, the spatial orientation new, and the responses slow, patience and encouragement are required, and it is best to set aside more time for these examinations. The effort to see is often tiring, and several trials may be needed to arrive at a consistent response. The chart should be well illuminated, the initial lens changes large, and the instructions clear and frequently repeated as the reduced acuity blunts discrimination. Letters, astigmatic dials, and cross cylinder powers should be commensurate with the reduced vision. It is conventional to repeat the refraction at weekly intervals until the consistency of the findings indicates a stabilized refraction, which may take weeks or months.

In binocular aphakes or in monocular aphakes fitted with a contact lens in anticipation of binocular vision with a phakic well-seeing eye, the motility requires attention so as to avoid diplopia. A residual superior rectus palsy due to the bridle suture generally disappears after several days. Motility balance is determined at distance and near with a Maddox rod, Lancaster red-green, von Graefe dissociating prism, or the Freeman or Bernell lantern. Since accommodation is absent in bilateral aphakia and almost always absent even in the phakic eye due to presbyopia, an exophoria or exotropia is common, resulting in an excessive convergence demand. Because the correcting lenses are high plus, any outward decentration of the optical centers will compound the problem, and it is useful to place the distance optical centers toward the near PD, thus providing some base-in effect. Additional decentration may be needed for near. Round segs here provide greater flexibility than flat-top segs. About 4 mm decentration is often required for fusion in binocular aphakia, and some decentration should be provided even in monocular cases so as not to interfere with habitual convergence for reading. Optical performance through bifocal segs may not be optimal because the segs are not aspheric. If inward decentration cannot be achieved for satisfactory binocular vision, a separate pair of reading glasses is a good alternative. The same considerations apply to vertical imbalances, which fortunately tend to disappear spontaneously without prism compensation.

MANAGEMENT OF APHAKIA

The management of aphakia begins before the operation (as in deciding whether the vision will be monocular or binocular); during the operation (as in selecting the corneal incision, number of sutures, and type of iridectomy); after the operation, (with an accurate refraction and choice of the most appropriate lens); and after the refraction (in teaching the patient to use his lens and his new vision most efficiently).

Monocular aphakia

The patient whose aphakic eye is the only one that sees is probably the simplest to care for. He most readily adapts to his new vision, if for no other reason than that there is no alternative, one advantage in delaying the operation until the acuity in the better eye drops to 20/60 or less. In younger patients earning a livelihood or alert older patients leading a life of active retirement, delay is not always possible or advisable, and such decisions must be made individually.

If the operated eye is the only seeing eye, early temporary lenses not only permit some reassuring vision useful in getting about in familiar surroundings, but the patient compares each subsequent lens change to the previous one and not to the precataractous vision. By making less precipitous changes, it gives more time and less trouble in adapting to the new magnification and spatial orientation. Depending on age and the available help at home, temporary lenses may be used for sedentary vision and limited mobile vision. Particular caution should be exercised in manipulating steps and curbs. Temporary single vision lenses rather than bifocals will significantly reduce the incidence of broken bones and torn ligaments.

An assortment of temporary lenses is available from various manufacturers. A plastic, aspheric, or full-field lenticular, preferably tinted lens in a sturdy frame is an obvious choice.* How closely it should resemble the final lens (full field or lenticular, single vision or bifocal, type of bifocal seg) depends on the vision, the stability of the refraction, the economic circumstances, or the adoption of a provisional lens plan. The greater the difference between temporary and final lens design the more likely the necessity to go through two adjustment periods.

In deciding on the timing of the final prescription, the vision and dominance of the unoperated eye are considered. If the vision is good enough to get about, one may encourage the use of this eye, reserving the aphakic correction for sedentary vision (movies, television, and reading). Frequently, the patient will help by indicating his own preference.

*For example, Ormaphax, Kramer-Tisher unit, Uniphakic, Unikat, Fresnel film over old glasses.

Unilateral aphakia and binocular vision

When the phakic eye is a well-seeing eye, an attempt may be made to restore binocular vision by correcting the aphakic eye with a contact lens. It is assumed, of course, that the patient had binocular vision before he developed his cataract and that it has not interfered with vision for a long time. The desirability of an attempt to achieve binocularity may be gauged by the patient's adjustment to monocular vision prior to surgery, the nature of his occupation or avocation, his motivation, and his age. Also to be considered is the presence of opacities in the phakic eye and the estimated rapidity of subsequent cataract development.

There is no clinical method to predict residual aniseikonia, and some patients make a satisfactory binocular adjustment to unilateral contact lenses even though they never really use both eyes together. The matter can only be decided by trial and error, especially if the occupation requires either wide side vision or stereopsis at some fixed distance. These patients frequently tolerate contact lenses well because their age reduces overall corneal sensitivity, and their surgery reduces it locally in the area of the incision.

Binocular aphakia

If the patient has had bilateral cataract extractions, he may be fitted with bilateral contact lenses or spectacles. Responses do not always follow optical theory, and some patients prefer bilateral spectacle lenses to bilateral contacts. Bilateral aphakia requires particular attention to centration and motility effects. The low convergence of these patients should not be further taxed by induced base-out prismatic effects. A certain time must also be allowed to relearn the use of both eyes. Because there is usually some residual anisometropia, a separate pair of reading glasses may be indicated.

Aphakia in children

The chief problem of unilateral cataracts in children is the possibility or probability of an associated amblyopia, depending on the cause and timing of the cataract and the degree of vision in the other eye. Whether a unilateral congenital cataract should be removed is a surgical problem we need not consider here. The refractive problem comes up not infrequently, however, in cases of traumatic cataracts in which the lens is either removed or has absorbed. In those cases in which good retinal function is maintained, every effort is made to restore binocular vision. Contact lens fitting may not be simple because of the associate corneal scar and pupil irregularities. But whether the aphakic eye is corrected or not, as in all cases of monocular vision the patient should be given safety glasses for constant wear.

The aphakic prescription

Intent and achievement are not necessarily twins; there is always a certain element of uncertainty. There will be fewer surprises if close cooperation with the optician is maintained. It is naive to hand the aphakic patient a prescription and assume variables such as vertex distance, pantoscopic tilt, lens design and construction, seg size and style, and frame size and shape will automatically fall into place. Most modern opticians are well-trained master craftsmen, a far cry from the "lens grinder wearing a clean coat," and are equally interested in obtaining a good visual result. But they cannot do it unless they know what you want, which means you must know what you want.

Spectacle fitting after cataract extraction requires much more dispensing skill than routine work and the highest cooperation between patient, doctor, and optician. Fitting the aphake is difficult; fitting without the necessary data is impossible. No elaborate ray tracing or sophisticated computation will guarantee a comfortable, cosmetically appealing frame and lens design. And unless the spectacles are properly fitted, the surgeon's skill is poorly expressed.

To translate the prescription into the optimal fit, consideration is given to optical, mechanical and cosmetic factors. The reduced weight, aspheric curve, and safety of plastic lenses explains their popularity. An aspheric surface is progressively shallower further from the center, computed zone by zone, so that most of the unwanted oblique astigmatism is eliminated. Not all aspheric lenses are identical; the design varies with power and base curve. A planoconvex lens does not permit the best aspheric design. Shallow base curves on the order of −3 D are commonly used because these lenses have minimal front curve (thus reducing shape magnification) and are less bulbous in appearance. The Volk "Catraconoid" is the only aspheric glass design and the only one aspheric over the bifocal seg.

All high plus lenses tend to produce a "bull's eye" appearance by magnifying the wearer's eye, and their weight in glass is heavy. Plastic lenses solve the weight problem but increase thickness, hence the advantage of reducing lens diameter by choosing the smallest compatible frame. Lenses with reduced aperture are termed "lenticular"; the smaller central "aperture" (usually 40 mm diameter) provides the power; the surrounding "margin" acts as a planocarrier for the effective aperture. Both full-field and lenticular lenses are available in glass or plastic, with or without aspheric curves (Fig. 21-3). (To check on aspheric curve, use a lens clock and note power drop variation in the center and periphery in the same meridian.) Aspheric bifocals can now be ordered with either flat-top or round segs in glass or plastic. Aspheric or full-field lenticular plastics offer obvious advantages, and although the price is higher, it seems inconceivable that the surgeon would recommend anything less than a lens that best reflects his operative dexterity. The lenticular aperture limits the field of view but keeps distortion to a minimum. They should not be used in patients with wide PDs or with large frame sizes because of the poached egg appearance. If the frame size can be kept small and not of unusual shape, the lenticular lens is the one of choice. For full-field lenses, specify the minimum effective diameter (MED) in which the optical center is decentered to provide equal thickness at the top and bottom lens edges.

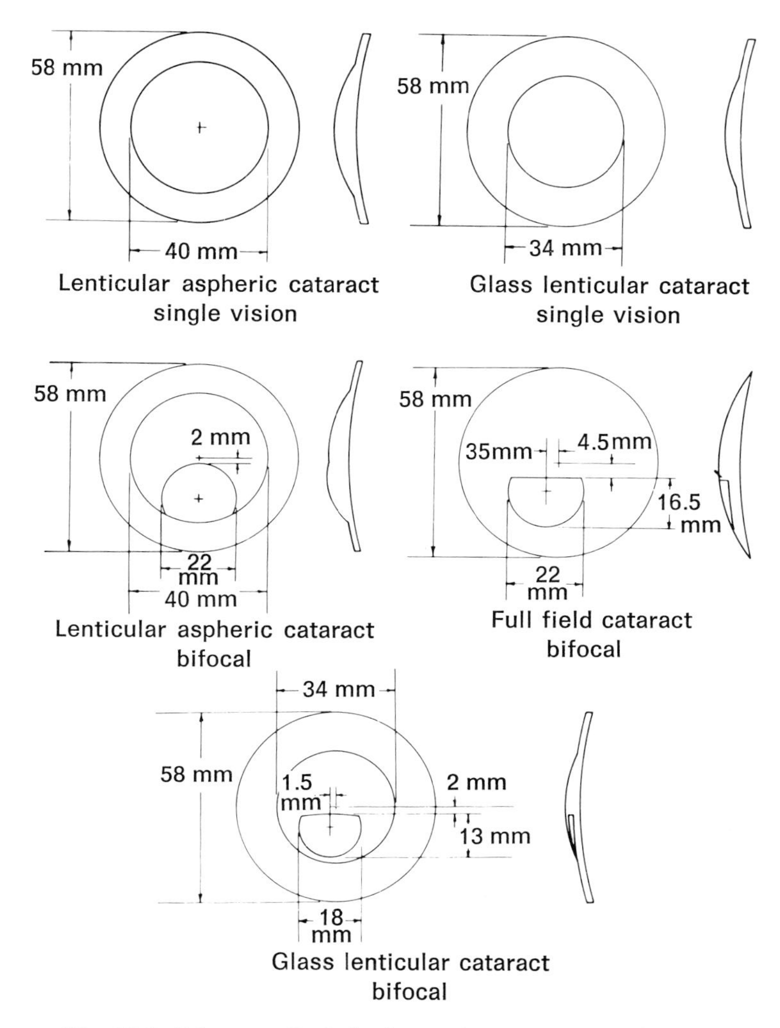

Fig. 21-3. Selection of aphakic lenses (American Optical Co.).

Above and beyond the lens design, the frame should be kept a minimal distance from the eye and the seg not too far below. When refracting with a trial frame, the vertex distance is often greater than the final fit, and the power winds up too weak (another reason for overcorrecting the aphake). Another frequent annoyance is a bifocal seg that is too low, partly the result of a poor choice in temple, which permits the frame to slip down the nose. The choice of reading seg varies with the type previously worn, the type worn over the seeing unoperated eye, or the adjustment the patient made to a temporary cataract bifocal. As for presbyopes, most aphakes prefer a flat-top seg, which allows the fixation axis to remain near the optical center and minimizes peripheral distortion and aberation. A variety of seg styles can be tried by pasting different shaped cutout Fresnel adds over the single vision temporary cataract lenses. Care should be taken in maintaining the same frame tilt throughout, since this is a major factor in introducing unwanted cylinder. A color coating (e.g., Cruxite) reduces glare, a nuisance especially with full iridectomies. The temporary cataract lenses should also preferably be dark (or a sunglass clip-on provided) until the initial photophobia disappears. Plastic lenses are dyed, so the color is evenly distributed. An antiglare coating is not required with plastic lenses.

In choosing a balancing lens the vision of the phakic eye is considered. If one anticipates operating this eye, the spherical aphakic equivalent will serve as a temporary lens and permit early binocular adjustment. If the aphakic eye is to wear the balancing lens, the prescription depends on ocular dominance and the possibility of mobilizing some peripheral vision. A separate aphakic correction can be ordered for occasional use.

A sturdy frame that does not readily get out of adjustment, with large nose pads to distribute weight and permit periodic minor adjustments, is strongly advisable. The clinician should make it a point to recommend a small frame, since patients are more likely to follow his advice, thus rendering the optician a real service. Not only do large frames increase weight, peripheral distortion, and adjustment problems, but lenticular lenses look unappealing. The importance of appearance should not be underestimated, and by proper choice of lens, frame, and fit, the best compromise between optics, mechanics, and cosmetics can be achieved. All requirements may not be met with the first aphakic correction, and the patient should be aware that periodic changes and adjustments will be necessary. In switching frames on a subsequent examination, especially in bilateral aphakia, the same vertex distance and tilt must be maintained, and replacement lenses should duplicate as far as feasible the previous base curves, color, and seg style.

Intraocular lenses

"For most of us," says Linksz, "the use of intraocular lenses are pioneering feats worthy of respectful admiration, if not emulation." One cannot help agreeing with this sentiment; the best place for a foreign body is outside the eye. The technique is inspiring when one considers how easy it would be to mess it up. But how calamitous the complicating consequences. Like most surgeons, I become more conservative with age and prefer unhindered sleep to the unhindered potential of binocular vision. Perhaps this is confusing fatigue with fatality; yet one would like to see more footage on the dangers than footnotes on the success of the procedure (dislocation, vitreous luxation, pressure on the chamber angle, secondary glaucoma, corneal edema, uveitis, fixed pupil, as well as interference with treatment of any subsequent uveitis or retinal detachment). Nor does an intraocular lens solve all the optical problems automatically, since the iseikonic design must be individualized, not an easy task without measuring axial length, corneal curvature, and refractive index.

Aphakic rehabilitation

With tact, encouragement, and specific instructions, a great deal may be done to help the aphakic patient adjust to his new vision. When teaching hand-eye, and foot-eye coordination, head movements rather than eye movements, the early use of temporary lenses, the trick of displacing single vision lenses for effectivity in reading, a walking stick, appropriate side vision precautions, emphasize that you are aware of the difficulties but that they are not unique to his surgical result. Sometimes a local university's department of physical medicine can be persuaded to undertake these rehabilitation measures. Where acuity is not optimum, a complete subnormal vision examination may suggest an aid that permits the best potential vision.

Periodic repeat refractions and lens adjustments should be scheduled. Although no major refractive changes are anticipated once the healing is completed, the exact timing cannot be predicted, and stronger or weaker lenses are often suggested by new visual habits.

REFERENCES

Benton, C., and Welsh, R. C.: Spectacles for aphakia, Springfield, Ill., 1964, Charles C Thomas, Publisher.

Binkhorst, R. D., Weinstein, G. W., and Troutman, R. C.: A weightless iseikonic intraocular lens, Am. J. Ophthal. **58**:73-78, 1964.

Cowan, A.: Aphakia, Am. J. Ophthal. **32**:419-424, 1949.

Cowan, A.: Monocular aphakia, Arch. Ophthal. **49**: 473-474, 1953.

Elenius, V., and Sopanen, V.: Power of the correcting lens of the aphakic eye as calculated from the keratometric measurement of the corneal radius and the ultrasonically measured axial length of the eye, Acta Ophthal. **41**:71-74, 1963.

Emsley, H. H.: The correction of unilateral aphakia with contact lenses, Optician **95**:283-284, 1938.

Enoch, J. M.: A spectacle–contact lens combination used as a reverse galilean telescope in unilateral aphakia, Am. J. Optom. **45**:231-240, 1968.

Floyd, G.: Changes in corneal curvature following cataract extraction, Am. J. Ophthal. **34**:1525-1533, 1951.

Gerard, J. P.: Stereoscopic vision and its measurement in monocular and bilateral aphakia corrected by contact lenses, Bull. Soc. Ophthal. Fr. **5**:347-355, 1961.

Gettes, B. C., and Ravdin, E. M.: Monocular aphakia and exotropia corrected by contact lenses, Am. J. Ophthal. **32**:850-851, 1949.

Goar, E. L.: The management of monocular cataracts, Arch. Ophthal. **54**:73-76, 1955.

Gordonson, L. C.: Postoperative cataract lenses, Arch. Ophthal. **84**:62, 1970.

Hirtenstein, A.: Contact lens in unilateral aphakia, Br. J. Ophthal. **34**:668-674, 1950.

Iliff, C. E., and Khodadoust, A.: Control of astigmatism in cataract surgery, Am. J. Ophthal. **65**:378-382, 1968.

Jaffe, N. S., and Light, D. S.: Vitreous changes produced by cataract surgery, Arch. Ophthal. **76**:541-553, 1966.

Kaplan, M. M.: Optical considerations of unilateral aphakia, Am. J. Optom. **46**:319-333, 1969.

Kirby, D. B.: Surgery of cataract, Philadelphia, 1950, J. B. Lippincott Co.

Kirsch, R. E.: Aphakic refraction, Int. Ophthal. Clin. **5**:403-411, 1965.

Liddy, B. S. L., Carr, K., and McCulloch, C.: Aphakic rehabilitation, Am. J. Ophthal. **63**:1793-1795, 1967.

Linksz, A.: Optical complications of aphakia, Int. Ophthal. Clin. **5**:271-308, 1965.

Lubkin, V., Linksz, A., and Chamby, G.: Stereopsis with spectacles in monocular aphakia, Am. J. Ophthal. **67**:547-553, 1969.

Lubkin, V., Stollerman, H., and Linksz, A.: Stereopsis in monocular aphakia with spectacle correction, Am. J. Ophthal. **61**:273-276, 1966.

Mann, W. A.: Optical correction of aphakia, Int. Ophthal. Clin. **1**:623-648, 1961.

Manschot, W. A.: Histopathology of eyes containing Binkhorst lenses, Am. J. Ophthal. **77**:865-871, 1974.

Michaels, D. D., and Zugsmith, G. S.: Optic neuropathy following cataract extraction, Ann. Ophthal. **5**:303-306, 1973.

Munchow, W.: History of intraocular correction of aphakia, Klin. Monatsbl. Augenheilkd. **145**:771-777, 1964.

Naylor, E. J.: Astigmatic difference in refractive errors, Br. J. Ophthal. **52**:422-425, 1968.

Ogle, K. N., Burian, H. M., and Bannon, R. E.: On the correction of unilateral aphakia with contact lenses, Arch. Ophthal. **59**:639-652, 1958.

Ridley, F.: Contact lenses in unilateral aphakia, Trans. Ophthal. Soc. U. K. **73**:373-386, 1953.

Ridley, H.: Intraocular acrylic lenses—past, present, and future, Trans. Ophthal. Soc. U. K. **84**:5-14, 1964.

Rosenbloom. A. A.: The correction of unilateral aphakia with corneal contact lenses, Am. J. Optom. **30**: 536-542, 1953.

Spaeth, E. B.: The treatment of monocular aphakia, Trans. Ophthal. Soc. U. K. **77**:517-532, 1957.

Spaeth, P., and O'Neill, P.: Functional results with contact lenses in unilateral congenital cataracts, high myopia and traumatic cataracts, Am. J. Ophthal. **49**:548-554, 1960.

Stokoe, N. L.: Binocular vision in bilateral aphakia, Proc. R. Soc. Med. **53**:192-195, 1960.

Town, A. E.: Contact glasses for correction of refractive errors in monocular aphakia: production of single binocular vision, Arch. Ophthal. **21**:1021-1926, 1939.

Troutman, R. C.: Microsurgical control of astigmatism in cataract and keratoplasty, Trans. Am. Acad. Ophthal. Otolaryngol. **77**:563-572, 1973.

Vail, D.: After results of vitreous loss, Int. Ophthal. Clin. **4**:789-813, 1964.

Welsh, R. C.: Postoperative—cataract spectacle lenses, Miami, 1961, Educational Press.

Welsh, R. C.: The roving ring scotoma, Am. J. Ophthal. **51**:1277-1281, 1961.

Welsh, R. C.: Aphakia: optical considerations, Am. J. Optom. **48**:852-859, 1971.

Welsh, R. C.: Remedies to inadvertent frequent-or-rare defects found in today's permanent spectacles for aphakia, Ann. Ophthal. **4**:432-435, 1972.

Welsh, R. C.: Prescribing single spectacles for distance in aphakia, Ann. Ophthal., pp. 1075-1076, Oct., 1973.

Welsh, R. C., Fall, E., and Ahorner, F.: Hand neutralization of aphakic—range corneal contact lenses, Arch. Ophthal. **73**:621-622, 1965.

Woods, A. C.: The adjustment to aphakia, Am. J. Ophthal. **55**:1268-1272, 1963.

Wyckoff, P.: Aphakia, Contact Lens J. **6**:12-15, 1972.

22

Contact lenses

The increasing number of people who wear contact lenses and the longer time they wear them comfortably while achieving good vision is a tribute to the advances in physiology and plastics technology. Originally intended only to protect the cornea or correct irregular refractive errors, contact lenses are now worn by any adventurous ametrope whose vision is capable of optical improvement. New lenses and new techniques afford not only cosmetic and optical advantages but also therapeutic, protective, and diagnostic benefits. Indeed the scope of the field is now so large and the advances so rapid that it is not feasible to cover more than general principles. But whether he fits contact lenses or delegates it, the clinician must be knowledgeable of the indications and contraindications, optical advantages and psychologic limitations, and anticipated changes and unexpected complications.

The conception of changing ocular curvature by a tube filled with water is described by Leonardo in his notebooks. Descartes theorized and Young proved that an eye immersed in water loses optical power.

The astronomer Herschel (1827), reading Airy's description of his own astigmatism correction, thought spectacles too indirect and worked out the optics more or less as it is known today. The first practical lens was a shell made by a German glass blower in 1887 to protect the only eye of a patient going blind from exposure keratitis after lid cancer. An attempt to correct vision with a contact lens was not made until the turn of the century by Fick (who coined the name). Fick made plaster molds of cadaver eyes, but his lens could only be worn 2 hours. The results led Abbe and Zeiss to devise blown and ground lenses for keratoconus. Blown glass lenses were more comfortable, and ground lenses gave better vision, but none achieved good tolerance. In 1929 Heine revived interest by using an afocal trial lens with a tear lens to correct refractive error. Later Dallos incorporated a cylinder for the correction of lenticular astigmatism. Feinbloom (1937) made the first partial plastic lens. A truly corneal contact lens was designed by Touhy in 1948, and the use of cross-linked hydrophilic polymers for contact lenses was first reported by Wichterle and Lim in 1960. Despite premature publicity and early gaseous claims, contact lenses are here to stay with electrifying potential not only for cosmetics but also for therapy and prevention of blindness.

EXAMINATION

Contact lenses may be indicated for cosmetic, optical, or therapeutic reasons, but between the supply of motivation and the demands of application, there is often a gulf of misconceptions. The most perfectly designed contact lens is useless to a patient who cannot tolerate it, insert it, gain useful vision with it, or maintain minimal hygienic standards and avoid complications. It is wise to balance apprehension with acceptability by a trial wear demonstration.

History

Motivation does not ensure successful contact lens wear, but its absence guarantees failure. The history helps identify the weeds of illusion in the garden of desire by balancing cosmetic enthusiasm with the vexation of spectacles. The interview should establish the intended use (optical, cosmetic, social, sports, therapeutic); the previous response to spectacles; the degree of resolution needed; the patient's mature willingness to make the adaptative effort and follow instructions; and his dexterity in handling insertion, removal, and hygienic precautions. Moreover motivation changes with time; if men make no passes at girls that wear glasses, marriage and progeny soon lead to contact lens apathy.

Past ocular history should investigate congenital disorders, recurrent herpes, uveitis, allergy, or blepharitis; rapid fluctuations in refractive error, extraocular muscle balance, accommodative amplitude; and previous surgical procedures on the lids, cornea, lens, or retina. The general health may adversely affect contact lens tolerance as in diabetes, thyroid dysfunction, or pregnancy. Various agents ranging from topical cosmetics, injectable antidiabetics, and oral contraceptives alter physiologic adaptation, and occupations involving smoke, dust, or flying particles increase the incidence of complications.

Ocular examination

Certain aspects of the ocular examination deserve special attention, and some require special equipment. The lids are examined for scars, deformities, ectropion, entropion, abnormal lashes, tumors, and infection. Evaluate the tight lids of the apprehensive "squeezer," the lax lids of senility, the drooping lids of ptosis, and the immobile lids of paralysis, which make contact lens motion and therefore tolerance impossible. A wide palpebral aperture, prominent eyes, and loose lids will usually require a larger diameter contact lens. The normal blink polishes the corneal epithelium, a function to be taken over by the contact lens. Nonblinkers fail to move the lens, tear exchange is poor, and corneal metabolism is impaired.

The accessory glands of the lid and the lacrimal gland maintain the precorneal film. Adequate tear flow is essential to corneal lubrication, transparency, and oxygenation. If tear flow is decreased or mucous balance altered, the resistance to lens movement is changed. A Schirmer test provides an indication of tear output in questionable cases. Older patients frequently show dry spots, that is, localized disappearance of the precorneal film causing burning and stinging. The reason is not known, but dry spots also occur with corneal injury, such as after cataract surgery, and their disappearance is a useful criterion of when to fit the postoperative aphakic patient. Tear secretion is reduced during sleep, one reason contact lenses should be removed upon retiring.

The conjunctiva is evaluated for folliculosis, hyperemia, edema, allergy, or infection. Pinguecula, pterygia, filtering blebs, and degenerative changes may interfere with lens tolerance. Infections should be treated, tumors removed, and xerosis eliminated before prescribing contact lenses.

The cornea is most sensitive with a natural intolerance to foreign bodies, hence one of the major obstacles to successful contact lens wear. Conversely, an anesthetic cornea or one with subnormal sensation is unable to register the usual warnings of a poor fit. Corneal sensation may be tested with a wisp of cotton or more elaborate graduated anesthesiometers. Sensitivity is normally greater in the center than peripherally and decreases with age and in the region of a surgical scar such as the incisional area for a cataract extraction. Corneal sensation may be diminished for several hours after contact lens removal in adapted wearers.

Corneal diameter determines lens size and is measured with a caliper or millimeter ruler (use a +8 D lens in the ophthalmoscope or the slit lamp for magnification). The average corneal diameter in the horizontal meridian is 11 to 12 mm and is smaller in children (about 9 to 10 mm in newborns). Average corneal thickness is 1 mm, but resilience varies, and some corneas are more easily deformed by external pressure than others. The anterior corneal surface is covered by epithelium whose remarkable healing powers adds a fortunate safety factor to contact lens adjustment.

The cornea is evaluated for congenital and dystrophic anomalies, scarring, opacification, and the neovascularization of previous injury or inflammation. Some opacities may be circumvented and some curvature irregularities replaced by a contact lens, which substitutes a more perfect refracting surface for an imperfect one. Keratoconus, bullous keratopathy, and even indolent ulcers can now be treated successfully, if not for visual improvement, at least for protection and the relief of pain.

The diameter, position, and responsiveness of the pupil are measured, the transparency of the media evaluated, and the fundus studied. Exophthalmos, nystagmus, albinism, strabismus, and glaucoma may influence contact lens tolerance.

Keratometry

The topography of the cornea is fundamental to a proper contact lens fit. The cornea is not spherical but toroidal and flattens peripherally. The central area where curvature is relatively constant (average diameter 6 mm) is the zone measured by keratometry. In Fig. 22-1, the virtual image h' of an object is formed at B'. Since the object distance l is large compared to the image distance l', the small distance x can be neglected; hence $Bf = BB' = b$. Because the focal length of a mirror is half the radius, $h'/h = f/b = r/2b$ or:

$$r = \frac{2b}{h} h'$$

In most clinical instruments, b is made constant by focusing the reticle through a short-focus telescope with a small depth of field. The object consists of illuminated mires carried on the telescope. Constant small eye movements make ordinary image size measurement impractical; so a doubling system is provided. When the eye moves, the doubled images move too but not relative to each other. Object size may be varied (fixed doubling itself varied (variable doubling), usually the latter. In variable doubling, object size is constant, and image size is directly proportional to the radius of the anterior surface of the cornea.

The power of the cornea is computed from the standard equation for single spherical refracting surfaces in air

$$F = \frac{n - 1}{r}$$

which in the case of the keratometer is based on an assumed index of 1.3375; hence $F_c = \frac{337.5}{r_c}$, where r_c is the corneal radius in millimeters and F_c the corneal power in diopters. Since the true index of the cornea is 1.376,

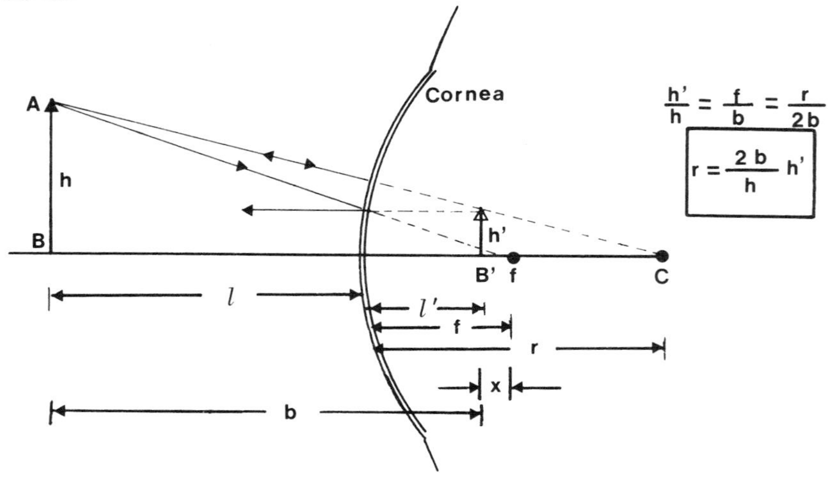

Fig. 22-1. Optics of the keratometer are illustrated in this schematic diagram. The virtual image h' of an object (mire) h is formed at B'. Since the object distance l is large, the image distance l' is practically equal to the focal length f and the small distance x can be ignored. Thus $Bf = BB' = b$. Since the focal length of a mirror is half its radius and $h'/h = f/b$, one obtains the equation for the keratometer: $r = \frac{2b}{h}(h')$. In clinical instruments, b is kept constant. In variable doubling instruments, h is constant, and h' varies according to the radius of the curvature.

the keratometer reading gives 337/376 or nine tenths of the power of the anterior surface. But since the rear corneal surface is not actually measured and its power is $\frac{1.333 - 1.376}{0.0068} = -5.88$ D, the reading approximates total corneal power.

Because the assumed keratometer index is nearly that of tears, the reading also gives the power of the rear surface of the liquid or "tear lens" (but of opposite sign). It follows that the keratometer measures only nine tenths of the astigmatism of the anterior surface. It is generally assumed that the posterior corneal surface parallels any toroidal curvature.

A topogometer is a device attached to the standard keratometer for peripheral measurements by controlling fixation. When the patient looks at a specified eccentric point, the paracentral corneal zone can be evaluated. Photoelectric keratometers photograph a series of concentric rings imaged by the cornea, and a computer converts the results into radii of curvature, the basis of lens fabrication. With marked corneal deformities, molding the eye followed by a plaster cast may be required.

Refraction

Contact lenses optically replace the anterior surface of the eye; hence corneal irregularities and astigmatism are mostly eliminated by the tear film. Contact lenses thus not only correct the same refractive errors as spectacles but also some anomalies that cannot be improved by other optical aids. Contact lenses change the optical characteristics of the eye as well as the incident light vergence, with secondary effects on accommodation and motility balance.

All candidates for contact lenses are refracted with and without cycloplegia as indicated. Reliance should not be placed on previous corrections that may be incorrect or out of date. A separate pair of spectacles is often required, and refractive findings also provide a base for ruling out later corneal warping. A high degree of astigmatism makes the fit more complicated or even impossible. Visual acuity is generally as good and sometimes better than with spectacles and refraction determines what resolution can be achieved. If the contact lens is fit parallel to the cornea, the spherical contact lens prescription will approximate the spectacle sphere, with all minus cylinders eliminated. Comparing the residual minus cylinder thus provides an indication of physical fit. Since contact lenses are fitted at the cornea, the vertex distance must be carefully measured, particularly for powers over 4 D. In aphakic refraction a trial frame is recommended, and vertex distance is measured with a stenopeic slit and metal ruler or one of the calipers commercially available. If no vertex distance is specified, the laboratory generally assumes an average 13 mm.

Contact lenses like spectacles alter the accommodative stimulus, but because of the new optical effectivity, the contact lens–corrected myope must accommodate more than he would with spectacles, and the contact lens–corrected hyperope must accommodate less. Although contact lenses are seldom prescribed exclusively for their motility effects, these changes must be anticipated. For example, a prepresbyopic myope switching from spectacles to contacts may find himself unable to read fine print and require reading glasses earlier, a symptom not conducive to successful adaptation. On the other hand, the esophoric hyperope gets some extra benefit from his contacts by a reduced accommodative, hence accommodative convergence, demand. These differences increase with ametropia and AC/A ratio and a change in the near point of accommodation is also expected. Refraction should therefore include phoria, fusional, and accommodative tests. Since the clinical effects may differ from those anticipated on a theoretical basis, it is best to measure these functions with and without the contact lens in place.

OPTICAL PRINCIPLES

When a contact lens is placed on the eye, the correcting optical system consists of the lens proper (made of glass or plastic) and the layer of tear fluid intervening between it and the cornea. The fluid acts as a second liquid or "tear" lens. The back vertex power of

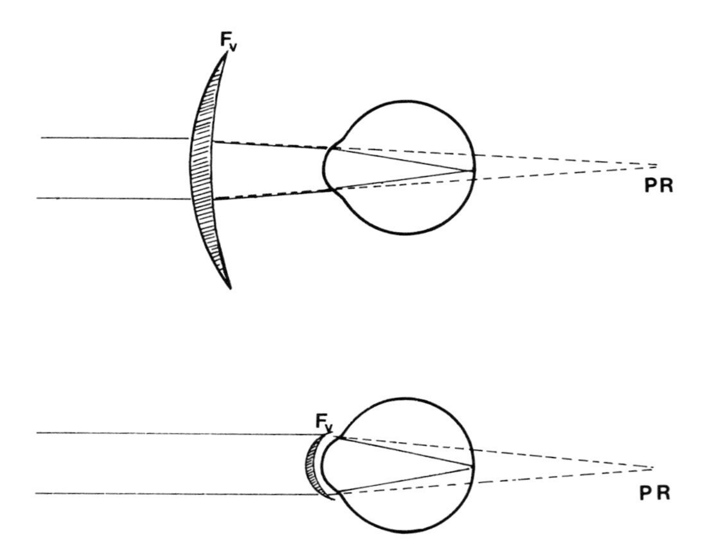

Fig. 22-2. The correcting optical system, whether a spectacle lens *(top)* or contact lens *(bottom)*, must have its back vertex focal point coincide with the far point *(PR)* of the eye.

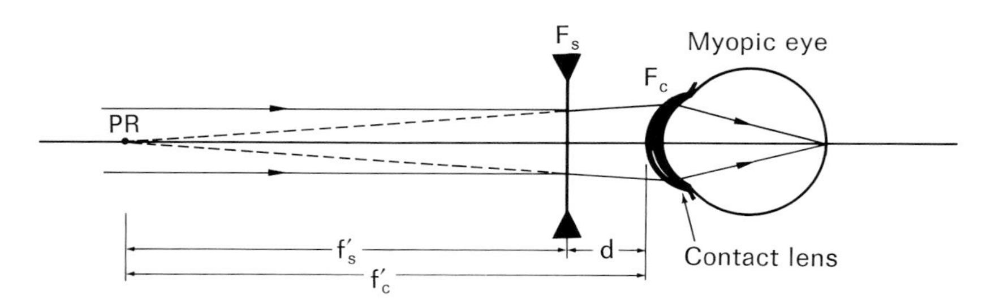

Fig. 22-3. Comparison of back vertex focal lengths of a contact lens *(f'_c)* and spectacle lens *(f'_s)* for a myopic eye. The far point *(PR)* of this eye must coincide with either lens for emmetropic correction. Note that the contact lens focal length is larger by the vertex distance *d;* hence it must have less minus power.

the system must equal the ocular refraction (Fig. 22-2).

The spectacle correction is ordinarily specified to be worn at the spectacle plane, some practical distance d from the cornea (or at a more theoretically correct distance from the principle plane of the eye, which is, however, not measurable). The contact lens, on the other hand, is worn in contact with the cornea, and the prescription changes according to the effectivity relation derived in Chapter 7:

$$F_o = \frac{F_s}{1 - dF_s}$$

where F_o is the ocular (or contact lens) prescription, F_s is the spectacle prescription, and d is the vertex distance. For example, if the spectacle refraction (F_s) at a vertex distance d = 12 mm is −5.00 D, the required back vertex power of the contact lens system (F_o) must be $\frac{-5}{1 - 0.012\ (5)} = -4.72$ D (Fig. 22-3). If the spectacle correction had been +5.00 D, the required contact lens prescription would be $\frac{+5}{1 - 0.012\ (+5)} = +5.32$ D. In myopia the required contact lens power is weaker and in hyperopia it is stronger than the indicated spectacle correction.

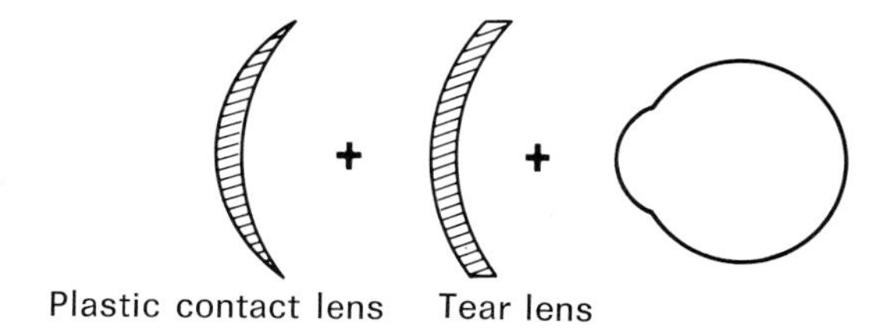

Fig. 22-4. A typical contact lens system consists of a plastic shell and the tear fluid separating it from the cornea. For purposes of clinical computation, it is convenient to consider each as a separate lens separated by an infinitely thin layer of air. Their dioptric power can then be added together directly.

It makes no difference how the required power in the contact lens system is divided. For example, in an older system of fitting, the contact lens itself was afocal, and the required correction was supplied by the fluid or tear lens, a principle still used with afocal trial contact lenses.

Ignoring for the moment the effect of thickness, the powers of any refracting surface can be computed from the standard equation: $F = \frac{n' - n}{r}$, where n′ and n are the refractive indices on the two sides of the single refracting boundary and r is its radius of curvature (in meters). By assuming that an infinitely thin layer of air separates the contact lens, tear lens, and cornea, each may be treated as an infinitely thin lens whose powers are directly additive (Fig. 22-4).

For example, if the ocular refraction is 2.75 D myopia and the keratometer reading is 45 D, the required correcting contact lens system might consist of an afocal plastic shell (n = 1.49) with radius 8 mm for each surface, giving a front surface power of $\frac{1.49 - 1}{0.008} = +61.25$ D and a rear surface power of $\frac{1 - 1.49}{0.008} = -61.25$ D, for a total of zero power. The tear lens (n = 1.33) would have a front surface power of $\frac{1.33 - 1}{0.008} = +41.25$ D and a rear surface power of $\frac{1 - 1.333}{0.0075} = -44.00$ D (7.5 mm is the radius of the cornea corresponding to a keratometer reading of 45 D when the instrument is calibrated for an index of 1.3375.) The power of tear lens is therefore −44.00 + 41.25 = −2.75 D, which is the required correction.

In practice the tear lens is no longer used to supply the correction because the required base curves are excessive and the physical fit necessarily poor. A hyperope would need a steeper fit and the myope a flatter fit. Modern fitting attempts to approximate the rear surface of the plastic lens to the corneal curvature and supply the correction by grinding it on the front surface of the contact lens proper. The tear lens now has parallel surfaces, and its optical contribution is negligible (about 0.12 D per 0.1 mm of thickness).

Although contact lenses are thin, their central thickness may not be negligible relative to their high curvature. The back vertex power (F'_v) of a thick lens is given by $F'_v = \frac{F_1}{1 - t/n\,F_1} + F_2$. For a plastic shell of radius 8 mm, as in the previous example, but center thickness of 0.5 mm, the power turns out to be $\frac{+61.25}{0.98} + (-61.25) = +1.25$ D as contrasted to the theoretical zero power if the lens is assumed to be thin.

A series of simple computations for average radii of corneal curvatures (e.g., $\frac{1.49 - 1}{0.0080} = 61.25$ D; $\frac{1.49 - 1}{0.0079} = 62$ D; $\frac{1.49 - 1}{0.0081} = 60.5$ D) reveals that every 0.1 mm change alters the surface power of the contact lens (n = 1.49) by about 0.75 D. Similar computation shows center thickness changes of 0.1 mm involves a power change of about 0.25 D.

Proper centering is as important in contact as in spectacle lenses, but unfortunately precise positioning is not as easy to achieve because the lens lags with lid movements. The fluid prismatic effect has a base in the same direction as lens displacement; if the lens power is minus, the prismatic effect is partly neutralized; if plus, the prismatic effect is exaggerated.

We have already seen that convex spectacle

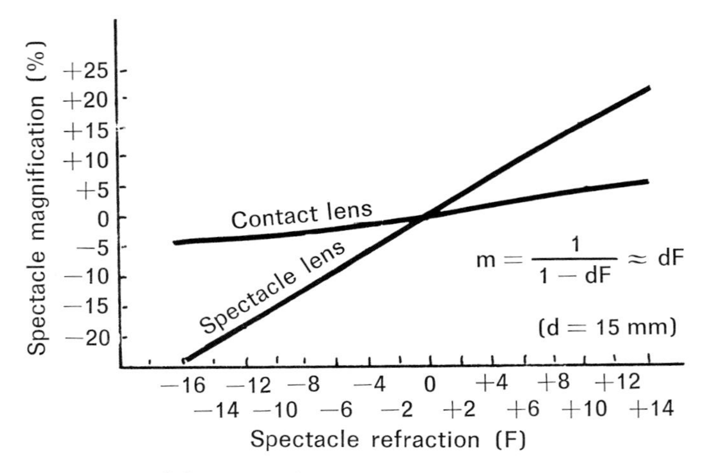

Fig. 22-5. Graphic comparison of the spectacle magnification (in percentages) for a range of spectacle refractions at an assumed vertex distance (d = 15 mm). The only reason the magnification by a contact lens differs from unity is that it is still 3 to 6 mm from the true optical reference point—the principal planes (or the entrance pupil) of the eye, not the cornea. The graph shows, for example, that for 10 D aphakia, the spectacle lens induces 15 mm + 3 = 18 mm = 1.8 cm; hence 1.8 (+10) = 18% magnification, contrasted to the contact lens, which induces 0.3 (+10) = 3% magnification. The first magnification is generally incompatible with binocular vision (although some rare exceptions occur).

lenses magnify and concave spectacle lenses minify the retinal image compared to the size without the lens in the same eye. An approximate formula for spectacle magnification (in percent) is dF, where d is the vertex distance in centimeters and F is the back vertex lens power. Theoretically contact lenses should cause no change, but since they are still 3 to 6 mm from the principal planes of the eye, they do produce minimal magnification. For example, a +10.00 D contact lens produces about 3% to 6% magnification as contrasted to the same power in a spectacle lens at a vertex distance of 15 mm, which gives 15 + 3 = 18 mm; 10 (1.8) = 18% magnification. Similarly, a −5.00 D contact lens causes 7.5% less magnification than an equivalent spectacle lens. A graphic comparison of spectacle and contact lens magnification is shown in Fig. 22-5.

Because of reduced magnification, the aphakic patient experiences less difficulty in his visuotactual coordination, the ring scotoma is reduced, the jack-in-the box effect is eliminated, and there is less distortion. Anisometropic patients may tolerate contact lenses better than spectacles because they diminish aniseikonia and anisophoria.

PATHOLOGIC PHYSIOLOGY

It is the general opinion of those practitioners experienced enough to have an opinion that a period of trial wear will do more than anything else to bring out the potential difficulties as well as benefits of contact lenses. For some patients the biggest hurdle is not building up tolerance but getting them in and out of the eye. Symptoms and signs of contact lens wear may be expected as an accompaniment of initial adaptation or from a poor fit of an improperly designed lens. Symptoms may also follow an unanticipated interference with normal ocular function.

The chief symptom when a contact lens is first placed on the eye is naturally a foreign body sensation. The cornea, plentifully supplied with nerve fibers almost reaching its surface, must adjust itself to the new weight, new movement, and new barrier to gaseous exchange. Although the cornea is more sensitive than the conjuctiva, the irritation to the inner lining of the upper lid is more annoying, and the patient finds relief by looking down or tilting his head back in the typical contact lens salute from which he may petulantly refuse to deviate. Lacrimation may be initally so excessive that the lens is floated out of position, and the patient experiences the sensation of loose lenses. Excess tearing generally disappears with adaptation; if at first you don't succeed, cry again. Conjunctival injection results from irritation usually caused by poor fit and may lead to microhemorrhages and neovascularization. Not only is the red eye esthetically unappealing, but also secretions are altered, leading to deposits on the lens. Blink rate may be increased and, if not, feels as if it were because of the greater awareness of each movement. The oxygen in the tears trapped under the contact lens is consumed in about 14 seconds; hence frequent lens movement and tear exchange is mandatory for comfortable wear. A reduced blink rate interferes with tear exchange, and these patients must be taught proper blinking by demonstration and exercises.

Blepharospasm and photophobia are common in the initial wearing period, the result of trigeminal irritation. If photophobia persists, it is probably due to improper fit or corneal injury and requires lens adjustment. Most patients benefit from a moderate tint, which does not unduly reduce retinal illumination and incidentally makes a lost lens easier to find.

Blurred vision in the initial stages may follow excessive lacrimation but, if it occurs regularly after several hours wear, is likely due to corneal edema. Edema is the single most important pathophysiologic obstacle to contact lens tolerance, and its presence in relation to wearing time is one of the most valuable objective signs of physical fit. There are gross or microscopic changes in all eyes, and the clinician must be aware of which are acceptable and which are not. Because of reduced evaporation and oxygenation, glycolysis leads to accumulation of lactate and depletion of glycogen. The surface epithelial cells are easily distorted or torn away, causing a foreign body sensation with lacrimation, photophobia, and apprehension. There may be congestion of limbal vessels and lid edema. Gradually the cornea adapts, and the sensitivity is reduced as wearing time increases. The mechanism of this toughening process is not known, although some corneal thickening is found, attributed to anoxia. Corneal sensitivity returns to normal within a week after contact lenses are discontinued. Mechanical movements of the lens cause fold in the surface, which may be outlined by fluorescein, and the formation of dimples that can trap air bubbles. A transient decrease in myopia is observed, lasting several hours, and may be noted as an improvement of vision without correction on arising. Two types of corneal edema are recognized; gross edema due to intraocular accumulation of fluid in the epithelium and microcystic edema, which may be diffuse or localized. Gross corneal edema is best seen with sclerotic scatter and low slit lamp magnification, appearing as a round, disclike, hazy area, usually following the wearing of an excessively steep contact lens. There may be keratometrically demonstrable changes in corneal curvature and distortions of the mires. Microcystic edema is best seen with high magnification using retroillumination and has the characteristic appearance of "bedewing" of acute congestive glaucoma. Corneal edema causes diffraction with decreased acuity, halos, and haze. Patient comfort is not a reliable index of edema because pain may be absent, although stinging, burning, and epiphora are common complaints. If the epithelium loses its integrity, the stroma imbibes water, and the epithelium may desquamate, leaving exposed nerve endings. Its most marked form is the overwearing syndrome, which occurs several hours after the lenses are removed. A 1:00 A.M. phone call from a distraught patient complaining of extreme pain, photophobia, blurred vision, hyperemia, lid edema, inability to open the eye, and excessive lacrimation is typical. The treatment is topical anesthesia, a tight bandage, and a more judicious wearing schedule.

Blurred vision may also result from uncorrected residual astigmatism, from the changes in accommodation and accommodative vergence (particularly in reading), and from simple alteration in the curvature of the cornea due to lens pressure, augmented by pressure from the upper lid. When the contact lens is removed, the previously acceptable spectacle correction is no longer adequate. This "spectacle blur" may last from minutes to days and is differentiated from corneal edema in that good acuity can be achieved by an optical change in spectacle blur but not in edema. Lenses that have been fitted too steeply often produce the worst spectacle blur. Occasionally it results in irreversible corneal warping, which may occur after years of comfortable wear. Finally, blurred vision may follow warping, scratches, deposits, or discoloration of the plastic lens material.

Flare, or glare, is caused by a lens edge that encroaches on the pupil, usually because the lens is too small, too heavy, or rides an eccentric corneal cap. When it is asymmetrical, the patient may note intermittent diplopia.

Neovascularization may follow prolonged overwear, with new vessels growing from the limbus on the side where the lens rests on the conjunctiva. This complication may occur without other symptoms in well-adapted wearers. The vessels are superficial or deep, and the change is irreversible.

Psychologic complications associated with apprehension and nervousness are not uncommon and disappear with the assurance of practice. Awareness of the lens immediately after insertion is normal, should not be painful, and gradually disappears. Reappearance of this symptom in an adapted wearer suggests reexamining the lens edge or surface for accumulated mucous deposits. The general optical and psychologic profile of the successful contact lens wearer is well known, but there are many individual differences, particularly with respect to age and maturity. Discussion of anticipated symptoms and written instructions regarding care and maintenance are helpful and always appreciated.

The optical complications of contact lenses include residual astigmatism, spectacle blur, improper magnification, prismatic effects and induced motility imbalances, increased accommodative demands in myopia with reduced reading range, and changes in refractive error due to altered corneal curvature. There is no convincing evidence that contact lenses stop or retard the progression of myopia; the apparent effect seems to be due to a natural plateau in axial growth because of the age at which contact lenses are fitted.

More serious complications include corneal abrasions caused by improper fit or improper insertion technique, abrasion or lacerations of the lid or conjunctiva, permanent changes in the physical contour of the cornea, allergic reactions, and reactivation of herpes, pterygia, or blepharitis. A foreign body sensation that persists may be due to a particle embedded in the cornea or abrasion. The most dreaded complication is secondary infection with ulceration, perforation, and endophthalmitis, especially with pseudomonas organisms. One horror story in the literature of the loss of both eyes after an abrasion caused by contact lenses teaches biologic humility in making the most of their optical utility.

TYPES OF CONTACT LENSES

New lenses and techniques are being introduced so rapidly that the clinician is obliged to flit from one to the next like a bee among the flowers just to keep abreast of changing developments. There are claims for lenses large and small, hard and soft, wet and dry, vented and unvented, spherical and toric, with or without ballast, all to be wetted, bathed, soaked, cleaned, cushioned, sterilized, and stored in containers whose shape and properties defy classification. Undoubtedly half the claims are right; we need only decide which half.

Most hard contact lenses are now made of plastic methylmethacrylate (Plexiglas), a hydrophobic material of low specific gravity, excellent light transmission, little or no toxicity, and a refractive index of 1.49. The lens is cut from a button about ½ inch in diameter and ¼ inch thick. The inside radius is lathe cut or cast and pitch polished; the outside radius, determined by lens power, is lathe cut. The size of the optical zone is controlled by the secondary curve. The lens is cut to size on a spindle, and finally, the edge is polished.

The *scleral* contact lens is designed with a posterior surface that conforms to cornea and sclera, with a central portion arching over the corneal apex. When the eye rotates, corneal touches may occur but disappear as it returns to the primary position. The interspace is filled with buffered solution or the patient's own tears. Scleral (or haptic) lenses were the earliest practical contacts, but wearing time and tolerance was poor. Flush-fitting fluidless scleral lenses used in the treatment of corneal burns, pemphigus, bullous keratopathy, as well as a cosmetic prosthesis, have been mostly replaced by soft lenses.

Corneal contact lenses were developed as a result of the difficulties in fitting scleral lenses. The most important feature of the corneal lens is its posterior surface, which must approximate the corneal topography while allowing sufficient movement for tear fluid to circulate. The posterior surface is divided into central, intermediate, and peripheral curves. The anterior surface, similarly classified, carries the prescription. It may be single cut or lenticular with a peripheral carrier. Most lenses have a central posterior curve ranging from 6 to 8 mm and a peripheral posterior zone of 0.2 to 0.8 mm, with or without an intermediate curve. Lens diameter is measured from edge to edge and may range from 6 to 14.5 mm, depending on the type of lens. The various posterior curves are blended to produce the best fit, or an aspheric curve may be ordered. The lens edge is rounded away from the corneal surface, and central thickness depends on refractive index and power.

A *toric* lens is asymmetrical on one or both surfaces (bitoric), indicated in high astigmatism to secure a better lens-cornea relationship, to promote adequate tear circulation or satisfy the optical requirements. More than 1.50 D of corneal astigmation suggests ordering a toric central posterior curve. The fitting principle is to align each lens meridian parallel to the primary meridians of the apical zone of the cornea. The power requirement is based on a cross cylindrical equivalent of the refraction. The chief difficulty is proper centering, controlled by ordering a lens that is not too large, prism ballast, and truncation.

Bifocal contact lenses are available for simultaneous or consecutive viewing. Simultaneous vision lenses consist of a central distance portion and a concentric reading portion. The patient looks through both at the same time and selects the one he needs. Visual efficiency is reduced, and some blur is always present. The consecutive lens has a fused reading segment similar to a spectacle bifocal. When the patient looks down, the lens is pushed up by the lower lid so that he utilizes the reading segment. The lens is constructed with a prism ballast to maintain proper centration and orientation. Bifocal contact lenses are difficult to fit and are potentially hazardous in walking and driving, but patients accustomed to single vision contact lenses may learn to accept them with proper motivation. A multirange bifocal is available using the pinhole principle, that is, giving an extensive depth of focus but at the price of reduced illumination.

Lenticular contact lenses follow the same principle as lenticular spectacle lenses; that is, they reduce weight and center thickness for a constant diameter. The size of the central optical zone is chosen according to whether there is a complete iridectomy. Single-cut aphakic lenses will be satisfactory for most patients if their diameter is kept small but pose problems in handling, especially removal. A minus carrier for a lenticular lens gives better lid grip to compensate for sagging.

The *silicone* lens is a rubbery polymer with the capacity for gaseous exchange. The lens is flexible, but its optics are stable, although in high powers it may become distorted with variable vision. The central thickness is about one third greater than hard lenses; the diameter averages 10 mm with an index of 1.44; and it moves on the cornea with a subjective sensation similar to hard lenses. Silicone lenses cannot be modified, and special cleansing, wetting, and soaking solutions are required.

Soft lenses are hydrophilic polymers that take up water to achieve the consistency of soft rubber without losing transparency. The lenses are equilibrated in normal saline, but pliability alone is not the chief factor in comfort; rather it is their moisture, which promotes a gliding lid action so wearing time can be increased more rapidly. Most hydrophilic lenses are larger than the cornea. There is less movement; hence tear exchange is not as effective as with hard lenses. Corneal anoxia is avoided because the polymer is permeable to oxygen and because the pumping action of the blink on the lens expedites tear replacement. Successful wear of soft lenses during sleep is attributed to unconscious eye movements, but individual variations suggest it is more prudent to remove them. Extreme stability allows repeated boiling for sterilization without adverse effects. The lens is fabricated by casting in a spinning mold; the power is altered by varying the size and curve of the mold, the amount of polymer injected, and the speed of the spin. The process is completely automated; hence the surfaces do not vary, and replacement is no problem. The posterior aspheric curve closely approximates the cornea, eliminating the need for blending secondary curves. The edges are subject to tearing and nicking, however, and require proper handling. Optical characteristics require fitting with a trial set, since lens shape and power depends on its position relative to the cornea. Since the lens is flexible, it molds itself to the cornea, and corneal astigmatism over 1.50 D cannot be corrected. The lens cannot be tinted. Currently available for optical correction is the Bausch & Lomb "Soflens."

Recent developments include a flexible contact lens (Flexinyl) that is thinner, lighter, and smaller than conventional hard lenses. This lens has excellent adaptability and stability and may be fitted to high astigmats. A new, lathe-cut hydrophilic lens (Hydrocurve) and a modification of the former Griffin lens (Softcon) is now available. A new gas permeable hard contact lens (RS-56) is in the developing stages. A chemical method of sterilizing soft lenses is soon likely.

FITTING PRINCIPLES

Details of contact lens fitting are outside the scope of this book, although the principles are simple in theory if somewhat more complex in execution. The lens should be light and thin, with an evenly distributed bearing surface and smoothed out peripheral curves. It should be tight enough to center properly, yet not so steep as to compromise corneal metabolism, and loose enough to allow free tear circulation without falling out of the eye. Above all the lens should be comfortable, physiologically innocuous, optically effective, and psychologically desirable.

Base curve

The base curve may be determined empirically by a trial lens, by keratometry, or preferably by a combination of the two methods. An additional advantage of the trial set method is that the patient can make his decision on the basis of actual experience.

The posterior curve may be flatter, steeper, or parallel to the cornea. In general, a lens parallel to the cornea floats on a precorneal tear layer with minimal pressure at any given point. Based on the keratometer reading, select a trial contact lens that approximates the flattest corneal meridian. For example, if the K-readings are 46.00-48.00, select a lens with a radius of $\frac{337.5}{46}$ = 7.34 mm. Trial lenses should, of course, be accurately fabricated and calibrated. If the lens is too flat, it will tend to sag with gravity or lag with each lid movement; a lens that is too steep tends to stick to the cornea with little or no movement. The lens should move about 1 to 2 mm with each blink. When the eye opens, the upper lid carries the lens with it and it then settles down again. A loose lens settles too rapidly; a tight one does not move. A slow recentering movement, usually with a slight rotary component, is optimum. Observation of lens lag and centering is best learned by clinical trial and is the criterion most frequently relied on by experienced fitters. The smaller the lens the less exacting the variables; that is, the greater the permissable error.

Lens diameter

Lens diameter depends on the size of the cornea, prominence of the eyes, width of the palpebral aperture, and tightness of the lids. A small lens gives better tear flow because it moves more easily, but it does not center as well and is more readily blinked out. Larger lenses cause more limbal irritation and tend to require longer adaptation periods. The ideal lens diameter has yet to be agreed on, but it should be small enough to avoid the limbus and large enough to cover the pupil. If the pupil is dilated, surgically enlarged, or eccentric, a larger lens is required. For an average-sized cornea (11 to 12 mm), begin with an 8 to 8.5 mm diameter lens. The cornea can be measured with a millimeter rule under slit lamp observation or directly, as in some of the newer keratometers, by a projected reticule grid. Use a larger lens for wider palpebral apertures, exophthalmos, and loose lids. When the eye is in the primary position, the lens should cover the pupil and center properly; when the eye turns down and the upper lid is raised, the lower lid should push the lens no more than 1 mm above the limbus; a lens that is too large will be pushed well onto the sclera, whereas one too small will remain below the superior limbus when looking down. A simple rule of thumb is to flatten the base curve radius by 0.1 mm for every 0.5 mm increase in optical zone size.

Peripheral curves

The optical zone is generally 1.5 mm smaller than the lens diameter, with 6 mm considered minimal. Secondary (or intermediate) curves are ground with a 1 mm longer radius (or 5 D flatter curvature); peripheral curves may be 10 D flatter than the base curve. If the peripheral curve is too small, it will feel tight, strangulate limbal circulation, and interfere with tear flow. The edge (or bevel) should be as thin as possible to allow the lid to move smoothly over the lens. The wider the separation between contact lens and corneal periphery the more readily tears enter the space between them, as in a funnel. The lacrimal film therefore renews itself more easily if the edge does not tightly approximate the cornea. The most frequently made office adjustments of contact lenses are polishing, smoothing, and opening peripheral curves.

Thickness

Thickness depends on lens diameter and power. For diameters 7.5 to 8.5 mm, the thickness of a plano contact lens is 0.16 mm. For lenses 9 mm in diameter, the thickness of a plano lens is 0.18 mm, and for lenses 9.5 mm in diameter, the thickness is 0.20 mm. Subtract 0.01 mm for every diopter of minus power, and add 0.02 mm for plus power. For example, a 9 mm diameter lens of +10 D power would have a thickness of 0.18 + 0.2 or 0.38 mm.

Slit lamp observation

Slit lamp observation of lens position with and without fluorescein, is performed by direct and indirect illumination. The depth of the tear layer between lens and cornea (i.e., the lacrimal lens) should be of even thickness with some increase in the apical zone. The entire surface is inspected, since only a single vertical section may be misleading in astigmatism. Low magnification sclerotic scatter is useful for detecting large areas of corneal edema; high magnification retroillumination, for detecting microbullae.

The five or six different methods of slit lamp examination are easier to use than name. Direct illumination is similar to an ordinary magnifying loupe, but the angulated light beam acts as an optical knife transecting the translucent optical media of the eye so that they may be seen in cross section.

In the indirect technique the light is focused on an area adjacent to the one being studied, and in retroillumination the posterior structures act as reflectors; for example, the light reflected from the iris is used to study the cornea. A special method of retroillumination is sclerotic scatter, in which the internal structure of the cornea is made to "glow" by throwing the beam onto the adjacent sclera.

All slit lamp examinations should be preceded by naked eye inspection; more than one expert has been fooled into diagnosing "uveitis" in a glass prosthesis.

Fluorescein pattern

Fluorescein patterns permit direct visualization of the geometric fit over the entire surface of the cornea. The principle is based on variations in thickness of the tear fluid between cornea and lens, made visible by the yellow-green fluorescence under violet light. After staining, the cornea is examined under the slit lamp in 1 or 2 minutes, and the intensity of the color depends on the thickness of the fluid layer. Normally a well-fitted lens on a spherical cornea shows vaulting of the corneal cap. Color is absent at points of contact: peripherally if the lens is too steep and centrally if too flat. Fluorescein patterns may be misleading in astigmatism, and appearance should always be correlated with symptoms. Many fitters no longer use fluorescein routinely, and it should never be used with soft lenses, which take up the stain. Fluorescein is best applied with paper-impregnated strips, which avoid the danger of contamination present with solutions.

Refraction

A cycloplegic and postcycloplegic refraction is performed according to age. In refracting over the contact lens, the simplest method is to use one that already has the approximate correcting power. If any additional spectacle power does not exceed 4 D, vertex effect is minimal, and the spectacle power can be added directly to the trial contact lens power. If the difference is larger than 4 D, translation of ocular from spectacle refraction is obligatory. A common source of error is improper translation.

Retinoscopy is a simple and rapid method of evaluating whether the base curve of the contact lens parallels the cornea. If the prescription is written in minus cylinder form, the contact lens eliminates the cylinder. Now if the lens is too flat, retinoscopy shows a need for more minus sphere as compared to the previously established refraction. A flat lens creates −0.50 to −0.75 D per 0.1 mm radius change in the tear lens; a steeper fit creates a plus tear lens power of 0.50 to 0.75 D.

Residual astigmatism

Astigmatism can be induced by warping, by fitting too high or too low, or by an inappropriate cylinder in the contact lens. A warped lens may show spherical power on the lensometer; hence a radiuscope reading of the central posterior curve is needed. Although tears neutralize a posterior toric lens surface, an anterior toricity is not compensated, and this unwanted cylinder becomes manifest. An analogous situation is created by a soft lens that accommodates itself to an astigmatic cornea. Flexing is also a problem with hard thin lenses, which mold themselves to a toroidal cornea. The symptom is intermittent blurred vision, confirmed by keratometry with the lens in place. A high- or low-riding lens can create astigmatism because of a cylindrical effect in eccentric gaze (oblique astigmatism). This problem occurs with small aphakic lenses that sag because of weight.

True residual astigmatism is usually lenticular and is present to some extent in all contact lens–corrected eyes, since only corneal toricity is neutralized. Significant residual astigmatism is present in 10% of contact lens wearers, most frequently when the spectacle correction indicates against-the-rule astigmatism (plus cylinder axis at or near 180°).

The best way of evaluating residual astigmatism is to refract the patient with the contact lens in situ. If the residual cylinder is less than 0.75 to 1.00 D, depending on acuity, the spherical equivalent is usually satisfactory. Larger degrees of residual astigmatism produce ghost images, reading difficulties, and distortions, with or without ocular discomfort and headaches, even when acuity is not markedly reduced. If the required cylinder exceeds 1.50 D, a front surface cylinder is incorporated in the contact lens or, more rarely, in a separate spectacle lens. To control contact lens rotation, 1Δ to 2Δ base down is added for ballast, or the lens is truncated 0.5 to 1 mm, or both.

Lens specification

Data sent to the laboratory should include the refractive prescription, vertex distance, corneal diameter, keratometer readings, lens specification (radius, diameter, number of peripheral curves, bevel, color, fenestration, and thickness).

In evaluating contact lens power, many laboratories specify front vertex power, not back vertex power as for spectacle lenses. Agreement should be reached as to which power aspect to specify, since the difference may exceed 1.50 D in aphakic prescriptions.

Posterior vertex power may be measured on the lensometer, providing a conical lens holder is adapted to the instrument to compensate for the sagittal depth error of smaller radii and higher powers of contact lenses. A better method of measuring contact lens curvature is based on the Drysdale principle incorporated in the radiuscope. The instrument is essentially a binocular microscope with a plus lens placed between a point source and the reflecting surface to be measured. The image is focused on the surface, and while maintaining a constant distance between lens and source, the microscope is adjusted until the target is again focused, that is, when the image is at the center of curvature. The distance the source has moved equals the radius of the reflecting surface. The instrument measures concave or convex surfaces (tolerance is 0.1 mm) and can also be used to detect surface irregularities and thickness.

Lens diameter is most easily measured by a V-shaped gauge (tolerance = ±0.1 mm). Thickness can be checked with a thickness gauge or Geneva lens measure by comparing the reading of the contact lens (concave surface up) to that of a known flat (each diopter displacement represents 0.1 mm). Thickness can be specified and the requirements fulfilled with a tolerance of 0.02 mm.

Surface quality should be checked for nicks, tears, scratches, discoloration, deposits, color match, as well as diameter, base curve, bevel, thickness, and blending of peripheral curves.

Fenestration or venting

Fenestration is the drilling of one or more holes in the contact lens to reduce weight, eliminate air bubbles, improve corneal metabolism in nonblinkers, and help to restore centering. The holes are 0.3 to 0.4 mm in diameter, and the procedure is performed in the office using a special lathe. Plastic burrs that scratch the cornea must be avoided. If the holes are too small, they become clogged; if excessive, they interfere with vision.

Wetting and soaking solutions

The purpose of the wetting solution is to alter the balance of force between the molecules of tear fluid for each other (cohesion) and for the plastic material of the lens (adhesion). Ordinary plastic is relatively hydrophobic, and wetting solutions increase the even spread of tears over the lens surface and provide a viscous protective coating and stabilization. Most wetting solutions contain a viscosity-increasing additive, a preservative, and a wetting agent. There may also be buffering systems and other ingredients.

Soaking solutions are used to increase the pliability of the plastic lens material by keeping it in a quasipermanent state of hydration, to leach out absorbed chemical compounds that may have accumulated, and to maintain sterility. Some patients prefer not to soak and avoid the larger container and extra expense; indeed soaking should be discouraged unless the patient can be relied on to change the soaking solution regularly and cleanse the container occasionally.

Contact lenses become contaminated with oil, grime, mucous, etc., and should be cleaned by rubbing, hydraulic pressure, or ultrasound after each wearing period. Under no circumstances should saliva ever be used to cleanse or wet lenses.

Insertion and removal

Instructions in techniques of insertion and removal are best given before the lenses are delivered. The lens is inserted by balancing it on the tip of the middle finger. The forefinger of the same holds the lower lid down and the forefinger or the middle finger of the other hand lifts the upper lid. The lens is placed upon the cornea while the ye is directed straight ahead. The same procedure can be repeated for the left eye. It is not necessary to change hands, but this may be done if the patient prefers.

Some find it more expedient to change hands and to separate the lids scissorslike with one hand while placing the lens on the eye, with the lens balanced on the ball of the forefinger of the other hand. A mirror may be useful in the early stages of learning lens insertion.

The method of removal preferred by most patients is to open the eye wide, placing one finger at the external canthus. As the lid is pulled to the side, the patient blinks and catches the lens in the palm of the hand. The principle of removal is to have the lid margins held tightly against the sclera above and below the superior and inferior margins of the contact lens. When the patient blinks, the lids go between the cornea and lens, forcing it out. Since the lens tends to adhere to the eye, the entire act of removal must be coordinated. In patients with poor vision or poor coordination, the lens may be floated out of the eye with an eye cup, or removed by a suction cup and magnifying mirror.

Insertion of soft lenses is best accomplished by having the patient lie down. Pinch the lens between thumb and forefinger and place it on the lower sclera as the patient looks up. The lens makes a sound as the air is released from underneath. When the patient looks up, the lens slips into place. To remove the lens, place the forefinger on the lens as the patient looks straight ahead and pull it down while he looks up. Then bring in your thumb and pinch the lens out of the eye. Make sure the lens is not inside-out when inserting by noting any eversion of the edges, and avoid touching the inside of the lens after it has been sterilized.

Follow-up examination

When the final lens is delivered, the patient's acuity and adaptation are checked after 4 to 24 hours. Optical and physical characteristics of the lenses are confirmed, and refraction is repeated to determine residual astigmatism. The blink rate is measured, fluorescein staining may be used to evaluate fit, slit lamp examination is performed to detect corneal edema or abrasions, keratometry is used to rule out corneal warping or deformation, and motility is investigated to establish binocularity.

A wearing time of 2 hours twice a day with a 30-minute to 1-hour increase on successive days is a reasonable schedule. Appropriate instructions are given regarding lens (and container) cleaning; ocular hygiene; wetting, soaking, cleaning, and cushioning solutions; precautions; and follow-up visits. Follow-up visits 1 or 2 weeks later are used to reevaluate acuity, fit, symptoms, and corneal integrity and to make necessary adjustment. The patient is seen two or three more times and every 6 months thereafter.

Fitting soft lenses

The advantages of soft lenses are greater comfort, less spectacle blur, less corneal staining or lens dislocation, and faster adaptation. The disadvantages are that acuity may be erratic, lens durability is poorer, and wet storage and repeated sterilization is required. The reasons for the erratic vision are that the extreme lens flexibility allows changes with each blink or eye movement and a small optical zone. In general, soft lenses should therefore be fitted flatter than hard lenses.

Soft lenses are fitted from a trial set because the shape of the lens changes when it is placed on the cornea. The refractive power of the lens therefore changes as it adheres to the cornea, and the lacrimal lens plays no role in the optical fit. Because the tear lens is absent to neutralize any toroidicity, an eye with more than 1.5 D of corneal astigmatism is rarely a candidate for soft lenses.

Soft lenses presently approved for optical use come in a series with a range of hydrated anterior surface curvatures (e.g., 9.08, 8.76,

8.24, 8.14). Each lens in a series has a range of powers according to the posterior surface. Central thickness is constant at 0.17 mm, and bevel width is about 0.75 mm. Steeper lenses are fitted to steeper corneas. For example, if the keratometer reading is 46.00-47.00 and the spectacle prescription is −4.00 − 1.00 × 180, the spherical equivalent is −4.50 D or, translated to the ocular refraction by a 13 mm vertex distance, −4.25 D (a conversion chart is available for average vertex distances and for most practical spectacle powers). Next, determine average corneal radius, that is, 46.50 (or 7.25 mm); hence the corneal power compensation to be added to the ocular refraction (from the manufacturer's chart) is +0.25 D. In this case the lens gains 0.25 D because it is bent less than the 45.00 D (zero) standard. The clinician now selects the −4.25 D lens from the appropriate series. The lens is hydrated with normal saline and inserted. The acuity is checked, and even if somewhat poor, the patient may be allowed to wear the lens for half an hour. If acuity is very poor, switch to another series. Overrefraction helps determine optimum acuity and further power changes.

The wearing schedule is flexible. A 3 hours on and 2 hours off is reasonable for the first few days, increasing 1 hour per day thereafter. The patient is checked after 1 day and 1 week. All hydrophilic lenses require daily sterilization in normal saline, although chemical methods are in the developing stages. Soft lenses often provide the opportunity to recuperate hard lens failures, but the two lenses should not be routinely compared in the same patient since the soft one is more comfortable, but the hard one may provide better optical correction.

SOME CLINICAL APPLICATIONS

In addition to their cosmetic advantages, contact lenses offer optical improvement that may not be achieved by any other means. The newer lens designs affort protection of fragile corneal epithelium, relief of pain, splinting of wounds, prevention of symblepharon, and delivery of pharmacologic agents and play a role in various clinical diagnostic techniques ranging from gonioscopy to radiologic localization of intraocular foreign bodies.

Refractive error

Ametropia is the most common indication for prescribing contact lenses, and myopes are the largest group who wear them successfully, probably because of high motivation. Mild to moderate myopia is easier to correct than more severe forms, which may be accompanied by high astigmatism and macular changes. In myopia, contact lenses increase the size of the retinal image compared to spectacles, the acuity is equivalent to spectacles, but the accommodative demand is increased. Cosmetic considerations should not be underestimated, particularly for public figures, actors, athletes, etc. Contact lenses are also used to change the color of the iris or the visibility of the corneal opacity.

Hyperopia is less often corrected with contact lenses because the motivation is not as marked. An exception is the accommodative esotrope or when the hyperopia is associated with corneal irregularities. Irregular astigmatism is a major indication for contact lenses, particularly since no other nonsurgical method may afford relief.

The advantages of contact lenses in anisometropia is that they make binocular vision possible and thus help eliminate the amblyopia frequently associated with it. Although contact lenses can also induce a vertical prismatic imbalance, they seldom do so because both move (or sag) symmetrically. In marked anisometropia, however, a difference in weight or edge thickness may cause difficulties.

Successful fitting of presbyopia is possible but not likely with presently available bifocal contact lenses. To utilize them, the patient must rely on an imperfect fit that permits a certain lens mobility with unhappy consequences for the corneal epithelium. Overcorrecting one eye and undercorrecting the other seems an unphysiologic substitution of monocular for binocular vision; few patients will accept black and blue elbows to avoid reading glasses. Moreover the eye is continuously changing, requiring several contact lens changes.

When nystagmus is associated with ametropia, an improvement in acuity is often achieved with contacts over that which can be gained by spectacles. More rarely, a pinhole contact lens helps subnormal vision or, in combination with a spectacle lens, serves as a Galilean telescope.

Aphakia

Successful contact lens fitting in monocular or binocular aphakia hinges on the elimination or distortion, aberration, magnification, and ring scotomas associated with cataract spectacles. A certain manual dexterity, motivation, and psychologic awareness is necessary to use them safely and effectively. Eyes with round pupils have less flare and glare and can tolerate a smaller lens. An against-the-rule astigmatic error is expected but is usually less than 2 D with modern suturing techniques. If acuity permits, the eye can be given a plus overcorrection to give some intermediate distance vision, although most patients will require separate spectacle lenses for reading.

Aphakic contact lenses should not be fitted until the refraction is stable, the surgical wound completely healed, the intraocular pressure normal, the anterior chamber clear, and the eye white. Refraction is done in the usual manner, utilizing a trial frame and measuring the vertex distance. In keratometry, care is needed to maintain proper fixation. A trial lens fitting will demonstrate patient reaction and ocular adaptability. Begin with a lens of 7 to 8.5 mm, single cut; refraction is then performed with the contact lens in situ. Centering is evaluated and lens diameter selected. It is generally advisable to order a tinted lens (light gray or amber) to replace the filtering effect of the crystalline lens. Particular attention is devoted to teaching insertion, removal, and hygienic handling. A magnifying mirror is often helpful in learning insertion, and a suction cup is helpful for removal. Follow-up examination is necessary to evaluate corneal reaction and detect edema. In selected cases, soft lenses give more immediate comfort and more rapid adaptation.

Younger children who have been operated on for congenital or traumatic cataract may tolerate contact lenses if fitted with care and the full cooperation of the parents. This may afford the only means or regaining useful vision in eyes that might otherwise be permanently impaired.

Keratoconus

Keratoconus is a rare (0.15%) dystrophic or degenerative corneal disease, frequently bilateral, in which allergy and eye rubbing have been implicated. The first eye affected generally has the poorer vision, but vision may stabilize. In addition to decreasing vision, myopia with astigmatism (usually oblique, against the rule, or irregular), distortion of the keratometric mires, and peculiar ophthalmoscopic and retinoscopic reflexes (scissors motion) develop. An indentation of the lower lid when looking down (Munson's sign) and thinning of the corneal apex with a faint gray to brown line at the base (Fleischer ring) may be seen. Scarring and curvature irregularity are common, and rupture of Descemet's membrane may lead to an acute stromal edema with marked visual loss. True keratoconus should be differentiated from the secondary scarring of interstitial keratitis, trachoma, trauma, herpes, or marginal dystrophy.

Keratoconus is corrected with spectacles as long as feasible by painstaking subjective refraction. Retinoscopy and keratometry are seldom helpful, but self-rotation of the trial cylinder or stenopeic slit may expedite the refraction, providing the target is large enough for the available resolution. When spectacles no longer afford useful vision, contact lenses may be tried, but increased scarring should be avoided, although this is not always possible. Motivation is a major factor in tolerance. Fitting is helped by a special trial set, based on graduated sagittal depths. Corneal lenses do not appear to stop progression and success parallels the severity of the disease. Scleral lenses have been used; soft lenses are helpful although disappointing in severe forms of the disease. The latter are best treated by penetrating keratoplasty.

Therapeutic uses

Soft contact lenses have greatly enhanced the treatment of corneal diseases. They have proved useful in bullous keratophy, epithelial erosion, indolent ulcers, conjunctival scarring, and keratitis sicca. Bandage lenses may replace sutures in certain perforating corneal injuries and flat chambers and protect the eye in neuoparalytic keratitis following trigeminal lesions. A remarkable property of soft lenses is the ability to take up and slowly release medications to the eye, thus prolonging and potentiating their effects. Soft lenses provide a simple, safe, and effective means of relieving pain and, in selected corneal diseases, of restoring vision. Moreover their use does not preclude other treatment or later surgery.

Miscellaneous uses

Contact lenses are used not only in the correction of refractive errors but also in gonioscopy, electroretinography, and diagnostic radiology. More controversial is their use in preventing or improving myopia with progressively shaped sequential fitting.

The use of contact lenses to change corneal curvature by nonsurgical means is termed "orthokeratology." In most cases, the corneal change is considered an undesirable side effect rather than a therapeutic goal. Although no severe corneal abnormalities have been reported in selected trial series, the method is not without risk and requires further evaluation before final judgment is given on its merits.

REFERENCES

Ackerly, E.: Clinical physiology of the eye with contact lenses, Int. Ophthal. Clin. **1**:311-325, 1961.

Alpern, M.: Accommodation and convergence with contact lenses, Am. J. Optom. **25**:379-387, 1949.

Bailey, I. L., and Carney, L. G.: Analyzing contact lens induced changes of the corneal curvature, Am. J. Optom. **47**:761-768, 1970.

Bennett, A. G.: Optics of contact lenses, London, 1969, Association of Dispensing Opticians.

Bier, N.: Myopia controlled by contact lenses; a preliminary report, Optician **135**:427, 1958.

Bier, N.: Prescribing for presbyopia with contact lenses, Am. J. Optom. **44**:687-710, 1967.

Boeder, P.: Power and magnification properties of contact lenses, Arch. Ophthal. **19**:54-67, 1938.

Bonnet, R., and El Hage, S. G.: Deformation of the cornea by wearing hard and gel contact lenses, Am. J. Optom. **45**:309-321, 1968.

Brauninger, G. E., Shah, D. O., and Kaufman, H. E.: Direct physical demonstration of oily layer on tear film surface, Am. J. Ophthal. **73**:132-134, 1972.

Camp, R. N.: Subjective symptoms and objective signs, Contact Lens J. **7**:15-22, 1973.

Cassady, J. R.: Corneal contact lenses; report of a series of 300 cases, Arch. Ophthal. **66**:356-361, 1961.

Cassady, J. R.: Correction of aphakia with corneal contact lenses, Am. J. Ophthal. **68**:319-323, 1969.

Charles, A. M., Callender, M., and Grosvenor, T.: Efficacy of chemical asepticizing system for soft contact lenses, Am. J. Optom. **50**:777-781, 1973.

Cochet, P., and Amiard, H.: Contact lenses, International Ophthalmology Clinics vol. 9, Boston, 1969, Little, Brown & Co.

Dickinson, F.: The value of microlenses in progressive myopia, Optician **133**:263-264, 1957.

Dixon, J. M.: Ocular changes due to contact lenses, Am. J. Opthal. **58**:424-444, 1964.

Dohlman, C. H., Ahmad, B., Carroll, J. M., and Refojo, M. F.: Contact lens glued to Bowman's membrane, a review, Am. J. Optom. **46**:434-439, 1969.

Fonda, D.: Complications of contact lens wearing, South. Med. J. **55**:126-128, 1962.

Freeman, M. H.: The measurement of contact lens curvatures, Am. J. Optom. **42**:693-701, 1965.

Gasset, A. R., and Bellows, R. T.: Hydrophilic contact lenses in the treatment of shallow or flat chambers, Ann. Ophthal. **6**:996-998, 1974.

Gasset, A. R., and Kaufman, H. E.: Bandage lenses in the treatment of bullous keratopathy, Am. J. Ophthal. **72**:376-380, 1971.

Gasset, A. R., and Lobo, L.: Corneal diseases and soft contact lenses, Ophthal. Digest **36**:19-27, 1974.

Girard, L. J., editor: Corneal contact lenses, ed. 2, St. Louis, 1970, The C. V. Mosby Co.

Girard, L., Soper, J. W., and Gunn, C., editors: Corneal and scleral contact lenses, St. Louis, 1967, The C. V. Mosby Co.

Grosvenor, T. P.: Contact lens theory and practice, Chicago, 1963, Professional Press.

Grosvenor, T., and Callender, M.: The N and N soft contact lens, Am. J. Optom. **50**:489-498, 1973.

Hartstein, J., and Becker, B.: Researches into the pathogenesis of keratoconus, Arch. Ophthal. **84**:728-729, 1970.

Heyman, L. S.: Cosmetic lenses, Contact Lens J. **6**: 16-18, 1972.

Hodd, F. A. B.: Changes in corneal shape induced by the use of alignment fitted contact lenses, Contacto. **9**:18-14, 1965.

Jenkin, L., and Tyler-Jones, R.: Contact lens instrumentation, Optician **144**:472-475, 1962.

Jessen, G. N.: The long-term effects on the cornea and the visual system with contact lenses, Contacto **5**: 137-145, 1961.

Jessen, G. N.: Contact lenses as a therapeutic device, Am. J. Optom. **47**:429-435, 1964.
Kelly, T. S., and Butler, D.: Preliminary report on corneal lenses in relation to myopia, Br. J. Physiol. Opt. **21**:175-186, 1964.
Levinson, A., and Ticho, U.: The use of contact lenses in children and infants, Am. J. Optom. **49**:59-64, 1972.
Mandell, R. B.: Contact lens practice; basic and advanced, Springfield, Ill., 1965, Charles C Thomas, Publisher.
Miller, D.: Contact lens-induced corneal curvature and thickness changes, Arch. Ophthal. **80**:430-432, 1968.
Millodot, M.: Variation of visual acuity with contact lenses, Arch. Ophthal. **82**:461-465, 1969.
Morrison, R. J.: Observations on contact lenses and the progression of myopia, Contacto **2**:20-25, 1958.
New Orleans Academy of Opthalmology: Symposium on contact lenses. St. Louis, 1973, The C. V. Mosby Co.
Pascal, J. I.: Form and power of contact lenses, Arch. Ophthal. **22**:399-402, 1939.
Raiford, M. B., editor: Contact lens management, International Ophthalmology Clinics, vol. 1, Boston, 1961, Little, Brown & Co.
Rengstorff, R. H.: A study of visual acuity loss after contact lens wear, Am. J. Optom. **43**:431-440, 1961.
Rengstorff, R. H.: The Fort Dix report, a longitudinal study of the effects of contact lenses, Am. J. Optom. **42**:153-163, 1965.
Rengstorff, R. H.: Variations in myopia measurements: an after-effect observed with habitual wearers of contact lenses, Am. J. Optom. **44**:149-161, 1967.
Richardson, B., and Camp, R.: A systematic approach to fitting soft lenses, J. Contact Lens Soc. Am. **7**:9-12, 1973.
Ridley, F.: Contact lenses in treatment of keratoconus, Br. J. Ophthal. **40**:295-304, 1956.
Robertson, D. M., Ogle, K. N., and Dyer, J. A.: Influence of contact lenses on accommodation, Am. J. Ophthal. **64**:860-871, 1967.
Rubin, M. L.: A tale of the warped cornea: a real-life melodrama, Arch. Ophthal. **77**:711-712, 1967.
Sampson, W. G., Hoefle, F. B., Gould, H. L., Dohlman, C. H., Baldone, J. A., Boyd, H. H., and Halberg, G. P.: Symposium: soft contact lenses, Trans. Am. Acad. Ophthal. Otolaryngol. **78**:838-424, 1974.
Sarver, M. D., Harris, M. G., Mandell, R. B., and Weissman, B. A.: Power of Bausch and Lomb Soflens contact lenses, Am. J. Optom. **50**:195-199, 1973.
Sattler, C. H.: Erfahrungen mit haftgläsern, Klin. Monatsbl. Augenheilkd. **100**:172-177, 1938.
Sears, M., and Guber, D.: The change in the stimulus AC/A ratio after surgery, Am. J. Ophthal. **64**:872-876, 1967.
Spaeth, P. G.: Contact lenses, Int. Ophthal. Clin. **1**: 277-290, 1961.
Swanson, K. V.: Report of the Z-80 subcommittee on contact lens standard, Contact Lens J. **5**:17, 21, 1971.
Tredici, T. J., and Shacklett, D. E.: Orthokeratology—help or hindrance, Trans. Am. Acad. Ophthal. Otolaryngol. **78**:425-432, 1974.
Wesley, N. K.: Progression of myopia and contact lenses, Contacto **10**:10-12, 1966.
Ziff, S.: Orthokeratology in relationship to existing corneal curvatures, South. J. Optom. **7**:9-19, 1965.

23

Subnormal vision

Second only to the fear of death is the fear of life without sight. Vision is the most important means of orientation and communication, but the amount of information the nervous system can glean from the retinal image depends on the quality of the optics, the quantity of light, the integrity of transmission, and the facility of interpretation. Although the darkness of vision can be compensated by the light of intelligence, visual loss is a disaster of the first magnitude. Fortunately the partially blind are also partially seeing, and the restoration of vision to its best potential is a clinical challenge of the highest order.

An estimated 2.5 million Americans have some visual impairment, resulting in legal blindness in half a million. The number of people with defective vision is growing faster than the population because of longer life span, and diseases previously fatal now achieve chronic status. Estimates of partial blindness are even more difficult to obtain. Many people are reluctant to admit their handicap and frequently conceal it even from themselves. As a world problem, the prevalence is illustrated by the fact that trachoma alone may affect 10% of the population in some communities. And to the intangible costs of suffering and deprivation must be added the enormous price of lost productivity, medical care, and rehabilitation.

PATHOLOGIC PHYSIOLOGY

Patients with impaired vision respond differently from those with ordinary refractive errors. To evaluate the clinical tests and select the most appropriate treatment it is necessary to consider the anatomic, physiologic, optical, and psychologic characteristics of the responsible ocular disease. Although the common denominator is poor vision, in many cases the mechanism is not evident because the eye suffers multiple abnormalities.

Poor vision may result because the light rays are not assembled into a proper focus, cannot reach the photoreceptors, or cause their excitation or because the neural message is imperfectly transmitted or interpreted. These mechanisms may be broadly grouped into optical, retinal, conductive, central, and psychologic disturbances of the visual apparatus.

Optical mechanisms

In treating refractive errors, spheres and cylinders reduce the size and alter the shape of blur circles; in irregularities of refractive components, these methods do not work because the light rays are refracted in skewed directions. Corneal distortions as in keratoconus, pterygium, trauma, or degeneration are particularly disturbing because the cornea is the most potent refractive interface. The diagnosis is made by keratometry, which shows the mire distortions, and biomicroscopy, which reveals the underlying pathology. Since the cornea is frequently involved by spread from adjacent structures, the conjunctiva, sclera, and anterior chamber may reveal the source.

Because of its relative exposure, the cornea is easily injured or infected, its avascularity makes healing slow, and its delicate metabolism soon leads to opacification, which occurs early, persists late, and is a leading cause of subnormal vision. Opacities are classified as slight (nebulae), moderate (maculae) or severe (leukomas) and follow any defect that extends below Bowman's membrane. If the opacity is due to a perforating injury, iris tissue may be incarcerated in the scar, resulting in an irregular pupil. If the scar is extensive, the tissue stretches into a staphyloma. Electron micrographs of opaque cornea show irregularities in the density of collagen and empty spaces, which scatter light. The latter cause the hazy appearance even though the tissue may look normal under the light microscope. Hydration and edema predispose to neovascularization, potentiated by the release of histamine—like vasoformative substances. The new vessels may be superficial or deep, but they remain forever as ghosts to haunt the vision.

Lens irregularities are less common and optically less important. Lenticonus and lentiglobus are very rare, subluxation is more common, and changes due to cataract are frequent. Sudden changes in lens position or refractive index result in transient, optically correctable decrease of vision. All lens opacities are technically cataracts and may range from an inconsequential Mittendorf dot to a dangerous Morgagnian sac. Although cataracts can usually be successfully removed, it is not always prudent, possible, or permitted. There may be complicating retinal disease, or the eye and patient is medically, perhaps psychologically, unsuited. Congenital cataracts, a leading cause of newborn blindness, are a surgical challenge but may sometimes be circumvented by an optical iridectomy. Not all congenital cataracts are stationary, however, and if bilateral, early treatment is necessary to prevent permanent visual impairment. Only surgeons who do not operate have no complications; so one must face up to the capsular remnants, updrawn pupils, vitreous adhesions, elevated tensions, and the other horror stories of the anterior chamber that disturb vision, whether they are one's own or whether the surgery was performed "elsewhere."

Laboratory research has attempted to define, as yet with limited success, the chemical and physical basis of senile cataract. Experimental cataracts have been produced by radiation, nutritional deficiencies, metabolic poisons, endocrine imbalances, and toxic drugs; hence cataracts occur in a variety of systemic diseases, as well as after trauma and the wear and tear of age. The acuity deteriorates, the field of vision diminishes, and the sensitivity to light decreases. At some stages, there may be double vision or polyopia, progressive myopia, or even entoptic visualization of the opacities.

Opacities in the anterior chamber are rare but dangerous; traumatic hyphema or a luxated lens. Vitreous opacities include protein, cholesterol, blood, tumor cells, new vessels or old connective tissue, quiescent congenital remnants, or active parasites. The particles scatter light and degrade the retinal image; larger aggregates reflect light, producing glare not unlike the dirty windshield of night driving. Central opacities particularly interfere with daylight vision because the pupil clamps down and chokes off retinal illumination. Vision may be totally obstructed with a vitreous hemorrhage, minimally affected in asteroid hyalitis, but most often annoyed by floaters. Retinal opacities are possible because light must travel through its substance before reaching the rods and cones. They include membranes, veils, fluid, blood, and occasionally myelinated nerve fibers. A rare kind of pseudo-opacity is a drooping lid (ptosis, tumor, or senile change). All forms of opacities are aggravated in older people by lenticular pigmentation, senile miosis, and vitreous degeneration.

High refractive errors may be missed and poor vision attributed to some other cause, particularly when the retinoscopic reflex is dim and subjective discrimination poor. Moreover long-standing uncorrected refractive errors may give rise to amblyopia and strabismus. Unusual and unsuspected causes of refractive changes are lid tumors, strabismus surgery, central serous retinopathy, and retinal detachment repairs.

Treatment of optical disturbances is based on the underlying mechanism. Corneal irregularity is improved by substituting a more perfect refracting surface with a contact lens, keratoplasty, or prosthokeratoplasty; by isolating the best optical meridians with a stenopeic slit or optical iridectomy; by reducing the size of the blur circles with a pinhole; by enhancing illumination to compensate for light lost by irregular refraction; by increasing the size and contrast of the retinal image; by reducing glare and stray light with shades, sunglasses, and masking; by the surgical removal or replacement of ocular opacities; or by circumventing optical opacities with mydriasis or iridectomy or masking them by cosmetic contact lenses or corneal tatooing. Whatever the cause of subnormal vision, refractive errors should be corrected and presbyopia compensated to achieve the best visual potential. A magnifier has limited value if it only enlarges an out-of-focus picture.

The first successful corneal graft is attributed to von Hippel (1888), but the procedure did not come into widespread use until the past few decades. The transfer of biologic tissue is no immediate threat to turn us into modular men, depending as it does on immunologic, pathologic, psychologic, and, in case of the eye, optical considerations. Light perception and normal projection are essential prerequisites. It is also desirable for the eye to have normal tension and no inflammation and the cornea to provide at least a healthy rim of tissue to nourish the graft. Keratoplasty may be penetrating or lamellar, optical or therapeutic, and preparatory or final. With proper indications, considerable success can be achieved; on the other hand, the patient may lose what sight he has.

Refractive keratoplasty (keratomileusis) has been used to correct high degrees of myopia and astigmatism but is essentially an esoteric procedure of academic interest. More promising are supporting scleral grafts in degenerative myopia.

Retinal mechanisms

The retina is the essential element of the eye, and the rods and cones are the essential elements of the retina. Diseases of this complex tissue are a major cause of poor vision. More than a collection of receptors, the retina is an embryonic outgrowth of the brain subject to similar deleterious hypoxia, tenuous terminal blood supply, mechanical disorganization, and limited functional recovery.

The neural organization of the retina is both vertical and horizontal (Fig. 23-1). The vertical arrangement is more important and consists of photoreceptors, bipolars, and ganglion cells. These are connected en masse and by private wire (midget) systems. The horizontal connection is provided by horizontal and amacrine cells. The rods and cones, which usually differ in form, always differ in chemistry and function, the rods at low and the cones at high levels of illumination. Some diseases selectively affect rods (e.g., congenital night blindness); some, cones (e.g., achromatopsia); and some, both rods and cones (e.g., retinitis pigmentosa). Other diseases interfere with their nutrition by involving the adjacent pigment epithelium (e.g., albinism and chloroquine toxicity), by destroying Bruch's membrane (e.g., colloid bodies, angioid streaks), or by diffuse involvement (e.g., choroideremia, gyrate atrophy, and high myopia).

Nocturnal animals such as the gecko have no cones; some diurnal reptiles have no rods. The correlation of structure and function was first noted by Max Schultze (1866) and has come to be called the "duplicity theory." The best evidence for it is the "break" in the curves of such diverse functions as acuity, flicker fusion, color discrimination, and dark adaptation when plotted against illumination.

Clinical measurements of dark adaptation help diagnose the night blinding disorders that may be congenital or follow severe nutritional deficiencies, progressive chorioretinopathies, and administration of such toxic drugs as quinine and thioridazine (surprisingly, not chloroquine). The inheritance of congenital stationary night blindness in one family has been traced back to 1637, perhaps the most fascinating genealogy in all medicine. Diminished night vision is common in older people for physiologic reasons and in those whose pupils are constricted by antiglaucoma drugs.

The capacity for detailed vision depends on the density of the cones. Since visual

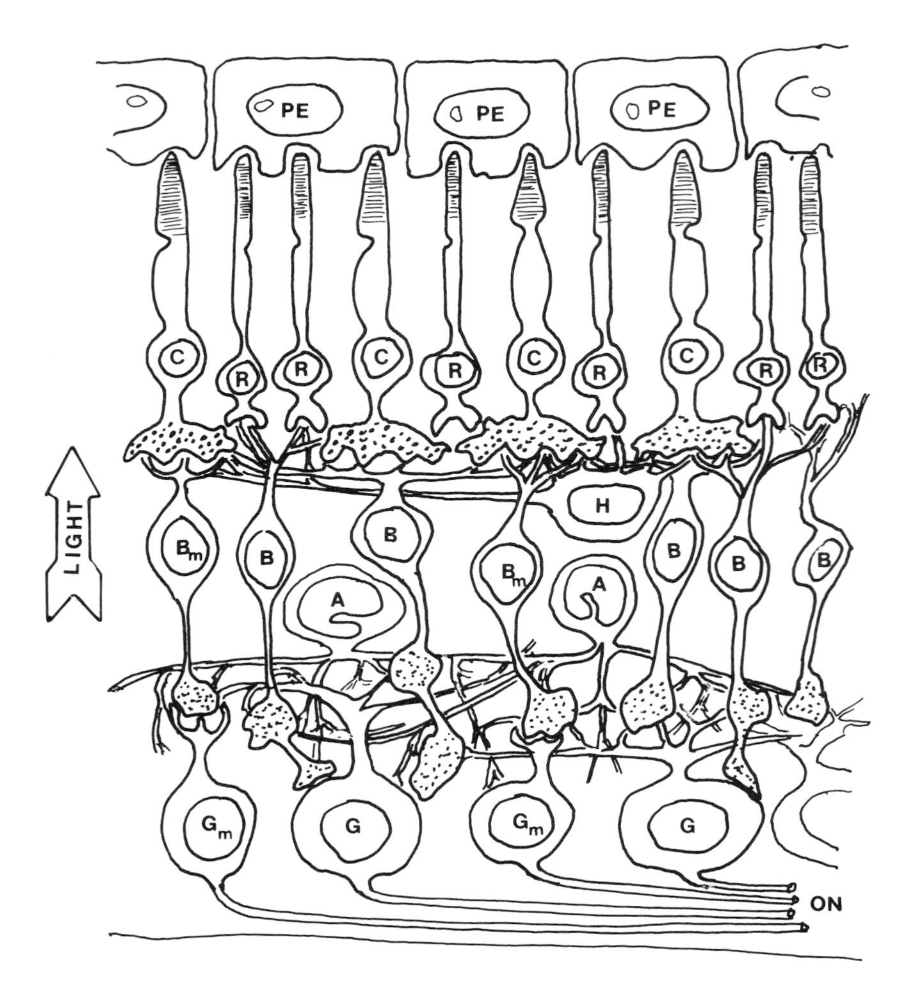

Fig. 23-1. Semischematic diagram of the retina showing the vertical connections of rods *(R)* and cones *(C)* by means of bipolars *(B)* and ganglion *(G)* cells. The private wire connections dominating the fovea are by midget bipolars *(B_m)* and ganglions *(G_m)*. Horizontal connections are provided by the horizontal *(H)* and amacrine *(A)* cells. The limiting retinal layers are the pigment epithelial cells *(PE)* on top and the internal limiting membrane below. The axons of the ganglion cells form the optic nerve fibers *(ON)*.

acuity is known to change with illumination, the number of elements per unit area also varies functionally through neurophysiologic interaction. So many cones are concentrated at the fovea that they look like rods, and even small lesions here are disastrous for resolution, although fortunately not for mobility if the peripheral retina is intact (and the patient is so informed for whatever consolation it is worth). Macular lesions occur secondarily to some peripheral retinal diseases (e.g., pars planitis, sickle cell retinopathy, Coat's disease) and although the loss of central vision dominates the clinical picture, they should be ruled out by indirect ophthalmoscopy with a dilated pupil. Most macular lesions can be seen with the ophthalmoscope; other tests of macular function are Haidinger brushes, Amsler grid, recovery following dazzling, color defects, and, of course, central acuity. Fluorescein angiography is useful in localizing vascular leaks, identifying hemangiomas, and differentiating blood from melanomas.

Acquired color defects in certain ocular diseases differ from congenital anomalies in that they are temporary, variable, and may affect one eye, and the patient, having had normal vision, can name the colors he actually sees. Kollner (1909) first formulated the generalization (not always true) that retinal disease affects blue-yellow vision, whereas optic nerve disease disturbs red-green vision. Although color defects are an ancillary finding, they are helpful in the diagnosis of retrobulbar neuritis, tobacco amblyopia, Leber's atrophy, pigment degenerations, myopic degeneration, and toxic retinopathies. In one form of juvenile macular degeneration (Stargardt), they precede all other findings and progress to total color blindness.

More exotic are electrophysiologic techniques of evaluating retinal function. The best of these is electroretinography (ERG), which measures the response of bipolars and photoreceptors (not ganglion cells) by a contact lens electrode relative to a skin electrode (an electrically shielded room is necessary). The chief difficulty is standardization; the parameters resemble those of perimetry, but the influence of stray light and media opacities is not known. In retinal anoxia or glaucoma, the ERG is normal because only ganglion cells are affected. The ERG is uniquely diagnostic in retinitis pigmentosa, congenital achromatopsia, and Leber's amaurosis; it is confirmatory in flecked retina syndromes and pigmented toxic retinopathies; and it is abnormal but not necessary for the diagnosis of retinal detachment.

The inner layers of the retina are nourished by the central retinal artery; the outer layers, by the choroid. Nutrition depends on a blood supply of adequate composition, pressure, flow, and viscosity; hence retinal insufficiency can occur in anemia, hypertension, diabetes, arteriosclerosis, and dysproteinemia. Of all the causes of visual loss, total occlusion of the central retinal artery is the most sudden, the most complete, and the most devastating. Only in the rare cases of a cilioretinal branch is some vision preserved. The most common diseases affecting the vascular tree are arteriosclerosis, hypertension, diabetes, embolization, collagen diseases, and vasculitis. Because of their number, collateral circulation, and large lumens, choriocapillary blockage is rare, although vision may deteriorate in some forms of choroidal sclerosis.

Central retinal artery pressure may be evaluated clinically by ophthalmodynamometry. In this method, the intraocular pressure is raised artificially by a calibrated spring-loaded plunger or by controlled suction. As the pressure rises, the ophthalmoscopically visible pulsations of the central retinal artery changes; the first point of collapsible pulsation is taken as the diastolic pressure; the point at which all pulsation disappears is the systolic pressure. There is no consistent correlation to the conventionally measured brachial pressure. The method is useful if not completely reliable in evaluating arterial insufficiency when there are symptoms of intermittent loss of vision and atheromatous disease of the ophthalmic or carotid arteries. A difference of 20% on the two sides is considered significant. (Ancillary procedures include palpating and auscultating the carotid vessels, radiologically visible calcification, angiography, and thermography.)

Throughout the retina there are two interconnecting capillary networks: a superficial one in the nerve fiber layer and a deeper one between the inner nuclear and outer plexiform layer. At the macula the capillary bed thins out, and only a single layer of capillaries is found in the perifoveal region. A capillary free zone is present around retinal arteries.

The superficial network is involved predominantly in diseases of the arterial circulation and manifests itself as flame-shaped or preretinal hemorrhages, cotton wool patches, retinal and optic nerve edema, and a macular star figure. The full-blown picture characterizes stage IV of the Keith-Wagener classification of hypertension and carries a poor prognosis (70% of patients are dead within a year.) Hypertension is a common cause of retinal vascular occlusion because of arterial narrowing, which does not improve despite otherwise therapeutic control of blood pressure.

Deep capillary network involvement tends to occur in diseases affecting the venous side of the circulation and is manifest by dot-and-blot hemorrhages, microaneurysms, and retinitis proliferans. These changes are most frequently seen in diabetic retinopathy, a leading cause of blindness; the more negligent the control of the disease the less negligible the ocular complications. Diabetes produces a selective degeneration of capillary cell walls, but the exact pathophysiology is not known, and the ocular prognosis in long-standing disease is poor despite systemic control. Poor responses to visual aids can be anticipated if there is neovasculari-

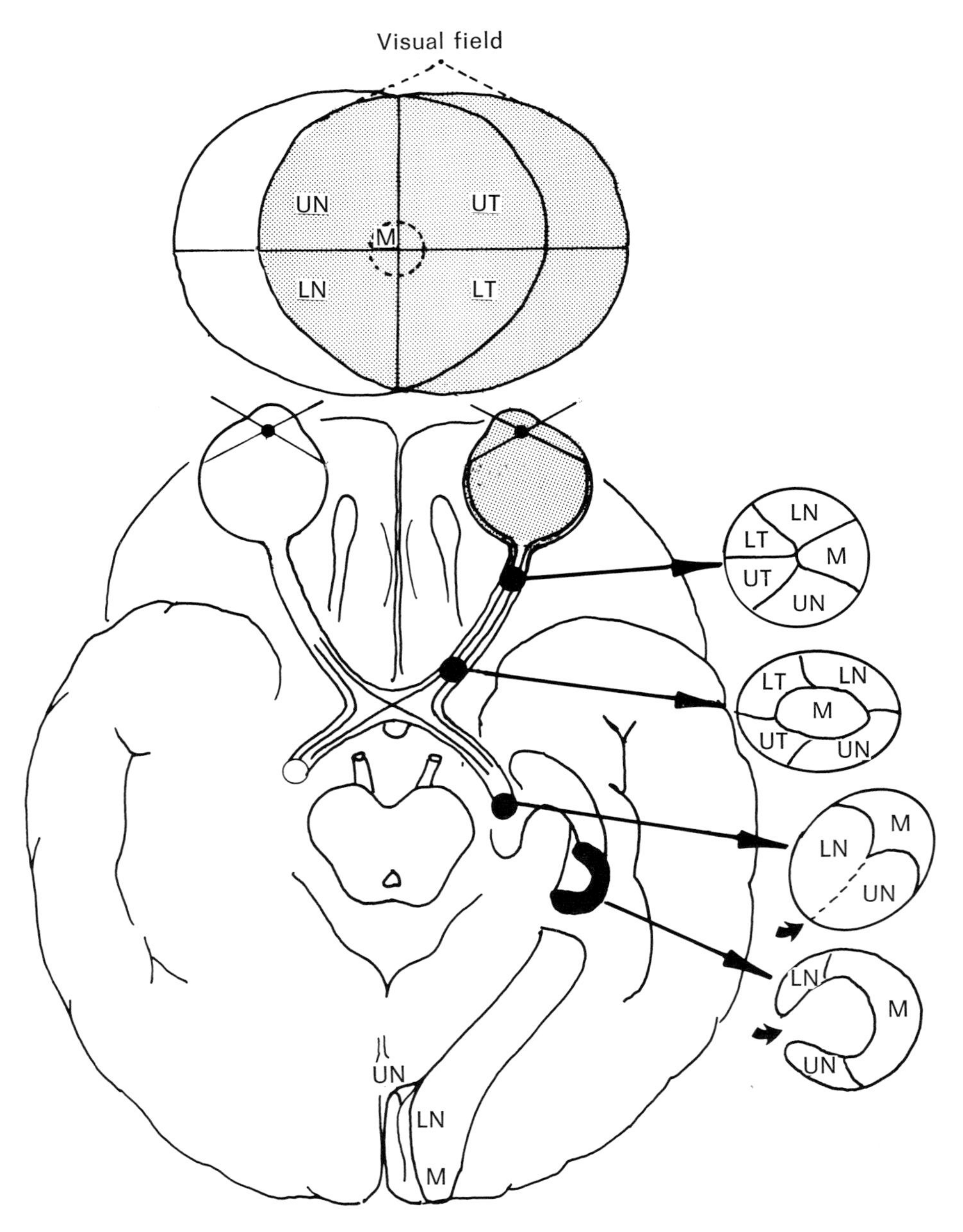

Fig. 23-2. The topography of retinal fibers in the visual pathway are indicated in this diagram according to the visual field: *UN*, upper nasal; *LN*, lower nasal; *UT*, upper temporal; and *LT*, lower temporal. Note the prominent area occupied by macular fibers *(M)*. Only the nasal field is represented past the chiasma, but fibers of both eyes participate. The optic tract undergoes a counterclockwise rotation compared to the optic nerve; the optic radiations "open up" along the dotted line as suggested by the solid arrows.

zation, proliferative retinopathy, and vitreous hemorrhages.

Conductive mechanisms

The optic nerve bridges the gap between ophthalmology and neurology. Increased intracranial pressure impedes the circulation at the disc, causing papilledema. The blind spot is enlarged, the disc swollen, and the venous pulse absent, but the vision is good unless prolonged. In optic neuritis, its chief mimic, vision is poor, there is a central scotoma, edema is minimal, and it is more likely to lead to optic atrophy especially after recurrent attacks typical of multiple sclerosis.

A pale, atrophic nerve head is a common finding in patients with poor vision, although appearance and vision are not always correlated. The causes are numerous; most important are glaucoma, multiple sclerosis, cerebrovasular accidents, brain tumors, or localized ocular disease. The field defect depends on the degree of nerve fiber damage and includes generalized constriction, central scotomas, or nerve fiber bundle defects.

Field defects are frequently a limiting factor in independent mobility, spatial orientation, and localization. An intact peripheral field is necessary to move about; 20/20 vision is of no help if confined to a central tunnel. Moreover any magnification throws the image into a nonseeing area, one reason patients with advanced glaucoma respond poorly to visual aids. Conversely, small central scotomas can be circumvented by magnification for the same reason one can read a partially obscured sentence. Lower altitudinal defects interfere with walking, and right hemianopia intrudes on left-to-right reading.

The functional evaluation of the field of vision, based on precise topographic representation of the retinal fibers in the visual path and cortex, is termed "perimetry." Typical defects help localize, if not identify, the cause of the interruption (Fig. 23-2). An isopter is the field boundary for a target of a given size, luminance, or color against a standardized background. Field defects may be partial or complete contractions or regional scotomas, which are classified according to size, shape, density, and whether visible or invisible

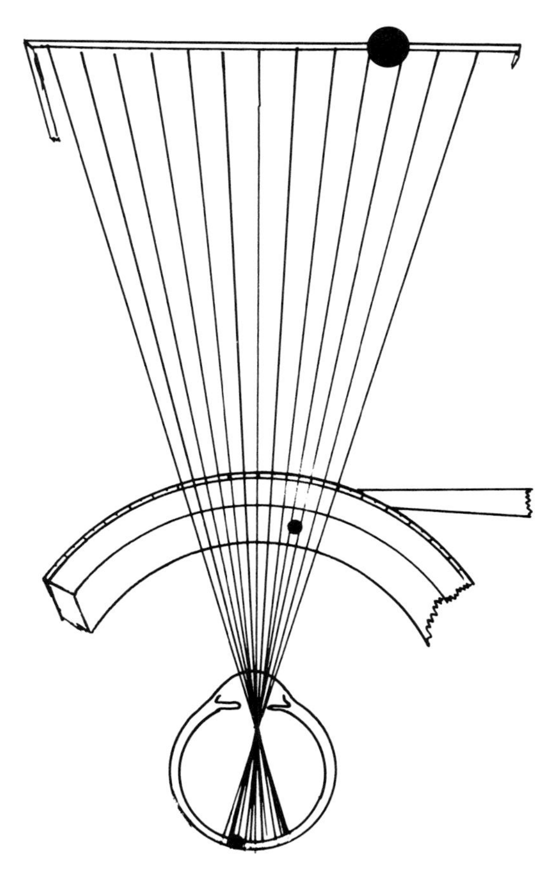

Fig. 23-3. Relative magnification of a scotoma (blind spot) by projecting it on a tangent screen, compared to a perimeter.

to the patient. Scotomas are best measured on a tangent screen that enlarges their appearance (Fig. 23-3).

The variety of defects produced by retinal lesions is ubiquitous, which helps in their diagnosis. Lesions of upper retinal fibers produce inferior field defects, and since they tend to remain up throughout their course, the rule is valid with some minor exceptions. The only location where a single lesion can produce a bitemporal defect is at the chiasma. It is now established that there is retrograde degeneration of ganglion cells with chiasmal lesions; yet despite a considerable reduction in ganglion population, acuity remains surprisingly good. Binasal hemianopia is so rare that it should probably be attributed to poor attention or poor technique until proved otherwise. All retrochiasmal lesions produce a homonymous field defect. Lesions of the tract and anterior radiation tend to be incongruous (i.e., nonsymmetrical); lesions of the posterior radiations and occipital lobe tend to be congruous. Several isopters must be plotted to demonstrate incongruity. Sparing of the macula is usually seen only with occipital lesions. A split macula can be confirmed by the disappearance of half the Haidinger brushes. Upper homonymous quadrantic field defects are characteristic of temporal lobe lesions; parietal tumors only produce defects after extensive disease; the main finding is anesthesia related to involvement of the postcentral gyrus. Occipital lesions, except for visual loss, are neurologically silent. Bilateral occipital lesions result in cortical blindness.

The turgidity of the eye, hence the stability of the refractive surfaces, is maintained by the intraocular pressure. A constant influx of aqueous is drained at an equivalent rate, a balance maintained by some unknown feedback mechanism. The delicate balance is vulnerable to malfunction, usually in the outflow mechanism. The result is glaucoma, a leading cause of blindness. The progressive cupping and associated visual field loss in untreated glaucoma are now attributed primarily to vascular insufficiency secondary to elevated intraocular pressure. Reduced blood flow to the optic nerve head results in relative ischemia with characteristic scimitar defects that progress in a predictable fashion. The reason for the vulnerability of the optic disc is related to its peculiar blood supply through the circle of Zinn, with shunting of blood and degeneration of glial-supporting tissue. Glaucoma occurs as a primary disease of unknown etiology or as a secondary complication to a variety of other ocular abnormalities (e.g., high myopia, central vein occlusion, retinitis pigmentosa, diabetes, retinal detachment, Fuchs dystrophy, traumatic hyphema) and may be induced by steroids, alpha-chymotrypsin, and as a surgical complication.

Central mechanisms

The age at which visual loss occurs, whether congenital, in early life, or in adulthood, determines what visual experiences have been established. Children who have never employed vision in their orientation and have no experience with visual images are likely to encounter greater difficulties in learning to read. Congenitally blind do not develop a system of thinking in word images and instead rely on verbal thoughts, which are usually closely linked to feeling. They may adapt mannerisms (frequently seen in mentally retarded blind children) involving rocking or handclapping to achieve sensory gratification. The person who loses vision later in life often develops normal patterns of visualization and continues in the same manner even after sight is lost.

Injury to the visual cortex produces a precise topographic defect. Injuries further forward produce reduction in general perceptual ability, such as the failure to perceive spatial relationships of objects or memory for visual events. Lesions affecting the intracerebral pathways are vascular (hemorrhage, ischemia, thrombosis, embolism, and aneurysms), tumors, infections, and trauma. The vascular lesions tend to sudden onset, with severe visual loss and variable resolution. Tumors tend to produce slowly progressive visual loss.

Flourens (1823) first established that vision depends on the integrity of the cerebral cortex. Panizza (1855) and later Hitzig (1874) localized it to the occipital cortex. Henschen (1890) and Wilbrand (1890), correlating human pathology, restricted it to the area striata. Differences in cerebral structural organization were first noted by Gennari, an Italian medical student in 1776, and Brodman (1909) used cytoarchitectural differences to map the brain into geographic areas. The primary visual area is area 17; the adjacent areas 18 and 19 are concerned with visual associations and perceptions, with input from other cortical areas. Visual disorganization in man resulting from lesions in these areas were first described by Holmes (1918).

Cortical blindness is characterized by loss of vision, with normal pupillary reflexes and a normal ophthalmoscopic picture. A striking feature of the condition is that the patient is blind to his blindness (Anton's syndrome), with visual hallucinations. Psycho-optical reflexes are absent despite normal ocular motility, such as no optokinetic responses. There may be a variable degree of recovery, especially when due to trauma or a vascular accident. The cause is generally some type of anoxia.

Agnosia is a failure to recognize sensory impressions, although the receptive mechanism is intact. Visual agnosia refers to the inability to recognize objects previously familiar despite the fact they are clearly seen or to appreciate their absolute or relative position in space. Generally, the cause is extensive brain damage, involving parietal, occipital, and temporal lobes.

The hysteric adopts blindness as a means of whistling in the graveyard of his anxieties. A rare condition, it is characterized by sudden onset and short duration. Unlike cortical blindness, the patient dodges obstacles and avoids collisions, indeed views his predicament with apparent indifference.

Micropsia and macropsia may be due to retinal disease (separation or crowding of functioning rods or cones) as seen in macular edema, or, more rarely, with cortical disturbances. Macropsia is frequently associated with convulsions or migraine. It may also occur as a transient phenomenon in drug intoxications (including metamorphopsia and visual hallucinations). Visual hallucinations are associated with epilepsy, recent blindness, and cortical disease and may arise from stimulation of visual fibers at any level. It is generally assumed that formed hallucinations arise as the result of temporal lobe involvement; unformed flashes, sparks, streaks, etc., originate in the occipital lobes.

A flash of light causes an electrical response of the visual cortex, but the results are masked by other brain waves unless extracted by a signal averaging system. The equipment for recording this visually evoked response (VER) is a standard electroencephalogram coupled to special digital computers. The record summates many responses, the light stimulus is an electronic flash, and the result reflects the function of the entire visual system from retina to cortex. The amplitude is greater for the photopic retina and the macula and hence can be used to measure acuity or visual fields objectively. It has even been used to measure refractive error. Obviously this technique is indicated only in special clinical situations in which subjective measurements are impossible. The chief technical problem is response variability, and more research is needed to evaluate its diagnostic potential.

In most blind people the eyes remain quiet unless voluntarily deviated from the midline, whereupon irregular, inconstant, slow "searching movements" occur. The mechanism appears to result from loss of sensory input to the visual system. A pendular nystagmus, sometimes jerky (with the fast component opposite to gaze direction), generally horizontal and conjugate, accompanies blindness acquired in early life.

Congenital nystagmus continues throughout life and is always bilateral, usually horizontal; pendular in character, becoming jerky on lateral gaze; and sometimes with compensatory head movements, which tend to become less pronounced. The cause is not known. Fortunately reading is generally possible if the head is held in a favored position (about 80% of these children can attend regular schools). Strabismus and refractive errors, often present, should be corrected.

A rare form of nystagmus with head nodding and torticollis in infants and young children is spasmus nutans. It never starts after the third year of life, carries an excellent prognosis, and requires no treatment except for associated refractive errors or strabismus.

Nystagmus may be physiologic but never in forward gaze. A nystagmus without vertigo is likely to be of central origin. Vertical nystagmus is frequently caused by drug intoxication. Positional nystagmus accompanies pathology of the labyrinth and may be encountered after head trauma or whiplash injury (along with convergence insufficiency).

Psychologic aspects

The anxieties of normal people relative to blindness are multiplied and exaggerated in those with partial visual loss. There may be fear of total blindness, of progressive incapacity, and of using the eyes for any task at any distance lest vision deteriorate further. After a sudden loss of vision there is a stage of depression, self-pity, recriminations, and hopelessness. During this period, the patient may blame the physician, the family, the employer, or the fates, followed by acceptance or denial of his blindness. Denial leads to rejection of any visual aid and a continuous search for a cure, so that little rehabilitation can be accomplished. Acceptance of the visual handicap despite its unpleasant realities sets the stage for more suitable optical rehabilitation. Sudden visual loss may provoke extreme anxiety in the ophthalmologist, as well as in the patient, but he should not blunt his own feelings of despair by holding out false hopes that can never be realized. Patients should be aware of the prognosis, at least whether the condition is reversible or progressive, and what future medical or surgical treatment is contemplated.

Younger patients with subnormal vision require professional guidance in their education. The decision whether it is better to have the child struggle with sight than commit him to a school for the blind or sight saving classes may require circumventing unenlightened official opinion and depends on the available resources as well as parental cooperation. The older view of sight conservation is gradually giving way to sight rehabilitation. Parents must be helped to balance guilt feelings and avoid overdependence. Teachers must balance independence with the frustration of keeping up with peers.

No patient need learn Braille if he can read type; conversely, Braille is indicated in severe visual loss not correctable by optical means or when progression is anticipated. Braille requires motivation, tactual facility, intelligence, and discrimination. Unfortunately the variety of occupational and reading material is severely limited. Finally, recall that a child who fails to flourish may suffer other neurologic, systemic, or behavioral anomalies.

Congenital ocular anomalies may be the sequel of abnormal genetics or intrauterine disturbances or occasionally both. Similar pictures follow different causes, depending on the type of anomaly and its specific time of action during organogenesis. Since birth is merely an incident in time, the term "congenital" has no etiologic significance. Some anomalies do not manifest themselves until long after birth, although they are due to maldevelopment. The fact that a disease is genetic in origin does not preclude effective treatment, such as congenital glaucoma. Appropriate genetic counseling can suggest cessation of propagating the noxious gene to future generations, and adequate environmental control helps promote maternal and fetal health during pregnancy (e.g., rubella) and after birth (e.g., retrolental fibroplasia).

If the handicap starts in middle life, the by-products are not only psychologic, but, in addition, there is loss of earning capacity and the need to redirect the career. Particular resentment is felt at the loss of independence, mobility, and the frequent isolation from former friends and acquaintances. In such instances it may be advisable to wait until the patient is prepared to make the necessary effort to use his reduced vision.

Older people make up the largest proportion of the visually handicapped, and debility, senility, arthritis, paresis, or tremors interfere with using a visual aid. Many develop an adaptational complacency and resist attempts to change their visual habits, especially if of long duration. Only a limited correction may be needed to satisfy their limited visual requirements. In contrast to children who cannot remember having good vision, the older patient is more likely to be resentful, despondent, and uncooperative.

OPTICAL PRINCIPLES

Few topics in optics are less clearly grasped, more dearly avoided, and most willingly forgotten than magnification. Magnification is indicated when vision with ordinary spectacles is so reduced that the patient finds it difficult or impossible to pursue desired tasks. Optically "magnification" means that the size of the image formed by a lens is larger than the object—the weaker the lens the further the image and the larger its size. Obviously, this is not what is meant by physiologic magnification, which is concerned with spreading the optical image over a greater number of photoreceptors—the stronger the lens the greater the visual angle and the higher the angular magnification. Clinically we care only about angular magnification. It may be increased by enlarging the target (by projection, television, or large type books) or by bringing it nearer. The simplest and greatest magnification is obtained by approaching the target. For example, if a target is moved 100 to 10 cm from

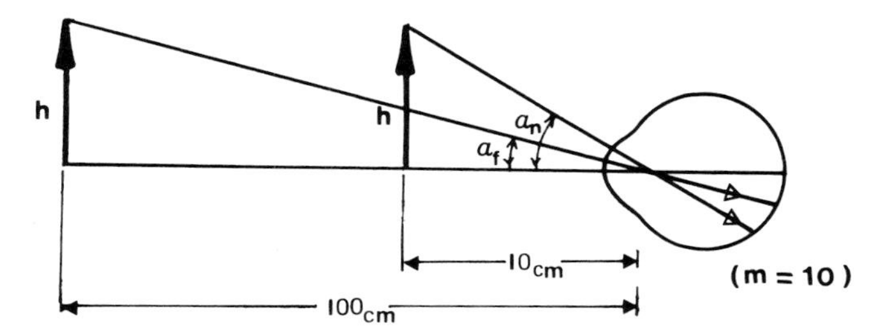

Fig. 23-4. Illustration (not to scale) showing the visual angle (α) is increased tenfold when the same target *(h)* is moved from 100 cm to 10 cm.

the eye, the angular magnification is increased tenfold (Fig. 23-4). It is only necessary for the eye to muster enough accommodation to see it clearly or for a lens to be supplied to make up for any accommodative insufficiency. It is true that the lens (by virtue of its power and shape) adds some magnification, but this is inconsequential compared to altering the distance. (Recall that a +10 D aphakic spectacle lens produces about 1.26 or 26% magnification of a distant object compared to the tenfold increase by reducing viewing distance one tenth.)

Children or adults with an adequate accommodative amplitude or high uncorrected myopia who can reduce their viewing distance seldom need an optical magnifier; presbyopes must be supplied with an appropriate add that may, of course, be considerably stronger than those for ordinary reading distances (e.g., a +10 D add for 10 cm, a +20 D add for 5 cm). By convention, adds stronger than +4 D are called microscopic lenses and may have special design characteristics such as aspheric curves, doublets, or even triplets to improve appearance, reduce weight, or eliminate aberrations, but their optical purpose is the same.

Strong reading adds are the most important means of clinical magnification, often preferred to all others when feasible. A big advantage is that, unlike a hand loupe, they leave both hands free and, since they are worn close to the eye, permit a wide field of view. Naturally there are limitations because as the target approaches the eye, binocular vision cannot be employed, it becomes progressively more difficult to illuminate the target properly, and the optical aberrations of strong lenses limit their performance. A minor optical problem but a major psychologic one is to get patients to hold the target close to the eye—they are not used to it, do not understand it, and do not like it, and some are afraid it may further damage their sight. Because the optics are strong, the depth of focus is reduced; so the patient without any residual accommodative flexibility is obliged to hold the target fairly steady at the required viewing distance. Thus a fundamental rule in supplying a visual aid is not only to choose the most appropriate optics but also to make it suit the patient's capability and requirements and to teach him how to use it.

Strong spectacle adds will not help the subnormal vision patient view a stage performance, discriminate street signs or a bus number, or recognize a price tag on a shelf. They also effectively prevent any work with the hands, which must necessarily be carried out at some reasonable distance (the same problem, incidentally, that the eye surgeon faces when he uses a "loupe" or microscope to achieve magnification at an operable distance). There is a tendency to confuse reading with near vision; many patients with poor vision will forego books and newspapers but not watching television, playing cards, signing checks, or inserting keys into doors. The necessary discrimination may be considerably less than for reading, but it cannot be carried out at a few inches from their face; so the required optical aid is a telescope (Fig. 23-5). Obviously it must be a Galilean system

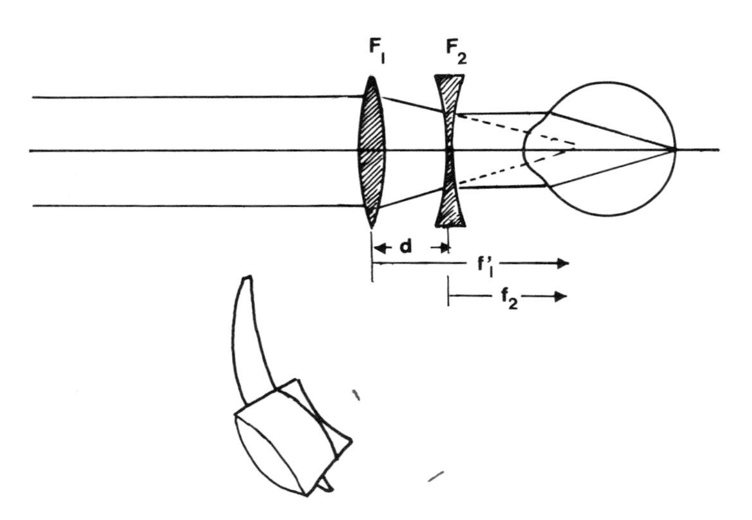

Fig. 23-5. A Galilean telescope consists of a plus and minus lens so arranged that the first focal point of the second lens coincides with the second focal point of the first lens. It produces a virtual, erect image, which is viewed by the eye. The magnification is in the ratio of the two focal lengths. A compact Galilean telescope can be constructed of solid glass or of individual lenses for insertion into a carrier spectacle lens (insert).

to provide an erect image. Since telescopes only change the visual angle, not the vergence, it is equally obvious that an ametropic eye will see enlarged blurs; so all refractive errors must be corrected. Alternatively the telescope is provided with a focusing adjustment (the myope reduces the distance and the hyperope increases the distance between the lenses of the Galilean system). The same adjustment or a reading cap permits its use for viewing objects nearer than optical infinity.

The magnification that can be practically achieved by clinical telescopes is limited to about 2× for several reasons. The patient cannot be expected to carry a long spyglass, so the lens system must be compact and light-weight; yet the optics cannot be so strong that aberrations limit performance. Even more important, the field of view depends on the diameter of the objective lens, which must be as large as possible. It is the reduced field of view that makes telescopes impractical for everyday wear (just as the glaucoma patient with only a central island of vision cannot move about freely). Their use is relatively restricted to stationary situations in which activity is confined to a small area (as in viewing television or a surgical field). Telescopes magnify movements as well as objects, and rapid head motion causes swimming. If the patient has a restricted visual field to begin with, putting a telescope before his eye will make him effectively blind; hence their use is limited to specific tasks. They may be worn as a kind of "add" in a spectacle (either for distance or near viewing), using the carrier for mobility, or carried in the pocket and brought out to see a distant object such as a bus, street, or house number. Telescopes are never worn while moving about, and their suggested use for driving under some circumstances is both ill-advised and dangerous—at any speed.

A hand-held convex lens is called a "simple magnifier." The simplicity is in its use (and price), not its optics, which is confusing because of several variables involved. Like a presbyopic add, a loupe reduces the need for accommodation by forming an image further from the eye than the original object. The image must be erect, so it is necessarily virtual. In fact, when the object is at the focal point of the lens, the image is at infinity, but it is now so far away that it can no longer be seen clearly. The use of a hand magnifier thus involves a compromise between bringing the object nearer without having to accommodate and the distance at which the virtual image is clear. The angle formed by the virtual image determines magnification.

Logically we compare the visual angle formed with the lens to the visual angle formed by the object without the lens. But where shall we place the object when viewed without the lens? By convention and on a purely arbitrary basis, the "standard" distance is set at 25 cm—the conventional "least distance of distinct vision." The angle formed by the virtual image (seen through the magnifier) is compared to the angle formed by the same object held at 25 cm. The standard eye is therefore not only assumed to have a maximum amplitude of 4 D but also to prefer reading at 25 cm. It matters not that most subnormal vision patients have a lesser amplitude or prefer a larger distance. The standard 25 cm seems established beyond hope of change at this writing.*

The principle is illustrated in Fig. 23-6. In the top sketch, an object *(h)* subtends an angle α at the entrance pupil of the eye. In the lower sketch, the object is viewed through a magnifying lens *(F)*, which forms an image *(h')* subtending an angle α'. The magnification (sometimes called "relative magnification") is as follows:

$$m = \alpha'/\alpha = \frac{h'/l' - d}{h/ -0.25}$$

In the special case where the target lies at the first focal plane of the lens, all the rays emerge parallel, and the distance of the lens from the eye does not affect the visual angle, which is now identical to that formed at the optical center of the lens. Hence:

$$m = \alpha'/\alpha = \frac{h/f}{h/ -0.25} = \frac{1}{4f} = \frac{F}{4}$$

The expression F/4 is the one used most often in specifying magnification; for example, a +8 D lens is said to have a magnification of 2×, a +16 D has a magnification of 4×. If in doubt, the lens should be placed in

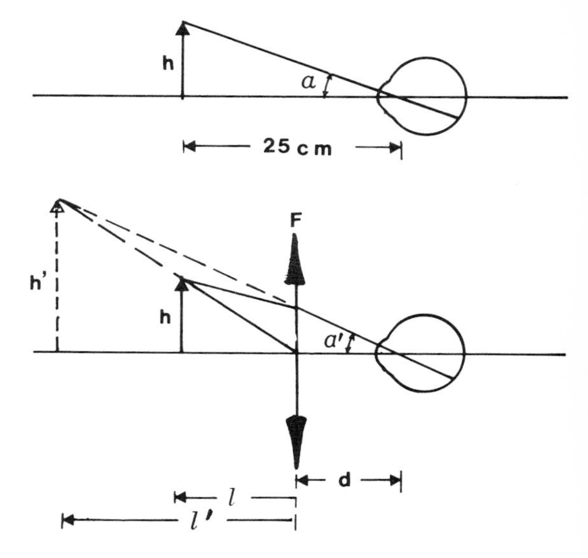

Fig. 23-6. Conventional magnification compares the angular magnification of a lens (α') *(bottom)* with the magnification (α) of an object at a distance of 25 cm *(top)*.

a lensometer and the power read off directly to determine what the magnification label attached to the device by the manufacturer means.

We said that when the object is held at the focal point of the lens the rays emerge parallel, and its distance from the eye has no influence on magnification. Fig. 23-7 shows that it does affect the field of view, however, since the shorter the eye-lens distance the larger the field. If the field of view is reduced too much, the patient sees only a word or perhaps a letter at a time, a situation not conducive to easy reading. Indeed, if reduced too much, the small field prevents localization completely, and the patient cannot find the target he wants to see (like looking for the moon through a high-powered telescope).

When the target is slightly inside the focal point of the magnifier, the virtual image is not at infinity but at some nearer distance, and adequate accommodation must be available to see it (or a presbyopic add supplied). Since the virtual image is seldom nearer than arm's length, the patient can wear his regular bifocals or reading glasses to use the magnifier. Most fixed loupes are in fact designed

*Southall (1933) proposed 1 meter, Ellerbrock (1954) and Sloan and Habel (1957) suggested 40 cm, and some "standard" near test cards are designed for viewing at 35 cm (14 inches).

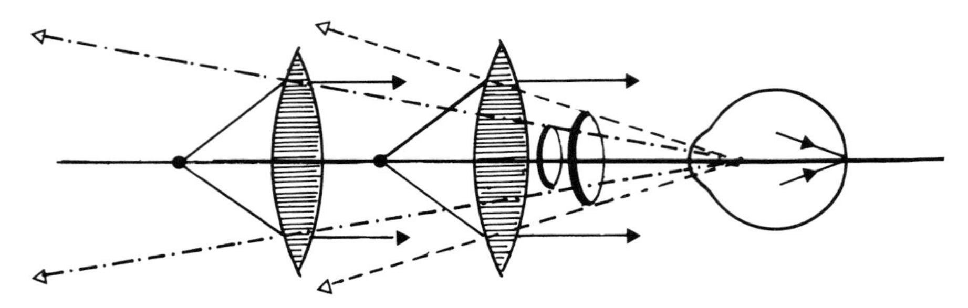

Fig. 23-7. Diagram illustrating that as long as the target occupies the focal plane of the lens, the emerging rays are parallel, and the focus is not influenced by the vertex distance. But the field of view is greatly reduced as the lens is moved farther from the eye.

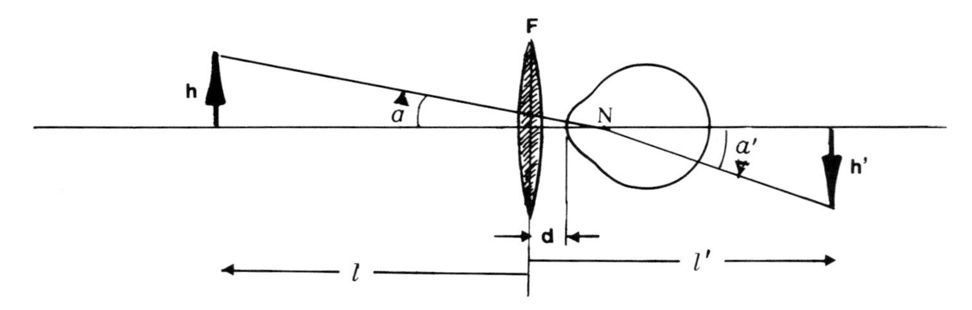

Fig. 23-8. Diagram used to derive the general magnification equation. For details see text.

to discharge divergent waves to compensate for the proximal accommodation induced by the near target.

Simple magnifiers are useful because they allow seeing an object at a convenient reading distance with moderate optical magnification but considerable enlargement resulting from the nearer distance.

Although magnification can be increased by stronger lenses, thus allowing nearer viewing distances, these reduce the field of view, increase aberrations, interfere with binocularity, and make illumination difficult. Hand magnifiers are suitable for brief reading, such as looking up phone numbers or checking price tags, but the fact that they must be held by older people subject to tremor, paresis, or arthritis makes the reduced depth of focus a serious handicap and limits such lenses to no more than +10 to +20 D. A stand magnifier is a useful substitute, since the target is automatically positioned at the focal plane, and allows binocular viewing but is restricted to an adequately illuminated flat surface, held perpendicular to the line of sight. The selection of these aids has been improved by a focusing adjustment that eliminates the need to maintain a critical object to lens distance, so the patient can hold the magnifier close to his eye to increase his field of view.

A general formula for magnification can be derived from Fig. 23-8, based on thin lens (i.e., paraxial) theory. In this diagram, a lens of power F is placed at a distance d from the entrance pupil (simplified to nodal point N). The angular size of the object h at N is α. The angular size of the image h' at N is α'. The object is at a distance l beyond the focal point of the lens, and the image is formed at a distance l' (and serves as the target for the eye).

Since the angles are small:

$$\alpha'/\alpha = \frac{h'/l' - d}{h/l + d} = \frac{h'(l + d)}{h(l' - d)}$$

Since $l'/l = \frac{h'}{h}$, substituting:

$$(l'/l)\left(\frac{l + d}{l' - d}\right)$$

Dividing by l'/l:

$$m = \frac{1 + d/l}{1 - d/l'}$$

Since $L = 1/l$ and $L' = 1/l'$ and $L = F - L'$:

$$m = \frac{1 + d\,L}{1 - d(F - L)}$$

The formula gives the magnification of any optical device by comparing the angular size of an object seen with and without the instrument.

When the target is at infinity, $L = 0$, and $m = \frac{1}{1 - dF}$, which is the standard magnification formula for thin spectacle lenses already derived in Chapter 7. When $F = L$, the object is at the focal plane, and the object seen with the lens is at the same distance as the object without the lens and $m = 1 + dF$. To compare angular size with the lens to the angular size at some specified distance (p): $d = (p - f)$ and $m = 1 - dF = 1 - (p - f)F = pF$. In this formula the target is held at a specified distance (p) from the eye, and the lens is held so that the target lies at its focal plane. If the specified distance is the conventional 25 cm, $p = 0.25$ and $m = F/4$.

If lens thickness is to be included, recall from Chapter 3 that the shape magnification is $m = \frac{1}{1 - t/n\,F_1}$, where t and n are the thickness and the index of the glass and F_1 the front surface power. The total magnification is the product of the power factor (i.e., the general formula just derived and the shape factor. F in the power factor is now the back vertex instead of the equivalent (thin) lens power.

An example will illustrate the application of the formulas. A target is held at 40 cm. If a +5.0 D (thin) lens at a vertex distance of 2 cm is now interposed and the target is brought to the focal point of the lens (20 cm), what is the magnification?

The reference distance in this example is 40 cm. Since the target is brought to the focal point of the lens, $F - L = 0$ and $m = 1 + dL$ or $1 + (0.02)(5) = 1.10$. This is the magnification assuming the target remains at the same distance, but there is a positional change from 40 cm to $20 + 2 = 22$ cm, yielding a magnification of $40/22 = 1.82$ and a total magnification (power and position) of $(1.1)(1.82) = 2.00$. Obviously the greatest magnification (82%) is brought about because the target can be moved nearer; only 10% is due to the lens per se.

The same result is obtained directly from $m = pF$, where (p) is the reference distance (40 cm in this example); $m = 0.4\,(+5) = 2.00$. If the conventional 25 cm distance had been used, the magnification would be $5/4 = 1.25$ because the positional change is only from 25 to 22 cm.

If we assume the lens has a thickness of 5 mm and a front surface power of +2.50 (i.e., equiconvex), the shape magnification is

$$m = \frac{1}{1 - t/n\,F_1} = \frac{1}{1 - \frac{0.005}{1.523}(+2.50)} = 1.008$$

and the total magnification of our example is $(2.00)(1.008) = 2.016$. It is evident that shape contributes negligible magnification for weak powers.

It is assumed in these examples that the patient can muster sufficient accommodation to see the near target clearly. If he can bring in additional accommodation, the nearest point of distinct vision will be closer and the magnification larger. (The power of the lens and of accommodation add as a lens system, separated by the vertex distance.)

EXAMINATION

Patients with subnormal vision form an appreciable percentage of every ophthalmic practice, and although their evaluation is an extension of routine methods, it takes more trouble, more time, and more tact. A significant proportion of patients with subnormal vision (up to 20%) may be helped by a proper refraction and ametropic or presbyopic correction, and even the patient waiting to have surgical treatment deserves the best vision possible in the meantime.

Clinical history

The history establishes whether the visual loss is recent or long-standing, whether the onset was sudden or gradual, the age at onset, whether the condition is progressive or stable, and the results of previous examinations or treatment. Determine the presence of associated diseases, particularly those which require control (hypertension, diabetes, glaucoma), the intake of medication, or the output of renal disease. The purpose of a potential visual aid is evaluated, that is, whether for school, industrial use, or occasional reading, walking, television, driving, etc., and the conditions under which it is to be used (working distance, size, contrast, color, illumination). The acceptance of a visual aid depends on the motivation to use it, coupled with cosmetic considerations and the fear of progressive visual loss from the use of the eyes. Where pertinent, the family history may throw light on genetic and congenital causes of poor vision.

The patient's bearing, mobility, head and eye posture, nystagmus, the use of an optical or nonoptical aid, tremors, palsies, and mental status can generally be established at a glance.

Visual acuity

Legal blindness does not mean there is no measurable vision; hence a patient who can see letters at some distance is a potential candidate for visual aids. Naturally an eye with better than 20/200 acuity has the best prognosis, whereas hand movement vision is seldom improved by optical devices. Much depends on central factors of interpretation, which allows getting the most out of a blurred image.

Visual aids are frequently selected on the basis of acuity; hence accurate measurements are important for more than medicolegal reasons. The patient should wear the optimum distance correction, with the targets appropriate for the age, of good contrast, and properly illuminated. The test distance is almost always closer than 20 feet (usually 10 feet and sometimes 5 feet). These distances give more choice and selection than conventional projected letters. Both single letter and multiple letters per line are tested. The patient should be allowed to tilt or turn his head for best vision. In converting from one test distance to another, the reciprocal is equivalent; thus 20/400, 10/200, 5/100 all represent 1/20 resolution. Each eye is tested and the results recorded separately. Generally, a visual aid will be prescribed for the better eye.

Acuity varies sharply with retinal eccentricity (Fig. 23-9); hence a small macular lesion may be devastating for central vision. Some patients learn and others must be taught to use an eccentric area for discrimination. In certain types of nystagmus, vision is best in a particular direction of gaze. Achieving the best acuity also depends on the educational background, the capacity to understand instructions, attention and fixation, motivation to achieve, and familiarity with the test situation. These factors are more disrupting with children than adults. Near vision is often incommensurate with distance acuity so that one cannot be predicted from the other in damaged eyes.

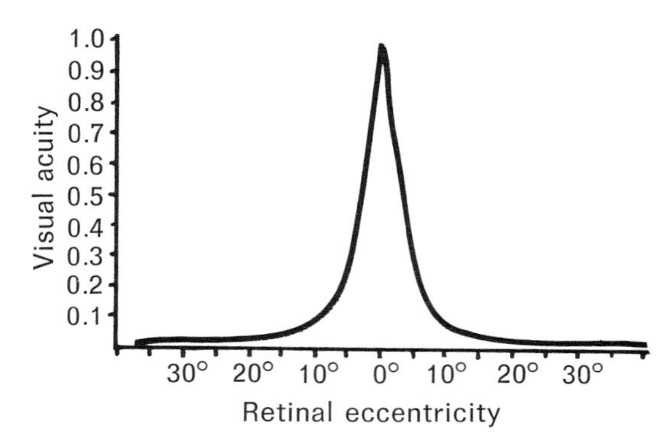

Fig. 23-9. Graph showing the rapid decrease in acuity with retinal eccentricity.

Reading vision

Reading with low vision aids is difficult even under favorable conditions because the letters are closer together, their legibility varies, the contrast depends on the paper and illumination, and reading ability changes with educational background. Whereas individual letters can be recognized by eccentric fixation in central scotomas, there is more interference in recognizing words and sentences. Continuous text therefore more nearly represents the visual task, and acuity alone does not indicate facility in reading.

Sloan and Brown (1963) devised samples of continuous text increasing in size from newsprint to print ten times larger. The cards are viewed from a reading distance of 40 cm, the patient wearing his distance correction and an appropriate presbyopic add. The text begins with larger targets and proceeds to smaller sizes until easy reading is no longer possible. Thus if the smallest size read is 5 M (i.e., five times the size of newsprint), one assumes he will be able to read newsprint at 1/5 the distance (i.e., 8 cm). The trial reading lens is selected on this basis (i.e.,

Table 23-1. Sloan print for reading acuity

Size of print in Sloan M units	*Acuity required at 40 cm*	*Equivalent distance acuity*	*Dioptric power of reading addition required to read 1 M print (assuming emmetropia and zero accommodation)*
1.0	40/100	20/50	+2.50 D
1.5	40/150	20/75	+3.75 D
2.0	40/200	20/100	+5.00 D
2.5	40/250	20/125	+6.25 D
3.0	40/300	20/150	+7.50 D
4.0	40/400	20/200	+10.00 D
5.0	40/500	20/250 or 16/200	+12.50 D
7.0	40/700	20/350 or 11.5/200	+15.00 D
10.0	40/1000	20/500 or 8/200	+25.00 D
14.0	40/1400	20/700 or 5.8/200	+30.00 D
20.0	40/2000	20/1000 or 4/200	+50.00 D

$\frac{100}{8}$ = +12.50 D). The equivalent print and M notation is given in Table 23-1.

There is often an abrupt transition from easy reading to slow or difficult recognition. Children and unskilled readers may read all print above a certain size but never manage easy reading. Such patients should be given a visual aid that correlates with their natural reading rate. Above a certain level of magnification, reading may decrease because the image falls into a paracentral scotoma or hemianopic area. Finally, no visual aid helps the patient who cannot or does not want to read.

Ocular examination

An eye with subnormal vision is almost always a diseased eye, requiring a medical diagnosis not only to evaluate the prognosis but the current status as well. It has been estimated that 10% are treatable, and many others require periodic monitoring to prevent progression. External examination, biomicroscopy, intraocular tension, gonioscopy, fundus studies, visual fields, adaptation, color vision tests, and selected ancillary procedures are used as indicated by the clinical problem.

Ophthalmoscopy is of particular importance and requires a dilated pupil. Indirect ophthalmoscopy affords better visualization when the media are hazy, and biomicroscopic fundus examination with a Hruby lens permits suitable enlargement and cross-sectional views of the posterior vitreous and fundus lesions (e.g., whether elevated or depressed). Vitreous opacities can be identified by ocular rotation and localized by parallax.

Refraction

A healthy eye is subject to refractive errors, a diseased eye even more so. Unless ametropia and presbyopia are properly corrected, the patient is unable to use a visual aid to best advantage. Age is a further handicap; discrimination is poor, and perceptual judgments take longer; hence it is best to set aside more time for these examinations.

Retinoscopy expedites the refraction but is not always possible because of irregularities and opacities of the ocular media. It may be necessary to approach within ophthalmoscopic distances (i.e., a few inches from the eye, so-called radical retinoscopy) to see a distinct reflex. In selected cases, retinoscopy is done with a contact lens in place. Keratometry is occasionally useful in high ametropia or when there are media opacities and if compared to other findings.

Subjective refraction is generally more reliable than objective methods and always used to confirm the results. The most elaborate retinoscopy is useless in optic atrophy; so a pinhole indicates the optical potential. A trial frame properly centered is used with targets appropriate for the age and distance requirements. The target is kept perpendicular to the line of sight and well illuminated. Begin with larger lens changes and letters easily recognized. Changes are made slowly with sufficient time for discrimination. A temporary telescope may be inserted to improve blur appreciation. Small astigmatic errors (less than 2 D) can frequently be ignored; high cylinders should be prescribed. A useful trick to find the axis is to allow the patient to rotate a stenopeic slit himself or to arbitrarily increase cylinder power and rotate the axis to best vision. Stronger lenses are placed in the trial cell nearest the eye and the total number of lenses kept to a minimum to avoid reflections and aberrations. The final back vertex power can be read off directly from the lensometer. Both distance and near refraction is determined. Occasionally a binocular correction is possible (about 20% of cases), so motility and fusion are checked.

TYPES OF VISUAL AIDS

Visual aids, properly selected* and fitted, may rehabilitate 50% to 75% of selected handicapped patients otherwise considered blind. Children may begin to resume their education; adults, to return to gainful employment; and older people, to fulfill a measure of independence.

In evaluating a visual aid, consideration is given to its power, magnification, field of view, aberrations, convenience of working distance, weight, size, portability, cosmetic appearance, provision for illumination, durability, mechanical construction, ease of fitting and adjustment, and cost.

Distance vision

Distance magnifiers are essentially Galilean telescopes, compacted, light in weight, of limited field of view (about 14°), monocular or binocular, and of moderate to high cost. Some units are designed for incorporating into spectacle frames; some may be clipped on or carried about (prism monocular). A telescopic system is possible with a strong minus corneal contact lens (e.g., −50 D) and a spectacle lens of +25 D in a spectacle frame at a vertex distance of (4 cm) + (−2 cm) = 2 cm. If the contact lens is designed with a weak carrier, the patient may use peripheral vision for getting about, but the practical difficulties are formidable.

Spectacle telescopic systems provide about 2× magnification (e.g., 1.8× and 2.2× in the Kollmorgen lens; 2.5× to 2.8× in the Selsi sport glasses; similarly for the Keller lenses, available in reduced aperture form with a carrier for peripheral vision). Monocular scopes provide magnification from 6× to 12× and are normally hand-held. A Selsi monocular can be clipped onto the regular spectacles (2.5×). A focusable, mass-produced, binocular spectacle scope may range from 1.5× to 4×. A compact Cassegrain (mirror telescope) was designed by Feinbloom and gives 3.5× magnification, with very limited field. A simplified double spectacle lens design may be used to form a Galilean system at surprisingly little cost. Patients with high irregular astigmatism may be fitted with standard contact lenses.

All nonfocusable telescopes are afocal, and the patient must wear his ametropic correction to use them. Their practical use is fairly limited to sedentary situations because of the restricted field and the magnification of motion. The headborne devices are indicated for more sustained tasks (television, movies, theater, blackboard), the monocular scopes are used intermittently (house numbers, street and bus signs).

For intermediate distance vision (reading music, price tags, certain restricted manual tasks involving a limited field and little movement) the telescopic system is fitted with a reading cap or focused for the appropriate distance. Naturally the depth of field is very limited, the chief drawback. The uncorrected aphakic patient may use a +3.00 D hand lens at about 10 inches to achieve 4× magnification for intermittent use (the uncorrected

*A catalogue of visual aids from various manufacturers is available from the National Society for the Prevention of Blindness (16 E. 40th St., New York, N. Y., 10016) and the New York Association for the Blind (111 E. 59th St., New York, N. Y., 10022).

aphakia, e.g., 12 D, acts as the minus component—hence 12/3 = 4×) as suggested by Fonda (1970). It is not unusual for patients to expect optical miracles from telescopic systems, at least before they have tried them; so a home trial should be arranged if possible to avoid inevitable disappointments.

Near vision

Near vision provides the greatest potential utility for subnormal vision aids. Recall that the child with high amplitude of accommodation or the adult with high myopia has a built-in reading magnifier preferable to most artificial aids (if astigmatism, corneal irregularities, or hazy media do not prevent it).

Near vision aids include strong spectacle adds (those above +4 D are called "microscopic lenses"; those above +8 D generally require some special design modification to reduce aberrations), hand-held magnifiers, and fixed focus and variable focus magnifiers.

Strong convex spectacle adds, either monocular or binocular, coupled with good illumination (which may be difficult at these close distances) provide good flexibility. The chief difficulty is the restricted field; it is difficult to read one word at a time—impossible to read one letter at a time. The magnification of the lens is F/4, and the lower powers should be planoconvex (up to +14 D) and biconvex for stronger powers. The lenses are available with aspheric curves in plastic (Jardon), as clip-on loupes comparable to the ordinary jeweler's loupe (Selsi, Ary), as microscopic lenses (Kestenbaum, Bechtold, Policoff, American Optical, Volk, Keeler, Feinbloom, and others), for single vision, or as bifocal inserts. Some of the microscopic lenses are systems (doublets or triplets) giving 8× to 22× magnification.

Simple magnifiers are probably the oldest known optical aid, their deserved popularity being due to simplicity of use, availability, cheap cost, and flexibility. Hand magnifiers are simply strong convex lenses mounted with a handle, for suspension around the neck (to keep both hands free), or even attached to the index finger by a small ring and swivel. They range in power from 4 D to 68 D, smaller powers (up to 10 D) for ordinary "readers." They are easily carried and brought out to look up phone numbers, labels, prices, notices, letters, etc. Hand magnifiers allow variation in the working distance, have good cosmetic acceptance, and are useful as an introduction to other types of visual aids. The patient must wear an appropriate presbyopic add when the virtual image is at some near distance, but the add must not be excessive (usually not over 5 D). Although maximum optical magnification occurs when the object is in the focal plane of the lens, the virtual image is now so far that the spreading magnification of the retinal image is reduced (for example, the angular magnification of an object 20 cm from the eye held at one half the focal length of a +10 D lens is not 10/4 = 2.5 but $\frac{2/25}{1/20}$ = 1.6, roughly half, and the needed accommodation is 4 D).* When the lens touches the target, the magnification is unity. Higher powered magnifiers reduce the field about 12° and increase aberrations. The chief disadvantage is maintaining the proper object-to-lens distance, especially by older people, and the reduced field makes reading difficult.

Stand magnifiers are designed to overcome the difficulty of maintaining an exact object-to-lens distance, critical with reduced depth of focus. A variable focus system (Sloan, Keeler) allows for ametropia and presbyopia. They often require accessory (built-in) illumination (e.g., Adisco). The available magnification ranges from 3.5× to 33 D, higher for the focusable magnifiers (up to 54 D). Large stand magnifiers may permit binocular vision.

*A simple rule (Linksz, 1956) by which about 2× angular magnification can be achieved with any loupe is to bring the target one half the focal length of the loupe nearer than the preferred reading distance and the loupe another one-half focal length from the target. For example, if the preferred reading distance is 40 cm and the loupe is a +10 D, the reading page is held at 35 cm and the loupe at 30 cm from the eye. The virtual image will be at 40 cm. It is assumed the patient has 2.5 D of accommodation or wears a +2.5 D add.

Nonoptical aids

Not all helpful visual aids are necessarily optical; a variety of simple devices may significantly improve vision. Pinholes (occasionally pinhole contact lenses) or stenopeic slits satisfy the optics in keratoconus or corneal scars. The number and diameter of the pinholes varies, and adequate light is essential. Filters, shades, visors, or hoods are useful to cut down glare, partly by promoting mydriasis. A rectangular slit cut in black cardboard, as suggested by Prentice, reduces haze and improves contrast, as in moderate cataracts. Projecting devices and closed circuit television are expensive and not portable and magnify the field as well as the target so that reading may not necessarily be faster but is at least achievable. Large type books, newspapers, and magazines are helpful but bulky and not always available. Bold ruled paper and felt-tipped pens are useful for letters and note-taking. A frequently neglected nonoptical aid is a high intensity lamp properly directed on the target. A reading stand maintains a constant distance and keeps the target steady and perpendicular to the visual axis.

PRESCRIBING FOR SUBNORMAL VISION

In selecting a visual aid, the chief factor is the purpose for which it will be used and ordering the weakest magnification to accomplish it. Improvement inconsequential to a normal eye may mean the difference between a life of quiet desperation and relative environmental independence. Aberrations intolerable to good vision may be entirely acceptable when discrimination is poor. No patient should therefore be dismissed as beyond help until a fair trial of visual aids, and trial should not be denied merely on the basis of ocular pathology or fundus appearance. The important thing is not what optical theory predicts but what the visual results are.

Such terms as "legally blind," "partially seeing," "amblyopic," or "almost blind" are poor descriptions of the visual potential. The best method is to test the vision at the required distance and for the required needs.

Except for corrective lenses, distance magnifiers are seldom feasible for walking about. When conventional or contact lenses do not improve distance vision, a telescope may bring the target optically closer and is the only means of improving distance vision. A trial telescopic lens set is therefore required (ranging from 1.8× to 2.5×), as well as some higher powered prism monoculars. The test distance is usually nearer than 20 feet, the target must be well lighted, and the full emmetropic correction is worn. The telescopic lenses are placed in the front cell of the trial frame as close to the eye as possible. Note the smallest line seen and increase the magnification as long as vision improves. Small refinements in the refractive correction may also be made at this point. The test is done monocularly on the better eye; repeat for the other eye in binocular cases. The patient must be trained to use the telescope, a procedure which generally requires several visits. Particularly annoying will be the reduced field, small depth of focus, and changes in spatial perception due to parallax. An attempt is next made to correct for intermediate distances (arm's length) by means of a reading cap or focusing adjustment. The least costly device is selected.

Near vision is the area of most useful application of a visual aid. Motivation to see and read are major factors in accepting any aid, and the patient who has forgotten how to read easily must be taught to do so again. A trial set for near vision testing includes selected standard trial lenses, microscopic lenses and systems, clip-ons, and hand or stand magnifiers. Some bifocal test lenses are also useful.

Kestenbaum as early as 1936 proposed a simple procedure for selecting the preliminary lens for reading on the basis of acuity. The basic formula is that the dioptric power required to read Jaeger 5 type equals the reciprocal of the distant acuity. Thus if distance vision is 20/400, the add required to read J 5 (about magazine type) is 20 D. To read smaller type, Kestenbaum's formula is modified by a factor (e.g., multiply by 1.4 for J 3 or telephone directory print and by 2.8 to read J 1). If the patient has some accommodative amplitude, the amount that can be comfortably sustained is subtracted from the required add (similarly for uncorrected myopia if no distance correction is worn).

In Sloan's method, the smallest continuous reading type is designated 1 M; the letters subtending 1 minute at the standard test distance of 40 cm requires 40/100 vision. The largest type is 10 M (ten times larger than 1 M). The patient wears his distance correction and an appropriate presbyopic add (2.50 D for absolute presbyopia). The strength of the reading add is determined by the distance the 1 M target must be moved. Thus if 5 M type can be read at 40 cm, 1 M type should be visible at one fifth the distance or 8 cm, requiring a +12.50 D add. Again the sustainable accommodation or uncorrected myopia may be subtracted from the indicated add.

In prescribing a near vision aid, most patients require continuous reassurance and encouragement to read at the indicated close distance from the eye. Proper illumination such as a goose-neck lamp are also essential, and the emmetropic and presbyopic correction must be worn. Recall that young patients or high myopes may require only minimal reading aids or none at all in order to approach the target and that bringing the target nearer the eye is the simplest, most flexible, and greatest magnification achievable. In teaching the patient to use a strong reading lens, it may be necessary to move the page across a restricted field of view. The patient may select his own preference. An add of up to 20 D can be prescribed in spectacle form; stronger adds have lower acceptance, require special lenses or systems, or alternatively a hand or stand magnifier. In doubtful cases, and most cases are, a trial period of home use should be arranged.

Subnormal vision aids require meticulous dispensing, and close cooperation with the optician is essential. Particular effort is devoted to centration, weight, comfort, and cosmetic appearance.

ADJUSTMENT AND REHABILITATION

No other investigation, wrote Gullstrand, has greater claim on scientists than a conscientious measurement of refraction and visual acuity. And no other field of refraction requires more physiologic, optical, and psychologic insight than the patient whose vision cannot be improved by conventional means.

The problems of subnormal vision and blindness are difficult to meet. There is an understandable reluctance to tell a patient a poor prognosis and a desire to postpone, procrastinate, and circumvent the magnitude of the disaster. Nevertheless, it is necessary to avoid false hope, which will delay adaptation by a vain search for a cure.

The eye is the first circle, wrote Emerson, the horizon which it forms the second; and throughout nature this primary picture is repeated without end. It is the highest emblem in the cipher of the world. Our life is an apprenticeship to the truth that around every circle, another can be drawn; that there is no end in nature, but every end is a beginning; that there is always another dawn risen, and under every deep a lower deep opens.

About 75% of patients with subnormal vision may achieve a successful outcome and carry out meaningful visual tasks at near, depending on motivation, adaptive training, and continued observation. A community low vision center, such as the Lighthouse for the Blind, and ancillary professional help (optometrist, optician, psychiatrist, social worker) can do much to relieve the fear, anxiety, and emotional depression of sudden visual loss. Low vision care requires regular follow-up usually at annual intervals, more frequently in cases of progressive diseases or active pathology. With modern technology, the outlook for the totally blind is also improving beyond the cane and seeing-eye dog.

Experimental use of electronic devices to help the blind are beginning to enter the clinical stage. Ultrasonic waves to replace light rays help orientation by echolocation, pulsed infrared photodetection, sonar sensors, closed circuit tactual television systems, and visual information through skin by tactual substitution are some of the devices and techniques currently under study. Perhaps the ultimate replacement system is implanted stimulators of the visual cortex.

APPENDIX

Sources of supply of subnormal vision aids

American Optical Co., Southbridge, Mass. 01550

Anco Wood Specialties, Inc., 71-08 80th St., Glendale, N. Y. 11227

Camalier & Buckley, Dept. HG 4, 1141 Connecticut Ave., Washington, D. C. 20036

Covington Plating Works, Inc., 330 Pike St., Covington, Ky. 41016

Edmund Scientific Co., Barrington, N. J. 08007

The Good-Lite Mfg. Co., 7638 W. Madison St., Forest Park, Ill. 60130

Hanover House, Hanover, Pa. 17331

B. Jadow & Sons, Inc., 53 W. 23rd St., New York, N. Y. 10010

Jesse Jones Box Corp., Box 5120, Philadelphia, Pa. 19143

Keeler Optical Co., 5536 Baltimore Ave., Philadelphia, Pa. 19143

McLeod Optical Co., 357 Westminster St., Providence, R. I. 02903

Miles Kimball Co., 41 W. 8th Ave., Oshkosh, Wisc. 54902

Charles Nusinov & Sons, Inc., 1404 E. Baltimore St., Baltimore, Md. 21231

Frederick Ravadge, Instruments, 731 Biddle Road, Glen Burnie, Md. 21061

Selsi Importing Co., 29 E. 22nd St., New York, N. Y. 10010

Shuron-Continental, Division of Textron, Inc., P.O. Box 831, Rochester, N. Y. 14603

Dr. Louise L. Sloan, Wilmer Institute, Johns Hopkins Hospital, Baltimore, Md. 21205

Spencer Gifts, 90 Spencer Bldg., Atlantic City, N. J. 08101

Sunset House, 122 Sunset Building, Beverly Hills, Calif. 90213

Superior Optical Co., 1500 S. Hope St., Los Angeles, Calif. 90015

Telesight, Inc., 1418 E. 88th St., Brooklyn, N. Y. 11236

REFERENCES

Anderson, D. R.: Optic atrophy, Am. J. Ophthal. **76:** 693-711, 1973.

Aulhorn, E., and Harms, H.: Visual perimetry, Handbook Sensory Physiol. **7**(4):102-145, 1972.

Barlow, J. S.: An electronic method for detecting evoked responses of the brain and for reproducing their average waveforms, Electroencephalogr. Clin. Neurophysiol. **9:**340-343, 1957.

Barlow, J. S.: Evoked responses in relation to visual perception and oculomotor reaction times in man, Ann. N. Y. Acad. Sci. **112:**432-467, 1964.

Batra, D. V., and Paul, S. D.: Macular illumination test in central serous retinopathy, Am. J. Ophthal. **63:**146-149, 1967.

Bedrossian, R. H.: The management of traumatic hyphema, Ann. Ophthal. **6:**1016-1021, 1974.

Berliner, M. L.: Biomicroscopy of the eye, 2 vols., New York, 1943-1949, Harper & Bros.

Berson, E. L., Rabin, A. R., and Mehaffey, L.: Advances in night vision technology, Arch. Ophthal. **90:**427-431, 1973.

Bier, N.: Correction of subnormal vision, London, 1960, Butterworth & Co.

Brazelton, F. H., Riley, L. H., Rosenbloom, A. A., Mehr, E. G., and Korb, D. R.: A symposium on the rehabilitation of the partially sighted, Am. J. Optom. **47:**585-628, 1970.

Campbell, C. J., Rittler, C., and Kramer, W. G.: A new projection adaptometer, Arch. Ophthal. **69:**564-570, 1963.

Cobb, W. A.: On the form and latency of the human cortical response to illumination of the retina, Electroencephalogr. Clin. Neurophysiol. **2:**104, 1950.

Consul, B. N., and Charan, H.: Macular function in macular diseases, Am. J. Ophthal. **58:**1007-1010, 1964.

Cushing, H.: The Harvey Cushing collection of manuscripts, New York, 1943, Henry Schuman.

Dawson, G. D.: A summation technique for the detection of small evoked potentials, Electroencephalogr. Clin. Neurophysiol. **6:**65-84, 1954.

DeVoe, R. G., Ripps, H., and Vaughan, H. G.: Cortical responses to stimulation of the human fovea, Vision Res. **8:**135-147, 1968.

Ellerbrock, V. J.: Magnification for near vision, Am. J. Optom. **31:**66-77, 1954.

Elstein, J. K., Sehgal, V. N., Kaplan, M. M., and Katzin, H. M.: Instrumentation and technique for refractive keratoplasty, Am. J. Ophthal. **68:**282-291, 1969.

Ernest, J. T., and Potts, A. M.: Pathophysiology of the distal portion of the optic nerve. 4. Local temperature as a measure of blood flow, Am. J. Ophthal. **72:** 435-444, 1971.

Faye, E. E.: The low vision patient, New York, 1970, Grune & Stratton, Inc.

Feinbloom, W.: Outline of the technique of examination of the partially blind patient with the clear image lens, Bulletin of the Optometric Society of the City of New York, 1953.

Fernald, R., and Chase, R.: An improved method for plotting retinal landmarks and focusing the eyes, Vision Res. **11:**95-96, 1971.

Fonda, G.: Report of five hundred patients examined for low vision, Arch. Ophthal. **56:**171-175, 1956.

Fonda, G.: Characteristics of low vision corrections in albinism, Arch. Ophthal. **68:**754-761, 1962.

Fonda, G.: Subnormal vision correction for aphakia, Am. J. Ophthal. **55:**247-255, 1963.

Fonda, G.: Management of the patient with subnormal vision, ed. 2, St. Louis, 1970, The C. V. Mosby Co.

Fonda, G.: Bioptic telescopic spectacles for driving a motor vehicle, Arch. Ophthal. **92:**348-349, 1974.

Francois, J.: Congenital cataracts, Springfield, Ill., 1963, Charles C Thomas, Publisher.

Francois, J.: The diagnosis of blindness in the infant, Ann. Ophthal., pp. 533-554, Sept., 1970.

Fraser, G. R., and Friedmann, A. L.: The causes of

blindness in childhood, Baltimore, 1967, The Johns Hopkins Press.

Freeman, E.: Optometric rehabilitation of the partially blind: a report of 175 cases, Am. J. Optom. **31:**230-239, 1954.

Gay, A. J.: Clinical ophthalmo-dynamometry, Int. Ophthal. Clin. **7:**729-744, 1967.

Genensky, S. M., Moshin, H. L., and Peterson, H. E.: Performance of partially sighted with Randsight I equipped with an X-Y platform, Am. J. Optom. **50:** 782-800, 1973.

Gettes, B. C.: Optical aids for low vision, Sight Saving Rev. **28:**81-83, 1958.

Goldmann, H.: Examination of the fundus of the cataractous eye, Am. J. Ophthal. **73:**309-320, 1972.

Gordon, D. M., and Ritter, C.: Magnification, Arch. Ophthal. **54:**704-716, 1955.

Gradle, H., and Stein, J.: Telescopic spectacles as aids to poor vision. In Transactions of the Section of Ophthalmology of the American Medical Association, 1924, p. 262.

Gunn, R. M.: Retrobulbar neuritis, Lancet, 1904.

Gunstensen, E.: Visual aids for the partially sighted, Br. J. Ophthal. **44:**672-678, 1960.

Harrington, D. O.: The visual fields, ed. 3, St. Louis, 1971, The C. V. Mosby Co.

Hiatt, R. L., Waddell, M. C., and Ward, R. J.: Evaluation of low-vision aids program, Am. J. Ophthal. **56:**596-602, 1963.

Hogan, M. J., and Zimmerman, L. E.: Ophthalmic pathology, ed. 2, London, 1962, W. B. Saunders Co.

Hollenhorst, R. W., Brown, J. R., Wagener, H. P., and Shick, R. M.: Neurologic aspects of temporal arteritis, Neurology **10:**490, 1960.

Hoover, R. E., Freiberger, H., Bliss, J. C., Bach-y-Rita, P., Brindley, G. S., and Rushton, D. N.: Symposium: prosthetic aids for the blind, Trans. Am. Acad. Ophthal. Otolaryngol. **78:**711-746, 1974.

Hoover, R., and Kupfer, C.: Low vision clinics: report, Am. J. Ophthal. **48:**177-187, 1959.

Huber, A.: Eye symptoms in brain tumors, translated by F. C. Blodi, ed. 2, St. Louis, 1971, The C. V. Mosby Co.

Jagerman, L. S.: Understanding magnification in ophthalmology, Rochester, 1970, American Academy of Ophthalmology and Otolaryngology.

Jayle, G. E., Ourgaud, A. C., Baisinger, L. F., and Holmes, W. J.: Night vision, Springfield, Ill., 1959, Charles C Thomas, Publisher.

Kaplan, H.: Refractive technique for the decreased vision problems of the aged, Am. J. Optom. **36:**511-519, 1959.

Keeler, C. H.: Helping the partially sighted, ed. 3, London, 1958, Keeler.

Keeney, A. H.: Field loss vs central magnification (editorial), Arch. Ophthal. **92:**273, 1974.

Kestenbaum, A.: Clinical methods of neuro-ophthalmologic examination, New York, 1946, Grune & Stratton, Inc.

Kestenbaum, A., and Sturman, R. M.: Reading glasses for patients with very poor vision, Arch. Ophthal. **56:** 451-470, 1956.

Klein, B. A.: Senile macular degeneration, Am. J. Ophthal. **58:**927-939, 1964.

Knox, D. L.: Examination of the cortically blind infant, Am. J. Ophthal. **58:**617-621, 1964.

Kollner, H.: Die Störung des Farbensinnes, Berlin, 1912, Karger.

Krill, A. J., and Klien, B. A.: Flecked retina syndrome, Arch. Ophthal. **74:**496, 1965.

Lebensohn, J.: Newer optical aids for children with low vision, Am. J. Ophthal. **46:**813-819, 1958.

Linksz, A.: Optical principles of loupe magnification, Am. J. Ophthal. **40:**831-840, 1955.

Linksz, A.: Visual aids for the partially sighted, Am. J. Ophthal. **41:**30-41, 1956.

Linksz, A.: Reflections, old and new, concerning acquired defects of color vision, Survey Ophthal. **17:** 229-240, 1973.

Mann, I.: Developmental abnormalities of the eye, Philadelphia, 1958, J. B. Lippincott Co.

Mann, I.: Researches into the regional distribution of eye disease, Am. J. Ophthal. **47:**134-144, 1959.

Michaels, D. D.: Projection campimetry and photographic recording, Am. J. Ophthal. **55:**107-115, 1963.

Milder, B.: Advantages of the optical aids clinic, Sight Saving Rev. **30:**78-84, 1960.

Miller, W. W.: Surgical treatment of degenerative myopia: scleral reinforcement, Trans. Am. Acad. Ophthal. Otolaryngol. **78:**896-910, 1974.

Moore, T. E., and Aronson, S. B.: The corneal graft, a multiple variable analysis of the penetrating keratoplasty, Am. J. Ophthal. **72:**205-293, 1971.

Newell, R. R., and Borley, W. E.: Roentgen measurement of visual acuity in cataractous eyes, Radiology **37:**59-61, 1961.

Newman, M.: Modern techniques in perimetry, Int. Ophthal. Clin. **7:**949-1003, 1967.

Pearlman, J. T., Owen, W. G., Brounley, D. W., and Sheppard, J. J.: Cone dystrophy with dominant inheritance, Am. J. Ophthal. **77:**293-303, 1974.

Potts, A. M., editor: The assessment of visual function, St. Louis, 1972, The C. V. Mosby Co.

Prince, J. H.: Completion of project 650 (special print for subnormal vision), Am. J. Ophthal. **48:**122-124, 1959.

Raiford, M. B.: Keeler low-vision aids, Am. J. Ophthal. **67:**542-546, 1969.

Reese, A. B.: Tumors of the eye, New York, 1963, Harper & Row, Publishers.

Rosenbloom, A.: Examining the partially blind patient, J. Am. Optom. Assoc. **29:**715, 1958.

Sandok, B. A., Trautman, J. C., Ramirez-Lassepas, M., Sundt, T. M., and Houser, O. W.: Clinical angiographic correlations in amaurosic fugax, Am. J. Ophthal. **78:**137-142, 1974.

Schultze, M.: Zur Anatomie und Physiologie der Retina, Arch. Mikrobiol. Anat. **2:**175-286, 1866.

Sehgal, V. N., Katzin, H. M., Kaplan, M. M., and

Elstein, J. K.: Refractive keratoplasty, Am. J. Ophthal. **70**:614-623, 1970.
Severin, S. L., Harper, J. Y., and Culver, J. F.: Photostress test for the evaluation of macular function, Arch. Ophthal. **70**:593-597, 1963.
Sloan, L. L.: New test charts for the measurement of visual acuity at far and near distances, Am. J. Ophthal. **48**:807-813, 1959.
Sloan, L.: New focusable stand magnifiers, Am. J. Ophthal. **58**:604-608, 1964.
Sloan, L.: Optical magnification for subnormal vision: historical survey, J. Opt. Soc. Am. **62**:162-168, 1972.
Sloan, L. L., and Brown, D.: Reading cards for selection of optical aids for the partially sighted, Am. J. Ophthal. **55**:1187-1199, 1963.
Sloan, L. L., and Brown, D. J.: Relative merits of headborne, hand, and stand magnifiers, Am. J. Ophthal. **58**:594-604, 1964.
Sloan, L., and Habel, A.: New methods of rating and prescribing magnifiers for the partially blind, J. Opt. Soc. Am. **47**:719-726, 1957.
Sloan, L. L., and Jablonski, M.: Reading aids for the partially blind: classification and measurement of more than 200 devices, Arch. Ophthal. **62**:465-484, 1959.
Snydacker, D.: The normal optic disc, Am. J. Ophthal. **58**:958-964, 1964.
Sorsby, A.: Genetics in ophthalmology, London, 1951, Butterworth & Co., Ltd.
Spivey, B. E., Pearlman, J. T., and Burian, H. M.: Electroretinographic findings (including flicker) in carriers of congenital X-linked achromatopsia, Doc. Ophthal. **18**:367, 1964.
Stimson, R. L.: Optical aids for low acuity, Los Angeles, 1957, Braille Institute of America.
Tillett, C. W.: Visual aids in office practice, Am. J. Ophthal. **46**:186-194, 1958.
Traquair, H. M.: An introduction to clinical perimetry, ed. 5, St. Louis, 1948, The C. V. Mosby Co.
Weale, R. A.: The duplicity theory of vision, Ann. R. Coll. Surg. Engl. **28**:16-35, 1961.
Weiss, S.: Optical aids for the partially sighted, Am. J. Ophthal. **55**:255-261, 1963.
Willets, G. S.: A survey of patients using spectacle magnifiers, Br. J. Ophthal. **44**:547-550, 1960.
Volk, D.: Conoid ophthalmic lenses in legal blindness, Am. J. Ophthal. **56**:195-203, 1963.
Zahl, P. A.: Blindness, Princeton, 1950, Princeton University Press.

24

Visual perception

Visual perception refers to seeing in the broadest sense. We perceive objects that appear to have extensity, solidity, and stability in a direction and at a distance from ourselves. We perceive ourselves in relation to them both spatially and psychologically. Objects may be fulfilling, threatening, useful, or desirable. If we are normal, we perceive our feelings and emotions and accept ourselves as independent, worthy observers. We "grasp" a problem, "perceive" a solution, "see" a meaning, and "visualize" a concept. The patient with an optical defect not only suffers blurred vision but also sees what there is incorrectly, or fails to see, or sees what there is not. Perceptual distortions may lead to inadequate understanding, incomplete knowledge, or inappropriate responses. Perception thus deals with the problems of gathering and assimilating information through the senses. Acting on this information leads to new perceptions and new actions in a chain that is the essence of behavior. Problems of perception have the dignity of age and the disaccord of multiple interests, including philosopher, physician, psychologist, physicist, photographer, and painter. The description of the modes of operation of perceptual systems is currently in a phase of expansion. Many details are clear; only the fundamentals have yet to be worked out. The purpose of this chapter is to provide a practical framework for interpreting the vast literature on perception, which impinges to a greater or lesser extent on clinical problems of vision.

SENSATION AND PERCEPTION

Sensation is the immediate, incommensurate response to receptor stimulation. Warmth, pain, pressure, hue, and brightness are modalities of sensations. All have, in addition, a spatial quality or extensity and an existence in time. Moreover visual sensations are external with a spatiality for which there seems to be no comprehensible stimulus equivalent in the two-dimensional retinal image. Indeed Sherrington classified receptors by their spatial character as exteroceptive, proprioceptive, and enteroceptive. To account for spatiality specifically and for the richness of experience despite the poverty of receptor messages generally, it has become traditional to distinguish between sensations and perceptions. Sensations in this view are the raw material; perceptions are the manufactured product. Sensations result from passive reception; perceptions are active operations of the mind, which collects, orders, differentiates, categorizes, and recalls them into meaningful ideas. Sensation is response; perception is knowledge. Psychologists differ on whether these operations of the mind are innate or learned, and some disagree on what is sensed and what is perceived. But whether learned or inborn, if perception is partly a construct of the mind, are we not separated from the real world by the filtering action of our senses? And since we can be deceived (e.g., by the bent stick in the water, the image behind the mirror, the distortions of aniseikonia), can our perceptual machinery be trusted to ever report faithfully and truthfully? This is the age-old dilemma to which every philosopher gives an answer, although few ask the question. It is a problem that attracts students of vision like flies to the flypaper with much the same result.

No understanding of perception is feasible without a glimpse, however brief, at the epistemologic flypaper and some of the glutinous problems to which psychology is heir. John Locke (1690), unimpressed by the wealth of ideas in the newborn, supposed the mind to be like blank paper, void of all character, on which experience writes. It was not a novel idea. (Plato had already rejected it because it could not account for error), but it seemed just right to Bishop Berkeley (1709) as a means of fighting the materialism of his age and incidentally glorifying the Deity. All knowledge, said Berkeley, is bundles of classified and interpreted sensations; matter is known only through our sensations; and material objects exist only through being perceived. Objects remain when we do not look at them because they persist in the mind of God forgetting that mind gives to objects their meaning, not their existence. But is not the mind itself but a bundle of feelings, ideas, and memories, argued Hume (1748), which we can know only through internalized perception? In one swoop, Hume abolished mind as Berkeley and Locke had dissolved matter.

Thus philosophers conject
The mind must have an architect,
Reach metaphysical perfection
With answers, if without the question.
This epistemologic dialectic
Not without its witty critic,
Spied in sensation undefined,
What was no matter could never mind
Summarizing his conclusion,
In theory, there's no solution.

It was Immanuel Kant (1781) who rescued philosophy from the ruins of its own making and saved science from skepticism. Although knowledge is derived from the senses, some is independent of sense experience. Experience can tell us only what is, not what ought to be. General truths must be independent, must exist before experience, and must remain true whatever subsequent experience. Mind is not a passive sheet, not merely internalized perceptions, but an active process that molds, selects, rejects, and coordinates, creating order out of the chaos of the senses. These necessary truths do not come from experience but derive their a- priori character from the natural and inherent way our minds work.

Spatial relations, according to Berkeley, are learned by correlating visual and tactual experiences and "by the perception of visible things in respect to one another, to make a sudden and true estimation of the situation of outward, tangible things corresponding to them." Granted we learn to attach verbal labels to directions, but the directions must exist before the labels and before eye movements, or we would not know which way to look. Moreover there is no intrinsic reason to accept the notion that the Mahomet of vision must come to the mountain of touch for its spatial education. For Kant, space and time are not sensations but modes of perceptions. Vision is spatial because it is inconveivable that any sensation, indeed any event, can be independent of space and time. The neurophysiologic correlate of sensation—the electrical activity of nerve cells in the eye or brain—occurs in space and has a temporal order. Only in dreams and fantasies can we normally escape the limitations of space and time and reconstruct the universe according to our own imagination.

Within the confines of experience, it is not possible to distinguish between the objective and subjective world, except historically. It is as if a counterfeiter had managed to duplicate an indistinguishable replica, detectable only if caught in the process of printing the forgery. Fortunately the joint testimony of the senses is usually in agreement and assures us that our perceptions square up with past experiences. Objects can never be known (although their existence may be inferred) apart from the process of knowing, yet we proceed as if they had an objective existence and not as if they were merely fabrications of our perceptual machinery. Nature is a phenomenon, not a noumenon, to which mind furnishes order and causality. Modern philosophy (e.g., Dewey, Bentley) has departed from Kant in so far that perception is not fixed and unalterable, but the perceiver himself changes in some respects with new habits and experiences. Perception thus becomes a dynamic series of transactions between the knower and the known.*

Despite a filial devotion to philosophy, the distinction between sensation and perception does not sit well with modern psychologists. Mental processes are so variable, so individual, and so subjective that they do not lend themselves to quantitative measurements. Moreover the Gestaltists view perception not as the sum of disparate sensations but as an integrated whole. Perception is

*The Roman poet Lucretius expressed a similar idea:

Nothing abides, but all things flow
Fragment to fragment clings, and grow
As we name them, fade
And are no more the things we know.

organized, three-dimensional, and precharged with meaning right from the beginning, although the meaning may later be revised, altered, or expanded. Perception begins with attention when stimuli become relevant to some current or derived organismic need and ends with cognition when evaluated and related to motivation and past experience. Cognition may then crystallize into adaptive behavior. The portion of the personality that utilizes information imparted by the senses is the ego, which contributes the interpretative, adaptive, and executive aspects. In psychoanalytic theory, the ego can no more be divorced from perception than walking can be made independent of gravity. By reducing the stimulus to its simplest elements, however, the behaviorist believes subjective concomitants of perception can be ignored or at least bypassed. In the meantime, the clinician cannot wait until all problems of behavior have been solved and works pragmatically, just as engineers may build bridges even though physics has yet to crack the riddle of the universe. The distinction continues in our thinking, as in the psychosomatic concept of symptom formation in which repressed tensions (mind) are channeled into bodily manifestations (matter).

METHODS AND MEASUREMENTS

Intensity, wavelength, exposure duration, and position of a retinal stimulus result in reasonably reproducible quantitative responses for all subjects with normal eyes and similar preadaptation. The stimulus is simple, and the isolated perceptual threshold response is common to all observers. Stimulus elements may be lines or points as in the classic approach or boundaries, surfaces, or textures as in more recent experimentation. It does not matter that there is no similarity between wavelength and hue or only a restricted parallelism between a triangular retinal image and a triangular percept. Stimulus and response are isomorphic, each belonging to its own sphere of categorization, but there is always some stimulus variable that corresponds to some property of the percept. Even complex percepts must have stimuli. Such is the basic assumption of psychophysics, which attempts to predict—at least statistically—the one from the other.

In psychophysics the specification of the stimulus is all important. Any object or environmental event is a potential stimulus: effective or ineffective; subliminal, threshold, or suprathreshold; fixed or variable; constant or intermittent; and unidimensional or polydimensional. Much effort is devoted to the measurement of absolute and differential thresholds because they define functional limits, circumscribe deficiencies, or stipulate the effects of learning and training. Attempts are then made to fit responses to a scale by equal intervals, equal ratios, or ordinal rating.

The visual stimulus is the retinal image. Since it makes no difference to the retina how the image is produced, an innumerable array of environmental events can produce the same image. Thus a retinal ellipse may represent an elliptic object or a circular one viewed obliquely. Solidity and depth have no obvious representation in the flat retinal image. How does vision come to have its precision for a community of observers? The difficulty lies in the term "image," the notion that the picture in the eye replicates the environment and is somehow transmitted to the brain where an homunculus looks at it as one would a painting in a gallery, perhaps comparing the picture sent by one eye to that of the other to "create" a third dimension through a form of mental machination. (The retinal image is a picture to an observer dissecting a scleral window, indeed a fairly good one as evidenced by photographs of it, but it is not a picture *to* the retina itself anymore than the cerebral fissure is a gulf to the mind.) In fact the retina's neural message is only a pattern of impulses; even the pattern is restricted by the all-or-none law. The retinal stimulus consists of gradients of illumination, color, pattern, size, shape, and texture. Gradients are not formed haphazardly but by lawful transformations to create boundaries, contours, surfaces, slants, and solidity.

The psychophysiologist goes one step further and attempts to relate perception to neurophysiologic mechanisms. For example, a plot of the scotopic *b* wave of the electroretinogram against log energy of the stimulus flash or visually evoked cortical potentials as a function of luminance are applicable to human experimentation. In other methods, usually confined to lower animals, a stimulus is presented and records obtained of neural activity of retinal cells, optic path, or visual cortex. Concomitant observations of reflex movements are made, or the results are correlated with simple discriminations or Pavlovian conditioning techniques. The classic experiments of Hecht on the clam *Mya* and the horseshoe crab *Limulus* illustrate this approach. There is a general feeling, at least among biologically trained workers, that perceptual phenomena for which a physiologic correlate can be found have a firmer if not more scientific footing. Much of the data on color and brightness perception falls into this category. Physiologic mechanisms most often invoked are variations in sensitivity and resolution (modulation transfer functions) through neural summation and inhibition.

To illustrate, all visual information passes through the retina, which acts as a transducer, converting light into electrical impulses. The retina communicates with the visual cortex by about a million nerve fibers, and the messages are coded by pattern and frequency. Thus the frequency but not the amplitude of nerve impulses in a single optic nerve fiber of *Limulus* increases with light intensity, varies with wavelength, obeys the time-intensity and area-intensity reciprocity laws, and changes as expected with retinal adaptation. The neurophysiologic results closely parallel the psychophysical ones. In the frog eye, however, only 20% of the optic nerve fibers respond as those of *Limulus*, with an initial burst followed by a continuous discharge. About 50% show an initial burst when the light appears and a final burst after the light goes off, with no discharge during steady illumination. The remaining 30% of the fibers show no response at all to illumination but give a vigorous burst after the light is turned off. Mammalian eyes show the types of discharge found in the frog, although the percentages differ. These results reflect the fundamental capacity of receptors to respond selectively to restricted ("adequate") stimuli and specifically to changes in stimuli. Thus the steady tick of the clock, the color of an optical filter, and the scent of a garden fade into the background after prolonged exposure.

How does the retina encode the messages? In 1938 Hartline succeeded in mapping receptive fields of a single optic nerve fiber by exploring the retina with a small spot of light. The stronger the light the larger the receptive field. Several weak spots could also effect a response, although each by itself was subliminal. Kuffler (1953) found that receptive fields of the cat's ganglion cells are organized concentrically, with a center surrounded by a disc-shaped outer zone. A spot of light in the central zone excites the on-response, whereas illumination of the surrounding area inhibits the discharge. In other fields, these characteristics are reversed. Simultaneous stimulation of center and periphery cancel each other and have little effect. Such responses are found in all vertebrate retinas studied, which contain five basic types of cells: photoreceptors, bipolars, and ganglions arranged in a vertical chain; and horizontal and amacrine cells, which provide horizontal interconnections. The horizontal chain allows modification of activity in neighboring vertical pathways. A small spot of light, illuminating only the receptor above the bipolar, actuates the vertical chain; illumination of the adjacent photoreceptors communicates with the bipolar through the horizontal cell but much more feebly. A similar arrangement holds for the bipolars and ganglions through amacrine cells. Thus the sustained-response ganglion appears to be driven directly mainly by the bipolar, whereas other ganglion cells, affected by amacrines, respond chiefly to changes in light stimulation. The lateral cells therefore have the capacity to modulate the signal transmitted along the vertical chain, coding the message to the brain according to whether the illumination is continuous or changing (Werblin, 1973). Finally, in a series of elegant experiments, Hubel and Wiesel (1962, 1967) found that some receptive fields of visual cortical cells in the cat and monkey are selectively sensitive to line stimuli. Size, shape, and orientation of these edges were critical for optimal response. Several different types of visual cortical cells have been identified: simple cells responding to stationary stimuli; complex cells directionally sensitive, which respond by continuous firing; and hypercomplex cells, possibly representing complex cells wired together in columns. Although there is less than complete clarity on cell types (e.g., too many cortical cells seem to be concerned with directional sensitivity, with predominance of oblique orientations, whereas human vision is oriented primarily along vertical-horizontal meridians), these are fascinating beginnings into the sensory basis of perception. The results explain how the retina can compress large amounts of information into a relatively simple code. Indeed the more complex the brain the less retinal processing is required. The neural messages differ from the optical retinal image but preserve its spatial and temporal character. This is the neurophysiologic significance of clinical refraction.

Intermediate between psychophysical and neurophysiologic studies are cybernetic models of perception in which the brain is compared to a computer. Using the vocabulary of engineering, perception is related to the detection of signals embedded in "noise." "Noise" is defined as any irregular interference relative to the signal of interest and may be external or internal. Unhappily in some cases, by imposing a classificatory structure on their terminology, the incremental parameters of the input result in a nonlinear ambivalence of the output, thus straining the put-put, restructuring a dichotomy of the cognitive paradigm, and making it hard to decipher what is being talked about.

No psychologic theory of perception can compare in scope and catalytic significance to the contributions of the Gestalt school. Wertheimer (1912) first showed that perception of movement cannot be understood as a series of retinal excitations; rather movement is perceived directly. Perceptions are not only separate sensory data but include qualities resulting from their mode of organization. The parts depend on the percept as a whole. A "square" is distinct from its line elements; a "melody" is more than the sum of its musical notes; each can be transposed without changing its character.

Instead of a blooming, buzzing confusion, the visual environment is composed of objects with form, color, brightness, texture, and solidity, organized into stable and meaningful wholes. According to the Gestaltists these organizing forces are innate, operating through external (retinal stimulation) and internal (cortical) forces. The internal forces tend to favor symmetry, simplicity, goodness of form, and closure, that is, stability, whereas the external forces tend to make the percept conform to the pattern of retinal stimulation. When the forces are weak (e.g., brief stimulation, defocusing, reduced intensity), the instability leads to perceptual fluctuations and illusions.

Although the Gestaltists stress the wholistic aspect of perception and the influence of that whole (context) on component parts, they do not deny the possibility of studying the parts. But they point out that the parts are abstractions and that the process destroys the essential nature of the configuration and may bear little relation to the original.

Whereas to the psychophysicist, suggestion, knowledge, emotion, and other subjective factors are vanishingly small when compared with the influences of physical and physiologic variables, to the phenomenologist only a naive observer would believe that his percepts are completely determined by external stimulus conditions and are exclusively referable to events outside himself.

Thus the psychopathology of perception, both diagnostic and therapeutic, remains phenomenologic and subjective. Only the patient can describe the perceptual distortions of his dreams, hallucinations, or delusions. It is the very subjectivity of perception that is the basis of the diagnostic Rorschach and Thematic Apperception tests.

Pope in his epistle to Sir Richard Temple summarized it neatly:

> Yet more; the difference is as great between
> The optics seeing, as the object seen.
> All manners take a tincture from our own;
> Or come discoulour'd thro' our Passions shown,
> Or Fancy's beam enlarges, multiplies,
> Contracts, inverts, and gives ten thousand dyes.

If psychology is a behavioral science, it is also a biologic science with anatomic, genetic, pharmacologic, and pathologic determinants. Anatomy, sometimes considered a dead subject, may not be able to tell us what happens, but it can rule out what is impossible. At least it will avoid such current statements in the psychologic literature as: "The belief that our 'perception' results exclusively from activity in the classic retino-geniculostriate pathways and that there is a point-to-point representation of the visual world on visual cortex is no longer tenable." More interesting are recent studies of the role of the superior colliculus in various aspects of visual attention and centering of the visual image on the cat retina, possibly also in human spatial vision and motion perception. The ontogenesis of perception and the measurement of visual capacities at birth (e.g., fixation, acuity, color, patterns, depth) and their subsequent maturation and differentiation provides insight into perceptual mechanisms, often under the category of the psychology of learning. Every drug has multiple sites of action; many are psychoactive by inducing neurochemical and neurophysiologic changes. Even aside from direct influences on brain excitability are the psychologic side effects of the pill-giving ritual—a placebo effect of which all physicians are aware. Neurobehavioral research based on drug responses have come to play an increasing role in diagnosis and treatment, although admittedly the problems of living are only temporarily dissolved by chemicals. Perceptual studies based on the effects of disease in humans, hereditary sensory defects, or extirpation studies in animals provide insights into localization of function and have replaced older notions of mental faculties. Perception does not occur in the brain; rather brain activity is perception. These methods, part of the field of physiologic psychology, are closely bound up with studies of autonomic, endocrine, and metabolic functions.

SOME PERCEPTUAL PROBLEMS

In clinical work, man is the measure of all things. The interrelation of function in terms of larger patterns of experience and behavior, however, does not preclude a natural desire to emphasize quantitative data. Reliable threshold measurements and the intelligent use of statistical methods summarize information, detect unsuspected relationships, and identify significant variables. Since mathematics is the most symbolic of languages, it is also the most universal, with the widest applications. Quantitative experiment is likely to remain the most direct method of evaluating perception just as perceptual problems are likely to remain a central issue in psychologic theorizing. In those cases in which behavior cannot be measured directly, one is forced to rely on indicants; for example, acuity as an indicator of ametropia and heterophoria as an indicator of fusional stress. The indicant is presumed to correlate to the underlying phenomenon, and the numerical values we assign to it are valid to the extent this correlation holds.

Perception of brightness

Brightness is the perceptual correlate of light intensity, the immediate response to adequate retinal illumination. The definition is at the outset psychophysical, and the experimental approach to how we discriminate brightness and brightness differences illustrates almost all the standard psychophysical methods and many psychophysiologic ones. The reason is that the stimulus-response relation is mostly straightforward. In those areas, such as brightness constancy, in which subjective factors play a larger role, theoretical speculations proliferate proportionately.

The immediate stimulus is retinal illumination, which depends on pupil size, ocular transmission, and the Stiles-Crawford effect. Attempts are made to keep these constant, and target luminance is changed (measured by instrumental photometry). Variables may be intensity proper, size, duration, region of the retina exposed, and wavelength composition. The probability of seeing depends not only on the stimulus but also on the observer's retinal adaptation, intrinsic stray light, and the basis on which the decision is made to respond.

Most experimental results can be explained by photochemistry and known retinal mechanisms of lateral inhibition and summation, types of receptors, their density, and convergence of photoreceptors on ganglion cells. Discrimination of brightness differences improves with increase in illumination, remains fairly constant at intermediate levels, and drops off again with dazzling. Discrimination improves both ways; light targets gets lighter and black ones blacker. Black, it should be noted, is not the absence of light; it takes light to see black. A gray ring that lies partially on a white and black field appears lighter on the black side, a phenomenon called "simultaneous contrast," probably on the basis of lateral retinal inhibition. Similar mechanisms are invoked to explain Mach bands, all designed to exaggerate the visibility of borders or contours. Adaptation is prevented by continuous small eye movements, as evidenced by fading when the retinal image is stablized. A rhythmic neural excitation (either photochemical or a neuron latent period fluctuation) probably accounts for the brightness overshoot of a briefly exposed flash (Broca-Sulzer phenomenon), the enhanced brightness that occurs with flickering lights (Brucke-Bartley phenomenon) and afterimages. The on-off pattern of neural discharge and the Broca-Sulzer phenomenon illustrate that the visual system is particularly sensitive to changes in luminance.

When the eye is exposed to a bright light followed by darkness, an afterimage is seen for some seconds. The initial brightness and color may be the same (positive afterimage), or the color and brightness are opposite (complimentary) to the original stimulus (negative afterimage). The latter appears when viewed against a light rather than a dark background as if the previous retinal excitation prevents the new stimulus from becoming effective. This type of afterimage is sometimes called "induced" or "consecutive." Afterimages can be prolonged by viewing them against an alternating light and dark surface, a technique useful in the clinical afterimage test for retinal correspondence. The afterimage is constant relative to the fixation axis and moves whenever the eye turns voluntarily (but not if displaced mechanically). The explanation for afterimages appears to be a residual local effect coupled with antagonistic processes (fatigue or inhibition). In one view, they are produced by the photoproducts of the primary light stimulus, their brightness being related to the amount of photoproduct present.

The temporal sequence of afterimages can be transformed into a spatial sequence by moving the initial stimulus across the retina, such as when a cigarette is quickly moved. The dark-adapted observer sees a consecutive cometlike effect (Bidwell's ghost).

Brightness contrast is as necessary for the perception of detail as angular separation. A receding target becomes smaller and loses detail because the eye is less sensitive to contrast differences for some separations than others. Contrast thresholds are measured by gratings of alternate black and white bars whose frequency is defined by the number of complete cycles that subtend 1° at the observer's eye. The patterns are produced by sine wave signals from a voltage generator on a cathode ray tube. For simple gratings the brightness varies in a sinusoidal manner, and the contrast is the variation of brightness around a mean level. Complex waveforms are built up by adding a number of sinusoidal wave forms. The best sensitivity to low contrast is found at a spatial frequency of three cycles per degree. The highest spatial frequency that can be perceived is about 50 cycles per degree but requires high contrast. Pattern detection may be reported verbally or recorded by electrical potentials evoked in the brain of man or animals. An analogous procedure using a photocell detector is used to measure the spatial modulation transfer function of ophthalmic lenses. In recent methods the optics of the eye is bypassed by laser interference fringes, useful in estimating acuity with cataracts. The threshold limits are set by the separation of cones, light scatter, lateral inhibition, and neural summation. The assumption is that the visual system evaluates spatial frequency in terms of its sine wave constituents, a procedure which lends itself to Fourier frequency analysis, a mathematic technique successfully used in electrical engineering.

When two stimuli are presented close to each other in time and space, resultant events are termed "visual masking." The masking stimulus may occur before, during, or after the target presentation. The effects are distinguished from simultaneous contrast in that the target or mask is of brief duration and the time interval small. In forward masking the mask precedes the target; in backward masking it follows the target. When the mask and target are adjacent to each other (e.g., a disc surrounded by an annulus or a rectangle flanked by two similar rectangles), the term "metacontrast" is used. Metacontrast can be obtained with a variety of shapes, providing the inner contours of the mask match the outer contours of the target. As the distance between target and mask increases, the masking effect decreases. The masking stimulus alters the sensitivity of the test stimulus (e.g., measured by brightness matching or threshold luminance), attributed to lateral retinal inhibition. Some contribution is made by the cortex, since interaction may occur between the two eyes.

When the general illumination of a field is changed, the apparent color and brightness of objects does not change proportionately but tends to remain constant, approximating the appearance under "average" illumination. Thus a white surface continues to look white and a black surface black even when the illumination is reduced to where the white paper reflects less light than the black paper did under higher illumination. The observer is apparently able to discriminate differences between changes in illumination and changes in the color or brightness of objects themselves. We tend to see objects as they are, not as the retinal image tells us. If objects are viewed through a reduction screen (an opaque screen with small aperture), the color and brightness separate from them and appear to float in the plane of the aperture. Under these circumstances, constancy breaks down. The classic experiment of the breakdown of constancy is the Gelb effect. When an observer views a circular black disc in a nearly dark room and a hidden spotlight is focused on the disc but not the background, the appearance is that of a white disc against a dark ground. If now a small piece

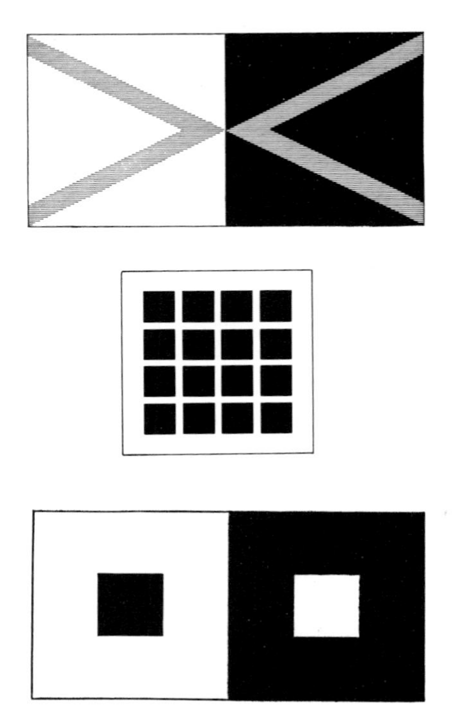

Fig. 24-1. Examples of simultaneous contrast. *Top,* The V on the black surface appears lighter than the one on the white background. The traditional explanation (Hering) is that the retinal elements responding to strong stimulation inhibit the responses of adjacent elements. The simplest example of this is shown at the *bottom. Middle,* Known as a Hermann grid (1870), this figure shows small gray areas at the intersections of the white bands. A physiologic explanation is that the white intersections inhibit the central core of receptive fields more effectively than in those areas where the black squares prevent activation of the concentric inhibitory annulus of the receptive field. This idea is used to measure the angular diameter of a receptor field; the white bands are narrowed until the gray patches disappear. The effect is demonstrable by moving the page closer or further from your eye.

of white paper is exposed next to the disc, it no longer appears white but black, even though luminance is unchanged. The presence of the white paper presumably allows a judgment comparison of the hidden light beam by the relative reflection. Helmholtz explained these brightness judgments by unconscious inference. Another explanation is that brightness is determined by the ratio of target luminance compared to the surround. In the Gelb experiment the comparison is made possible by introducing the white paper. The ratio explanation, however, only applies to a limited range, and the surround may induce simultaneous contrast, some illustrations of which are shown in Fig. 24-1. The theoretical problems are further discussed by Hurvich and Jameson (1966), and a retinal (i.e., physiologic) explanation is advanced by Cornsweet (1970). The instructions given in brightness matching tests, however, also influence the results, implying some nonretinal processes.

Although brightness constancy appears to violate simple correspondence between stimulus and response, behavior would be impos-

Fig. 24-2. Examples of reversible figures. The Necker cube *(top left)* illustrates spatial reversal due to ambiguous perspective. The Rubin vase or faces *(top right)* illustrate reversal of figure-ground organization. The central figure may be seen as a young woman or old hag, presumably due to context. The lower figure, an example of "op-art," shows a peculiar wavelike motion.

sible without it. The stability of the visual world hinges on objects maintaining a relatively uniform appearance despite variations in illumination, distance, or viewing conditions.

Perception of pattern and form

Visual objects exhibit attributes of size and shape for which the immediate stimulus is the size and shape of the retinal image when all other factors are held constant. Nevertheless, objects tend to look the same size irrespective of distance and maintain their shape despite variations in the angle of view. The traditional explanation is that we compensate and correct our perceptions according to past experience. The Gestaltists suggest that form perception follows organizing excitations occurring in the central nervous system. The psychophysicist maintains that the rigidity of shape has a correlate in the retinal image. The physiologist accounts for shape in terms of borders and contours enhanced by retinal neural mechanisms.

That past experience influences form perception is evident to any clinician who has measured visual acuity with Snellen letters compared to Landolt rings. Reading and mathematics clearly represent examples of perception inconceivable without prior knowledge of symbolic meanings. To the empiricist, perception is a kind of problem solving, the ability to read nonoptical reality from optical images in the eye. What we see goes beyond sensory evidence; the latter is used in conjunction with an internal repertoire of hypothesis to build an adequate knowledge of reality. In so far as the visual field is structured, objects are differentiated into figure and ground. Color, brightness, contour, and texture distinguish objects from their surroundings. That figure-ground organization is a dynamic process is evident in reversible perceptions (Fig. 24-2). Eye movements or blinking can help induce a change in the reversible appearance, but alternation also occurs spontaneously. The mechanism is poorly understood, possibly based on selective interpretation of ambiguous perspective. Some examples of these factors are shown in Fig. 24-3.

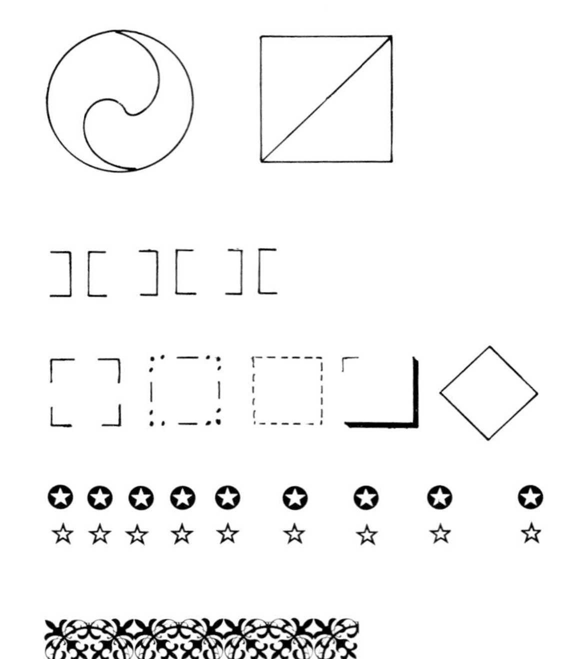

Fig. 24-3. Examples of perceptual organization. The first row illustrates that a circle and square are "better" figures than two S-shaped patterns or two triangles. The second row illustrates grouping by proximity. The third row shows that a "square" is independent of its component elements. When the square is rotated 45°, however, it is perceived as a diamond. Similar stars in the next row are grouped together; as the separation increases, association by proximity takes over. The bottom pattern, although composed of figures, is seen as a design, possibly because its overall symmetry dominates the individual elements.

Form perception in the Gestalt view proceeds in terms of the whole. The angles of a Necker cube appear as right angles, although some are acute, others are obtuse, and none are 90°. We see "things" and not the spaces between them unless the form is repetitive (patterns or designs) or structureless (clouds, haze, leaves, ground texture). The whole influences perception by providing a framework according to needs, moods, mental set, and sociocultural context. These psychologic forces influence the perceptual field like the lines in the Wundt, Hering, or Ehrenstein illusion distort apparent parallelism (Fig. 24-4).

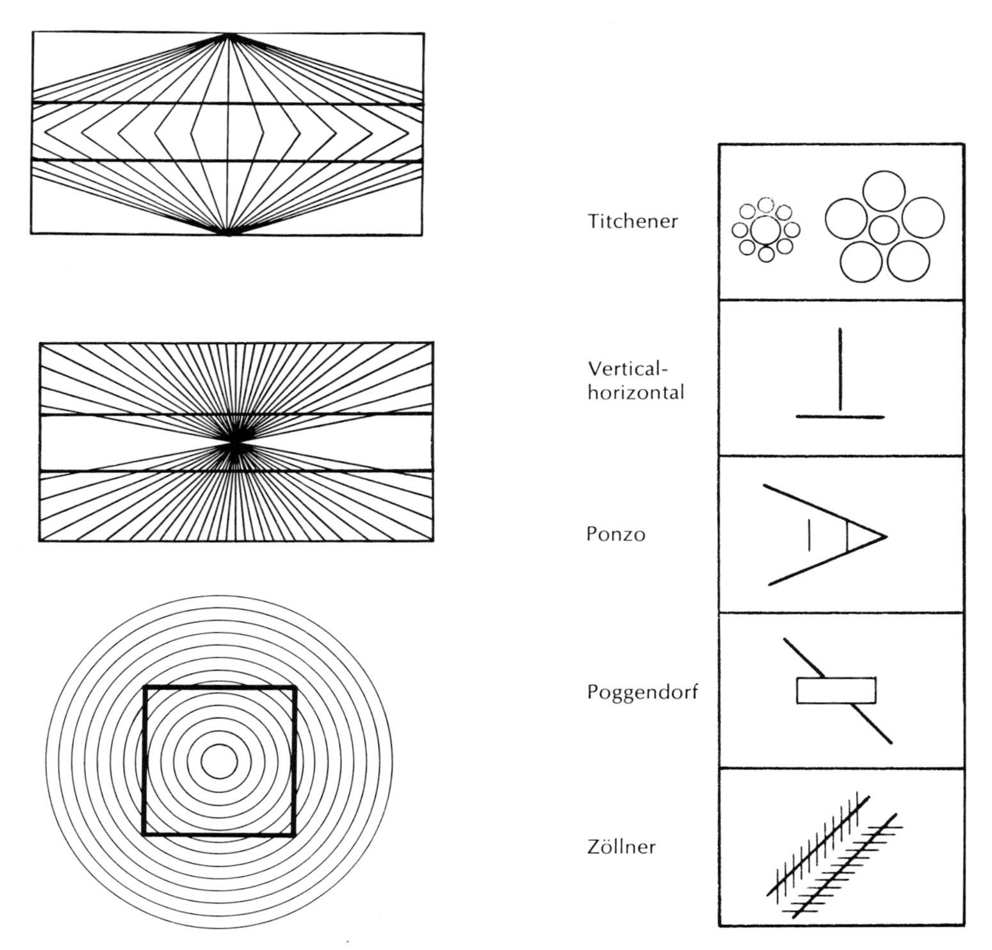

Fig. 24-4. Three classic geometric illusions of shape: Wundt *(top)*, Hering *(middle)*, and Ehrenstein *(bottom)*. The effect of surrounding field forces, in this case the background lines, on the appearance of the parallel lines is emphasized by the Gestalt school. In the field of social psychology, these forces are taken symbolically rather than geometrically.

Fig. 24-5. Some representative illusions of shape, size, and depth. They probably arise for a number of reasons, one of which is retinal inhibition, since they tend to break down when presented binocularly. Other theories emphasize the effect of the total perceptual field, which provides a confusing frame of reference.

Illusions, of unending fascination to the experimental psychologist, should not be considered oddities but special cases of general perceptual principles. The so-called optical illusions, for example, show that experienced form may disagree with measurable conditions with respect to direction, length, size, shape, or number of units (Fig. 24-5). An immense amount of effort has been spent in attempts to explain them (optical aberrations, retinal inhibition, cortical satiation, psychic fatigue), modified by age, set, or motivation.

In addition to simultaneous illusions are those produced by successive stimulation, for example, following the viewing of a rotating spiral (Fig. 24-6), a waterfall, a moving scene from a train window, or the sequential illumination of a light as in "moving" neon signs. No comprehensive explanation exists for some of these figural aftereffects, although the trend is to attribute them to neurophysiologic mechanisms.

Illusions are not perversions of the mind but have a reality indistinguishable from other percepts, hence the philosopher's difficulty with the problem of veridical sensory information. Because we cannot distinguish percepts whose implications are erroneous from those which are not, we call them "illusory." The imperfection is not in the brain but in insufficient information; the

Fig. 24-6. Illusions produced by sequential stimulation are termed "figural aftereffects." The first motion aftereffect was discovered by Purkinje after watching a cavalry parade. When the illustrated Plateau spiral is rotated, it appears to shrink or expand. A stationary figure observed subsequently expands in the opposite direction. A similar effect is obtained after observing a waterfall. The mechanism is unknown, but size and accurate fixation strongly influence the result. Its strength may be measured by a compensating rotation of the subsequent field.

Fig. 24-7. Only a few contours are sufficient to recognize Einstein's features, yet identification is lost when the figure is inverted. The criteria used to identify complex shapes such as faces are poorly understood, one reason for the difficulty in designing computers that can handle compound forms.

context or framework to which the illusory percept is related is inappropriate. A similar perceptual reality characterizes dreams, eidetic images, hypnosis, early stages of anesthesia, and certain pathologic conditions. Although illusions are rare under ordinary circumstances, they are not random but occur under specific conditions, which may differ for different kinds of illusions.

Recognition of similar forms implies the revival of previous sensory experiences, possibly similar to template matching, although the attributes may be symbolic rather than geometric. Few significant contours are necessary to recognize the face of a specific individual (Fig. 24-7), yet identification is lost when the figure is inverted. A related example is that a tilted square is seen as a diamond, not a rotation. The orientation of forms also depends on other sensory information such as postural reflexes, since in many cases, identification involves the assignment of top, bottom, and sides. For example, the impression of symmetry relies more on objects symmetrical about a vertical axis rather than a horizontal axis. All this implies that form perception depends to a greater extent on cognitive processes than simple stimulus-response relations imply.

The retinal image undergoes continuous changes as we turn our eyes or head or move about. Despite these deformations, objects retain their characteristic shape, a stability known as "shape constancy." When a circular disc is rotated about its vertical axis, we tend to continue seeing a circular disc. Although constancy is seldom complete, it is higher than the optical distortion of the retinal image leads one to suppose. If the disc is presented in a darkened room, it appears elliptic instead of foreshortened. Constancy thus requires sufficient information to perceive the slanted surface (e.g., compression of texture, variance in brightness) coupled to perspective transformation. Under normal conditions, motion parallax and binocular disparity help prevent our being fooled. In this sense,

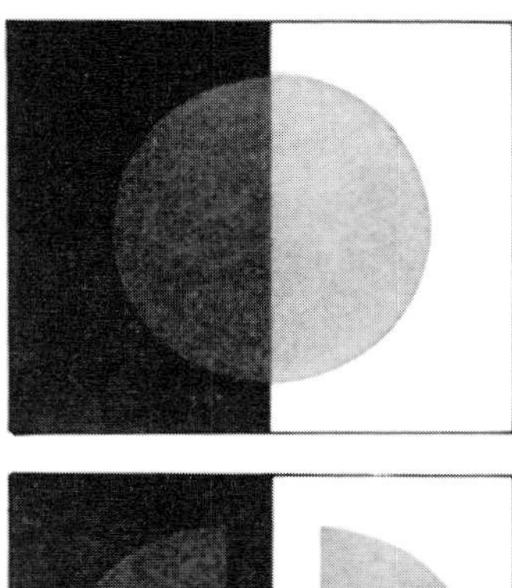

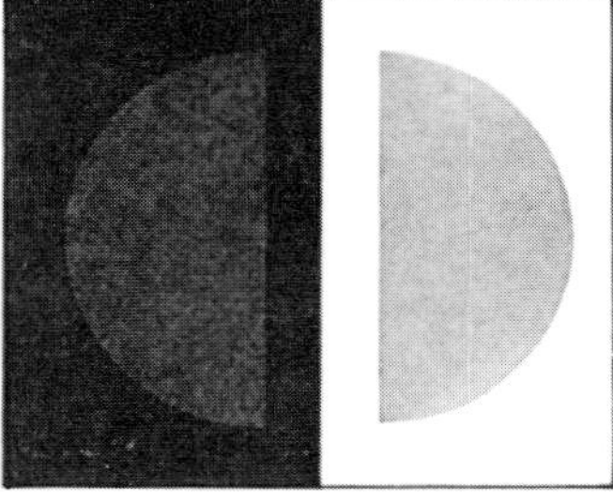

Fig. 24-8. Illustration showing that one of the factors in perceiving a surface as transparent is a continuous contour. Very little is known about the perception of transparency and translucency.

shape constancy is attributable to psychophysical correlates in the retinal image. Another factor, however, is undoubtedly the attitude adopted by the observer. The approach of the painter duplicating a landscape differs from the naive observer. The painter draws the perceptually parallel road perspectively convergent; the house in the distance smaller than the man in the foreground; the mountains blued by the intervening atmosphere; and the shadows colored and textured by the surfaces on which they fall. Although his retinal image, like the photograph or camera lucida drawing, always yields correct perspective, the artist discounts perceptual constancy, deducts past experience, and marks down a copy of the optical projection. But the results are often disappointing, so he may exaggerate, modify; or simplify his painting according to his esthetic, not perspective vision. Conversely, pictures drawn with high constancy, such as ancient Egyptian art, appear flat and lifeless. Seldom measured is the perception of transparency, which seems to depend on brightness differences at adjacent but not separated contours (Fig. 24-8).

Perception of motion

The perception of movement is basic, probably of fundamental importance to survival. The lowest velocity at which movement can be perceived is on the order of displacements of minutes of arc as a function of time in seconds (1 to 2 minutes of arc per second), somewhat influenced by the background against which the target is seen. The basic conditions for the perception of motion is a displacement of the object against a stable environment, in which case portions of the background are successively covered, or movement of the observer (eyes, head, body), in which case the entire array is transformed. An efferent copy of proprioceptive feedback is utilized in the second instance to maintain the stable appearance of the visual field. Additional factors are the displacement of the edge of the visible field as the head is rotated and perceived motion of the observer's own extremities.

Contrasted to real movement is the apparent movement of sequential retinal stimulation (phi phenomenon). The motion of motion pictures, television, and "moving" neon signs is of this type. Perceived motion depends on the interval between stimulation (for example, a long interval between two lights turned on and off is seen sequentially; with briefer intervals a single light appears to move smoothly from the first location to the second, and finally, one sees an objectless pure movement). It is not clear whether the same physiologic mechanism underlies both real and apparent movement.

Closely related to the perception of motion is the perception of time. We perceive events separated by temporal units, but the manner in which we estimate them is not clear despite much experimentation. Although there is some evidence that we judge time by the motivational state (e.g., the anticipation of pleasure or pain; activity or inactivity; attention or boredom), the experimental results are conflicting.

Perception of symbols

Perception is sometimes defined as the process of extracting information, and the perception and manipulation of symbols is probably the most distinguishing feature of human behavior. Through language we communicate with each other across space and time. It is the channel by which the genius of the past teaches its lessons to the present. Without language, thinking would be primitive and civilization impossible. It is in the nature of the human mind to think mostly with words, to some extent only about things for which we know words, and as the advertiser knows, words also shape our thinking. Language began, perhaps, when some prehistoric man pointed toward food or danger and grunted if he had both hands full. After some indeterminate period, he learned to transmit more lasting messages by drawing pictures on cave walls, clay tablets, or papyrus. Through use and the mechanics of manual duplication, these ideograms were transformed into the symbols of the alphabet. The symbols are collected into words whose meaning is defined by common consent.

A child learns that the symbols "CAT" represents a small, furry, four-legged animal. The French child learns to call it "CHAT" and the German child, "KATZE." Unlike the retinal image of a cat, which follows universal laws of optics, symbols are arbitrary—an accident of birth. Pictures, unless they are symbolic, are always specific and concrete and cannot represent such concepts as democracy, beauty, love, duty, or faith, for which there are no retinal equivalents. Only symbols can free our minds from the tyranny of the senses. But we must not conclude that thinking is nothing more than the act of naming. The relation between the object, its image, and the word that denotes it are logically connected. We can change the name from *cat* to *chat* or *katze* without affecting the relation, providing we use whatever language we have consistently. Thinking is generally about real objects; the words are necessary instruments of expression. Ambiguity is the price man pays for abstract thinking. A story goes that a brilliant Englishwoman once offered a prize of one thousand pounds to any philosopher submitting documentary evidence that he knows what he means, knows what anyone else means, knows what anything means, means what everyone else means, or can express what he means. No one claimed the prize. Because we tend to fill in the incomplete and rationalize the inconsistent, we may be misled by inaudible articulations, meaningless cliches, poisonous propaganda, and other forms of contextual goulashes. To guard against such verbal dysentery we need to nail down concrete referents—the war cry of the general semanticist.

Although thinking is not restricted to words or even the meaning of words, it is closely connected to their correct usage. Sentences are not random strings of words but ordered patterns. "This triangle has four sides" is perfectly good grammar but geometric nonsense. "The dog bit the man" is prosaic; reverse the order of the words, and the result is headline news.

To perceive symbols is to recognize their meaning; the meaning is brought to them by past experience and spoken associations. The cat's fur looks soft because we have handled it, and its softness becomes part of our mental idea of the cat. At least so goes the classic association theory of learning to which American psychology is much attached. To talk about sensations, images, and feelings is one thing; to describe a physiologic mechanism for their linkage is something else. It is easier to define memory traces, synaptic modifications, neurochemical imprints, or reverberating circuitry than to demonstrate them. There is no theory of learning so inconceivable that some psychologic theory has not conceived it or indeed for which some evidence cannot be mustered. The notion of unconscious inference, that is, thinking as internal trial and error, is plausible enough until we try to specify who makes the inference: an homunculus, a nerve cell, a group of nerve cells, or a cybernetic computer? And to whom would the computer report? Moreover our emotional predispositions perversely make some associations impossible no matter how strong the evidence in their favor, and the artist sees esthetic associations where the rest of us see only the commonplace. Some forms of learning unquestionably result from conditioning; others result from spontaneous reorganization or modification of memory traces.

Each person builds his own storehouse of meaningful symbols. The growth of language in children from early babbling through reduplicated monosyllables to the ability to follow simple commands generally occurs in the first 10 months. At about 1 year, the motor skill to produce sounds catches up with perceptive development. These sound-speak reflexes mark the beginning of parental teaching. The parent points and says the word; the child hears and repeats it. By the time he begins school, a considerable repertoire of spoken words has been built up. The advantage of the phonetic method of teaching reading is that it mobilizes this storehouse by applying it to the ordered arrangement of letters. By some perceptual miracle, the words eventually come to transcend the configuration of their component letters, and the meaning of a sentence transcends individual words. The child learns to read contextually, hardly aware of the symbols themselves unless confronted with a foreign language. He becomes a linguistic mammal. It is not clear why some children fail to achieve reading proficiency but continue beyond the normal age to make mistakes all children make when first learning to read (e.g., confusing letter orientation and the vector sequence of symbols; p for q, and "saw" for "was"). Apparently they are not deficient in certain neural figural analyzers because they do not have greater difficulty reading symbols in specific orientations. Nor is the dyslectic's visual system malfunctioning by transforming what other eyes leave alone, since they do no better with transformed text than normal readers. Rather than a peculiarity of the visual system, poor performance is most likely due to poor teaching or inattention, since the dyslectic generates internal language much as the normal reader does (Kolers, 1972).

Through teaching, ability to learn, and social reinforcement, new verbal habits are established. Old associations are replaced and weak ones extinguished. Their strength was first measured by Francis Galton (1879) by "word association tests" which "lay bare more about a man's thought and exhibit his mental anatomy more vividly than he would probably care to publish to the world." As new associations are built through discrimination of similarities, differences, and classification, vocabulary grows. Indeed the number of words we recognize and use at any given age is the single most dependable indicator psychologists use to measure intelligence. (The vocabulary of an average college student is estimated at some 156,000 words.) To a considerable extent therefore learning to think is learning words with which to think. The larger our verbal currency the bigger our mental checking account, the process accumulating interest through reading, writing, and speaking. The number and type of words we use, the proportion of nouns or adjectives, the length of our sentences, and the employment of similes or metaphors constitute our verbal style. It is by verbal style that we recognize the personality of the speaker or writer. Language most showeth the man, says Ben Johnson; speak that I may see thee.

Perception of spatial relationships

Of all the senses, vision predominates in the accurate localization of objects relative to each other and to ourselves. Olfaction and audition, the other telesthetic senses, come in a poor second both in speed and precision. This is the ultimate biologic value of vision, clearly recognized by Berkeley: "For this end the visive sense seems to have been bestowed on animals, to wit, that they may be able to foresee the danger or benefit of distant bodies, which foresight is necessary for preservation."

The spatial attributes of an object are as immediate as the perception itself. A pressure on the skin, a sound, or a light all have an immediate spatial quality. It may differ as to "where" or "how far" but not about "somewhere" and "sometime." The mind apprehends all experiences through Kantian categories of space and time; they are not things perceived but forms of perception. The intrinsic character of space and time is acknowledged by both nativists and empiricists, although they may disagree on how direction or distance come to be associated with a particular pattern or sequence of stimulation. We can imagine space in the abstract, or invent spaceless Euclidean "points," but these are concepts, not percepts. Space per se is not perceived; the relationship of objects to each other, to ourselves, and to a continuous background define experienced, subjective, visual space as distinct from the engineer's physical space of houses, bridges, and streets. Although visual space is subjective, subject to personal constants, it is not erratic; its stability is due to the unique, highly precise, anatomic construction of the visual system. If the incongruities between physical and visual space are widely espoused in theory, they are wisely neglected in practice. We can usually rely on the other driver to stay on his side of the street, and we hardly ever cross perceptual bridges that have no physical counterpart.

The stimulus for vision is retinal excition; yet the visual impression is not in the eye but external to ourselves. The reason, we are told, is that images of all objects in space falling on our retina are mentally projected back into space. This extrasomatic mental feat does not appear to be uniquely human because all animals with vision seem to manage it. The choice of "image" as the article to be projected (rather than percept, idea, notion, impression, sentiment, or simply object) dates back to the Stoic philosophers and was solemnly set down by Leonardo in his notebook:

> The eye transmits its own image through the air to all objects which are in front of it, and receives them into itself, that is on its surface, whence the understanding considers them, and such as it finds pleasing, commits to the memory. So I hold that the invisible powers of the images in the eye may project themselves forth to the object as do the images of the objects to the eye.

That da Vinci confused the action of mirrors with the retina is not nearly as remarkable than that the muddle should have persisted for nearly five centuries. If we really projected retinal "images," objects would change shape with every new angle of view; size, with every distance; and brightness, with every illumination. Although we would all have perfect knowledge of perspective, there could be no constancy, no stability, no consistency, and no meaning. When the retina is stimulated by pressure, no phosphene should be visible, since there is no optical image to be projected; an afterimage would be trapped within the eye by closing the lids; and prodding the visual cortex directly should cause us to see inside our heads. Presumably the space we project back into is subjective, empty, and ready to receive the images; from whence did it come if empty space cannot be perceived? And if projection is a process, as the verb implies, conceivably it could not develop or default in disease; would we then fail to project? Do we learn to project, and if so, can we be taught to see objects up when they are down and left when they are right as indeed some psychologists suggest? Thus the difficulties pile up by confusing physical and subjective events, the reversability of light rays with visual perception. From a grain of optical fact there has grown a mountain of psychologic fable, a pyramid of absurdities, and all to implement a nonexistent process.

All perceptions, visual, auditory, or tactual, are external to the ego. Although all sensory excitations occur in the physical body, it has no subjective reality except as it is somesthetically, proprioceptively, or visually perceived. My stomach may be upset, but it is "I" not my stomach that does not feel up to par. The perceived body, or body image, is the point of reference for absolute distance judgments. Vision is a distance sense modality in that objects are at some distance from the visual ego. The visual ego may be located by pointing a ruler at "yourself"; generally it is directed at the dominant eye or somewhere between the two eyes toward the center of the head. "Projection" is required only if we assume that bodily processes causing sensation must also be their seat. But vision does not occur in the retina, and the brain contains no images. There is nothing to be projected and no space into which to project. We are misled by the coincidence that a triangular retinal image also arouses a triangular percept, but this is only because the eye is an optical instrument tuned to discriminate light patterns. The cortical processes are no more required to be triangular than "red," "angry," or "painful." It is only necessary that they preserve a spatial and temporal isomorphism with the stimuli arousing them.

All visual percepts, in addition to being intrinsically spatial, also have a direction. How does the excitation of a particular retinal point come to be associated with a particular visual direction? Each retinal element, it seems, achieves a spatial value, a "local sign" by which it announces itself as here or there. The totality of this arrangement is called the "law of visual direction" or, in the case of touch, the "law of eccentric projection." It is easy to assume, and many people do, that visual directions are somehow tied to the directions of light rays forming the optical retinal image. Thus for Le Conte, visual impressions are referred back along the path of the principal light ray (even when the light enters the pupil eccentrically, and there is no central pupillary ray); Hyslop assumes the visual impression is referred along a line perpendicular to the retinal point stimulated; still others suppose that the percept is an extension of the longitudinal axis of the excited rod or cone. Visual directions are even referred to the nodal point (an optical fiction)* or, according to Wundt, the center of rotation of the eyeball (a geometric fiction). All explanations relating direction to light rays are, of course, invalidated by nonoptical pressure phosphenes, afterimages, or retinal vessel shadows. It makes no difference how the receptor is stimulated, and since the retina cannot detect perpendiculars, such "explanations" are tantamount to stating that seeing is explained by seeing (Müller, 1826).

Local signs were invented by Lotze (1852) in connection with touch and later applied to the retina and vision. Lotze, following Berkeley, believed local signs were learned by an association of the excitation of a given retinal (or dermal) point with an actual or intended muscular movement in a particular direction. The intensity of the movement or intended movement defines the spatial quality. In the case of vision this association need only be established initially. Subsequently all awareness of ocular movements fade from consciousness, leaving only the local signs, which now operate even without the benefit of such movements. Hence, says Lotze anticipating later nativists, "the orientation of the perceptual image *seems*, to us, to be something simply given."

Lotze's view, in fact, is a considerable advance from Berkeley who had also tackled the problem of visual direction in connection with retinal inversion. Ignoring the confusion between directions and the labels we must admittedly learn to attach to them, Berkeley had proposed that vision receives its spatial education through the sense of touch, transferring the primary problem from one to the other. In addition, however, he also offered the principle of directional relativity (still a favorite among modern thinkers). Although all parts of the optical retinal image are inverted relative to the physical object, the relations of top and bottom *within* the image are preserved. Since the image of the man has his head toward the image of the sky and his feet toward the image of the ground, and since the mind knows only what is on the retina and cannot compare the image to the object, up-down and left-right have no meaning, and retinal inversion is a pseudoproblem. Plausible as this sounds, it will not do because, as Lotze noted, when looking

*Recall that there are two nodal points in the schematic eye for mathematic simplification. The real light rays (insofar as a "ray" can be called "real") are refracted by each of the ocular media in turn. Each retinal point is thus required not only to identify the direction of the ray that strikes it but its antecedent history, thus arriving at a "mean" direction instantaneously, a rather remarkable optical feat. Hering's "lines of direction" are just such conceptual geometric constructions to help us relate the *stimulus* to its optical image on the retina. But the term is so similar to "visual directions" (sometimes the qualification "subjective" is added, as if they could be anything else) that the two are frequently confused (e.g., which does "principal direction" refer to?). Why English writers insist on translating German terminology, never famous for its clarity, is a mystery. The terms "oculocentric" and "egocentric directions" are more intelligible than "relative visual directions" (principal and secondary) or "absolute subjective visual directions" (common or otherwise).

through an astronomic telescope, although everything (object and background) is inverted, we are immediately aware of the transformation—or any other transformation—and before it is confirmed by the contradictory evidence from other senses. The inversion of the retinal image is therefore a necessary prerequisite to upright vision, necessary for appropriate ocular fixation movements and for the established spatial cooperation with touch and kinesthesis.

Hering (1861) adopted local signs but considered them native, that is, given. Indeed he endowed each retinal element with three spatial coordinates at once; horizontal, vertical, and depth. This unnecessary introduction of distance has confused the issue of visual directions ever since. Hering had to assume an anatomic isomorphism between retina and visual cortex; Helmholtz, supporting Lotze's empiricism, supposed the relation is learned. A tangible basis for nativism (Walls, 1951) was not discovered until the present century. We now know that there is a precise topographic representation of the retina in the lateral geniculate, superior colliculus, and area 17 of the visual cortex.* Local signs, it turns out, are not retinal but cortical, not "psychic" but anatomic, and not local but directional. Indeed the neurophysiologic evidence of the past decade provides ample support for the exquisite directional sensitivity of visual cortical cells. It is not necessary therefore to assume, for example, that each of seven odd million retinal cones must acquire its own local sign. It is sufficient that each has a different *position* in the retina and that this difference has its spatial counterpart in the visual cortex. Of course, cones do differ in their neural interconnections, perhaps in chemical composition, but these differences subserve intensity and wavelength discrimination, not direction.

*Such precision does not extend to other sensory systems where brain equipotentially, mass action, and cortical field theory are still the vogue. But the freedom to speculate that the brain is so plastic that anything may be assumed does not extend to the visual system. The anatomy is precise and immutable as any field defect will demonstrate.

In referring to local signs, most writers are careful to speak of retinal elements, not photoreceptors. This is because directional discrimination exceeds the diameter of a single cone (e.g., vernier acuity is on the order of *seconds* of arc). The basis is probably enhancement of physiologic excitation peaks through lateral summation and inhibition. Thus directional discrimination, although based on retinal receptor position, is neurophysiologically extended beyond their population density just as resolution acuity exceeds cone separation. Local signs cannot therefore be educated by ocular movements, if for no other reason than that the amplitude of the smallest movements is on the order of *minutes* of arc. Moreover fixation reflexes would be impossible unless the direction of the object to be fixated was given in advance.

The prewired totality of retinocortical local signs constitute the oculocentric field. Oculocentric directions are intrinsic properties based on the spatial position of retinal elements projecting to specific cortical cells in area 17, each cerebral hemisphere receiving the fibers from both retinas corresponding to the same half of the visual field. All visual perceptions occur within the oculocentric field; all visual impressions have a relative direction. The oculocentric directions define position relative to the fovea, which serves as the polar zero. When the eye turns voluntarily, the oculocentric directions move with it. As I fixate a target moving to the right, it remains imaged on the fovea, and its oculocentric direction does not change. In lower animals the pursuit reflexes (i.e., the maintenance of fixation on a moving target) are mediated by the superior colliculi (which are, therefore, anatomic maps of the retina), but in man the control is taken over by the visual cortex in close apposition to the receiving sensory cells. This sensorimotor sequence depends on and can only be mediated by an optically inverted retinal image, that is, by inherently fixed local signs. This is equivalent to saying that each retinal element signals a specific retinomotor sequence, which increases with its distance from the fovea. The fovea has a retinomotor value of zero with respect to optomotor reflexes. Although the oculocentric field moves with the eye, the direction of objects relative to "me," that is, egocentric directions, depend also on the innervation pattern sent to the extraocular muscles. This integration between oculocentric directions and the record of efferent innervation probably occurs in area 19. When my fovea remains glued to an object moving to the right, the "right" refers to egocentric direction. It the eye sweeps a stationary panorama, no object moves; egocentric directions remain stable, although the oculocentric field moves. But the fovea discriminates successive details with sharper resolution. If I displace the eye passively by a mechanical push, no correcting efferent innervation is available, and the panorama is also displaced egocentrically. In a muscle palsy, the perceptual mechanism operates according to the innervation record, which is now inadequate, and egocentric directions are fallacious, contradicting the sense of touch. (Only the labrynthine reflexes do not enter the perceptual record, hence the swimming of the egocentric field with caloric stimulation.) When the eye is in the primary position, the egocentric and oculocentric directions coincide; under these circumstances the fovea signals "straight-ahead." Retinal elements in the two eyes that share a common oculocentric direction are said to be "corresponding"; their relation is discussed in Chapter 25.

Discrepancies between objective and visual space are generally so small that special conditions are required to demonstrate them. Some are due to inconsistencies in the perception of vertical due to body tilt. The body image is somewhat plastic; for example, there is normally no visceral representation, and an arm or leg that has "fallen asleep" may feel temporarily detached. Visual cues may contradict those elicited by gravity (as in distorted rooms or tilted mirrors), although orientation is eventually reestablished (e.g., by astronauts on space walks). Some discrepancies are due to retinal asymmetries as in the Kundt partition (temporal size segment of a horizontal line is overestimated); some are anatomic (differences in interpupillary distance, refractive power, or axial length); some are psychologic (attitude, technical ability, visuomotor skill); and some are pathologic (retinal metamorphopsia, hemianopia, agnosia). These discrepancies suggest stability may be affected when the stimulus situation is ambiguous or conflicting. Various attempts to alter local signs by wearing reversing prisms, result in new sensorimotor readaptations, not new directions (despite the confused literature). When the optic nerves of salamanders, fishes, frogs, or toads are severed, they regenerate with visual recovery, restoring orderly communication to the brain. If the eye is turned upside down, regeneration is inevitably followed by reversed vision; no experience, no trial and error, and no learning could reverse the result that when bait was presented above, the animal lunges for it below (Sperry, 1956).

Not only are objects seen in particular directions but at different distances relative to each other and to ourselves. All objects are seen at some distance whether we observe them with monocular or binocular vision. There is no question that relative distances and near absolute distances are discriminated better with binocular vision, but this does not mean the monocular world is flat. Monocular patients, especially those who have never developed binocularity, achieve a high degree of depth discrimination and manage to make satisfactory behavioral adjustments in the three-dimensional world. The advantages of binocularity should not blind us to the fact that a great many visual activities, (e.g., reading, watching movies or television, looking at photographs and paintings) do not require two eyes. Indeed the introduction of stereopsis into 3-D movies was a financial if not perceptual flop. The factors involved in distance perception with monocular vision are considered here; binocular factors, in the next chapter.

The accepted theory of monocular distance perception is based on "cues," a perversion of "clue" to avoid the latter's explicit implication of a reasoning process. Either term, however, has its historical roots in empiricism (e.g., Berkeley). Cues may be defined as raw sensory events on which the mind acts analytically to arrive at knowledge. In this sense, the term is a criterion; the mind, a logician; and seeing, a judgment. Or cues may denote some aspect of the retinal image (perspective, light and shade, overlap, etc.), in which case the term is a stimulus; the mind, an organizer; and seeing, a response. The psychologist may also use cue for such subjective factors as motives, set, and past experience; here the term is a tension, the mind is consciousness, and seeing is a resolution of psychic forces. Or a cue may be a signal; the mind, a computer; and seeing, the extraction of information. The terminology varies with one's theoretical approach. Stereopsis has never been classified; the term is used both as a binocular cue for distance and as a perceptual response to retinal disparity. Cues are also classified as primary (i.e., binocular) and secondary (i.e., monocular), although no painter would agree with the relegation of his efforts to secondary status. The notion that binocular cues are primary has resulted in their theoretical overemphasis at the experimental expense of secondary cues.

The traditional cues for monocular distance discrimination are well known (and are well described in Leonardo's notebooks). They include size, geometric and aerial perspective, overlap, light and shade, brightness, clarity, density, horizon level, accommodation, shape, and motion parallax. All cues interact and reinforce each other in normal seeing, along with binocular cues and past experiences. Interesting and ambiguous results are obtained when one factor is pitted against another, the basis of many depth illusions (Fig. 24-9). When the information is ambiguous as in the Ames distorted rooms, the perceiver must rely on what to him yields the greatest degree of consistency and stability (hence variations with age and attitude). If monocular cues are reduced sufficiently, one ends up with a point of light in a dark room for which depth discrimination is almost nonexistent or a homogeneously illuminated unstructured fog for which depth is almost nondescribable.

When monocular cues collaborate, as in a well-executed realistic painting, the effect is so vivid that it is labeled a trick (trompe l'oeil). On the portrait of Cardinal Sauli by del Piombo in the National Gallery there is a painted fly of such realism that visitors frequently attempt to brush it away. A fur by

Fig. 24-9. Perspective illusion. The breakdown of size constancy is attributed to a misjudgment induced by the monocular depth cues. The arrow indicates the direction of the vanishing point *(VP)*, which lies on the horizon. The horizon determines the perspective "eye-level."

Rembrandt, a satin by Velasquez, or the skin of a Rubens have an almost palpable quality, and some of the frescoes painted on ceilings of Italian renaissance villas are a veritable furor perspecticus. The effect of light and shade and interposition in a photograph are as hard to discount as those of the actual scene, except by a Picasso or a Miro.

Despite the immediate and persistent impression of depth in pictures, it is never as striking as the real scene. The granularity of the surface, the visibility of brush strokes, and, most important, the invariance with different angles of view destroys the illusion. The visual field unlike the picture is not bounded by a framework but is oriented with reference to gravity and the body image; shapes are transformed as the head moves, and various portions spring into focus as accommodation and foveal fixation sweeps the panorama. But the analysis of pictures does show that everyday nonsymbolic visual perception is a close correlate of the retinal image as emphasized by Gibson (1950). There is a lawful optical projection of objects onto the retina producing gradients of size, tex-

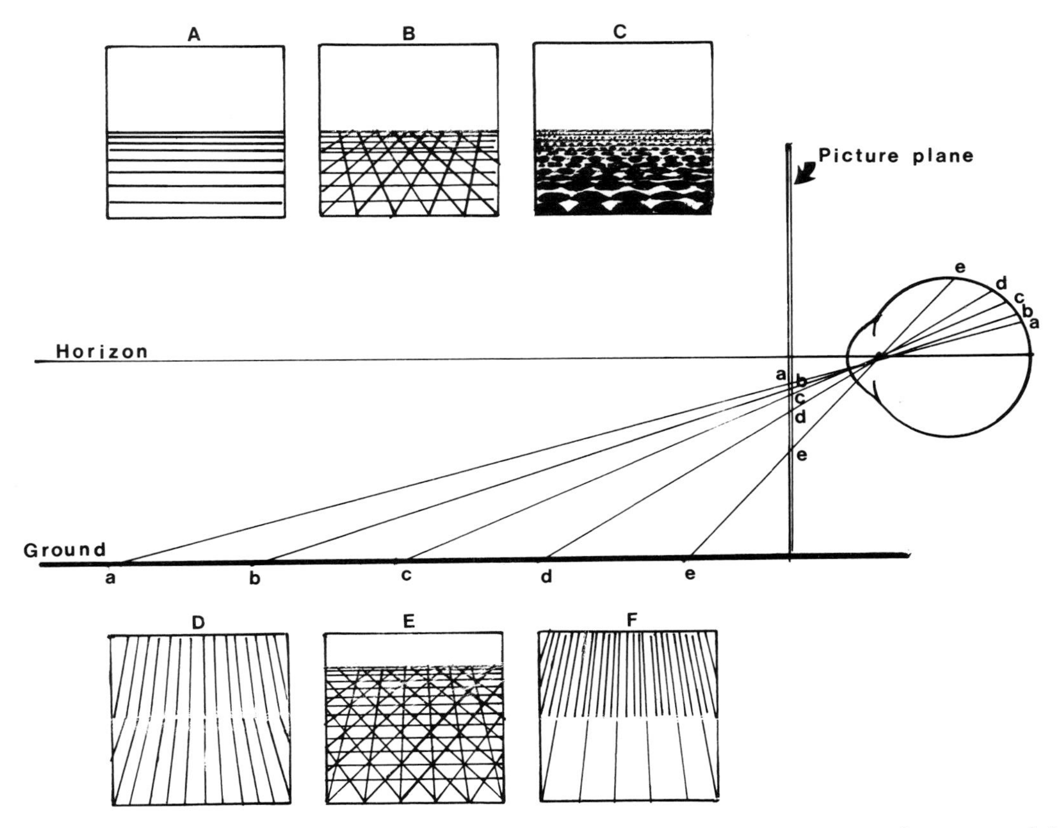

Fig. 24-10. Equal segments on the ground are transformed into unequal segments on the retina and the perpendicular picture plane. This gradient of horizontal spacing is illustrated in *A*. *B* shows gradients of geometric perspective; *C*, gradients of texture (detail and size); *D*, a change in gradient corresponding to a corner; *E*, a change in the horizon line or eye level; and *F*, a jump in gradients corresponding to an edge.

ture, density, and illumination as objects recede or approach the observer.

According to Gibson, the seeming poverty of the retinal image is exaggerated. Even static, momentary images arouse depth and distance without the need of a special mental process to supplement them. For example, the retinal image is related to the visual scene by a lawful (i.e., predictable) transformation of size, texture, density, and illumination. These transformations may be studied and varied by plane projections. Fig. 24-10 shows such a projection and illustrates gradients of vertical spacing, perspective, and texture. Changes in gradient correspond to a corner, a jump between gradients corresponds to an edge, and gradients of illumination (chiaroscuro) yield depth. A moving observer has available gradients of image deformation, a flow of perspective (e.g., position of horizon line), and changing overlap (e.g., motion parallax). The traditional cues for depth are thus replaced by retinal image gradients.

SOME CLINICAL IMPLICATIONS

It is fashionable to point out (e.g., Carr, 1935; Boring, 1942) that the nativism-empiricism is a psychologic dead issue or ought to be. Yet its clinical implications are far from obsolete. To the extent seeing is learned, it is subject to relearning, adaptation, and training; to the extent visual perception is prewired, it will not permit any outside tampering. Awareness of which is what will allow us to predict what kinds of spatial adaptations a patient may make to his new spectacles, evaluate the symptoms of a child who cannot read efficiently, or prescribe orthoptics with a success to match our conviction and persistence. It is not enough to vaguely indicate that vision is an affair of the mind. Stock analysts can turn vagueness into a science; astrologers, into an art; prophets, into a poem, but the clinician's reputation hangs on his results. He must know if not all that may happen at least what is unachievable.

Clinical refraction employs psychophysical and phenomenologic techniques. We observe a retinoscopic reflex, a fusional refixation movement, or a pupillary response; we ask the patient to discriminate differences in color, brightness, size, and distance; and when the results are bad, the patient is back with introspective lamentations. When we change the size, color, contrast, or orientation of a target, we anticipate predictable responses. We must know to what extent they result from physical, physiologic, or psychologic variables. There are spatial distortions from anisometropia, retinal edema, conversion syndromes, and malingering, each with its differential diagnostic features. There are changes brought about by neurophysiologic modification, by improved functional capacity, and by better utilization of psychic potential, and there are the improved scores resulting simply from clearer instructions. There are global changes in perception and those confined to a specific and limited task; some permanent, some temporary. There are changes from learning and from maturation. If "improved visual efficiency" is to be more than a muddled verbal puddle, we must know which changes we are talking about; whether the new spectacles brought about improved utilization of optical information, ameliorated a motor defect, or fulfilled a subconscious drive.

It is misleading to call some tests perceptual; all vision tests measure visual perception. The isolation of some stimulus aspect for diagnosis or for training a specific response pattern does not transmute us into educators, psychologists, or epibiologists, nor do the results necessarily provide new insights into the perceptual process.

REFERENCES

Adrian, E. D.: The basis of sensation, New York, 1928, Norton.

Allport, F. H.: Theories of perception and the concept of structure, New York, 1955, John Wiley & Sons, Inc.

Alpern, M.: Metacontrast, J. Opt. Soc. Am. **43**:648-657, 1953.

Ames, A.: Visual perception and the rotating trapezoidal window, Psychol. Monogr. **65**:324, 1951.

Asch, S. E., and Witkin, H. A.: Studies in space orientation, J. Exp. Psychol. **38**:325-337, 1948.

Asher, H.: Contrast in eye and brain, Br. J. Psychol. **40**:187-194, 1950.

Attneave, F.: Some informational aspects of visual perception, Psychol. Rev. **61**:183-193, 1954.

Aubert, H.: Physiologio der Netzhaut, Breslau, 1865, E. Morgenstein.

Ayer, A. J., and Winch, R., editors: British empirical philosophers, New York, 1968, Simon & Schuster, Inc.

Ayres, J. J., and Harcum, E. R.: Directional response-bias in reproducing brief visual patterns, Percept. Mot. Skills **14**:155-165, 1962.

Baird, J. C.: Psychophysical analysis of visual space. In Eysenk, H., editor: International monograph in experimental psychology, New York, 1970, Pergamon Press.

Bartlett, F. C.: An experimental study of some problems of perceiving and imagining, Br. J. Psychol. **8**:222-266, 1916.

Bartley, S. H.: The principles of perception, New York, 1969, Harper & Row, Publishers.

Bauermeister, M.: Effect of body tilt on apparent verticality, apparent body position, and their relation, J. Exp. Psychol. **67**:142-147, 1964.

Begelman, D. A.: The role of retinal orientation in the egocentric organization of a visual stimulus, J. Gen. Psychol. **79**:283-289, 1968.

Berkeley, G.: An essay towards a new theory of vision, 1709, New York, 1910, Dutton.

Bishop, P. O., and Henry, G. H.: Spatial vision, Ann. Rev. Psychol. **22**:119-160, 1971.

Boring, E. G.: Sensation and perception in the history of experimental psychology, New York, 1942, D. Appleton-Century Co.

Bourdon, B.: La perception visuelle de l'espace, Paris, 1902, Schleicher Freres.

Boynton, R. M.: Spatial vision, Ann. Rev. Psychol. **13**: 171-200, 1962.

Brown, J. F.: The visual perception of velocity, Psychol. Forsch. **14**:199-232, 1931.

Brown, J. L.: Afterimages. In Graham, C. H., (editor): Vision and visual perception, New York, 1965, John Wiley & Sons, Inc.

Brucke, E.: Vorlesungen über Physiologie, Vienna, 1884, Braumueller.

Buswell, G. T.: How people look at pictures, Chicago, 1935, University of Chicago Press.

Campbell, F. W., and Maffei, L.: Electrophysiological evidence for the existence of orientation and size detectors in the human visual system, J. Physiol. **207**:635-652, 1970.

Campbell, F. W., and Maffei, L.: Contrast and spatial frequency, Sci. Am. **231**:106-114, 1974.

Campbell, F. W., and Robson, J. G.: Application of Fourier analysis to the visibility of gratings, J. Physiol. (London) **197**:551-566, 1968.

Carr, H. A.: An introduction to space perception, New York, 1935, Longmans, Green & Co.

Clarke, F. J. J.: A study of Troxler's effect, Opt. Acta **7**:219-236, 1960.

Cornsweet, T. N.: Visual perception, New York, 1970, Academic Press, Inc.

Day, R. H.: Application of statistical theory to form perception, Psychol. Rev. **63**:139-148, 1956.

Dember, W. N.: The psychology of perception, New York, 1960, Henry Holt & Co.

Deregowski, J. B.: Pictorial perception and culture, Sci. Am. **227**:82-88, 1972.

De Silva, H. R.: An analysis of the visual perception of movement, Br. J. Psychol. **19**:268-305, 1929.

Dewey, J.: Experience and nature, New York, 1958, Dover Publications, Inc.

Dews, P. B., and Wiesel, T. N.: Consequences of monocular deprivation on visual behaviour in kittens, J. Physiol. (London) **206**:437-455, 1970.

Djang, S.: The role of past experience in the visual perception of masked forms, J. Exp. Psychol. **20**:29-59, 1937.

Dowling, J. E., and Boycott, B. B.: Organization of the primate retina: electron microscopy, Proc. R. Soc. [Biol.] **166**:80-111, 1966.

Ewert, J. P.: The neural basis of visually guided behavior, Sci. Am. **230**:34-42, 1974.

Ewert, P. H.: Factors in space localization during inverted vision, Psychol. Rev. **43**:522-546, 1936.

Fantz, R. L.: Pattern vision in young infants, Psychol. Rec. **8**:43-47, 1958.

Fieandt, K. von, and Wertheimer, M.: Perception, Ann. Rev. Psychol. **20**:159-192, 1969.

Flock, H. R.: Three theoretical views of slant perception, Psychol. Bull. **62**:110-121, 1964.

Forgus, R. H.: Perception, New York, 1966, McGraw-Hill Book Co.

Fry, G. A., and Bartley, S. H.: The effect of one border in the visual field upon the threshold of another, Am. J. Physiol. **112**:414-421, 1935.

Ganz, L.: Lateral inhibition and the location of visual contours; an analysis of visual aftereffects, Vision Res. **4**:465-481, 1964.

Gelb, A.: Die Farbenkonstanz der Sehdinge. In Handbuch d. normal. und pathol. physiologie, Berlin, 1929, J. Springer.

Geldard, F. A.: The human senses, New York, 1972, John Wiley & Sons, Inc.

Gibson, E. J.: Principles of perceptual learning and development, New York, 1969, Appleton-Century-Crofts.

Gibson, E. J., Bergman, R., and Purdy, J.: The effect of prior training with a scale of distance on absolute and relative judgments of distance over ground, J. Exp. Psychol. **50**:97-105, 1955.

Gibson, J. J.: The perception of the visual world, Boston, 1950, Houghton Mifflin Co.

Gibson, J. J.: The senses considered as perceptual systems, Boston, 1966, Houghton Mifflin Co.

Gibson, J. J.: What gives rise to the perception of motion, Psychol. Rev. **75**:335-346, 1968.

Gilinsky, A. S.: Perceived size and distance in visual space, Psychol. Rev. **58**:460-482, 1951.

Gombrich, E. H.: The visual image, Sci. Am. **227**: 82-96, 1972.

Gordon, B.: The superior colliculus of the brain, Sci. Am. **227**:72-83, 1972.

Gottschaldt, K.: The influence of past experience on the perception of figures, Psychol. Forsch. **8**:261-317, 1926.

Granit, R.: Receptors and sensory perception, New Haven, 1955, Yale University Press.

Green, D. G.: Testing the vision of cataract patients by means of laser-generated interference fringes, Science **168**:1240-1242, 1970.

Gregory, R. L.: Eye and brain, New York, 1971, McGraw-Hill Book Co.

Gruber, H. E.: The relation of perceived size to perceived distance, Am. J. Psychol. **67**:411-426, 1954.

Harris, C. S.: Perceptual adaptation to inverted, reversed, and displaced vision, Psychol. Rev. **72**:419-444, 1965.

Hartline, H. K.: The response of single optic nerve fibers of the vertebrate eye to illumination of the retina, Am. J. Physiol. **121**:400-415, 1938.

Hartline, H. K.: The nerve messages in the fibers of the visual pathway, J. Opt. Soc. Am. **30**:239-247, 1940.

Hartline, H. K.: The receptive fields of optic nerve fibers, J. Physiol. **130**:690-699, 1940.

Hartline, H. K., and Graham, C. H.: Nerve impulses from single receptors in the eye, J. Cell. Comp. Physiol. **1**:277-295, 1932.

Hebb, D. O.: The organization of behavior, New York, 1949, John Wiley & Sons, Inc.

Hecht, S.: The nature of the photoreceptor process. In Murchison, C. A., editor: Handbook of experimental psychology, Worcester, 1934, Clark University Press.

Held, R., and Hein, A.: Adaptation of disarranged hand-eye coordination contingent upon re-afferent stimulation, Percept. Mot. Skills **8**:87-90, 1958.

Helson, H.: Current trends and issues in adaptation-level theory, Am. Psychol. **19**:26-38, 1964.

Hering, E.: Eine Methode zur Beobachtung des Simultankontrastes, Pfluegers Arch. Gesamte Physiol. **47**: 236-242, 1890.

Hermans, T. G.: The perception of size in binocular, monocular, and pin-hole vision, J. Exp. Psychol. **27:** 203-207, 1940.
Hirsch, H. V. B., and Spinelli, D. N.: Visual experience modifies distribution of horizontally and vertically oriented receptive fields in cats, Science **168:** 869-871, 1970.
Hochberg, J.: Perception, Englewood Cliffs, 1964, Prentice-Hall Co.
Holland, H. C.: The spiral aftereffect, London, 1965, Pergamon Press.
Howard, I. P., and Templeton, W. B.: Human spatial orientation, New York, 1966, John Wiley & Sons, Inc.
Hubel, D. H., and Wiesel, T. N.: Receptive fields, binocular interaction, and functional architecture in the cat's visual cortex, J. Physiol. (London) **160:**106-154, 1962.
Hubel, D. H., and Wiesel, T. N.: Cortical and callosal connections concerned with the vertical meridian of visual fields in the cat, J. Neurophysiol. **30:**1561-1573, 1967.
Huey, E. B.: The psychology and pedagogy of reading, Boston, 1968, The M. I. T. Press.
Humphrey, N. K., and Weiskrantz, L.: Vision in monkeys after removal of the striate cortex, Nature **215:** 595-597, 1967.
Hurvich, L. N., and Jameson, D.: The perception of brightness and darkness, Boston, 1966, Allyn & Bacon, Inc.
Hyslop, J. H.: Upright vision, Psychol. Rev. **4:**71-73, 142-163, 1897.
Ittelson, W. H.: Size as a cue to distance, Am. J. Psychol. **64:**54-67, 1951.
Ittelson, W. H.: The Ames demonstrations in perception, Princeton, 1952, Princeton University Press.
James, W.: Principles of psychology, New York, 1891, Henry Holt & Co.
Jones, R. C.: How images are detected, Sci. Am. **219:** 110-117, 1968.
Katz, D.: The world of colour, London, 1935, Kegan Paul, Trench, Trubner & Co.
Kaufman, L.: On the spread of suppression and binocular rivalry, Vision Res. **3:**401-415, 1963.
Kohler, W.: Unsolved problems in the field of figural after-effects, Psychol. Rec. **15:**63-83, 1965.
Kolers, P. A.: Experiments in reading, Sci. Am. **227:** 84-91, 1972.
Korzybski, A.: Science and sanity, Lancaster, 1933, Science Press.
Kuffler, S. W.: Discharge patterns and functional organization of mammalian retina, J. Neurophysiol. **16:**37-68, 1953.
Lashley, S. K.: The problem of cerebral localization in vision. In Kluver, H., editor: Biological Symposia, vol. 7, Lancaster, 1942, Jacques Cattell Press.
Le Conte, J.: Sight, New York, 1895, D. Appleton & Co.
Leibowitz, H. L., and Harvey, L. O.: Perception, Ann. Rev. Psychol. **24:**207-229, 1973.
Linksz, A.: Physiology of the eye; vision, vol. 2, New York, 1952, Grune & Stratton, Inc.
Lotze, R. H.: Psychologie oder Physiologie der Seele, Leipzig, 1852, Weidmann.
Lowenstein, W. R.: Biological transducers, Sci. Am. **203:**98-104, 1960.
Luckiesh, M.: Visual illusions; their cause, characteristics and applications, New York, 1965, Dover Publications, Inc.
Mach, E.: The analysis of sensations, translated by C. M. Williams, New York, 1959, Dover Publications, Inc.
Magnus, R.: Korperstellung, Berlin, 1924, J. Springer.
Marchbanks, G., and Levin, H.: Cues by which children recognize words, J. Educ. Psychol. **56:**57-61, 1965.
Martin, L. C.: The photometric matching field. Proc. R. Soc. (London) **104:**302-315, 1923.
Matin, L.: Eye movements and perceived visual direction, Handbook Sensory Physiol. **7:**331-380, New York, 1972, Springer.
McCurdy, E.: Leonardo da Vinci's note-books, New York, 1923, Empire State Book Co.
McDougall, W.: The sensations excited by a single momentary stimulation of the eye, Br. J. Psychol. **1:**78-113, 1904.
McIlwain, J. T.: Central vision: visual cortex and superior colliculus, Ann. Rev. Physiol. **34:**291-314, 1972.
Metelli, F.: The perception of transparency, Sci. Am. **230:**91-98, 1974.
Michael, C. R.: Receptive fields of single optic nerve fibers in a mammal with an all-cone retina, J. Neurophysiol. **31:**249-256, 1968.
Michotte, A.: The perception of causality, London, 1963, Methuen.
Miller, G. A.: Language and communication, New York, 1951, McGraw-Hill Book Co.
Miller, G. A.: The magical number seven, plus or minus two: some limitations on our capacity for processing information, Psychol. Rev. **63:**81-97, 1956.
Milner, P. M.: Physiological psychology, New York, 1970, Holt, Rinehart & Winston, Inc.
Müller, J.: Zur vergleichenden Physiologie des Gesichtssinnes des Menschen und der Thiere, Leipzig, 1826, Knobloch.
Murch, G. M.: Visual and auditory perception, Indianapolis, 1973, Bobbs-Merrill Co., Inc.
Murphy, G.: An historical introduction to modern psychology, New York, 1932, Harcourt, Brace & Co.
Noton, D., and Stark, L.: Scanpaths in eye movements during pattern perception, Science **171:**308-311, 1971.
Ogden, C. K., and Richards, I. A.: The meaning of meaning, New York, 1947, Harcourt, Brace & Co.
Ogle, K. N.: The optical space sense. In Davson, H., editor: The eye, vol. 4, London, 1962, Academic Press, Inc., pp. 211-417
Osterberg, G.: Topography of the layer of rods and cones in the human retina, Acta. Ophthal. (supp.) **6:**1-106, 1935.

Piaget, J.: The child's construction of reality, London, 1955, Routledge & Kegan Paul.
Pick, H. L., and Ryan, S. M.: Perception, Ann. Rev. Psychol. **22**:161-192, 1971.
Pierce, J. R.: Symbols, signals, and noise, New York, 1961, Harper & Brothers.
Pratt, C. C.: The role of past experience in visual perception, J. Psychol. **30**:85-107, 1950.
Purdy, D. M.: The Bezold-Brucke phenomenon and contours for constant hue, Am. J. Psychol. **49**:313-315, 1937.
Raab, D. H.: Backward masking, Psychol. Bull. **60**: 118-129, 1963.
Rapaport, D., Gill, M. M., and Schafer, R.: Diagnostic psychological testing, Chicago, 1945, Year Book Publishers.
Ratliff, F., and Hartline, H. K.: The responses of *Limulus* optic nerve fibers to patterns of illumination on the retinal mosaic, J. Gen. Physiol. **42**:1241-1255, 1959.
Redlich, F. C., and Freedman, D. X.: The theory and practice of psychiatry, New York, 1966, Basic Books, Inc.
Riggs, L. A., Ratliff, F., Cornsweet, J. C., and Cornsweet, T. N.: The disappearance of steadily fixated visual test objects, J. Opt. Soc. Am. **43**:495-501, 1953.
Rock, I.: The nature of perceptual adaptation, New York, 1966, Basic Books, Inc.
Rock, I.: The perception of disoriented figures, Sci. Am. **230**:78-85, 1974.
Rock, I., and Victor, J.: Vision and touch: an experimentally created conflict between the two senses, Science **143**:594-596, 1964.
Rosenblith, W. A., editor: Sensory communication, New York, 1961, John Wiley & Sons, Inc.
Rubin, E.: Visuell wahrgenonmene Figuren, Copenhagen, 1921, Gyldendalska.
Sanford, F. H.: Speech and personality, Psychol. Bull. **39**:811-845, 1942.
Sekuler, R.: Spatial vision, Ann. Rev. Psychol. **25**: 195-232, 1974.
Skinner, C. E., editor: Educational psychology, New York, 1947, Prentice-Hall, Inc.
Smith, K. V., and Smith, W. M.: Perception and motion, Philadelphia, 1962, W. B. Saunders Co.
Spearling, G.: The information available in brief visual presentations, Psychol. Monogr. **74**:11, 1960.
Sperry, R. W.: The eye and the brain, Sci. Am. **194**: 48-52, 1956.
Stevens, S. S., editor: Handbook of Experimental Psychology, New York, 1951, John Wiley & Sons, Inc.
Stevens, S. S.: The surprising simplicity of sensory metrics, Am. Psychol. **17**:29-39, 1962.
Swets, J. A.: Detection theory and psychophysics: a review, Psychometrika **26**:49-63, 1961.
Taylor, J. G.: The behavioral basis of perception, New Haven, 1962, Yale University Press.
Thouless, R. H.: Phenomenal regression to the real object, Br. J. Psychol. **21**:338-359, 1931.
Titchener, E. G.: Textbook of psychology, New York, 1910, The Macmillan Co.
Uhr, L. M.: Pattern recognition, New York, 1966, John Wiley & Sons, Inc.
van de Geer, J., and Jaspars, J. M.: Cognitive functions, Ann. Rev. Psychol. **17**:145-176, 1966.
Vernon, M. D.: A further study of visual perception, Cambridge, 1952, Cambridge University Press.
Walk, R. D., and Gibson, E. J.: A comparative and analytical study of visual depth perception, Psychol. Monogr. **75**:1-44, 1961.
Walls, G. L.: The problem of visual direction, Am. J. Optom. **28**:55-83, 1951.
Weale, R.: After images, Br. J. Ophthal. **34**:190-192, 1950.
Weisstein, N.: Metacontrast, Handbook Sensory Physiol. **7**:233-272, 1972.
Werblin, F. S.: The control of sensitivity in the retina, Sci. Am. **228**:70-79, 1973.
Werner, H.: Studies in contour, Am. J. Psychol. **47**: 40-64, 1935.
Werner, H., and Wapner, S.: Toward a general theory of perception, Psychol. Rev. **59**:324-338, 1952.
Wertheimer, M.: Experimentelle Studien über das Sehen von Bewegung, Z. Psychol. **61**:161-265, 1912.
Willey, R., and Gyr, J. W.: Motion parallax and projective similarity as factors in slant perception, J. Exp. Psychol. **79**:525-532, 1969.
Wundt, W.: Beiträge zur Theorie der Sinneswahrnehmung, Z. Ration. Med. **3**:12, 1862.

25

Binocular vision

Binocular vision depends on a sequence of harmoniously favorable factors: sensory and motor, spatial and temporal, psychophysical and psychophysiologic. The sequence may be interrupted by a variety of obstacles, ranging from anisometropia and excessive tonic vergence to deficient fusional amplitudes or intrinsic anomalies of the retina and visual pathway. The consequences range from motility imbalances in which binocularity is inefficient or uncomfortable to strabismus in which binocular vision is not possible. To maintain useful visual orientation, there follow certain inevitable concessions and adaptations: suppression, amblyopia, eccentric fixation, and spatial reorganization. The differential diagnosis is made by tracing the etiologic obstacle and its effects in the total clinical picture.

The clinical evaluation of binocular vision is often based on the use of stereoscopes (or haploscopes). These instruments merely reproduce conditions that would have been obtained if a real three-dimensional object or scene were observed. In interpreting the results, it is always useful and often necessary to think in terms of the spatial geometry—the direction and distance of real objects—not only the horizontal separation of stereograms in an artificial setting. The parents of a strabismic child want to know whether the anomaly or its treatment affects learning, reading, driving, and similar activities, not how much depth can be seen in a stereoscope.

It would be an error, however, to think of binocular anomalies only in terms of strabismus. Every refractive condition and spectacle correction has its motor and spatial consequences that may shift the binocular balance in a favorable or unfavorable direction. The presbyope whose new reading glasses may precipitate an intermittent exotropia; the pseudomyope with a convergence spasm; the accommodative esotrope; and the patient with anisometropia, anisophoria, aphakia, or aniseikonia are each, to a variable extent, binocular cripples. Conversely, an optical correction or spectacle modification is often the primary and always a supplementary means of treating manifest ocular deviations.

CLINICAL PHYSIOLOGY

Binocular vision is one of those subjects (like color vision) about which it is not possible to say a little without saying a great deal that is misleading. Vision like any other sensation is a psychophysical process; to analyze the response one must specify the stimulus. As in elementary optics, the binocular stimulus analysis is simplified by a series of assumptions, which, although only approximately true, are true enough for a first-order clinical approach.

In Fig. 25-1, two reduced eyes are substituted for the compound optics of the real eyes. Each eye is assumed to be in focus (to permit use of the nodal point instead of the

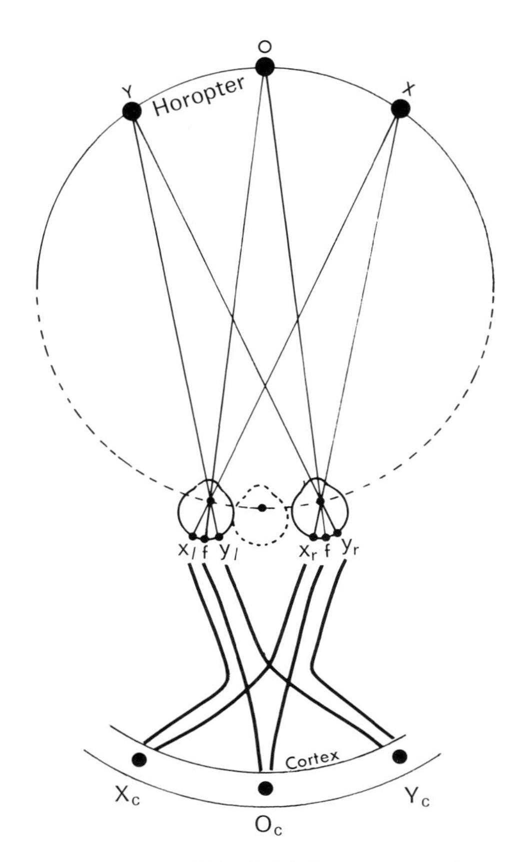

Fig. 25-1. Basis of Vieth-Muller horopter (not to scale). For details see text.

exit pupil). Each retina is further assumed to be perfectly spherical, and each eye is supposed to pivot about a fixed center of rotation that coincides with the nodal point. The two eyes fixate a target *(O)* in the median plane. The lines connecting the object and the two foveas represent the hypothetical, most direct path of a light ray (actual rays might enter the pupil eccentrically but would end up at the same retinal point). These hypothetical paths pass through each nodal point (or center of rotation) and are termed "lines of direction." The two foveas are arbitrarily defined as corresponding points—"corresponding" because their topographic neural representation is a single cortical locus and "arbitrarily" because this is not a clinical criterion by which correspondence can be measured. We further assume that two retinal points x_l and x_r, which have identical angular displacements at the nodal point, also have identical cortical representation at X_c, hence are corresponding; and similarly for the retinal points y_l and y_r at Y_c. It is evident that to stimulate the points x_l and x_r, an object must be located at a position X, and similarly to stimulate points y_l and y_r, an object would be required at Y. Connecting the object positions X, Y, and O yields an arc, part of the Vieth-Muller horopter. The horopter thus represents *potential* target positions that would stimulate all possible corresponding points in the horizontal meridian for a given fixation distance. The circle is incomplete because the binocular field does not extend 360°, but its extension passes through the nodal point of each reduced eye (by the assumed definition of correspondence). If the fixation distance increases, the horopter circle gets larger. If the horopter is to represent corresponding points in meridians other than the horizontal, the Vieth-Muller circle becomes a partial sphere (spherical because the retinas were assumed hemispherical and partial because the binocular field is limited). The Vieth-Muller "circle" is simply a cross section of the total horopter "sphere." The diameter of the circle depends on the fixation distance, that is, the convergence of the eyes. But it is independent of versions, since for a constant angle of convergence, any other horopter point fixated does not alter the correspondence relation (assuming symmetrical distribution of corresponding elements and a center of rotation coincident with the nodal point). Most textbook diagrams show only the horopter circle because it fits the plane of the page. We see later that the choice of the horizontal plane has a physiologic as well as a pictorial advantage.

Everything shown in Fig. 25-1, the two reduced eyes, the lines of direction, the horopter, and the cortical loci, exist in objective space. The purpose of the diagram is to define the position of the visual *stimulus* albeit in a simplified and theoretical fashion. Indeed we can carry it one step further. Since the distribution of corresponding points is assumed to be symmetrical, the two reduced eyes can be superimposed at a single position midway between them. The horopter circle naturally passes through the nodal point of this equivalent or "cyclopean eye." The cyclopean eye is therefore an imaginary construct in objective space by which to explain to ourselves the relative location of retinal stimuli. A pin stuck through its "fundus" automatically passes through corresponding points, since they cover each other by definition. It is therefore a simple matter to identify the position of retinal stimuli (e.g., with respect to the foveas) in a case of muscle paralysis to analyze the anticipated type of diplopia (as in the red glass test). The validity of such applications depends on the soundness of the assumptions on which the cyclopean eye is based.

But the cyclopean eye is not the seat of a psychic homunculus that "projects" visual sensations through its nodal point. In fact it has nothing whatever to do with subjective direction or the visual ego. The question arises, however: why does it "work" for those who confuse the two and assume subjective percepts follow objective lines of direction? And, of course, it does work most of the time (or we would not use it to interpret the red glass test results). Visual orientation would indeed be chaotic if the most direct line between object and fovea did not parallel the direction in which we see it. The biologic advantage of light as an adequate stimulus would be lost if the visual system did not make use of the very linearity of light trajectories. And what could be more linear than a line of direction? The most elementary knowledge of vision points to a psychophysical parallelism between lines of direction and visual directions so precise that it staggers the imagination. Hence it is easy to stumble onto the notion that they are identical (particularly when we represent both by convenient lines drawn on paper). Nevertheless, the two are no more identical than color and wavelength, light intensity and brightness, visceral distention and pain, or any other psychophysical relation. It is true that as regards visual directions, special conditions are required to demonstrate the discrepancies between visual and objective space,* but this does not entitle us to confuse the two in our thinking. The parallelism is a fact of nature, not a result of projection that might fail to develop or fall short if the retina is stimulated mechanically with no lines of direction to guide the percept out of the eye.

An essential requirement for binocular vision is the superimposition of the two monocular fields of vision. Frontal eye position has developed several times independently in fishes, birds, and mammals. Within each series, frontality is closely tied to predatory habits. Animals that are obliged to detect danger from all directions tend to retain lateral eyes and panoramic vision. If a binocular field is present at all, it is narrow or even limited to a single point—the striking distance—determined by the two temporal foveas. Various degrees of frontality represent a compromise between the instincts for hunting and self-preservation. The primates, on the other hand, have parallel visual axes not because they are predatory but to exploit their manipulative skills. Only mammals with central foveas have totally conjugated, not just coordinated, ocular movements, and only they examine objects binocularly. These voluntary movements are supplemented by static and statokinetic reflexes, especially in the horizontal plane—the plane of greatest biologic utility.

In man the superimposed foveas occupy the center of his binocular field of vision, which extends about 200° horizontally and 130° vertically. Beyond the binocular field is a monocular crescent whose size varies with the prominence of the bridge of the nose. Any object in the binocular field necessarily stimulates both eyes yet is seen in a single direction. How this comes about is a central problem in the physiology of binocular vision.

*This is not to imply that each person builds up his own private space world that could conceivably differ from everyone else. Solipsism is a theoretical position reserved only for philosophers; anyone else holding such a view is promptly locked away. Spatial orientation must square up with reality, or behavior would be impossible. Evolution has taken care of the physiologic solipsists long ago; the rest of us have the same equipment and necessarily respond in the same sensorial way to identical stimulus conditions (including illusions). Color vision is actually a poor analogy in this respect. Color blindness is compatible with behavior; spatial disorientation is not.

Binocular perception of direction

Visual directions are subjective; lines of direction are objective. Each belongs to its own sphere of categorization. We saw in the previous chapter that every retinal element has a subjective local sign by virtue of its position relative to the fovea. This individuality is anatomically preserved throughout the geniculocalcarine path and even enhanced by cortical magnification (neuronal arborization) to counteract the funnel effect of multiple photorecepters connecting to one ganglion cell. The cortex thus distinguishes the individual direction of objects by the angular separation of their images, a process that would not be feasible if the retina were a continuum instead of a mosaic. The totality of local signs of each eye defines its "*oculocentric directions.*" The fovea is the polar zero, polar because all oculocentric directions are relative to it and zero because it has a zero retinomotor (fixation) value. Such terms as "primary" (or principal) for the foveal oculocentric direction are best avoided. They not only tend to confuse oculocentric and egocentric direction but also imply "straight ahead," true only when the eye is in the primary position (because the primary position is defined that way). An object A is to the left of another object B, and both are to the left of the fovea whether I turn my eye voluntarily, displace the globe mechanically, or interpose a prism or a magnifier. When the eye follows a moving object, its image remains on the fovea, its oculocentric direction is constant, and it continues to be seen with maximum resolution. Oculocentric directions are therefore independent of eye movements.

If my eye follows a moving target to the right or I change my foveal gaze from a target in front of me to another on the right, the "right" obviously refers to something other than oculocentric direction. It refers to its direction relative to "me." The "me" does not mean my body in objective space but my visual ego. The visual ego is the polar zero of "*egocentric directions.*"

Although there is a set of oculocentric directions for each eye, there is normally only one set of egocentric directions whose origin is a unitary locus in the mind (not necessarily at a specific anatomic position in the brain). But the visual ego, so to speak, views the world through both eyes. If the eye is compared to a television camera and the visual ego to the receiver, the latter sees the scene *as if* it were inside the camera. If the camera is aimed downhill, the receiver registers that direction whether it sits on the floor, on a table, or on a high shelf. Similarly, the visual ego perceives directions *as if* it were stationed inside the eye (at a point variously called the "station point," "vantage point," "sighting center," or "perspective center"). Operationally therefore egocentric directions originate (clearly they can never cross each other) as if at some point in the eye, probably the entrance pupil (in analogy with the camera). The reason the visual ego perceives directions according to which way the eye is pointing is not, of course, that it really sits in the eye but that the brain is informed of its movements by the innervation sent and of the retinal points stimulated by their local signs.

In binocular vision each eye has equal claim to be the seat of the visual ego. The preference is given to the dominant or "sighting" eye. Indeed this preference is the only legitimate definition of ocular dominance. When only one eye is available, it is automatically the sighting eye. If there is no clear dominance, the site of the visual ego is indeterminately somewhere between the eyes (but not from a quasisubjective counterpart of the cyclopean eye, as in Tschermak's often reproduced diagram). There is nothing wrong, of course, with a diagram showing visual directions as lines drawn on paper; indeed they are extremely useful (e.g., in explaining physiologic diplopia), but they must not be confused with lines of direction. (Directions, in fact, are not lines but points on the compass or an angle made with some axis of reference; the lines representing *visual directions* are really pointers, whereas *lines of directions* represent, at least hypothetically, the actual trajectories of photons.)

Experimental methods to locate the visual egocenter turn out to be modified techniques of measuring the sighting eye. And the results are less than clear-cut. Pointing a binocularly seen rod at "yourself" is basically a sighting task. One can then close one eye, and if the rod is still aligned, the open eye is the dominant one. A more precise point can be achieved by triangulation, that is, by aiming the rod from various directions and distances. The result generally coincides with the entrance pupil of the dominant eye. Variations occur if the target is asymmetrical, with successive fixations, or with changes in head or body posture.

The egocentric "straight ahead" is easier to define than to measure. I fixate the page in front of me. When I now turn my eyes, my head, or my body to the right, at what point does the page acquire an egocentric direction "to my left"? The criterion clearly depends on the total visual framework (the page remains centered on my desk). If you place a dissociating vertical prism before the eyes, you see two visual scenes. Which one is straight ahead? When a squinting child is asked whether an ophthalmoscopically projected star on its fundus is seen straight ahead, is his frame of reference to be his eye, head, or body as he feels them; the examiner or instrument as he sees them; or his median plane as he conceives it?

Egocentric directions are, in fact, independent of vision. Objects are felt or heard to the left or right of me whether I look at them or have both eyes closed. Such nonvisual egocentric directions depend on the effects of gravity, proprioception, and innervation sense. (An innervation sense is not unique for the visual system; an anesthesized arm can be pointed with fair accuracy even when it is totally insensitive to passive movements.) Normally visual and "other" egocentric directions coincide—we see objects where we feel or hear them. But they can be dissociated, as in watching a film of a roller coaster ride spectacular enough to make our stomach flip by visual cues alone; or conversely, when the visual vertical is altered by tilting the head. The change in the visually perceived vertical is seldom more than 10° to 15° regardless of how much the head is tilted. Since ocular torsion cannot compensate for more than 10° of head tilt, other compensatory mechanisms operate (e.g., afferent impulses from the muscles of the head, neck, and labyrinth). Proprioception from eye muscles, on the other hand, are ineffective; we can seldom identify the position of the eye by its "feel," and most people are unaware that the eye rolls up when the lids are closed.

When my eye follows a moving target as contrasted to sweeping a stationary environment, a relative retinal displacement of object and background occurs in the first case but not in the second. Is this difference in retinal displacement sufficient to distinguish between a moving and stable environment? No. A mechanical displacement of the globe by finger push alters egocentric directions despite total displacement of the retinal field. Moreover an afterimage maintains the same egocentric direction regardless of such mechanical displacement but moves when I will my eye to turn. Some sort of extraretinal signal is therefore required by which shifting oculocentric directions are transformed into stable target perception during voluntary saccades and shifting target perception during pursuit movements. This is the "innervation sense" (also called "innervation factor," "willed movement," "copy of efferent innervation," and, perhaps most aptly by von Kries' original term, "adjustment factor").

Recent studies (Matin and others) have focused on quantitative characteristics of extraretinal signals that modify oculocentric directions. In these experiments, subjects reported on the visual direction of a brief flash presented at various times either before, during, or after a voluntary saccade relative to the visual direction of a fixation target viewed and extinguished prior to the saccade. The stimuli were presented in complete darkness, thus eliminating the influence of visual context. It was found that the mapping of visual direction changes in the time associated with the saccade. This change occurs in the absence of ongoing retinal stimulation and is attributed to the growth of an extraretinal signal whose size is related to the size of local sign shift. Since the growth of the extraretinal signal is much slower and more prolonged than the saccade, it suggests that the extraretinal signal is not the sole factor responsible for perceived directional stability unless sufficient time separates the double stimuli.

Unlike saccades, during which vision is in abeyance (probably through backward masking), the perception of target movement during tracking (pursuit) eye movements is continuous. The stimulus for pursuit movements is some sort of retinal error or slip. During such movements, visual directions are modified by an extraretinal signal, which is not opposed by a shift in the position of the retinal image.

The nature and character of the extraretinal signal associated with involuntary eye movements remains unknown but is partly controlled by retinal image displacement. No clear picture emerges about extraretinal signals, if any, associated with cyclofusional movements, accommodation, or pupil size changes. There is some evidence that convergence initiates a signal that produces changes in apparent size in a direction favoring size constancy.

If oculocentric directions are transmuted into egocentric directions by an efferent innervation factor supported by proprioceptive feedback from muscular contraction (acting chiefly in a metabolic capacity), what role if any does the visual feedback from the retina play? The current conception is that it serves as the final fine adjustment of ocular movements—to correct any overshoot or undershoot of the intended movement. This is in accord with the other definition of the polar zero of the fovea, that is, its zero retinomotor value. Under normal conditions therefore there is an interplay between retinal image displacement, intended movements, and actual movements of the eye in the elaboration of egocentric directions.

Visual directions under binocular viewing conditions are defined in terms of corresponding points. Unfortunately there are about a dozen different criteria of correspondence, some of them mutually exclusive. In reading the literature, it is not always easy to determine which criterion a writer is using, and difficulties are compounded by a certain vagueness in terminology. The same difficulties inevitably carry over into such clinical areas as "anomalous correspondence." There is probably no definition of correspondence either normal or abnormal that is entirely satisfactory from all points of view.*

Corresponding points are said to be those elements in each eye that have a symmetrical geometric distribution, subtend the same angle at the nodal point, have the same oculocentric direction, project to the same cortical locus, or share a common subjective visual direction; whose identical simultaneous stimulation gives rise to fusion or does not give rise to fusional movements; that have zero depth value; or whose stimulation gives, for a constant fixation distance, the appearance of an apparent frontoparallel plane, equidistant plane, or equiradial surface. Some of these criteria can be dismissed outright as arbitrary simplifications (e.g., symmetrical distribution and angular equality); some have not been proved (e.g., a common cortical locus); some do not involve correspondence except by assumption (e.g., the frontoparallel or equidistant plane); some suffer from semantic afflictions (e.g., shared common directions); and some cannot be measured at all (e.g., failure to elicit fusional movements).

The definition of corresponding points as those retinal elements that have the same oculocentric direction is clear enough, providing the visual axes are parallel. In convergence on a near target, however, the foveas, which are corresponding points "par excellence," have directions inclined toward each other. Their directions cannot be the *same* because they originate from different eyes. The correct term, suggested by Walls, is "homologous"; each point signals the same spatial locus under normal-seeing conditions. The tacit implication is that this locus has two coordinates: direction and distance.

If binocular oculocentric directions cannot be identical (except for fixation at infinity), it is hard to see how they can come to have a "common" visual direction. "It must be," writes Ogle, "that for any given retinal element of one eye there is an element of the other eye, which, when stimulated separately, would give rise to the same primary visual direction. By definition, such elements are said to be corresponding retinal points."* Although primary visual direction is never explained, it must mean the same egocentric direction (primary or otherwise). Images falling on both foveas, irrespective of their actual angular positions relative to the observer, are

*A background on these differences of opinion will be obtained by reading Helmholtz's *Treatise on Physiologic Optics*, vol. 3, translated by J. P. C. Southall, 1925, Optical Society of America, pp. 403-493; Verhoeff, F. H.: Arch. Ophthal. **19**:663-699, 1938; and Burian, H. M.: Arch. Ophthal. **37**:336-368, 504-533, 618-648, 1947.

*Ogle, K. N.: Spatial localization according to direction. In Davson, H., editor: The eye, vol. 4, New York, 1962, Academic Press, Inc., p. 223.

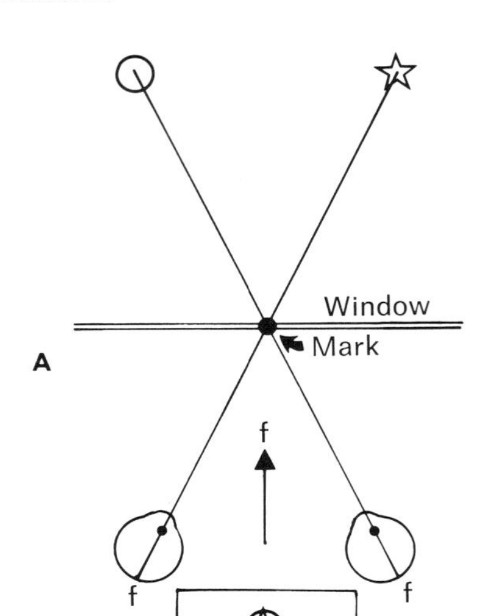

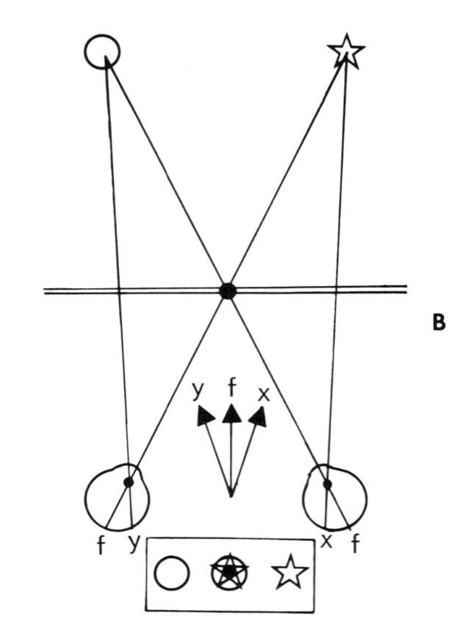

Fig. 25-2. Hering's experiment "proving" the law of identical visual directions. **A** is the incomplete version and **B**, the complete version. For details see text.

then supposed to give rise to the same subjective (principal) visual direction.

The last sentence is often given as a law—the law of identical visual directions—based on a classic experiment by Hering. An observer is to stand about two feet from a window through which he views a distant scene. Holding the head steady, he closes one eye and sights along the open eye, aligning some prominent distant target (e.g., a star) with a mark on the window. Now closing this eye and opening the other, he aligns another target (e.g., a circle) with the window marker. When both eyes are opened and fixed on the window marker, the star and circle appear in the "same visual direction" as the window mark (Fig. 25-2, *A*).

Although Hering's experiment is easy to reproduce, the results are often misrepresented. In fact, as soon as both eyes fixate the window marker, one sees two stars and two circles in physiologic diplopia, of which the innermost pair are superimposed. The observer thus sees not only a star-circle aligned with the window marker (more or less straight ahead if convergence is symmetrical) but also another star on his egocentric right and a circle on his egocentric left (Fig. 25-2, *B*). There is nothing in these three egocentric directions by which one could claim preference over the others. Indeed most people would avoid any judgment, since the chief characteristics of the distant targets is confusion. Not only are they blurred (because seen in indirect vision) and smaller (the micropsia of overconvergence), but the superimposed star and circle rival with each other while the lateral targets tend to merge with the background (hence are often overlooked). But if a judgment *had* to be made, common sense and past experience would suggest the two lateral targets (because there *are* two lateral targets), not one superimposed whose direction is obviously fallacious. At any rate, the result is rather equivocal for a law of binocular vision and a fundamental law at that.

Another experiment based on afterimages is no more convincing. If an afterimage is created on the retina of one eye, it appears in the binocular field of view. There is no basis for the additional conclusion that it is also seen in a common visual direction with its unstimulated partner in the other eye. It is simply seen in a single egocentric direction—period, and the direction depends on which way the eye is pointing (whether the motor innervation is voluntarily initiated by the stimulated or unstimulated eye). Since it is usually im-

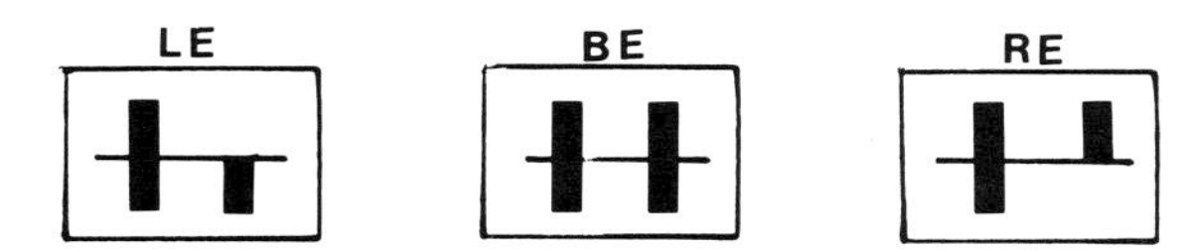

Fig. 25-3. Volkmann's vernier method of locating corresponding points. For details see text.

possible to tell which eye carries the afterimage, the egocentric direction is generally given to the dominant eye just as it is in viewing any other single target with both eyes. The same results apply if an afterimage is created simultaneously in both eyes.

The difficulty of defining corresponding points by the criterion of common directionality is also evident in the most exact vernier (or nonius) experimental method, first suggested by Volkmann (1859). A vertical line crossing a horizontal axis is presented as a fixation point for both eyes in a stereoscope. Two additional, half-vertical lines at the same distance from the observer, each seen by one eye, are viewed by indirect vision (Fig. 25-3). The offset vertical lines are movable until they appear aligned, that is, in the same visual direction. The image of these lines is now *assumed* to fall on corresponding points.

For reliable results, the measurements must be confined to within a few degrees of the binocular fixation line (to avoid diplopia, suppression, and the blind spot). It is evident that the vernier alignment does not involve simultaneous binocular vision much less corresponding points. The vernier targets must lie on the horopter to stimulate corresponding points; yet this technique is used to measure the horopter. Finally, the results are invalidated by fixation disparity. It is not difficult to appreciate Ogle's conclusion: "The existence of corresponding elements is established by definition. The central problem related to this existence pertains to their function and stability."*

In summary, corresponding points are retinal elements in each eye that have homologous (not identical) oculocentric directions and that usually mediate unified spatial vision. A "homologous direction" means the same locality reference in subjective space. They usually give rise to fusion because in normal vision we fixate real objects that present similar if not identical contours and colors to each eye. This unified percept is seen in a particular egocentric direction, originating at the visual ego situated in the dominant eye and influenced by the position of that eye (innervation factor). This egocentric direction is not a "common" direction because egocentric directions do not originate from each eye, nor is it "the same as" anything. Only in abnormal situations, such as a paralytic squint or prismatic dissociation, is the visual ego in a quandary as to which of the doubled egocentric directions is real.* And the result is disorientation. Even here, the egocentric directions do not originate from each eye simultaneously but from one or the other position of the visual reference point. Obviously this is not the case in concomitant squints, whose directions are not doubled and whose behavior is not confused.

*From Ogle, K. N.: Spatial localization according to direction. In Davson, H., editor: The eye, vol. 4, New York, 1962, Academic Press, Inc., p. 224.

Fusion and retinal rivalry

The term "fusion" has been used to describe two different aspects of binocular vision. On the one hand, there are the group of neuromuscular reflexes by which each eye is turned in response to limited egocentric disparity (i.e., horizontal, vertical, and cyclofusional movements), and on the other, there

*There is an element of confusion in terminology here also. In the phenomenon of past-pointing with an acutely palsied muscle, the neurologist (correctly) calls the localization "false" because the patient reaches for an object at a different place than where he sees it. The effect is monocular and has nothing to do with correspondence or oculocentric directions (the localization of objects relative to each other is undisturbed). The orthoptist, however, may refer to the result as "normal" because there is no anomalous correspondence (i.e., there is no "false macula").

is the process of so-called perceptual union, or unification of visual excitations from corresponding points (or areas) into a single visual perception. The first is also called "motor fusion" and the second, "sensory fusion." One of the unsolved problems in the physiology of binocular vision is whether there is in fact a process to which the term "sensory fusion" can legitimately apply.

We touch one object with both hands, hear one sound with both ears, and see one target with both eyes. Since semidecussation characterizes only the visual apparatus, this anatomic fact cannot mediate unitary perception. Semidecussation is not the reason for frontal eye position. In all vertebrates except mammals, the optic fibers decussate completely, and no animals are known to have totally uncrossed optic nerves. Only mammals, however, have semidecussating optic nerves, and the relative proportion of uncrossed fibers parallels the degree of frontality reaching 50% in primates. Since only mammals cannot move their eyes independently, the biologic "purpose" of semidecussation is most reasonably interpreted as a means of bringing the sensory afferent visual impulses into the closest relation with motor efferent cells. Only in this way can the vision of one eye control the movements of both eyes. Semidecussation is not the reason for sensory fusion (unless we are to believe that birds see double), nor does it preserve the spatial sequence of the visual field (as Cajal believed but Ovio disproved).

If a single object point is bifixated in normal binocular vision, its image falls on corresponding points, which in turn project to the same cortical locus. Another object point further or nearer stimulates noncorresponding or disparate retinal elements and is generally seen double. This doubling is inevitable if the target separation exceeds certain limits and is called "physiologic diplopia." If the target is beyond the point of fixation, the diplopia is uncrossed; if nearer than the fixation point, the diplopia is crossed.

The term "diplopia" is actually misleading, since few people are ever aware of it. Special attention and a strongly contoured figure against a homogeneous background are required to notice it. Even trained observers may confuse it with blurredness or fail to notice that the more distant target is smaller or that physiologic diplopia can also be seen in flat stereograms. Indeed there is a theory that attributes Panum's limits to blur rather than diplopia. The blur effect is well illustrated by viewing a scene through an iconoscope (the opposite of a range finder), which optically gives each eye the same view. In this form of binocular vision without disparity, everything appears *sharper*, although only physiologic diplopia has been eliminated.

If a single solid (i.e., three-dimensional) object is bifixated, all parts of it cannot be imaged on corresponding points. If the two views of it obtained by each eye are not too dissimilar, some slightly disparate elements are stimulated, and a single three-dimensional percept results. The limits of retinal disparity that do not give rise to diplopia define Panum's area. In terms of the horopter surface, there is therefore a range beyond and inside of which no diplopia occurs. This range gives the horopter "thickness"; the classic horopter surface is more or less at the midpoint of this range. Since Panum's areas increase in size toward the peripheral retina, the horopter is also thicker peripherally and has its thinnest portion near the point of fixation (Fig. 25-4). It should be noted that each element within Panum's area retains its local sign. Although the solid object is fused into a single percept, different portions of it are in a different direction relative to each other and from myself. The shift in direction that occurs in viewing stereograms results from alternately comparing two monocular targets relative to each other at the same distance (and which involves no correspondence) with a binocularly seen target at a *different* distance from me. Why this shift should be interpreted as "a change of uniocular directions into a compromise mean binocular direction" is a puzzle.

The advantage of horopter thickness is evident by considering a plane surface such as the center of a door now rotated about its vertical axis. The most peripheral image of

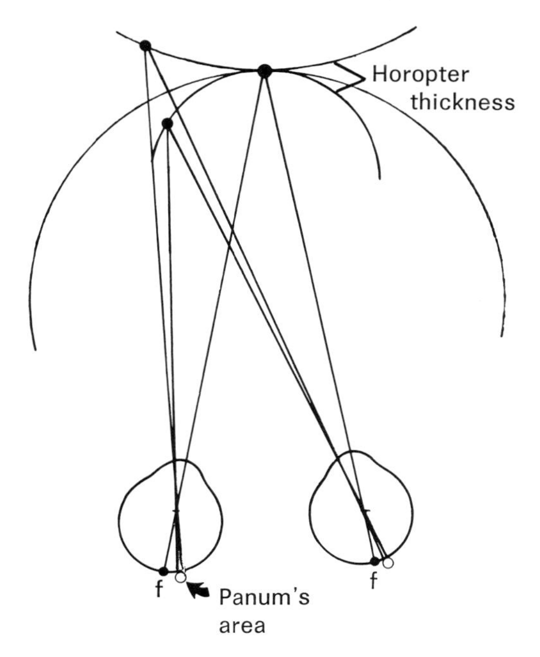

Fig. 25-4. Horopter thickness and Panum's areas (not to scale). For details see text.

the door, having the largest disparity, falls on retinal areas of the greatest tolerance to fusion (larger Panum's areas). Of course, some portions of any large target may eventually exceed even these, and the farthest and nearest edge of the door is then seen double, whereas its middle remains single. Fortunately this does not interfere either with seeing a continuous door or its depth (although the depth-despite-doubling should not be called "stereopsis," since it exceeds the disparity limits).

Single vision has been said to be the hallmark of correspondence. It is true that single binocular vision is not possible with disparate stimulation (outside Panum's area), but it is not the inevitable outcome of corresponding stimulation. Correspondence is one prerequisite for sensory fusion; the two retinal stimuli must also be reasonably similar in size, shape, luminance, color, and contrast. This is generally the case with real objects. Unequal stimulation of corresponding points normally can only be achieved by stereograms and abnormally in anisometropia, aniseikonia, or strabismus. If the two stimuli are not similar, the result is suppression (if there is marked dissimilarity) or "retinal rivalry" (if each ocular stimulus has equal claim to attention).

Normal suppression is demonstrated by several striking experiments. If a short tube is held before one eye and the open hand before the other, the scene viewed with the tube appears as if through a hole in the hand. In another illustration, the two index fingers are held horizontally before both eyes. When the two tips are separated, the bright space between them suppresses the dark corresponding area when fixating beyond the fingers, resulting in a disembodied finger that appears to float free in space.

Retinal rivalry is a misnomer; the process is cortical not retinal, although retinal factors can modify it. Retinal rivalry presupposes fusion and conversely. Whenever an object in a real visual scene is bifixated, there are always *dissimilar* targets stimulating corresponding points other than the foveas (since not all objects can lie on the horopter). The alternate suppression of such stimuli is the essence of rivalry. There is no need for rivalry if two percepts are seen in different directions. Rivalry is the reason we are unaware of physiologic diplopia (the disparate stimuli do not rival with each other; rather each alternates with the background that stimulates its corresponding point in the other eye).

Rivalry is readily demonstrated by stereoscopically presenting each eye contours of equal dominance, such as a set of diagonal black lines running in opposite oblique directions (Fig. 25-5, *A*). One sees a patchwork or mosaic of one set of lines or the other (they never actually cross), and this pattern fluctuates with a mystical rhythm more or less independent of volition. Rivalry is influenced, however, by intensity, contour, contrast, clarity and meaning. The effect of contour is mutually exclusive (Fig. 25-5, *B*), the pattern with stronger contour predominating (Fig. 25-5, *C*). If the uniocular patterns are geometrically identical but differ in luminance, such as a black and white square, the result is binocular "lustre." This is a peculiar appearance, neither gray nor black or white, resembling that produced by viewing a metal

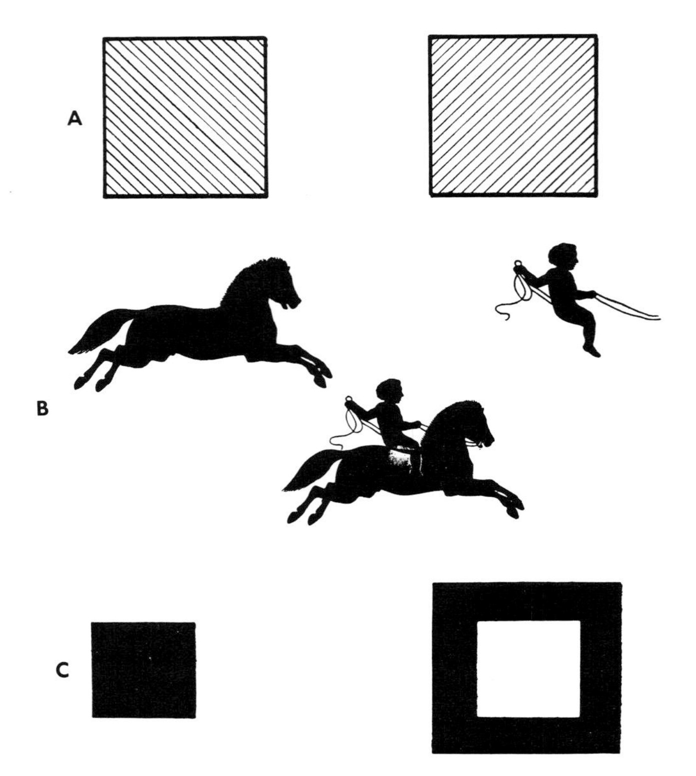

Fig. 25-5. Some aspects of "retinal" rivalry. Rivalry is induced by similar but opposite contours (**A**); but each figure tends to retain a complete form (**B**); and the predominance of contour favors the hollow square (**C**).

surface obliquely in which the same area reflects different amounts of light to each eye. Unlike rivalry for contours, the rivalry for colors is all or none; one color wipes off the other. Or the colors may fuse (e.g., red and green into binocular yellow). If a figure makes sense and fits into a meaningful context (e.g., an upright vs inverted face), the one with meaning predominates. In this sense, rivalry varies with "fluctuating attention."

A peculiarity of sensory fusion is that there is no way to prove that it is taking place when the two uniocular images are exactly alike. It is necessary to make each stimulus different in some way to demonstrate that the binocular result is something more (or even less) than each monocular percept. Of course, one cannot say fusion results *because* corresponding points are stimulated if correspondence is defined by the absence of diplopia, without ending up with a circular definition.

Sherrington had already concluded that binocular vision cannot represent a neurophysiologic common path but that two independently elaborated monocular images are "psychically" unified. Indeed it is not difficult to demonstrate that the binocular response is not a simple summative effect of monocular stimulation. If one eye is mechanically displaced by the finger, the two uniocular fields are entirely comparable in brightness, and when the doubling is allowed to slip back into registry, the brightness does not increase. Periodically Fechner's paradox is rediscovered, as in viewing the sky through two paper tubes with their monocular fields partially overlapping. Here the binocular field does look brighter but only because the corresponding monocular area is inhibited by the diminished illumination (from the inside of the tube) in the other eye. Similarly, binocular acuity is not twice as good as monocular, although it is somewhat better. This is attributed to chance summation, ocular dominance, and the operation of Fechner's paradox by occluding one eye.

Various attempts have been made to solve the binocular puzzle of how one plus one can still equal one. Unquestionably, the simplest and most satisfying is the assumption of a neurophysiologic union of monocular impulses from corresponding elements at the same cortical cell with an inhibitory mechanism to prevent actual summation. Such an effect could result from the operation of binocular neuron circuits as contrasted to monocular circuits. The arrangement of visual cortical cells in columns responding to similar edge orientations provides some support for this idea. This three-dimensional cortical arrangement is tangible evidence for Linksz's view (subsequently elaborated by Ronne) of a quasicyclopean cortical retina in which disparate stimulation results in a three-dimensional neural excitation. Unfortunately it leaves no clear explanation for depth perception mediated by nondisparate stimulation.

Binocular perception of two different monocular views into a unitary three-dimensional percept is often cited as the most obvious evidence of sensory fusion. For example, if two parallel vertical lines are presented to the left eye, and two identical parallel lines are presented closer together to the right, the stereoscopic perception is two lines, the right one nearer to the observer than the left. (The nearer line incidentally also appears smaller, although no one has yet proposed a "common" mean binocular *size* to account for it.) The fusion of the lines has been described as a unique sensation because it is not suggested by anything in monocular vision. This idea is not entirely convincing. Color is not suggested by wavelength or motion parallax by head movement; sensations do not have to resemble the stimulus. The binocular percept does not differ qualitatively (i.e., it has the same texture, color, brightness, and sharpness) except for the absence of diplopia. The essential difference is that the binocular percept has a different spatial locus (direction and distance), which is hardly unique. The uniqueness of stereopsis is that retinal disparity can provide this signal of relative spatial discrimination even when all other cues are absent.

For these and other reasons, a true mergence of uniocular stimuli has been denied. Of the many nonfusion (or suppression) theories of binocular vision the best known is that of Verhoeff. In this view, binocular percepts are made up of bits and pieces of uniocular vision. In this mosaic, which resembles contour rivalry but at a necessarily faster rate (since we are always aware of rivalry), no part of the puzzle is ever missing, no part ever represents both eyes, and all parts are constantly changing (on a subconscious level) in shape, size, and color. This view like that of mental amalgamation is entirely immune from being disproved—or proved. So are the theories that we see one target because we "project" to the locus where the lines of direction normally intersect, or we see singly because the sensorimotor conditions for diplopia are absent. All these are only rephrasings of the phenomenon to be explained. Perhaps the current explosion of neurophysiologic experimentation will eventually provide an answer. If there is an answer, this is the only direction from which it can come.

Recent studies by Barlow, Blakemore, Bishop, and Pettigrew, and their associates have provided new insights into the neurophysiology of binocular vision. In the lateral geniculate body the input from the two eyes are segregated into alternate layers. Although there are inhibitory effects between them, a cell in one layer receives impulses from only one eye. In the visual cortex (e.g., of the cat), individual cortical cells respond to the input from both eyes. For every cortical cell, two retinal areas can be defined, driven by a moving oblique slit in a specific (within 10°) orientation. These areas (about 1° across) are roughly corresponding (they are separated by the exotropia resulting from ocular muscle paralysis of the experiment). Pettigrew noted that it is not possible to superimpose all response fields of the two eyes simultaneously and postulated a specific disparity detection mechanism. Working out the formidable experimental conditions, he was able to show that simple cortical cells generally respond weakly from the ipsilateral eye alone, strongly from the contralateral side, and optimally when binocularly driven. When the response fields partially overlap (achieved by rotary prisms before the cat's eyes), there is a decrease in binocular firing. Some neurons could detect a disparity as small as 2 minutes of arc. Moreover there is almost total suppression of the strong response of the contralateral eye when the image in the ipsilateral eye is inappropriately located, an inhibition that persists for more than 1° in either direction of the optimal position. An oblique moving slit can thus stimulate the contralateral response fields of two neurons, exciting one and inhibiting the other. Complex cortical neurons (probably representing a chain of simple neurons in a column) were found, which retained a high degree of disparity specificity (i.e., 2 minutes) in spite of large response fields (e.g., 6°). Other complex binocular neurons responded over a wide range of disparity, as well as visual fields. The total range of disparity variation demonstrated the expected increased threshold for eccentric (peripheral) retinal stimulation. The total range found was 6° of horizontal disparity and 2° of vertical disparity; the vertical disparity range thus allows for some fixational imprecision, although they do not participate in stereoscopic vision.

Binocular perception of distance

We perceive the distance of objects relative to each other and their absolute distance from us. For example, a house may be unequivocally stated to be nearer than a mountain, although the judgment of the absolute distance of either (or their metric separation) may blunder grossly, especially on a hazy day. Relative distance perception, which involves stereopsis, is considerably more efficient than absolute distance judgments in which stereopsis plays little role. Indeed one can dispense with binocular vision without great loss of absolute distance accuracy (we may even close one eye in aiming a rifle), and the poor correlation of certain visual tasks with stereoacuity makes more sense if this distinction is kept in mind.

One of the less obvious differences between viewing a stereogram and normal binocular vision is that in the latter we perceive parts of the body (e.g., the nose) in the visual scene, which helps relative orientation. The awareness of the self, the impression of being here versus objects being there, also has its psychopathologic consequences (e.g., the multifaceted perspective of Van Gogh's later paintings). It should be recalled that monocular as well as binocular cues operate in distance judgments, the basis of including a human figure in a landscape photograph to distinguish hills from mountains. The perspective of retinal imagery, for example, sets the stage of the horizon and "point of view" to which stereocues can add but not subtract. Contradictions are almost invariably resolved in favor of monocular cues, as in ignoring disparity of a concave facemask. Absolute distance judgments depend on the total visual framework and on the set and past experience of the observer. Normally all cues cooperate harmoniously, simultaneously, and effectively because we view real scenes and objects, not atomized points and lines in haploscopes. The depthiness of a movie or television screen are acceptable without (or with constant) binocular cues. Only when the various cues are systematically eliminated or artificially perverted does depth perception become inaccurate. We may then confuse the identification light of an airplane with a distant star in the night sky.

Judging from experiments by Gibson and associates, depth perception is present early in human infants (as well as chicks, turtles, rats, lambs, kids, pigs, kittens, and dogs). When animals are placed on the unique "visual cliff" in which parts of the ground seen through a glass plate are optically more than a foot below the surface, they immediately freeze into a crouching, defensive position. "Despite repeated experience of the tactual solidity of the glass, the animals never learned to function without optical support. Their sense of security or danger continued to depend on the visual cues that give them their perception of depth."* Conditioning experiments by Bower (1966) confirm that human infants can recognize shape (which depends on slant and therefore distance) and orientation of objects, although their ability to process the information is limited. For example, the infant may misreach not because of poor depth perception but because he simply does not know how long his arm is. Information processing capacity presumably increases with maturation. The results suggest that because depth discrimination has survival value, it develops early (and is

*From Gibson, E. J., and Walk, R. D.: The visual cliff, Sci. Am. **202**:54-71, 1960.

established by the time the infant or animal can move independently).

Among the binocular cues for depth perception are convergence, retinal disparity without fusion (i.e., physiologic diplopia), and retinal disparity with fusion (i.e., stereopsis). Convergence was at one time (e.g., Wundt) held to be the primary cue for spatial localization either through proprioception or innervation sense. Despite numerous attempts to define its contribution (including the micropsia of overconvergence), few studies have differentiated between the effects of various convergence components. For example, although accommodation is a binocular function, it is delegated to the status of a monocular cue because it *can* operate under monocular conditions.* But so can accommodative convergence; indeed all but fusional convergence may operate in one-eyed vision (although we now call it "adduction"). Individual differences in proximal convergence are known to be large, which may explain why some people are more proficient in using stereoscopic instruments. Whatever the mechanism, convergence effects are limited to within a few meters and hence play a negligible role in absolute or relative distance judgments. It is doubtful therefore that convergence alone sets a useful metric for absolute spatial judgments, although it may define the plane of reference by which objects are judged to be nearer or further.

Since physiologic diplopia may be crossed or uncrossed, this has been postulated as a binocular cue for relative distance perception. But the value of a cue that is seldom noticed is doubtful, except as a vague depthy "background" for the sharply seen foveal "figure." It is therefore difficult to follow the rationale of training people to become aware of physiologic diplopia or to understand what benefit could conceivably derive from such an exercise.

The basis of stereoscopic vision is illustrated in Fig. 25-6. Two points *A* and *B* at unequal distances from the observer are viewed in binocular vision. Their angular separation results in a retinal disparity d′ − d. If the disparity is large, fixation on one target results in physiologic diplopia for the other. As the disparity is reduced by bringing *A* nearer to *B*, a critical threshold value is reached where both targets are perceived singly at once, yet one further than the other. Wheatstone discovered that this small disparity is the immediate and sufficient stimulus for stereopsis; that is, no other cue is needed for the observer to become aware of depth. If the disparity is reduced still further, a point is reached at which it is no longer possible to discriminate one target as nearer or further than the other. This amount of disparity, usually expressed in seconds of arc ($\alpha - \beta$), is the stereoscopic threshold, and its reciprocal is the stereoscopic acuity. The range through which the separation of the targets extends from diplopia at one end to equidistance at the other

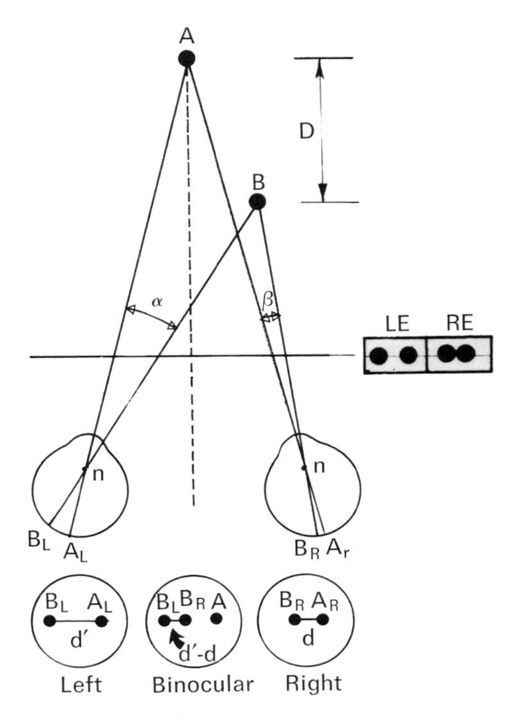

Fig. 25-6. Optical basis of stereoscopic vision. For details see text.

*Conversely, such a monocular cue as "overlap" has its binocular consequences, since the area of a further target covered by a nearer one differs according to the point of view of each eye.

is the stereoscopic range. If one target continues to be brought nearer, a diplopia of opposite disparity eventually results. The total range from diplopia with crossed disparity to diplopia with uncrossed disparity is called the "range of single binocular vision." This range is an expression of the width of Panum's area.

The stereoscopic response is practically instantaneous, independent of eye movements or the direction of fixation. But stereopsis is not the primary factor in space perception as any one-eyed individual will confirm. Nor is it the strongest single depth clue (overlap is more definitive); nor does it cause more "vivid" depth (unless all other cues are absent as in a stereoscope); nor is it the final human culmination of an evolutionary process (the hawk does better). Motion parallax, for example, compares favorably to binocular parallax in terms of threshold and quality of response. But one is consecutive; the other, simultaneous. Since stereopsis can also be obtained by consecutive stimuli (one may lag behind the other by up to 70 msec), there is no absolute distinction. The biologic "purpose" of stereopsis is not only to lower the depth threshold in less time but to provide a useful signal when all other cues are absent. One has only to think of camouflage, the protective coloring of animals, or the beautiful computer-generated stereograms* of Julesz to appreciate the advantage of recognizing objects binocularly that are invisible under monocular condition.

Stereoscopic vision mediates relative, not

*Computer-generated stereograms reveal no information when viewed monocularly but readily demonstrate a depth pattern when viewed in a stereoscope. They are not, of course, "random dot" stereograms as usually described, since the pattern is carefully generated—the disparity is confined to microstructure. Nor does it "prove" that contextual clues are unnecessary for stereopsis, since no one ever maintained they were (there is no reasoning or experience that makes us see one line in an ordinary stereogram in front or behind the other). But they do demonstrate disparity effects of texture and brightness to a greater extent than can be realized with ordinary contours and forms. The notion of using computer-generated patterns to produce an ideal camouflage is intriguing—if we could limit the enemy to one-eyed pilots.

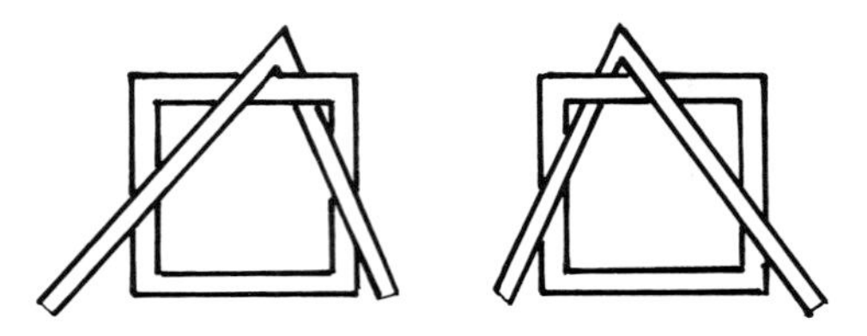

Fig. 25-7. One of Schriever's nonsense stereograms in which disparity and overlap contradict each other. The monocular cue predominates, and the binocular perception is never very comfortable.

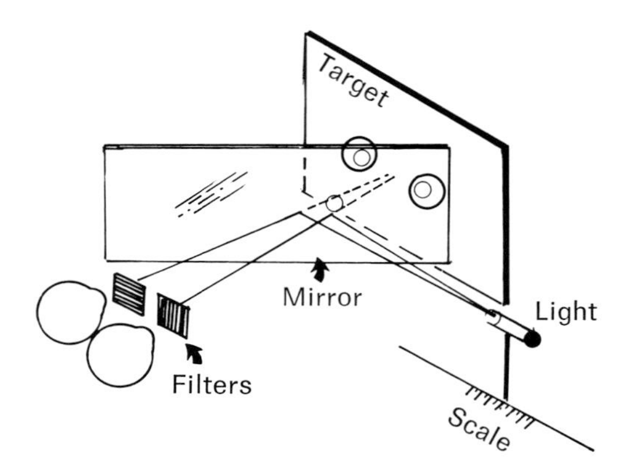

Fig. 25-8. Apparatus to gauge the depth seen in stereograms. The eyes view the vectographic targets through the angulated half-silvered mirror. The light source can be moved along the scale so that it appears to float in front of the stereogram at the same distance as the three-dimensional ball.

absolute distance discrimination. When a stereogram is viewed and there is no other visible object to which its absolute distance can be compared, there is considerable uncertainty as to how much one target "floats" in front of the other for a given amount of disparity. Thus a 3-D ball appears in greater depth against a homogeneous background than when it is part of a column attached to a building. And disparity may yield a perverted or no stereoscopic effect if other cues contradict it as in Schrievers nonsense stereograms (Fig. 25-7).

Within horizontal Panum's limits, the larger the disparity the greater the stereoeffect. Contextual and individual variations can be demonstrated by an apparatus such as in Fig. 25-8. This permits actual measurements of the perceived separation of stereotargets.

Panum's limits themselves are not precise as will be evident if one tries to measure them. The distinction between seeing a target double or single is not easy to make and the end points are established by psychophysical techniques (e.g., probability of doubling). The reliability (i.e., duplication) of the measurements is practically useless beyond 10° to 15° from the fixation point. Naturally depth discrimination may persist outside Panum's areas because of other cues, and this should not be called "stereopsis" (by definition). Ogle, for example, differentiated between patent (or obligatory) stereopsis and qualitative (or empirical) stereopsis for the depth effect outside Panum's area.

The rule for making stereograms is also illustrated in Fig. 25-6. The targets must be so drawn that the image they present to each eye is identical to what would have been seen if a real solid object or three-dimensional scene had been viewed. Many theoretical misinterpretations about binocular vision will be avoided by keeping this in mind. Depth differences in the real object that are easily perceived in relief are also seen in stereograms; and if the disparity differences exceed the limits of binocular vision, they are doubled, whether viewed in the original or in the stereogram.

A positive relationship exists between stereoscopic and resolution acuity. A fair amount of blur difference must exist between the two eyes to interfere with depth perception, depending on the complexity of the stimulus. Because of their high correlation, there is a strong probability that stereoscopic and vernier acuity have a similar physiologic basis (the discrimination of visual directions). Stereopsis is also affected by differences in image luminance, the classic example being the Pulfrich phenomenon. When an oscillating pendulum in the frontal plane is viewed binocularly with a neutral density filter before one eye, it appears to describe an elliptic path, extending toward and away from the observer. The apparent displacements are attributed to differences in the visual latent periods for the two eyes. Its clinical implications are erroneous rotations from the frontal plane and errors in spatial relationships (anisopia), although not all investigations have confirmed the effect. Both base-in and base-out prism may impair stereoacuity in some cases, partly the result of chromatic dispersion and of accommodation induced by vergence changes that throw the eyes out of focus. If a vertical disparity (up to 30 minutes of arc) is added to an existing horizontal disparity, stereopsis is maintained. The explanation is not known, but the results are in line with the fact that patients with anisometropia can enjoy good stereoacuity despite the vertical prismatic effects induced by their spectacles. The changes in stereoacuity in dark adaptation are consistent with the changes in resolution acuity. Heterophoria has little or no effect on stereoscopic depth perception. There is no consistent relationship between stereopsis and retinal rivalry.

The nature of stereopsis is still poorly understood. Current opinion holds that it is a sensory phenomenon with its own physiologic mechanism. Hering believed that each disparate retinal element has a depth value, similar to its local sign of height and breadth —in short, that stereopsis is an innate capacity. The theory is unacceptable in this form because there is a limit beyond which disparity is ineffective. It is not clear how the eyes can distinguish between crossed and uncrossed disparity or in what manner disparity is made inoperative when other cues contradict it. Nevertheless, recent neurophysiologic evidence (see discussion on p. 473) does suggest that there are cortical cells which can discriminate disparity, if not in terms of depth, at least in terms of direction of contours, as well as the inhibitory mechanism necessary to differentiate one type of disparity from another. Still unresolved is at what point retinal disparity within Panum's area signals a fusional movement (the evidence is that any disparity does) and how retinal elements may fuse on the one hand yet maintain their directional individuality on the other.

Although it appears that stereopsis cannot be created by training if it is absent, it may be improved by practice. It is not known to what degree such practice effects result from using stereoscopic instruments more efficiently, better interpretation of ancillary cues, or actual learning of skills capable of transfer to other visual tasks.

The horopter

The ideal horopter is defined as a surface in space any point of which has images in the two eyes falling on corresponding points for a particular fixation distance. The Vieth-Muller horopter is a geometric fiction because corresponding points are not symmetrically distributed, the retinas are not perfect hemispheres, and the eyes do not rotate about the nodal point. Moreover any near target not in the median plane cannot stimulate corresponding points in the vertical and horizontal meridians simultaneously because it is necessarily nearer one eye than to the other. Even if the horopter is limited to the horizontal plane, the eyes undergo cyclofusional movements in convergence, and the horizontal meridians of the retinas are inclined toward each other.

For these and other reasons, the experimentally determined horopter in the horizontal plane (longitudinal horopter) differs from the idealized Vieth-Muller circle. Based on the previously listed differences in the criteria of defining corresponding points, there are differences of opinion as to what the horopter really measures. In the usual experimental set-up, the targets are short vertical rods or threads, evenly illuminated against a homogeneous background, whose observation is confined by a horizontal slit (or multiple slits). Measurements are limited to within 12° to 15° on either side of the fixation point, based on one of several criteria. The criterion of equating oculocentric directions is based on Volkmann's vernier method. The principle is that two vertical, eccentrically viewed rods in vernier alignment are seen in the same direction. Although this is considered the most direct and theoretically correct method, both eyes never actually see the vernier targets simultaneously; each eye views its own target (only the fixation point is seen binocularly). That corresponding points are being measured is therefore an a priori assumption of their definition according to direction. In the criterion of apparent frontoparallel plane, the rods are aligned so as to appear in the same frontal plane as the fixation point. The principle, originated by Hering, is that when the rods appear in the frontal plane they cannot stimulate disparate points (which would result in their perception in depth); hence the stimulated points are corresponding. This method is limited to a near test distance, and since stereopsis falls off rapidly beyond a few degrees of fixation, peripheral settings can be made with one eye closed. In aniseikonia one retinal image is larger than the other, and the frontoparallel plane appears rotated further on that side. The objective settings of the rods are opposite to compensate. Indeed the eikonometer is basically a horopter apparatus adapted for clinical use. A variant of the frontal plane horopter is the criterion of equidistance, subject to the same limitations.

Object points that are perceived in a frontoparallel plane are theoretically tangent to the Vieth-Muller horopter for symmetrical convergence. To stimulate corresponding points, however, there must be a disparity in the angle subtended at the nodal point of each eye by any eccentric object point simply because it is outside the Vieth-Muller circle. This was experimentally confirmed by Hering and Hillebrand and hence has come to be known as the Hering-Hillebrand deviation. A geometric consequence is that for a constant angle of disparity, the locus of object points tends to approximate the Vieth-Muller circle for a near fixation distance, becoming tangent to it at about 2 meters, then convex toward the observer's eyes beyond 2 meters. A particular object point for changing convergence describes a hyperbola, and the family of such hyperbolas become practically coincident, the subjective lines of direction originating from a single locus.

In all the previously discussed criteria, the horopter is defined as a surface, and deviations from this surface are taken to be errors of measurements. In the criterion of the zone of single binocular vision, the horopter has thickness, and deviations define Panum's areas. Corresponding *points* are not stimulated at all, and the horopter "surface" is taken as the statistical midpoint of this zone.

The horizontal (or longitudinal) horopter is the one usually determined. Only horizontal disparities lead to stereopsis, and the perspective horizon (set by the height of the observer's head above the ground) is a key factor in spatial orientation. Unfortunately different methods of measuring the longitudinal horopter lead to data so variable that they cannot be correlated. Under similar test conditions, however, these psychophysical thresholds are fairly constant for the same individual (interpupillary distance and other personal constants). For the same individual therefore it is possible to develop an equation that expresses the deviation of his subjective from objective visual space. Aside from its considerable theoretical importance, however, the horopter has little clinical value, and its measurement for practical diagnosis and treatment would serve no purpose.

Ocular dominance

The term "eye dominance" is purely descriptive. Nothing in one eye per se determines sighting dominance in over 90% of the population. The concept of retinal receptor dominance has been suggested but never seriously. Yet it still appears in the form of a minor variant for those who insist that visual acuity is somehow the determinant of eye dominance, even though it has been shown over and over again that the dominant eye may or may not have the better vision. Such terms as "sighting eye," "master eye," "preferred eye," or "fixing eye" were originally applied to the eye used to aim a gun (indeed the early studies were undertaken purely in the interest of better marksmanship).

The simplest method to determine the dominant eye is the Miles test. A folded cardboard cone is pressed open with both hands. The smaller distal opening ensures monocular vision while the sighting end is large enough to accommodate both eyes. Other criteria are the eye used to align two targets, to look through a tube or monocular microscope, the eye that continues to fixate after the "break" in the near point of convergence test, the eye that does not deviate in monocular congenital squint, or the eye that actually fixates in fixation disparity.

There is little doubt that sighting dominance is a motor characteristic. Walls (1951) proposed to call it "directional dominance"; that is, it is the eye that determines egocentric directions—the apparent locus of the visual ego. His thesis is that in any individual who has a dominant eye, an innervation record is filed only for the muscles of that eye and that this is true whether one is seeing binocularly or monocularly with either eye. This innervation record is one-sided with respect to the eye, not the brain, since oculomotor control for each eye comes from both hemispheres. (Unilateral cerebral lesions produce conjugate, not uniocular deviations.)

Some believe eye dominance is learned, for example, because reading material is held by hand. This implies that handedness is not learned and that left-handed people would learn to become left eye dominant, which is not true. The evidence is that sighting dominance is native or established early and maintains itself throughout life. For those who feel that eye-hand cross dominance is a factor in incoordination or dyslexia (e.g., Orton's "strephosymbolia"), the question arises whether training the eye or hand reestablishes an originally uniform cerebral dominance that somehow went wrong in development. If dominance is a maturational process, genetically predetermined, any changes therapeutically imposed would be as bad as the changes induced by social pressure,* The implication is that cross dominance as

*Parents, teachers, and circumstances generally conspire against the left-handed child to get him to switch—at least for certain tasks. This is supposed to be deplorable but does not appear to have any effect on dexterity in other tasks. For example, a horse must be mounted from the left—it is entirely immaterial which of the rider's feet is dominant. And woe to the army recruit who does not step out with his left foot. The most efficient individual may well be the one who most readily adapts himself to these altered requirements.

a physiologic anomaly resides in the brain and that eye, hand, foot, etc., dominance are cerebrally determined. If this cannot be proved, at least it cannot readily be disproved. Not enough is known about cerebral dominance, except for the speech center, to allow its identification by any simple test for eyedness or handedness. Animal studies do not help, since dominance does not occur in infrahuman species, and, as everybody knows, they have no speech.

The old kind of dominance is generally called "motor" or "sighting dominance." Currently another type of dominance is described under such headings as "sensory dominance," "laterality dominance," "laterality difference in perception," and "controlling eye." The chief criterion of this new kind of dominance is the prevalence of one eye's pattern in a rivalry situation. Let us clearly understand the difference. In over 90% of the people, one eye is motor dominant; in rivalry there is continuous alternation. It is as if the right hand were dominant on Mondays, Wednesdays, and Fridays, and the left hand on Tuesdays, Thursdays, and Saturdays—hardly a reasonable criterion of dominance. Recall that the essence of motor dominance is the preference of the visual ego for one eye. Its uniqueness is that it determines egocentric directions in binocular vision, perhaps because a copy of the efferent innervation is filed in the brain only for the muscles of the dominant eye (the steering-gear theory of dominance). In rivalry, the predominance of one pattern can be altered at will by simple making one contour more complex or more meaningful.

Could cerebral dominance still be somehow related to one eye's preference in rivalry? No, at least not in the anatomic sense of brain laterality, since what is represented in each hemisphere is not each eye but each half of the visual field. Various attempts have been made to relate field dominance to brain dominance, but the results are inconclusive. What is conclusive is that attempts to alter dominance are of dubious benefit to children with reading or learning problems. Even if we could all agree on what dyslexia is (which we cannot), directional confusion does not characterize poor readers any more than its absence distinguishes good ones (Bettman and associates, 1967).

Fixation disparity

It is generally assumed that during binocular fixation the visual axes intersect at the object of regard. In fact they may cross in front or behind the plane of the target although the object continues to be seen singly. This phenomenon represents an additional elasticity in the functioning of the binocular mechanism, a tolerance allowing some imprecision in fixation. The theory of ocular motor dominance moreover suggests that the fixational imprecision be delegated to the nondominant eye so as to avoid any falsification of egocentric localization. When both eyes are open but only the acuity of one is being tested (as in a stereogram where the other eye sees dummy targets), the acuity of the nondominant eye will be improved by occluding the dominant eye. The fixational imprecision is now abolished because the visual point of reference actually changes. This probably accounts for the clinical phenomenon of the "controlling eye" described by Berner and Berner (1938).

Fixation disparity (also called "retinal slip," "fusional disparity," or "associated phoria")

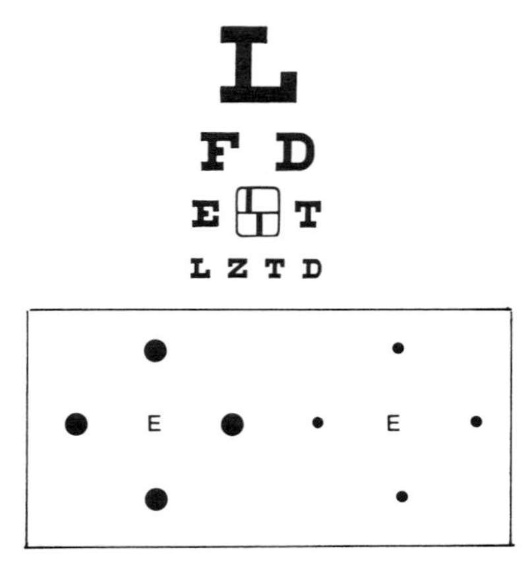

Fig. 25-9. Targets to demonstrate fixation disparity, having peripheral fusion *(top)* and central fusion *(bottom)*.

can be demonstrated with a stereoscope or polarization. Each eye is presented its own target, which contains partly fusible and partly nonfusible elements (e.g., presenting a pair of playing cards such as the four of diamond to one eye and the five of diamond to the other). Although the peripheral elements are seen singly, the central diamond may be found displaced to one side or the other, or up and down. The displacement is usually but not invariably in the direction of an existing heterophoria. It makes no difference whether the fusible elements are central or peripheral (Fig. 25-9). Ames and Gliddon (1928) first showed that the amount of prism (e.g., base out or base in) necessary to reduce the disparity to zero is the same for either type of target, and a strong correlation exists between the prism required and the subject's phoria.

Ogle and associates (1967) have extensively elaborated the variables involved in fixation disparity and refined the techniques of measuring it. They arranged two monocularly seen lines in the center of a binocular field. The lines could be displaced relative to each other and the angular value of the displacement read off in minutes of arc at that viewing distance (Fig. 25-10). Various amounts of convergence and divergence were induced by base-out and base-in prism. As expected, they found the fixation disparity altered, although the relationship was not the same for all subjects.

A typical result is shown in Fig. 25-11 and is characteristic of the majority of (but not all) patients. The ideal subject illustrated is orthophoric and has no initial fixation disparity. As the eyes are made to diverge in response to a base-in prism, the visual axes lag more and more behind, leaving it to Panum's area to make up the fusional deficiency. Eventually Panum's limit is exhausted, and fusion breaks. The converse situation obtains with base-out prism, but here the response is adequate longer because convergence exceeds divergence. The two break limits define the horizontal dimensions of Panum's area, measured in minutes of arc. It will be evident that such angular values are much too small to be detected by the cover-uncover test; hence fixation disparity should not be confused with microstrabismus. Nor is it related to anomalous correspondence, foveal suppression, poor stereopsis, or aniseikonia.

Results similar to those obtained with prisms can be induced by plus and minus lenses. A comparison of the two graphs per-

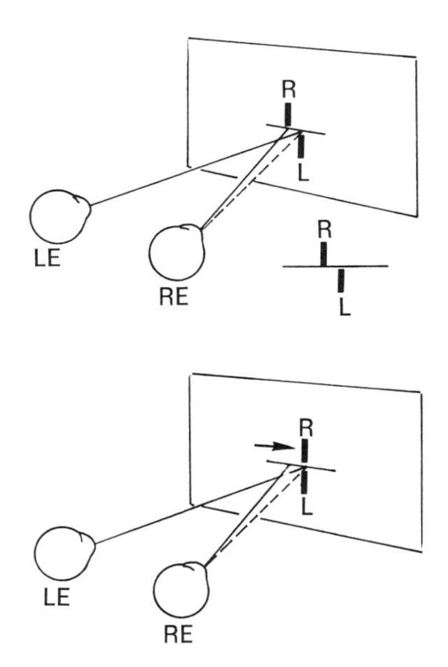

Fig. 25-10. Vernier method of measuring fixation disparity.

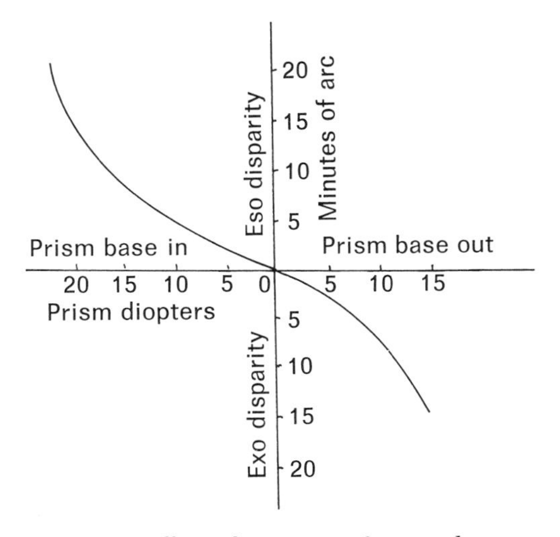

Fig. 25-11. Effect of prisms on fixation disparity. For details see text.

mits computation of the AC/A ratio as described in Chapter 18. An interesting observation is that less fixation disparity is demonstrable when the fusional target itself has some disparity and is seen stereoscopically.

The practical determination of fixation disparity is easily made with the polarized targets available for the acuity projector. Its exact clinical significance, however, as a cause for symptoms remains unclear. In some patients, at least, it is better correlated with complaints than the phoria itself. A change in fixation disparity induced by anisometropic prescriptions can be demonstrated over a period of time, suggesting some form of adaptation, more so in some patients than others. This adaptation differs from the innervational factors responsible for "natural" heterophoria, which never show adaptation. Finally, fixation disparity illustrates that the oculomotor imbalance in heterophoria persists during binocular fusion even though it is not generally measured under "associated" conditions.

PATHOLOGIC PHYSIOLOGY

Functional disturbances of binocular vision are enormously complex because they involve not only monocular defects but also anomalies peculiar to the cooperation of the eyes. The understanding of the pathophysiology is limited by the lack of unifying principles, a "leitmotif" to tie the diverse physiologic strands together. Only recently have myographic and neurophysiologic techniques become available sophisticated enough to provide new insights into an area generally restricted by subjective methods and the absence of experimental models. It is likely that the next decade will see a change in approach, perhaps even provide pharmacologic treatment for binocular motor defects.

Incidence and pathogenesis

The incidence of binocular anomalies is not established, but the obvious cosmetic blemish of squint is estimated to affect about 2% of the world's population, whereas the prevalence of amblyopia ranges from 1% to 3%. Obstacles to binocular vision have been traced to various innervational levels and to mechanical, muscular, and genetic factors. Without assembling a laundry list of theoretical possibilities, the causes are basically optical, motor, and sensory. Probably there is more than one cause, with an interplay of etiologic mechanisms. A key issue still unresolved is whether motor defects induce sensory adaptations or sensory obstacles lead to motor anomalies. Of current interest is the adaptability of function after muscle transplants in the monkey, the apparent contradiction of neurophysiologic and psychophysical results in measuring ocular movements, the genetic defects of optic nerve fiber topography in albino and cross-eyed animals, the neurophysiologic evaluation of cortical cell function in binocular disparity discrimination, and the extraretinal signals that transmute oculocentric directions into a stable egocentric environment.

Suppression and amblyopia

"Suppression" is a descriptive term applied to both a normal phenomenon in which two unequal ocular stimuli are resolved in favor of one predominating in contour or brightness and an abnormal phenomenon in which equal ocular stimuli do not elicit the anticipated binocular response. Whether these are both accomplished by the same mechanism is not known. But it seems an oversimplification to consider suppression as a purposive, voluntary effort to avoid diplopia. It differs from the voluntary "suspension" of vision in one eye (e.g., by a binocular individual using a monocular microscope), which is always temporary and does not carry over into other binocular tasks. Moreover suspension is a consequence of ocular dominance, whereas pathologic suppression is transferred from one eye to the other, depending on which one is fixating. It also differs from normal suppression operating to eliminate physiologic diplopia, which never involves the foveas. And of course it differs from amblyopia (sometimes considered an extension of suppression) in that suppression is manifest only under binocular conditions.

There is in fact some evidence that the suppression of squinters has a distinct pathophysiologic basis, based on reduced visually evoked response amplitudes elicited by the nonfixating eye. Most workers have not been able to demonstrate a similar result in retinal rivalry. The work with electroretinography, on the other hand, is not convincing, since this is a mass response that does not differentiate the relative contribution of the fovea.

Some evidence comes from, of all places, cross-eyed cats. It is known that Siamese cats are often cross-eyed. It has now been demonstrated that they have a peculiar defect in the representation of their optic nerve fibers in the lateral geniculate body. Some fibers are misdirected, ending up on the same instead of the opposite side, although in the correct layering. Although nonstrabismic cats also have this defect, it is quantitatively greater in the strabismic ones. Other albino species (e.g., tigers, rats, guinea pigs, mice, and mink) also demonstrate this neural defect, although the role of tyrosine metabolism is still a mystery.

The geniculate misarrangement in the cat would lead to a partial reversal of spatial patterns, inconceivable for normal behavior. The cats solve their dilemma in two ways: in one, the cat's brain simply ignores the input from the abnormal geniculate cells; in the other, there is a compensatory rearrangement of visual cortical cells to accommodate the abnormal geniculate ones.

Pertinent to the problems of suppression is that some esotropic cats (presumably those without cortical compensation) act as if they cannot see with their temporal retinas in monocular vision but appear normal if both eyes are open. It is not yet clear whether there is also an actual deficit in binocularly driven cortical cells from the affected retinal areas. (Guillery, 1974; Hubel and Wiesel, 1971; Sanderson and others, 1969).

In monocular strabismus the target of fixation stimulates a noncorresponding area in the deviated eye. In a normal binocular individual the result would be diplopia, but here there is suppression. First, the area stimulated is some distance from the fovea, hence has lower resolution; second, its cortical representation is not neurophysiologically enhanced as is the fovea of the fixing eye (indeed its cortical receptors may be deficient in irreversible amblyopia); third, the area corresponding to it in the fixing eye has not only figural and resolution dominance but also is the zero retinomotor point, providing visual feedback of precise ocular alignment; and finally, the fixing eye is the dominant eye—the nondominant deviated eye is suppressed by whatever mechanism makes *it* nondominant even in a normal patient. One could hardly expect anything but suppression from such a horse and ox combination.

If the fixing eye is now occluded and the deviated eye has good resolution and equal dominance, its fovea takes up fixation. It is not known whether the alternating squint patient is born with no dominance (the incidence of alternating heterotropia and alternating dominance is about the same, roughly 10%) or whether it occurs purely because each eye has the same acuity. The switch in egocentric localization from one eye to the other accounts for the total suppression characteristic of alternating exotropia.

If the deviated eye has poor acuity and the other is highly dominant, it is not difficult to see why its fixation is erratic and inconstant when the sound eye is occluded. It provides no effective retinal feedback for precise motor alignment, and its diminished resolving power makes it a poor reference for the all-important business of spatial localization. Coupled with this, there is a massive translatory slippage in innervation to the muscles of both eyes (massive enough to produce a manifest squint, translatory in that it is concomitant). If therefore the brain files only the innervation sent to the sound eye (although now covered), the usually deviated eye makes no refixation movement. The area used for monocular fixation is the same area stimulated when both eyes are open. This area acquires zero retinomotor value (it becomes the so-called false macula) within the limits of its erratic fixation. It sees the object where it actually is in space, although it is now said to have "false projection" or "anomalous correspondence." On the other hand, if the brain accepts the innervation record sent to the usually deviated eye, that is, if its true fovea retains zero retinomotor value, the localization may be as the normal-but-dissociated patient sees (i.e., "normal projection") or depending on the motor slippage and dominance, somewhere in between (so-called subharmonious correspondence).

If both eyes are open, the stimulus to the deviated eye is naturally suppressed. But under experimental conditions (e.g., visuscopy, afterimages, or a bright flashing light in a synoptophore), the squinter can be made aware of double targets whose relative position would follow one of the above three previously described possibilities. Of course, he could never see both targets simultaneously long enough to obtain fusion or stereopsis.

No one actually knows how the squinter sees, but it seems to me more plausible that it is tied to a primary motor (not sensory) defect and a change in egocentric (not oculocentric) directions. The motor defect is self-evident (considering the high incidence of anisometropia, there would otherwise be many more squinters on the basis of this sensory obstacle alone). Oculocentric directions are never disturbed in strabismus (the patient sees objects to the left or right of each other whichever eye he uses, if he can see them at all). Alternating vision (or alternating suppression) can be easily demonstrated in monocular squinters (e.g., by the Bagolini striated glass test). And a change in egocentric direction is the inevitable consequence of any motor defect, which is why a surgical attack is effective and other treatment methods frequently fail. Whatever the true nature of the squinter's vision, unlike the binocular individual, normal correspondence is absent. But it is not necessary to suppose that he must now acquire an anomalous new oculocentric relationship.

The squinting eye must cope with one additional problem. *Its* fovea is stimulated by some pattern that is not the object of attention. Under normal circumstances this pattern *being* foveal should have equal claim to cortical attention as does the fovea of the opposite eye, yet here it is suppressed to avoid the conflict (or "confusion") of seeing two different objects (superimposed or in rivalry with each other) at the same spatial locus. The mechanism is not known. Presumably it is physiologic at least initially, since amblyopia is reversible during early years of life. It may be simply an extension of the same suppression mechanism that makes the other eye dominant to begin with.

Eccentric fixation

When the dominant or fixing eye in monocular strabismus is occluded, the usually deviating eye must take up fixation. If its fovea is incapacitated by amblyopia or suppression, the fixation is likely to be eccentric. The area used for fixation is often unstable, the "scatter" depending on the severity of the visual loss and how far it is from the fovea. Grossly wandering fixation can be detected by observing the corneal reflection as the patient follows a moving light. Smaller fixational anomalies can be documented with the Visuscope. Fixation may be paramacular or peripheral or even paradoxic in some post-surgical cases (e.g., esotropes who fixate with an area temporal to the fovea). Such terms as "false fovea" or "false macula" for the eccentric fixation point are now happily being abandoned. The displaced fixation area never has the anatomic or physiologic characteristics of the true fovea. Its acuity is never better than the eccentric photoreceptor population permits, although it may have the best resolution under the circumstances. Fixation anomalies may also develop under monocular conditions, for example, after occlusion in a formerly dominant eye.

The pathogenesis of eccentric fixation is not known. There is evidence both for and against the theories that disturbed fixation is responsible for poor vision in amblyopia and that amblyopia causes poor fixation. That the two are somehow related is the basis for pleoptic therapy. Cuppers believed that eccentric fixation follows anomalous retinal correspondence; that is, the abnormal monocular fixation behavior is maintained because a peripheral retinal element becomes associated with the "principal visual direction." This is in contrast to the view that the fixation anomaly is a direct result of deep suppression scotomas (Bangerter, 1953). In recent carefully documented studies, von Noorden could not confirm Cupper's theory. The relationship between the angle of anomaly and the area used for eccentric fixation occurs only in a limited number of cases. At this writing, we are left with no satisfactory theory to explain the clinical phenomenon.

Anomalous correspondence

The conventional conception of anomalous correspondence can be described as follows:

It represents a sensory adaptation of the eye in certain types of squints, particularly those that occur early in life, and where the angle of deviation is fairly constant. There is a "rearrangement" of the common visual directions of the retinal elements of the two eyes, in which corresponding points lose their common visual direction and disparate elements acquire a common visual direction. For example, the two foveas may acquire two different visual directions. In contrast to normal correspondence, anomalous correspondence is acquired by usage, requiring time and individual adaptability. Anomalous correspondence may fully or partially replace normal correspondence in squinters, indeed in some cases resulting in the use of both eyes in a way approaching normal binocular vision. Anomalous correspondence may be superficial or "deep-rooted," and this determines how easily normal correspondence can be elicited in various clinical tests. The subjective angle between the visual directions of the foveas may be equal or less than the objective angle of deviation (angle of anomaly) and is then classified as harmonious or subharmonious. Harmonious anomalous correspondence represents a complete adaptation but can develop only under favorable circumstances. Frequent changes in the angle of squint make complete adaptation impossible, and the best the organism can do is to choose an average common visual direction, that is, an unharmonious adaptation. Coexisting with anomalous correspondence are areas of selective suppression in the deviated eye, involving the macula and the extramacular area on which the image of the fixation point impinges. A special type of unharmonious correspondence is paradoxic diplopia, explained by the persistence of anomalous retinal relationships that have not yet had time to catch up with changed motor conditions (surgery, glasses, or prisms). The physiologic basis of anomalous correspondence is an extension of the "assimilation of visual directions of the two eyes within Panum's areas which normally take place in sensory fusion." It is conjectured that the mechanism by which normal correspondence is "loosened" and a new relation "created" is a change in the synaptic threshold of unknown cortical cells, facilitating the transmission of impulses at some unknown points and impeding it at others, thus redirecting the flow of impulses.

The theory of anomalous correspondence is a fascinating intellectual construct. There is, however, no physiologic evidence for a flexibility of retinocortical relationships, and its clinical inconsistencies and theoretical artificiality throw a kind of general improbability over it. The theoretical and practical difficulties of measuring normal correspondence have already been described. The obstacles are compounded in squinters, usually children whose attention is circumscribed and descriptive vocabulary limited. We have only to cover one eye and uncover it to demonstrate that the direction of a fixated object is not altered. Nothing is assimilated in normal people, much less rearranged in squinters. Although the visual directions supposedly altered are oculocentric, no such change can be demonstrated in monocular vision. Moreover any such shift must be translatory and involve the entire sheaf of oculocentric directions of one eye relative to the other. This means that there are insufficient retinal elements in the nonfoveal area of the deviated eye (because there are fewer photoreceptors) to match up with the high-density, foveal population of the fixing eye. The result should lead to "gaps" in the percept, not to speak of a new functional blind spot (in addition to the real one) and an enormous aniseikonia. Even if such a new coupling could be achieved, it should include the surrounding Panum area; yet we never hear of anomalous corresponding areas, and indeed only rudimentary (peripherally induced) fusional movements can be elicited in restricted cases. Moreover, if a new correspondence is established, previously corresponding elements are now disparate, yet their stimulation does not give rise to stereopsis. If anomalous correspondence depends on usage, on what basis is the conditioning established, since *all* squint angles change with every fixation distance. And in exotropia the cortical neuron arborizations would need be extensive, since the impulses end up in the opposite hemisphere. If anomalous correspondence takes time to develop, how is it so rapidly altered by surgery but not by lenses, prisms, or orthoptics? And indeed why should one attempt to alter it at all if it represents the best adaptation to squint, since treatment seldom reestablishes normal correspondence? What is the biologic value of an adaptation that only partially accomplishes its purpose as in nonharmonious correspondence, and why should amblyopia or suppression develop at the so-called pseudomacula in harmonious adaptation? Why does this adaptation never occur in congenital macular disease, hemianopia, or paralytic squint or never result in stereoscopic vision when it is supposedly present in concomitant strabismus? Monocular diplopia is always cited as a crucial bit of evidence for the theory of anomalous correspondence; yet it is extremely rare, occurring only after surgery, and there are other causes and other interpretations. Finally and perhaps most important, the theory of anomalous correspondence leads to no useful clinical consequences. It does not alter the indications for surgery or other treatment, it does not change the prognosis, and even if it did, it is generally agreed that anomalous correspondence cannot be modified by nonsurgical means.

Many if not all the results obtained by tests that presumably demonstrate anomalous correspondence can be equally well interpreted as manifestations of suppression and eccentric fixation, changes in the perception of distance as well as direction (probably the chief factor in the tests done in "free space"), and modifications in the innervation factor, which lead to changes in egocentric (not oculocentric) directions, depending on ocular dominance. It must be kept in mind that all tests for anomalous correspondence are subjective, designed to elicit double vision in squinters unaccustomed to it, and hence intrinsically artificial. The chief characteristic of the results is that they are universally inconstant and variable not only on the same test but on different tests and not necessarily because anomalous correspondence is "deep-rooted," which implies a quantification to eliminating double vision when in fact the test conditions simply differ. The interpretation of these tests rests on the ability to measure the objective angle exactly, maintain fixation precisely, and compare it to the subjective angle simultaneously. None of these conditions can be said to exist in clinical testing. For these and other reasons, although the theory of anomalous correspondence is indeed classic, it can hardly be said to be generally accepted. Dissenting opinions have been filed by Chavasse, Linksz, Walls, Verhoeff, Boeder, Morgan, LeGrand, and others.

Spatial disturbances

Spatial disturbances in anisometropia and aniseikonia were considered in Chapter 19. In strabismus the patient must utilize monocular cues for depth perception, and his effectiveness is of practical clinical concern. It can be evaluated by the ability to pass a pencil through a small ring or strike a nail with a hammer. A certain degree of learning is involved, as evidenced by improved capacity with time in acquired squint or after the loss of one eye. The evidence suggests that as regards depth perception (but not stereoacuity) in everyday activities, most congenital squinters do about as well as subjects with normal binocular vision.

One of the more fascinating unsolved problems in the pathophysiology of binocular vision is to what extent the deviated eye contributes to the vision of the squinter, even if only in a subsidiary fashion. Is there in fact some form of rudimentary binocularity at some distance or for some directions of gaze? Although a number of orthoptic philosophies are based on this idea, there are no clear answers to this question.

CLINICAL EVALUATION

The clinical evaluation of binocular anomalies is based on an analysis of signs and symptoms. As usual, the symptoms are polymorphic, and their critical assessment requires insight, instinct, and integration with physiologic principles. Complaints may range from an inability to sustain vision without discomfort, intermittent diplopia, spatial distortions and perversions, poor acuity, faulty depth perception, and failure to read efficiently and coordinate proficiently to the psychologic and cosmetic blemish of a manifest squint. Optical and motor obstacles and signs have been described in previous chapters. It remains to consider some clinical tests for the sensory cooperation of the eyes. Unfortunately their number is exceeded only by the diversity of diagnostic criteria while their use diminishes in proportion to therapeutic disappointments. Fortunately there is no shortage of theories and new hypotheses appear at regular intervals trumpeted with all the delicate cacophony of agitated pots and pans. An ideal evaluation of binocular vision should include tests that are simple to use, easy to understand, consistent when repeated, noninstrumental if possible, and practical in their diagnostic and therapeutic implications.

Tests for acuity and amblyopia

The evaluation of monocular and binocular acuity in children and adults was described in Chapter 10. Methods suitable for the diagnosis of amblyopia based on its pathophysiology were given in Chapter 15. Promoting and preserving good vision in each eye is the predominant goal of any therapeutic effort, whether it be correcting refractive error, improving subnormal vision, or preventing amblyopia. Moreover all sensorial tests of binocular vision depend on subjective discrimination (and attentive cooperation). For example, the fixation of a severly amblyopic eye is so erratic that subjective responses are unreliable, whereas haploscopic techniques require a degree of maturity that is often unobtainable.

Tests for correspondence and fixation

A patient who demonstrates fusion and stereopsis may be safely assumed to have normal correspondence. This assumption can be confirmed by a fusional vergence movement to eliminate foveal double vision when a low prism is interposed before one or both eyes. Measuring the horopter by whatever criteria is clearly unsuitable for clinical purposes.

The simplest test for fixation is the objective observation of the corneal reflex in following a moving light. In older children, several other techniques are available. The most objective of these is the "Visuscope," a modified ophthalmoscope that projects a fixation target (e.g., a star) on to the patient's fundus. The patient is instructed to look at the star with the opposite eye occluded, and the examiner simultaneously observes its position relative to the foveal reflex. This position can be documented by serial fundus photography (Mackensen and von Noorden, 1962). Fixation may be central, parafoveal, peripheral, or paradoxic (e.g., in surgical overcorrection). If the area used for fixation is not the fovea, its position and stability (or scatter) are noted. The patient is asked to indicate whether he has the impression or feeling of looking "at" the star or "past" the star. Fixation is classified as eccentric in the first case and as eccentric viewing in the second. In eccentric viewing, the patient responds to a fundus displacement of the star by a motor adjustment of the eye in which the star is placed first on the fovea and then eccentrically. (A similar effect can be induced in a normal observer by a temporarily induced, electronic flash, foveal scotoma.)

The target may also be projected on the fovea of the deviated eye while the sound eye fixates a light on a calibrated screen (or Maddox cross) at 1 meter through an angulated mirror. The patient indicates the position of the star relative to the fixation light (Cuppers' bifoveal correspondence test).

A monocular afterimage may be induced

on the fovea of the sound eye, and the patient fixates a light on a calibrated screen with the deviated eye. The patient indicates the relative position of the afterimage and fixation light (Brock and Givner afterimage test).

The absence of normal correspondence may be diagnosed by several useful tests. In the Hering-Bielschowsky afterimage test, each eye fixates the center (blocked-out area) of a linear light streak (horizontal for one eye, vertical for the other). If fixation is normal, the afterimages should form a cross in normal correspondence. A displacement suggests the absence of normal correspondence (anomalous localization). In the Bagolini striated glass test, the patient views a small light source through striated plano lenses, which act as modified Maddox rods without otherwise altering normal-seeing conditions. In the absence of a manifest deviation and if normal fixation is present, the streaks are seen to cross at the light source if normal correspondence is present. If one or the other streak is absent, there is suppression. If the streaks cross but there is a deviation on the cover test, normal correspondence is absent. Both the direction and distance of each streak relative to the observer and to each other are noted.

It will be evident that all tests for the absence of normal correspondence are based on the subjective localization of double targets, a diplopia to which the patient is unaccustomed and which is inherently confusing. The interpretation depends on the assumption that there is simultaneous (not alternating) vision, that the fixation pattern is consistent, that anomalous localization involves only direction and not relative distance, and that there is no suppression. The presence of anomalous localization does not mean, however, that a new correspondence has been established. It is conceivable, for example, that variations in response depend on whether localization is referred to the dominant eye, the usually deviated eye, or neither, depending on the efferent innervation pattern (even if the innervation induces no overt change in ocular posture).

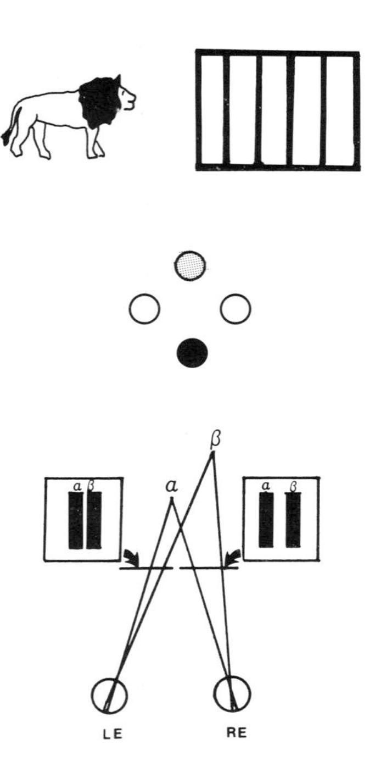

Fig. 25-12. Worth classification of fusion into first *(top)*, second *(middle)*, and third degree *(bottom)*. For details see text.

Tests for fusion and suppression

Whenever the two monocular retinal images are reasonably similar in size, position, brightness, color, shape, and configuration, the normal result is a unitary percept. If under these test conditions there is diplopia or alternating or permanent suppression, sensory fusion is said to be absent. The amount by which the two monocular stimuli can differ one from the other in any of the just listed characteristics may be said to define the quality of the fusional response.

The traditional Worth classification (Fig. 25-12) of fusion into first, second, and third degree is gradually and properly being abandoned. First-degree fusion or simultaneous macular perception obviously involves no fusion at all. Separate and different targets are presented to each eye and their relative positions compared. The standard lion in the cage targets are not seen simultaneously but undergo a patchwork rivalry, and their position depends on heterophoria and fixation disparity. Second-degree fusion, or the Worth four dot test, involves looking at different colored lights (one red, two green, and one white) through red-green glasses. The glasses alone are sufficient to break fusion in some individuals, and the size and distance of the lights have never been standardized. An abnormal response will also be obtained if vision alternates rapidly, and a normal response can be anticipated if target size exceeds the diameter of a suppression scotoma. Third-degree fusion, or stereopsis, is basically diagnosed by the presence of depth discrimination. Unfortunately for the interpretation of the test, some patients with otherwise normal binocular vision have very poor stereoacuity, whereas depth perception can be elicited even in the presence of double vision.

It follows that so-called tests for fusion are really tests for the absence of suppression. The simplest (and most discriminating because the most natural) is observation of a polarized line of printed letters (or pictures) through polarizing filters. Some targets are seen by one eye, some by the other, and some by both. The patient with suppression simply skips the targets seen by the involved eye. The test is easily quantified as to the size of the suppression area by the separation of targets and the fixation distance.

Any of the tests for correspondence may also be used to diagnose suppression. In selected cases the size of the suppression scotoma can be mapped by a stereocampimeter, the Travers screen test, the Lancaster red-green projection test, or any modification of these. Perhaps the simplest is the 4Δ base-out prism test. The prism is placed over one eye while the patient fixates a small light source, and the examiner observes the movement of the *other* eye. A fusional movement occurs in the absence of suppression; no such movement occurs if the retinal image has been shifted onto a nonfunctioning suppression area. If there is a suppression area, the eye without the prism executes a (outward) version movement only, but the secondary fusional inward movement is absent. It is assumed, of course, that the 4Δ prism is not sufficient to dissociate the eyes.

Tests for depth perception and stereopsis

The three-dimensional world is imaged on a two-dimensional retinal surface. The image is not haphazard, however, but a lawful transformation; near objects are larger and further from the horizon with better detail, clearer texture, and more color saturation and luminosity, with outlines organized about one or more vanishing points. Although the retinal image is not a literal copy of the world, it corresponds to it in those very laws of perspective that transforms one to the other but not to a different relation. The greater the number of monocular cues the less ambiguity in the transformation and the more accurate relative and absolute distance perception. This is even more true when binocular vision works efficiently. Monocular cues are not absent with both eyes open, however, and hence more or less influence all tests of binocular depth discrimination.

Most monocular cues are not easily subject to quantification. It may be necessary to learn that when one object overlays another it means nearer, but once learned, the performance cannot be further improved. The size of known objects and therefore their distance, the awareness of aerial perspective, shadows, and motion parallax can undoubtedly be utilized more efficiently by a trained artist or anyone else who has lost the vision of one eye by trial and error. The artistic inability to reproduce perspective does not preclude its effective use in everyday vision. Various simple tests can be used to evaluate this performance, such as threading a needle, passing a pencil through a ring, hitting a nail, reaching for objects at different distances, etc. These are the kinds of tasks the congenital squinter, the patient with macular degeneration, or the monocular aphake is obliged to deal with—the last despite magnification. Still poorly understood are the modifying variables that influence monocular depth when the observer himself is moving.

Many tests are available for measuring depth perception under binocular conditions. In some instances one wants to determine depth discrimination with all available cues operating; in others, only the stereoscopic threshold (with all monocular factors eliminated as far as possible). Failure to make this distinction may result in equating stereoacuity with depth perception, which would certainly be fallacious for those individuals whose capacity to utilize monocular cues is more than adequate for the tasks they perform. One might employ the first type of test in designing an industrial job profile and the second type in analyzing the functional cooperation of the eyes.

Stereothreshold tests may be based on three-dimensional objects (e.g., Howard-Dolman test, Hering sphere test, Verhoeff Stereopter, etc.) or their two-dimensional equivalents (stereoscopic, anaglyph, or polarized targets). More sophisticated tests may measure spatial discrimination or distortion (horopter, eikonometer, leaf room, or tilting plane). It is obviously easier to eliminate monocular cues in two-dimensional tests, but the eyes must necessarily be dissociated so each sees its own target. Two-dimensional tests therefore usually give higher thresholds. Poor visual acuity, eccentric fixation, aniseikonia, suppression, or amblyopia lower stereoacuity on all tests. Three dimensional object tests involve such monocular factors as size, blur, and convergence; two-dimensional tests involve individual differences in the ability to use a stereoscope, proximal convergence and accommodation, fusional dissociation (by red-green or polarizing filters), and the monocular detection of target offset. As might be expected, the correlation between tests varies considerably. In those tests in which uniocular cues operate, their contribution potentiates binocular stereoacuity even though their sensitivity, acting alone, is poor.

The *Howard-Dolman test* involves the relative distance discrimination of two black thin rods (their thickness and separation have never been standardized) viewed through a limiting aperture. Time is not limited. The subject adjusts the movable rod to equidistance with the stationary one by a cord at 6 meters. Average thresholds range from 2 to 10 seconds of arc (for a 60 mm interpupillary distance, each centimeter separation of the rods is equivalent to about 3.5 seconds of arc). The monocular threshold is about six times larger than the binocular. The test is available as a portable model through Bernell Corporation.

The *Hering sphere test* involves the discrimination of the relative distance of falling spheres compared to a vertical thread, viewed through a limiting aperture or tube. The threshold is higher because of the time limit. The test is not commercially available but can be easily constructed.

The *Verhoeff Stereopter* is a convenient hand-held instrument consisting of four preset vertical bars viewed through a limiting aperture against a diffusely illuminated background. The bar thickness is varied to introduce misleading monocular size cues. The test is done at the reading distance. The patient determines which of the bars in the four test situations is nearer or farther. Threshold scatter is greater than for the Howard-Dolman test and larger. Uncorrected myopes may do better on the Verhoeff test. The test is commercially available through various optical supply houses.

Stereoscopic tests include an infinite variety of targets (circles, lines, pictures) whose contour, color, and background influence the result despite controlled disparity. Stereograms are available from Keystone, Titmus, Baush & Lomb, and other suppliers. Recently available are computer-generated stereograms, which provide no visible monocularly seen targets but present simple geometric forms if observed binocularly (Storz). These should be particularly useful for children once standardized. Anaglyph targets present difficulties to the color-blind and require careful matching of filter to target color. The instrument (stereoscope, haploscope, synoptophere, etc.) must naturally be adjusted for the viewing distance for which the stereograms are designed.

Polarized or vectographic tests are available (Titmus, American Optical) for both gross stereopsis (e.g., the "Stereo-fly") and quantative measurements (pictures, circles). They may be used at distance (American Optical polarized acuity slide) and for near testing (Titmus circles and American Optical Vectographs). The principle is the same for all.

It is evident that there is no one "best" test for stereoacuity; the clinical choice depends on what one wants to measure and the maturity of the subject to be evaluated.

BINOCULAR MOTILITY IMBALANCES

A certain number of patients complain of headaches, discomfort when reading, blurred vision, or intermittent diplopia despite adequate refractive corrections or low ametropia. Such symptoms are sometimes due to binocular motility imbalances. Duane (1896) first classified them on a rational basis. He assembled phorias, fusional vergences, and the near point of convergence into four diagnostic syndromes: convergence excess or insufficiency and divergence excess or insufficiency. Convergence anomalies manifest themselves primarily at the reading distance; divergence anomalies cause symptoms with distance vision. Thus convergence insufficiency is characterized by minimal or no distance heterophoria and a large exophoria at near; convergence excess manifests little or no heterophoria at distance and esophoria at near. Divergence insufficiency is characterized by esophoria at far and minimal or no heterophoria at near; divergence excess shows little heterophoria at near, but exophoria at far. Since many patients have large phorias without symptoms, the near point of convergence (NPC) was added to the syndromes. A remote NPC, presumably characteristic of low positive fusional amplitude, would indicate inadequate convergence compensation for exophoria, whereas a normal NPC would suggest comfortable reading vision.

It was soon found that some patients did not fit into clear-cut categories. Secondary effects were postulated, such as convergence spasm secondary to divergence insufficiency or divergence excess secondary to convergence spasm. Since there was no agreement on how large a phoria (or NPC) was abnormal, the diagnostic criteria became blurred. Some authors spoke of convergence insufficiency as an entity without exophoria, as secondary to primary divergent strabismus, or associated with "pseudoheterophoria" (whatever that means). One group reported an incidence of divergence excess as high as 2%, although Bielschowsky stated he had seen only one case in fifty years of practice. Because the symptoms were characteristically uncharacteristic and the pathophysiology poorly understood, it was hard to decide which was cause and which was effect, a chicken-egg game still played in literature.

A muscular etiology for binocular motility imbalances was postulated by von Graefe but abandoned for lack of evidence. Heterophorias do not change significantly with angle of gaze but do vary predictably with accommodation; hence it is not likely that the cause is anatomic (anomalous check ligaments, adhesions, etc.) except as a postsurgical complication. Whether a distance phoria is called an "anomalous position of rest" or an "anomalous tonic innervation" is more semantic than clinically verifiable. There is a shift in the midpoint of the fusional amplitudes only because the measured values are altered by the heterophoria (i.e., there is a difference between the measured and true amplitude).

Duane's conception of oculomotor imbalance as a defect in the supranuclear innervation mechanism was thus a considerable step forward. But although an active convergence mechanism was accepted, the required divergence action was denied. There is now abundant evidence that divergence is an active process, not merely convergence inhibition or the pull of orbital elasticity. For example, when prism base in is placed before the eyes and a fusional divergence elicited, increased firing of both lateral recti can be demonstrated. This increase is maintained throughout the response. Unfortunately this is not evidence for a divergence "center," much less a center that is overexuberant in some and sluggish in others. No one has yet demonstrated such misfiring in patients with motility syndromes. Nor is there neuroanatomic evidence for a divergence center—or even a convergence center.

In fact the entire conception of motor centers is open to reexamination. Much of the evidence is based on clinical neurologic lesions, vaguely circumscribed and seldom subject to systematic examination of oculomotor function or prolonged periods of observation. Experimental electrical stimulation with and without combined ablation of restricted cortical and collicular areas in monkeys indicates no specific oculomotor centers (not even in Brodman's area 8 or the superior colliculus). For example, a normal pattern of eye movements can still be elicited by stimulation of the preoccipital cortex following collicular destruction. Thus even the evolutionary concept of progressive encephalization of oculomotor functions is in question. Eye movement control seems to be widely distributed over the cortex, probably consequent to extensive afferent impulses. The cortex exerts a predominantly contralateral effect, but no particular region is indispensable for the function of another (Pasik and Pasik, 1964). Obviously it is more difficult to ignore clinical custom than break neurologic laws; so "centers" remain, at least in our speculative thinking.

Convergence insufficiency

Symptoms of convergence insufficiency are associated with reading or other close work. They may include headaches, discomfort, blurred vision, occasional diplopia, or an intermittent outward deviation of which the patient is unaware and that causes a cosmetic blemish.

The signs are a minimal distance heterophoria (normal range 1 eso to 3 exo), and a near exophoria (larger than 9 exo at 40 cm). The near point of convergence tends to be more remote than average (normal 2 to 4 inches from the root of the nose). When the positive fusional amplitude is less than 20Δ true value (i.e., measured value plus the exophoria), there may be occasional diplopia. Since the near point of convergence is made up primarily of accommodative and fusional vergence, it is remote. It is known that barbituates inhibit fusional vergence and accommodative vergence; hence central nervous system depression (e.g., lack of sleep, overwork, general illness) exaggerates the symptoms. Ultimately, the correlation of symptoms and signs depends on the response to treatment.

A low AC/A ratio reduces the accommodative stimulus and increases the near exophoria. Thus uncorrected or undercorrected myopia, overcorrected hyperopia, or presbyopia reduces the accommodative demand, hence the accommodative convergence. Retinal blur, whether due to astigmatism, amblyopia, or anisometropia, lowers accommodative innervation. Fatigue and CNS depression may lower the compensating fusional vergence. Cycloplegia or intraocular inflammation producing accommodative paralysis may lead to intermittent exotropia. More rarely, accommodation is relaxed through inattention (or to gain attention) and results in decompensation.

The differential diagnosis includes convergence insufficiency of exophthalmos and Grave's disease (Moebius' sign), amaurosis and severe amblyopia, and the true paresis of organic intracranial disease of sudden onset (encephalitis, disseminated sclerosis, tabes, head trauma). A horizontal diplopia also occurs in narcolepsy and cataplexy.

The treatment of convergence insufficiency is gratifying because almost everything works, at least at the beginning. The refractive prescription may be modified by giving less plus or more minus; the exophoria may be relieved by base-in prism; the fusional vergences improved by base-out exercises.

Complications of convergence insufficiency are related to decompensation, resulting in diplopia or suppression with manifest deviation. Symptoms occur in the early age group, characterized by intermittent exotropia on near fixation, usually with suppression and no diplopia (although the child may close one eye in bright light). The angle of deviation and the frequency of deviation increase, spread to distance fixation, and may eventually result in a constant exotropia. The most likely mechanism in most cases is the gradual decrease of hyperopia or development of myopia. This further embarrasses accommodation and accounts for the high familial incidence. A decrease in hyperopia is as significant as actual myopia (hence the lack of correlation with the degree of myopia). Any factors (e.g., anisometropia or astigmatism) that blur the retinal image of one eye serve as an obstacle to fusion. Suppression may be assumed if there is no diplopia. The sensory status is notoriously difficult to evaluate because of the variability of the fusion-free position. The fusional potential (or quality) can be estimated, however, by the frequency of the deviation, the ease with which fusion is broken (e.g., red glass test), and how promptly binocularity is regained on the cover-uncover test. A fairly good prognosis as far as progression is possible if treatment is started early (correction of myopia, overcorrection with minus lenses, orthoptic antisuppression, and base-out exercises). A significant percentage (about 50%) improve spontaneously by age 15 years. Surgical treatment should not be undertaken unless there is a manifest deviation at both distance and near. The correction of a near exotropia with minimal distance heterophoria may result in a permanent distance esotropia. Large horizontal deviations are often associated with vertical imbalances (but require no separate treatment). The procedure frequently recommended is bilateral medial rectus resection or unilateral resection-recession of the deviating eye.

Convergence excess

The symptoms of convergence excess are similar to insufficiency: discomfort, headaches, intermittent (uncrossed) diplopia, and blurred vision, precipitated by close work. The signs are minimal heterophoria at distance and esophoria at near. The near point of convergence is within average limits, and fusional amplitudes may be normal, or the negative fusional convergence at near is diminished. The tonic convergence is normal, and the AC/A ratio is high. Symptoms are aggravated by excessive accommodation (uncorrected hyperopia, accommodative spasm, overcorrected myopia), and treatment is directed along these lines. Base-in exercises may be used to build up fusional divergence amplitude.

The differential diagnosis includes convergence spasm with paralysis of upward gaze and retraction nystagmus. I have seen two cases, both in teen-age girls, associated with a mild febrile upper respiratory disease, in which the accommodative spasm and esotropia responded to atropine cycloplegia, and which resolved spontaneously after 3 or 4 weeks. There was no evidence of hysteria or neurosis. If fusional reflexes fail to develop or are inadequate, the patient manifests the typical findings of accommodative esotropia. A bifocal correction coupled with temporary cycloplegia or miotics are useful.

Convergence spasm may accompany accommodative spasm in pseudomyopia.

Divergence insufficiency

Divergence insufficiency is a rare condition whose existence as a clinical entity has been denied. It is characterized by an esodeviation at distance and heterophoria with fusion at near. It may occur following central nervous system disease (e.g., trauma, multiple sclerosis, head trauma, increased intracranial pressure) with a sudden onset of uncrossed diplopia. The diplopia has a constant displacement regardless of the direction of gaze. There is apparent weakness of the lateral rectus muscle. Under these circumstances it may represent a stage in the development of bilateral sixth nerve palsy. The symptoms may subside spontaneously, but if symptoms persist, a bilateral rectus resection may be considered.

Divergence excess

Characteristic findings are a larger exophoria at distance than near; the near phoria is within normal limits or even in the eso direction. Symptoms include blurred distance vision (overaccommodation to help align the visual axes), occasional diplopia, or asymptomatic exodeviations. Diplopia may be avoided by closing one eye. Unlike convergence insufficiency, the deviation occurs with distance vision, although the eyes may deviate with inattention (relaxation of accommodation). The deviation may also be increased by anger or fright. The near point of convergence is normal or remote (depending on the amplitude of positive fusional vergence). As the child grows older, the exodeviation may increase, suggesting a progressive disease. The pathophysiology is not clear. An excessive tonic divergence has been postulated, but the evidence is not convincing. For example, although active firing of both lateral recti can be demonstrated during fusional divergence, there is no electromyographic bilateral activity when the eyes deviate *from* the fusion position. If the near phoria is repeated through binocular +3 D spheres, the patient naturally manifests the distance exophoria (except for proximal convergence). This is not necessarily a separate clinical entity (although sometimes called "simulated divergence excess"). The significance of anatomic or mechanical factors in the etiology is speculative. Sensory adaptations such as amblyopia are uncommon. The AC/A ratio is normal or high. Treatment consists of undercorrecting hyperopia, correcting anisometropia and astigmatism, base-in relieving prisms, and base-out fusional amplitude exercises. Prisms seldom work well because they disrupt near vision. If the exodeviation increases and fusion cannot be maintained because of accommodative insufficiency, myopia, or a low AC/A ratio, surgical treatment is required. The greater the myopia (or decrease in hyperopia) and the larger the angle of deviation (over 25Δ), the worse the prognosis. Recommended surgical treatment is a bilateral lateral rectus recession.

Summary

In summary, the treatment of motility imbalances include modification of the spectacle correction, prismatic relief, orthoptics, or surgery. Modifying the prescription depends on the AC/A ratio. Prescribing prisms is always a delicate balance between solving old complaints and creating new difficulties. They frequently cause an increase in the deviation, which has been interpreted either as an adaptational return to the status quo or as the "uncovering" of more deviation analogous to latent hyperopia. Depending on one's point of view, the prism is likely to be reduced in the former and increased in the latter.

Orthoptic management of motility imbalances is aimed at gaining or restoring comfortable binocular vision by supplementing optical or surgical treatment. It may range from antisuppression therapy to prism exercises to improve convergence amplitude.

Most clinicians are understandably reluctant to operate on binocular motility problems with uncertain symptoms, equivocal test findings, and speculative pathophysiology and where poor results may precipitate a permanent squint with amblyopia. Surgical correction may be considered when fusion is absent and intermittent or constant diplopia interferes with the patient's work, endangers his mobility, or when the deviation is a cosmetic blemish. In all cases one attempts to achieve useful binocular vision or, when this is not possible, comfortable monocular vision.

REFRACTIVE MANAGEMENT OF STRABISMUS

In normal binocular vision, all convergence components operate; in strabismus the fusional component is missing or defective. Each of the remaining components contributes to the pathophysiology of squint; for example, tonic innervation has long been recognized as a factor in the basic deviation to be differentiated from the accommodative component. Since surgery is usually indicated for the former but rarely in the latter, the treatment of strabismus begins in the refracting room.

Convergence and divergence are believed to be antagonistic tonic innervations normally held in balance. In congenital strabismus an imbalance follows some sensory or motor obstacle. Excess tonic convergence may cause esotropia, and the converse dynamics influence the relative position of the eyes in exodeviations. Because divergence is more active with distance fixation, exotropias are often larger at far, whereas esotropias are usually larger at near. To these tonic effects are added mechanical and anatomic orbital factors, which, in the nature of things, tend to play a greater role in exotropias than in esotropias. Little is known of the disturbed neurology, possibly subcortical and supranuclear (to explain the stability of the squint, the involvement of both eyes, and the lack of consistent variation with accommodation). That nonaccommodative heterotropia may be due to an anomaly of versions rather than vergences has been suggested because of nasal displacement of the midpoint of horizontal excursions, and a mechanical factor has been postulated to account for this shift. It is not clear, however, whether this is a primary cause or a secondary muscle contracture. Despite its practical importance, no overall clarifying theory of the pathophysiology of congenital heterotropia is available, and even the same clinical and experimental findings are subject to varying interpretation. A still unsolved puzzle in strabismus is the failure of fusional reflexes to overcome a misalignment. Although rudimentary fusional movements can be demonstrated in some squinters, Alpern points out that "it is an obvious over-simplification to imagine that such a patient continues to be a squinter because two of the types of vergences are not in harmony (i.e., because, for example, his amplitude of negative fusional vergence is not sufficient to overcome the rather large amount of positive tonic convergence)."*

The significance of accommodative vergence in strabismus was first emphasized by Donders who noted the common association of hyperopia and esotropia. Thus an esodeviation results in cases of high hyperopia with a normal AC/A ratio (refractive esotropia) and in those patients with little or no hyperopia but an abnormally high AC/A ratio (accommodative esotropia). Neither refractive error nor AC/A ratio can be implicated as the cause of esotropia, however, but only as the reason for the deviation.

Accommodative esotropia may accompany a high hyperopia with a normal AC/A ratio or relatively little ametropia with an excessive AC/A ratio. What constitutes a high ratio is speculative and rather arbitrary. Thus a difference of 10Δ or less on alternate cover measurements at 6 meters and ⅓ meter is considered normal; a ratio greater than 10Δ is taken as abnormal. The difficulty is evident when considering two patients, PD = 60 mm, each without a distant tropia but one with 10 exo and the other with 10 eso near deviation. The first has an AC/A ratio of 8/3 or 2.67; the second, 28/3 or 9.33. The first ratio is actually lower than average.

In refractive esotropia, correcting the ametropia reduces the deviation at all fixation distances. In accommodative esotropia, the hyperopic correction leaves a residual near point esodeviation, which can only be eliminated by further inhibiting accommodation. In fact the diagnosis is frequently made by the response of the near deviation to an additional +3 D lens or to cycloplegics. The accommodative esotrope is the ideal candidate for bifocals, pharmacologic therapy, or both. Although these children tolerate bifocals well, their purpose should be explained to the parents (and distinguished from their more questionable use in controlling myopia). Bifocals are preferably of the straight-across type, with the seg line splitting the pupil. The prescribed add is the minimum needed to straighten the eyes and may range from +1.00 D to +3.00 D. Also effective are anticholinesterases such as 0.125% echothiophate (Phospholine) iodide, 1 drop instilled in both eyes at bedtime for 1 to 3 months. Iris cysts are avoided by the simultaneous use of 5% to 10% phenylephrine drops. Neither drugs nor lenses "cure" the strabismus but only improve the angle of deviation. Additional orthoptic techniques are required in older children to teach or enhance fusion. In the younger squinter, the nervous system may occasionally be plastic enough to permit spontaneous development of a normal binocular pattern, although the correction often remains only cosmetic. Lenses and drugs do not "normalize" an abnormal AC/A ratio but may help compensate for it. Surgery alters the AC/A ratio only by changing the mechanical degree of rotation achieved per unit muscular contraction.

When treatment of accommodative esotropia is delayed, it becomes unresponsive to ciliary activity, and the angle of deviation can no longer be easily changed. But a therapeutic trial of plus overcorrection (regardless of refractive findings) coupled with cycloplegics or anticholinesterases is worth a trial. Premature surgical management in patients with a hidden accommodative component are especially liable to overcorrection as the child grows older. It is best to make the diagnosis of accommodative esotropia early before the secondary innervational and perhaps structural changes compound the therapeutic difficulties.

Exotropia may occur with unequal refractive errors, astigmatism, or even in hyperopia if clear vision is unattainable. Myopic exodeviations are the obverse of the Donders mechanism—the reduced accommodation lowers convergence beyond the capacity of positive fusional reflexes. Only half the exotropes are myopes, however, and since accommodative convergence reduces divergent deviations, the eye may wander out only when the child is tired or inattentive. This history, obtained from the parents, is usually reliable. Exotropic children have as few subjective complaints as their esotropic counterparts,

*From Alpern, M.: Muscular mechanisms. In Davson, H., editor: Physiology of the eye, vol. 3, New York, 1969, Academic Press, Inc., p. 211.

since diplopia and confusion are avoided by sensory adaptations. The diagnosis depends on the ease with which binocular vision is sustained, as well as the magnitude of the deviation when it is given up. Prism cover test may therefore fail to give a true picture of the phoria-tropia dissociation. Moreover the phoria is seldom as large as the tropia in these intermittent cases and occasionally are in the opposite direction. The exotropia angle is usually smaller at near than at far. This may be reversed with high amblyopia, excess tonicity, a large proximal factor, and secondary muscle contractions. A "normal" correspondence has been described in the phoria stage and an "abnormal" one in the tropia stage. But the progression from exophoria to constant exotropia via intermittent exotropia is neither common nor inevitable. The significance of the AC/A ratio in exodeviations remains speculative; an abnormal ratio has been implicated both as a cause and as an adaptive mechanism. A high AC/A ratio in patients with exodeviations suggests minus spheres (hyperopic undercorrection or myopic overcorrection) in conjunction with other forms of treatment to reduce the deviation. Up to -3.00 D can be given, which mobilizes 18Δ of accommodative vergence in a patient with an AC/A ratio of 6/1. This may be enough to move the eye into the range of available fusional reserves or turn an esotropic surgical overcorrection into a cosmetic success.

The surgical treatment of strabismus is outside the scope of this book. In principle, surgery may be performed for functional or cosmetic reasons, preferably both. The mechanical effect does not alter the innervational mechanisms except indirectly and hence is best confined to the static (e.g., nonaccommodative component) angle of deviation. Based on an adequate diagnosis, the surgeon must make three decisions: when to operate, which muscle to operate, and how much to operate. If a proper diagnosis can be established, early operation is better than late. There is no definite rule on symmetrical versus asymmetrical procedures; routine methods should be avoided. The surgical result depends on the experience of the operator, and the procedure that gives him the most reproducible effect is the most useful. This includes recessions and resections and their various modifications. The ideal operation is of such magnitude as to achieve a perfect alignment of the eyes. Frequently, this cannot be accomplished by one operation or by operating on the muscles of one eye, and it may be necessary to titrate the surgical effect. All other factors being equal, more delicate handling of tissue will lead to better and more predictable results, with fewer complications. Whenever feasible, surgery is combined with refractive, orthoptic, and pharmacologic treatment in a coordinated attempt to achieve functional binocularity.

Orthoptics

In the broadest sense, orthoptics (or vision training) refers to the education and rehabilitation of proper visual habits. It includes not only the management of obvious binocular defects but the promotion of comfortable and efficient ocular coordination; the rehabilitation of amblyopia, aphakia, and subnormal vision; and the prevention of recurrences or further visual deterioration. Orthoptics may involve a few minutes showing one patient how to use his new aphakic spectacles to spending months on another with amblyopia. Instrumentation may range from an ordinary eye patch to a sophisticated haploscope. Techniques vary from simple finger to nose exercises to complex pleoptic therapy requiring the expert assistance of a well-trained orthoptist. Vision training may be brilliantly successful in indicated cases or a dismal failure in attempts to achieve what is not possible. Obviously little can be accomplished without an adequate diagnosis and effective rapport. It is absurd to expect the orthoptist to achieve success if her efforts are restricted only to those cases in which everything else failed. Treatment should be a cooperative effort in which optical, pharmacologic, surgical, and rehabilitation techniques are integrated and coordinated. Orthoptics is a reeducation process, and the fundamental principles of learning apply. Its quintessence is a knowledgeable teacher whose enthusiasm sparks the motivation of the learner.

Scientific orthoptics began in the midnineteenth century as a fortuitous combination of surgical overenthusiasm and a familial squint. One victim of what in later years he was to call the "massacre of the internal recti" was Emile Javal's father, whose convergent squint was operated on by Desmarres. When his sister Sophie became afflicted by the family squint, Javal dropped a career in engineering and took up ophthalmology, determined to cure strabismus by a nonsurgical method. His inaugural thesis on strabismus won a prize in 1868. Javal devoted a considerable part of his professional life to orthoptics, which he supervised himself, often without payment. In later years his enthusiasm dimmed, and he came to agree with von Graefe's comment: *les gens ne sont pas dignes de tant de peine*. Javal's main contribution was in stimulating interest in the pathophysiology of squint, developing orthoptic techniques, and emphasizing differential diagnosis so that potential rewards can be balanced by the time, trouble, and expense involved.

REFERENCES

Allen, J. H., editor: Strabismus ophthalmic symposium, St. Louis, 1950, The C. V. Mosby Co.

Allen, J. H., editor: Strabismus ophthalmic symposium II, St. Louis, 1958, The C. V. Mosby Co.

Alpern, M.: Muscular mechanisms. In Davson, H., editor: The eye, vol. 3, New York, 1969, Academic Press, Inc.

Ames, A., and Gliddon, G. H.: Ocular measurements, Transactions of the Section on Ophthalmology of the American Medical Association, 1928, pp. 102-169.

Ames, A., Ogle, K. N., and Gliddon, G. H.: Corresponding retinal points, the horopter, and size and shape of ocular images, J. Opt. Soc. Am. **22**:538-632, 1933.

Andersen, E. E., and Weymouth, F. W.: Visual perception and the retinal mosiac, Am. J. Physiol. **64**: 561-594, 1923.

Arruga, A.: The use of space diagnostic methods and prismotherapy in the treatment of sensory alterations of convergent squint. In Transactions of the First International Congress on Orthoptics, London, 1967, Henry Kimpton Publishers.

Arruga, A., editor: International strabismus symposium, Basel, 1968, S. Karger.

Asher, H.: The suppression theory of binocular vision, Am. J. Optom. **31**:246-251, 1954.

Azar, R. F.: Postoperative paradoxical diplopia, Am. Orthopt. J. **15**:64-71, 1965.

Bagolini, B.: Anomalous correspondence: definition and diagnostic methods, Doc. Ophthal. **23**:346-386, 1967.

Bagolini, B., and Capobianco, N. M.: Subjective space in comitant squint, Am. J. Ophthal. **59**:430-442, 1965.

Bangerter, A.: Amblyopiebehandeung, New York, 1953, S. Karger.

Barlow, H. B., Blakemore, C., and Pettigrew, J. D.: The neural mechanism of binocular depth perception, J. Physiol. (London) **193**:327-342, 1967.

Beale, J. P.: Pleoptics: an interim report of double-blind studies of the treatment of amblyopia with eccentric fixation, Trans. Am. Acad. Ophthal. Otolaryngol. **67**:815-821, 1963.

Bedrossian, E. H.: Surgical results after recession-resection operation for intermittent exotropia. In Manley, D. R., editor: Symposium on horizontal ocular deviations, St. Louis, 1971, The C. V. Mosby Co.

Bennett, A. G.: The optics of the prism stereoscope as a clinical instrument. In Transactions of the International Optical Congress, London, 1972, The British Optical Association.

Berens, C., Hardy Le G. H., and Stark, E. K.: Divergence excess; its incidence, its correlation with refraction, and the value of orthoptic treatment, Trans. Am. Ophthal. Soc. **27**:263, 1929.

Berlucchi, G., and Rizzolatti, G.: Binocularly driven neurons in visual cortex of split-chiasm cats, Science **195**:308-310, 1968.

Berner, G. F., and Berner, D. E.: Relations of ocular dominance, handedness, and the controlling eye in binocular vision, Arch. Ophthal. **50**:603-608, 1953.

Bettman, J. W., Stern, E. L., Whitsell, L. J., and Gofman, H. F.: Cerebral dominance in developmental dyslexia, Arch. Ophthal. **78**:722-729, 1967.

Bielschowsky, A.: Divergence excess, Arch. Ophthal. **12**:157, 1934.

Bielschowsky, A.: Application of after-image test in investigation of squint, Arch. Ophthal. **17**:408-419, 1937.

Bielschowsky, A.: Lectures on motor anomalies, Am. J. Ophthal. **21**:843-854, 1938.

Binder, H. F., and Arndt, C. L.: Binocularity in anomalous retinal correspondence, Acta Ophthal. **41**: 653-658, 1963.

Binder, H. F., Engel, D., Ede, M. L., and Loon, L.: The red filter treatment of eccentric fixation, Am. Orthopt. J. **13**:64-69, 1963.

Blakemore, C.: The range and scope of binocular depth discrimination in man, J. Physiol. (London) **211**: 599-622, 1970.

Blakemore, C., Carpenter, R. H., and Georgeson, M. A.: Lateral inhibition between orientation detectors in the human visual system, Nature **228**:37-39, 1970.

Blakemore, C., and Julesz, B.: Stereoscopic depth aftereffect produced without monocular cues, Science **171**:286-288, 1971.

Blank, A. A.: Analysis of experiments in binocular space perception, J. Opt. Soc. Am. **48**:911-925, 1958.

Blumenfeld, W.: Untersuchunger über die schienbare Grösse in Sehraum, Z. Psychol. **63**:241-404, 1913.

Boeder, P.: Anomalous retinal correspondence refuted, Am. J. Ophthal. **58**:366-373, 1964.

Boeder, P.: Single binocular vision in strabismus, Am. J. Ophthal. **61**:78-86, 1966.

Boeder, P.: The response shift, Doc. Ophthal. **23**:88-100, 1967.

Bower, T. G. R.: Morphogenetic problems in space perception. In Proceedings of the Association for Research of Nervous and Mental Diseases, Stanford, 1968, Stanford University Press.

Braunstein, M. L.: Sensitivity of the observer to transformations of the visual field, J. Exp. Psychol. **72**: 683-689, 1966.

Brecher, G. A.: Form and extent of Panum's areas in foveal vision, Arch. Gesamte Physiol. **246**:315-328, 1942.

Bredemeyer, H. G., and Bullock, K.: Orthoptics theory and practice, St. Louis, 1968, The C. V. Mosby Co.

Breese, B. B.: Binocular rivalry, Psychol. Monogr. **3**:18-21, 44-48, 59-60, 1899.

Brenner, R. L., Charles, S. T., and Flynn, J. T.: Pupillary responses in rivalry and amblyopia, Arch. Ophthal. **82**:23-29, 1969.

Brewster, A.: A treatise on optics, London, 1831.

Brindley, G. S., and Lewin, W. S.: The sensations produced by electrical stimulation of the visual cortex, J. Physiol. (London) **196**:479-493, 1968.

Brock, F. W., and Givner, I.: Fixation anomalies in amblyopia, Arch. Ophthal. **47**:775-786, 1952.

Brown, D. R., and Owen, D. H.: The metrics of visual form, Psychol. Bull. **68**:243-259, 1967.

Brown, J. P., Ogle, K. N., and Reiher, L.: Stereoscopic acuity and observation distance, Invest. Ophthal. **4**:894-900, 1965.

Bruse, G. M.: Ocular divergence; its physiology and pathology, Arch. Ophthal. **13**:639, 1935.

Burian, H. M.: Fusional movements in permanent strabismus: a study of the role of central and peripheral retinal regions in the act of binocular vision in squint, Arch. Ophthal. **26**:626-650, 1941.

Burian, H. M.: Sensorial retinal relationship in concomitant strabismus, Arch. Ophthal. **37**:336-368, 504-533, 618-648, 1947.

Burian, H. M.: Anomalous retinal correspondence; its essence and its significance in diagnosis and treatment, Am. J. Ophthal. **34**:237-253, 1951.

Burian, H. M.: The behavior of the amblyopic eye under reduced illumination and the theory of functional amblyopia, Doc. Ophthal. **23**:189-202, 1967.

Burian, H. M.: Comment on the determination of retinal correspondence, Arch. Ophthal. **80**:146-147, 1968.

Burian, H. M.: Pathophysiologic basis of amblyopia and of its treatment, Am. J. Ophthal. **67**:1-12, 1969.

Burian, H. M., and Capobianco, N. M.: Monocular diplopia (binocular triplopia) in concomitant strabismus, Arch. Ophthal. **47**:23-30, 1952.

Burian, H. M., and von Noorden, G. K.: Binocular vision and ocular motility, St. Louis, 1974, The C. V. Mosby Co.

Burri, C.: Process of learning simultaneous binocular vision, Arch. Ophthal. **28**:235-244, 1942.

Byron, H. M.: Results of pleoptics in the management of amblyopia, Arch. Ophthal. **63**:675-680, 1960.

Carter, D. B.: Studies in fixation disparity, Am. J. Optom. **37**:408-419, 1960.

Carter, D. B.: Fixation disparity and heterophoria following prolonged wearing of prisms, Am. J. Optom. **42**:141-152, 1965.

Cass, E. E.: Monocular diplopia occurring in cases of squint, Br. J. Ophthal. **25**:565-577, 1941.

Chamberlain, W.: Cyclic esotropia, Am. Orthopt. J. **18**:31-34, 1968.

Chamberlain, W.: Diagnostic techniques in the selection of surgery for intermittent exotropia. In Manley, D. R., editor: Symposium on horizontal ocular deviations, St. Louis, 1971, The C. V. Mosby Co.

Chandler, P. A.: Practical considerations concerning choice of operation in convergent squint, Am. J. Ophthal. **34**:375-381, 1951.

Charnwood, J. R. B.: An essay on binocular vision, London, 1950, Hatton Press, Ltd.

Chow, K. L., Lindsley, D. F., and Gollender, M.: Modification of response patterns of lateral geniculate neurons after paired stimulation of contralateral and ipsilateral eyes, J. Neurophysiol. **31**:729-739, 1968.

Cibis, P.: Faulty depth perception caused by cyclotorsion, Arch. Ophthal. **47**:31-42, 1952.

Clark, W. E. Le Gros: The anatomy of cortical vision, Trans. Ophthal. Soc. U. K. **79**:455-461, 1959.

Cobb, W. A., Ettlinger, G., and Morton, H. B.: Cerebral potentials evoked in man by pattern reversal and their suppression in visual rivalry, J. Physiol. (London) **195**:33-34, 1968.

Cobb, W. A., Morton, H. B., and Ettlinger, G.: Cerebral potentials evoked by pattern reversal and their suppression in visual rivalry, Nature **216**:1123-1125, 1967.

Colenbrander, M. C.: What do we see straight, Ophthalmologica **132**:291-294, 1956.

Cooper, C. F.: Orthoptic appraisal, Am. Orthopt. J. **13**:127-134, 1963.

Costenbader, F. D., and O'Neill, J. F.: Secondary surgery of esotropia. In Manley, D. R., editor: Symposium on horizontal ocular deviations, St. Louis, 1971, The C. V. Mosby Co.

Cowan, L. J., Bennett, M. M., and Ogg, D. K.: Diagnostic and therapeutic uses of filters in orthoptics and pleoptics, Am. Orthopt. J. **16**:24-29, 1966.

Crone, R. A.: The kinetic and static function of binocular disparity, Invest. Ophthal. **8**:557-560, 1969.

Cuppers, C.: Moderne Schielbehandlung, Klin. Monatsol. Augenheilkd. **129**:579-604, 1956.

Cuppers, K.: Egozentrische Lokalisationswandel, Inst. Barraguer Second Curso Int. Oftal. **1**:431-452, 1958.

Cuppers, C.: Some reflections on the possibility of influencing the pathologic fixation act, Ann. R. Coll. Surg. (Engl.) **38**:308-325, 1966.

Day, R. H.: On interocular transfer and the central origin of visual after effects, Am. J. Psychol. **71**:784-789, 1958.

Descartes, R.: Discours de la methode plus la dioptrique et les meteors, Paris, 1658, Henry Le Gras.

Douglas, A. A.: Value of orthoptics in the treatment of squint, Br. J. Ophthal. **36**:169-200, 1952.

Dove, H. W.: Ueber Stereoskopie, Ann. Physiol. (series 2) **110**:494-498, 1841.

Duane, A.: Projection and double vision; some new viewpoints, Arch. Ophthal. **54**:233-251, 1925.

Duane, A.: Binocular vision and projection, Arch. Ophthal. **5**:734-753, 1931.

Duane, A.: Diplopia and other disorders of binocular projection, Arch. Ophthal. **7**:187-210, 1932.

Duke-Elder, S., and Wybar, K.: System of ophthal-

mology: ocular motility and strabismus, Vol. 6, St. Louis, 1973, The C. V. Mosby Co.
Dunlap, E. A.: The role of strabismus in reading problems, Am. Orthopt. J. **16**:44-49, 1966.
Dunlap, E. A.: Vertical displacement of horizontal recti. In New Orleans Academy of Ophthalmology: Symposium on strabismus, St. Louis, 1971, The C. V. Mosby Co.
Dunnington, J. H.: Divergence excess; its diagnosis and treatment, Arch. Ophthal. **56**:344, 1927.
Dyer, J. A.: Atlas of extraocular muscle surgery, Philadelphia, 1970, W. B. Saunders Co.
Efron, R.: Stereoscopic vision: effect of binocular temporal summation, Br. J. Ophthal. **41**:709-730, 1963.
Ellerbrock, V. J.: The prescription of visual training by a graphical method, Am. J. Optom. **30**:559-568, 1953.
Enoksson, P.: Binocular rivalry and stereoscopic acuity, Acta Ophthal. **42**:165-178, 1964.
Fells, P., editor: The First Congress of the International Strabismological Association, St. Louis, 1971, The C. V. Mosby Co.
Fender, D. H., and Julese, B.: Extension of Panum's fusional area in binocularly stabilized vision, J. Opt. Soc. Am. **57**:819-830, 1967.
Fender, D. H., and Saint-Cyr, A. J.: Information transfer between the oculomotor systems of the two eyes during tracking tasks, J. Opt. Soc. Am. **59**:512-516, 1969.
Fletcher, M. C.: Biostatistical approach to learning disabilities, Am. Orthopt. J. **17**:73-77, 1967.
Fletcher, M. C.: Natural history of idiopathic strabismus. In New Orleans Academy of Ophthalmology: Symposium on strabismus, St. Louis, 1971, The C. V. Mosby Co.
Fletcher, M. C., Silverman, S., and Abbott, W. P.: Strabismus: reliability of recorded data as used in clinical research, Am. J. Ophthal. **60**:1047-1055, 1965.
Flom, M. C.: Treatment of binocular anomalies of vision. In Hirsch, M. J., and Wick, R. E., editors: Vision of children, Philadelphia, 1963, Chilton Books.
Flom, M. C., Heath, G. G., and Takahashi, E.: Contour interaction and visual resolution; contralateral effects, Science **142**:979-980, 1963.
Flom, M. C., and Kerr, E.: Determination of retinal correspondence; multiple-testing results and the depth of anomaly concept, Arch. Ophthal. **77**:200-213, 1967.
Flom, M. C., and Weymouth, F. W.: Retinal correspondency and the horopter in anomalous correspondence, Nature **189**:34-36, 1961.
Flynn, J. T.: Receptive field function in stimulus deprivation amblyopia, Am. Orthopt. J. **21**:7-14, 1971.
Flynn, J. T., Grundmann, S., and Mashikian, M.: Binocular suppression scotoma; its role in phorias and intermittent tropias, Am. Orthopt. J. **20**:54-67, 1970.
Folk, E. R., and Welched, M. C.: The effect of the correction of refractive errors on nonparalytic esotropia, Am. J. Ophthal. **40**:232-236, 1955.
Fry, G. A.: Visual space perception, Am. J. Optom. **27**:531-553, 1950.
Fry, G. A.: Comments on Luneburg's analysis of binocular vision, Am. J. Optom. **29**:3-11, 1952.
Fry, G. A., and Kent, P. R.: The effects of base-in and base-out prism on stereo-acuity, Am. J. Optom. Monograph 4, 1944.
Gerebtzoff, M. A.: Essai de localisation anatomique du siege de l'amblyopie, Doc. Ophthal. **23**:109-127, 1967.
Gibson, E. J., and Walk, R. D.: The visual cliff, Sci. Am. **202**:54-71, 1960.
Gibson, G. G.: Surgical principles of concomitant strabismus, Am. J. Ophthal. **34**:1431-1437, 1951.
Gibson, H. W.: Textbook of orthoptics. London, 1955, Hatton Press, Ltd.
Gilbert, D. S.: Monocular estimates on distance and direction with stabilized and non-stabilized retinal images, Vision Res. **9**:103-115, 1969.
Goldstein, J. H.: The role of miotics in strabismus, Survey Ophthal. **13**:31-46, 1968.
Goldstein, J. H., Hornblass, A., and Clahane, A. C.: Effect of peripheral vision on the binocular relationship; a further study, Am. J. Ophthal. **66**:86-89, 1968.
Graham, C. H.: Depth and movement, Am. Psychol. **23**:18-26, 1968.
Gregory, R. L.: Eye movements and the stability of the visual world, Nature **182**:1214-1216, 1958.
Guillery, R. W.: Visual pathways in albinos, Sci. Am. **230**:44-54, 1974.
Guillery, R. W., and Kaas, J. H.: A study of normal and congenitally abnormal retinogeniculate projections in cats, J. Comp. Neurol. **143**:73-100, 1971.
Haessler, F. H.: The divergence impulse, Arch. Ophthal. **26**:293-308, 1941.
Hallden, U.: Fusional phenomena in anomalous correspondence, Acta Ophthal. (supp.) **37**:5-93, 1952.
Hallden, U.: An optical explanation of Hering-Hillebrand's horopter deviation, Arch. Ophthal. **55**:830-835, 1956.
Hamburger, F. A.: Binocular vision, stereoscopic vision, and stereoscopic visual acuity, Klin. Monatsbl. Augenheilkd. **141**:321-335, 1962.
Hardesty, H.: Treatment of recurrent intermittent exotropia, Am. J. Ophthal. **60**:1036-1046, 1965.
Harley, R. D.: A and V patterns in horizontal deviations. In Manley, D. R., editor: Symposium on horizontal ocular deviations, St. Louis, 1971, The C. V. Mosby Co.
Harms, H.: Ort und Wesen der Bildhemmung bei Schielenden, Graefe's Arch. Ophthal. **138**:149-210, 1937.
Hebbard, F. W.: Foveal fixation disparity measurements and their use in determining the relationship between accommodative convergence and accommodation, Am. J. Optom. **37**:3-26, 1960.
Hebbard, F. W.: Comparison of subjective and objective measurements of fixation disparity, J. Opt. Soc. Am. **52**:706-712, 1962.
Heinemann, E. G., Tulving, E., and Nachmias, J.:

The effect of oculomotor adjustments on apparent size, Am. J. Psychol. **72**:32-45, 1969.

Helveston, E. M., and von Noorden, G. K.: Microtropia; a newly defined entity. Arch. Ophthal. **78**:272-281, 1967.

Henry, G. H., Bishop, P. O., and Coombs, J. S.: Inhibitory and sub-liminal excitatory receptive fields of simple units in cat striate cortex, Vision Res. **9**:1289-1296, 1969.

Hepler, R. S.: Neurologic disturbances of gaze movements, Am. Orthopt. J. **19**:63-68, 1969.

Hering, E.: Das Gesetz der identischen Sehrichtungen, Arch. Anat. Physiol. Wiss. Med. 1864.

Hering, E.: Spatial sense and movements of the eye, (translated by C. Radde), Baltimore, 1942, American Academy of Optometry.

Hermann, J. S., and Priestley, B. S.: Statistical methods in pleoptics, Am. J. Ophthal. **57**:129-131, 1964.

Herzau, W.: On the horopter in oblique pointing of the eye, Graefe's Arch. Ophthal. **121**:756, 1929.

Herzau, W., and Ogle, K. N.: Difference in size of ocular images of both eyes in a symmetrical convergence and its significance for binocular vision; contribution to problem of aniseikonia, Arch. Ophthal. **137**:327-363, 1937.

Hess, A., and Pilar, G.: Slow fibers in the extraocular muscles of the cat, J. Physiol. (London) **196**:780-798, 1963.

Hillebrand, F.: Lehre von den Gesichtsempfindungen, Vienna, 1929, Springer.

Hofstetter, H. W.: Orthoptics specification by a graphical method, Am. J. Optom. **26**:439-444, 1949.

Holway, A. H., and Boring, E. G.: Determinants of apparent visual size with distance variant, Am. J. Psychol. **54**:21-37, 1941.

Horn, G., and Hill, R. M.: Modifications of receptive fields of cells in the visual cortex occurring spontaneously and association with body tilt, Nature **221**:186-188, 1969.

Howard, H. J.: A test for the judgment of distance, Am. J. Ophthal. **2**:656-576, 1919.

Hubel, D. H., and Wiesel, T. N.: Aberrant visual projections in the Siamese cat, J. Physiol. (London) **218**:33-62, 1971.

Hugonnier, R., and Hugonnier, S. C.: Strabismus, heterophoria, ocular motor paralysis, translated by S. V. Troutman, St. Louis, 1969, The C. V. Mosby Co.

Irvine, S. R.: A simple test for binocular vision; clinical application useful in the appraisal of ocular dominance, amblyopia ex anopsia, minimal strabismus, and malingering, Am. J. Ophthal. **27**:740-746, 1944.

Jampolsky, A.: Differential diagnostic characteristics of intermittent exotropia and true exophoria, Am. Orthopt. J. **4**:48-55, 1954.

Jampolsky, A.: Esotropia and convergent fixation disparity, Am. J. Ophthal. **41**:825-833, 1956.

Jampolsky, A.: Management of small degree esodeviations. In New Orleans Academy of Ophthalmology: Strabismus, St. Louis, 1962, The C. V. Mosby Co.

Jampolsky, A.: Some uses and abuses of orthoptics—the present status. In New Orleans Academy of Ophthalmology: Symposium on strabismus, St. Louis, 1971, The C. V. Mosby Co.

Jampolsky, A., and Thorson, J. C.: Membrane fresnel prisms: a new therapeutic device. In Fells, P., editor: The First Congress of the International Strabismological Association, St. Louis, 1971, The C. V. Mosby Co.

Javal, E.: Manuel theoretique et practique der strabisme, Paris, 1896, G. Masson.

Jones, R.: On the origin of changes in the horopter deviation, Vision Res. **14**:1047-1049, 1974.

Jones, R., and Kerr, K. E.: Motor responses to conflicting asymmetrical vergence stimulus information, Am. J. Optom. **48**:989-1000, 1971.

Joshua, D. E., and Bishop, P. O.: Binocular single vision and depth discrimination: receptive field disparities for central and peripheral vision and binocular interaction on peripheral single units in cat striate cortex, Exp. Brain Res. **10**:389-416, 1969.

Julesz, B.: Binocular depth perception without familiarity cues, Science **145**:356-362, 1964.

Julesz, B.: Foundations of cyclopean perception, Chicago, 1971, The University of Chicago Press.

Kahmann, H.: Ueber das Fovealsehen der Wirbeltiere, Arch. Ophthal. **135**:265-276, 1936.

Kaufman, L.: Suppression and fusion in viewing complex stereograms, Am. J. Psychol. **77**:193-205, 1964.

Kaufman, L.: On the nature of binocular disparity, Am. J. Psychol. **77**:393-402, 1964.

Kertesz, A. E., and Jones, R. W.: Human cyclofusional response, Vision Res. **10**:891-896, 1970.

Kimura, D.: Cerebral dominance and perception of verbal stimuli, Can. J. Psychol. **15**:166-174, 1961.

Knapp, P.: Management of exotropia. In New Orleans Academy of Ophthalmology: Symposium on strabismus, St. Louis, 1971, The C. V. Mosby Co.

Krauskopf, J., Consweet, T. N., and Riggs, L. A.: Analysis of eye movements during monocular and binocular fixation, J. Opt. Soc. Am. **50**:572-578, 1960.

Krimsky, E.: The stereoscope in theory and practice, also a new precision type stereoscope, Br. J. Ophthal. **21**:161-197, 1937.

Lancaster, J. E.: The learning process in orthoptics, Am. J. Ophthal. **32**:1577-1585, 1949.

Lancaster, J. E.: A manual of orthopics, Springfield, Ill., 1951, Charles C Thomas, Publisher.

Land, E. H.: Vectographs, J. Opt. Soc. Am. **30**:230-238, 1940.

Lang, J.: Practical importance of abnormal retinal correspondence, Ophthalmologica **133**:215-217, 1957.

Lang, J.: Which cases of strabismus can be cured? Ophthalmologica **156**:190-196, 1968.

Lang, J.: Microtropia, Arch. Ophthal. **81**:758-762, 1969.

Lau, E.: Neue Untersuchungen über das Tiefen and Ebenensehen, Z. Psychol. Sinn. (Abst. 2) **53**:1-35, 1921.

Lee, D. N.: Binocular stereopsis without spatial disparity, Percept. Psychophy. **9**:216-221, 1971.

LeGrand, Y.: Form and space vision, translated by

M. Millodot and G. G. Heath, Bloomington, 1967, University of Indiana Press.

Leibowitz, H., and Moore, D.: Role of changes in accommodation and convergence in the perception of size, J. Opt. Soc. Am. **56:**1120-1123, 1966.

Levelt, W. J. M.: Binocular brightness averaging and contour information, Br. J. Psychol. **56:**1-13, 1965.

Levelt, W. J. M.: On binocular rivalry, Soesterberg, 1965, Institute for Perception.

Levinge, M.: Value of anomalous retinal correspondence in binocular vision, Br. J. Ophthal. **38:**332-344, 1954.

Limpaecher, E., and Apt, L.: Prognosis in pleoptic therapy, Am. Orthopt. J. **24:**42-46, 1974.

Ling, B. C.: A genetic study of sustained visual fixation and associated behavior in the human infant from birth to six months, J. Genet. Psychol. **61:**227-277, 1942.

Linksz, A.: The stereoscope as an orthoptic instrument, Arch. Ophthal. **26:**389-407, 1941.

Linksz, A.: Physiology of the eye: vision, vol. 2, New York, 1952, Grune & Stratton, Inc.

Linksz, A.: Pleoptics, Trans. Am. Acad. Ophthal. Otolaryngol. **65:**548-562, 1961.

Linksz, A.: Thoughts of a nonstrabologist on ocular motility and the pathophysiology of binocular vision, Int. Ophthal. Clin. **11:**1-22, 1971.

Lit, A.: The magnitude of the Pulfrich stereophenomenon as a function of binocular differences of intensity at various levels of illumination, Am. J. Psychol. **62:** 159-181, 1949.

Lit, A.: Depth-discrimination thresholds as a function of binocular differences in retinal illuminance at scotopic and photopic levels, J. Opt. Soc. Am. **49:**746-752, 1959.

Lowe, S. W., and Ogle, K. N.: Dynamics of the pupil during binocular rivalry, Arch. Ophthal. **75:**395-403, 1966.

Ludlam, W. M., and Kleinman, B. I.: The long range results of orthoptic treatment of strabismus, Am. J. Optom. **42:**647-684, 1965.

Ludvigh, E.: Control of ocular movements and visual interpretation of environment, Arch. Ophthal. **48:**442-448, 1952.

Ludvigh, E.: Direction sense of the eye, Am. J. Ophthal. **36:**139-143, 1953.

Luneburg, R. K.: The metric of binocular visual space, J. Opt. Soc. Am. **40:**627-642, 1950.

Lyle, T. K.: The present position of orthoptics in treatment of manifest and latent strabismus, Trans. Ophthal. Soc. U. K. **70:**19-20, 1950.

Lyle, T. K., and Foley, J.: Subnormal binocular vision with special reference to peripheral fusion, Br. J. Ophthal. **39:**474-487, 1955.

Lyle, T. K., and Wybar, K. C.: Practical orthoptics in the treatment of squint, London, 1967, H. K. Lewis & Co., Ltd.

Mackensen, G., and von Noorden, G. K.: Zur Phenomenologie und Pathogenese der exzentric Fixation bei der Schielamblyopic, Graefe's Arch. Ophthal. **164:** 235-273, 1962.

Maddox, E. E.: The bearing of stereoscopes on the relation between accommodation and convergence, Br. J. Ophthal. **11:**330-337, 1927.

Manas, L.: The constancy of the AC/A ratio, Am. J. Optom. **32:**304-315, 1955.

Manley, D. R., editor: Symposium on horizontal ocular deviations, St. Louis, 1971, The C. V. Mosby Co.

Marlow, F. W.: Tentative interpretation of findings of prolonged occlusion test on evolutionary basis, Arch. Ophthal. **19:**194-204, 1938.

Matin, L.: Eye movements and perceived visual direction. In Jameson, D., and Hurvich, L. M., editors: Visual psychophysics, vol. 7, New York, 1972, Springer Verlag.

Matteucci, P., Basino, L., and Maraini, G.: Desequilibre moteur et vision binoculaire a normale dans l'espace, Doc. Ophthal. **23:**399-422, 1967.

McLaughlin, S. C.: Visual perception in strabismus and amblyopia, Psychol. Monogr. **78:**1-23, 1964.

Mead, L. C.: The influence of size of test stimuli, interpupillary distance, and age on stereoscopic depth perception, J. Exp. Psychol. **33:**148-158, 1943.

Michaels, D. D.: Ocular dominance, Survey Ophthal. **17:**151-163, 1972.

Miles, P. W.: Limits of stereopsis due to physiologic diplopia, Am. J. Ophthal. **32:**1567-1573, 1949.

Miles, W. R.: Ocular dominance demonstrated by unconscious sighting, J. Exp. Psychol. **12:**113-126, 1929.

Mitchell, D. E.: Qualitative depth localization with diplopic images of dissimilar shape, Vision Res. **9:**991-993, 1969.

Mitchell, D. E.: Properties of stimuli eliciting vergence eye movements and stereopsis, Vision Res. **10:**145-162, 1970.

Monje, M.: New method for testing acuity of stereoscopic vision, Z. Sinn. Physiol. **69:**73-90, 1940.

Morgan, M. W.: An investigation of the use of stereoscopic targets in orthoptics, Am. J. Optom. Monograph. 31, 1947.

Morgan, M. W.: Methods used in the treatment of squint, Am. J. Optom. **25:**57-74, 1948.

Morgan, M. W.: Anomalous correspondence interpreted as a motor phenomenon, Am. J. Optom. **38:**131-148, 1961.

Mountcastle, V. B., editor: Interhemispheric relations and cerebral dominance, Baltimore, 1962, Johns Hopkins Press.

Naylor, E. J., Shannon, T. E., and Stanworth, A.: Stereopsis and depth perception after treatment for convergent squint, Br. J. Ophthal. **40:**641-651, 1956.

Nordlow, W.: Spontaneous changes in refraction and angle of squint together with the state of retinal correspondence and visual acuity in concomitant convergent strabismus during the years of growth, Acta Ophthal. **29:**383-421, 1951.

Ogle, K. N.: Analytical treatment of the longitudinal horopter, J. Opt. Soc. Am. **22:**665-728, 1932.

Ogle, K. N.: Researches in binocular vision, Philadelphia, 1950, W. B. Saunders Co.

Ogle, K. N.: Precision and validity of stereoscopic

depth perception from double images, J. Opt. Soc. Am. **43**:906-913, 1953.

Ogle, K. N.: A consideration of the horopter, Invest. Ophthal. **1**:446-461, 1962.

Ogle, K. N.: Present status of our knowledge of stereoscopic vision, Arch. Ophthal. **60**:755-774, 1958.

Ogle, K. N.: The optical space sense. In Davson, H., editor: The eye, vol. 4, New York, 1962, Academic Press, Inc.

Ogle, K. N., and Boeder, P.: Distortion of stereoscopic spatial localization, J. Opt. Soc. Am. **38**:723-733, 1948.

Ogle, K. N., and Groch, J.: Stereopsis and unequal luminosities of the images in the two eyes, Arch. Ophthal. **56**:878-895, 1956.

Ogle, K. N., Martens, T. G., and Dyer, J. A.: Oculomotor imbalance in binocular vision and fixation disparity, Philadelphia, 1967, Lea & Febiger.

Ogle, K. N., Mussey, F., and Prangen, A. de H.: Fixation disparity and the fusional processes in binocular single vision, Am. J. Ophthal. **32**:1069-1087, 1949.

Orton, S. T.: Word-blindness in school children, Arch. Neurol. Psychiatr. **14**:581-615, 1925.

Ovio, J.: Anatomie et physiologie de l'oeil dans la serie animale, Paris, 1927, Librarie Felix Alcan.

Panum, P. L.: Physiologische Untersuchungen über das Sehen mit zwei Augen, Kiel, 1858.

Parks, M. M.: Management of eccentric fixation and ARC in esotropia. In Manley, D. R., editor: Symposium on horizontal oculur deviations, St. Louis, 1971, The C. V. Mosby Co.

Parks, M. M.: Sensory adaptations in strabismus. In New Orleans Academy of Ophthalmology: Symposium on strabismus, St. Louis, 1971, The C. V. Mosby Co.

Parks, M. M., and Pullen, R. M.: Recent developments in sensory testing, Am. Orthopt. J. **15**:85-94, 1965.

Pascal, J. I.: Parallactic angle in binocular depth perception, Arch. Ophthal. **28**:258-262, 1942.

Pearlman, J. T.: Stereoscopic vision testing: a review, Am. Orthopt. J. **19**:78-86, 1969.

Pettigrew, J. D.: The neurophysiology of binocular vision, Sci. Am. **227**:84-95, 1972.

Pettigrew, J. D., Nikara, T., and Bishop, P. O.: Binocular interaction on single units in cat striate cortex: simultaneous stimulation by single moving slit with receptive fields in correspondence, Exp. Brain Res. **6**:391-410, 1968.

Priestley, B. S., Hermann, J. S., and Nutter, A. H.: Home pleoptics, Arch. Ophthal. **70**:616-624, 1963.

Pugh, M.: Amblyopia and the retina, Br. J. Ophthal. **46**:193-211, 1962.

Pulfrich, D.: Die Stereoskopie im Dienste der isochromen und heterochromen Photometrie, Naturwissenschaften **10**:533-564, 569-601, 714-722, 735-743, 751-761, 1922.

Revell, M. J.: Strabismus; a history of orthoptic techniques, London, 1971, Barrie & Jenkins, Ltd.

Richards, W.: Independence of Panum's near and far limits, Am. J. Optom. **48**:103-108, 1971.

Richards, W., and Kaye, M. G.: Local versus global stereopsis, Vision Res. **14**:1345-1347, 1974.

Riesen, A.: Interocular transfer of habits in cats after alternating monocular visual experience, J. Comp. Physiol. Psychol. **49**:516-520, 1956.

Riggs, L. A.: Curvature as a feature of pattern vision, Science **181**:1070-1072, 1973.

Riggs, L. A.: Responses of the visual system to fluctuating patterns, Am. J. Optom. Physiol. Opt. **51**:725-735, 1974.

Rochon-Duvigneaud, A.: Recherches sur l'oeil et la vision chez les vertebres, Laval, 1933, Imprimerie Barneoud.

Romano, P. E., and von Noorden, G. K.: Atypical responses to the four-diopter prism test, Am. J. Ophthal. **67**:935-941, 1969.

Romano, P. E., von Noorden, G. K., and Awaya, S.: A reevaluation of diagnostic methods for retinal correspondence, Am. Orthopt. J. **20**:13-21, 1970.

Ronne, G.: The physiological basis of sensory fusion, Acta Ophthal. **34**:1-26, 1956.

Roth, E.: Which squints respond best to orthoptic treatment, Am. J. Ophthal. **30**:748-752, 1947.

Sanderson, K. J., Darian-Smith, I., and Bishop, P. O.: Binocular corresponding receptive fields of single units in the cat dorsal lateral geniculate nucleus, Vision Res. **9**:1297-1303, 1969.

Schlossberg, H.: Stereoscopic depth from single pictures, Am. J. Psychol. **54**:601-605, 1941.

Schriever, W.: Experimentelle Studien ueber stereoskopische Schen, Z. Psychol. **95**:113-170, 1924.

Schubert, G.: Problem of aniseikonia, Arch. Ophthal. **140**:55-60, 1939.

Schubert, G.: Die Physiologie des Binocularsehens, Doc. Ophthal. **23**:1, 1967.

Scott, A. B.: Neurologic aspects of oculomotor disorders, Int. Ophthal. Clin. **4**:705-773, 1964.

Scott, W. E., and Mash, J.: Stereoacuity in normal individuals, Ann. Ophthal., pp. 99-101, Feb., 1974.

Seiff, S. S.: Foveal slip, Am. Orthopt. J. **19**:120-124, 1969.

Sharpe, M. R.: Living in space: the astronaut and his environment, New York, 1969, Doubleday & Co.

Shepard, R. N., and Metzler, J.: Mental rotation of three-dimensional objects, Science **171**:701-703, 1971.

Shepherd, J. S.: A study of the relationship between fixation disparity and target size, Am. J. Optom. **28**:391-404, 1951.

Shipley, T., and Rawlings, S. C.: The nonius horopter. 1. History and theory, Vision Res. **10**:1225-1263, 1970.

Shipley, T., and Rawlings, S. C.: The nonius horopter. 2. An experimental report, Vision Res. **10**:1263-1299, 1970.

Skavenski, A. A., and Steinman, R. M.: Control of eye position in the dark, Vision Res. **10**:193-203, 1970.

Sloan, L. L., and Altman, A.: Factors involved in several tests of binocular depth perception, Arch. Ophthal. **52**:524-544, 1954.

Sloane, A. E., and Gallagher, J. R.: Evaluation of stereopsis: a comparison of the Howard-Dolman and the Verhoeff test, Arch. Ophthal. **34**:357-359, 1945.

Spekreijse, H., van der Twell, H., and Regan, D.: In-

terocular sustained suppression; correlations with evoked potential amplitude and distribution, Vision Res. **12**:521-526, 1972.
Sperling, G.: Binocular vision; a physical and neural theory, J. Am. Psychol. **83**:461-534, 1970.
Squires, P. C.: Luneburg theory of visual geodesics in binocular space perception, Arch. Ophthal. **65**:288-297, 1956.
Stark, E. K.: Historic American contributions to orthoptics: 1784-1930, Am. Orthopt. J. **21**:127-135, 1971.
Sterling, P., and Wickelgren, B. G.: Visual receptive fields in the superior colliculus of the cat, J. Neurophysiol. **32**:1-15, 1969.
Stoll, M. R., and Boeder, P.: The stereoscope as a training instrument, Arch. Ophthal. **39**:27-36, 1948.
Stromeyer, C. F., and Psotka, J.: The detailed texture of eidetic images, Nature **225**:346-349, 1970.
Swan, K. C.: The clinical physiology of the fusional vergences, Am. Orthopt. J. **15**:6-13, 1965.
ten Doesschate, G.: Results of an investigation of depth perception at a distance of 50 metres, Ophthalmologica **129**:56-57, 1955.
ten Doesschate, G., and Fischer, F. P.: Physiologic and psychologic factors in perspective, Ophthalmologica **97**:1-19, 1939.
Tomlinson, E. A.: Treatment of anomalous retinal correspondence employing retinal after image, Am. Orthopt. J. **15**:59-63, 1965.
Travers, T., a'B: Suppression of vision in squint and its association with retinal correspondence and amblyopia, Br. J. Ophthal. **22**:577-604, 1938.
Travers, T. a'B: Origin of abnormal retinal correspondence, Br. J. Ophthal. **24**:58-64, 1940.
Tschermak-Seysenegg, A. von: Introduction to physiological optics, translated by P. Boeder, Springfield, Ill., 1952, Charles C Thomas, Publisher.
Tscherning, M.: Physiologic optics, translated by Weiland, C., Philadelphia, 1904, Keystone Press.
Urist, M. J.: Horizontal squint with secondary vertical deviations, Arch. Ophthal. **46**:245-267, 1951.
Verhoeff, F. W.: A new theory of binocular vision, Arch. Ophthal. **13**:151-175, 1935.
Verhoeff, F. H.: Anomalous projection and other visual phenomena associated with strabismus, Arch. Ophthal. **19**:663-699, 1938.
Verhoeff, F. H.: Improved kinetic test for binocular stereopsis, Am. J. Ophthal. **23**:320-321, 1940.
Verhoeff, F. H.: Fixation disparity, Am. J. Ophthal. **48**:339-341, 1959.
Vieth, G. A.: Ueber die Richtung der Augen, Ann. Physiol. **58**:233-251, 1818.
Volkmann, A. W.: Die stereoskopischen Erscheinungen in ihrer Beziehung zu der Lehre von den identischen Netzhautpunkten, Arch. Ophthal. **5**(2):1-100, 1859.
von Noorden, G. K.: Pathogenesis of eccentric fixation, Am. J. Ophthal. **61**:399-422, 1966.
von Noorden, G. K.: The etiology and pathogenesis of fixation anomalies in strabismus, Trans. Am. Ophthal. Soc. **67**:698-751, 1969.
von Noorden, G. K., and Mackensen, G. K.: Phenomenology of eccentric fixation, Am. J. Ophthal. **53**: 642-661, 1962.
von Noorden, G. K., and Maumenee, A. E.: Atlas of strabismus, ed. 2, St. Louis, 1973, The C. V. Mosby Co.
Vukovich, V., and Schubert, G.: Fixationsschwankungen und binokulares Einfachsenen, Graefe's Arch. Ophthal. **149**:706-718, 1949.
Wald, G., and Burian, H. M.: The dissociation of form vision and light perception in strabismic amblyopia, Am. J. Ophthal. **27**:950-963, 1944.
Walker, R. Y.: Differences in judgment of depth perception between stationary and moving objects, J. Aviation Med. **12**:218-225, 1941.
Wallach, H., and Zuckerman, C.: The constancy of Stereoscopic depth, Am. J. Psychol. **76**:404-412, 1963.
Walls, G. L.: The vertebrate eye and its adaptive radiation, Cranbrook, 1942, Cranbrook Institute of Science.
Walls, G. L.: A theory of ocular dominance, Arch. Ophthal. **45**:387-412, 1951.
Walls, G. L.: The common-sense horopter, Am. J. Optom. **29**:460-477, 1952.
Wells, D. W.: The stereoscope in ophthalmology, with special reference to treatment of heterophoria and heterotropia, Boston, 1926, E. F. Mahady Co.
Werner, H.: Studies on contour, Am. J. Psychol. **47**:40-64, 1935.
Werner, H.: Binocular vision—normal and abnormal, Arch. Ophthal. **28**:834-844, 1942.
Wiesel, T. N., and Hubel, D. H.: Extent of recovery from the effects of visual deprivation in kittens, J. Neurophysiol. **28**:1060-1072, 1965.
Williams, T. D.: Vertical disparity in depth perception, Am. J. Optom. **47**:339-344, 1970.
Winkelman, J. E.: The significance of peripheral fusion, Ophthalmologica **115**:62-64, 1948.
Worth, C.: Squint; its causes, pathology and treatment, ed. 4, Philadelphia, 1915, Blakiston Co.
Zajac, J. L.: Convergence, accommodation and visual angle as factors in perception of size and distance, Am. J. Psychol. **73**:142-146, 1960.

Index

N

O

P